Figures available in three downloadable sizes (resolutions)

Citations in text link to references in bibliography

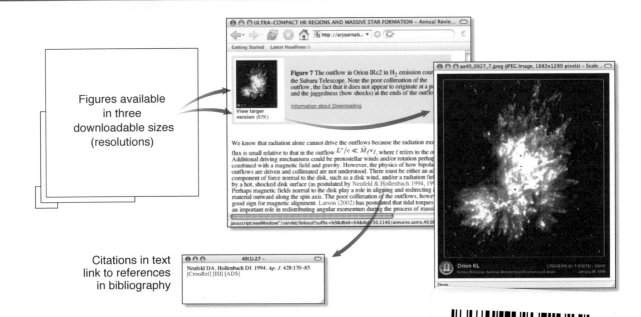

References in Annual Reviews chapter bibliography link out to sources of cited articles online

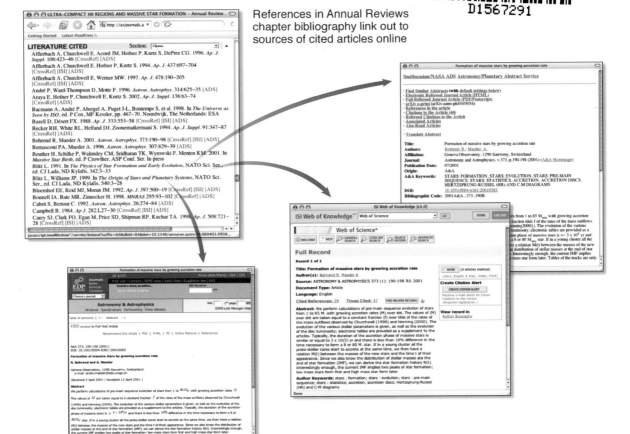

Editorial Committee (2006)

Roger Blandford, Stanford University
John Carlstrom, University of Chicago
David Jewitt, University of Hawaii
John Kormendy, University of Texas, Austin
Rolf Kudritzki, University of Hawaii
Robert Lin, University of California, Berkeley
Marcia J. Rieke, University of Arizona
Ewine van Dishoeck, Leiden Observatory

Responsible for the Organization of Volume 44 (Editorial Committee, 2004)

Roger Blandford
Geoffrey Burbidge
Hugh Hudson
John Kormendy
Kenneth I. Kellermann
Marcia J. Rieke
Allan Sandage
Ewine van Dishoeck
Mike Norman (Guest)
George Rieke (Guest)
Wallace Tucker (Guest)
Ben Zuckerman (Guest)

Production Editor: Roselyn Lowe-Webb
Bibliographic Quality Control: Mary A. Glass
Electronic Content Coordinator: Suzanne K. Moses
Illustration Editor: Douglas Beckner
Subject Indexer: Cheri Walsh

Annual Review of Astronomy and Astrophysics

Volume 44, 2006

Roger Blandford, *Editor*
Stanford University

John Kormendy, *Associate Editor*
University of Texas, Austin

Ewine van Dishoeck, *Associate Editor*
Leiden Observatory

www.annualreviews.org • science@annualreviews.org • 650-493-4400

Annual Reviews
4139 El Camino Way • P.O. Box 10139 • Palo Alto, California 94303-0139

Annual Reviews
Palo Alto, California, USA

COPYRIGHT © 2006 BY ANNUAL REVIEWS, PALO ALTO, CALIFORNIA, USA. ALL RIGHTS RESERVED. The appearance of the code at the bottom of the first page of an article in this serial indicates the copyright owner's consent that copies of the article may be made for personal or internal use, or for the personal or internal use of specific clients. This consent is given on the condition that the copier pay the stated per-copy fee of $20.00 per article through the Copyright Clearance Center, Inc. (222 Rosewood Drive, Danvers, MA 01923) for copying beyond that permitted by Section 107 or 108 of the U.S. Copyright Law. The per-copy fee of $20.00 per article also applies to the copying, under the stated conditions, of articles published in any *Annual Review* serial before January 1, 1978. Individual readers, and nonprofit libraries acting for them, are permitted to make a single copy of an article without charge for use in research or teaching. This consent does not extend to other kinds of copying, such as copying for general distribution, for advertising or promotional purposes, for creating new collective works, or for resale. For such uses, written permission is required. Write to Permissions Dept., Annual Reviews, 4139 El Camino Way, P.O. Box 10139, Palo Alto, CA 94303-0139 USA.

International Standard Serial Number: 0066-4146
International Standard Book Number: 0-8243-0944-8
Library of Congress Catalog Card Number: 63-8846

All Annual Reviews and publication titles are registered trademarks of Annual Reviews.

∞ The paper used in this publication meets the minimum requirements of American National Standards for Information Sciences—Permanence of Paper for Printed Library Materials, ANSI Z39.48-1992.

Annual Reviews and the Editors of its publications assume no responsibility for the statements expressed by the contributors to this *Annual Review*.

TYPESET BY TECHBOOKS, FALLS CHURCH, VIRGINIA
PRINTED AND BOUND BY FRIESENS CORPORATIONS, ALTONA, MANITOBA, CANADA

Preface

The Editorial Committee meeting for Volume 44 was held in San Diego on August 21, 2004. In attendance were Geoff Burbidge (editor), Roger Blandford, John Kormendy (Associate Editors), Hugh Hudson, Kenneth I. Kellermann, Marcia R. Rieke, and Ewine van Dishoeck (Members of the Editorial Board). In addition we were joined by guests: Mike Norman (UCSD), George Rieke (UA, Tucson), Wallace Tucker (Harvard-Smithsonian CfA), and Ben Zuckerman (UCLA). Representing Annual Reviews were Sam Gubins and Roselyn Lowe-Webb. Some 20 reviews were selected, of which 12 appear in this volume. After two unusually large volumes, the size of this volume is closer to the size of *ARAA* a decade ago, as well as the size of most other Annual Reviews.

Although the number of reviews appearing in a volume is subject to large fluctuations, there is an unmistakable trend to fewer, longer, individual reviews. Only one review this year is within the traditional page limit! We have accepted the inevitable and started to commission reviews of different lengths, dictated by quantity of research papers being reviewed. We will, however, continue to use editorial control to promote concision over prolixity.

Volume 44 also represents the first adopting a new Annual Reviews-wide layout. Reaction to change is always mixed but we hope that you appreciate its merits.

Another issue is the relationship of *ARAA* to the *Annual Review of Earth and Planetary Sciences* and the *Annual Review of Nuclear and Particle Science*. We will continue to publish reviews on topics covered primarily by these two other series as long as the content is deemed to be of considerable interest and relevance to astronomers and astrophysicists.

We hope that you enjoy Volume 44 and find it useful. We would be pleased to receive your general comments on the content and format of the *Annual Review of Astronomy and Astrophysics*.

Roger Blandford
John Kormendy
Ewine van Dishoeck

Contents

Annual Review
of Astronomy
and Astrophysics

Volume 44, 2006

An Engineer Becomes an Astronomer
Bernard Mills 1

The Evolution and Structure of Pulsar Wind Nebulae
Bryan M. Gaensler and Patrick O. Slane 17

X-Ray Properties of Black-Hole Binaries
Ronald A. Remillard and Jeffrey E. McClintock 49

Absolute Magnitude Calibrations of Population I and II Cepheids and
Other Pulsating Variables in the Instability Strip of the
Hertzsprung-Russell Diagram
Allan Sandage and Gustav A. Tammann 93

Stellar Population Diagnostics of Elliptical Galaxy Formation
Alvio Renzini 141

Extragalactic Globular Clusters and Galaxy Formation
Jean P. Brodie and Jay Strader 193

First Fruits of the *Spitzer Space Telescope*: Galactic
and Solar System Studies
*Michael Werner, Giovanni Fazio, George Rieke, Thomas L. Roellig,
and Dan M. Watson* 269

Populations of X-Ray Sources in Galaxies
G. Fabbiano 323

Diffuse Atomic and Molecular Clouds
Theodore P. Snow and Benjamin J. McCall 367

Observational Constraints on Cosmic Reionization
Xiaohui Fan, C.L. Carilli, and B. Keating 415

X-Ray Emission from Extragalactic Jets
D.E. Harris and Henric Krawczynski 463

The Supernova–Gamma-Ray Burst Connection
S.E. Woosley and J.S. Bloom 507

Indexes

Subject Index 557

Cumulative Index of Contributing Authors, Volumes 33–44 567

Cumulative Index of Chapter Titles, Volumes 33–44 570

Errata

An online log of corrections to *Annual Review of Astronomy and Astrophysics* chapters (if any, 1997 to the present) may be found at http://astro.annualreviews.org/errata.shtml

Related Articles

From the *Annual Review of Earth and Planetary Sciences*, Volume 34 (2006)

Binary Minor Planets
Derek C. Richardson and Kevin N. Lornez

Planetesimals to Brown Dwarfs: What is a Planet?
Gibor Basri and Michael E. Brown

Cosmic Dust Collection in Aerogel
Mark J. Burchell, Giles Graham, and Anton Kearsley

From the *Annual Review of Nuclear and Particle Science*, Volume 55 (2005)

Leptogenesis as the Origin of Matter
W. Buchmüller, R.D. Peccei, and T. Yanagida

Ascertaining the Core Collapse Supernova Mechanism: The State of the Art and the Road Ahead
Anthony Mezzacappa

Annual Reviews is a nonprofit scientific publisher established to promote the advancement of the sciences. Beginning in 1932 with the *Annual Review of Biochemistry*, the Company has pursued as its principal function the publication of high-quality, reasonably priced *Annual Review* volumes. The volumes are organized by Editors and Editorial Committees who invite qualified authors to contribute critical articles reviewing significant developments within each major discipline. The Editor-in-Chief invites those interested in serving as future Editorial Committee members to communicate directly with him. Annual Reviews is administered by a Board of Directors, whose members serve without compensation.

2006 Board of Directors, Annual Reviews

Richard N. Zare, *Chairman of Annual Reviews, Marguerite Blake Wilbur Professor of Chemistry, Stanford University*
John I. Brauman, *J.G. Jackson–C.J. Wood Professor of Chemistry, Stanford University*
Peter F. Carpenter, *Founder, Mission and Values Institute, Atherton, California*
Sandra M. Faber, *Professor of Astronomy and Astronomer at Lick Observatory, University of California at Santa Cruz*
Susan T. Fiske, *Professor of Psychology, Princeton University*
Eugene Garfield, *Publisher,* The Scientist
Samuel Gubins, *President and Editor-in-Chief, Annual Reviews*
Steven E. Hyman, *Provost, Harvard University*
Daniel E. Koshland Jr., *Professor of Biochemistry, University of California at Berkeley*
Joshua Lederberg, *University Professor, The Rockefeller University*
Sharon R. Long, *Professor of Biological Sciences, Stanford University*
J. Boyce Nute, *Palo Alto, California*
Michael E. Peskin, *Professor of Theoretical Physics, Stanford Linear Accelerator Center*
Harriet A. Zuckerman, *Vice President, The Andrew W. Mellon Foundation*

Management of Annual Reviews

Samuel Gubins, President and Editor-in-Chief
Richard L. Burke, Director for Production
Paul J. Calvi Jr., Director of Information Technology
Steven J. Castro, Chief Financial Officer and Director of Marketing & Sales
Jeanne M. Kunz, Human Resources Manager and Secretary to the Board

Annual Reviews of

Anthropology
Astronomy and Astrophysics
Biochemistry
Biomedical Engineering
Biophysics and Biomolecular Structure
Cell and Developmental Biology
Clinical Psychology
Earth and Planetary Sciences
Ecology, Evolution, and Systematics
Entomology
Environment and Resources

Fluid Mechanics
Genetics
Genomics and Human Genetics
Immunology
Law and Social Science
Materials Research
Medicine
Microbiology
Neuroscience
Nuclear and Particle Science
Nutrition
Pathology: Mechanisms of Disease

Pharmacology and Toxicology
Physical Chemistry
Physiology
Phytopathology
Plant Biology
Political Science
Psychology
Public Health
Sociology

SPECIAL PUBLICATIONS
Excitement and Fascination of Science, Vols. 1, 2, 3, and 4

Bernard Mills

An Engineer Becomes an Astronomer

Bernard Mills

Key Words

radio astronomy, radio telescopes, relativity

Abstract

The early days of radio astronomy in Australia are revisited. The evolution of ideas and the way they led to various instrumental developments and some of the results of these developments are presented. Besides these personal reminiscences, an indication of the political background that sometimes influenced developments is given and, as a coda, an account of a different approach to relativity through the so-called twin paradox.

The explosive development of radio astronomy following World War II was, in Australia, centered in the Radio Physics Laboratory of the then Council for Scientific and Industrial Research, now the Division of Radiophysics in CSIRO (the Commonwealth Scientific and Industrial Research Organization). This laboratory had been established by the Australian Government in 1939 to develop radar for the military services and a general account of its wartime activities may be found in MacLeod (1999). The laboratory was located in the grounds of the University of Sydney and it became customary under wartime regulations for the final year Electrical Engineering Honors students to be assigned to the laboratory as Assistant Research Officers, once their course work had been completed.

It was thus, in December 1942, that my professional career began as an engineer, well versed in structural design and electrical engineering but with my only knowledge of astronomy, or even twentieth-century physics, derived from reading popular science and science fiction. The possibility of a career in astronomy would have never entered my head. I was placed in the receiver group at the laboratory and over the next few years became expert in the design and general development of receivers and display systems. This is just what was needed later, but a lot had to happen before I became involved in the emerging field of radio astronomy.

Shortly after the end of World War II, the activities of the laboratory were changing and I was assigned to the development of an X-ray machine using a resonant cavity powered by a magnetron and producing voltages around 1 MV. This was my first exposure to physics that went beyond Maxwell's equations, a practical introduction to relativity, and even some quantum physics. The challenge was fascinating and it eventually led to a major publication (Mills 1950). I then spent some time assisting with the development of Australia's first digital computer until, in 1948, Joe Pawsey gave me the choice of continuing this work on the computer or joining his group, which had recently made some remarkable astronomical discoveries. The choice was easy, not only because the mechanics of computation did not really interest me, but because of the opportunity to study some completely new phenomena that were apparently not understood by anyone, and also to be working with Pawsey, whom I had come to know well and respect. By this time the study of radio astronomy was well established in the laboratory with various groups of people involved in its different aspects. I have already given an overall account of the development of radio astronomy in Australia (Mills 1988) and here I will limit myself to some of the matters in which I was directly involved, particularly in the evolution of early ideas.

My initiation began through assisting Chris Christiansen with observations of the solar eclipse of November 1, 1948, at a wavelength of 50 cm. I undertook the local observations while he and Don Yabsley departed to distant locations to provide three independent records of the eclipse, which allowed us to precisely locate the hot spots of solar emission. This led to my first paper in radio astronomy, albeit as junior author (Christiansen, Yabsley & Mills 1949).

Following these observations, I had a long discussion with Pawsey about the technical problems facing the group and what project I could adopt as my own. He had two suggestions, either to begin a program on the discrete sources making use of a 97 MHz swept-lobe interferometer, which he had initiated to study the apparently fast moving bursts of radiation from the Sun (Little & Payne-Scott 1951), or else to attempt to detect the H I 21-cm line that had been predicted by van de Hulst. If I had been a trained astronomer and therefore aware of the possible great importance of the H line no doubt this would have been my choice, but I looked on it as merely a technical challenge, whereas I was intrigued by the mystery of the discrete sources and had no hesitation in choosing this option. This did ensure some friction within the group as John Bolton had made discrete sources his own, following his use of the cliff-top interferometer to discover the first such source (Bolton & Stanley 1948) and to establish the existence of this class of object by finding several others. However, Pawsey knew that the future lay with the use of horizontal baselines and Bolton was still making effective use of the interferometer that had proved so successful for him previously.

I began my project by modifying the swept-lobe interferometer for the detection of faint signals, and then, together with fellow engineer Adin Thomas, began a series of observations of Cygnus A, which had been the first and brightest source to be found by Bolton. The primary aim was to determine its position as accurately as possible, and this involved a thorough study of the intricacies of astronomical position measurement as well as the general study of astronomy that I was undertaking at the time.

Eventually we determined a position with probable errors of some two minutes of arc. A photograph of the general area of the Cygnus source had been sent to Bolton by Rudolph Minkowski and on the photograph, after much trouble, we found a small nebulous object within our positional uncertainties. We were confident that this was the source and that it was a Galactic nebulosity because of its location so close to the Galactic plane. However, before publishing I wrote to Minkowski to seek his interpretation, naturally expecting that he would confirm our identification. He did not; the object was a distant galaxy quite similar to others nearby and was certainly not the source that was most probably a nearby star of peculiar properties. This was all very disappointing but we accepted the judgement of such an eminent astronomer. Consequently I decided against immediate publication and spent a long time investigating the violent fluctuations in intensity of the source and their relation to the ionosphere; the final paper made only a passing reference to the galaxy and rejected the identification (Mills & Thomas 1951). Shortly after this paper eventually appeared Graham Smith, who had been carrying out a series of observations on this and other sources at Cambridge, published his final position in *Nature*. This induced Baade & Minkowski (1954) to examine the region with the Hale telescope and led to identification with the same galaxy that we had found; they also described the object as two galaxies in collision and this became the standard explanation for most radio galaxies during the next decade. I could not really blame Minkowski for dismissing our result, because three positions of the source had been published previously, all in wild disagreement and in disagreement with ours; the radio measurements of obviously ignorant newcomers were clearly not taken seriously, especially when in conflict with the conventional wisdom. However, the whole episode marked the beginning of my development of a healthy scepticism toward authoritative pronouncements and the confidence to rely on my own judgment, although I did accept without question the colliding galaxy story because it was a very plausible explanation for the extreme abnormality of a whole galaxy.

Our paper also demonstrated that we were all in unfamiliar territory. In discussing the various corrections necessary to determine the true position of the source, I stated that no correction was needed for aberration, an interesting if elementary mistake that went unnoticed by colleagues who read the paper. It arose because the textbook discussion of aberration was then based on the passage of light down the tube of a telescope, whereas interferometers obviously required a different approach. I thought that relativity would be needed and should have consulted an appropriate textbook, but worked it out for myself employing both inertial frames, thus concluding that there would be no change in the interference fringes with changes of velocity. This was rather like Dingle's famous blunder of some years later, if at a more elementary level. However, it did not matter because in those days the errors of observation were larger than the aberration.

From the beginning of our observations of Cygnus A it was clear that the sensitivity was inadequate for a general study of the discrete sources and that antennas very much larger than the simple Yagis of the swept-lobe interferometer would be required, preferably located in an area less subject to electrical interference. I found a suitable area some 50 km away and, as soon as our observations ceased, set about constructing a new interferometer. This presented no problems as the Radio Physics Laboratory (now the Division of Radiophysics in CSIRO) had excellent workshop and design facilities stemming from its wartime activities. Apart from larger antennas, the interferometer was similar to the earlier one and operated at nearly the same wavelength, close to 3 m. Also, provision was made for the simultaneous recording of the output of two interferometer spacings from the three antennas disposed along an east-west line; this enabled identification of the central lobe of the interference pattern and, as an unexpected bonus, measurement of some angular sizes.

Although the new interferometer was designed for accurate positional measurements, I began with a general sky survey, finding 77 sources with only rough positions. This survey immediately generated some surprising and puzzling results (Mills 1952). It was then believed that the discrete sources were stars having radio emissions many orders of magnitude greater than normal stars and that they were randomly distributed over the sky, the general radio emission from the Galaxy being the integrated emission from such radio stars, analogous to the optical situation. The identifications with nebulae suggested by Bolton were believed to be either wrong or else the nebulae were not representative of the population of discrete sources; Bolton himself appeared to believe the latter alternative.

I began as far as possible without assumptions and an operational definition, which stated that "discrete sources are defined in terms of a particular pattern on the pen recorder." The survey then showed that the distribution of such discrete sources was not completely random and the strongest sources were distributed along the plane of the Galaxy, with only the weaker appearing to be randomly distributed over the remainder of the sky. Clearly, this indicated a population of strong emitters distributed thinly through the Galaxy, possibly like the Crab Nebula, but what were the others, radio stars or radio galaxies or both? It was easy to show that the integrated emission from the population of Galactic sources was insufficient to explain the total Galactic radio emission and another origin was required. It seemed more plausible that the intense nonthermal radiation was associated with stars rather than the tenuous interstellar medium, but identifications with nebulae, both Galactic and extragalactic, suggested the latter and the survey had also produced evidence that some of the sources detected had angular sizes in the region of half a degree.

All identifications were based on positional coincidences, but to decide with certainty whether the randomly distributed sources were galaxies or stars it seemed necessary to measure angular sizes with a resolution of better than a minute of arc. So I added a small portable antenna connected to the interferometer system by a radio link, thus permitting flexibility in orientation and spacings of many kilometers. The principal technical problem of equalizing the propagation time from each interferometer antenna to the mixing point was overcome by the use of adjustable mercury-filled acoustic delay lines borrowed from the computing group where I had been working previously, and where they were used as memory devices.

Some results were needed quickly because the 1952 General Assembly of L'Union Radio-Scientifique Internationale (URSI) was to be held in Sydney with a strong emphasis on radio astronomy, and it would be nice to have something to say about this contentious issue. Accordingly, I undertook a series of observations on the four strongest sources visible from Sydney (Cygnus A, Taurus A, Virgo A, and Centaurus A) using four east-west spacings chosen from available locations to give definite answers as quickly as possible. The sources were all resolved with indicated sizes consistent with the size of the identified nebulae, although the limited spacings caused me to miss the complexity for Cygnus A and, hence, the double nature of the source. It turned out that the Cambridge and Manchester groups had also been engaged in measuring angular sizes using different techniques and also reported the results at the URSI meeting; they had measured the sources Cygnus A and Cassiopeia A. All results were similar although not identical, and we arranged to publish them simultaneously in *Nature*, a rare example of collaboration at the time (see Sullivan 1982). Later I made further observations at three different azimuths (Mills 1953) and constructed simple two-dimensional models, which were consistent with the observations but unfortunately gave very misleading information for Centaurus A because of the complexity of the source, the limited number of observations, and the absence of phase information. This was also the first attempt at gray scale presentation of radio brightness distributions, although prepared by darkroom manipulation rather than computer printout.

Continuing on, from the survey results and what I had gleaned from talking to other workers in the field (whom I had met for the first time at the URSI meeting), I had begun to think seriously of future needs. Simple interferometers were clearly inadequate to deal with the complexity that was showing up. Of course, everyone was aware that Fourier synthesis enabled complex distributions to be reconstructed but the technical problems were daunting. The phase between antennas had to be maintained accurately over long periods and at different locations. More seriously, the only way to perform Fourier transforms in a reasonable time was to use the Radiophysics computer, which was still under development and hardly to be considered as an essential part of an observing program. Perhaps just as important for me was the long time that the actual observations would take; Fourier synthesis had no appeal, and I did not really consider it seriously.

Early work of Hanbury Brown at Manchester had demonstrated the virtues of a simple parabolic reflector and there were now plans to build the large steerable reflector at Jodrell Bank. First thoughts were also being given for a similar instrument in Australia. The need for such a general purpose instrument was obvious but the cost ruled it out at that time and, like many others in the Laboratory, I began to think of cheaper ways of making an instrument that could directly image the sky with a reasonably high resolution. One thought, quickly dismissed, envisaged an enormous half Luneberg lens sitting on a horizontal plane reflector, but what could be used for the dielectric construction material? A helpful suggestion was that a mixture of beer bottles and sand would be appropriate for an Australian telescope!

By now it was clear that the spectrum of nonthermal emission was such that at low frequencies the performance of a very large antenna would be dominated by resolution rather than sensitivity. Accordingly I began thinking of constructing partially filled antennas such as rings, squares, and crosses, but all suffered from severe problems with unwanted responses. A solution occurred to me after discussing the imaging problem with Christiansen who was using two grating arrays along the sides of a reservoir to produce maps of the Sun by the first application of earth rotation synthesis. However, fast imaging was really needed because of the variable solar emission, quite apart from the inconvenience of carrying out Fourier transforms when no computer was available. With my thoughts concentrated on linear arrays I soon realized that a solution to both our needs was an antenna in the form of a symmetrical cross, with the outputs of the arms combined through a phase reversing switch as then used in my interferometer systems. Only the signals received in the overlapping area of the fan beams would produce a modulated signal that could be picked out with a phase-sensitive detector to produce a simple pencil beam response or, in the case of grating arrays, an array of pencil beams. This process effectively multiplied the two antenna responses.

There was opposition to the idea in the Laboratory for some technical reason that I never really understood, and perhaps a political reason, which I could well understand. However, Pawsey supported me and gave approval for the construction of a small experimental model to explore the technique. He also assigned the laboratory's brightest young Technical Officer, Alec Little, to help and this was the beginning of a long and fruitful association. A small "Cross" of a mere 36 m was quickly constructed and confirmed all expectations, even detecting continuum radiation from the Large Magellanic Cloud for the first time (Mills & Little 1953). After this successful demonstration, manpower and funds were made available for the construction and operation of a full scale Cross, and I set about establishing the basic design and finding a suitable site. Fortunately there was a disused airstrip not far from my existing interferometer that provided an adequate area of flat ground, which we were able to lease; construction began there in 1953 and this became the Fleurs Radio Observatory.

During all this activity I received an invitation to spend six months in the United States visiting the California Institute of Technology and the Department of

Terrestrial Magnetism, which was developing a program in radio astronomy. The invitation came at an awkward time, but to decline was unthinkable and I had no qualms about leaving the supervision of construction in the capable hands of Alec Little. This visit was well worthwhile as the few months spent at Cal Tech marked a turning point in my grasp of astronomy and astrophysics. Discussions with some of the leading astronomers and astrophysicists of the day (particularly the iconoclastic Fritz Zwicky), attendance at colloquia, and even a postgraduate course on stellar structure all helped to fill in some of the numerous gaps in the knowledge I had managed to acquire. I returned home in early 1954 with my mind full of plans for observational programs.

Physical construction of the Cross was now essentially complete and there only remained its adjustment and the completion of the associated electronics, all of which took but a few months. A full description is given in the radio astronomy issue of the *P.I.R.E* (Mills et al. 1958). Briefly, it operated at a wavelength of 3.5 m and had a beamwidth of 48 arcmin at the zenith; the sensitivity approached 1 Jy under ideal conditions, which was close to the level at which I expected resolution would become a problem. Provision was made for the rapid switching of the north-south array to provide five outputs separated by half a beamwidth for the production of isophotes and for fast surveys, although with reduced sensitivity.

The next few years saw frenetic activity, exploring and making sense of the radio universe and what it contained. Much of this work is described in three review articles (Mills 1959a, 1964, 1984). The last describes in considerable detail our well-known exposure of the errors in the Cambridge 2C catalog of radio sources, and only a few comments are needed here.

I had just begun a systematic sky survey, together with Bruce Slee, when, as the result of a letter from Fred Hoyle, I first heard of the amazing cosmological claims made by Ryle, which were based on the very large number of faint radio sources included in the Cambridge 2C catalog. It was clear from the survey we had begun that such numbers were impossibly high and it was equally clear that the primary resolution of the Cambridge interferometer was insufficient to separate out so many sources. On obtaining a copy of the catalog we found gross discrepancies with our results in a limited area we had surveyed, but in correspondence Ryle appeared unable or unwilling to understand that he had made such a grievous error and dismissed our results. Accordingly, we published a formal criticism of the catalog and of the conclusion that Ryle had drawn, namely that the radio sources displayed very strong evolutionary effects and therefore the Steady State cosmology could be ruled out (Mills & Slee 1957). Although we found a small excess of faint and presumably distant sources, or perhaps more likely a deficiency of nearby strong sources, we showed that instrumental effects tended to produce this result and stated that "there is no clear evidence for any effect of cosmological importance in the source counts, but there is some evidence for significant clustering of the radio sources, which may be indicative of metagalactic structure." Completion of our surveys over the next several years gave no reason to change these opinions, although the evidence for large-scale clustering of the sources was never strong enough to make definite claims. However, evidence for small-scale clusterings, which caused blending of sources in the antenna beam, continued to be found at a significant level.

Although the Cambridge 2C catalog was largely useless, an important result had been obtained from an analysis of the statistics of the interferometer output (e.g., Scheuer 1957). This showed that the observed radio emission could not have originated in a population of unresolved discrete radio sources randomly distributed throughout a nonevolving universe. My view expressed at the time was that many of the stronger sources would have been resolved by the interferometer, producing smaller output deflections, and it seemed likely that the distribution of the sources was not random, so that nothing could be said directly about evolution.

I have never been particularly optimistic about cosmological research because the observational uncertainties and necessary, but often unrecognized, assumptions lead to great difficulties in coming to meaningful conclusions. Nevertheless the subject fascinated me and I did undertake two more activities; an attempt to derive a luminosity function for radio galaxies (Mills 1960), and the installation of a remote antenna connected by radio link to the Cross, in the hope that crude angular size data would provide useful additional information (Goddard et al. 1960). However, the first program was not very significant without the yet to be discovered quasars and I had left CSIRO before a useful amount of angular size information had been obtained in the second program.

The most effective use of the Cross turned out to be the study of the closer objects such as nearby galaxies, the Galaxy itself, and some of its components. This period also saw the solution to a problem that had been perplexing me from the beginning, the physical process responsible for nonthermal radio emission. I was aware of suggestions that synchrotron radiation might be the mechanism for the hypothetical radio stars but had somehow missed a less publicized suggestion by Kiepenheuer proposing the same mechanism for the interstellar medium. However, everything fell into place with the confirmation of Shklovskii's conclusion that the optical emission from the Crab Nebula was synchrotron radiation and should be polarized. Work in the Soviet Union on the subject then came to light and there seemed no difficulties in ascribing all the nonthermal radio emission to this process. In particular I found that early Cross observations fitted quite well the model of Galactic emission advanced by Shklovskii (1952) of a near spherical distribution combined with a disk, although this disk was nonthermal rather than thermal as proposed by Shklovskii. I even went so far as to predict a maximum in the spectrum of Galactic emission at a frequency around 5 MHz as a result of equipartition considerations, and finally laid to rest the radio star hypothesis on finding that there was no detectable radiation from bright globular clusters (Mills 1955). Baade later told me that this last result finally convinced him, and all this induced me to take a course in reading scientific Russian.

Observations continued on emission nebulae, seen both in emission and absorption, and on supernova remnants. A survey of a strip along the Galactic plane and the preparation of isophotes were also completed, thus ending the most obvious programs possible with the Cross.

By now I was a fully fledged astronomer and attended the 1958 IAU General Assembly in Moscow (my first) and also presented many of the Cross results at the associated Paris Symposium on Radio Astronomy. In one paper I dealt with discrete radio sources and the current problems, and in another I presented new results on Galactic structure (Mills 1959b); the latter involved analyses to determine the distribution of high latitude radiation and disk radiation. When observed with the resolution of the Cross, the high latitude radiation was very complex and could not be modelled directly, but there was clear evidence for an irregular, near spherical, distribution and I estimated its gross parameters as well as I could, but with little confidence in the details. For the disk component, however, things were much clearer and, after subtracting discrete sources and background radiation, the longitude distribution displayed a series of marked "steps," some of which corresponded with the tangential directions of known spiral arms, indicating an association. Moreover, all the eight steps found corresponded with all the tangential directions of a two-arm equiangular spiral with an arm inclination of $6.°8$, suggesting that it represented the basic spiral form of the Galaxy. Needless to say this stirred the H line fraternity and there was much debate.

I returned from my first IAU meetings and my first visit to Europe well pleased with the reception of our work and with making contact with so many astronomers, but the next two years were to present a multitude of problems and difficult decisions. The very basic questions of a decade earlier had now been answered and further development would clearly involve great expense. In-house funding had so far served us well but the much larger and more expensive instruments required for further advances meant that it would no longer suffice. External funding had already been obtained for the 64-m steerable reflector that was under construction at Parkes (see **Figure 1**) and would meet requirements for high frequency and spectral studies, but which had limitations at the lower frequencies. I believed that, for the study of radio sources and nonthermal emission generally, a large Cross would far outperform the Parkes radio telescope and could be constructed at a reasonable cost, leaving the Parkes instrument free to be used for programs that were more appropriate. However, there seemed little prospect of funding for a Cross, particularly as the other branch of radio astronomy, solar astronomy, had an imposing and expensive program to be funded.

This was also a period of great turbulence in the radio astronomy group because of the undermining of Pawsey's authority, as Taffy Bowen, the Chief of the Division of Radiophysics and the architect of the Parkes reflector, took increasing control. Pawsey was planning to leave CSIRO before his sudden illness and death in 1962.

Figure 1

The first working party at Parkes—fixing the location of the radio telescope. From the left: Bernard Mills; the owner of the property; a technician; Chris Christiansen.

Jack Piddington transferred to another CSIRO division and Christiansen took up the Chair of Electrical Engineering at Sydney University. My own solution to funding problems came with an offer made by Harry Messel, the new Professor of Physics at Sydney University to undertake the funding for a Cross of an appropriate size. Messel was revitalizing a very run down department and had demonstrated an extraordinary ability to raise money for various scientific projects, including support for the stellar intensity interferometer of Hanbury Brown who would shortly be joining the School of Physics. I accepted Messel's offer with relief, but also with some trepidation because I would no longer have access to the excellent design and workshop facilities of Radiophysics. Indeed, there was opposition from Bowen who could not tolerate what he saw as a rival establishment setting up shop next door.

In June 1960, I took up the position of Reader in Physics at Sydney University, but for the next few years became an engineer again and, after working out a basic design for the "One Mile Cross" and receiving major funding from the NSF (I believe their first and largest foreign grant), I found myself manager of a big engineering project. It was not an enjoyable job but there was no one else to do it and I was much helped by my engineering contacts, stretching back in some cases to student days. Also, the new Professor of Electrical Engineering, Chris Christiansen, took responsibility for development and construction of the receiving system, which, unlike the first Cross, was based on recently developed solid state technology, now essential because of the complexity of the system. I was also joined by two Radiophysics colleagues, first Arthur Watkinson and later Alec Little, and an international flavor was given by the appointments of Bruce McAdam and Tony Turtle from Cambridge and Michael Large from Manchester.

A location for the Cross had caused some problems as no suitable area could be found close enough to the University to permit daily visits for observations and maintenance; a remote self-contained observatory was necessary. Here my contacts with Radiophysics colleagues proved helpful. A thorough search had been made for possible locations for the 64-m radio telescope and I was given access to the files, which immediately turned up an ideal site some 200 km from Sydney but not far from the Australian National University in Canberra and the associated Mt. Stromlo Observatory. For us this location was the next best thing to a location near Sydney, so Messel set about acquiring the site, with success. The Molonglo River, which flows through Canberra, is nearby and hence it was named the Molonglo Radio Observatory. Construction began there in late 1962; observations with the east-west arm commenced in 1965 and with the full Cross in 1967.

The Cross is described in the radio astronomy issue of the Proc. I.R.E. (Australia) by Mills et al. (1963). The operating frequency was 408 MHz and the beamwidth just under 3 arcmin at the zenith; multi-beaming was provided as a matter of course. An additional lower frequency, described in the reference, was restricted to the east-west arm because it was realized that the process of installing this frequency for the full Cross would have hampered the initial observing programs, and it was probably unnecessary anyway because of developments at Radiophysics.

A general account of the first twenty-five years of operation at Molonglo has been presented by several authors in an issue of the *Australian Journal of Physics* (1991), and here I will only present some of the background, with references limited to personal highlights.

Besides having a new radio telescope, my change of employer signaled a great change in activities. With this appointment went all the usual pursuits of university life: lecturing, supervision of research students, and general university responsibilities, as well as much less time for personal research. A new department in the School of Physics was created in 1965, the Astrophysics Department, with me as Professor and Head; this complemented the existing Astronomy Department under Hanbury Brown and formalized the arrangements under which I had been working.

During the next decade things proceeded much as planned. By now the mysteries of the radio sources had been largely solved but many questions remained and here the Cross was very effective with accurate position measurements leading to optical identifications, and for plotting low level brightness distributions. Also it was outstanding for rapid surveys of the southern sky, eventually producing the Molonglo Reference Catalogue (Large et al. 1981) and detailed surveys of many regions such as the southern Galaxy and the Magellanic Clouds. The discovery of pulsars provided another avenue of research, and the Cross proved to be an excellent search instrument, discovering more than half of the known pulsars by the end of its life and also providing the first association of a pulsar with a supernova remnant.

I was satisfied with the performance of the Cross and the use that had been made of it but, after nearly 10 years operation, the raw observational data for the whole southern sky had been recorded for posterity on magnetic tape (and indeed work has recently begun on these archival data, now available on CD). It was clear that a higher frequency and better resolution would be needed to break new ground. This situation had been anticipated and provision had been made in the original specifications for a structural accuracy sufficient to permit operation at 1420 MHz. There were originally plans for conversion of the Cross to this frequency, but the development of the Fleurs Synthesis Telescope, also operating at 1420 MHz with sensitivity comparable to that of the projected Cross, led me to reconsider. Two radio telescopes in the one university operating at the same frequency and with similar sensitivity, even if vastly different speeds, hardly seemed an efficient arrangement.

Future needs appeared to be largely directed toward the mapping of particular objects or regions of the sky with the highest possible sensitivity and resolution. Perhaps now was the time to consider synthesis, using the tiltable east-west arm alone and a phased feed for this arm to produce the equivalent of an alt-alt mounting. Earth rotation synthesis could then produce fully sampled images of sufficiently southern regions in a single twelve-hour observation and useful images of more northerly regions for which observation times were less. The phased feed system using the cheap circularly polarized antennas that I had devised for the planned 1420 MHz Cross (my last piece of real engineering) was ideal for this purpose. Fortunately I immediately saw the best way to carry out synthesis and was not tempted to spend time on other possibilities. Again the multiplication of the outputs of two long antennas was the required solution, although this time the antennas were collinear, the east and west arms of the Cross (Mills, Little & Joss 1976).

By now, plans for 1420 MHz operation had been abandoned, primarily because the cost would have been much more than could be provided by the Australian Research Grants Committee, our source of operating funds. External funding would be needed but we would be competing with CSIRO, which had a much more sophisticated synthesis telescope on the drawing board (eventually the Australia Telescope). The result of such competition could be easily foreseen, so a more modest plan was adopted using a lower frequency for which the existing reflector mesh was adequate and which corresponded with one of the resonant frequencies of the existing waveguide feed structures, which could be retained. The only mechanical changes involved were the provision of the circularly polarized antennas and their driving mechanism, which reduced the cost of conversion to manageable proportions. Several frequencies were physically possible and the final choice of 843 MHz was made after discussions with those in control of frequency allocations; it was deemed to be the least likely to suffer interference in the future.

With funding assured, the Cross was closed down in 1978 and conversion to the Molonglo Observatory Synthesis Telescope or the MOST began; the first synthesis maps were obtained in 1981. Our celebrations were marred, however, by the death of Alec Little who suffered a heart attack while working at the observatory. Alec, the Director of the Observatory, had contributed mightily to the success of MOST, first by carrying out the development of the phased feed system and then by supervising the actual conversion. He had an encyclopedic knowledge of all aspects of the observatory and was sorely missed. I personally lost an old and valued colleague who, for some thirty years, had been deeply involved in instrumental developments with me. It is sad that he did not have the opportunity to see the full results of his work.

My own departure from the scene was also approaching, owing to the then mandatory retiring age of sixty-five and the University policy of quickly dispensing with emeritus professors. However, I did manage to see through one more large observational program, a study of the supernova remnants in the Magellanic Clouds and a comparison with the conventional wisdom derived from Galactic remnants. There were some anomalies (Mills et al. 1984).

So ended my life as an active astronomer: it had been an absorbing search for what the radio spectrum can tell us of the constituents of the Universe and I learned to be grateful for my engineering background, which provided the know-how for development of the necessary instrumentation. It was certainly a privilege to be involved in some of the immense advances that occurred during the early times, and for this I have always acknowledged a debt of gratitude to Joe Pawsey who started me off and provided support and encouragement.

After retirement and with time to contemplate fundamental issues, my interests gradually changed from the very large to the very small, and I will conclude with an account of some of this postretirement meditation. Since the days when the Dingle fiasco focused attention on the basis of relativity, I had known that an absolute framework is not actually excluded, just unnecessary and inconvenient (e.g., Builder 1958), and I began to speculate that perhaps the quantum world, which is not directly observable, is an absolute world with peculiar and possibly inconceivable properties that produce the observable world we know. The current successful model of the quantum world is necessarily relativistic as it is based on familiar things, and that is the way the familiar world works; just as mankind has created gods in his own image, so he has created a quantum world in the image of the world he knows. With these thoughts in mind I contemplated ways of testing my speculations and thought I had found a logical flaw in the relativistic viewpoint, but this was due to a simple mistake that was pointed out to me in time to save further embarrassment. Then I realized that instead I had stumbled on a new and very simple way of removing the paradox from the clock paradox. Besides being interesting, this approach could well have a role in the presentation of relativity and, as a coda, I give a brief description illustrating the basic idea.

The clock paradox is now known as the twin paradox, although the actual paradox has everything to do with clocks and nothing to do with twins. I will adopt the usual scenario involving twins in which one remains in an inertial frame and the other takes off, traveling with constant velocity for a long time then returning at the same speed, eventually coming to rest beside his twin. However, to ensure that each twin knows the time recorded by the other they will now be provided with identical clocks, transmitters for the time signals and receivers to record the signals transmitted by the other twin. It will be assumed that no time signals are lost in the exchange and that the time spent traveling at constant velocity is vastly greater than the time spent accelerating so that time signals sent and received during periods of acceleration may be ignored.

Two postulates are needed, both based on the postulates of relativity, but weaker as they do not directly involve the velocity of light.

1. Each twin finds the same ratio between the rates of time signals received and transmitted.
2. Signals are received in the same order they are transmitted, without overlap.

The second is now well established (e.g., binary pulsars) and therefore need no longer be a postulate, but I am not aware if there is similar direct evidence for the first.

To proceed, let the ratio of the rate of received to transmitted signals between the twins be written as $r/t = s$ when separating and $r/t = a$ when approaching and take the total numbers of time signals transmitted and received by the stay-at-home (twin #1) as T_1 and R_1 and the corresponding numbers for the traveler (twin #2) as T_2 and R_2.

Twin #2 transmits an equal number of signals, $T_2/2$, when separating and approaching because equal distances are covered with equal speed. During these periods the number of signals he receives will therefore be $s\, T_2/2$ and $a\, T_2/2$, whence

$$R_2 = T_2(s + a)/2.$$

Because all transmitted signals are received, we therefore have for twin #1

$$T_1 = R_2 = T_2(s + a)/2.$$

But twin #1 receives the equal number of signals, $T_2/2$, transmitted by twin #2 when separating and approaching and, while receiving these signals, will be transmitting the numbers of $(T_2/2)/s$ and $(T_2/2)/a$. So we may also write

$$T_1 = T_2(1/s + 1/a)/2, \quad \text{whence}$$
$$s + a = 1/s + 1/a \quad \text{and} \quad s \cdot a = 1$$

Thus, s and a are reciprocals and the sum of reciprocals is greater than 2, whence $T_1 > T_2$.

The difference in the times recorded arises because red-shifted and blue-shifted signals are received for different times as measured by the clock of twin #1 but for the same time as measured by the clock of twin #2. There is no symmetry and therefore no paradox in the resulting time difference. The so-called paradox arises because the usual verbal description is incomplete.

This approach has made no use of velocities and so yields no measure of the actual time difference, for which we need the crucial postulate that the velocity of light is the same in all inertial frames and the postulate that accompanies it, the apparent isotropy of space in all inertial frames. The ratios s and a are the Doppler shifts for propagation of light between the twins and will require the same velocity of light, c, for each twin. As the elapsed times for the twins are different, an extra factor needs to be introduced that must also be the same for both twins and independent of the direction of motion; let us call it δ. Taking the velocity of twin #2 as v, the Doppler shifts may then be written as $s = \delta/(1 + v/c)$ and $a = \delta/(1 - v/c)$, and because $s \cdot a = 1$, then $\delta = \sqrt{1 - v^2/c^2}$. This familiar result may then be applied to give the Doppler shifts, s and a, yielding the final result, $T_1 = T_2/\sqrt{1 - v^2/c^2}$.

From the result for δ, and employing the usual kind of argument in reverse, we may also derive the Lorentz transformation and hence the rest of Special Relativity.

LITERATURE CITED

Aust. J. Phys. 1991. Papers presented at the Molonglo Observatory 25th Anniversary Symposium, November 1990. *Aust. J. Phys.* 44:719–803
Baade W, Minkowski R. 1954. *Ap. J.* 119:206–14
Bolton JG, Stanley GJ. 1948. *Nature* 161:312–13
Builder G. 1958. *Aust. J. Phys.* 11:279–97
Christiansen WN, Yabsley DE, Mills BY. 1949. *Aust. J. Sci. Res. A* 2:506–23
Goddard BR, Watkinson AR, Mills BY. 1960. *Aust. J. Phys.* 14:497–507
Large MI, Mills BY, Little AG, Durdin JM, Kesteven MJ. 1981. *MNRAS* 194:693–704
Little AG, Payne-Scott R. 1951. *Aust. J. Sci. Res. A* 4:489–507
MacLeod R, ed. 1999. *Hist. Rec. Aust. Sci.* 12:411–93
Mills BY. 1950. *JIEE* 97(Part III):425–37
Mills BY. 1952. *Aust. J. Sci. Res. A* 5:266–87
Mills BY. 1953. *Aust. J. Phys.* 6:452–70
Mills BY. 1955. *Aust. J. Phys.* 8:368–89
Mills BY. 1959a. *Handb. Phys.* 53:239–74
Mills BY. 1959b. *Paris Symposium on Radio Astronomy, IAU Symp. 9/ URSI Symp.1*, 1958, ed. R Bracewell, pp. 431–46. Stanford, CA: Stanford Univ. Press. 612 pp.

Mills BY. 1960. *Aust. J. Phys.* 13:550–77
Mills BY. 1964. *Annu. Rev. Astron. Astrophys.* 2:185–212
Mills BY. 1984. *The Early Years of Radio Astronomy—Reflections Fifty Years After Jansky's Discovery*, ed. WT Sullivan III, pp. 147–65. Cambridge, UK: Cambridge Univ. Press. 421 pp.
Mills BY. 1988. *J. Elect. Electron. Eng. Aust.* 8:12–23
Mills BY, Aitchison RE, Little AG, McAdam WB. 1963. *Proc. IRE (Aust.)* 24:156–65
Mills BY, Little AG. 1953. *Aust. J. Phys.* 6:272–78
Mills BY, Little AG, Joss GH. 1976. *Proc. Astron. Soc. Aust.* 3:33–34
Mills BY, Little AG, Sheridan KV, Slee OB. 1958. *Proc. IRE* 46:67–84
Mills BY, Slee OB. 1957. *Aust. J. Phys.* 10:162–94
Mills BY, Thomas AB. 1951. *Aust. J. Sci. Res. A* 4:158–71
Mills BY, Turtle AJ, Little AG, Durdin JM. 1984. *Aust. J. Phys.* 37:321–57
Scheuer PAG. 1957. *Proc. Camb. Philos. Soc.* 53:764–73
Shklovskii IS. 1952. *Astron. J. USSR* 29:418–49
Sullivan WT III. 1982. *Classics in Radio Astronomy*, pp. 275–90. Dordrecht, Holland: Reidel. 348 pp.

The Evolution and Structure of Pulsar Wind Nebulae

Bryan M. Gaensler[1] and Patrick O. Slane

Harvard-Smithsonian Center for Astrophysics, 60 Garden Street, Cambridge, Massachusetts 02138; email: bgaensler@cfa.harvard.edu, pslane@cfa.harvard.edu

Key Words

acceleration of particles, magnetic fields, shock waves, supernova remnants, winds and outflows

Abstract

Pulsars steadily dissipate their rotational energy via relativistic winds. Confinement of these outflows generates luminous pulsar wind nebulae, seen across the electromagnetic spectrum in synchrotron and inverse Compton emission and in optical emission lines when they shock the surrounding medium. These sources act as important probes of relativistic shocks, particle acceleration, and interstellar gas. We review the many recent advances in the study of pulsar wind nebulae, with particular focus on the evolutionary stages through which these objects progress as they expand into their surroundings, and on morphological structures within these nebulae that directly trace the physical processes of particle acceleration and outflow. We conclude by considering some exciting new probes of pulsar wind nebulae, including the study of TeV gamma-ray emission from these sources, and observations of pulsar winds in close binary systems.

[1] Current address: School of Physics, University of Sydney, NSW 2006, Australia

1. INTRODUCTION

The Crab Nebula (**Figure 1**) is almost certainly associated with a supernova explosion observed in 1054 CE (Stephenson & Green 2002, and references therein). However, this source differs substantially from what is now seen at the sites of other recent SNe, in that the Crab Nebula is centrally filled at all wavelengths, whereas sources such as Tycho's and Kepler's supernova remnants show a shell morphology. This and other simple observations show that the Crab Nebula is anomalous, its energetics dominated by continuous injection of magnetic fields and relativistic particles from a central source.

SN(e): supernova(e)

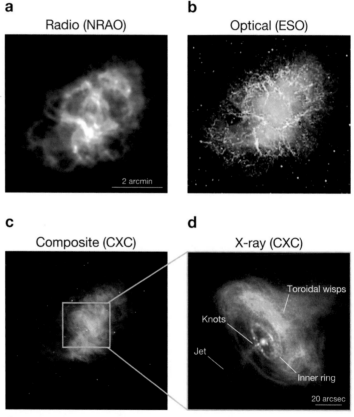

Figure 1

Images of the Crab Nebula (G184.6–5.8). (*a*) Radio synchrotron emission from the confined wind, with enhancements along filaments. (*b*) Optical synchrotron emission (*blue-green*) surrounded by emission lines from filaments (*red*). (*c*) Composite image of radio (*red*), optical (*green*), and X-ray (*blue*) emission. (*d*) X-ray synchrotron emission from jets and wind downstream of the termination shock, marked by the inner ring. Note the decreasing size of the synchrotron nebula going from the radio to the X-ray band. Each image is oriented with north up and east to the left. The scale is indicated by the 2 arcmin scale bar, except for panel (*d*), where the 20 arcsec scale bar applies.

A 16th magnitude star embedded in the Crab Nebula was long presumed to be the stellar remnant and central engine (Minkowski 1942; Pacini 1967). This was confirmed when 33-ms optical and radio pulsations were detected from this star in the late 1960s (Staelin & Reifenstein 1968; Cocke, Disney & Taylor 1969), and these pulsations were then shown to be slowing down at a rate of 36 ns per day (Richards & Comella 1969). The conclusion was quickly reached that the Crab Nebula contains a rapidly rotating young neutron star, or "pulsar," formed in the SN of 1054 CE. The observed rate of spin down implies that kinetic energy is being dissipated at a rate of $\sim 5 \times 10^{38}$ ergs s^{-1}, a value similar to the inferred rate at which energy is being supplied to the nebula (Gold 1969). Following this discovery, a theoretical understanding was soon developed in which the central pulsar generates a magnetized particle wind, whose ultrarelativistic electrons and positrons radiate synchrotron emission across the electromagnetic spectrum (Pacini & Salvati 1973; Rees & Gunn 1974). The pulsar has steadily released about a third of its total reservoir of $\sim 5 \times 10^{49}$ ergs of rotational energy into its surrounding nebula over the last 950 years. This is in sharp contrast to shell-like SNRs, in which the dominant energy source is the $\sim 10^{51}$ ergs of kinetic energy released at the moment of the original SN explosion.

SNR: supernova remnant

Neutron star: a compact degenerate stellar remnant, formed in the core-collapse of a massive star

Pulsar: a rapidly rotating, highly-magnetized neutron star, which generates coherent beams of radiation along its magnetic poles

Pulsar wind nebula(e) [PWN(e)]: a bubble of shocked relativistic particles, produced when a pulsar's relativistic wind interacts with its environment

Observations over the past several decades have identified 40 to 50 further sources, in both our own Galaxy and in the Magellanic Clouds, with properties similar to those of the Crab Nebula (Kaspi, Roberts & Harding 2006; Green 2004)—these sources are known as "pulsar wind nebulae."[1] Sometimes a PWN is surrounded by a shell-like SNR, and the system is termed "composite" (see **Figure 2**). In other cases, best typified by the Crab itself, no surrounding shell is seen.[2]

More recently, an additional category of PWNe has been identified, in which pulsars with high space velocities produce nebulae with cometary or bow shock morphologies as they move through the interstellar medium at supersonic speeds. The sample of such sources is currently small, but high spatial resolution observations, especially in the X-ray band, are rapidly adding to this group.

Because PWNe have a well-defined central energy source and are close enough to be spatially resolved, they act as a marvelous testing ground for studying both relativistic flows and the shocks that result when these winds collide with their surroundings. Studies of PWNe, particularly the spectacular images now being taken by the *Chandra X-ray Observatory*, allow us to resolve details of the interaction of relativistic flows with their surroundings that may never be possible in other classes of source, and can provide the physical foundation for understanding a wide range of astrophysical problems.

We here review current understanding of the structure and evolution of PWNe, with an emphasis on the explosion of new data and new ideas that have emerged in the past few years. Our focus is primarily observational; theoretical considerations

---

[1] PWNe are also often referred to as "plerions." However, given this term's obscure origin (Weiler & Panagia 1978; Shakeshaft 1979), we avoid using this terminology here.

[2] The absence of a shell around the Crab Nebula is presumably because it has not yet interacted with sufficient surrounding gas (Frail et al., 1995; Seward, Gorenstein & Smith 2006).

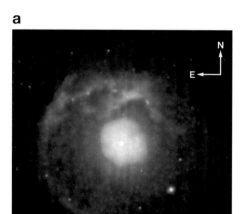
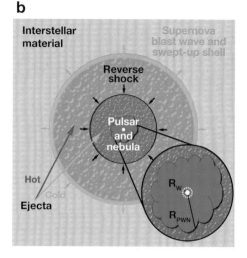

Figure 2

(*a*) A deep *Chandra* X-ray image of the composite SNR G21.5–0.9 (Matheson & Safi-Harb 2005). A circular supernova remnant (SNR) of diameter ≈5′ surrounds a symmetric pulsar wind nebula (PWN) of diameter ≈1′.5, with the young pulsar J1833-1034 at the center (Gutpa et al., 2005; Camilo et al., 2006). The central location of the pulsar and PWN and the symmetric appearance of the PWN and SNR both argue for a relatively unevolved system in which the PWN expands freely and symmetrically into the unshocked interior of the SNR.
(*b*) A schematic diagram of a composite SNR showing the swept-up interstellar medium shell, hot and cold ejecta separated by the reverse shock, and the central pulsar and its nebula. The expanded PWN view shows the wind termination shock. Note that this diagram does not correspond directly to G21.5–0.9, in that a significant reverse shock has probably yet to form in this young SNR.

have been recently discussed by van der Swaluw, Downes & Keegan (2004), Melatos (2004), and Chevalier (2005).

The outline of this review is as follows: in Section 2 we explain the basic observational properties of pulsars and their nebulae; in Section 3 we review current understanding of the evolutionary sequence spanned by the observed population of PWNe; in Section 4 we discuss observations of PWNe around young pulsars, which represent the most luminous and most intensively studied component of the population; in Section 5 we consider the properties of the bow shocks produced by high velocity pulsars; and in Section 6 we briefly describe other recent and interesting results in this field.

2. OVERALL PROPERTIES

2.1. Pulsar Spin Down

Because a pulsar's rotational energy, E_{rot}, is the source for most of the emission seen from PWNe, we first consider the spin evolution of young neutron stars.

2.1.1. Spin-down luminosity, age, and magnetic field.
An isolated pulsar has a spin period, P, and a period derivative with respect to time, $\dot{P} \equiv dP/dt$, both of which can be determined from observations of the pulsed signal.

The "spin down luminosity" of the pulsar, $\dot{E} = -dE_{rot}/dt$, is the rate at which rotational kinetic energy is dissipated, and is thus given by the equation:

$$\dot{E} \equiv 4\pi^2 I \frac{\dot{P}}{P^3}, \qquad (1)$$

where I is the neutron star's moment of inertia, which for a mass of $1.4\,M_\odot$ and a radius of 10 km has the value 10^{45} g cm^{-2}. Values of $\dot{E}$ for the observed pulsar population range between $\approx 5 \times 10^{38}$ ergs s^{-1} for the Crab pulsar and PSR J0537-6910, down to 3×10^{28} ergs s^{-1} for the slowest known pulsar, PSR J2144-3933 (Manchester et al., 2005). Typically only pulsars with $\dot{E} \gtrsim 4 \times 10^{36}$ ergs s^{-1} (of which $\sim$15 are currently known) produce prominent PWNe (Gotthelf 2004).

The age and surface magnetic field strength of a neutron star can be inferred from P and $\dot{P}$, subject to certain assumptions. If a pulsar spins down from an initial spin period P_0 such that $\dot{\Omega} = -k\Omega^n$ (where $\Omega = 2\pi/P$ and n is the "braking index"), then the age of the system is (Manchester & Taylor 1977)

$$\tau = \frac{P}{(n-1)\dot{P}}\left[1 - \left(\frac{P_0}{P}\right)^{n-1}\right], \qquad (2)$$

where we have assumed k to be a constant and $n \neq 1$. The braking index, n, has only been confidently measured for four pulsars (Livingstone, Kaspi & Gavriil 2005, and references therein), in each case falling in the range $2 < n < 3$.

If for the rest of the population we assume $n = 3$ (corresponding to spin down via magnetic dipole radiation) and $P_0 \ll P$, Equation 2 reduces to the expression for the "characteristic age" of a pulsar,

$$\tau_c \equiv \frac{P}{2\dot{P}}. \qquad (3)$$

Equation 3 often overestimates the true age of the system, indicating that P_0 is not much smaller than P (e.g., Migliazzo et al., 2002). PWNe resembling the Crab Nebula tend to be observed only for pulsars younger than about 20,000 years (see Section 4); older pulsars with high space velocities can power bow-shock PWNe (see Section 5).

In the case of a dipole magnetic field, we find $k = 2M_\perp^2/3Ic^3$, where $M_\perp$ is the component of the magnetic dipole moment orthogonal to the rotation axis. We can thus estimate an equatorial surface magnetic field strength:

$$B_p \equiv 3.2 \times 10^{19}(P\dot{P})^{1/2}\,\mathrm{G}, \qquad (4)$$

where P is in seconds. Magnetic field strengths inferred from Equation 4 range between 10^8 G for recycled (or "millisecond") pulsars up to $>10^{15}$ G for "magnetars." Most pulsars with prominent PWNe have inferred magnetic fields in the range 1×10^{12} to 5×10^{13} G.

2.1.2. Time evolution of $\dot{E}$ and P.

A pulsar begins its life with an initial spin-down luminosity, $\dot{E}_0$. If n is constant, its spin-down luminosity then evolves with time, t, as (e.g., Pacini & Salvati 1973):

$$\dot{E} = \dot{E}_0 \left(1 + \frac{t}{\tau_0}\right)^{-\frac{(n+1)}{(n-1)}}, \tag{5}$$

where

$$\tau_0 \equiv \frac{P_0}{(n-1)\dot{P}_0} = \frac{2\tau_c}{n-1} - t \tag{6}$$

is the initial spin-down timescale of the pulsar. The pulsar thus has roughly constant energy output until a time τ_0, beyond which $\dot{E} \propto t^{-(n+1)/(n-1)}$. The spin period evolves similarly:

$$P = P_0 \left(1 + \frac{t}{\tau_0}\right)^{\frac{1}{n-1}}, \tag{7}$$

so that $P \approx P_0$ for $t \ll \tau_0$, but at later times $P \propto t^{1/(n-1)}$.

2.2. Radio and X-Ray Emission from Pulsar Wind Nebulae

As discussed in Section 4.1, the resultant deposition of energy with time generates a population of energetic electrons and positrons, which in turn powers a synchrotron-emitting nebula. Radio synchrotron emission is characterized by a power-law distribution of flux, such that $S_\nu \propto \nu^\alpha$, where S_ν is the observed flux density at frequency ν, and α is the source's "spectral index." At X-ray energies, the emission is often described as a power-law distribution of photons, such that $N_E \propto E^{-\Gamma}$, where N_E is the number of photons emitted between energies E and $E + dE$, and $\Gamma \equiv 1 - \alpha$ is the "photon index." Typical indices for PWNe are $-0.3 \lesssim \alpha \lesssim 0$ in the radio band, and ($\Gamma \approx 2$) in the X-ray band. This steepening of the spectrum implies one or more spectral breaks between these two wavebands, as discussed in Section 4.6.

If the distance to a PWN is known, the radio and X-ray luminosities, L_R and L_X, respectively, can be calculated over appropriate wavelength ranges. Typical ranges are 100 MHz to 100 GHz for L_R and 0.5–10 keV for L_X. Observed values for L_R and L_X span many orders of magnitude, but representative values might be $L_R \sim 10^{34}$ ergs s^{-1} and $L_X \sim 10^{35}$ ergs s^{-1}. The efficiency of conversion of spin-down luminosity into synchrotron emission is defined by efficiency factors $\eta_R \equiv L_R/\dot{E}$ and $\eta_X \equiv L_X/\dot{E}$. Typical values are $\eta_R \approx 10^{-4}$ and $\eta_X \approx 10^{-3}$ (Frail & Scharringhausen 1997; Becker & Trümper 1997), although wide excursions from this are observed. Note that if the synchrotron lifetime of emitting particles is a significant fraction of the PWN age (as is almost always the case at radio wavelengths, and sometimes also in X rays), then the PWN emission represents an integrated history of the pulsar's spin down, and η_R and η_X are not true instantaneous efficiency factors.

3. PULSAR WIND NEBULA EVOLUTION

We now consider the phases of evolution that govern the overall observational properties of PWNe. The detailed theoretical underpinning for these evolutionary phases is given in studies by Reynolds & Chevalier (1984), Chevalier (1998 2005), Blondin, Chevalier & Frierson (2001), Bucciantini et al. (2003), and van der Swaluw, Downes & Keegan (2004).

We defer a discussion of the details of how the wind is generated to Section 4, and here simply assume that the pulsar's continuous energy injection ultimately results in an outflowing wind that generates synchrotron emission.

ISM: interstellar medium

3.1. Expansion into Unshocked Ejecta

Because a pulsar is formed in a SN explosion, the star and its PWN are initially surrounded by an expanding SNR. The SNR blast wave at first moves outward freely at a speed $> (5 - 10) \times 10^3$ km s^{-1}, while asymmetry in the SN explosion gives the pulsar a random space velocity of typical magnitude 400–500 km s^{-1}. At early times the pulsar is thus located near the SNR's center.

The pulsar is embedded in slowly moving unshocked ejecta from the explosion and, as $t \ll \tau_0$, has constant energy output so that $\dot{E} \approx \dot{E}_0$ (see Equation 5). The pulsar wind is highly over-pressured with respect to its environment, and the PWN thus expands rapidly, moving supersonically and driving a shock into the ejecta. In the spherically symmetric case, the PWN evolves as (Chevalier 1977):

$$R_{PWN} \approx 1.5 \dot{E}_0^{1/5} E_{SN}^{3/10} M_{ej}^{-1/2} t^{6/5}$$

$$= 1.1 \text{pc} \left(\frac{\dot{E}_0}{10^{38} \text{ ergs s}^{-1}} \right)^{1/5} \left(\frac{E_{SN}}{10^{51} \text{ ergs}} \right)^{3/10} \left(\frac{M_{ej}}{10 \text{ M}_\odot} \right)^{-1/2} \left(\frac{t}{10^3 \text{ years}} \right)^{6/5}, \tag{8}$$

where R_{PWN} is the radius of the PWN's forward shock at time t, and E_{SN} and M_{ej} are the kinetic energy and ejected mass, respectively, released in the SN.

Because the PWN expansion velocity is steadily increasing, and the sound speed[3] in the relativistic fluid in the nebular interior is $c/\sqrt{3}$, the PWN remains centered on the pulsar. Observationally, we expect to see a rapidly expanding SNR, with a reasonably symmetric PWN near its center, and a young pulsar near the center of the PWN. A good example of a system at this stage of evolution is the recently discovered pulsar J1833-1034, which powers a bright X-ray and radio PWN, which in turn lies at the center of the young SNR G21.5–0.9 (**Figure 2a**; Matheson & Safi-Harb 2005; Gupta et al., 2005; Camilo et al., 2006). This system is estimated to be ∼1000 years old.

3.2. Interaction with the Supernova Remnant Reverse Shock

As the expanding SNR sweeps up significant mass from the ISM or circumstellar medium, it begins to evolve into the "Sedov-Taylor" phase, in which the total energy

---

[3]This assumes that the bulk of the PWN is particle dominated (see Section 4.1).

is conserved and is partitioned equally between kinetic and thermal contributions (see Truelove & McKee 1999, for a detailed discussion).

The region of interaction between the SNR and its surroundings now takes on a more complex structure, consisting of a forward shock where ambient gas is compressed and heated, and a reverse shock where ejecta are decelerated. The two shocks are separated by a contact discontinuity at which instabilities can form. The reverse shock at first expands outward behind the forward shock, but eventually moves inward. In the absence of a central pulsar or PWN, and assuming that the SNR is expanding into a constant density medium (which, given the effects of progenitor mass loss by stellar winds, may not be the case; see Chevalier 2005), the reverse shock reaches the SNR center in a time (Reynolds & Chevalier 1984):

$$t_{Sedov} \approx 7 \left(\frac{M_{ej}}{10~M_\odot} \right)^{5/6} \left(\frac{E_{SN}}{10^{51}~\text{ergs}} \right)^{-1/2} \left(\frac{n_0}{1~\text{cm}^{-3}} \right)^{-1/3} \text{kyr}, \qquad (9)$$

where n_0 is the number density of ambient gas. At this point the SNR interior is entirely filled with shock-heated ejecta, and the SNR is in a fully self-similar state that can be completely described by a small set of simple equations (Cox 1972). The radius of the shell's forward shock now evolves as $R_{SNR} \propto t^{2/5}$.

In the presence of a young pulsar, the inward moving SNR reverse shock collides with the outward moving PWN forward shock after a time $t_{coll} < t_{Sedov}$, typically a few thousand years (van der Swaluw et al., 2001; Blondin, Chevalier & Frierson 2001). Even in the simplest case of a stationary pulsar, an isotropic wind, and a spherical SNR, the evolution is complex. The reverse shock compresses the PWN by a large factor, which responds with an increase in pressure and a sudden expansion. The system reverberates several times, resulting in oscillation of the nebula on a timescale of several thousand years and a sudden increase in the nebular magnetic field that serves to burn off the highest energy electrons (Reynolds & Chevalier 1984; van der Swaluw et al., 2001; Bucciantini et al., 2003). The crushing of the PWN produces Rayleigh-Taylor instabilities, which can produce a chaotic, filamentary structure and mixing of thermal and nonthermal material within the PWN (Chevalier 1998; Blondin, Chevalier & Frierson 2001).

In a more realistic situation, the pulsar's motion carries it away from the SNR's center by the time the reverse shock collides with the PWN. Furthermore, if the SNR has expanded asymmetrically, then the reverse shock moves inward faster on some sides than on others. This results in a complicated three-dimensional interaction, spread over a significant time interval, during and after which the PWN can take on a highly distorted morphology and be significantly displaced from the pulsar position (Chevalier 1998; van der Swaluw, Downes & Keegan 2004). An example of such a system is the Vela SNR, shown in **Figure 3a**.

3.3. A Pulsar Wind Nebula Inside a Sedov Supernova Remnant

Once the reverberations between the PWN and the SNR reverse shock have faded, the pulsar can again power a steadily expanding bubble. However, the PWN now

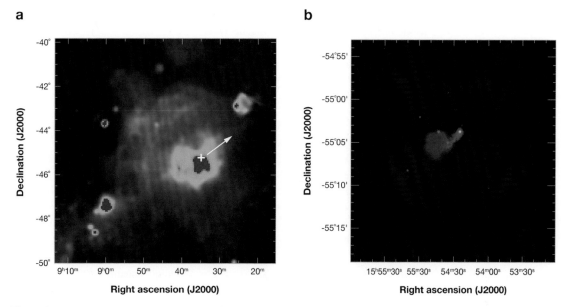

Figure 3

(*a*) A 2.4-GHz Parkes map of the Vela supernova remnant (SNR) (G263.9−3.3) (Duncan et al., 1996). A limb-brightened shell and a central radio pulsar wind nebula (PWN) can both be seen. The cross indicates the location of the associated pulsar B0833-45, whereas the white arrow indicates its direction of motion (Dodson et al., 2003). The fact that the pulsar is neither at nor moving away from the PWN's center indicates that a reverse shock interaction has taken place. (*b*) The composite SNR G327.1−1.1. An 843 MHz Molonglo image is shown in red (Whiteoak & Green 1996), and a 0.2–12 keV *XMM-Newton* image is in blue. The radio morphology consists of a faint shell enclosing a central PWN. The peak of X-ray emission indicates the likely position of an (as yet undetected) pulsar. The offset between the X-ray and radio nebulae indicates that the radio nebula is a "relic PWN" as discussed in Section 3.3. The pulsar is likely to be still moving subsonically through the SNR interior, generating a new PWN as it moves away from its birthsite.

expands into hot, shocked, ejecta at subsonic speeds, and no longer accelerates. In the spherically symmetric case, there are two solutions, depending on whether $t < \tau_0$ or $t > \tau_0$ (see Equation 6). In the former case, $\dot{E}$ is approximately constant and the PWN radius evolves as $R_{PWN} \propto t^{11/15}$ (van der Swaluw et al., 2001), whereas for the latter situation $\dot{E}$ is decaying, and (for $n = 3$) we expect $R_{PWN} \propto t^{3/10}$ (Reynolds & Chevalier 1984).

At this point, the distance traveled by the pulsar from the explosion site can become comparable to or even larger than the radius of an equivalent spherical PWN around a stationary pulsar. The pulsar thus escapes from its original wind bubble, leaving behind a "relic PWN," and generating a new, smaller PWN around its current position (van der Swaluw, Downes & Keegan 2004). Observationally, this appears as a central, possibly distorted radio PWN, showing little corresponding X-ray emission. The pulsar is to one side of or outside this region, with a bridge of radio and X-ray emission linking it to the main body of the nebula. An example is the PWN in the SNR G327.1–1.1, shown in **Figure 3b**.

The sound speed in the shocked ejecta drops as the pulsar moves from the center to the edge of the SNR. Eventually the pulsar's space motion becomes supersonic, and it now drives a bow shock through the SNR interior (Chevalier 1998; van der Swaluw, Achterberg & Gallant 1998). The ram pressure resulting from the pulsar's motion tightly confines the PWN, so that the nebula's extent is small, $\lesssim 1$ pc. Furthermore, the pulsar wind is in pressure equilibrium with its surroundings, so that the PWN no longer expands steadily with time.

For a SNR in the Sedov phase, the transition to a bow shock takes place when the pulsar has moved 68% of the distance between the center and the forward shock of the SNR (van der Swaluw, Achterberg & Gallant 1998; van der Swaluw et al., 2003). The pulsar is now surrounded by a Mach cone, and the PWN takes on a cometary appearance at X-ray and radio wavelengths. An example of such a system is PSR B1853+01 in the SNR W44, as shown in **Figure 4a**.

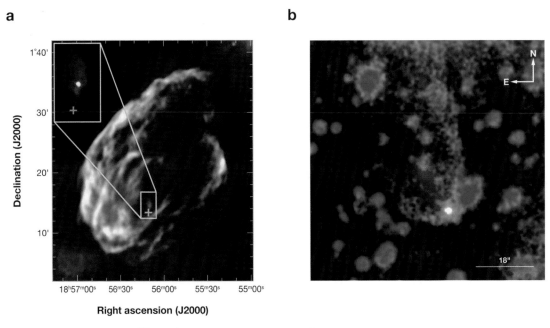

Figure 4

(*a*) The supernova remnant (SNR) W44 (G34.7–0.4). The main panel shows a 1.4 GHz VLA image of the SNR (Giacani et al., 1997), while the inset shows 8.4 GHz VLA data on the region surrounding the associated young pulsar B1853+01 (Frail et al., 1996), whose position is marked by a cross. The pulsar is nearing the edge of the SNR, and now drives a small bow-shock pulsar wind nebula as a result of its supersonic motion. (*b*) The recycled "Black Widow" pulsar (PSR B1957+20) and its bow shock (Stappers et al., 2003). The green shows Hα emission imaged with the Anglo-Australian Observatory, and the red shows X-ray emission observed with *Chandra* (the *blue* emission indicates background stars). The pulsar is moving at a position angle of 212° (north through east).

A pulsar will typically cross its SNR shell after ~40,000 years (see Equation 15 in Section 5). If the SNR is still in the Sedov phase, the bow shock has a Mach number $\mathcal{M} \approx 3.1$ at this point (van der Swaluw et al., 2003). The injection of energy from the pulsar may brighten and re-energize the SNR shell during its passage (Shull, Fesen & Saken 1989; van der Swaluw, Achterberg & Gallant 2002).

3.4. A Pulsar in Interstellar Gas

Once outside its SNR, a pulsar's motion is often highly supersonic in interstellar gas. A bow-shock PWN results, with a potentially large Mach number, $\mathcal{M} \gg 1$.

In cases where the pulsar propagates through neutral gas, the forward shock driven by the PWN is visible, in the form of Hα emission produced by shock excitation and charge exchange (see Section 5.2). The shocked wind also produces synchrotron emission, resulting in a bright head and cometary tail, both best seen in radio and X rays (see Section 5.3). An example of an interstellar bow shock is the structure seen around PSR B1957+20, show in **Figure 4b**.

As the pulsar now moves through the Galaxy, its $\dot{E}$ drops, and its motion carries it away from the denser gas in the Galactic plane where most neutron stars are born. Eventually most pulsars will end up with low spin-down luminosities in low density regions, where they may no longer be moving supersonically and their energy output is insufficient to power an observable synchrotron nebula. In this final stage, a pulsar is surrounded by a static or slowly expanding cavity of relativistic material with a radius $\gg 1$ pc and confined by the thermal pressure of the ISM (Blandford et al., 1973; Arons 1983). Deep searches have yet to detect such "ghost nebulae" (e.g., Gaensler et al., 2000).

An alternate evolutionary path may take place for old pulsars in binary systems, which can eventually be spun up via accretion from a companion. This produces a recycled pulsar with a low value of $\dot{P}$ but a very rapid spin period, $P \sim 1 - 10$ ms. Such pulsars can have spin-down luminosities as high as $\dot{E} \approx 10^{34} - 10^{35}$ ergs s^{-1}, which is sufficient to generate observable bow-shock nebulae, as shown in **Figure 4b**.

4. YOUNG PULSAR WIND NEBULAE

4.1. Pulsar Winds

Despite more than 35 years of work on the formation of pulsar winds, there are still large gaps in our understanding. The basic picture is that a charge-filled magnetosphere surrounds the pulsar, and that particle acceleration occurs in the collapse of charge-separated gaps either near the pulsar polar caps or in outer regions that extend to the light cylinder (i.e., to $R_{\rm LC} = c/\Omega$). The maximum potential generated by the rotating magnetic field has been calculated for the case of an aligned rotator (i.e., with the magnetic and spin axes co-aligned) by Goldreich & Julian (1969) as:

$$\Delta\Phi \approx \frac{B_p \Omega^2 R_{NS}^3}{2c} \approx 6 \times 10^{12} \left(\frac{B_p}{10^{12}\ {\rm G}}\right)\left(\frac{R_{NS}}{10\ {\rm km}}\right)^3 \left(\frac{P}{1\ {\rm s}}\right)^{-2} {\rm V}, \qquad (10)$$

where R_{NS} is the neutron star radius. The associated particle current is $\dot{N}_{GJ} \approx (\Omega^2 B_p R_{NS}^3)/Zec$, where Ze is the ion charge. This current, although considerably modified in subsequent models, provides the basis for the pulsar wind. In virtually all models, the wind leaving the pulsar magnetosphere is dominated by the Poynting flux, $F_{E \times B}$, with a much smaller contribution from the particle energy flux, F_{particle}. The magnetization parameter, σ, is defined as (e.g., Kennel & Coroniti 1984):

$$\sigma \equiv \frac{F_{E \times B}}{F_{\text{particle}}} = \frac{B^2}{4\pi\rho\gamma c^2}, \qquad (11)$$

where B, ρ, and γ are the magnetic field, mass density of particles, and Lorentz factor, in the wind, respectively. As the wind flows from the pulsar light cylinder, typical values of $\sigma > 10^4$ are obtained (see Arons 2002). However, models for the structure of the Crab Nebula (Rees & Gunn 1974; Kennel & Coroniti 1984a) require $\sigma \ll 1$ just behind the termination shock (see Section 4.3), in order to meet flow and pressure boundary conditions at the outer edge of the PWN. The high ratio of the synchrotron luminosity to the total spin-down power also requires a particle-dominated wind (Kennel & Coroniti 1984b), and implies $\gamma \sim 10^6$, a value considerably higher than that expected in the freely expanding wind (Arons 2002). Between the pulsar light cylinder and the position of the wind termination shock, the nature of the wind must thus change dramatically, although the mechanism for this transition is as yet unclear (see Melatos 1998; Arons 2002).

The loss of electrons from the polar regions of the star represents a net current that needs to be replenished to maintain charge neutrality. This may occur through ion outflow in equatorial regions. As discussed in Section 4.4, ions may contribute to nonthermal electron acceleration in the inner regions of the PWN.

4.2. Observed Properties of Young Pulsar Wind Nebulae

The deceleration of a pulsar-driven wind as it expands into the confines of cold, slowly expanding supernova ejecta produces a wind termination shock, at which electron/positron pairs are accelerated to ultrarelativistic energies (see Section 4.3). As these particles move through the wound-up magnetic field that comprises the PWN, they produce synchrotron radiation from radio wavelengths to beyond the X-ray band. For a power-law electron spectrum, the constant injection of particles plus a finite synchrotron-emitting lifetime lead to a spectral break at a frequency (Ginzburg & Syrovatskii 1965):

$$\nu_b = 10^{21} \left(\frac{B_{\text{PWN}}}{10^{-6}\,\text{G}}\right)^{-3} \left(\frac{t}{10^3\,\text{years}}\right)^{-2}\,\text{Hz}, \qquad (12)$$

where B_{PWN} is the nebular magnetic field strength. Particles radiating at frequencies beyond ν_b reach the outer portions of the PWN in ever-diminishing numbers; most radiate their energy before they are able to travel that far. The result is that the size of the PWN decreases with increasing frequency, as is clearly observed in the Crab Nebula (**Figure 1**). For PWNe with low magnetic fields, the synchrotron loss times

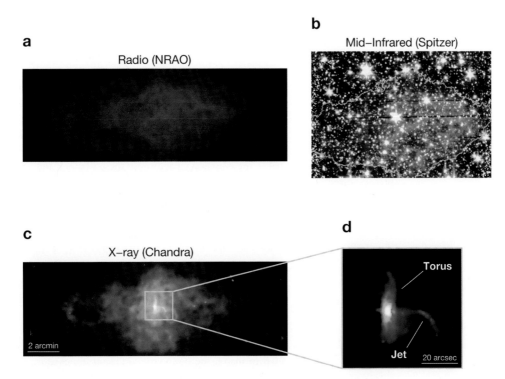

Figure 5

Images of the pulsar wind nebula 3C 58 (G130.7+3.1). (*a*) Radio synchrotron emission from the confined wind, with filamentary structure (Reynolds & Aller 1988). (*b*) Infrared synchrotron emission with morphology similar to the radio nebula (whose outer contour is shown in *green*). (*c*) X-ray synchrotron emission (*blue*), thermal emission (*red*) from shock-heated ejecta, and central torus/jet structure, shown expanded in panel *d* (Slane et al., 2004). Images are shown with north up and east to the left. The scale for the figures is indicated by the 2 arcmin scale bar except for panel *d*, where the 20 arcsec scale bar applies.

are longer and there may not be a significant difference in size between the radio and higher frequency bands (e.g., 3C 58 in **Figure 5**).

The morphology of a young PWN is often elongated along the pulsar spin axis due to the higher equatorial pressure associated with the toroidal magnetic field (Begelman & Li 1992; van der Swaluw 2003). This effect is seen clearly in many PWNe (e.g., **Figures 1** and **5**) and allows one to infer the likely projected orientation of the pulsar. As the nebula expands (see Section 3.1), Rayleigh-Taylor instabilities form as the fast-moving relativistic fluid encounters and accelerates slower-moving unshocked supernova ejecta. These form dense, finger-like filamentary structures that suffer photoionization from the surrounding synchrotron emission and radiate recombination lines in the optical and ultraviolet (UV) bands (**Figure 1***b*; Hester et al., 1996). The increased density compresses the magnetic field around the filaments, causing enhanced synchrotron emission. One thus observes radio structures that correspond to the optical/UV filaments.

At the core of the PWN lies the pulsar itself. As its free-flowing equatorial wind encounters the more slowly-expanding nebula, a termination shock is formed. Particles accelerated at the shock form a toroidal structure, while some of the flow is collimated at speeds of $\sim 0.5c$ along the rotation axis, possibly contributing to the formation of jet-like structures (Bogovalov et al., 2005). These structures generate synchrotron radiation that is observed most readily in the X-ray band (**Figure 1d**), although a toroid is also observed in optical images of the Crab Nebula (Hester et al., 1995). The emission pattern from jets or ring-like structures, as well as the larger scale geometry of the PWN, thus provides an indication of the pulsar's orientation. As we discuss in Section 4.4, the emission structures in the postshock and jet regions provide direct insight on particle acceleration, magnetic collimation, and the magnetization properties of the winds in PWNe. In addition, for pulsars whose proper motion is known, constraints on mechanisms for producing this population's high-velocity birth kicks can be derived based on the degree of alignment between the pulsar spin axis and the direction of motion (Lai, Chernoff & Cordes 2001; Ng & Romani 2004).

4.3. The Wind Termination Shock

The highly relativistic pulsar wind and its wound-up toroidal magnetic field inflate an expanding bubble whose outer edge is confined by the expanding shell of SN ejecta. As the wind is decelerated to match the boundary condition imposed by the more slowly-expanding ambient material at the nebula radius, a wind termination shock is formed at the radius, R_w, at which the ram pressure of the wind is balanced by the internal pressure of the PWN (**Figure 2b**):

$$R_w = \sqrt{\dot{E}/(4\pi\omega c \mathcal{P}_{\rm PWN})}, \tag{13}$$

where ω is the equivalent filling factor for an isotropic wind, and $\mathcal{P}_{\rm PWN}$ is the total pressure in the shocked nebular interior. Upstream of the termination shock, the particles do not radiate, but flow relativistically along with the frozen-in magnetic field. At the shock, particles are thermalized and reaccelerated, producing synchrotron emission in the downstream flow. From estimates of the field strength, the observed X-ray emission implies Lorentz factors $\gamma \gtrsim 10^6$ in the shock.

A reasonable pressure estimate can be obtained by integrating the broad-band spectrum of the PWN, using standard synchrotron theory (Ginzburg & Syrovatskii 1965), and assuming equipartition between particles and the magnetic field. Typical $\mathcal{P}_{\rm PWN}$ and $\dot{E}$ values yield termination shock radii of order 0.1 pc, implying an angular size of several arcsec at distances of a few kiloparsecs. For the Crab Nebula, the equipartition field is $B_{\rm PWN} \approx 300$ μG (Trimble 1982), and the pressure reaches equipartition at a radius of $\sim(5-20)R_w$ (Kennel & Coroniti 1984a). The associated spin-down luminosity, $\dot{E} = 5 \times 10^{38}$ ergs s^{-1}, yields $R_w \sim 4 \times 10^{17}$ cm, consistent with the position of the optical wisps and the radius of the X-ray ring seen in **Figure 1d**. Similar calculations indicate a much weaker field, $B_{\rm PWN} \approx 80$ μG in 3C 58 (Green & Scheuer 1992), yielding a termination shock radius similar to the

Crab, $R_w \sim 6 \times 10^{17}$ cm, given the smaller $\dot{E}$ of the pulsar. Indeed structure that possibly corresponds to the termination shock is seen at this separation from the pulsar (Slane, Helfand & Murray 2002). This lower field strength is also consistent with the fact that the observed size of 3C 58 is similar in the radio and X-ray bands (**Figure 5**).

It must be noted that high resolution X-ray observations of 3C 58 (Slane, Helfand & Murray 2002), G21.5–0.9 (Camilo et al., 2006), G292.0+1.8 (Hughes et al., 2001), and many other young PWNe and SNRs do not reveal directly the ring-like emission that is observed just outside the termination shock in the Crab pulsar, as seen in **Figure 1d**. Rather, the compact emission around the pulsar appears slightly extended (see **Figure 5d**), possibly originating from regions similar to the Crab Nebula's torus, downstream from the termination shock. However, the extent of such emission still provides a lower limit on $\mathcal{P}_{PWN}$, as well as an indication of the pulsar orientation.

4.4. Formation of Tori, Jets, and Wisps

The geometry implied by the X-ray morphology of the Crab Nebula (see **Figure 1d**) is a tilted torus, with a jet of material that flows along the toroid axis, extending nearly 0.25 pc from the pulsar. A faint counter-jet accompanies the structure, and the X-ray emission is significantly enhanced along one edge of the torus. Both effects are presumably the result of Doppler beaming of the outflowing material, whereby the X-ray intensity, I, varies with viewing angle as (Pelling et al., 1987):

$$\frac{I}{I_0} = \left[\frac{\sqrt{1-\beta^2}}{1-\beta\cos\phi}\right]^{\Gamma+1}, \qquad (14)$$

where βc is the flow speed immediately downstream of the termination shock, ϕ is the angle of the flow to the line of sight, and I_0 is the unbeamed intensity.

Similar geometric structures are observed in G54.1+0.3, for which *Chandra* observations reveal a central point-like source surrounded by an X-ray ring whose geometry suggests an inclination angle of about 45° (Lu et al., 2002). The X-ray emission is brightest along the eastern limb of the ring. If interpreted as the result of Doppler boosting, this implies a postshock flow velocity of $\sim 0.4c$. The ring is accompanied by faint bipolar elongations aligned with the projected axis of the ring, consistent with the notion that these are jets of shocked material flowing along the pulsar rotation axis. The total luminosity of these structures is similar to that of the central ring. This is to be contrasted with the Crab and 3C 58 (**Figures 1** and **5**), for which the torus outshines the jets by a large factor. Moreover, for G54.1+0.3 the brighter portion of the outflow lies on the same side of the pulsar as the brightest portion of the ring, which is inconsistent with Doppler boosting. A similarly troubling observation is that the brightness distribution around the inner ring of the Crab also fails to show Doppler brightening consistent with that seen in its surrounding torus; the brightness is reasonably uniform except for some compact structures that vary in position and brightness with time (see Section 4.5).

The formation of these jet/torus structures can be understood as follows. Outside the pulsar magnetosphere, the particle flow is radial. The rotation of the pulsar forms an expanding, toroidal, alternating magnetic field for which the Poynting flux varies as $\sin^2 \psi$, where ψ is the angle from the rotation axis. Conservation of energy flux along flow lines leads to a latitude dependence of the Lorentz factor of the wind of the form $\gamma = \gamma_0 + \gamma_m \sin^2 \psi$ (Bogovalov & Khangoulyan 2002), where γ_0 is the wind Lorentz factor just outside the light cylinder ($\gamma_0 \sim 10^2$ in standard models for pulsar winds; e.g., Daugherty & Harding 1982), and γ_m is the maximum Lorentz factor of the preshock wind particles ($\gamma_m \sim 10^6$ near the termination shock; Kennel & Coroniti 1984a). From Equation 11, we see that this corresponds to a latitude variation in the magnetization parameter also, with σ much larger at the poles than at the equator. This anisotropy results in the toroidal structure of the downstream wind. Moreover, modeling of the flow conditions across the shock shows that magnetic collimation produces jet-like flows along the rotation axis (Komissarov & Lyubarsky 2004; Bogovalov et al., 2005). This collimation is highly dependent on the magnetization of the wind. For $\sigma \gtrsim 0.01$, magnetic hoop stresses are sufficient to divert the toroidal flow back toward the pulsar spin axis, collimating and accelerating the flow to speeds of $\sim 0.5c$ (Del Zanna, Amato & Bucciantini 2004); smaller values of σ lead to an increase in the radius at which the flow is diverted. The launching points of the jets observed in the Crab Nebula, which appear to form much closer to the pulsar than the observed equatorial ring, apparently reflect such variations in the value of σ. Near the poles σ is large, resulting in a small termination shock radius and strong collimation, while near the equator σ is much smaller and the termination shock extends to larger radii (Bogovalov & Khangoulyan 2002).

Because γ is smaller at high latitudes, models for the brightness of the jet-like flows produced by the collimation process fall short of what is observed. However, kink instabilities in the toroidal field may transform magnetic energy into particle energy, thus accelerating particles and brightening the jet (Bogovalov et al., 2005). Such instabilities limit the duration of the collimation (Begelman 1998). Evidence for the effects of kink instabilities is seen in the curved nature of many PWN jets, particularly for the Vela PWN where the jet morphology is observed to change dramatically on timescales of months (Pavlov et al., 2003). There appears to be a wide variation in the fraction of $\dot{E}$ channeled into PWN jets, ranging from $\sim 2.5 \times 10^{-5}$ in 3C 58 to nearly 10^{-3} for the PWN powered by PSR B1509-58, based on their synchrotron spectra (Gaensler et al., 2002; Slane et al., 2004). This apparently indicates considerable differences in the efficiency with which additional acceleration occurs along the jets.

The torus surrounding the Crab pulsar is characterized by the presence of wisp-like structures (see **Figure 1d**), whose position and brightness vary with time in the optical, infrared, and X-ray bands, and which emanate from the termination shock and move outward through the torus with inferred outflow speeds of $\sim 0.5c$ (Hester et al., 2002; Melatos et al., 2005). The exact nature of these structures is not fully understood. Hester et al. (2002) suggest that they are formed by synchrotron instabilities. However, the position of the arc-like structure surrounding PSR B1509-58 seems inconsistent with this hypothesis given the much lower magnetic field in the PWN (Gaensler et al., 2002). An alternative suggestion is that the wisps are the sites of compression of the electron/positron pair plasma on scales of the cyclotron gyration radius of ions in the outflow (∼0.15 pc in the Crab Nebula), which can also explain the radius of the X-ray arc seen around PSR B1509-58 (Gallant & Arons 1994; Spitkovsky & Arons 2004, and references therein).

It is worth noting that VLA observations of the Crab Nebula show variable radio structures very similar to the optical and X-ray wisps, indicating that acceleration of the associated particles must be occurring in the same region as for the X-ray–emitting population (Bietenholz et al., 2004).

4.5. Filamentary and Compact Structures in Pulsar Wind Nebulae

In the Crab Nebula, an extensive network of filaments is observed in Hα, [O$_{\rm III}$] and other optical lines, surrounding the nonthermal optical emission from the PWN (**Figure 1b**). The detailed morphology and ionization structure of these filaments indicate that they form from Rayleigh-Taylor instabilities as the expanding relativistic bubble sweeps up and accelerates slower moving ejecta (Hester et al., 1996). Magnetohydrodynamic simulations support this picture, indicating that 60–75% of the swept-up mass can be concentrated in such filaments (Jun 1998; Bucciantini et al., 2004). As the expanding PWN encounters these filaments, compression increases the density and magnetic field strength, forming sheaths of enhanced synchrotron emission observed as a corresponding shell of radio filaments (Reynolds 1988). X-ray observations reveal no such filaments in the Crab Nebula (Weisskopf et al., 2000). This is presumably because the higher energy electrons required to produce the X-ray emission suffer synchrotron losses before reaching the outer regions of the PWN, consistent with the smaller extent of the nebula in X rays than in the radio.

A different picture is presented by X-ray observations of 3C 58, which reveal a complex of loop-like filaments most prominent near the central regions of the PWN, but evident throughout the nebula (**Figure 5c**; Slane et al., 2004). These nonthermal X-ray structures align with filaments observed in the radio band (Reynolds & Aller 1988). Optical observations of 3C 58 also reveal faint filaments (van den Bergh 1978), whose origin is presumably similar to those in the Crab Nebula. However, a detailed X-ray/optical comparison shows that most of the X-ray filaments do not have optical counterparts. Although comparisons with deeper optical images are clearly needed, the fact that many of the X-ray features without optical counterparts are brighter than average suggests that these may actually arise from a different mechanism. Slane et al. (2004) propose that the bulk of the discrete structures seen in the X-ray and radio images of 3C 58 are magnetic loops torn from the toroidal field by kink instabilities. In the inner nebula, the loop sizes are similar to the size of the termination shock radius (∼0.1 pc), as suggested by Begelman (1998). As the structures expand, they enlarge slightly as a consequence of decreasing pressure.

There is also considerable loop-like filamentary structure evident in high resolution X-ray images of the Crab Nebula (**Figure 1d**; Weisskopf et al., 2000). These filaments appear to wrap around the torus, perpendicular to the toroidal plane, and may be signatures of kink instabilities in the termination shock region.

In some PWNe, compact knot-like structures are observed close to the pulsar, which dissipate and reappear on timescales of order months (Hester et al., 2002; Melatos et al., 2005). Examples can be seen for the Crab in **Figure 1d**, and similar structures are also seen for PSR B1509-58 (Gaensler et al., 2002). Some of these features appear in projection inside the termination shock region. However, it is believed that they actually correspond to unstable, quasi-stationary shocks in the region just outside the termination shock, at high latitudes where the shock radius is small due to larger values of σ (e.g., Komissarov & Lyubarsky 2004).

4.6. Pulsar Wind Nebula Spectra

As noted in Section 2.2, the spectra of PWNe are characterized by a flat power law index at radio wavelengths ($\alpha \approx -0.3$) and a considerably steeper index in X rays ($\Gamma \approx 2$). (Recall that $\Gamma \equiv 1 - \alpha$.) The nature of this spectral steepening is not understood. Simple assumptions of a power-law particle spectrum injected by the pulsar would predict a power-law synchrotron spectrum, with a break associated with the aging of the particles (see Equation 12); the expected increase in spectral index is $\Delta \alpha = 0.5$, which is smaller than what is typically observed (Woltjer et al., 1997). Moreover, for many PWNe a change in spectral index is inferred at low frequencies that would imply unrealistically high magnetic fields (e.g., Green & Scheuer 1992). Relic breaks in the spectrum can be produced by a rapid decline in the pulsar output over time, and these breaks propagate to lower frequencies as the PWN ages (Pacini & Salvati 1973). The inherent spectrum of the injected particles, which may deviate from a simple power law, as well as modifications from discrete acceleration sites all contribute to a complicated integrated spectrum. As a result, the interpretation of spectral steepening as being due to synchrotron losses can lead to drastically wrong conclusions about PWN properties.

At frequencies for which the synchrotron lifetime is shorter than the flow time to the edge of the PWN, a steepening of the spectrum with radius is expected. Radial steepening is indeed observed in the X-ray spectra of the Crab Nebula, 3C 58, and G21.5–0.9 (Slane et al., 2000; Willingale et al., 2001; Slane et al., 2004), but the spectra steepen rather uniformly with radius, whereas generalizations of the Kennel & Coroniti (1984b) model predict a much more rapid steepening near the outer regions (Reynolds 2003). Some mixing of electrons of different ages at each radius seems to be required, perhaps due to diffusion processes in the PWN.

5. BOW SHOCKS AROUND HIGH VELOCITY PULSARS

Many pulsars are born with high space velocities, typically $V_{PSR} = 400–500$ km s^{-1}, but for some sources exceeding 1000 km s^{-1}. These high velocities are almost certainly the result of kicks given to the star during or shortly after core collapse (Lai 2004). Young pulsars thus have the highest velocities of any stellar population, and many have sufficient speeds to eventually escape the Galaxy.

As discussed in Section 3.3, a pulsar's ballistic motion allows it to eventually escape its original PWN, and to propagate through the shocked ejecta in the SNR interior.

At first the pulsar's motion will be subsonic in this hot gas, but by the time the pulsar nears the edges of the SNR, the sound speed drops sufficiently for the pulsar's motion to be supersonic. In the simplest situation of a spherical SNR in the Sedov phase, expanding into a uniform medium, this transition occurs at half the crossing time (given in Equation 15 below), at which point the pulsar has traveled 68% of the distance from the center of the SNR to its edge (van der Swaluw, Downes & Keegan 2004). This simple result is independent of V_{PSR}, n_0, or E_{SN}. The pulsar's supersonic motion now produces a PWN with a bow shock morphology.

Because the SNR is decelerating, the pulsar ultimately penetrates and then escapes the shell. A pulsar moving at $V_{PSR} \gtrsim 650$ km s^{-1} will escape while the SNR is still in the Sedov phase, after a time (van der Swaluw et al., 2003):

MHD: magnetohydrodynamic

$$t_{cross} = 44 \left(\frac{E_{SN}}{10^{51} \text{ergs}} \right)^{1/3} \left(\frac{n_0}{1 \text{ cm}^{-3}} \right)^{-1/3} \left(\frac{V_{PSR}}{500 \text{ km s}^{-1}} \right)^{-5/3} \text{kyr}. \quad (15)$$

At times $t > t_{cross}$, a pulsar proceeds to move through the ambient ISM.

The speed of sound in interstellar gas is a function of temperature: typical values are approximately 1, 10, and 100 km s^{-1} for the cold, warm, and hot components of the ISM, respectively. Thus, except in the case of a particularly slow moving pulsar moving through coronal gas, a pulsar will move supersonically and drive a bow shock through the ISM.

5.1. Theoretical Expectations

The pulsar wind in a bow shock is decelerated at a termination shock, just as for younger PWNe (see Section 4.3). However, the external source of pressure balance is now ram pressure from the pulsar's motion, rather than the internal pressure of the shocked wind. Furthermore, because ram pressure is not isotropic, the termination shock radius varies as a function of angle with respect to the pulsar's velocity vector. In the direction of the star's motion, the termination shock radius is referred to as the "stand-off distance," R_{w0}, and is defined by (cf. Equation 13):

$$\frac{\dot{E}}{4\pi \omega R_{w0}^2 c} = \rho_0 V_{PSR}^2, \quad (16)$$

where ρ_0 is the ambient density. If the wind is isotropic and $\omega = 1$, then at a polar angle θ with respect to the bow shock's symmetry axis, the analytic solution for the termination shock radius as a function of position is (Wilkin 1996):

$$R_w(\theta) = R_{w0} \csc \theta \sqrt{3(1 - \theta \cot \theta)}. \quad (17)$$

It is important to note that the above solution assumes an efficiently cooled thin-layer shock, in contrast to the double shock expected for pulsar bow shocks. Full hydrodynamic and MHD simulations show that Equation 17 is a reasonable approximation in regions near the apex ($\theta \lesssim \frac{\pi}{2}$), but performs more poorly further downstream (Bucciantini 2002a; van der Swaluw et al., 2003).

A result of such simulations is shown in **Figure 6a**. The double-shock structure is clearly apparent, consisting of a forward shock where the ISM is heated, plus the

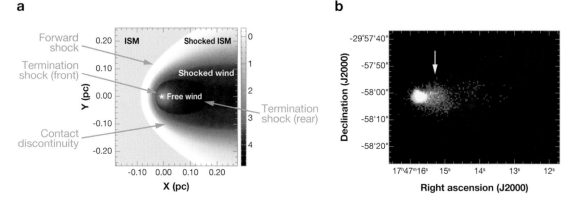

Figure 6

(*a*) A hydrodynamic simulation of a pulsar bow shock, adapted from Gaensler et al. (2004). The pulsar, whose position is marked with an asterisk, is moving from right to left with a Mach number $\mathcal{M} = 60$. The intensity in the image and the scale bar indicate density, in units of $\log_{10}(\rho_0/10^{-24}\ \mathrm{g\ cm^{-3}})$. (*b*) *Chandra* X-ray (*blue*) and VLA radio (*red*) images of G359.23−0.82 ("the Mouse"), the bow shock associated with PSR J1747-2958 (Gaensler et al., 2004). The white arrow marks a bright compact region of X-ray emission behind the apex, which possibly corresponds to the surface of the termination shock.

termination shock where the pulsar's wind decelerates. As expected, the termination shock is not of uniform radius around the pulsar: specifically, for low Mach numbers, $\mathcal{M} \sim 1 - 3$ [as may be appropriate for pulsars traveling supersonically inside their SNRs; see Section 3.3 and van der Swaluw, Downes & Keegan (2004)], the ratio of termination shock radii between polar angles $\theta = \pi$ and $\theta = 0$ is approximately $\mathcal{M}$ (Bucciantini 2002a; van der Swaluw et al., 2003), but for $\mathcal{M} \gg 1$ (typical of bow shocks in the ambient ISM; see Section 3.4), this ratio approaches a limit of $\sim 5 - 6$ (Gaensler et al., 2004; Bucciantini, Amato & Del Zanna 2005).

5.2. Observations: Forward Shock

If a pulsar drives a bow shock through neutral gas, then collisional excitation and charge exchange occur at the forward shock, generating optical emission in the Balmer lines (Bucciantini & Bandiera 2001; Bucciantini 2002b). Indeed several pulsar bow shocks have been identified in the 656-nm Hα line, associated with (in order of discovery) B1957+20 (**Figure 4*b***; Kulkarni & Hester 1988), B2224+65 ("the Guitar"; Cordes, Romani & Lundgren 1993), J0437-4715 (Bell et al., 1995), RX J1856.5-3754 (van Kerkwijk & Kulkarni 2001), B0740-28 (Jones, Stappers & Gaensler 2002), and J2124-3358 (Gaensler, Jones & Stappers 2002).

If the distance to the system is known, R_{w0} can be directly measured, provided that one adopts a scaling factor of $\sim 0.4 - 0.6$ to translate between the observed radius of the forward shock to that of the termination shock (Bucciantini 2002a; van der Swaluw et al., 2003). If $\dot{E}$ and V_{PSR} have been measured, Equation 16 can then be applied to yield ρ_0. This is an approximation, because $\omega = 1$ is usually assumed, and because of the unknown inclination of the pulsar's motion to the line of sight,[4] but certainly suggests ambient number densities ~ 0.1 cm^{-3}, as expected for warm neutral ambient gas (Chatterjee & Cordes 2002; Gaensler, Jones & Stappers 2002). For pulsars for which V_{PSR} is not known, one can write:

$$\rho_0 V_{\text{PSR}}^2 = \gamma_1 \mathcal{M}^2 \mathcal{P}_{\text{ISM}}, \qquad (18)$$

where $\gamma_1 = 5/3$ and $\mathcal{P}_{\text{ISM}}$ are the adiabatic coefficient and pressure of the ISM, respectively. Estimates for $\mathcal{P}_{\text{ISM}}$ and ω in Equations 16 and 18 then yield $\mathcal{M}$.

For the bow shocks associated with RX J1856.5-3754, PSR J0437-4715, and PSR B1957+20 (**Figure 4b**), the shape of the forward shock is a good match to the solution predicted by Equation 17 and simulated in **Figure 6a**. However, the optical emission around PSRs B2224+65, J2124-3358, and B0740-28 all show strong deviations from the expected shape in that there are abrupt kinks and inflection points in the Hα profile (Jones, Stappers & Gaensler 2002; Chatterjee & Cordes 2002). Furthermore, in the case of PSR J2124-3358 there is an apparent rotational offset between the symmetry axis of the bow shock and the velocity vector of the pulsar (Gaensler, Jones & Stappers 2002). These systems imply the presence of some combination of anisotropies in the pulsar wind (as are observed in young pulsars; see Section 4.4), gradients and fluctuations in the density of the ISM, or a bulk velocity of ambient gas with respect to the pulsar's local standard of rest. Extensions of Equation 17 to account for these effects have been presented by Bandiera (1993) and by Wilkin (2000), and have been applied to interpret the morphology of PSR J2124-3358 by Gaensler, Jones & Stappers (2002).

5.3. Observations: Termination Shock

Just as for the PWNe around the youngest pulsars discussed in Section 4.3, particles in the pulsar wind inside a bow shock will be accelerated at the termination shock, producing nonthermal synchrotron emission that can be potentially observed in the radio and X-ray bands. Indeed cometary radio and X-ray PWNe aligned with the direction of motion have been now identified around many pulsars, convincing examples of which include PSRs B1853+01 (**Figure 4a**; Frail et al., 1996; Petre, Kuntz & Shelton 2002), B1957+20 (**Figure 4b**; Stappers et al., 2003), and B1757-24 ("the Duck"; Frail & Kulkarni 1991; Kaspi et al., 2001).

The most spectacular example of this class is G359.23–0.82, the X-ray/radio bow shock powered by PSR J1747-2958 ("the Mouse"; Yusef-Zadeh & Bally 1987; Gaensler et al., 2004), multi-wavelength observations of which are shown in **Figure 6b**. The extent of the radio trail in this system is larger than in X rays, reflecting the difference in synchrotron lifetimes between these bands (cf. Section 4.2).

The X-ray morphology of the Mouse in **Figure 6b** appears to consist of two components: a bright compact region extending ~ 0.2 pc from the pulsar, superimposed

---

[4]Note that the correction factor for motion inclined to the line of sight cannot be derived through simple trigonometry (Gaensler, Jones & Stappers 2002).

on a larger fainter component extending ∼1 pc from the pulsar. Comparison with the hydrodynamic simulation in **Figure 6a** suggests that the bright component of the X-ray emission corresponds to the surface of the wind termination shock, while fainter, more extended X-ray emission originates from the shocked wind (Gaensler et al., 2004). In this identification, the bright X-ray component of this bow shock corresponds to the inner toroidal rings discussed in Section 4.4, but are elongated due to the pulsar motion. This interpretation is supported by the fact that the extent of the termination shock region along the symmetry axis (in units of R_{w0}) should be a function of Mach number, as was discussed in Section 5.1. Indeed for the Mouse, with $\mathcal{M} \sim 60$ (Gaensler et al., 2004), this bright region is about twice as long (relative to R_{w0}) as for the pulsar bow shock seen inside the SNR IC 443 (Olbert et al., 2001), which must have $\mathcal{M} \lesssim 3$ because it is moving through shocked gas in the SNR interior (van der Swaluw et al., 2003).

In PWNe for PSRs B1757-24 (Kaspi et al., 2001) and B1957+20 (**Figure 4b**; Stappers et al., 2003), only a short (∼0.1 pc) narrow X-ray trail is apparent. Such features have been interpreted as a rapid back-flow or nozzle, which transports particles downstream (Wang, Li & Begelman 1993; Kaspi et al., 2001). However, **Figure 6** suggests that the short trail seen behind PSRs B1757-24 and B1957+20 is the surface of the termination shock, and that emission from the post-shock wind further downstream is too faint to see (Gvaramadze 2004; Gaensler et al., 2004). Deeper X-ray observations are required to test this possibility.

Hydrodynamic models predict that the pulsar's motion should divide the post-shock emitting region of a bow shock into two distinct zones: a highly magnetized broad tail originating from material shocked at $\theta \lesssim \frac{\pi}{2}$, plus a more weakly magnetized, narrow, collimated tail, produced by material flowing along the axis $\theta \approx \pi$ (Bucciantini 2002a; Romanova, Chulsky & Lovelace 2005). These two structures are both apparent in the Mouse (Gaensler et al., 2004). Through relativistic MHD simulation, this and other issues related to the structure of the post-shock flow are now being explored (e.g., Bucciantini, Amato & Del Zanna 2005).

6. OTHER TOPICS AND RECENT RESULTS

6.1. Pulsars and Pulsar Wind Nebulae in Very Young Supernova Remnants

The youngest known Galactic PWNe are the Crab Nebula and 3C 58, powered by pulsars thought to correspond to the SNe of 1054 CE and 1181 CE, respectively.[5] It would be of great interest to identify PWNe at earlier evolutionary stages.

SN 1987A formed a neutron star, but deep searches have failed to detect a central object, down to a luminosity $\lesssim 10^{34}$ ergs s^{-1} (e.g., Graves et al., 2005). This is well below the luminosity of the Crab Nebula, and may indicate that the central neutron

---

[5]The association of 3C 58 with 1181 CE is not completely secure; see Stephenson & Green (2002) and Chevalier (2005), and references therein.

IC: inverse Compton

star has collapsed further into a black hole, accretes from fall-back material, or does not generate a wind (e.g., Fryer, Colgate & Pinto 1999).

Searches for PWNe in other extragalactic SNe and SNRs have generally not produced any convincing candidates (Reynolds & Fix 1987; Bartel & Bietenholz 2005). However, recent high resolution radio images have revealed the gradual turn-on of a central flat-spectrum radio nebula in SN 1986J, which may correspond to emission from a very young PWN (Bietenholz, Bartel & Rupen 2004). New wide-field radio and X-ray images of other galaxies may lead to further identification of young PWNe, whereas optical and UV spectroscopy of recent SNe may identify PWNe through emission lines that broaden with time as gas is swept up by the expanding pulsar wind (Chevalier & Fransson 1992).

6.2. Winds from Highly Magnetized Neutron Stars

A growing population of neutron stars have surface magnetic fields (inferred via Equation 4) above the quantum critical limit of $B_p = 4.4 \times 10^{13}$ G. The properties of these stars indicate that they are comprised of two apparently distinct populations: the high-field radio pulsars (McLaughlin et al., 2003, and references therein), and the exotic magnetars (Woods & Thompson 2006). The winds and PWNe of these sources potentially provide a view of different spin-down processes than those seen in normal pulsars.

Most of the high-field radio pulsars have $\dot{E} \approx 10^{32} - 10^{34}$ ergs s^{-1}. For typical efficiency factors $\eta_R, \eta_X \lesssim 10^{-3}$, this implies PWNe too faint to be detectable. However, the very young pulsars J1846-0258 ($B_p = 4.9 \times 10^{13}$ G) and J1119-6127 ($B_p = 4.1 \times 10^{14}$ G) have high spin-down luminosities ($\dot{E} > 10^{36}$ ergs s^{-1}) and are near the centers of SNRs. In both cases PWNe are detected, although with very different properties. PSR J1846-0258 puts a large fraction ($\eta_X \sim 0.2$) of its spin-down power into a luminous X-ray PWN $\sim$2 pc in extent (Helfand, Collins & Gotthelf 2003), whereas PSR J1119-6127 powers an under-luminous ($\eta_X \sim 2 \times 10^{-4}$) and small ($\sim$0.2 pc) X-ray nebula (Gonzalez & Safi-Harb 2003). Clearly PWN properties are dominated by factors such as age, environment, and evolutionary state (see Section 3) rather than the associated pulsar's surface magnetic field.

Magnetars also spin down, albeit in some cases not smoothly (Woods et al., 2002). Just as for radio pulsars, this rotational energy output is thought to go into a relativistic wind, but traditional PWNe have not been detected around magnetars,[6] presumably because they have time-averaged spin down luminosities $\dot{E} \lesssim 10^{34}$ ergs s^{-1} for all these sources. Magnetars likely experience an enhanced torque over the dipole spin-down presumed to act in radio pulsars, as a result of either Alfvén waves and outflowing relativistic particles driven by seismic activity, or by a large-scale twist of the external magnetic field (Harding, Contopoulos & Kazanas 1999; Thompson, Lyutikov & Kulkarni 2002). Under either circumstance, the spin-down behavior deviates from that described in Equations 2 to 7 (e.g., Thompson et al., 2000).

---

[6]The magnetar SGR 1806–20 was originally presumed to power the radio nebula G10.0–0.3, but a revision in the position of this neutron star now makes this unlikely (Hurley et al., 1999).

A transient radio PWN was proposed to account for the short-lived radio nebula seen in 1998 following the giant flare from the magnetar SGR 1900+14 (Frail, Kulkarni & Bloom 1999). However, recent observations of a radio nebula in the aftermath of a giant flare from SGR 1806-20 suggest that the synchrotron emission from these nebulae is powered by ejected baryonic material, so that these sources are more analogous to SNRs than to PWNe (Gaensler et al., 2005).[7]

6.3. TeV Observations of Pulsar Wind Nebulae

The Crab Nebula is a well-known source of TeV gamma-rays (Weekes et al., 1989, and references therein). This emission is well explained as inverse Compton emission, the relativistic particles in the shocked wind acting as scattering centers for the synchrotron photons that they themselves emit at lower energies (Atoyan & Aharonian 1996). Under this interpretation, the emitted spectrum can be modeled to provide the mean and spatial distribution of the nebular magnetic field strength, and hence the PWN's particle content, the time-averaged injection rate of particles, and an independent estimate of the magnetization parameter, σ (e.g., de Jager et al., 1996). The estimated values of B_{PWN} and σ are in good agreement with those derived from the MHD model of Kennel & Coroniti (1984a).

The new generation of ground-based Čerenkov detectors (most notably the High Energy Stereoscopic System, HESS) has now begun to detect other PWNe in the TeV band. These detections indicate that acceleration of particles to considerable energies must have occurred and provide estimates of the nebular magnetic field strength, which can be used in modeling and interpreting the other nebular structures discussed in Section 4. The much lower synchrotron luminosities of these other sources compared to the Crab Nebula imply that the seed photons for IC scattering are in these cases primarily external, originating from a combination of the cosmic microwave background and a local contribution from dust and starlight. Recent TeV detections of PWNe by HESS include G0.9+0.1 and G320.4–1.2/PSR B1509-58 (Aharonian et al., 2005a, b). As shown in **Figure 7**, the latter is spatially resolved by HESS and has a TeV morphology that is a good match to the X-ray synchrotron nebula. Such observations can potentially provide direct measurements of spatial variations in the magnetic fields of PWNe.

6.4. Pulsar Winds in Binary Systems

The recently discovered dual-line double pulsar PSR J0737-3039 consists of a 23-ms pulsar ("A") and a 2.8-s pulsar ("B") in a 2.4-hour orbit, viewed virtually edge-on (Lyne et al., 2004). This system is proving to be a remarkable new probe of pulsars and their winds, providing information at much closer separations to the pulsar than is possible for the sources discussed in Sections 4 and 5.

---

[7]A possibly transient radio source has also recently been identified coincident with the flaring magnetar XTE J1810-197, but the nature of this source is as yet unclear (Halpern et al., 2005).

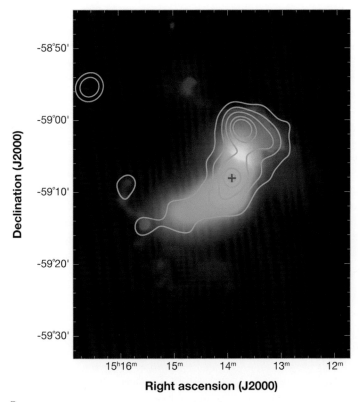

Figure 7

Multiwavelength images of the pulsar wind nebula (PWN) powered by the young pulsar B1509-58. *ROSAT* PSPC data (in *blue contours* at levels of 5%, 10%, 20%, 40%, and 60% of the peak) show the extent of the X-ray PWN (Trussoni et al., 1996), while 843 MHz Molonglo data (in *red*) correspond to the surrounding SNR G320.4–1.2 (Whiteoak & Green 1996). TeV emission from the High Energy Stereoscopic System is shown in green (Aharonian et al., 2005b). The cross marks the position of PSR B1509-58.

The line of sight to pulsar A passes within ≈0.01 light-seconds of pulsar B, well within the unperturbed light cylinder radius of the slower pulsar. For ≈30 s at conjunction, the pulse-averaged radio emission from A is modulated with a complicated time-dependence, showing intermittent periodicities at both 50% and 100% of the rotational period of pulsar B (McLaughlin et al., 2004b). Detailed modeling shows that this behavior can be interpreted as synchrotron absorption from a relativistic plasma confined by a dipolar magnetic field, providing direct evidence for the field geometry commonly adopted in pulsar electrodynamics (Rafikov & Goldreich 2005; Lyutikov & Thompson 2005).

Drifting subpulses in the pulsed emission from pulsar B are also observed, with fluctuations at the beat frequency between the periods of A and B, and with a separation between drifting features corresponding to the period of A (McLaughlin et al., 2004a). This provides clear evidence that pulsed radiation from A is interacting with the magnetosphere of B, supporting the conclusion that pulsar winds are magnetically dominated ($\sigma \gg 1$) in their inner regions, as discussed in Section 4.1.

Other binary systems also provide information on conditions very close to the pulsar. PSR B1957+20 is in a circular 9.2-hour orbit around a ~ 0.025 M$_\odot$ companion, interaction with which produces a termination shock just 1.5×10^{11} cm from the pulsar[8] (Phinney et al., 1988). The X-ray flux from this nebula shows possible modulation at the orbital period (Stappers et al., 2003), as should result from Doppler boosting of the flow around the companion (Arons & Tavani 1993). PSR B1259-63 is in a highly eccentric 3.4-year orbit around a Be star. Near periastron, the pulsar is subject to a time-varying external pressure, producing a transient X-ray/radio synchrotron nebula plus TeV emission from IC scattering of light from the companion star (e.g., Johnston et al., 2005; Aharonian et al., 2005c). At future periastra of this system, coordinated X- and γ-ray observations with INTEGRAL, GLAST, and HESS can directly probe particle acceleration in this pulsar's wind (Tavani & Arons 1997; Kirk, Ball & Skjæraasen 1999).

SUMMARY POINTS

1. A magnetized relativistic wind is the main reservoir for a pulsar's rotational energy loss. The termination of this wind due to surrounding pressure produces a pulsar wind nebula (PWN), usually observed as a synchrotron nebula. A high velocity pulsar can also produce a line-emitting optical bow shock where the pulsar wind shocks surrounding gas.

2. PWNe move through a series of distinct evolutionary states, moderated by the pulsar's location (inside a supernova remnant versus in interstellar gas), the ambient conditions (cold ejecta versus shocked ejecta versus interstellar medium) and the Mach number of the pulsar (subsonic versus mildly supersonic versus highly supersonic).

3. High resolution X-ray observations of young pulsars reveal the imprint of the rotation axis on the morphology of the surrounding PWN, in the form of equatorial tori, polar jets, and overall elongation of the nebula. Using these structures, one can locate the wind termination shock and can infer the composition, flow speed, and geometry of the pulsar's wind.

4. An increasing number of bow shocks are being found around high-velocity pulsars. These systems impose a second axis of symmetry on the PWN, providing additional probes of the pulsar's wind and environment.

ACKNOWLEDGMENTS

We thank NASA for generous support through the LTSA and General Observer programs, and the Radcliffe Institute for hosting a stimulating PWN workshop in

---

[8]On much larger scales, PSR B1957+20 also powers a bow shock, as shown in **Figure 4b**.

2005. We also acknowledge support from an Alfred P. Sloan Research Fellowship (BMG) and from NASA Contract NAS8-03060 (POS). We have used images provided by Tracey DeLaney, Bruno Khélifi, CXC/SAO, NRAO/AUI, and ESO. We thank Shami Chatterjee, Roger Chevalier, Maxim Lyutikov and Steve Reynolds for useful discussions and suggestions. Finally, we thank our other enthusiastic collaborators on PWNe, most notably Jon Arons, Rino Bandiera, Fernando Camilo, Yosi Gelfand, David Helfand, Jack Hughes, Vicky Kaspi, Fred Seward, Ben Stappers, and Eric van der Swaluw.

LITERATURE CITED

Aharonian F, Akhperjanian AG, Aye KM, Bazer-Bachi AR, Beilicke M, et al. 2005a. *Astron. Astrophys.* 432:L25–29

Aharonian F, Akhperjanian AG, Aye KM, Bazer-Bachi AR, Beilicke M, et al. 2005b. *Astron. Astrophys.* 435:L17–20

Aharonian F, Akhperjanian AG, Aye KM, Bazer-Bachi AR, Beilicke M, et al. 2005c. *Astron. Astrophys.* 442:1–10

Arons J. 1983. *Nature* 302:301–5

Arons J. 2002. In *Neutron Stars in Supernova Remnants*, ed. PO Slane, BM Gaensler, pp. 71–80. San Francisco: Astronomical Society of the Pacific

Arons J, Tavani M. 1993. *Ap. J.* 403:249–55

Atoyan AM, Aharonian FA. 1996. *Astron. Astrophys. Suppl.* 120:453–56

Bandiera R. 1993. *Astron. Astrophys.* 276:648–54

Bartel N, Bietenholz MF. 2005. *Adv. Space Res.* 35:1057–64

Becker W, Trümper J. 1997. *Astron. Astrophys.* 326:682–91

Begelman MC. 1998. *Ap. J.* 493:291–300

Begelman MC, Li ZY. 1992. *Ap. J.* 397:187–95

Bell JF, Bailes M, Manchester RN, Weisberg JM, Lyne AG. 1995. *Ap. J.* 440:L81–83

Bietenholz MF, Bartel N, Rupen MP. 2004. *Science* 304:1947–49

Bietenholz MF, Hester JJ, Frail DA, Bartel N. 2004. *Ap. J.* 615:794–804

Blandford RD, Ostriker JP, Pacini F, Rees MJ. 1973. *Astron. Astrophys.* 23:145–46

Blondin JM, Chevalier RA, Frierson DM. 2001. *Ap. J.* 563:806–15

Bogovalov SV, Chechetkin VM, Koldoba AV, Ustyugova GV. 2005. *MNRAS* 358:705–14

Bogovalov SV, Khangoulyan DV. 2002. *Astron. Lett.* 28:373–85

Bucciantini N. 2002a. *Astron. Astrophys.* 387:1066–73

Bucciantini N. 2002b. *Astron. Astrophys.* 393:629–35

Bucciantini N, Amato E, Bandiera R, Blondin JM, Del Zanna L. 2004. *Astron. Astrophys.* 423:253–65

Bucciantini N, Amato E, Del Zanna L. 2005. *Astron. Astrophys.* 434:189–99

Bucciantini N, Bandiera R. 2001. *Astron. Astrophys.* 375:1032–39

Bucciantini N, Blondin JM, Del Zanna L, Amato E. 2003. *Astron. Astrophys.* 405:617–26

Camilo F, Ransom SM, Gaensler BM, Slane PO, Lorimer DR, et al. 2006. *Ap. J.* 637:456–65

Chatterjee S, Cordes JM. 2002. *Ap. J.* 575:407–18
Chevalier RA. 1977. In *Supernovae*, ed. DN Schramm, pp. 53–61. Dordrecht: Reidel
Chevalier RA. 1998. *Mem. Soc. Astron. It.* 69:977–87
Chevalier RA. 2005. *Ap. J.* 619:839–55
Chevalier RA, Fransson C. 1992. *Ap. J.* 395:540–52
Cocke WJ, Disney MJ, Taylor DJ. 1969. *Nature* 221:525–27
Cordes JM, Romani RW, Lundgren SC. 1993. *Nature* 362:133–35
Cox DP. 1972. *Ap. J.* 178:159–68
Daugherty JK, Harding AK. 1982. *Ap. J.* 252:337–47
de Jager OC, Harding AK, Michelson PF, Nel HI, Nolan PL, et al. 1996. *Ap. J.* 457:253–66
Del Zanna L, Amato E, Bucciantini N. 2004. *Astron. Astrophys.* 421:1063–73
Dodson R, Legge D, Reynolds JE, McCulloch PM. 2003. *Ap. J.* 596:1137–41
Duncan AR, Stewart RT, Haynes RF, Jones KL. 1996. *MNRAS* 280:252–66
Frail DA, Giacani EB, Goss WM, Dubner G. 1996. *Ap. J.* 464:L165–68
Frail DA, Kassim NE, Cornwell TJ, Goss WM. 1995. *Ap. J.* 454:L129–32
Frail DA, Kulkarni SR. 1991. *Nature* 352:785–87
Frail DA, Kulkarni SR, Bloom JS. 1999. *Nature* 398:127–29
Frail DA, Scharringhausen BR. 1997. *Ap. J.* 480:364–70
Fryer CL, Colgate SA, Pinto PA. 1999. *Ap. J.* 511:885–95
Gaensler BM, Arons J, Kaspi VM, Pivovaroff MJ, Kawai N, Tamura K. 2002. *Ap. J.* 569:878–93
Gaensler BM, Jones DH, Stappers BW. 2002. *Ap. J.* 580:L137–41
Gaensler BM, Kouveliotou C, Gelfand JD, Taylor GB, Eichler D, et al. 2005. *Nature* 434:1104–6
Gaensler BM, Stappers BW, Frail DA, Moffett DA, Johnston S, Chatterjee S. 2000. *MNRAS* 318:58–66
Gaensler BM, van der Swaluw E, Camilo F, Kaspi VM, Baganoff FK, et al. 2004. *Ap. J.* 616:383–402
Gallant YA, Arons J. 1994. *Ap. J.* 435:230–60
Giacani EB, Dubner GM, Kassim NE, Frail DA, Goss WM, et al. 1997. *Ap. J.* 113:1379–90
Ginzburg VL, Syrovatskii SI. 1965. *Annu. Rev. Astron. Astrophys.* 3:297–350
Gold T. 1969. *Nature* 221:25–27
Goldreich P, Julian WH. 1969. *Ap. J.* 157:869–80
Gonzalez M, Safi-Harb S. 2003. *Ap. J.* 591:L143–46
Gotthelf EV. 2004. In *Young Neutron Stars and Their Environments*, ed. F. Camilo, BM. Gaensler, pp. 225–28. San Francisco: Astron. Soc. Pac.
Graves GJM, Challis PM, Chevalier RA, Crotts A, Filippenko AV, et al. 2005. *Ap. J.* 629:944–59
Green DA. 2004. *Bull. Astron. Soc. India* 32:335–70
Green DA, Scheuer PAG. 1992. *MNRAS* 258:833–40
Gupta Y, Mitra D, Green DA, Acharyya A. 2005. *Curr. Sci.* 89:853–56
Gvaramadze VV. 2004. *Astron. Astrophys.* 415:1073–78
Halpern JP, Gotthelf EV, Becker RH, Helfand DJ, White RL. 2005. *Ap. J.* 632:L29–32

Harding AK, Contopoulos I, Kazanas D. 1999. *Ap. J.* 525:L125–28
Helfand DJ, Collins BF, Gotthelf EV. 2003. *Ap. J.* 582:783–92
Hester JJ, Mori K, Burrows D, Gallagher JS, Graham JR, et al. 2002. *Ap. J.* 577:L49–52
Hester JJ, Scowen PA, Sankrit R, Burrows CJ, Gallagher III JS, et al. 1995. *Ap. J.* 448:240–63
Hester JJ, Stone JM, Scowen PA, Jun B, Gallagher JS, et al. 1996. *Ap. J.* 456:225–33
Hughes JP, Slane PO, Burrows DN, Garmire G, Nousek JA, et al. 2001. *Ap. J.* 559:L153–56
Hurley K, Kouveliotou C, Cline T, Mazets E, Golenetskii S, et al. 1999. *Ap. J.* 523:L37–40
Johnston S, Ball L, Wang N, Manchester RN. 2005. *MNRAS* 358:1069–75
Jones DH, Stappers BW, Gaensler BM. 2002. *Astron. Astrophys.* 389:L1–5
Jun BI. 1998. *Ap. J.* 499:282–93
Kaspi VM, Gotthelf EV, Gaensler BM, Lyutikov M. 2001. *Ap. J.* 562:L163–66
Kaspi VM, Roberts MSE, Harding AK. 2006. In *Compact Stellar X-ray Sources*, ed. WHG Lewin, M van der Klis. Cambridge: CUP In press. **http://arxiv.org/abs/astro-ph/0402136**
Kennel CF, Coroniti FV. 1984a. *Ap. J.* 283:694–709
Kennel CF, Coroniti FV. 1984b. *Ap. J.* 283:710–30
Kirk J, Ball L, Skjæraasen O. 1999. *Astropart. Phys.* 10:31–45
Komissarov SS, Lyubarsky YE. 2004. *MNRAS* 349:779–92
Kulkarni SR, Hester JJ. 1988. *Nature* 335:801–3
Lai D. 2004. In *Cosmic Explosions in Three Dimensions: Asymmetries in Supernovae and Gamma-Ray Bursts*, ed. P Höflich, P Kumar, JC Wheeler, pp. 276–84. Cambridge: CUP
Lai D, Chernoff DF, Cordes JM. 2001. *Ap. J.* 549:1111–18
Livingstone MA, Kaspi VM, Gavriil FP. 2005. *Ap. J.* 633:1095–100
Lu FJ, Wang QD, Aschenbach B, Durouchoux P, Song LM. 2002. *Ap. J.* 568:L49–52
Lyne AG, Burgay M, Kramer M, Possenti A, Manchester RN, et al. 2004. *Science* 303:1153–57
Lyutikov M, Thompson C. 2005. *Ap. J.* 634:1223–41
Manchester RN, Hobbs GB, Teoh A, Hobbs M. 2005. *Astron. J.* 129:1993–2006
Manchester RN, Taylor JH. 1977. *Pulsars* San Francisco: Freeman
Matheson H, Safi-Harb S. 2005. *Adv. Space Res.* 35:1099–1105
McLaughlin MA, Kramer M, Lyne AG, Lorimer DR, Stairs IH, et al. 2004a. *Ap. J.* 613:L57–60
McLaughlin MA, Lyne AG, Lorimer DR, Possenti A, Manchester RN, et al. 2004b. *Ap. J.* 616:L131–34
McLaughlin MA, Stairs IH, Kaspi VM, Lorimer DR, Kramer M, et al. 2003. *Ap. J.* 591:L135–38
Melatos A. 1998. *Mem. Soc. Astron. It.* 69:1009–15
Melatos A. 2004. In *Young Neutron Stars and Their Environments*, ed. F. Camilo, BM. Gaensler, pp. 143–50. San Francisco: Astron. Soc. Pac.
Melatos A, Scheltus D, Whiting MT, Eikenberry SS, Romani RW, et al. 2005. *Ap. J.* 633:931–40

Migliazzo JM, Gaensler BM, Backer DC, Stappers BW, van der Swaluw E, Strom RG. 2002. *Ap. J.* 567:L141–44
Minkowski R. 1942. *Ap. J.* 96:199–213
Ng CY, Romani RW. 2004. *Ap. J.* 601:479–84
Olbert CM, Clearfield CR, Williams NE, Keohane JW, Frail DA. 2001. *Ap. J.* 554:L205–8
Pacini F. 1967. *Nature* 216:567–68
Pacini F, Salvati M. 1973. *Ap. J.* 186:249–65
Pavlov GG, Teter MA, Kargaltsev O, Sanwal D. 2003. *Ap. J.* 591:1157–71
Pelling RM, Paciesas WS, Peterson LE, Makishima K, Oda M, et al. 1987. *Ap. J.* 319:416–25
Petre R, Kuntz KD, Shelton RL. 2002. *Ap. J.* 579:404–10
Phinney ES, Evans CR, Blandford RD, Kulkarni SR. 1988. *Nature* 333:832–34
Rafikov RR, Goldreich P. 2005. *Ap. J.* 631:488–94
Rees MJ, Gunn JE. 1974. *MNRAS* 167:1–12
Reynolds SP. 1988. *Ap. J.* 327:853–58
Reynolds SP. 2003. In *Cosmic Explosions: On the 10th Anniversary of SN 1993J (IAU Colloquium 192)* Suppl., ed. JM Marcaide, KW Weiler, pp. 161–65. Springer: Berlin. **http://arxiv.org/abs/astro-ph/0308483**
Reynolds SP, Aller HD. 1988. *Ap. J.* 327:845–52
Reynolds SP, Chevalier RA. 1984. *Ap. J.* 278:630–48
Reynolds SP, Fix JD. 1987. *Ap. J.* 322:673–80
Richards DW, Comella JM. 1969. *Nature* 222:551–52
Romanova MM, Chulsky GA, Lovelace RVE. 2005. *Ap. J.* 630:1020–28
Seward FD, Gorenstein P, Smith RK. 2006. *Ap. J.* 636:873–80
Shakeshaft JR. 1979. *Astron. Astrophys.* 72:L9
Shull JM, Fesen RA, Saken JM. 1989. *Ap. J.* 346:860–68
Slane P, Chen Y, Schulz NS, Seward FD, Hughes JP, Gaensler BM. 2000. *Ap. J.* 533:L29–32
Slane P, Helfand DJ, van der Swaluw E, Murray SS. 2004. *Ap. J.* 616:403–13
Slane PO, Helfand DJ, Murray SS. 2002. *Ap. J.* 571:L45–49
Spitkovsky A, Arons J. 2004. *Ap. J.* 603:669–81
Staelin DH, Reifenstein EC III. 1968. *Science* 162:1481–83
Stappers BW, Gaensler BM, Kaspi VM, van der Klis M, Lewin WHG. 2003. *Science* 299:1372–74
Stephenson FR, Green DA. 2002. *Historical Supernovae and Their Remnants*. Oxford: Oxford University
Tavani M, Arons J. 1997. *Ap. J.* 477:439–64
Thompson C, Duncan RC, Woods PM, Kouveliotou C, Finger MH, van Paradijs J. 2000. *Ap. J.* 543:340–50
Thompson C, Lyutikov M, Kulkarni SR. 2002. *Ap. J.* 574:332–55
Trimble V. 1982. *Rev. Mod. Phys.* 54:1183–1224
Truelove JK, McKee CF. 1999. *Ap. J. Suppl.* 120:299–326
Trussoni E, Massaglia S, Caucino S, Brinkmann W, Aschenbach B. 1996. *Astron. Astrophys.* 306:581–86

van den Bergh S. 1978. *Ap. J.* 220:L9–10
van der Swaluw E. 2003. *Astron. Astrophys.* 404:939–47
van der Swaluw E, Achterberg A, Gallant YA. 1998. *Mem. Soc. Astron. It.* 69:1017–22
van der Swaluw E, Achterberg A, Gallant YA. 2002. In *Neutron Stars in Supernova Remnants*, ed. PO Slane, BM Gaensler, pp. 135–40. San Francisco: Astron. Soc. Pac.
van der Swaluw E, Achterberg A, Gallant YA, Downes TP, Keppens R. 2003. *Astron. Astrophys.* 397:913–20
van der Swaluw E, Achterberg A, Gallant YA, Tóth G. 2001. *Astron. Astrophys.* 380:309–17
van der Swaluw E, Downes TP, Keegan R. 2004. *Astron. Astrophys.* 420:937–44
van Kerkwijk MH, Kulkarni SR. 2001. *Astron. Astrophys.* 380:221–37
Wang QD, Li ZY, Begelman MC. 1993. *Nature* 364:127–29
Weekes TC, Cawley MF, Fegan DJ, Gibbs KG, Hillas AM, et al. 1989. *Ap. J.* 342:379–95
Weiler KW, Panagia N. 1978. *Astron. Astrophys.* 70:419–22
Weisskopf MC, Hester JJ, Tennant AF, Elsner RF, Schulz NS, et al. 2000. *Ap. J.* 536:L81–84
Whiteoak JBZ, Green AJ. 1996. *Astron. Astrophys. Suppl.* 118:329–43
Wilkin FP. 1996. *Ap. J.* 459:L31–34
Wilkin FP. 2000. *Ap. J.* 532:400–14
Willingale R, Aschenbach B, Griffiths RG, Sembay S, Warwick RS, et al. 2001. *Astron. Astrophys.* 365:L212–17
Woltjer L, Salvati M, Pacini F, Bandiera R. 1997. *Astron. Astrophys.* 325:295–99
Woods PM, Kouveliotou C, Göğüş E, Finger MH, Swank J, et al. 2002. *Ap. J.* 576:381–90
Woods PM, Thompson C. 2006. In *Compact Stellar X-ray Sources*, ed. WHG Lewin, M van der Klis. Cambridge: CUP In press. **http://arxiv.org/abs/astro-ph/0406133**
Yusef-Zadeh F, Bally J. 1987. *Nature* 330:455–58

RELATED RESOURCES

Davidson K, Fesen RA. 1985. *Annu. Rev. Astron. Astrophys.* 23:119–46
Weiler KW, Sramek RA. 1988. *Annu. Rev. Astron. Astrophys.* 26:295–341
Woltjer L. 1972. *Annu. Rev. Astron. Astrophys.* 10:129:58
Internet supernova remnant catalog: **http://www.mrao.cam.ac.uk/surveys/snrs/**
Internet pulsar catalog: **http://www.atnf.csiro.au/research/pulsar/psrcat/**
Internet pulsar wind nebula catalog: **http://www.physics.mcgill.ca/~pulsar/pwncat.html**

X-Ray Properties of Black-Hole Binaries

Ronald A. Remillard[1] and Jeffrey E. McClintock[2]

[1]Kavli Institute for Astrophysics and Space Research, MIT, Cambridge, Massachusetts 02139; email: rr@space.mit.edu

[2]Harvard-Smithsonian Center for Astrophysics, Cambridge, Massachusetts 02138; email: jem@cfa.harvard.edu

Key Words

accretion physics, black holes, general relativity, X-ray sources

Abstract

We review the properties and behavior of 20 X-ray binaries that contain a dynamically-confirmed black hole, 17 of which are transient systems. During the past decade, many of these transient sources were observed daily throughout the course of their typically year-long outburst cycles using the large-area timing detector aboard the *Rossi X-Ray Timing Explorer*. The evolution of these transient sources is complex. Nevertheless, there are behavior patterns common to all of them as we show in a comprehensive comparison of six selected systems. Central to this comparison are three X-ray states of accretion, which are reviewed and defined quantitatively. We discuss phenomena that arise in strong gravitational fields, including relativistically-broadened Fe lines, high-frequency quasi-periodic oscillations (100–450 Hz), and relativistic radio and X-ray jets. Such phenomena show us how a black hole interacts with its environment, thereby complementing the picture of black holes that gravitational wave detectors will provide. We sketch a scenario for the potential impact of timing/spectral studies of accreting black holes on physics and discuss a current frontier topic, namely, the measurement of black hole spin.

1. INTRODUCTION

BH: black hole
BHB: black-hole binary
GR: general relativity
ISCO: innermost stable circular orbit

Oppenheimer & Snyder (1939) made the first rigorous calculation describing the formation of a black hole (BH). The first strong evidence for such an object came from X-ray and optical observations of the X-ray binary Cygnus X–1 (Bolton 1972, Webster & Murdin 1972). Today, a total of 20 similar X-ray binary systems are known that contain a compact object believed to be too massive to be a neutron star or a degenerate star of any kind (i.e., $M > 3\ M_\odot$). These systems, which we refer to as black-hole binaries (BHBs), are the focus of this review.

These 20 dynamical BHs are the most visible representatives of an estimated $\sim 10^8$ to 10^9 stellar-mass BHs that are believed to exist in the Galaxy (e.g., Brown & Bethe 1994; Timmes, Woosley & Weaver 1996). Stellar-mass BHs are important to astronomy in numerous ways. For example, they are one endpoint of stellar evolution for massive stars, and the collapse of their progenitor stars enriches the universe with heavy elements (Woosley, Heger & Weaver 2002). Also, the measured mass distribution for even the small sample featured here is used to constrain models of BH formation and binary evolution (e.g., Fryer & Kalogera 2001; Podsiadlowski, Rappaport & Han 2003). Lastly, some BHBs appear to be linked to the hypernovae believed to power gamma-ray bursts (Israelian, Rebolo & Basri 1999; Brown et al. 2000; Orosz et al. 2001).

The BHBs featured here are mass-exchange binaries that contain an accreting BH primary and a nondegenerate secondary star. For background on X-ray binaries, see Psaltis (2005). For comprehensive reviews on BHBs, see McClintock & Remillard (2005), Tanaka & Shibazaki (1996) and Tanaka & Lewin (1995). X-ray observations of BHBs allow us to gain a better understanding of BH properties and accretion physics. In this review, we emphasize those results that challenge us to apply the predictions of general relativity (GR) in strong gravity (Section 8). Throughout, we make extensive use of the extraordinary database amassed since January 1996 by NASA's *Rossi X-Ray Timing Explorer* (RXTE; Swank 1998).

In an astrophysical environment, a BH is completely specified in GR by two numbers, its mass M and its specific angular momentum or spin $a = J/cM$, where J is the BH angular momentum and c is the speed of light. The spin value is conveniently expressed in terms of a dimensionless spin parameter, $a_* = a/R_g$, where the gravitational radius is $R_g \equiv GM/c^2$. The mass simply supplies a scale, whereas the spin changes the geometry. The value of a_* lies between 0 for a Schwarzschild hole and 1 for a maximally-rotating Kerr hole. A defining property of a BH is its event horizon, the immaterial surface that bounds the interior region of space-time that cannot communicate with the external universe. The event horizon, the existence of an innermost stable circular orbit (ISCO), and other properties of BHs are discussed in many texts (e.g., Shapiro & Teukolsky 1983; Kato, Fukue & Mineshige 1998). The radius of the event horizon of a Schwarzschild BH ($a_* = 0$) is $R_S = 2R_g = 30$ km $(M/10M_\odot)$, the ISCO lies at $R_{ISCO} = 6R_g$, and the corresponding maximum orbital frequency is $\nu_{ISCO} = 220$ Hz $(M/10M_\odot)^{-1}$. For an extreme Kerr BH ($a_* = 1$), the radii of both the event horizon and the ISCO (prograde orbits) are identical, $R_K = R_{ISCO} = R_g$, and the maximum orbital frequency is $\nu_{ISCO} = 1615$ Hz $(M/10 M_\odot)^{-1}$.

2. A CENSUS OF BLACK-HOLE BINARIES AND BLACK-HOLE CANDIDATES

Following the discovery of Cygnus X–1, the second BHB to be identified was LMC X–3 (Cowley et al. 1983). Both sources are persistently bright in X-rays, and their secondaries are massive O/B-type stars (White, Nagase & Parmar 1995). The third identified BHB, A 0620–00, is markedly different (McClintock & Remillard 1986). A 0620–00 was discovered as an X-ray nova in 1975 when it suddenly brightened to an intensity of 50 Crab[1] to become the brightest nonsolar X-ray source ever observed (Elvis et al. 1975). Then, over the course of a year, the X-ray nova decayed back into quiescence to become a feeble (1 μCrab) source (McClintock, Horne & Remillard 1995). Similarly, the optical counterpart faded from outburst maximum by $\Delta V \approx 7.1$ mags to $V \approx 18.3$ in quiescence, thereby revealing the optical spectrum of a K-dwarf secondary.

As of this writing, there are a total of 20 confirmed BHBs and, remarkably, 17 of them are X-ray novae like A 0620–00. They are ordered in the top half of **Table 1** by right ascension (column 1). Column 2 gives the common name of the source (e.g., LMC X–3) or the prefix to the coordinate name that identifies the discovery mission (e.g., XTE J, where a "J" indicates that the coordinate epoch is J2000). For X-ray novae, the third column gives the year of discovery and the number of outbursts that have been observed. The spectral type of the secondary star is given in column 4. Extensive optical observations of this star yield the key dynamical data summarized respectively in the last three columns: the orbital period, the mass function, and the BH mass. Additional data on BHBs are given in tables 4.1 and 4.2 of McClintock & Remillard (2005).

An observational quantity of special interest is the mass function, $f(M) \equiv P_{orb} K_2^3 / 2\pi G = M_1 \sin^3 i / (1 + q)^2$ (see **Table 1**, column 6). The observables on the left side of this equation are the orbital period P_{orb} and the half-amplitude of the velocity curve of the secondary K_2. On the right, the quantity of greatest interest is M_1, the mass of the BH primary (given in column 7); the other parameters are the orbital inclination angle i and the mass ratio $q \equiv M_2/M_1$, where M_2 is the mass of the secondary. The value of $f(M)$ can be determined by simply measuring the radial velocity curve of the secondary star, and it corresponds to the absolute minimum allowable mass of the compact object.

An inspection of **Table 1** shows that 15 of the 20 X-ray sources have values of $f(M)$ that require a compact object with a mass $\gtrsim 3 M_\odot$. This is a widely agreed limit for the maximum stable mass of a neutron star in GR (e.g., Kalogera & Baym 1996). For the remaining five systems, some additional data are required to make the case

---

[1] 1 Crab $= 2.43 \times 10^{-9}$ erg cm^{-2} s^{-1} keV^{-1} = 1.00 mJy (averaged over 2–11 keV) for a Crab-like spectrum with photon index $\Gamma = 2.08$; Koyama et al. (1984).

Table 1 Twenty confirmed black holes and twenty black-hole candidates[a]

Coordinate Name	Common[b] Name/Prefix	Year[c]	Spec.	P_{orb} (hr)	$f(M)$ ($M_\odot$)	M_1 ($M_\odot$)
0422+32	(GRO J)	1992/1	M2V	5.1	1.19±0.02	3.7–5.0
0538–641	LMC X–3	–	B3V	40.9	2.3±0.3	5.9–9.2
0540–697	LMC X–1	–	O7III	93.8[d]	0.13±0.05[d]	4.0–10.0:[e]
0620–003	(A)	1975/1[f]	K4V	7.8	2.72±0.06	8.7–12.9
1009–45	(GRS)	1993/1	K7/M0V	6.8	3.17±0.12	3.6–4.7:[e]
1118+480	(XTE J)	2000/2	K5/M0V	4.1	6.1±0.3	6.5–7.2
1124–684	Nova Mus 91	1991/1	K3/K5V	10.4	3.01±0.15	6.5–8.2
1354–64[g]	(GS)	1987/2	GIV	61.1[g]	5.75±0.30	–
1543–475	(4U)	1971/4	A2V	26.8	0.25±0.01	8.4–10.4
1550–564	(XTE J)	1998/5	G8/K8IV	37.0	6.86±0.71	8.4–10.8
1650–500[h]	(XTE J)	2001/1	K4V	7.7	2.73±0.56	–
1655–40	(GRO J)	1994/3	F3/F5IV	62.9	2.73±0.09	6.0–6.6
1659–487	GX 339–4	1972/10[i]	–	42.1[j,k]	5.8±0.5	–
1705–250	Nova Oph 77	1977/1	K3/7V	12.5	4.86±0.13	5.6–8.3
1819.3–2525	V4641 Sgr	1999/4	B9III	67.6	3.13±0.13	6.8–7.4
1859+226	(XTE J)	1999/1	–	9.2:[e]	7.4±1.1:[e]	7.6–12.0:[e]
1915+105	(GRS)	1992/Q[l]	K/MIII	804.0	9.5±3.0	10.0–18.0
1956+350	Cyg X–1	–	O9.7Iab	134.4	0.244±0.005	6.8–13.3
2000+251	(GS)	1988/1	K3/K7V	8.3	5.01±0.12	7.1–7.8
2023+338	V404 Cyg	1989/1[f]	K0III	155.3	6.08±0.06	10.1–13.4
1524–617	(A)	1974/2	–	–	–	–
1630–472	(4U)	1971/15	–	–	–	–
1711.6–3808	(SAX J)	2001/1	–	–	–	–
1716–249	(GRS)	1993/1	–	14.9	–	–
1720–318	(XTE J)	2002/1	–	–	–	–
1730–312	(KS)	1994/1	–	–	–	–
1737–31	(GRS)	1997/1	–	–	–	–
1739–278	(GRS)	1996/1	–	–	–	–
1740.7–2942	(1E)	–	–	–	–	–
1743–322	(H)	1977/4	–	–	–	–
1742–289	(A)	1975/1	–	–	–	–
1746–331	(SLX)	1990/2	–	–	–	–
1748–288	(XTE J)	1998/1	–	–	–	–
1755–324	(XTE J)	1997/1	–	–	–	–
1755–338	(4U)	1971/Q[l]	–	4.5	–	–
1758–258	(GRS)	1990/Q[l]	–	–	–	–

(*Continued*)

Table 1 Twenty confirmed black holes and twenty black-hole candidates[a]

Coordinate Name	Common[b] Name/Prefix	Year[c]	Spec.	P_{orb} (hr)	$f(M)$ ($M_\odot$)	M_1 ($M_\odot$)
1846−031	(EXO)	1985/1	−	−	−	−
1908+094	(XTE J)	2002/1	−	−	−	−
1957+115	(4U)	−	−	9.3	−	−
2012+381	(XTE J)	1998/1	−	−	−	−

[a]See McClintock & Remillard (2005, and references therein) for columns 3–5, Orosz (2003) for columns 6–7, plus additional references given below.
[b]A prefix to a coordinate name is enclosed in parentheses. The presence/absence of a "J" indicates that the epoch of the coordinates is J2000/B1950.
[c]Year of initial X-ray outburst/total number of X-ray outbursts.
[d]Period and f(M) corrections by A.M. Levine and D. Lin, private communication.
[e]Colon denotes uncertain value or range.
[f]Additional outbursts in optical archives: A 0620(1917) and V404 Cyg (1938, 1956).
[g]Casares et al. 2004; possible alias period of 61.5 h.
[h]Orosz et al. 2004.
[i]Estimated by Kong et al. 2002.
[j]Hynes et al. 2003.
[k]Period confirmed by A.M. Levine and D. Lin, private communication.
[l]"Q" denotes quasi-persistent intervals (e.g., decades), rather than typical outburst.

for a BH (Charles & Coe 2005, McClintock & Remillard 2005). Historically, the best available evidence for the existence of BHs is dynamical, and the evidence for these 20 systems is generally very strong, with cautions for only two cases: LMC X–1 and XTE J1859+226 (see McClintock & Remillard 2005). Thus, assuming that GR is valid in the strong-field limit, we choose to refer to these compact primaries as BHs, rather than as BH candidates.

BHC: black-hole candidate

Figure 1 is a schematic sketch of 16 Milky Way BHBs with reasonably accurate dynamical data. Their diversity is evident: There are long-period systems containing hot and cool supergiants (Cyg X–1 and GRS 1915+105) and many compact systems containing K-dwarf secondaries. Considering all 20 BHBs listed in **Table 1**, only 3 are persistently bright X-ray sources (Cyg X–1, LMC X–1, and LMC X–3). The 17 transient sources include 2 that are unusual. GRS 1915+105 has remained bright for more than a decade since its first known eruption in August 1992. GX 339–4 undergoes frequent outbursts followed by very faint states, but it has never been observed to fully reach the quiescent state (Hynes et al. 2003).

The second half of **Table 1** lists 20 X-ray binaries that lack radial velocity data. In fact, most of them even lack an optical counterpart, and only three have known orbital periods. Nevertheless, they are considered black-hole candidates (BHCs) because they closely resemble BHBs in their X-ray spectral and temporal behavior (Tanaka & Lewin 1995, McClintock & Remillard 2005). Some X-ray and radio characteristics of the BHCs are given in table 4.3 in McClintock & Remillard (2005); also given there is a subjective grade (A, B, or C) indicating the likelihood that a particular system does contain a BH primary. The seven A-grade BHCs are: A 1524–617, 4U 1630–47, GRS 1739–278, 1E 1740.7–2942, H 1743–322, XTE J1748–288, and GRS 1758–258.

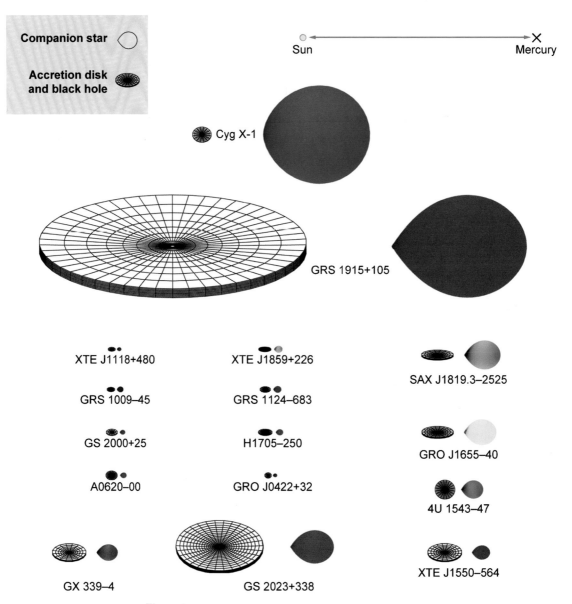

Figure 1

Scale drawings of 16 black-hole binaries in the Milky Way (courtesy of J. Orosz). The Sun–Mercury distance (0.4 AU) is shown at the top. The estimated binary inclination is indicated by the tilt of the accretion disk. The color of the companion star roughly indicates its surface temperature.

3. X-RAY OBSERVATIONS OF BLACK-HOLE BINARIES

We first discuss X-ray light curves obtained by wide-angle X-ray cameras that are used to discover X-ray novae and to monitor hundreds of sources on a daily basis. We then discuss timing and spectral analyses of data obtained in pointed observations that reveal, in detail, the properties of accreting BHs. Throughout this work we feature RXTE results derived from the huge and growing archive of data amassed since 1996 by the All-Sky Monitor (ASM) and the large-area Proportional Counter Array (PCA) detector (Swank 1998).

PDS: power density spectrum

3.1. X-Ray Light Curves of Black-Hole Binaries in Outburst

Nearly all BHBs are X-ray novae (see Section 2 and **Table 1**) that are discovered when they first go into outburst. Their discovery and subsequent daily monitoring are largely dependent on wide-field X-ray cameras on orbiting satellites. The light curves of all 20 BHBs and many BHCs (**Table 1**) can be found either in McClintock & Remillard (2005) or in a review paper on pre-RXTE X-ray novae by Chen, Shrader & Livio (1997). These researchers discuss the striking morphological diversity among these light curves, which show broad distributions in their timescales for rise and decay.

For X-ray outbursts that last between ~20 days and many months, the generally accepted cause of the outburst cycle is an instability that arises in the accretion disk. When the accretion rate from the donor star is not sufficient to support continuous viscous flow to the compact object, matter fills the outer disk until a critical surface density is reached and an outburst is triggered. This model was developed initially for dwarf novae (e.g., Smak 1971; Lasota 2001) and extended to X-ray novae (e.g., Dubus, Hameury & Lasota 2001).

This model predicts recurrent outbursts; indeed, half of the BHBs are now known to recur on timescales of 1 to 60 years (**Table 1**). Outbursts on much shorter or longer timescales do occur, but these are not understood in terms of the disk instability model. Sources such as GRS 1915+105 and 4U 1755 − 338 exhibit "on" and "off" states that can persist for $\gtrsim 10$ years. The behavior of the companion star may play a role in causing these long-term changes in the accretion rate.

After a decade of continuous operation, the ASM aboard RXTE continues to scan most of the celestial sphere several times per day (Levine et al. 1996). It has discovered 8 BHB/BHC X-ray novae and an additional 15 recurrent outbursts. Detailed X-ray light curves have been archived for each of these sources, and complete outbursts for six such systems are shown in Section 5.

3.2. X-Ray Timing

Our most important resource for examining the near-vicinity of a BH is the rapid variations in X-ray intensity that are so often observed (McClintock & Remillard 2005, van der Klis 2005). The analysis tool commonly used for probing fast variability is the power-density spectrum (PDS; e.g., Leahy et al. 1983). Related techniques for computing coherence and phase lag functions are reviewed by Vaughan & Nowak (1997). PDSs are interpreted with the presumption that the source variations are a locally stationary process. More generalized considerations of time series analyses are given by Scargle (1981), whereas recent topics in nonlinear processes are well described by Gliozzi, Papadakis & Räth (2005).

The PDS is used extensively in this work. The continuum power in the PDS is of interest for both its shape and its integrated amplitude (e.g., 0.1–10 Hz), which is usually expressed in units of rms fluctuations scaled to the mean count rate. PDSs of BHBs also exhibit transient, discrete features known as quasi-periodic oscillations (QPOs) that may range in frequency from 0.01 Hz to 450 Hz. QPOs are generally modeled with Lorentzian profiles, and they are distinguished from broad power peaks using a coherence parameter, $Q = \nu/FWHM \gtrsim 2$ (Nowak 2000, van der Klis 2005). PDSs are frequently computed for a number of energy intervals. This is an important step in linking oscillations to an individual component in the X-ray spectrum.

QPO: quasi-periodic oscillation

PL: power law

3.3. X-Ray Spectra

It has been known for decades that the energy spectra of BHBs often exhibit a composite shape consisting of both a thermal and a nonthermal component. Furthermore, BHBs display transitions in which one or the other of these components may dominate the X-ray luminosity (see Tanaka & Lewin 1995; McClintock & Remillard 2005). The thermal component is well modeled by a multitemperature blackbody, which originates in the inner accretion disk and often shows a characteristic temperature near 1 keV (see Section 7). The nonthermal component is usually modeled as a power law (PL). It is characterized by a photon index Γ, where the photon spectrum is $N(E) \propto E^{-\Gamma}$. The PL generally extends to much higher photon energies (E) than does the thermal component, and sometimes the PL suffers a break or an exponential cutoff at high energy.

X-ray spectra of BHBs may also exhibit an Fe Kα emission line that is often relativistically broadened (Section 8.2.3). In some BHBs, particularly those with inclinations that allow us to view the disk largely face-on, the spectral model requires the addition of a disk reflection component (e.g., Done & Nayakshin 2001). In this case, the X-ray PL is reflected by the accretion disk and produces a spectral bump at roughly 10 keV to 30 keV. Finally, high-resolution grating spectra of BHBs sometimes reveal hot gas that is local to the binary system (e.g., Lee et al. 2002). Such features may eventually help us to interpret X-ray states, but at present there are too few results to support any firm conclusions.

4. EMISSION STATES OF BLACK-HOLE BINARIES

4.1. Historical Notes on X-Ray States

The concept of X-ray states was born when Tananbaum et al. (1972) observed a global spectral change in Cyg X–1 in which the soft X-ray flux (2–6 keV) decreased by a factor of 4, the hard flux (10–20 keV) increased by a factor of 2, and the radio counterpart turned on. Thereafter, a similar X-ray transition was seen in A 0620–00 (Coe, Engel & Quenby 1976) and in many other sources as well. The soft state, which was commonly described as ~1 keV thermal emission, was usually observed when the source was bright, thereby prompting the name "high/soft state." The hard state, with a typical photon index $\Gamma \sim 1.7$, was generally seen when the source was faint, hence the name "low/hard state." In this state, the disk was either not observed above 2 keV or it appeared much cooler and withdrawn from the BH. An additional X-ray state of BHBs was identified in the *Ginga* era (Miyamoto & Kitamoto 1991, Miyamoto et al. 1993). It was characterized by the appearance of several-Hz X-ray QPOs, a relatively high luminosity (e.g., $>0.1 L_{\rm Edd}$), and a spectrum comprised of both a thermal component and a PL component that was steeper ($\Gamma \sim 2.5$) than the hard PL. This state was named the "very high" state.

Rapid observational developments in the RXTE era challenged the prevailing views of X-ray states in BHBs. First, it was shown that the soft state of Cyg X–1 is not consistent with a thermal interpretation (Zhang et al. 1997); instead, the spectrum is dominated by a steep PL component ($\Gamma \sim 2.5$). Thus, Cyg X–1 is not a useful prototype for the high/soft state that it helped to define. Secondly, the spectra of BH transients near maximum luminosity were often found to exhibit a steep PL spectrum, rather than a thermal spectrum (McClintock & Remillard 2005). Thirdly, a number of different QPO types were commonly observed over a wide range of luminosities, (e.g., Morgan, Remillard & Greiner 1997; Sobczak et al. 2000a; Homan et al. 2001). These findings attracted great interest in the nature of the very high state.

During this period, Gamma-ray observations (~40–500 keV) of seven BHBs brought clarity to the distinction between the soft and hard types of X-ray PL components (Grove et al. 1998; Tomsick et al. 1999). Sources in the low/hard state ($\Gamma \sim 1.7$) were found to suffer an exponential cutoff near 100 keV, whereas sources with soft X-ray spectra ($\Gamma \sim 2.5$) maintained a steep, strong, and unbroken PL component out to the sensitivity limit of the gamma-ray detectors (~1 MeV).

More recently, radio observations have cemented the association of the low/hard state with the presence of a compact and quasi-steady radio jet (see Fender 2005; McClintock & Remillard 2005). In brief, the evidence for the association includes the following: (*a*) the presence of compact jets in VLBI images of two BH sources (Dhawan, Mirabel & Rodriguez 2000; Stirling et al. 2001); (*b*) correlated X-ray and radio intensities and/or the presence of flat or inverted radio spectra (e.g., Gallo, Fender & Pooley 2003), which allow the jet's presence to be inferred even in the absence of VLBI images; (*c*) a 2% linear radio polarization at nearly constant position angle observed for GX 339–4 (Corbel et al. 2000); and (*d*) the frequently-observed quenching of the persistent radio emission that occurs when a BHB switches from the low/hard state to the high/soft state (e.g., Fender et al. 1999).

4.2. A Quantitative Three-State Description for Active Accretion

In McClintock & Remillard (2005), a new framework was used to define X-ray states that built on the preceding developments and the very extensive RXTE data archive for BHBs. In **Figure 2**, we illustrate the character of each state by showing examples of

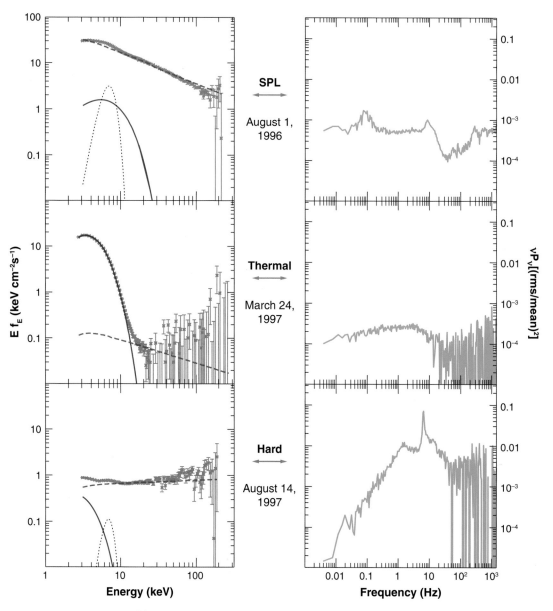

Figure 2

Sample spectra of black-hole binary GRO J1655–40 illustrating the three outburst states: steep power law, thermal, and hard. Each state is characterized by a pair of panels. Left panels show the spectral energy distribution decomposed into three components: thermal (*red, solid line*), power-law (*blue, dashed line*), and a relativistically broadened Fe Kα line (*black, dotted line*). Right panels show the PDSs plotted as $\log(\nu \times P_\nu)$ versus $\log \nu$.

PDSs and energy spectra for the BHB GRO J1655–40. The relevance of X-ray states fundamentally rests on the large differences in the energy spectra and PDSs that can be seen in a comparison of any two of these states. For thorough discussions, many illustrative spectra, and detailed references, see McClintock & Remillard (2005). Discussions of physical models for these states are given in Section 7.

The sharpest point of departure from the old description of states is that luminosity is abandoned as a criterion for defining the state of a source. In defining states, McClintock & Remillard (2005) adopted the pragmatic and generic strategy of utilizing a spectral model consisting of a multitemperature accretion disk and a PL component (with a possibile break near 15 keV or an exponential cutoff in the range of 30–100 keV). When required, an Fe emission line or a reflection component was included. The model also included photoelectric absorption by neutral gas. McClintock & Remillard (2005) used four parameters to define X-ray states: (*a*) the disk fraction *f*, which is the ratio of the disk flux to the total flux (both unabsorbed) at 2–20 keV; (*b*) the PL photon index (Γ) at energies below any break or cutoff; (*c*) the rms power, *r*, in the PDS integrated from 0.1–10 Hz, expressed as a fraction of the average source count rate; and (*d*) the integrated rms amplitude, *a*, of any QPO detected in the range 0.1–30 Hz. PDS criteria, i.e., *a* and *r*, utilize a broad energy range, e.g., the bandwidth of the RXTE PCA instrument, which is effectively 2–30 keV. Quantitative definitions of the three states are given in **Table 2**.

In the thermal state [formerly high/soft state, and "thermal dominant" state in McClintock & Remillard (2005)], the flux is dominated by the heat radiation from the inner accretion disk, the integrated power continuum is faint, and QPOs are absent or very weak (see **Table 2**). There is usually a second, nonthermal component in the spectrum, but its contribution is limited to <25% of the flux at 2–20 keV. The

Table 2 Outburst states of black holes: nomenclature and definitions

New State Name (Old State Name)	Definition of X-Ray State[a]
Thermal (High/Soft)	Disk fraction $f^b > 75\%$ QPOs absent or very weak: $a^c_{max} < 0.005$ Power continuum level $r^d < 0.075^e$
Hard (Low/Hard)	Disk fraction $f^b < 20\%$ (i.e., Power-law fraction > 80%) $1.4^f < \Gamma < 2.1$ Power continuum level $r^d > 0.1$
Steep Power Law (SPL) (Very high)	Presence of power-law component with $\Gamma > 2.4$ Power continuum level $r^d < 0.15$ Either $f^b < 0.8$ and 0.1–30 Hz QPOs present with $a^c > 0.01$ or disk fraction $f^b < 50\%$ with no QPOs

[a] 2–20 keV band.
[b] Fraction of the total 2–20 keV unabsorbed flux.
[c] QPO amplitude (rms).
[d] Total rms power integrated over 0.1–10 Hz.
[e] Formerly 0.06 in McClintock & Remillard (2005).
[f] Formerly 1.5 in McClintock & Remillard (2005).

state is illustrated in the middle row of panels in **Figure 2**. The spectral deconvolution shows that the thermal component (*red line*) is much stronger than the PL component (*blue, dashed line*) for $E \lesssim 10$ keV. The PDS (*right panel*), which is plotted in terms of $\log(\nu \times P_\nu)$ versus $\log \nu$, appears featureless. Similar displays of paired energy spectra and PDSs for nine other BHBs/BHCs in the thermal state are shown in McClintock & Remillard (2005).

The hard state (formerly low/hard state) is characterized by a hard PL component ($\Gamma \sim 1.7$) that contributes $\geq 80\%$ of the 2–20 keV flux (**Table 2**). The power continuum is bright with $r > 0.1$ and QPOs may be either present or absent. A hard state osbervation of GRO J1655–40 is shown in the bottom row of panels in **Figure 2**. The accretion disk appears to be faint and cool compared to the thermal state. As noted previously (Section 4.1), the hard state is associated with the presence of a quasi-steady radio jet, and clear correlations between the radio and X-ray intensities are observed.

The hallmark of the steep power law (SPL) state (formerly the very high state) is a strong PL component with $\Gamma \sim 2.5$. In some sources, this PL has been detected without a break to energies of ~ 1 MeV or higher (Section 4.1). This state is also characterized by the presence of a sizable thermal component and the frequent presence of X-ray QPOs (see **Table 2**). An example of the SPL state is shown in the top row of panels in **Figure 2**, and many additional illustrations are displayed in McClintock & Remillard (2005). There are similarities between the SPL state and the thermal state; both show a thermal component and a steep PL component. However, in the thermal state the PL is faint and has a more variable photon index, while the SPL state is plainly distinguished by its powerful PL component and the commonly-occurring QPOs. The SPL state tends to dominate BHB spectra as the luminosity approaches the Eddington limit, and it is this state that is associated with high-frequency QPOs (Section 6.2; McClintock & Remillard 2005).

Intermediate states and state transitions are another important aspect of BHB studies. The three states defined by McClintock & Remillard (2005) represent an attempt to define spectral and timing conditions that are quasi-stable and that appear to have distinct physical origins. There are gaps in the parameter ranges used to define these states (**Table 2**), and this gives rise to intermediate states. The hybrid of the hard and SPL states is particularly interesting for its correlations with radio properties (Sections 4.3 and 5) and also with disk properties (Section 6.1).

4.3. The Unified Model for Radio Jets

Many researchers investigate the spectral evolution of BHBs using a hardness intensity diagram (HID), which is a plot of X-ray intensity versus a "hardness ratio" (*HR*), i.e., the ratio of detector counts in two energy bands (e.g., Homan et al. 2001, Belloni 2004, van der Klis 2005). This diagram is widely used in tracking the behavior of accreting neutron stars. Compared to the spectral-fitting approach described above, the HID approach has the advantage that it is model independent and the disadvantage that it is difficult to relate the results to physical quantities. Interpretations of variations in the HID depend on the particular energy bands chosen to define *HR* in a given

---

HID: hardness intensity diagram

HR: hardness ratio

study. If both bands are above ∼5 keV, then the *HR* value effectively tracks the slope of the PL component (i.e., lower *HR* means a steeper PL). Softer energy bands admit a mixture of thermal and nonthermal components, and interpretations are then more complicated.

The HID is also used in illustrating the "unified model for radio jets" proposed by Fender, Belloni & Gallo (2004). **Figure 3** shows their schematic for the relationships between jets and X-ray states, where the state of an observation is distinguished simply by the value of *HR*. The figure shows qualitatively how the jet Lorentz factor (*lower panel*) and the morphology of the jet (sketches *i–iv*) evolve with changes in the X-ray state. Tracks for state transitions of an X-ray source in the HID are also shown (*top panel*). The solid vertical line running through both panels in **Figure 3** is the "jet line." To the right of the jet line the X-ray spectrum is relatively hard and a steady

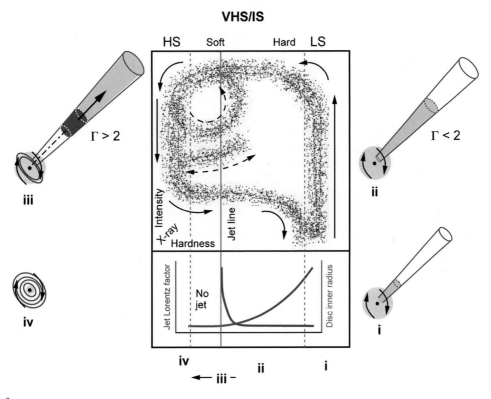

Figure 3

A schematic representation of the model for disk-jet coupling in black-hole binaries from Fender, Belloni & Gallo (2004). The top panel shows evolutionary tracks in a HID, which is a plot of X-ray intensity versus X-ray hardness. (These quantities increase upward and to the right, respectively.) The bottom panel gives a qualitative impression of how the jet's bulk Lorentz factor (*blue curve*) and the inner disk radius (*red curve*) vary with X-ray hardness. The X-ray states are labeled at the top in the old nomenclature (i.e., HS = high/soft state, VHS/IS = very high and intermediate states, and LS = low/hard state).

radio jet is present, and to the left the spectrum is soft and the jet is quenched. The jet line also marks an instability strip where violent ejections of matter may occur (see sketch *iii*), as indicated by the spike in the Lorentz factor (*lower panel*). The tracks for state evolution in the HID are influenced by observations of GX 339–4 (Belloni et al. 2005). In the Belloni et al. study *HR* was defined as the ratio of source counts at 6.3–10.5 keV to the counts at 3.8–7.3 keV.

The vertical source track on the right side of **Figure 3** corresponds to the low/hard or hard state with $HR > 0.8$ in the case of GX 339–4 (see Belloni et al. 2005). The vertical track on the far left (and most observations of GX 339–4 with $HR < 0.2$) are in the high/soft or thermal state. Thus, the HID state classifications and the McClintock & Remillard (2005) classifications agree very well in these two regimes. However, for the intermediate values of *HR*, which fall between the dashed lines, the states are described differently. The states in this region are further divided by Belloni et al. (2005) into "soft intermediate" and "hard intermediate" states, based on the *HR* values and the properties of the power continuum in the X-ray PDS. This completes a description of the four X-ray states defined in the unified jet model. Further comparisons of the HID state classifications and the McClintock & Remillard (2005) classifications are given in Section 5.

The unified model for X-ray states and radio jets provides opportunities to study the disk-jet coupling explicitly, and the HID format is very easy to apply to observations. On the other hand, the state definitions of McClintock & Remillard (2005) are more quantitative, and they provide spectral information that is more directly applicable to physical models. In Section 5, we present outburst data for selected BHB/BHC systems using both depictions for X-ray states.

4.4. Quiescent State

The quiescent state corresponds to luminosities that are three or more orders of magnitude below the levels of the active states described above. The typical BHB with transient outbursts spends most of its time in a quiescent state that is characterized by an extraordinarily faint luminosity ($L_x = 10^{30.5} - 10^{33.5}$ erg s^{-1}) and a spectrum that is distinctly nonthermal and hard ($\Gamma = 1.5 - 2.1$). The quiescent state is particularly important in two ways: (*a*) It enables firm dynamical measurements to be made because the optical spectrum of the secondary star becomes prominent and is negligibly affected by X-ray heating (van Paradijs & McClintock 1995); and (*b*) its inefficient radiation mechanism underpins a strong argument for the event horizon (Section 8.1).

For a thorough review of the X-ray properties of this state and discussions of physical models, see McClintock & Remillard (2005). As commonly remarked, it is possible that the hard and quiescent states represent a single mechanism that operates over several orders of magnitude in X-ray luminosity. However, this question remains controversial. Corbel, Tomsick & Kaaret (2006) have recently derived precise spectral parameters for XTE J1550 – 564. They conclude that the quiescent spectrum of this source (and a few others) is softer than the spectrum in the much more luminous hard state.

5. X-RAY OVERVIEW OF STATE EVOLUTION AND ENERGETICS

5.1. Overview Plots for Six Individual Sources

In Section 4, we described two ways of defining X-ray states: a quantitative method based on generic X-ray spectral modeling and PDS analysis (McClintock & Remillard 2005), and another based on radio properties, X-ray PDSs, and HIDs. Here we synthesize the two approaches to provide the reader with a comprehensive picture of the behavior of an accreting BH.

In **Figures 4–9** respectively, we show detailed overviews of the behavior of six X-ray novae: five BHBs and one BHC (H 1743–322). Each figure contains seven panels that review the data for a single source. The overview plots can be used in two ways. One can focus on the figure (7 panels) for a particular BHB and examine each color-coded state for many aspects of the behavior of the source. The key science questions addressed by each panel are: (a, b) How do states and luminosity vary with time?; (c) How does the radiation energy divide between thermal and nonthermal components (2–20 keV)?; (d) How do the states of McClintock & Remillard (2005) relate to the states of Fender, Belloni, & Gallo (2004), which are presented in a HID?; and (e, f, g) How do three key X-ray properties—PL index, disk fraction, and rms power—vary as a function of either HR or the X-ray state? Alternatively, one can choose a particular panel and compare the behavior of the six sources to draw general conclusions about common behavior patterns in BHBs as well as their differences. Such conclusions are discussed at the end of this Section and in Section 5.2.

The following comments pertain to all six figures. As an inspection of the various ASM light curves (*panel a*) shows, the data cover outbursts observed with RXTE during the time interval 1996–2004. The remaining panels (*b–g*) show results derived from RXTE pointed observations. These latter panels display several different kinds of data, but in every panel there is a common use of symbol type and color to denote the state of the source [McClintock & Remillard (2005) definitions]: thermal (*red "x"*), hard (*blue square*), SPL (*green triangle*), and any intermediate type (*black circle*).

Panel b, which mimics the ASM light curve, shows the 2–20 keV X-ray flux derived from the fitted spectral model (see Section 3.3). Panel c, which is also based on the fitting results, shows how the energy is divided between the thermal (accretion disk) component and the nonthermal (PL) component; we refer to this plot as the energy-division diagram. Panel d, which is based on raw count rates, shows how the states are distributed in the hardness intensity diagram (HID). We use this panel to examine how the McClintock & Remillard (2005) states are distributed in the HID and how they correspond to the states that are defined by the unified jet model (Section 4.3). Finally, in panels e–g we plot spectral hardness on the x-axis versus three of the parameters that are used to define BH states, namely, the PL index Γ, the disk fraction, and the integrated rms power in the PDS (0.1–10 Hz). The HIDs use the normalization scheme and hard color definition ($HR = 8.6$–18.0 keV counts/5.0–8.6 keV counts) given by Muno, Remillard & Chakrabarty (2002), for which the Crab Nebula yields 2500 c s^{-1} PCU^{-1} and $HR = 0.68$. The RXTE pointed observations

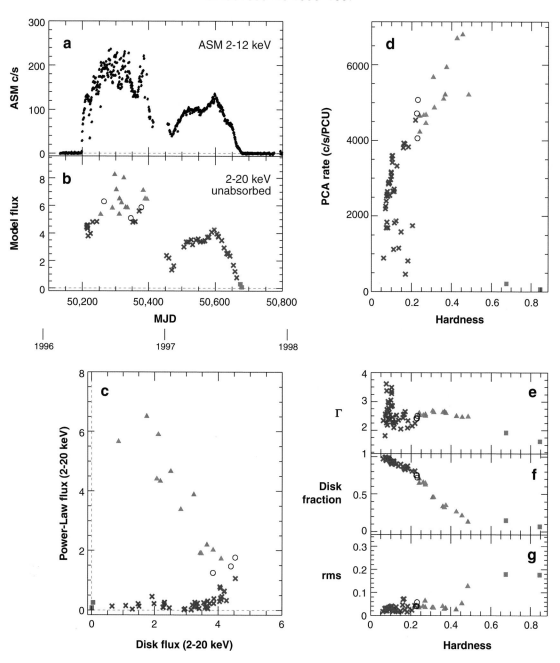

are selected to have a minimum exposure time of 500 s and a minimum source flux of 10^{-10} erg cm^{-2} s^{-1} (i.e., 4 mCrab) at 2–20 keV.

We begin our discussion of individual sources with GRO J1655–40, the BHB used in Section 4.2 to illustrate the BH states (McClintock & Remillard 2005). In **Figure 4** we show the overview for the 1996–1997 outburst. The data are mostly derived from publications (Remillard et al. 1999, Sobczak et al. 1999, Remillard et al. 2002a), with some supplementary results added for completeness. GRO J1655–40 shows an orderly and monotonic evolution of states along the arcs displayed in the energy-division diagram (*panel c*) and the HID (*panel d*). In the HID, the McClintock & Remillard (2005) states are cleanly sorted along the hardness axis. Furthermore, panels *e–g* show clear correlations between the key state parameters and the hardness ratio. Quite similar behavior is seen in the X-ray overview for the next BHB, 4U 1543–47 (**Figure 5**). Here we have used the results derived by Park et al. (2004), while adding results from similar analyses to extend the coverage to the end of the outburst. As these two overviews show (i.e., **Figures 5** and **6**), both sources favor the thermal state, and they only enter the hard state at low luminosity.

Figure 6 exhibits four outbursts of XTE J1550–564 with successively shorter durations and decreasing maxima (*panels a and b*). The results shown in panels *c–g* are dominated by the first outburst (1998–1999), which provides 202 of 309 state assignments and most of the points at high flux levels. Several authors have noted the complex behavior of this source (e.g., Sobczak et al. 2000b, Homan et al. 2001). The energy-division diagram (*panel c*) and the HID (*panel d*) display many branches. The thermal branch (*red "x" symbols*) in the energy diagram covers a wide range of luminosity. Note that there are examples of both high luminosity hard-state observations and low luminosity SPL-state observations. One must conclude that luminosity does not drive a simple progression of X-ray states (as implied by the old state names: low/hard, high/soft, and very high). Thus, the X-ray state must depend on some unknown and important variable(s) in addition to the BH mass and the mass accretion rate (Homan et al. 2001).

Similar patterns of behavior are shown in the next two examples. The overview for H 1743–322 (**Figure 7**) shows many state transitions (*panel b*), and the complex tracks in panels *c* and *d* are reminiscent of XTE J1550–564 (**Figure 6**). The hysteresis in transitions to and from the hard state are especially evident in the HID (*panel d*; see also Maccarone & Coppi 2003). The overview for XTE J1859+226 (**Figure 8**) is similar to that of H 1743–322. All the of brightest observations are found in the SPL state, and the vertical track in the HID (*green triangles* in **Figure 8d**) is a consequence

---

Figure 4

X-ray overview of GRO J1655–40 during its 1996–1997 outburst. The All-Sky Monitor light curve is displayed in panel *a*. All other data are derived from Rossi X-Ray Timing Explorer pointed observations organized into 62 time intervals. A variety of parameters are displayed (see text). The symbol color and type denote the X-ray state (see **Table 2**): thermal (*red "x"*), steep power law (SPL) (*green triangle*), hard (*blue square*), and any intermediate type (*black circle*). This outburst shows simple patterns of evolution that favor the thermal and SPL states.

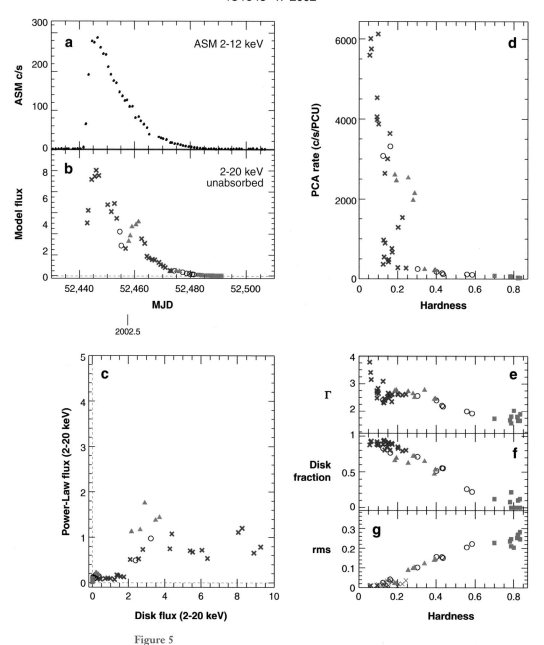

Figure 5

X-ray overview of 4U 1543–47 during its 2002 outburst. The presentation format and the state-coded plotting symbols follow the conventions of **Figure 4**. The Rossi X-Ray Timing Explorer pointed observations (49 time intervals) show relatively simple patterns of state evolution, and the thermal state is prevalent when the source is bright.

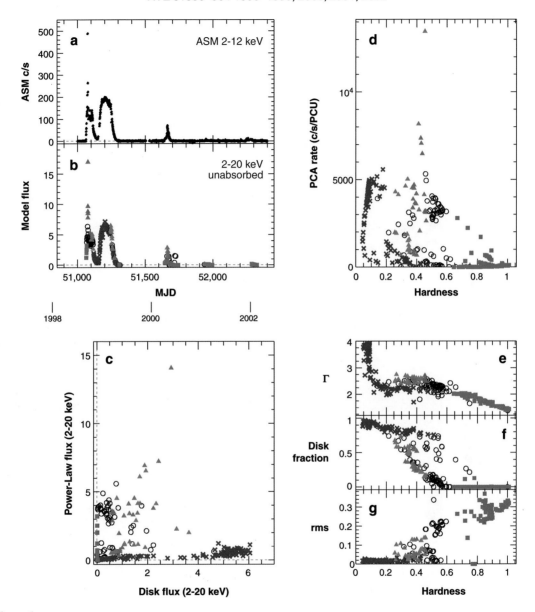

Figure 6

X-ray overview (309 time intervals) of XTE J1550–564 that includes a series of four outbursts with decreasing maxima. The presentation format and the state-coded plotting symbols follow the conventions of **Figure 4**. The two brighter outbursts (1998–1999 and 2000) show great complexity in the temporal evolution of states and the energy division between thermal and nonthermal components. In contrast, the subsequent pair of faint outbursts were confined to the hard state.

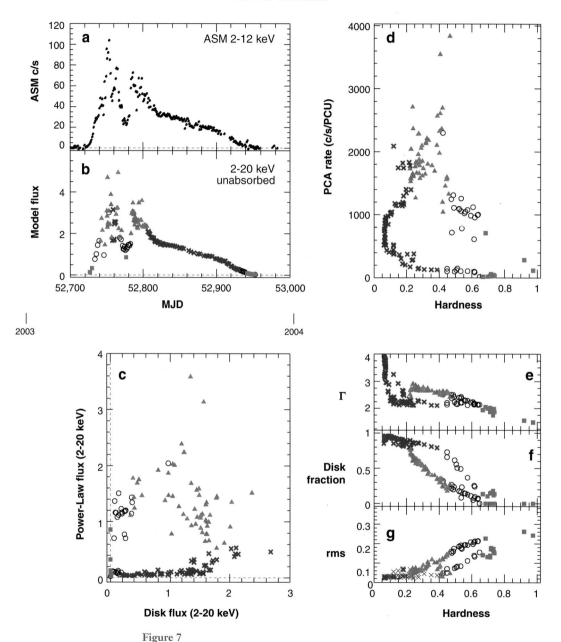

Figure 7

X-ray overview of H 1743−322 during its 2003 outburst. The presentation format and the state-coded plotting symbols follow the conventions of **Figure 4**. The Rossi X-Ray Timing Explorer pointed observations (170 time intervals) show complex state evolution with similarities to XTE J1550−564.

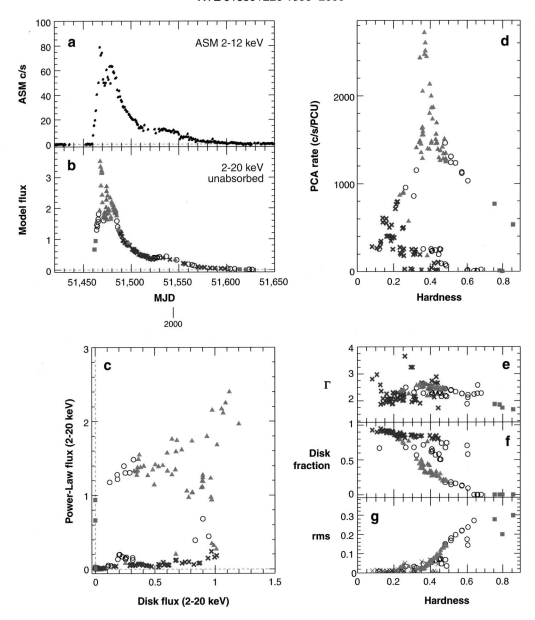

Figure 8

X-ray overview of XTE J1859+226 during its 1999–2000 outburst. The presentation format and the state-coded plotting symbols follow the conventions of **Figure 4**. The Rossi X-Ray Timing Explorer pointed observations (130 time intervals) show complex behavior very similar to that seen for H 1743−322. In panels b, c, and d it is apparent that the steep power law state is prevalent when the source is bright.

of constant values of Γ in the SPL state. These complex results reinforce the need to probe deeply in attempting to identify the underlying variables that govern the emission of radiation from BHBs.

Finally, we show the overview of GX 339–4 in **Figure 9**. This source is known to produce frequent X-ray outbursts and to remain for long intervals in the hard state, as shown in panel *b*. It is therefore a frequent target for radio studies (e.g., Fender et al. 1999; Corbel et al. 2000; Gallo, Fender & Pooley 2003; Belloni et al. 2005). There is an apparent similarity between the HID for GX 339–4 (*panel d*) and the schematic sketch in **Figure 3** that illustrates the unified model for jets. The HID shows four branches for the hard state, one of which peaks at 75% of the maximum count rate in the thermal and SPL states. Despite the wide range in luminosity for each of the three states, the plots of the key parameters (*panels e–g*) appear orderly and well correlated with the hardness ratio. Thus, the behavior of GX 339–4 supports the theme that recurs throughout this work, namely, that X-ray states are important, although they are plainly not a simple function of source luminosity.

On the one hand, the six X-ray overviews presented in **Figures 4–9** show that BHB outbursts can be very complex. They typically begin and end in the hard state, but between those times there is common disorder in the temporal evolution of luminosity (*b* panels), and in the division of radiation energy between thermal and nonthermal states (*c* panels). On the other hand, there are clear correlations involving the key spectral and timing properties, examined versus either the *HR* value or the color-coded state symbols (*e*, *f*, and *g* panels). These results, which are robust for a variety of BHBs, including sources displaying multiple outbursts, confirm the prevailing wisdom that complex BHB behavior can be productively organized and studied within the framework of X-ray states. Further discussions of the overview figures are continued in the following section.

5.2. Conclusions from Overviews of X-Ray States

Several conclusions can be drawn from the set of overview plots shown in Section 5. First, when the state assignments of McClintock & Remillard (2005) are examined in the HID format (i.e., sorting symbol color/type versus hardness in the *d* panels of **Figures 4–9**), it is apparent that the state prescriptions of McClintock & Remillard (2005) and unified-jet model overlap significantly. For example, for the particular *HR* defined in Section 5.1, observations with $HR < 0.2$ would be assigned to the high/soft state in the unified-jet model, and nearly all of these same observations are here assigned to the thermal state of McClintock & Remillard (2005). Similarly, observations with $HR > 0.65$ correspond to the hard state in both prescriptions. For the intermediate states in the HID, there is a divergence between the two approaches, and this is especially obvious for sources that show complex behavior, e.g., XTE J1550–564. For this source, the McClintock & Remillard (2005) states are strongly disordered in the interval $0.20 < HR < 0.65$ (see **Figure 6**, panel *d*).

We have emphasized that source luminosity is not a criterion used for identifying X-ray states in either prescription (Sections 4.2 and 4.3). This is evident in **Figures 4–9**, because there are lines of constant flux that intercept several different McClintock

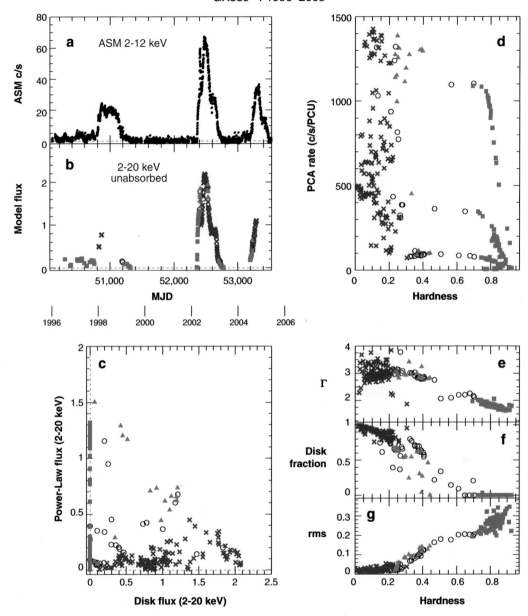

Figure 9

X-ray overview of GX 339–4 (274 time intervals). This source exhibits frequent outbursts and long intervals in the hard state at low luminosity. The presentation format and the state-coded plotting symbols follow the conventions of **Figure 4**. The state evolution shown here was used as the prototype for making the connection between the inner disk and a steady radio jet (compare panel *d* and **Figure 3**).

& Remillard (2005) states (*b* panels), and because there are lines of constant intensity that intercept observations with very different *HR* values (*d* panels). However, it is clear that there is still some degree of correlation between X-ray states and source brightness. For example, all of the sources show transitions to the hard state when the source becomes faint during its decay phase (see the *b* and *d* panels in **Figures 4–9**). At the opposite extreme, the highest luminosity observations for four of the six sources occur in the SPL state (**Figures 4** and **6–8**).

The energy-division diagrams (2–20 keV; *c* panels of **Figures 4–9**) routinely show vertical tracks for the hard state (*blue squares*), whereas horizontal tracks with gentle curvature are seen for the thermal state (*red x's*). These tracks indicate a free flow of energy into the hard PL spectrum during the hard state and into the accretion disk during the thermal state. This point was first made by Muno, Morgan & Remillard (1999) in describing the behavior of GRS 1915+105. The curvature of the thermal tracks (*c* panels, **Figures 4–9**) can be interpreted as being due to increased Comptonization with increasing thermal luminosity. Tracing smooth lines through these tracks would constrain Comptonization to a maximum of <20% of the flux (2–20 keV) at the peak of the thermal state. The energy-division tracks for the SPL state (*green triangles*) are not well defined, and the tracks of different sources do not resemble each other. The energy-division diagrams are more effective than the HIDs in delineating intrinsic differences between the SPL and thermal states.

Comparisons between the SPL and thermal states in terms of the photon index (Γ; *e* panels of **Figures 4–9**) are of interest, as these states are sometimes presumed to share a common PL mechanism that is relatively muted during the thermal state. Generally, we find broad distributions in Γ during the thermal state, and in some cases the thermal tracks are shifted from the SPL tracks (e.g., **Figures 6*e*, 7*e*, 8*e***). We find no compelling evidence that the PL mechanisms for the thermal and SPL states are the same.

In the following section, we consider one additional observational topic—X-ray QPOs—before discussing the ongoing efforts to relate X-ray states to physical models for BH accretion.

6. X-RAY QUASI-PERIODIC OSCILLATIONS

X-ray QPOs are specialized and extraordinarily important avenues for the study of accreting BHs. They are transient phenomena associated with the nonthermal states and state transitions. For definitions of QPOs and analysis techniques, see van der Klis (2005) and some details in Section 3.2.

QPOs play an essential role in several key science areas, such as probing regions of strong field (Sections 6.2 and 8.2.4) and defining the physical processes that distinguish X-ray states. Thus far in this review, however, QPOs have been considered only tangentially (e.g., Section 4.2). In this section we focus on the significance of QPOs, their subtypes, and spectral/temporal correlation studies that involve the QPO frequency. Following the literature, we divide the discussions of QPOs into low-frequency and high-frequency groups. In doing so, we disregard the additional, infrequent appearances of very low frequency QPOs (e.g., the QPO below 0.1 Hz in the upper-right panel of **Figure 2**) that are not understood. Physical models for QPOs are briefly discussed in Section 7, and the importance of high-frequency QPOs for probing strong gravity is highlighted in Section 8.2.4.

6.1. Low-Frequency Quasi-Periodic Oscillations

Low-frequency QPOs (LFQPOs; roughly 0.1–30 Hz) have been detected on one or more occasions for 14 of the 18 BHBs considered in **Table 4.2** of McClintock & Remillard (2005). They are important for several reasons. LFQPOs can have high amplitude (integrated rms/mean values of $a > 0.15$) and high coherence (often $Q > 10$), and their frequencies and amplitudes are generally correlated with the spectral parameters for both the thermal and PL components (e.g., Muno, Morgan & Remillard 1999; Revnivtsev, Trudolyubov & Borozdin 2000; Sobczak et al. 2000a; Vignarca et al. 2003). With the exception of Cyg X–1, QPOs generally appear whenever the SPL contributes more than 20% of the flux at 2–20 keV (Sobczak et al. 2000a), which is one component of the definition of the SPL state (**Table 2**; McClintock & Remillard 2005).

LFQPOs can vary in frequency on timescales as short as one minute (e.g., Morgan, Remillard & Greiner 1997). On the other hand, LFQPOs can also remain relatively stable and persistent. For example, in the case of GRS 1915+105, a 2.0–4.5 Hz QPO is evident in every one of the 30 RXTE observations conducted over a 6-month interval in 1996–1997 (Muno et al. 2001). This degree of stability suggests that LFQPOs are tied to the flow of matter in the accretion disk. However, the frequencies of these QPOs are much lower than the Keplerian frequencies for the inner disk. For example, a 3 Hz orbital frequency around a Schwarzschild BH of 10 $M_\odot$ corresponds to a radius near 100 R_g, while the expected range of radii for X-ray emission in the accretion disk is $\lesssim 10\ R_g$.

LFQPOs are seen in the SPL state, the hard:SPL intermediate state, and in some hard states, particularly when the X-ray luminosity is high (Rossi et al. 2004; McClintock & Remillard 2005). The rms amplitude generally peaks at photon energies $\gtrsim 10$ keV (e.g., Rodriguez et al. 2002; Vignarca et al. 2003), and detections have been made at energies above 60 keV (Tomsick & Kaaret 2001). This behavior clearly ties LFQPOs to the nonthermal component of the X-ray spectrum. However, in principle, the oscillation could still originate in the accretion disk if the PL mechanism is inverse Compton scattering of disk photons and the coherence of the original oscillation is not destroyed by the scattering geometry. Alternatively, the mechanism that creates the energetic electrons required for Comptonization could be an oscillatory type of instability. In this case, the disk temperature or the thermal energy flux might control the QPO properties.

These concepts regarding QPO origins motivate correlation studies that compare QPO frequencies and amplitudes with various spectral parameters. In **Figure 10** we show an example of such a correlation between LFQPO frequency and disk flux for two sources. The X-ray states are symbol coded, using the conventions adopted in Section 5. The QPO frequencies in the hard and intermediate states are highly correlated with disk flux, but this is not true for the QPOs at higher frequency in

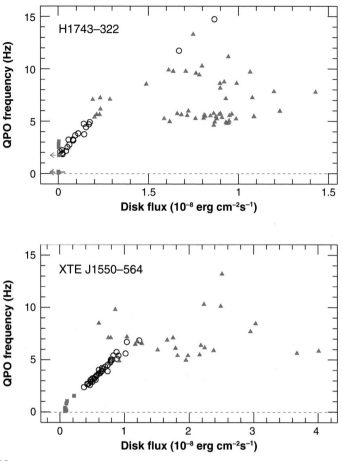

Figure 10

Quasi-periodic oscillation (QPO) frequency versus disk flux for XTE J1550–564 and H 1743–322. See **Figure 4** for definitions of the symbol types. The QPO frequencies in the hard and especially the hard:SPL intermediate state are highly correlated with disk flux; these are C-type low-frequency QPOs (LFQPOs) based on their phase lags (see text). The steep power law QPOs are not correlated with disk flux, and these are A and B type LFQPOs.

the SPL state. Presently, there is no explanation for this result. Possibly, the QPO mechanisms could differ between the hard and SPL states, or there could be a common mechanism that exhibits some type of dynamical saturation as the source moves into the SPL state.

Another avenue for QPO investigations is the study of phase lags and coherence functions that compare two different energy bands, e.g., 2–6 versus 13–30 keV. Such analyses have been conducted for the QPOs in XTE J1550-564. Unexpectedly, the phase lag measurements showed groups distinguished by positive, negative, and ∼zero lags defining, respectively, LFQPO types A, B, and C (Wijnands, Homan & van der Klis 1999; Cui, Zhang & Chen 2000; Remillard et al. 2002b; Casella et al. 2004).

These details are surely complicated, but the ramifications are very significant. The A and B types are associated with the SPL state and the presence of high-frequency QPOs. On the other hand, C-type LFQPOs mostly occur in the intermediate and hard states, and they are responsible for the frequency versus disk flux correlation shown in **Figure 10**.

Given the relatively high amplitudes of all LFQPOs above 6 keV, it is clear that the C-type oscillations are well connected to both the thermal and PL components in the X-ray spectrum. In this sense, QPOs can provide insights regarding the origin of the PL spectrum. Futhermore, we now have a comprehensive archive of accurate LFQPO measurements for a wide range of disk conditions. These data are available for testing any detailed models that are proposed to explain the nonthermal states in BHBs.

6.2. High-Frequency Quasi-Periodic Oscillations

High-frequency QPOs (HFQPOs; 40–450 Hz) have been detected in seven sources (5 BHBs and 2 BHCs). These oscillations are transient and subtle ($a \sim 0.01$), and they attract interest primarily because their frequencies are in the expected range for matter in orbit near the ISCO for a $\sim 10\ M_\odot$ BH.

As an aside, we briefly note that broad power peaks ($Q < 1$) have been reported at high frequencies in a few cases (e.g., Cui et al. 2000; Homan et al. 2003; Klein-Wolt, Homan & van der Klis 2004). At the present time, these broad PDS features do not impact the field significantly, because they are relatively rare and poorly understood. Consequently, we do not consider these broad features further.

The entire sample of HFQPOs with strong detections ($>4\sigma$) is shown in **Figure 11**. Three sources have exhibited single oscillations (Cui et al. 2000, Homan et al. 2003, Remillard 2004). The other four sources display pairs of HFQPOs with frequencies that scale in a 3:2 ratio. Most often, these pairs of QPOs are not detected simultaneously. The four sources are GRO J1655–40 (300, 450 Hz: Remillard et al. 1999, 2002a; Strohmayer 2001a; Homan et al. 2005a), XTE J1550–564 (184, 276 Hz: Homan et al. 2001, Miller et al. 2001, Remillard et al. 2002a), GRS 1915+105 (113, 168 Hz: Remillard 2004), and H 1743–322 (165, 241 Hz: Homan et al. 2005b, Remillard et al. 2006b). GRS 1915+105 also has a second pair of HFQPOs with frequencies that are not in a 3:2 ratio (41, 67 Hz: Morgan, Remillard & Greiner 1997; Strohmayer 2001b).

HFQPOs are of further interest because they do not shift freely in frequency in response to sizable luminosity changes (factors of 3–4; Remillard et al. 2002a; Remillard et al. 2006b). There is evidence of frequency shifts in the HFQPO at lower frequency (refering to the 3:2 pairing), but such variations are limited to 15% (Remillard et al. 2002a; Homan et al. 2005a). This is an important difference between these BHB HFQPOs and the variable-frequency kHz QPOs seen in accreting neutron stars, where both peaks can shift in frequency by a factor of two (van der Klis 2005). Overall, BHB HFQPOs appear to be a stable and identifying "voice-print" that may depend only on the mass and spin of the BH (see Section 8.2.4).

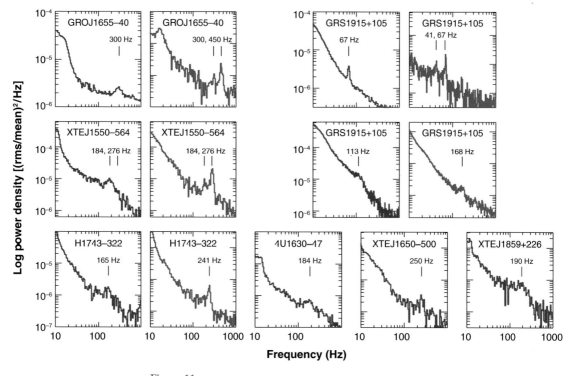

Figure 11

High-frequency quasi-periodic oscillations observed in black-hole binary and black-hole candidate systems. The traces in blue show power density spectrums (PDSs) for the range 13–30 keV. Red traces indicate PDSs with a broader energy range, which may be either 2–30 keV or 6–30 keV.

All of the strong detections ($>4\sigma$) above 100 Hz occur in the SPL state. In three of the sources that exhibit HFQPOs with a 3:2 frequency ratio, the $2\nu_0$ QPO appears when the PL flux is very strong, whereas $3\nu_0$ appears when the PL flux is weaker (Remillard et al. 2002a, 2005b). Currently, there is no explanation for this result.

The commensurate frequencies of HFQPOs suggest that these oscillations are driven by some type of resonance condition. Abramowicz & Kluzniak (2001) proposed that orbiting blobs of accreting matter could generate the harmonic frequencies via a resonance between a pair of the coordinate frequencies given by GR. Earlier work had used GR coordinate frequencies and associated beat frequencies to explain fast QPOs in both neutron-star and BH systems (Stella, Vietri & Morsink 1999), but without invoking a resonance condition. Current work on resonances as a means of explaining HFQPOs includes more realistic models for fluid flow in the Kerr metric. Resonance models are considered in more detail in Section 8.2.4.

7. PHYSICAL MODELS FOR X-RAY STATES

We briefly describe some physical models for the three active emission states (Section 4) and QPO phenomena (Section 6). Our focus is on basic principles and the current interface between observations and theory.

7.1. Thermal State

For the thermal state there is a satisfactory paradigm, namely, thermal emission from the inner regions of an accretion disk. Observations and magnetohydrodynamic (MHD) simulations continue to increase our understanding of accretion disks. Also, fully-relativistic models of disk spectra have recently become publicly available, and results from this advance are discussed below.

The best-known hydrodynamic model of a radiating gas orbiting in the gravitational potential of a compact object is the steady-state, thin accretion disk model (Shakura & Sunyaev 1973, Pringle 1981). A central problem for this model is a prescription for the viscosity that is required to drive matter inward and heat it while transporting angular momentum outward. Initially, the viscosity was modeled using an ad hoc scaling assumption (Shakura & Sunyaev 1973). This model leads to a temperature profile $T(R) \propto R^{-3/4}$ and the conclusion that the inner annulus in the disk dominates the thermal spectrum, because $2\pi R dR \sigma T^4 \propto L(R) \propto R^{-2}$. This result has a striking observational consequence: X-ray astronomy is the window of choice for probing strong gravity near the horizon of an accreting stellar-mass BH.

The cardinal importance of the inner disk region highlights the need for an accurate model for the radiation emitted near the inner disk boundary associated with the ISCO (see Section 1). The ISCO lies at $6R_g$ for a Schwarzschild BH ($a_* = 0$), decreasing toward $1R_g$ as a_* approaches 1. Observationally, the thermal-state spectra of BHBs are well fitted using the classical model for a multitemperature accretion disk (Mitsuda et al. 1984, Makishima et al. 1986, Kubota & Makishima 2004, Kubota et al. 2005). However, the derived spectral parameters (i.e., the temperature and radius of the inner disk) cannot be interpreted literally for several reasons. The model neglects the physically-motivated torque-free boundary condition at the ISCO (see Gierliński, Maciolek-Niedzwiecki & Ebisawa 2001; Zimmerman et al. 2005). Furthermore, the classical model ignores the sizable effects due to GR and radiative transfer (e.g., see Zhang, Cui & Chen 1997). Fortunately, accretion disk models for Kerr BHs have recently become publicly available (Li et al. 2005; Dovčiak, Karas & Yaqoob 2005), and there now exists a fully-relativistic treatment of the effects of spectral hardening (Davis et al. 2005). Applications of these models are discussed in Section 8.

In parallel with these developments, MHD simulations have advanced our understanding of the nature of viscosity in accretion disks. The magnetorotational instability (MRI) has been shown to be a source of turbulent viscosity (Balbus & Hawley 1991), a result that has been confirmed by several global GR MHD simulations (e.g., McKinney & Gammie 2004; DeVilliers, Hawley & Krolik 2003; Matsumoto, Machida & Nakamura 2004). Investigators are now considering how MRI and MHD turbulence influence disk structure (e.g., Gammie 2004), the emerging thermal spectrum (e.g., Merloni 2003), and Comptonization effects (e.g., Socrates, Davis & Blaes 2004). MHD simulations of accretion disks are currently three-dimensional, global, and based on the Kerr metric. They will soon include dissipative processes (i.e., radiation; see Johnson & Gammie 2003), and it is hoped that they will then connect more directly with observation.

7.2. Hard State

The association of the hard state with the presence of a steady radio jet marked a substantial advance. Indirect signatures of this jet can now be recognized in the X-ray data (Sections 4–5). However, the relationship between the disk and jet components and the origin of the X-ray properties of the hard state remain uncertain.

Difficulties in understanding the hard state are illustrated by results obtained for XTE J1118+480, a BHB with an extraordinarily small interstellar attenuation (e.g., only 30% at 0.3 keV) and a display of weak outbursts confined to the hard state. This source provides the best direct determination of the apparent temperature and radius of the inner disk in the hard state. Using simultaneous HST, EUVE, and *Chandra* observations (McClintock et al. 2001), the disk was found to be unusually large ($\sim 100 R_g$) and cool (~ 0.024 keV). Slightly higher temperatures (≈ 0.035–0.052 keV) were inferred from observations with BeppoSAX (Frontera et al. 2003). Though it seems clear that the blackbody radiation is truncated at a large radius, the physical condition of material within this radius remains uncertain. Alternative scenarios include a thermal advection-dominated accretion flow (ADAF; Esin et al. 2001), a radiative transition to synchrotron emission in a relativistic flow that is entrained in a jet (Markoff, Falcke & Fender 2001), and a radiative transition to a Compton corona (Frontera et al. 2003), which must then be sufficiently optically thick to mask the ~ 1 keV thermal component normally seen from the disk. Such a corona might be a hot wind leaving the disk (Blandford & Begelman 1999, 2004).

Recent investigations of other BHBs in the hard state suggest that both synchrotron and Compton components contribute to the broadband spectrum; the Compton emission is presumed to originate at the base of the jet (Kalemci et al. 2005; Markoff, Nowak & Wilms 2005). It is also possible that the jet is supplied by hot gas from a surrounding ADAF flow (Yuan, Cui & Narayan 2005).

Guidance in understanding the accretion flow in this inner region may eventually come from other types of investigations, such as the study of correlated optical/X-ray variability (Malzac et al. 2003). Also promising are spectral analyses that focus on features that indicate densities higher than that expected for an optically thin flow, such as the ADAF mentioned above. One such feature is the broad Fe emission line (e.g., Miller et al. 2002b, 2002c). The Fe line profile can provide information on the Keplerian flow pattern and constrain the inner disk radius (see Section 8.2.3). Another diagnostic spectral feature is an X-ray reflection component (Done & Nayakshin 2001). In one study of the reflection component of Cyg X–1, the hard-state disk appeared to be truncated at a few tens of Schwarzschild radii (Done & Zycki 1999).

7.3. Steep Power Law

The physical origin of the SPL state remains one of the outstanding problems in high-energy astrophysics. It is crucial that we gain an understanding of this state, which is capable of generating HFQPOs, extremely high luminosity, and spectra that extend to $\gtrsim 1$ MeV.

Most models for the SPL state invoke inverse Compton scattering as the operant radiation mechanism (see Zdziarski & Gierliński 2004). The MeV photons suggest that the scattering occurs in a nonthermal corona, which may be a simple slab operating on seed photons from the underlying disk (e.g., Zdziarski et al. 2005). Efforts to define the origin of the Comptonizing electrons have led to models with more complicated geometry and with feedback mechanisms, such as flare regions that erupt from magnetic instabilities in the accretion disk (Poutanen & Fabian 1999). There are alternative models of the SPL state. For example, bulk motion Comptonization has been proposed in the context of a converging sub-Keplerian flow within $50R_g$ of the BH (Titarchuk & Shrader 2002; Turolla, Zane & Titarchuk 2002).

An analysis of extensive RXTE spectral observations of GRO J1655–40 and XTE J1550–564 shows that as the PL component becomes stronger and steeper, the disk luminosity and radius appear to decrease while the temperature remains high. These results can be interpreted as an observational confirmation of strong Comptonization of disk photons in the SPL state (Kubota & Makishima 2004).

7.4. Quasi-Periodic Oscillation Mechanisms

In addition to spectral observations, it is also necessary to explain timing observations of LFQPOs (0.1–30 Hz), which are commonly seen in the SPL state. There are now a large number of proposed LFQPO mechanisms in the literature, and we mention only a few examples here. The models are driven by the need to account for both the observed range of frequencies and the fact that the oscillations are strongest at photon energies above 6 keV, i.e., where the PL component completely dominates over the disk component. The models include global disk oscillations (Titarchuk & Osherovich 2000), radial oscillations of accretion structures such as shock fronts (Chakrabarti & Manickam 2000), and oscillations in a transition layer between the disk and a hotter Comptonizing region (Nobili et al. 2000). Another alternative, known as the accretion-ejection instability model, invokes spiral waves in a magnetized disk (Tagger & Pellat 1999) with a transfer of energy out to the radius where material corotates with the spiral wave. This model thereby combines magnetic instabilities with Keplerian motion to explain the observed QPO amplitudes and stability.

The behavior of the SPL state is complex and challenging. Nevertheless, we have much to work with, such as the exquisite quality of the data for this (usually) bright state, the regularities in behavior among various sources (**Figures 4–9**), and the remarkable couplings between the timing and spectral data (e.g., **Figure 10**). It appears that a successful model must allow for a highly dynamical interplay between thermal and nonthermal processes and involve mechanisms that can operate over a wide range of luminosity. Finally, we note that a physical understanding of the SPL state is required as a foundation for building any complete model of the HFQPO mechanisms, a topic considered in further detail in Section 8.2.4.

8. ACCRETING BLACK HOLES AS PROBES OF STRONG GRAVITY

The continuing development of gravitational wave astronomy is central to the exploration of BHs. In particular, we can reasonably expect that LIGO and LISA will provide us with intimate knowledge concerning the behavior of space-time under the most extreme conditions. Nevertheless, gravitational wave detectors are unlikely to provide us with direct information on the formation of relativistic jets, on strong-field relativistic MHD accretion flows, or on the origin of high-frequency QPOs or broadened Fe lines. Accreting BHs—whether they be stellar-mass, supermassive, or intermediate mass—promise to provide detailed information on all of these topics and more. In short, accreting BHs show us uniquely how a BH interacts with its environment. In this section, we first sketch a scenario for the potential impact of BHBs on physics, and we then discuss a current frontier topic, namely, the measurement of BH spin.

8.1. Black Holes Binaries: The Journey from Astrophysics to Physics

Astrophysics has a long history of impacting physics: e.g., Newton's and Einstein's theories of gravity, the ongoing research on dark matter and dark energy, the equation of state at supranuclear densities, and the solar neutrino puzzle. Likewise, astrophysical BHs have the potential to revolutionize classical BH physics; after all, the only real BHs we know, or are likely ever to know, are astrophysical BHs. But how can BHBs contribute to the study of BH physics? Very roughly, we envisage a five-stage evolutionary program that is presently well underway.

Stage I—*Identify Dynamical BH Candidates*: This effort is already quite advanced (see Section 1 and **Table 1**) and represents an important step, because mass is the most fundamental property of a BH. Nevertheless, the dynamical data do not probe any of the effects of strong gravity, and therefore we curtail the discussion of Stage I.

Stage II—*Confirm that the Candidates are True Black Holes*: Ideally, in order to show that a dynamical BH candidate (i.e., a massive compact object) is a genuine BH, one would hope to demonstrate that the candidate has an event horizon—the defining characteristic of a BH. Strong evidence has been obtained for the reality of the event horizon from observations that compare BHBs with very similar neutron-star binaries. These latter systems show signatures of the hard surface of a neutron star that are absent for the BH systems. For example, X-ray observations in quiescence show that the BH systems are about 100 times fainter than the nearly identical neutron-star binaries (Narayan, Garcia & McClintock 1997; Garcia et al. 2001). The ADAF model (Section 7) provides a natural explanation for the faintness of the BHs, namely, the low radiative efficiency of the accretion flow allows a BH to "hide" most of its accretion energy behind its event horizon (e.g, Narayan, Garcia & McClintock 2002).

In quiescence, one also observes that BHs lack a soft thermal component of emission that is very prevalent in the spectra of neutron stars and can be ascribed to surface emission (McClintock, Narayan & Rybicki 2004). During outburst, the presence of a surface for an accreting neutron star likewise gives rise to distinctive phenomena that are absent in BHBs: (*a*) type I thermonuclear bursts (Narayan & Heyl 2002, Tournear et al. 2003, Remillard et al. 2006a), (*b*) high-frequency timing noise (Sunyaev & Revnivtsev 2000), and (*c*) a distinctive spectral component from a boundary layer at the stellar surface (Done & Gierliński 2003).

Of course, all approaches to this subject can provide only indirect evidence for the event horizon because it is quite impossible to detect directly any radiation from this immaterial surface of infinite redshift. Nevertheless, barring appeals to very exotic physics, the body of evidence just considered makes a strong case that dynamical BH candidates possess an event horizon.

Stage III—*Measure the Spins of Black Holes*: An astrophysical BH is described by two parameters, its mass M and its dimensionless spin parameter a_*. Because the masses of 20 BHs have already been measured or constrained (see Stage I and Section 2), the next obvious goal is to measure spin. Indeed, several methods to measure spin have been described in the literature, and various estimates of a_* have been published, although few results thus far can be described as credible. We consider this stage to be a central and active frontier in BH research. Consequently we return to this subject below, where we discuss four approaches to measuring spin and some recent results for two BHBs.

Stage IV—*Relate Black Hole Spin to the Penrose Process and Other Phenomena*: A number of phenomena observed in astrophysical BHs have been argued to be associated with BH spin. The most notable examples are the explosive and relativistic radio jets associated with the hard-to-soft X-ray transition that occurs near the jet line (Section 4.3). Such ejections have been observed for at least eight BHBs and BHCs (Mirabel & Rodriguez 1999; Fender & Belloni 2004; McClintock & Remillard 2006, and references therein). Also, large-scale relativistic X-ray jets have been reported for XTE J1550-564 (Hannikainen et al. 2001, Corbel et al. 2002) and H 1743–322 (Corbel, Tomsick & Kaaret 2005). For many years, scientists have speculated that these jets are powered by BH spin via something like the Penrose (1969) process, which allows energy to be milked from a spinning BH. Detailed models generally invoke magnetic fields (Blandford & Znajek 1977, Hawley & Balbus 2002, Meier 2003, McKinney & Gammie 2004). A number of beautiful ideas have been published along these lines, but there has been no way of testing or confirming them. Recently, however, some progress has been made on measuring BH spin, and it may soon be possible to attack the jet-spin/Penrose-process connection in earnest.

Stage V—*Carry out Quantitative Tests of the Kerr Metric*: One of the most remarkable predictions of BH physics is that the space-time surrounding a stationary rotating BH is described by the Kerr metric, which is completely specified by just two parameters, M and a_*. Testing this prediction is the most important contribution astrophysics can make to BH physics. Obviously, in order to carry out such a test, one must first measure M and a_* with sufficient precision (Stage III). Once suitable measurements of M and a_* have been amassed for a number of BHs, we presume that astronomers will be strongly motivated to devise ways of testing the metric, a topic which is beyond the scope of this work.

8.2. Measuring Black Hole Spin: A Current Frontier

We now elaborate on Stage III by discussing four avenues for measuring BH spin. These include (*a*) X-ray polarimetry, which appears very promising but thus far has not been incorporated into any contemporary X-ray mission; (*b*) X-ray continuum fitting, which is already producing useful results; (*c*) the Fe K line profile, which has also yielded results, although the method is hampered by significant uncertainties; and (*d*) high-frequency QPOs, which arguably offer the most reliable measurement of spin once a model is established. We now consider each of these in turn.

8.2.1. Polarimetry.
As pointed out by Lightman & Shapiro (1975) and Meszaros et al. (1988), polarimetric information (direction and degree) would increase the parameter space used to investigate compact objects from the current two (spectra and time variability) to four independent parameters that models need to satisfy. Such constraints are likely to be crucial in our attempts to model the hard state with its radio jet and the SPL state. However, because of the complexities of the accretion flows associated with these states (Sections 4, 5, and 7), it appears unlikely that their study will soon provide quantitative probes of strong gravity. We therefore focus on disk emission in the thermal state.

The polarization features of BH disk radiation can be affected strongly by GR effects (Connors, Piran & Stark 1980). The crucial requirement for a simple interpretation is that higher energy photons come from smaller disk radii, as they are predicted to do in conventional disk models (Section 7). If this requirement is met, then as the photon energy increases from 1 keV to 30 keV, the plane of linear polarization swings smoothly through an angle of about 40° for a $9M_\odot$ Schwarzschild BH and 70° for an extreme Kerr BH (Connors, Piran & Stark 1980). The effect is due to the strong gravitational bending of light rays. In the Newtonian approximation, on the other hand, the polarization angle does not vary with energy, except for the possibility of a sudden 90° jump (Lightman & Shapiro 1976). Thus, a gradual change of the plane of polarization with energy is a purely relativistic effect, and the magnitude of the change can give a direct measure of a_*.

A model is now available in XSPEC that allows one to compute the Stokes parameters of a polarized accretion disk spectrum (Dovčiak, Karas & Yaqoob 2004). While the theoretical picture is bright, and very sensitive instruments can be built (e.g., Kaaret et al. 1994; Costa et al. 2001), unfortunately, results to date are meager and there are no mission opportunities on the horizon. Important advances in this promising area could be made by a relatively modest mission given that BHBs in the thermal state are bright.

8.2.2. Continuum fitting.
Pioneering work in fitting the spectrum of the X-ray continuum to measure spin was carried out by Zhang, Cui & Chen (1997), and the method was advanced further by Gierliński, Maciolek-Niedzwiecki & Ebisawa

(2001). Very recently, two developments have allowed this method to be applied more widely and with some confidence, namely: (*a*) models of thin accretion disks are now publicly available in XSPEC ("KERRBB," Li et al. 2005; "KY," Dovčiak, Karas & Yaqoob 2004) that include all relativistic effects and allow one to fit for a_*; and (*b*) sophisticated disk atmosphere models now exist for computing the spectral hardening factor, $f_{col} = T_{col}/T_{eff}$ as a function of the Eddington-scaled luminosity of the disk (Davis et al. 2005).

This method of measuring a_* depends on the properties of thin accretion disks described in Section 7 and is most convincing when it is applied to BHBs in the thermal state (Sections 4, 5, and 7). Effectively, in this technique, one determines the radius R_{in} of the inner edge of the accretion disk and assumes that this radius corresponds to R_{ISCO} (see Section 1). Because R_{ISCO}/R_g is a monotonic function of a_*, a measurement of R_{in} and M directly gives a_*. Provided that (*a*) *i* and *D* are sufficiently well known, (*b*) the X-ray flux and spectral temperature are measured from well-calibrated X-ray data in the thermal state, and (*c*) the disk radiates as a blackbody, it is clear that R_{in} can be estimated. A major complication is that the disk emission is not a true blackbody but a modified blackbody with a spectral hardening factor f_{col}. Therefore, the observations only give the quantity R_{in}/f_{col}^2, and one needs an independent estimate of f_{col} in order to estimate a_* (Shimura & Takahara 1995; Davis et al. 2005). A second caveat is that the orbital inclination may differ significantly from the inclination of the BH's spin axis (Maccarone 2002).

Using the new disk models mentioned above, Shafee et al. (2006) fitted ASCA and RXTE spectral data on the BHB GRO J1655–40 and found $a_* \sim 0.65$–0.75. For 4U 1543–47, they found $a_* \sim 0.75$–0.85, although this result is based only on RXTE data. The authors consider it unlikely that either BH has a spin close to the theoretical maximum, $a_* = 1$. On the other hand, in the case of 4U 1543–47, the estimated spin appears too high to be explained by spin up due to accretion, which suggests that their measurements are sensitive to the BH's natal spin.

8.2.3. Fe K line profile. The first broad Fe Kα line observed for either a BHB or an AGN was reported in the spectrum of Cyg X–1 based on EXOSAT data (Barr, White & Page 1985). Since then, the line has been widely studied in the spectra of both BHBs and AGN. The Fe K fluorescence line is thought to be generated through the irradiation of the cold (weakly-ionized) disk by a source of hard X-rays (likely an optically-thin, Comptonizing corona). Relativistic beaming and gravitational redshifts in the inner disk region can serve to create an asymmetric line profile (for a review, see Reynolds & Nowak 2003).

The line has been modeled in the spectra of several BHBs. In some systems the inner disk radius deduced from the line profile is consistent with the $6R_g$ radius of the ISCO of a Schwarzschild BH, suggesting that rapid spin is not required (e.g., GRS 1915+105, Martocchia et al. 2002; V4641 Sgr, Miller et al. 2002a). On the other hand, fits for GX 339–4 indicate that the inner disk likely extends inward to $(2 - 3)R_g$, implying $a_* \geq 0.8$–0.9 (Miller et al. 2004). XTE J1650–500 is the most extreme case with the inner edge located at $\approx 2R_g$, which suggests nearly maximal spin (Miller et al. 2002c; Miniutti, Fabian & Miller 2004). Large values of a_* have also been reported for XTE J1655–40 and XTE J1550–564 (Miller et al. 2005). Sources of uncertainty in the method include the placement of the continuum, the model of the fluorescing source, and the ionization state of the disk (Reynolds & Nowak 2003). Also, thus far the analyses have been done using the LAOR model in XSPEC, which fixes the spin parameter at $a_* = 0.998$ (Laor 1991). A reanalysis of archival data using new XSPEC models that allow one to fit for a_* may prove useful (KY, Dovčiak, Karas & Yaqoob 2004; KD, Beckwith & Done 2004).

8.2.4. High-frequency quasi-periodic oscillations. Arguably, HFQPOs (see Section 6.2) are likely to offer the most reliable measurement of spin once the correct model is known. Typical frequencies of these fast QPOs, e.g., 150–450 Hz, correspond respectively to the frequency at the ISCO for Schwarzschild BHs with masses of 15–5 $M_\odot$, which in turn closely matches the range of observed masses (**Table 1**). As noted in Section 6.2, these QPO frequencies (single or pairs) do not vary significantly despite sizable changes in the X-ray luminosity. This suggests that the frequencies are primarily dependent on the mass and spin of the BH. Those BHs that show HFQPOs and have well-constrained masses are the best prospects for contraining the value of the BH spin (a_*).

The four sources that exhibit harmonic pairs of frequencies in a 3:2 ratio (see Section 6.2) suggest that HFQPOs arise from some type of resonance mechanism (Abramowicz & Kluźniak 2001; Remillard et al. 2002a). Resonances were first discussed in terms of specific radii where particle orbits have oscillation frequencies in GR (see Merloni et al. 1999) that scale with a 3:1 or a 3:2 ratio. Current resonance concepts now consider accretion flows in a more realistic context. For example, the "parametric resonance" concept (Abramowicz et al. 2003; Kluźniak, Abramowicz & Lee 2004; Török et al. 2005) describes oscillations rooted in fluid flow where there is a coupling between the radial and polar coordinate frequencies. As a second example, one recent MHD simulation provides evidence for resonant oscillations in the inner disk (Kato 2004); in this case, however, the coupling relation involves the azimuthal and radial coordinate frequencies. If radiating blobs do congregate at a resonance radius for some reason, then ray tracing calculations have shown that GR effects can cause measurable features in the X-ray power spectrum (Schnittman & Bertschinger 2004).

Other models utilize variations in the geometry of the accretion flow. For example, in one model the resonance is tied to an asymmetric structure (e.g., a spiral wave) in the inner accretion disk (Kato 2005). In an alternative model, state changes are invoked that thicken the inner disk into a torus; the normal modes (with or without a resonance condition) can yield oscillations with a 3:2 frequency ratio (Rezzolla et al. 2003; Fragile 2005). All of this research is still in a developmental state, and these proposed explanations for HFQPOs are basically dynamical models that lack radiation mechanisms and fail to fully consider the spectral properties of HFQPOs described in Section 6.2.

Theoretical work aimed at explaining HFQPOs is motivated by the following empirical result that is based on a very small sample of three sources: XTE J1550–564, GRO J1655–40, and GRS 1915+105. These sources are presently the only ones that both exhibit harmonic (3:2) HFQPOs and have measured BH masses.

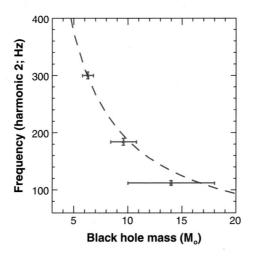

Figure 12
Relationship between high-frequency quasi-periodic oscillation (HFQPO) frequency and BH mass for XTE J1550–564, GRO J1655–40, and GRS 1915+105. These three systems display a pair of HFQPOs with a 3:2 frequency ratio. The frequencies are plotted for the stronger QPO that represents $2 \times \nu_0$. The fundamental is generally not seen in the power spectra. The dashed line shows a relation, ν_0 (Hz) $= 931 \, (M/M_\odot)^{-1}$, that fits these data.

As shown in **Figure 12**, their frequencies appear to scale inversely with mass (McClintock & Remillard 2005), which is the dependence expected for coordinate frequencies (see Merloni et al. 1999) or for diskoseismic modes in the inner accretion disk (Wagoner 1999, Kato 2001). If these HFQPOs are indeed GR oscillations, then **Figure 12** further suggests that the three BHs have similar values of the spin parameter a_*. Obviously it is of great importance to attempt to confirm this result by obtaining the requisite frequency and mass measurements for additional sources.

8.3. Critique of Methods for Measuring Spin

In short, there are four avenues for measuring spin—polarimetry, continuum fitting, the Fe K line, and HFQPOs. Because spin is such a critical parameter, it is important to attempt to measure it by as many of these methods as possible, as this will provide arguably the best possible check on our results. The best current method, continuum fitting, has the drawback that its application requires accurate estimates of BH mass (M), disk inclination (i), and distance. In contrast, assuming we have a well-tested model, QPO observations require knowledge of only M to provide a spin estimate. Broadened Fe K lines and polarimetry data do not even require M, although knowledge of i is useful in order to avoid having to include that parameter in the fit. On the other hand, the Fe-line and HFQPO methods are not well-enough developed to provide dependable results, and the required polarimetry data are not available, whereas the continuum method, despite its limitations, is already delivering results.

9. BLACK-HOLE BINARIES YESTERDAY, TODAY, AND TOMORROW

Three historic milestones in stellar BH research are separated by roughly one human generation, namely, Einstein's 1915 paper on GR, Oppenheimer & Snyder's 1939 paper on gravitational collapse, and the 1972 identification of BH Cyg X–1 (see Section 1). Now, one more generation later, we have measured or constrained the masses of 20 BHBs and obtained spin estimates for two of them. As described herein, a rich X-ray data archive, as well as ongoing observations, are providing an intimate look at the behavior of transient BHs as they vary in X-ray luminosity by 5–8 orders of magnitude. We have obtained strong circumstantial evidence for the existence of the event horizon and observed harmonic pairs of HFQPOs and relativistically-broadened Fe lines emanating from near the ISCO. However, the RXTE detectors are unable to resolve the Fe K line, and the HFQPOs are all near the limiting sensitivity of the mission. Further major advances will require a new timing mission with order-of-magnitude increases in detector area, telememetry capability, and spectral resolving power. It is important to press ahead with a new timing mission soon in order to complement effectively the vigorous programs underway in gravitational wave astronomy and in observational studies of supermassive and intermediate-mass BHs.

SUMMARY POINTS

1. The topics of black-hole binaries and candidates are introduced, along with the perspective of general relativity on black hole mass and spin.

2. The presence of a black hole is deduced from dynamical measurements of its binary companion. Twenty such systems are identified.

3. Data analysis techniques are summarized for the X-ray observations that characterize black-hole binaries.

4. Accreting black holes exhibit different X-ray states that are seen as distinct and very different combinations of X-ray energy spectra and power-density spectra.

5. The temporal evolution of X-ray states and the manner in which states are related to primary spectral and timing properties are illustrated for six selected sources.

6. Quasi-periodic oscillations occur in some states. They span a wide range in frequency, and they impose requirements on physical models.

7. Physical models are briefly reviewed for X-ray states and quasi-periodic oscillations.

8. We discuss the present and future for efforts to utilize black-hole binaries as a test bed for applications of general relativity.

ACKNOWLEDGEMENTS

We are grateful to Ramesh Narayan, Jerry Orosz, Rob Fender, and Alan Levine for contributing material for this review. We also thank Marek Abramowicz, Wlodek Kluźniak, Omer Blaes, Jeroen Homan, Phil Kaaret, and Jon Miller for helpful disussions while this paper was in preparation. This work was funded in part by the NASA contract to MIT for support of RXTE and also by NASA Grant NNG-05GB31G to JEM.

LITERATURE CITED

Abramowicz MA, Karas V, Kluźniak W, Lee WH, Rebusco P. 2003. *Publ. Astron. Soc. Jpn.* 55:467–71
Abramowicz MA, Kluźniak W. 2001. *Astron. Astrophys.* 374:L19–20
Agol E, Krolik JH. 2000. *Ap. J.* 528:161–70
Balbus SA, Hawley JF. 1991. *Ap. J.* 376:214–22
Barr P, White NE, Page CG. 1985. *MNRAS* 216:65P–70P
Beckwith K, Done C. 2004. *MNRAS* 352:353–62
Belloni T. 2004. *Nucl. Phys. B-Proc. Suppl.* 132:337–45
Belloni T, Homan J, Casella P, van der Klis M, Nespoli E, et al. 2005. *Astron. Astrophys.* 440:207–22
Blandford RD, Begelman C. 1999. *MNRAS* 303:L1–5
Blandford RD, Begelman C. 2004. *MNRAS* 349:68–86
Blandford RD, Znajek RL. 1977. *MNRAS* 179:433–56
Bolton CT. 1972. *Nature* 235:271–73
Brown GE, Bethe HA. 1994. *Ap. J.* 423:659–64
Brown GE, Lee C-H, Wijers RAM, Lee HK, Israelian G, Bethe HA. 2000. *New Astron.* 5:191–210
Casares J, Zurita C, Shahbaz T, Charles PA, Fender RP. 2004. *Ap. J.* 613:L133–36
Casella P, Belloni T, Homan J, Stella L. 2004. *Astron. Astrophys.* 426:587–600
Chakrabarti SK, Manickam SG. 2000. *Ap. J.* 531:L41–44
Charles PA, Coe MJ. 2006. In *Compact Stellar X-ray Sources*, ed. WHG Lewin, M van der Klis, pp. 215–66. Cambridge: Cambridge Univ.
Chen W, Shrader CR, Livio M. 1997. *Ap. J.* 491:312–38
Coe MJ, Engel AR, Quenby JJ. 1976. *Nature* 259:544–45
Connors PA, Piran RF, Stark T. 1980. *Ap. J.* 235:224–44
Corbel S, Fender RP, Tzioumis AK, Nowak M, McIntyre V, et al. 2000. *Astron. Astrophys.* 359:251–68
Corbel S, Fender RP, Tzioumis AK, Tomsick JA, Orosz JA, et al. 2002. *Science* 298:196–99
Corbel S, Kaaret P, Fender RP, Tzioumis AK, Tomsick JA, Orosz JA. 2005. *Ap. J.* 632:504–13
Corbel S, Tomsick JA, Kaaret P. 2006. *Ap. J.* 636:971–78
Costa E, Soffitta P, Bellazzini R, Brez A, Lumb N, et al. 2001. *Nature* 411:662–65
Cowley AP, Crampton D, Hutchings JB, Remillard R, Penfold JE. 1983. *Ap. J.* 272:118–22

Cui W, Shrader CR, Haswell CA, Hynes RI. 2000. *Ap. J.* 535:L123–27
Cui W, Zhang SN, Chen W. 2000. *Ap. J.* 531:L45–48
Davis SW, Blaes OM, Hubeny I, Turner NJ. 2005. *Ap. J.* 621:372–98
DeVilliers J-P, Hawley JF, Krolik JH. 2003. *Ap. J.* 599:1238–53
Dhawan V, Mirabel IF, Rodriguez LF. 2000. *Ap. J.* 543:373–85
Done C, Gierliński M. 2003. *MNRAS* 342:1041–55
Done C, Nayakshin S. 2001. *MNRAS* 328:616–22
Done C, Zycki PT. 1999. *MNRAS* 305:457–68
Dovčiak M, Karas V, Yaqoob T. 2004. *Ap. J. Suppl.* 153:205–21
Dubus G, Hameury J-M, Lasota J-P. 2001. *Astron. Astrophys.* 373:251–71
Elvis M, Page CG, Pounds KA, Ricketts MJ, Turner MJL. 1975. *Nature* 257:656–57
Esin AA, McClintock JE, Drake JJ, Garcia MR, Haswell CA, et al. 2001. *Ap. J.* 555:483–88
Fender RP. 2006. In *Compact Stellar X-ray Sources*, ed. WHG Lewin, M van der Klis, pp. 381–420. Cambridge: Cambridge Univ.
Fender RP, Belloni T. 2004. *Annu. Rev. Astron. Astrophys.* 42:317–64
Fender RP, Belloni T, Gallo E. 2004. *MNRAS* 355:1105–18
Fender R, Corbel S, Tzioumis T, McIntyre V, Campbell-Wilson D, et al. 1999. *Ap. J.* 519:L165–68
Fragile CP. 2005. In *Proceedings of 22nd Texas Symposium on Relativistic Astrophysics at Stanford*, ed. P Chen, E Bloom, G Madejski, V Patrosian, econf/C041213 (astro-ph/0503305)
Frontera F, Amati L, Zdziarski AA, Belloni T, Del Sordo S, et al. 2003. *Ap. J.* 592:1110–18
Fryer C, Kalogera V. 2001. *Ap. J.* 554:548–60
Gallo E, Fender RP, Pooley GG. 2003. *MNRAS* 344:60–72
Gammie CF. 2004 *Ap. J.* 614:309–13
Garcia MR, McClintock JE, Narayan R, Callanan P, Barret D. 2001. *Ap. J.* 553:L47–50
Gierliński M, Maciolek–Niedzwiecki A, Ebisawa K. 2001. *MNRAS* 325:1253–65
Gliozzi M, Papadakis IE, Räth C. 2006. *Astron. Astrophys.* In press (astro-ph/0512026)
Grove JE, Johnson WN, Kroeger RA, McNaron-Brown K, Skibo JG, Phlips BF. 1998. *Ap. J.* 500:899–908
Hannikainen D, Campbell-Wilson D, Hunstead R, McIntyre V, Lovell J, et al. 2001. *Astrophys. Space Sci.* 276:45–48
Hawley, JF, Balbus SA. 2002. *Ap. J.* 573:738–48
Homan J, Klein–Wolt M, Rossi S, Miller JM, Wijnands R, et al. 2003. *Ap. J.* 586:1262–67
Homan J, Miller JM, Wijnands R, Lewin WHG. 2005a. *BAAS* 207:1331–1331
Homan J, Miller JM, Wijnands R, van der Klis M, Belloni T, et al. 2005b. *Ap. J.* 623:383–91
Homan J, Wijnands R, van der Klis M, Belloni T, van Paradijs J, et al. 2001. *Ap. J. Suppl.* 132:377–402
Hynes RI, Steeghs D, Casares J, Charles PA, O'Brien K. 2003. *Ap. J.* 583:L95–98
Israelian G, Rebolo R, Basri G. 1999. *Nature* 401:142–44

Johnson BM, Gammie CF. 2003. *Ap. J.* 597:131–41
Kaaret PE, Schwartz J, Soffitta P, Dwyer J, Shaw PS, et al. 1994 *SPIE* 2010:22–27
Kalemci E, Tomsick JA, Buxton MM, Rothschild RE, Pottschmidt K, et al. 2005. *Ap. J.* 622:508–19
Kalogera V, Baym G. 1996. *Ap. J.* 470:L61–64
Kato S. 2001. *Publ. Astron. Soc. Jpn.* 53:1–24
Kato S. 2005. *Publ. Astron. Soc. Jpn.* 57:L17–20
Kato S, Fukue J, Mineshige S. 1998. *Black–Hole Accretion Disks*, Kyoto: Kyoto Univ. 594 pp.
Kato Y. 2004. *Publ. Astron. Soc. Jpn.* 56:931–37
Klein-Wolt M, Homan J, van der Klis M. 2004. *Nucl. Phys. B-Proc. Suppl.* 132:381–86
Kluźniak W, Abramowicz MA, Lee W. 2004. *AIPC* 714:379–82
Kong AKH, Charles PA, Kuulkers E, Kitamoto S. 2002. *MNRAS* 329:588–96
Koyama K, Ikegami T, Inoue H, Kawai N, Makishima K, et al. 1984. *Publ. Astron. Soc. Jpn.* 36:659–66
Kubota A, Ebisawa K, Makishima K, Nakazawa K. 2005. *Ap. J.* 631:1062–71
Kubota A, Makishima K. 2004. *Ap. J.* 601:428–38
Laor A. 1991. *Ap. J.* 376:90–94
Lasota J-P. 2001. *New Astron. Rev.* 45:449–508
Leahy DA, Darbro W, Elsner RF, Weisskopf MC, Sutherland PG, et al. 1983. *Ap. J.* 266:160–70
Lee JC, Reynolds CS, Remillard RA, Schulz NS, Blackman EG, Fabian AC. 2002. *Ap. J.* 567:1102–11
Levine AM, Bradt H, Cui W, Jernigan JG, Morgan EH, et al. 1996. *Ap. J.* 469:L33–36
Li L-X, Zimmerman ER, Narayan R, McClintock JE. 2005. *Ap. J. Suppl.* 157:335–70
Lightman AP, Shapiro SL. 1975. *Ap. J.* 198:L73–75
Lightman AP, Shapiro SL. 1976. *Ap. J.* 203:701–3
Maccarone TJ. 2002. *MNRAS* 336:1371–76
Maccarone TJ, Coppi PS. 2003. *MNRAS* 338:189–96
Makishima K, Maejima Y, Mitsuday K, Bradt HV, Remillard RA, et al. 1986. *Ap. J.* 308:635–43
Malzac J, Belloni T, Spruit HC, Kanbach G. 2003. *Astron. Astrophys.* 407:335–45
Markoff S, Falcke H, Fender R. 2001. *Astron. Astrophys.* 372:L25–28
Markoff S, Nowak MA, Wilms J. 2005. *Ap. J.* 635:1203–16
Martocchia A, Matt G, Karas V, Belloni T, Feroci M. 2002. *Astron. Astrophys.* 387:215–21
Matsumoto R, Machida M, Nakamura K. 2004. *Prog. Theor. Phys. Suppl.* 155:121–31
McClintock JE, Haswell CA, Garcia MR, Drake JJ, Hynes RI, et al. 2001. *Ap. J.* 555:477–82
McClintock JE, Horne K, Remillard RA. 1995. *Ap. J.* 442:358–65
McClintock JE, Narayan R, Rybicki GB. 2004. *Ap. J.* 615:402–15
McClintock JE, Remillard RA. 1986. *Ap. J.* 308:110–22
McClintock JE, Remillard RA. 2006. In *Compact Stellar X-ray Sources*, ed. WHG Lewin, M van der Klis, pp. 157–214. Cambridge: Cambridge Univ.
McKinney JC, Gammie CF. 2004. *Ap. J.* 611:977–95

Meier DL. 2003 *New Astron. Rev.* 47:667–72
Merloni A. 2003. *MNRAS* 341:1051–56
Merloni A, Vietri M, Stella L, Bini D. 1999. *MNRAS* 304:155–59
Meszaros P, Novick R, Szentgyorgyi A, Chanan GA, Weisskopf MC. 1988. *Ap. J.* 324:1056–67
Miller JM, Fabian AC, in't Zand JJM, Reynolds CS, Wijnands R, et al. 2002a. *Ap. J.* 577:L15–18
Miller JM, Fabian AC, Nowak MA, Lewin W. 2005. *Procs., Tenth Marcel Grossmann Meeting on General Relativity*, ed. M Novello, S Perez-Bergliaffa, R Ruffini. In press. Singapore: World Scientific. (astro-ph/0402101)
Miller JM, Fabian AC, Reynolds CS, Nowak MA, Homan J, et al. 2004. *Ap. J.* 606:L131–34
Miller JM, Fabian AC, Wijnands R, Remillard RA, Wojdowski P, et al. 2002b. *Ap. J.* 578:348–56
Miller JM, Fabian AC, Wijnands R, Reynolds CS, Ehle M, et al. 2002c. *Ap. J.* 570:L69–73
Miller JM, Wijnands R, Homan J, Belloni T, Pooley GG, et al. 2001. *Ap. J.* 563:928–33
Miniutti G, Fabian AC, Miller JM. 2004. *MNRAS* 351:466–72
Mirabel IF, Rodriguez LF. 1999. *Annu. Rev. Astron. Astrophys.* 37:409–43
Mitsuda K, Inoue H, Koyama K, Makishima K, Matsuoka M, et al. 1984. *Publ. Astron. Soc. Jpn.* 36:741–59
Miyamoto S, Iga S, Kitamoto S, Kamado Y. 1993. *Ap. J.* 403:L39–42
Miyamoto S, Kitamoto S. 1991. *Ap. J.* 374:741–43
Morgan EH, Remillard RA, Greiner J. 1997. *Ap. J.* 482:993–1010
Muno MP, Morgan EH, Remillard RA. 1999. *Ap. J.* 527:321–40
Muno MP, Remillard RA, Morgan EH, Waltman EB, Dhawan V, et al. 2001. *Ap. J.* 556:515–32
Muno MP, Remillard RA, Chakrabarty D. 2002. *Ap. J.* 568:L35–39
Narayan R, Garcia MR, McClintock JE. 1997. *Ap. J.* 478:L79–82
Narayan R, Garcia MR, McClintock JE. 2002. In *Proc. Ninth Marcel Grossmann Meeting*, ed. VG Gurzadyan, et al. pp. 405–25. Singapore: World Scientific
Narayan R, Heyl JS. 2002. *Ap. J.* 574:L139–42
Nobili L, Turolla R, Zampieri L, Belloni T. 2000. *Ap. J.* 538:L137–40
Nowak MA. 2000. *MNRAS* 318:361–67
Oppenheimer JR, Snyder H. 1939. *Phys. Rev.* 56:455–59
Orosz JA. 2003. *A Massive Star Odyssey: From Main Sequence to Supernova*, ed. KA van der Hucht, A Herraro, C Esteban, pp. 365–71. San Francisco: Astron. Soc. Pac.
Orosz JA, Kuulkers E, van der Klis M, McClintock JE, Garcia MR, et al. 2001. *Ap. J.* 555:489–503
Orosz JA, McClintock JE, Remillard RA, Corbel S. 2004. *Ap. J.* 616:376–82
Park SQ, Miller JM, McClintock JE, Remillard RA, Orosz JA, et al. 2004. *Ap. J.* 610:378–89
Penrose R. 1969. *Riv. Nuovo Cim.* 1:252–76
Podsiadlowski P, Rappaport S, Han A. 2003. *MNRAS* 341:385–404

Poutanen J, Fabian AC. 1999. *MNRAS* 306:L31–37
Pringle JE. 1981. *Annu. Rev. Astron. Astrophys.* 19:137–62
Psaltis D. 2006. In *Compact Stellar X-ray Sources*, ed. WHG Lewin, M van der Klis, pp. 1–38. Cambridge: Cambridge Univ.
Remillard RA. 2004. *AIPC* 714:13–20
Remillard RA, Lin D, Cooper R, Narayan R. 2006a. *Ap. J.* 646: In press (astro-ph/0509758)
Remillard RA, McClintock JE, Orosz JA, Levine AM. 2006b. *Ap. J.* 637:1002–9
Remillard RA, Morgan EH, McClintock JE, Bailyn CD, Orosz JA. 1999. *Ap. J.* 522:397–412
Remillard RA, Muno MP, McClintock JE, Orosz JA. 2002a. *Ap. J.* 580:1030–42
Remillard RA, Sobczak GJ, Muno MP, McClintock JE. 2002b. *Ap. J.* 564:962–73
Revnivtsev M, Trudolyubov SP, Borozdin KN. 2000. *MNRAS* 312:151–58
Reynolds CS, Nowak MA. 2003. *Phys. Rept.* 377:389–466
Rezzolla L, Yoshida S'i, Maccarone TJ, Zanotti O. 2003. *MNRAS* 344:L37–41
Rodriguez J, Durouchoux P, Mirabel IF, Ueda Y, Tagger M, Yamaoka K. 2002. *Astron. Astrophys.* 386:271–79
Rossi S, Homan J, Miller JM, Belloni T. 2004. *Nucl. Phys. B-Proc. Suppl.* 132:416–19
Scargle JD. 1981. *Ap. J. Suppl.* 45:1–71
Schnittman JD, Bertschinger E. 2004. *Ap. J.* 606:1098–1111
Shafee R, McClintock JE, Narayan R, Davis SW, Li L-X, Remillard RA. 2006. *Ap. J.* 636:L113–16
Shakura NI, Sunyaev RA. 1973. *Astron. Astrophys.* 24:337–66
Shapiro SL, Teukolsky SA. 1983. *Black Holes White Dwarfs and Neutron Stars: The Physics of Compact Objects*. New York: Wiley. 645 pp.
Shimura T, Takahara F. 1995. *Ap. J.* 445:780–88
Smak J. 1971. *Acta Astron.* 21:15–21
Sobczak GJ, McClintock JE, Remillard RA, Bailyn CD, Orosz JA. 1999. *Ap. J.* 520:776–87
Sobczak GJ, McClintock JE, Remillard RA, Cui W, Levine AM, et al. 2000a. *Ap. J.* 531:537–45
Sobczak GJ, McClintock JE, Remillard RA, Cui W, Levine AM, et al. 2000b. *Ap. J.* 544:993–1015
Socrates A, Davis SE, Blaes O. 2004 *Ap. J.* 601:405–13
Stella L, Vietri M, Morsink SM. 1999. *Ap. J.* 524:L63–66
Stirling AM, Spencer RE, de la Force CJ, Garrett MA, Fender RP, Ogley RN. 2001. *MNRAS* 327:1273–78
Strohmayer TE. 2001a. *Ap. J.* 552:L49–53
Strohmayer TE. 2001b. *Ap. J.* 554:L169–72
Sunyaev R, Revnivtsev M. 2000. *Astron. Astrophys.* 358:617–23
Swank J. 1998. *Nucl. Phys. B-Proc. Suppl.* 69:12–19
Tagger M, Pellat R. 1999. *Astron. Astrophys.* 349:1003–16
Tanaka Y, Lewin WHG. 1995. In *X-ray Binaries*, ed. WHG Lewin, J van Paradijs, EPJ van den Heuvel, pp. 126–174. Cambridge: Cambridge Univ.
Tanaka Y, Shibazaki N. 1996. *Annu. Rev. Astron. Astrophys.* 34:607–44

Tananbaum H, Gursky H, Kellogg E, Giacconi R, Jones C. 1972. *Ap. J.* 177:L5–10
Timmes FX, Woosley SE, Weaver TA. 1996. *Ap. J.* 457:834–43
Titarchuk L, Osherovich V. 2000. *Ap. J.* 542:L111–14
Titarchuk L, Shrader C. 2002. *Ap. J.* 567:1057–66
Tomsick JA, Kaaret P. 2001. *Ap. J.* 548:401–9
Tomsick JA, Kaaret P, Kroeger RA, Remillard RA. 1999. *Ap. J.* 512:892–900
Török G, Abramowicz MA, Kluźniak W, Stuchlík Z. 2005. *Astron. Astrophys.* 436:1–8
Tournear D, Raffauf E, Bloom ED, Focke W, Giebels B, et al. 2003. *Ap. J.* 595:1058–65
Turolla R, Zane S, Titarchuk L. 2002. *Ap. J.* 576:349–56
van der Klis M. 2006. In *Compact Stellar X-ray Sources*, ed. WHG Lewin, M van der Klis, pp. 39–112. Cambridge: Cambridge Univ.
van Paradijs J, McClintock JE. 1995. In *X-ray Binaries*, ed. WHG Lewin, J van Paradijs, EPJ van den Heuvel, pp. 58–125. Cambridge: Cambridge Univ.
Vaughan BA, Nowak MA. 1997. *Ap. J.* 474:L43–46
Vignarca F, Migliari S, Belloni T, Psaltis D, van der Klis M. 2003. *Astron. Astrophys.* 397:729–38
Wagoner RV. 1999. *Phys. Rept.* 311:259–69
Webster BL, Murdin P. 1972. *Nature* 235:37–38
White NE, Nagase F, Parmar AN. 1995. In *X-ray Binaries*, ed. WHG Lewin, J van Paradijs, EPJ van den Heuvel, pp. 1–57. Cambridge: Cambridge Univ.
Wijnands R, Homan J, van der Klis M. 1999. *Ap. J.* 526:L33–36
Woosley SE, Heger A, Weaver TA. 2002. *Rev. Mod. Phys.* 74:1015–71
Yuan F, Cui W, Narayan R. 2005. *Ap. J.* 620:905–14
Zdziarski AA, Gierliński M. 2004. *Prog. Theor. Phys. Suppl.* 155:99–119
Zdziarski AA, Gierliński M, Rao AR, Vadawale SV, Mikolajewska J. 2005. *MNRAS* 360:825–38
Zhang SN, Cui W, Chen W. 1997. *Ap. J.* 482:L155–158
Zhang SN, Cui W, Harmon BA, Paciesas WS, Remillard RE, van Paradijs J. 1997. *Ap. J.* 477:L95–98
Zimmerman ER, Narayan R, McClintock JE, Miller JM. 2005. *Ap. J.* 618:832–44

Absolute Magnitude Calibrations of Population I and II Cepheids and Other Pulsating Variables in the Instability Strip of the Hertzsprung-Russell Diagram

Allan Sandage[1] and Gustav A. Tammann[2]

[1]The Observatories of the Carnegie Institution of Washington, 813 Santa Barbara Street, Pasadena, California 91101

[2]Astronomisches Institut der Universität Basel, Venusstrasse 7, 4102 Binningen, Switzerland; email: G-A.Tammann@unibas.ch

Key Words

stars, variable, absolute magnitude calibrations, Cepheids, RR Lyraes, SX Phe stars, AHB1 stars, anomalous Cepheids

Abstract

The status of the absolute magnitude calibrations is reviewed for the long period Cepheids of population I and II, RR Lyrae stars, evolved "above horizontal branch" (AHB1) variables (periods 0.8 to 3 days), dwarf Cepheids of both populations (the Delta Scuti and SX Phoenicus variables), and the anomalous Cepheids (AC). Evidence shows that the period-color and period-luminosity (P-L) relations for population I Cepheids in the Galaxy and in the Large and Small Magellanic Clouds have different slopes and zero points. This greatly complicates use of Cepheids for the extragalactic distance scale. Strategies are discussed to patch the problem. A consensus exists for the long distance scale for RR Lyrae stars whose calibrations favor $\langle M_V(RR) \rangle = 0.52$ at $[Fe/H] = -1.5$. Exceptions exist for "second parameter" clusters where the variation of the morphology of the horizontal branch with metallicity is anomalous, the most blatant being NGC 6388 and NGC 6441. The status and calibrations of ABH1 and AC show that different evolutionary paths and masses explain the difference P-L relations for them. AC appear predominantly in the dwarf spheroidal galaxies, but are almost absent in Galactic globular clusters. AHB1 stars are absent in dwarf spheroidals but are present in globular clusters. The difference may be used to study the formation of the remote Galactic halo if it is partially made by tidal disruption of companion dwarf spheroidals.

1. INTRODUCTION

Generally, the more deeply a scientific problem is studied, the more complex becomes its solution. Although the first approximations made at the beginning can scout out a territory, as the database expands, first approximations must often be replaced. The use of Cepheid variables and RR Lyrae stars as distance indicators has now reached this point.

Recent observations have almost certainly shown that the period-luminosity (P-L) relation of classical Cepheid variables in the Galaxy has a different slope and zero point than that of Cepheids in the Large Magellanic Cloud (LMC) and Small Magellanic Cloud (SMC). Equivalent expressions of these differences are the observed offsets of the period-color relations between the Galaxy and the Clouds (due to Fraunhoffer blanketing differences plus the effects of real temperature differences in the position of the edges of the instability strip), and different slopes of the lines of constant period in the Hertzsprung-Russell (HR) diagrams of the three galaxies.

The discovery of the color differences was made 40 years ago by Gascoigne & Kron (1965) who found SMC Cepheids to be bluer than Galactic ones of the same period. The color difference was in the correct direction to be due to the known metal weakness in the SMC but was greater than could be accounted for by only the technical effect of line blanketing on the colors. However, that conclusion depended on the accuracy of the reddening corrections to the Galactic Cepheids, which were uncertain at the time.

The discovery was followed by the strong conclusion by Laney & Stobie (1986) that the temperature of the ridge-line locus of the instability strip for Cepheids in the SMC is indeed hotter than the ridge-line locus for Cepheids in the LMC and the Galaxy. Their value for the Galaxy-SMC temperature difference was 210 ± 80 K.

More recent analyses (Tammann, Sandage & Reindl 2003; Sandage, Tammann & Reindl 2004) confirm their conclusion, although the Galaxy-LMC temperature difference of 150 K is smaller than that of Laney & Stobie, and it depends on period (see figure 3 of Sandage, Tammann & Reindl 2004).

From an intricate analysis, Sandage, Tammann & Reindl (2004) concluded that the modern reddening values of Galactic Cepheids by Fernie (1990, 1994), and by 15 others reduced to the Fernie system by Fernie et al. (1995), are highly reliable after a small scale correction, removing the previous doubt for real temperature differences based on early Galactic reddening values.

Clearly, differences in the slope of the period-color relations between the Galaxy and the Clouds require that the slope of the P-L relations must also differ in the various colors. Even if the P-L relations were the same in one photometric band, they would differ in all other bands. Furthermore, if real temperature differences do exist, it is likely that the luminosity zero point of the individual P-L relations must also differ, unless unlikely compensation exists in the mass-luminosity relations such that the luminosity from the pulsation equation does not change.

This review addresses this and other problems concerning the uniqueness of the Cepheid P-L relation from galaxy to galaxy.

Since the last reviews in these pages and elsewhere about RR Lyrae (RRL) luminosities as a function of metallicity, nasty problems have also been discovered in their use as distance indicators. Until 1997, most data and analyses of period-luminosity-metallicity correlations of RRL were consistent, with their absolute magnitudes being tightly correlated with metallicity, such that if the [Fe/H] metallicity was known, then the RRL luminosity was also believed to be known to within narrow limits. The effects of evolution off the zero age horizontal branch (ZAHB) complicated the correlation, but a nearly unique period-amplitude-metallicity relation for unevolved stars (Sandage 1981a,b; Carney, Storm & Jones 1992, their equation 16; Alcock et al. 1998; Alcock 2000) was believed to be able to flag stars in this evolutionary state, permitting correction of the nonevolved calibration of the M_V(RR)-[Fe/H] relation. However, this assumption has now been challenged by the discovery of large-amplitude RRL variables with abnormally long periods in the metal rich ([Fe/H] ~ -0.5) globular clusters NGC 6388 and NGC 6441 (Rich et al. 1997; Pritzl et al. 2000, 2001). It is believed that the abnormal RRL period distribution in these clusters is due to an anomalously bright RRL HB luminosity (at the level of about 0.25 mag) compared with the canonical M_V-([Fe/H]) calibration known to be valid before 1997. This discovery showed that not all RRL follow the same luminosity-metallicity relation. There must be an unknown component to the pulsation physics (abnormal He core mass or some parameter like it) that complicates and spreads the HB zero-point luminosity values at given [Fe/H] values.

This review addresses these and other problems in the calibration and use of Cepheids and RRL variables as distance indicators. Earlier reviews of these problems include Feast & Walker (1987) and Reid (1997, 1999) in these *Annual Review* pages, and, among others, Carney, Storm & Jones (1992), Gieren, Fouque & Gomez (1998), Feast (1999, updated in 2003), Cacciari & Clementini (2003), Bono (2003), Tammann, Sandage & Reindl (2003), Fouque, Storm & Gieren (2003), and Sandage, Tammann & Reindl (2004), each with references to others to which the reader is referred. Many of the intricate details are not given again here.

Progress has also been made in testing the luminosity calibration of the Galactic classical Cepheids by (*a*) the method of angular radius measurements using stellar optical interferometers, (*b*) direct *Hipparcos* trigonometric parallaxes, (*c*) added data for the Baade (1926)-Becker (1940)-Wesselink (1946) moving atmospheres method, and (*d*) new data for the open cluster and association main sequence fittings.

New luminosity calibrations of RRL variables both in globular clusters and the general field have also been made by diverse methods including (*a*) luminosities calculated from a pulsation equation using observed inputs on period-metallicity correlations and temperature-metallicity correlations at the fundamental blue edge (FBE) of the instability strip, (*b*) main sequence fittings in the clusters, (*c*) Baade-Becker-Wesselink (BBW) moving atmosphere parallaxes for field variables, (*d*) the statistical parallax method using new radial velocities and updated proper motions, and (*e*) use of the newly discovered Delta Scuti variables (periods 0.05 to 0.25 days) in globular clusters (i.e., the SX Phe population II subclass) as stepping stones from their known trigonometric absolute magnitudes between M_V of +2 and +4 to the much brighter HB that contains the RRLs.

Progress has also been made in the luminosity calibration of the population II long period Cepheids (periods 13 to 30 days) and for the 1–3 day "above horizontal branch" (AHB1) variables in globular clusters and in the local dwarf spheroidal galaxies. The status is also reviewed of the "anomalous Cepheids" (AC) being discovered in an expanding data base.

This review contains the following: The instability strip for Cepheids is defined in Section 2 with a short history of its discovery and placement in the HR diagram. Modern period-color relations are in Section 3, based on data obtained by Berdnikov, Dambis & Voziakova (2000) for the Galaxy and by Udalski et al. (1999a,b) for the LMC and SMC. New slope and zero-point calibrations for the Cepheid P-L relations are set out in Section 4, based on a summary of data made available since Feast & Walker (1987) and Feast (1999). Comparison is made of the derived calibrations from main sequence fittings and the Baade-Becker-Wesselink moving atmosphere method with the absolute magnitudes from *Hipparcos* trigonometric parallaxes and from optical interferometer diameters of Cepheids.

New RRL luminosity calibrations are in Section 5, stressing the need for, and the evidence supporting, a nonlinear M_V-metallicity relation, both for the ZAHB for different metallicities and for evolved HB configurations. The nonuniqueness problem posed by the second parameter effect in its exaggerated form that has been signaled by the metal-rich clusters NGC 6388 and NGC 6441 is also discussed.

The status of the AHB1 variables in the field and in clusters is discussed in Section 6 as a prelude to a review of the modern calibration of the population II Cepheids relative to the cluster RRLs in Section 7.

Advances in the understanding of the anomalous Cepheids and their absolute magnitude calibration are in Section 8.

2. THE CLASSICAL CEPHEID INSTABILITY STRIP IN THE HR DIAGRAM

In the late 1800s, August Ritter (1879) began the study of the pulsation of stars by showing that the period of pulsation must be related to the mean density of a pulsating star by $P(\rho)^{1/2} = Q(P)$, where Q is a very weak function of period if the restoring force after displacement from the equilibrium position is gravity. The relation follows directly from dimensional analysis using Newtonian inverse square attraction and the law of inertia.

Cepheid variables had been identified as a homogeneous class in the last decade of the nineteenth century based on the similarity of light curves (see Fernie 1969 for a review), but had been shown to be pulsating stars only when Shapley (1914) gave a series of convincing arguments. An understanding of why a relatively tight P-L relation exists for Cepheids had then become clear when the Cepheids were discovered to have a highly restricted range of spectral type (temperature) at a given period (Adams & Joy 1927, Shapley 1927a). **Figure 1** shows this important conclusion from the paper by Adams & Joy. This temperature restriction (spectral type changed to temperature) at a given period, put together with the P-L relation, gives a temperature-luminosity locus in the HR diagram. This is the instability strip.

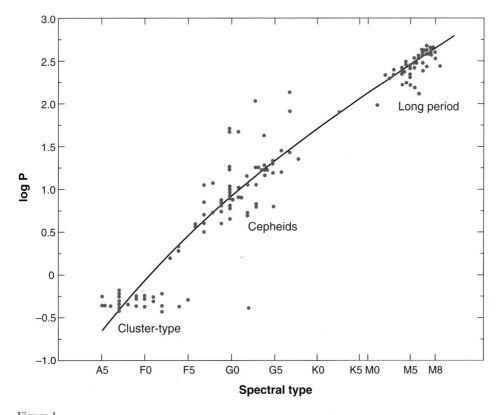

Figure 1
The relation by Adams & Joy (1927) between spectral type at maximum light and period for RR Lyrae-type variables, Cepheids, and long period Miras based on the first 15 years of sidereal observations made at Mount Wilson. The scatter at a given period is real due to the finite width of the instability strip. A similar relation using spectral type at midlight is by Shapley (1927a).

The isolation of the strip in the HR diagram was implicit in the attempts by Shapley (1927b) and Russell (1927) to understand the slope of the P-L relation. However, their papers were not particularly transparent in the modern language of the subject. They did realize that the Ritter P $(\rho)^{1/2}$ = Q(P) requirement would not restrict the luminosity at a given period unless there was a temperature restriction to the pulsation condition. Said differently, lines of constant period can be drawn in the HR diagram using the Ritter law. They slant over the diagram faintward and toward lower temperatures. This produces a large range of periods at a given luminosity if the temperatures of the variables can vary over an appreciable part of the HR diagram. However, the temperature restriction (the tight spectral-type, period correlation of **Figure 1**) cuts these constant period lines at discrete luminosities, producing a tight P-L relation.

Either because both the Shapley and Russell papers are semiopaque, or because interest in the problem settled on other aspects of it, these two central papers became almost lost in the archives of history between 1927 and 1950. However, the problem they solved was rediscovered once the evolutionary tracks of Cepheids across the HR diagram settled the problem of the Cepheid masses, at least in principle. Also, photoelectric observations of the Galactic Cepheids began to be obtained by Eggen (1951), permitting a color-magnitude HR diagram to be plotted (Eggen's figures 42 and 43). The language of an "instability strip" became explicit (Sandage 1958b) with its intrinsic width and lines of constant period, leading to an intrinsic scatter in the P-L relation that varies with wavelength (Sandage 1958b, 1972; Sandage & Tammann 1968, 1969).

The instability strip of the RRL variables is an extension of the Cepheid strip to fainter magnitudes and shorter periods. The RRL star position in this fainter part of the strip near $M_V = 0.5$ was discovered by Schwarzschild (1940), who showed that the RRL stars in the globular cluster M3 were confined to a narrow range of color along the cluster's HB. It was later shown that the same is true for all globular clusters and that the color boundaries of the RRL instability "gap" are a continuation of the Cepheid strip. The 1927 period-spectral relation of Adams & Joy in **Figure 1** shows the continuity of the period-spectral type relation of the RRL variables with the later spectral types of the long period Cepheids.

Hence, we are dealing with a single instability strip in the HR diagram (see Cox 1974, figure 1 for a review of the location of the various classes of variable stars in the HR diagram) that can be extended to the main sequence to include the Delta Scuti variables ultra short period Cepheids (USPC), to be discussed in Section 5.4, and even faintward to $M_V = +10$ to include the ZZ Ceti pulsating white dwarfs (figure 1 of Gautschy & Saio 1995).

2.1. Period-Color Relations for Cepheids in the Galaxy, LMC, and SMC

Since the last reviews cited earlier, two major advances have been made in the observations that permit nearly definitive determinations of the period-color and color-luminosity (i.e., the HR diagram) relations for the classical Cepheids in the Galaxy, the LMC, and the SMC.

Berdnikov, Dambis & Voziakova (2000) have published accurate B,V,I photometry on the Cape (Cousins) system as realized by Landolt (1983, 1992) for hundreds of Galactic Cepheids for which Fernie (1990, 1994) and Fernie et al. (1995) have determined E(B-V) color excesses. Tammann, Sandage & Reindl (2003) have made slight corrections to the color excess values on the Fernie system to remove a mild correlation of the initial Fernie color excess with residuals from the period-color relation; i.e., the Cepheids with large color excess on the initial Fernie system are systematically redder in their derived intrinsic colors than Cepheids of the same period with smaller excess values. The Tammann, Sandage & Reindl corrections produce the period-color relations in B-V and V-I for Galactic Cepheids of

$$(B-V)^\circ = (0.366 \pm 0.015)\log P + (0.361 \pm 0.013) \quad (1)$$

from 321 stars, and

$$(V\text{-}I)^\circ = (0.256 \pm 0.017)\log P + (0.497 \pm 0.016) \qquad (2)$$

from 250 stars.

These can be compared with similar period-color relations for Cepheids in the LMC and SMC from the extensive photometry by Udalski et al. (1999a,b). Analyzed in the same way as for the Galactic Cepheids, Tammann, Sandage & Reindl obtained period-color relations for both the LMC and SMC that differ from Equations 1 and 2. The differences are similar to those found by both Gascoigne & Kron and Laney & Stobie mentioned earlier. The comparisons in $(B\text{-}V)^\circ$ are shown in **Figure 2**.

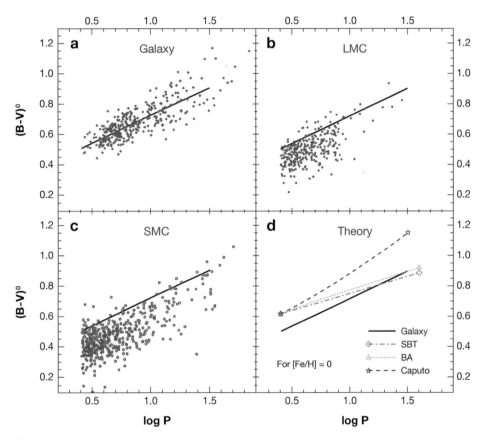

Figure 2

Comparison of the period-intrinsic $(B\text{-}V)^\circ$ color relations for the Galaxy, LMC, and SMC from the photometry of Berdnikov, Dambis & Voziakova (2000) for the Galaxy, and of Udalski et al. (1999a,b) for the LMC and SMC. The paucity of points with log P > 1.0 for the LMC show a selection effect in the data by Udalski et al. due to photometric saturation of the detectors used for the observations. The least squares fit to the Galaxy data in the upper left panel is repeated as the solid black line in the other three panels. The offsets discovered by Gascoigne & Kron (1965) and by Laney & Stobie (1986) are evident.

Figure 2a gives the color-period data for the Galaxy from Berdnikov, Dambis & Voziakova (2000). A linear least squares correlation line is drawn. This line is repeated in the other three panels to illustrate the color offset using the data for the LMC and SMC from Udalski et al. (1999a,b). The paucity of points with log P > 1.0 for the LMC shows a selection effect in the Udalski et al. data due to photometric saturation of the brightest Cepheids in the LMC, hence their rejection from the Udalski et al. database.

Figure 2d also shows predicted period-color ridge lines from three model calculations by Sandage, Bell & Tripicco (1999), Caputo, Marconi & Musella (2000b), and Baraffe & Alibert (2001), showing that the theoretical models, with their adopted transformations of temperature to B-V color, differ by ~0.2 mag at short periods but, except for the predictions of Caputo et al., agree to better than 0.05 mag in color for log P > 1.0.

Comparison of the LMC data with the Galactic Cepheids is shown in **Figure 3**. Closed circles are from Berdnikov, Dambis & Voziakova (2000). Open circles are

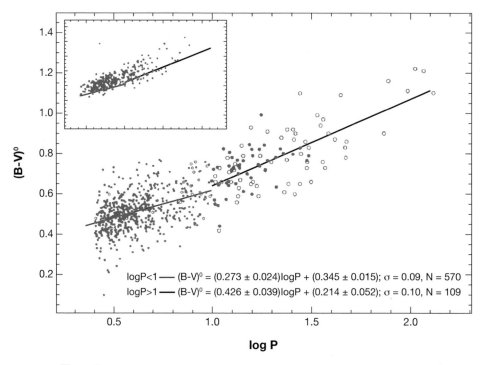

Figure 3

Comparison of the period-color relation of the Large Magellanic Cloud (LMC) with that of the Galaxy, showing the break at the 10-day period in the LMC but not in the Galaxy, and the difference in slope at all periods. The blue dots are from Udalski et al. (1999a). The open blue circles are additional data from the literature. Note the difference in the equations for the LMC (in the interior of the figure) with that of the Galaxy from Equation 1 here. The insert shows the individual Galaxy data compared with the two LMC mean lines. The difference is at the 4 sigma level. Diagram is from figure 1a of Sandage, Tammann & Reindl (2004).

additional longer period Cepheids from the literature. The difference in slope and zero point between the Galaxy and LMC is evident (shown in the insert of **Figure 3**). A similar difference exists in the period-(V-I)° relation from Sandage, Tammann & Reindl (2004), not shown here.

These color differences translate to differences in the positions of the instability strips in the HR diagrams for the Galaxy, LMC, and SMC. These can be constructed once the absolute magnitude scale is chosen. The result is shown in **Figure 4** using the true distance moduli shown in each panel, justified in Tammann, Sandage & Reindl (2003).

A more detailed comparison between the Galaxy and LMC in (B-V)° and (V-I)° is in **Figure 5**, taken from an updated analysis of the Udalski data (Sandage, Tammann & Reindl 2004, their figure 8). Four lines of constant period are shown, labeled by their log P values. The break at 10 days for the LMC is explicit. The insert sets out the individual data points for the Galaxy Cepheids, showing the offsets from the LMC ridge and the blue and red color boundaries as the lines in the insert diagrams.

The color data for the individual LMC Cepheids in **Figure 5** have been changed to temperatures by the color-temperature relations in Sandage, Bell & Tripicco (1999, their table 6) with the result shown in **Figure 6**. The two solid lines in both **Figures 5** and **6** near the middle of the strip for log P smaller and larger than 1.0 are drawn, as are the blue and red color boundaries of the strip, shown as dashed lines. Five lines of constant period in both diagrams are the sloping dotted lines crossing the instability strip, marked with their log P values. The ridge line of the Galaxy instability strip in **Figure 6** is transferred from **Figure 4** and changed to temperature.

The main conclusion from **Figure 6** is that there is a real temperature difference between the instability strips of the Galaxy and the LMC. The consequences, of course, are (*a*) there must be a difference in the zero points of the P-L relation between the Galaxy and the LMC (seen by the different log L values at the intersection of the lines of constant period with the strip boundaries and the ridge lines), and (*b*) the slopes of the P-L relations must also differ because the Galaxy and the LMC ridge lines are not parallel in **Figure 6**.

These are dire results, boding ill for the use of Cepheids as precision distance indicators at the 0.3-mag level unless corrections from one P-L relation to the other can be made. However, one wants to be convinced that the color and temperature differences shown in **Figures 2–6** are real.

The early evidence for color differences between the Galaxy and the LMC Cepheids by Gascoigne & Kron and by Laney & Stobie has been made stronger by the reddening values derived by Fernie that are confirmed by Tammann, Sandage & Reindl (2003), as discussed earlier. That evidence has been made even stronger by the recent analysis of Ngeow & Kanbur (2005).

The break in the Cepheid LMC period-color relation at 10 days and its absence in the Galaxy is the other principal difference between the two galaxies. The original suggestion of the color break by Tammann & Reindl (2002) and Tammann et al. (2002), supported by Kanbur & Ngeow (2002) and later argued by them in more detail (Kanbur & Ngeow 2004, Ngeow & Kanbur 2005, Ngeow et al. 2005), supports the reality of the difference.

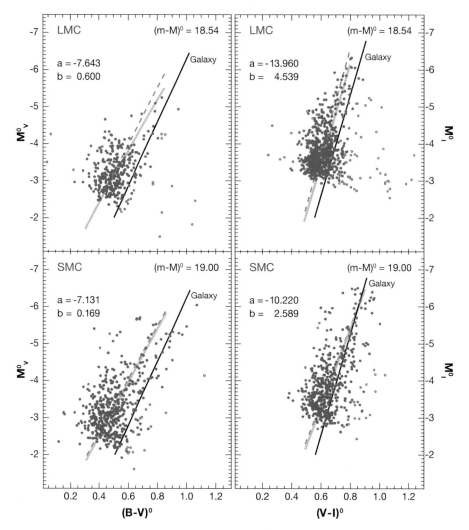

Figure 4

The instability strip in color for the Galaxy, LMC, and SMC. The ridge line for the Galaxy from Tammann, Sandage & Reindl (2003, their figure 16) is compared with the individual LMC and SMC data points showing the offset in both B-V and V-I at given $M_{V,I}$ values. The coefficients of $M_{V,I} = a(\text{color}) + b$ for the least squares fits are in each panel. The green dashed and light blue dotted lines are for two subsamples of the Udalski et al. data.

A most telling independent piece of evidence is by Tanvir et al. (2005) using Fourier decomposition of light curves into principal components to conclude that there are systematic differences in the light curve shapes at the same period between the Galaxy, LMC, and SMC.

Of course, the color differences require different slopes to the P-L relations, and the color break in the LMC (but not in the Galaxy) at ∼10 days also produces a break

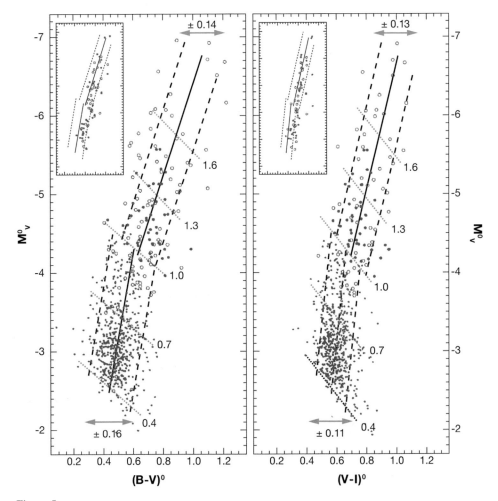

Figure 5
The instability strips for the LMC in (B-V)° and (V-I)° compared in the insert with the individual data for the Galactic Cepheids. Four lines of constant period, labeled by their log P values, are shown. The individual data in the main panels are for the LMC. Diagram is from Sandage, Tammann & Reindl (2004, their figure 8).

in the LMC P-L relation, but again, not in the Galaxy. It was at this point that a search for slope and zero point differences in the P-L relations became crucial.

3. CALIBRATIONS OF THE P-L RELATIONS FOR THE GALAXY, LMC, AND SMC

Before the reddening and absolute magnitude data for Galactic Cepheids became known with sufficient accuracy, it was common practice to adopt the P-L slope from Cepheids in the LMC and then to set the zero point in some way, such as by main

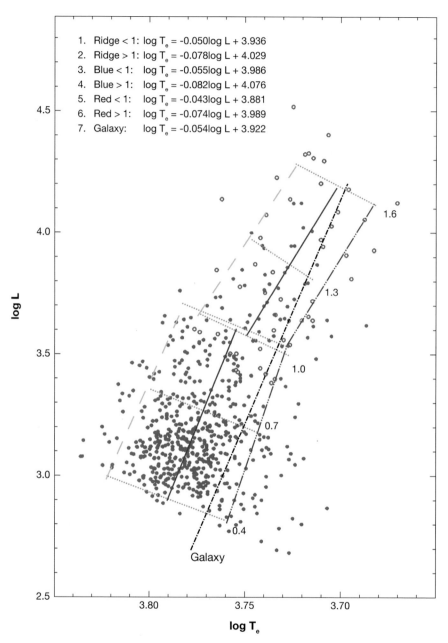

Figure 6

The instability strip in log L, log T_e for the LMC compared with the ridge-line relation for the Galaxy, shown as the black, dashed-dot line. Five lines of constant period (log P of 1.6, 1.3, 1.0, 0.7, and 0.4) are shown. The hot and cold boundary lines to the strip are the same lines as in **Figure 5** transformed to temperature. Diagram is from figure 20 of Sandage, Tammann & Reindl (2004).

sequence fittings or by some other means (Sandage & Tammann 1968). Obviously, it was impossible in this way to discover differences, if they exist, in the slopes and zero points of the Galactic and the LMC/SMC P-L relations that must be present if the color differences are real.

A fundamental advance on the problem was made beginning in the 1990s when the absolute magnitudes and interstellar absorptions of many Cepheids in the Galaxy had been determined by both main sequence fittings in clusters and associations as summarized by Feast & Walker (1987) and Feast (1999, 2003), and by Gieren et al. (1998) using a variation of the BBW moving atmospheres method introduced by Barnes & Evans (1976). Comparison of the results from the two methods were summarized in 2003 by Tammann, Sandage & Reindl. They were revised with additions to each list in a new summary by Sandage, Tammann & Reindl (2004, their tables 3 and 4). The additions to the data for the main sequence fittings since Feast (1999) are from Turner & Burke (2002) and Hoyle, Shanks & Tanvir (2003). The fiducial main sequence to which the cluster data were fitted uses a zero point based on a Pleiades modulus of $(m-M)^\circ$ = 5.61 from Stello & Nissen (2001). This is within 0.06 mag of the photometric distance (5.60) by Pinsonneault et al. (1998), the moving cluster distance (5.58) of Narayanan & Gould (1999), and the latest *Hipparcos* direct parallax distance (5.55) by Makarov (2002).

The 1998 BBW listings by Gieren et al. (2005), used by Tammann, Sandage & Reindl in 2003, were replaced by Sandage, Tammann & Reindl (2004) using the revised BBW distances of Fouque et al. (2003), to which new BBW distances by Barnes et al. (2003) were added, plus the interferometer distances from angular diameter measurements as summarized by Kervella et al. (2004).

The two new independent sets of Galactic calibrators (main sequence fittings and BBW) used by Sandage, Tammann & Reindl (2004) agree in their P-L relations to within 0.05, 0.07, and 0.10 mag in B, V, and I at P = 10 days. This is highly satisfactory, and Sandage, Tammann & Reindl (2004) combined the data to form a mean Galactic calibration.

The results are in **Figure 7**, taken from figure 5 of Sandage, Tammann & Reindl (2004). The open circles are from the BBW method; the closed circles are from main sequence fittings. The dashed lines are the calculated envelope boundaries to the P-L ridge lines whose width is given by the product of the width of the instability strip and the slope of the constant period lines. There is no break in slope in the 10-day period in these Galactic data, differing in that respect from that in LMC. The B, V, and I equations of the combined Galactic P-L relations are:

$$M_B = -(2.692 \pm 0.093)\log P - (0.575 \pm 0.107), \qquad (3)$$

with rms = 0.25 mag per star,

$$M_V = -(3.087 \pm 0.085)\log P - (0.914 \pm 0.098), \qquad (4)$$

rms = 0.23 mag per star, and

$$M_I = -(3.348 \pm 0.083)\log P - (1.429 \pm 0.097), \qquad (5)$$

rms = 0.23 mag per star.

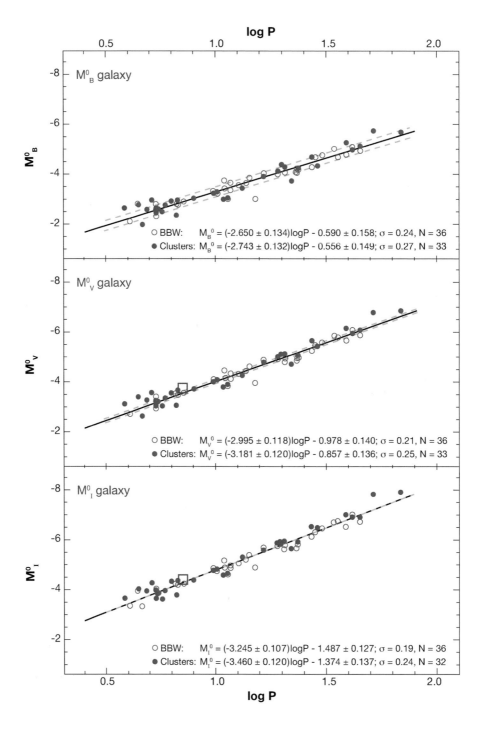

The P-L data in B,V, and I for the LMC, transformed to absolute magnitude using an LMC modulus of $(m-M)^\circ = 18.54$ that is independent of Cepheid data (summarized in table 6 of Tammann, Sandage & Reindl 2003), are shown in **Figure 8**, taken from figure 4 of Sandage, Tammann & Reindl (2004). The data for $\log P < 1.5$ are from Udalski et al. (1999). Data for longer period Cepheids, not measured by Udalski et al. because of saturation of the CCD chips for such bright Cepheids, are from previous photoelectric measurements published in the literature, summarized in Tammann, Sandage & Reindl (2003).

Figure 8 shows the change of slope at $P = 10$ days for the LMC Cepheids. The lower right panel shows the difference between the P-L relations in the Galaxy and the LMC. The individual data points in that panel are for the Galaxian Cepheids. The lines are from the other three panels. The equations for the LMC P-L relations for periods smaller and larger than 10 days are shown within the borders of **Figure 8**.

The absolute V magnitudes of Galactic Cepheids according to Equation 4 and the LMC Cepheids from the equations in **Figure 8** differ significantly at the short and long period ends of the P-L relation. At $\log P = 0.5$ ($P = 3$ days) Cepheids in the Galaxy are 0.36 mag fainter than in the LMC in V. At $\log P = 1.7$ ($P = 50$ days), Galactic Cepheids are 0.16 mag brighter in V. The consequences of these differences in using Cepheid distances to determine the Hubble constant are profound.

The evidence in the lower right panel of **Figure 8** for the reality of the difference is strong if the Galactic Cepheid data are not plagued by systematic error. That both the main sequence fitting method and the BBW method used in **Figure 7** give the same slope and zero point to better than 1 sigma of the statistics gives support to the reality of a difference. Furthermore, there is no break in the Galactic data at 10 days, but it is clearly evident in the LMC data.

In an attempt to reconcile the Galactic and the LMC data, Gieren et al. suggested at the 2005 Rome conference on pulsating variables and in the literature (Gieren et al. 2005) that a break in his Galactic P-L relation at 10 days could be produced if the velocity projection factor, p, necessary in the BBW method, is not a constant but varies with period. Although that might be made to reconcile the Galaxy and the LMC P-L slopes, no such fix is possible in the main sequence fitting method, and the excellent agreement of the two methods for Galactic Cepheids militates against such a fix. A refutation of the Gieren et al. (2005) conclusion has been made by Ngeow et al. (2005).

Nevertheless, the value of the p projection factor is a central component in the BBW method. In a discussion on its determination and its uncertainties, Fernley

---

Figure 7

The Galactic P-L relations for 36 Cepheids with cluster distances (*blue dots*) and 33 with BBW moving atmosphere distances (*open blue circles*). The mean midpoint ridge lines for the combined data are expressed by Equations 3, 4, and 5 of the text. Separate equations for the main sequence fittings and the BBW distances are in the borders of the figure. The open red square is the *Hipparcos* calibration by Groenewegen & Oudmaijer (2000). The light blue dashed lines are the expected envelope lines due to the finite width of the instability strip. They are calculated assuming strip widths of 0.13 mag in B-V and 0.10 mag in V-I and the slopes of the lines of constant period derived by Sandage, Tammann & Reindl (2004).

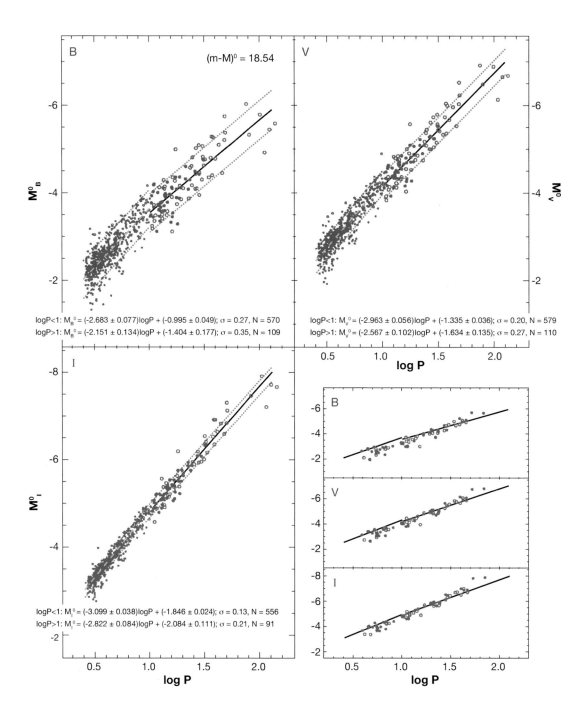

(1994) reviews the reasons that the factor is expected to be a function of (*a*) the Fraunhofer line strengths for the lines used to measure the pulsation velocities, (*b*) the velocity amplitude of the pulsation, and (*c*) the temperature of the star at each observation. A single factor for all temperatures and at all phases of the cycle is not theoretically correct. Hence, systematic errors at the level of 0.1 mag still seem possible in the BBW method due to mishandling of the p factor alone. The problem in another context has also been discussed by Gieren et al. (2005), but, again, the main objection for a p-factor revision comes from the excellent agreement of the slopes in B, V, and I of the BBW unrevised results with those from main sequence fittings. In addition, the differences still remain between the Galaxy and the LMC—the temperature difference in the instability strip with its different slope, and in the break in the P-L relation at 10 days.

However, problems in the main sequence fitting method are also present at the 0.1-mag level if the reddening corrections are known only to 0.02 mag, because the main sequence slope at the fitting point is about $dM_V/d(B-V) = 6$. However, random errors in the reddening can only increase the scatter in the P-L relation, not the slope. The method of removing a slight scale error in the previous Galactic reddenings by Tammann, Sandage & Reindl (2003) has been discussed above.

The error in the Pleiades main sequence placement in M_V may still be at the 0.04-mag level. Added to this uncertainty is the effect of variation of the metallicity of the individual Cepheids and of their parent clusters and associations. The adopted absolute magnitudes from the main sequence fitting method are based on making no correction for variations of the metallicities from that of the Pleiades. Feast & Walker (1987) discuss this potential problem, concluding that, for their list of clusters and associations containing Cepheids, "there is little direct evidence that the calibrations clusters in general have [Fe/H] significantly different from" that of the Pleiades to which the main sequence fit has been made. And "when zero points for each calibrating Cepheid were recalculated assuming a galactic metallicity gradient of delta [Fe/H] = -0.1 kpc^{-1}, the change in the Cepheid luminosities was insignificant." Nevertheless, the change of main sequence position with metallicity is steep near the solar metallicity of [Fe/H] = 0. A summary (Sandage & Cacciari 1990, their figure 4) of the determinations of main sequence position in the HR diagram using models prior to 1990 gave the gradient of dM_V/d[Fe/H] ~ 0.1, showing that an error of 0.1 mag in M_V would be made by neglecting a metallicity change of 0.1 dex.

However, the Feast & Walker test for the absence of an effect of variations in [Fe/H] on the luminosities of the cluster main sequences is supported by other

---

Figure 8

The period-luminosity (P-L) relations in B, V, and I for Cepheids in the LMC fitted to two linear regressions broken at 10 days. The dashed envelope lines have the same meaning as in **Figure 7**. Comparisons with the calibration for Galactic Cepheids from Equations 3–5 are shown in the lower right panel. The lines in the insert are the regressions for the LMC from the equations in the borders of the B, V, and I main panels. The individual Galactic Cepheids are shown as the data points from **Figure 7**. The differences in the P-L slopes are evident. The diagram is from figure 4 of Sandage, Tammann & Reindl (2004).

evidence. Consider first a straw-man possibility of a set of circumstances suggested by J. Kormendy (private communication) that could artificially give an incorrect slope and zero point to the Cepheid P-L relation determined from main sequence fittings in the presence of a Galaxian [Fe/H] gradient with geocentric distance.

If long period Cepheids, being brighter than those of short period, have larger geocentric distances on average, then, because of the Galactic metallicity gradient, a systematic error could occur in the derived main sequence fitted absolute magnitudes that is progressive with period. This, in fact, is the test made by Feast & Walker. They recalculated the absolute magnitude of each calibrating Cepheid using a metallicity gradient of $d[Fe/H]/dR(gal) = -0.1$ kpc^{-1}, and found no mean effect. This is because there is, in fact, no systematic difference in R(gal) with period for the Feast/Walker calibrators. But what about the zero point?

[Fe/H] values are available for 15 of the 25 calibrating Cepheids from Fry & Carney (1997), Andrievsky et al. (2002), and Luck (2003). There is overlap between the three lists, and, although [Fe/H] scatters considerably, the mean is $\langle[Fe/H]\rangle = -0.02 \pm 0.02$ (rms = 0.09) and there is no significant trend of [Fe/H] with period, which is the crucial point. Hence, the mean [Fe/H] of the Cepheid-bearing clusters is solar on average. If this is correct also for the Pleiades, which sets the zero point of the Cepheid scale, the main sequence fitting does not introduce a systematic metallicity related error in either the slope or the zero point of the Galaxian P-L relation.

Furthermore, the Galaxian gradient of -0.1 kpc^{-1} assumed by Feast & Walker in their test is a worst case value. Modern values are smaller at -0.06 kpc^{-1} (Luck et al. 2003; Kovtyukh, Wallerstein & Andrievsky 2005; Chen & Hou 2005).

There is still a further test for the absence of a main sequence fitting error. If such an error exists, and if there is an appreciable range of [Fe/H] for the calibrating Cepheids, the rms scatter of the P-L relation for cluster-fitted Cepheids (*blue dots* in **Figure 7**) should be larger than for the BBW calibrators (*open blue circles* in **Figure 7**) where no error due to [Fe/H] variations exists. However, the rms scatter about the P-L relation in Tammann, Sandage & Reindl (2003, their tables 3 and 4) is nearly identical for the main sequence fittings (rms = 0.259 mag) and the BBW data (rms = 0.245 mag) in the V band. Furthermore, these values are what is expected from the intrinsic spread due to the finite width of the instability strip.

Finally, the agreement in the P-L slopes and zero points between the blue dots and the open circles in **Figure 7**, based on two totally independent methods, is evidence in favor of an absence of systematic errors in each.

Potentially useful are the angular diameter measurements of Cepheids during their cycle, and their subsequent use to determine distances and therefore absolute magnitudes. A review of the results to 2004 by Kervella et al. (2004) gives absolute V and K magnitudes for seven long period Cepheids determined by the optical interferometric method. At this writing, the large statistical errors of the interferometer absolute magnitudes confirm the Equation 4 zero point, but only at the level of 0.3 mag. But great promise is believed to be in store for the method eventually.

Feast (2003) again discusses the problem of the Galactic calibration, including the *Hipparcos* trigonometric parallaxes discussed first by Feast & Catchpole (1997), statistical parallaxes, pulsation parallaxes, and water maser parallax for NGC 4258, which is also discussed by Saha et al. (2005). His conclusion is that the mean calibration for the Galactic Cepheids is

$$M_V = -2.81 \log P - 1.35 \pm 0.05. \qquad (6)$$

This differs from Equation 4 both in slope and zero point, because Feast uses the slope of the LMC to set the slope of the Galactic Cepheids, as was the tradition before 2003. Equation 6 cannot be correct if the slopes are indeed different as indicated by **Figures 7** and **8**. However, at $\log P = 1.5$ ($P = 32$ days) where many of the Cepheids in the HST Cepheid programs lie, Equations 4 and 6 are the same within 0.02 mag. But the analysis by Feast of the calibration of the Galactic Cepheids should be redone using the slope of $dM_V/d\log P = -3.087$ as in Equation 4 rather than adopting the slope of the LMC Cepheids, and zero-pointing it by the Galactic data.

If the slopes of the LMC and Galactic Cepheids do indeed differ, the conclusion of Feast (2003) "that the calibration presented [using the LMC slope] is valid to about 0.1 mag (rms) at least for Cepheids near solar metallicity" is not correct. Equation 6 differs from Equation 4 by 0.3 mag at $\log P = 0.5$ and by 0.12 mag in the opposite direction at $\log P = 2.0$. Feast emphasizes, "To reduce the uncertainty substantially below $\sim$0.1 mag will require extensive work on metallicity effects."

To this end, the problem should become clearer when the P-L relations in SMC, IC 1613, and other dwarf galaxies of low and intermediate metal abundance are analyzed in the same way as was done in Tammann, Sandage & Reindl (2003) and Sandage, Tammann & Reindl (2004) for the Galaxy and LMC. In addition, if there are metallicity gradients across the faces of giant-to-intermediate luminosity galaxies, as in M101, M33, and perhaps NGC 300, an analysis of the deviations from one another of separate P-L relations for separate spatial regions in these galaxies can be expected to clarify the role of metallicity in the P-L differences.

Early attempts to make this test are by Freedman & Madore (1990) in M31, and by Kennicutt et al. (1998) in M101. The difficulty of such a test is to separate the effects of metallicity on the P-L scatter from the severe differential absorption that will dominate the scatter. To date, the efforts to extract the metallicity effect from the observed P-L data have not been convincing at the necessary level of accuracy. However, the attempt to be made by Pietrzynski et al. (2002) in NGC 300, a galaxy with small internal absorption, is anticipated.

It can also be expected that the DIRECT program to use eclipsing binaries and Cepheids to refine the distances of M31 and M33 may be capable of making a more definitive test. A description of the program is by Macri (2004) with references to the first nine papers of the series.

4. USE OF CEPHEIDS FOR THE EXTRAGALACTIC DISTANCE SCALE

The differences in the slopes and zero points of the P-L relations in the Galaxy and LMC greatly complicate the use of Cepheids to determine Cepheid distances to galaxies using HST, and now also complicate the use of the large ground-based telescopes using adaptive optics (e.g., Thim et al. 2003 for NGC 5236 = M83). Until the reason is understood, effective correction methods to overcome the differences cannot be established in any definitive manner. Yet, if Cepheids are to be used to gauge cosmic distances in this new era of uncertainty of what P-L relation to use, we must attempt to understand, even at the 0.3-mag level in distance moduli.

At this point, the most reasonable hypothesis is that the difference in slope in the P-L relations between the Galaxy and LMC is related to metallicity differences, because the difference in metallicity between the Galaxy and LMC is well established to be about delta [Fe/H] = 0.4 dex. It is also known that the instability strip in the LMC is hotter on average than that for the Galaxy (see **Figure 6**, from figure 20 of Sandage, Tammann & Reindl 2004) and that the slopes of the boundary lines of the strips differ (**Figure 8**, *lower right panel*), requiring, from the pulsation equation, that the P-L slopes must also differ.

That these differences are due to differences in the chemical compositions is supported by theoretical models of the positions of the fundamental blue edges of the instability strip as functions of the abundances of hydrogen, helium, and the metals. These variations were first calculated by Iben & Tuggle (1975), and later by many others, among whom are Chiosi et al. (1992) and Saio & Gautschy (1998). A review by Sandage, Bell & Tripicco (1999) discussed the effect of variations in the Y and Z chemical abundance variations on the position of the fundamental blue edge. Indeed, changes in the temperature of the edge are predicted by the theoretical models for the edge position for reasonable changes of Y and Z. Hence, the principle is established that changes in chemical composition can change the position of the instability strip, and hence the slope and zero point of the P-L relation, but we must proceed empirically to avoid making decisions between the various conflicting models.

For any application of the P-L relations to Cepheid data, we must decide which P-L relation to use—that for the Galaxy or the LMC, or some interpolation between them for intermediate metallicities, or extrapolations for metallicities weaker than for the LMC, on the assumption that metallicity change is the cause of the differences.

Or, alternatively, one can proceed as done in Gibson et al. (2000) and Freedman et al. (2001) by adopting a single fiducial P-L relation (that, as usual before 2003, uses the LMC slope and Galactic calibrators) and then by correcting for metallicity variations by a Kennicutt et al. (1998)-like correction that is linear in delta [Fe/H], or by a Sakai et al. (2004)-like correction in delta [O/H]. The 2002–2005 literature contains both approaches.

The resulting distance scale by Freedman et al. (2001) differs from distances determined by interpolating between the Galaxy and LMC according to the measured mean metallicity of the parent galaxy. The details of this interpolation method have been developed in detail by Saha et al. (2005) and applied there to the same galaxies used by Gibson that were included in the original program galaxies in the HST SNe Ia calibration campaign led in a consortium by Tammann (Saha et al. 2005, Sandage et al. 2006).

Comparison of the two Cepheid distance scales differ from each other at the 0.2-mag level. The distance moduli of the eight supernovae calibrating galaxies measured by the Tammann consortium and 26 other galaxies in common with Gibson et al. (2000) average 0.20 mag more distant. The detailed comparison set out in Sandage et al. (2006) shows that part of this is due to the different treatments of the P-L relations, and part due to the different treatment of the correction of the Cepheid data for internal absorption in the parent galaxies, showing the extreme difficulties in photometry, even with HST, at $V = 25$ (Saha et al. 2005 for details). Because this review is not concerned with the extragalactic distance scale per se, but rather only with the Cepheid P-L relation, readers interested in the detailed reasons for the difference between the Freedman et al. (2001) and the Sandage et al. (2006) distance scales are referred to the literature references just cited.

Clearly, much work lies ahead. We are only at the beginning of a new era in distance determinations using Cepheids. It can be expected that much will be discovered and illuminated in the years to come.

5. ABSOLUTE MAGNITUDE CALIBRATIONS OF RR LYRAE STARS IN GLOBULAR CLUSTERS AND IN THE FIELD

5.1. Linear Correlations of $M_V(RR)$ with Metallicity

When it was discovered that the division of globular clusters into two groups with mean periods of the RRL stars differing at $\langle P \rangle = 0.55$ and $\langle P \rangle = 0.65$ days (Grosse 1932; Hachenberg 1939; Oosterhoff 1939, 1944) was also a division by metal abundance (Arp 1955, verified by Kinmann 1959), it became obvious that a difference in mean luminosity between the two groups of about 0.2 mag could explain the observations. Temperature differences between the two Oosterhoff period groups were not sufficient (Sandage 1958). It was further found that the Oosterhoff period difference with metallicity was also present in the field RRL stars and that the correlation of mean period with metallicity was a continuum rather than a division into two discrete period groups (Preston 1959). It was further found that the period shifts in globular clusters existed star by star when the cluster variables were compared between clusters of different metallicity and where the RRL parameters were read at the same amplitude, or the same temperature, or the same rise time (Sandage 1981b, 1982). Hence, the Oosterhoff effect is not caused by differences in ensemble averages over different distributions of periods in different clusters, which was an early usual way of discussing the Oosterhoff effect, but is a star-by-star effect. Further evidence of the star-by-star explanation is in two general summaries (Sandage 1990a,b, 1993a,b).

It became customary to assume a linear relation between absolute magnitude and metallicity of the form $M = a + b[Fe/H]$. The calibration problem then reduced to finding a and b by whatever calibration method was used. The three most popular have been (*a*) the traditional way through statistical parallaxes, (*b*) the BBW moving atmosphere method, and (*c*) main sequence fitting.

The abundant literature on the first two methods had been reviewed by Smith (1995) in his textbook on RRL stars, and by Cacciari & Clementini (2003) using the extensive later literature. The results are listed in their table 6.2.

They give $\langle M_V(RR) \rangle = 0.78 \pm 0.12$ at $[Fe/H] = -1.5$ from the statistical parallax method, based primarily on the modern analysis of new radial velocities and proper motions by Layden et al. (1996) and by Gould & Popowski (1998). These statistical parallax values are near the faint end of the new calibrations. They are representative of the short distance scale for the RRLs, being about 0.25 mag fainter than the calibrations defining the long RRL distance scale, among which are as follows.

The Cacciari/Clementini (2003) listing for the BBW moving atmospheres method is $M_V = 0.55 \pm 0.12$ at [Fe/H] = −1.5, based on a high weight determination of RR Cet by Clementini et al. (1995) and Fernley et al. (1998a). This differs from the summary review by Fernley et al. (1998b) who derive the calibration of $M_V = (0.98 \pm 0.05) + (0.20 \pm 0.04)[Fe/H]$ from many other BBW determinations, giving $\langle M_V \rangle = 0.68$ at [Fe/H] = −1.5, midway to the short distance scale. The difference between these two values using the BBW method (0.56 mag versus 0.68 mag) illustrates the level of systematic differences in different applications of the BBW method for RRLs, which is at about the 0.15-mag level, although in better agreement are many calibrations favoring the long distance scale to be set out below. But consider first the value of b.

A large literature exists on the measurement of b. Among the largest values of b is from a linear pulsation calibration by Sandage (1993b) who used an empirical correlation of log P with [Fe/H] and adopted variations of mass and T_e with [Fe/H] to give $M_V = 0.94 + 0.30[Fe/H]$. This gives $M_V = 0.49$ at $M_V = -1.5$. This is among the brightest of the calibrations but is supported by McNamara (1997a,b) who derived $M_V(RR) = 0.96 + 0.29[Fe/H]$ from his application of the BBW method, giving $M_V = 0.53$ at [Fe/H] = −1.5. McNamara (1999) also derived $M_V = 1.00 + 0.31[Fe/H]$ using RRLs of different metallicities in the Galactic bulge from data by Alcock et al. (1998) on the period-amplitude-metallicity relation, giving $M_V = 0.54$ at [Fe/H] = −1.5. A high value of b was also obtained by Feast (1997) who gave $M_V = 1.13 + 0.37[Fe/H]$, or $M_V = 0.58$ at $M_V = -1.5$. In a rediscussion of Fernley's (1993, 1994) BBW results, McNamara (1999) derived $1.06 + 0.32[Fe/H]$, also giving $M_V = 0.58$ at [Fe/H] = −1.5.

Smaller values of b are more common. Fernley (1993) derived $b = 0.19$ using (V-R)° colors. He later obtained $b = 0.21$ from his assessment (Fernley 1994), mentioned earlier, of the velocity projection factor, p, needed in the BBW method. Fernley et al. (1998a) later derived $b = 0.18$ from new BBW data.

Carretta et al. (2000) derived $M_V(RR) = 0.74 + 0.18[Fe/H]$, from main sequence fitting (see below), giving $M_V(RR) = 0.47$ at [Fe/H] = −1.5.

From a study of eight clusters in M31 that have RRL photometric data, Fusi Pecci et al. (1996) obtained the shallow slope of $b = 0.13$. They later revised the value to 0.22 (Rich et al. 2001) from a larger sample of M31 clusters.

The most secure determination to date is by Clementini et al. (2003) from their LMC sample of RRL where they derive $b = 0.214 \pm 0.047$, and a zero point of $M_V = 0.52$ at [Fe/H] = −1.5. Their linear calibration is $M_V = 0.84 + 0.214[Fe/H]$, zero pointed using an LMC modulus of $(m-M)° = 18.54$, which can be made independent of any long period Cepheid data (Tammann, Sandage & Reindl 2003, their table 6).

Jones et al. (1992) and Carney et al. (1992) derived $b = 0.16$ in reviews of the BBW results to 1992.

> **COMPARISON OF CEPHEIDS AND RR LYRAE VARIABLES IN GALAXIES WHERE BOTH APPEAR TOGETHER**
>
> It has often been hoped that the Cepheid and RR Lyrae (RRL) calibrations can be tested relative to one another by comparing the apparent magnitude levels of each in galaxies that contain them both. This seemed straightforward and powerful until the Cepheid P-L relation was shown to be variable from galaxy to galaxy by the arguments set out in Section 2. Nevertheless, it can still be expected that comparisons between the RRL and Cepheids in a given galaxy will yet prove to be useful, perhaps by turning the problem around to determine the Cepheid P-L relation using RRL, if the second parameter problem of the RRL can be solved (see below). Recent use of the method of joint comparisons, made before the results of Section 2 or the blatant second parameter problem of the RRL from NGC 6388 and NGC 6441 was fully appreciated, has been made by Fusi-Pecci et al. (1996) for M31; van den Bergh (1995) and Sandage, Bell & Tripicco (1999) for M31, the LMC, the SMC, and IC 1613; Smith et al. (1992) and Walker & Mack (1988) for SMC; and Dolphin et al. (2001) for IC 1613.

Comprehensive reviews of many of the extant linear calibrations are by Gratton et al. (1997), Gratton (1998), and Carretta et al. (2000).

Many of the calibrations cited earlier for RRL are consistent with $\langle M_V \rangle = 0.82 + 0.20[Fe/H]$, giving $M_V = 0.52$ at $[Fe/H] = -1.5$, which fits most of the cited calibrations to within about 0.05 mag.

However, the evidence is strong that the luminosity-metallicity relation is not linear, but rather that $\underline{b}$ is a function of [Fe/H] in the sense of being larger at the metal-rich end than at the metal-poor end of the distribution. The evidence is from (*a*) theoretical models of the HB (both at zero age and in an evolved state), (*b*) the pulsation equation using the observed input parameters of mass-[Fe/H] and $\log T_e$ (from colors) as functions of metallicity relations, (*c*) and from semiempirical use of observational data.

5.2. Nonlinear Calibrations of $M_V(RR)/([Fe/H])$

Most theoretical HB models made after ~1990 predict a nonlinear relation for M_V with [Fe/H] for the ZAHB. Examples are the models of Lee, Demarque & Zinn (1990); Castellani, Chieffi & Pulone (1991); Bencivenni et al. (1991); Dorman (1992); Caputo et al. (1993); Caloi, D'Antona & Mazzitelli (1997); Salaris, Degl'Innocenti & Weiss (1997); Cassisi et al. (1999); Ferraro et al. (1999); Demarque et al. (2000); VandenBerg et al. (2000, their figures 2, 3, and 20); and Catelan, Pritzl & Smith (2004). A graphical summary showing several of these calibrations is given by Cacciari & Clementini (2003). To illustrate the theme of all these models we show the prediction of the VandenBerg et al. (2000) models (their figure 3) in **Figure 9**, and a summary

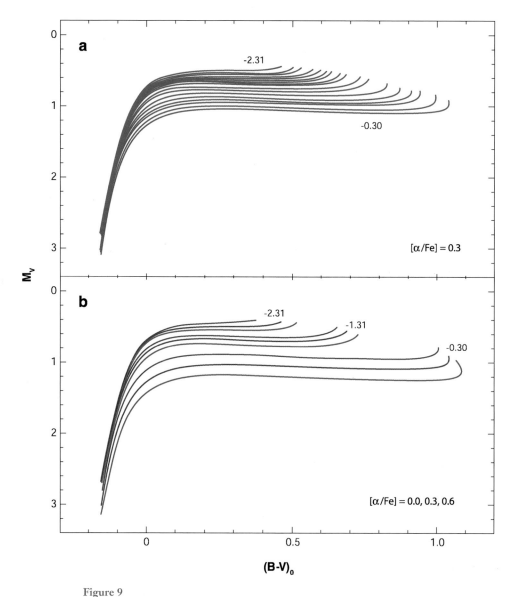

Figure 9

(*a*) Predicted level of the zero age horizontal branch (ZAHB) by VandenBerg et al. (2000) for 17 different metallicities in steps of 0.15 in [Fe/H] for an alpha element enhancement of 0.3 dex over solar. (*b*) The sensitivity of the horizontal branch level to variations of the alpha element enhancements for three values of [alpha/Fe]. The higher the alpha element enhancement, the fainter is the ZAHB. Diagram is from figure 3 of VandenBerg et al. (2000).

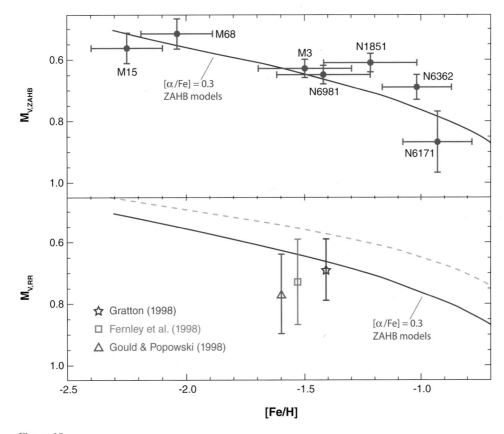

Figure 10

Top panel is the zero age horizontal branch (ZAHB) level in M_V from the analysis of DeSantis & Cassisi (1999) of individual data in selected globular clusters, based on the pulsation equation and their adopted input parameters of mass and temperature. The curve is the predicted calibration by VandenBerg et al. (2000) for an alpha element enhancement of [alpha/Fe] = 0.3 over solar. The bottom panel shows this prediction for the ZAHB as the solid red line. The effect of evolution, producing the mean position of the horizontal branch (HB) due to the vertical structure of the HB (Sandage 1990a), is the dashed light blue line. Three of the empirical calibrations cited in the text are shown in the bottom panel as the open symbols. They are fainter by ~0.2 mag than the final calibration given later here (Equation 8). The diagram is from figure 20 of VandenBerg et al. (2000).

given by VandenBerg et al. (their figure 20) of a selection of the observational evidence shown here as **Figure 10**.

The theoretical models of VandenBerg et al. (2000) in M_V, B-V in the top panel of **Figure 9** show the sensitivity of the position of the unevolved HB for 17 values of [Fe/H] ranging from −0.3 to −2.31, computed for an alpha element enhancement of a factor of 2 over solar (i.e., [alpha/Fe] = 0.3). The variation of the level of the absolute magnitudes with [Fe/H] is evident in **Figure 9** for the ZABH (i.e., without evolution).

The lower panel shows the effect on the luminosity level for alpha element enhancements of 0.0, 0.3, and 0.6 in the log (i.e., for factors of 1, 2, and 4 in the alpha/Fe numerical ratios) for each of the three values of [Fe/H]. The lowest luminosity level for each [Fe/H] family is for the highest (0.6) alpha element enhancement.

Figure 10 (*top*) compares the predictions from **Figure 9** for the ZAHB using [alpha/Fe] = 0.3 with the absolute magnitudes derived from a number of globular clusters by DeSantis & Cassisi (1999) from semitheoretical considerations from the pulsation equation. The lower panel repeats the solid line from the top panel and also shows three observational calibrations using different empirical methods. The cross from Gratton (1998) is from using *Hipparcos* trigonometric parallaxes giving $M_V = 0.69 \pm 0.1$ at $\langle[\text{Fe/H}]\rangle = -1.41$. The open square by Fernley et al. (1998a) at $M_V = 0.73 \pm 0.14$ at $\langle\text{Fe/H}\rangle = -1.53$ is from the BBW method averaged from many pre-1998 investigations. The triangle by Gould & Popowski (1998) at $M_V = 0.77 \pm 0.13$ at $\langle[\text{Fe/H}]\rangle = -1.68$ is from statistical parallaxes.

Because these determinations are from the observational data, they refer to the mean luminosity after evolution from the ZAHB. The effect of evolution is shown in the bottom panel by the dashed line (Sandage 1990a), which is 0.09 mag brighter than the level of the ZAHB.

Clearly, the three observational determinations in the bottom panel are ∼0.2 mag fainter than the level of the dashed line. Either the VandenBerg models are too bright, or these three points have systematic errors of ∼0.2 mag too faint. We show later in this section (Equation 8) that we favor the second possibility and therefore that the VandenBerg et al. (2000) models would then be close to reality.

Other observational and/or semitheoretical calibrations also show the nonlinearity, among which are the studies by Caputo (1997); Gratton et al. (1997); DeSantis & Cassisi (1999, their figure 15); Caputo et al. (2000); McNamara et al. (2004); and Sandage (1993b, 2006).

Caputo et al. (2000a) combine a pulsation equation that relates period, luminosity, temperature, and mass with observational data on the periods of RRLs at the blue edge of the instability strip for first overtone variables and the red edge of the fundamental mode in a number of clusters. They determined the distance modulus of each cluster by comparing the observed period-apparent magnitude data with new HB models used by Caputo et al., from which the absolute magnitude of the cluster variables can be determined. **Figure 11** shows the result taken from figure 2 of Caputo et al. (2000a). Five earlier calibrations from the literature are drawn. The upper panel shows the Caputo et al. analysis of their data for an [alpha/Fe] overabundance ratio of $f = 3$. The lower panel is for the solar [alpha/Fe] value.

The calibration is nonlinear, shown as **Figure 12** here as given by Caputo et al. 2000a. Two linear relations are fitted to the data giving $M_V(\text{RR}) = 0.71 + 0.17[\text{Fe/H}] + 0.03 f$ for $[\text{Fe/H}] < -1.5$, and $M_V(\text{RR}) = 0.92 + 0.27[\text{Fe/H}] + 0.03 f$ for $[\text{Fe/H}] > -1.5$, where again $\underline{f}$ is the overabundance ratio ($\log f = [\text{alpha/Fe}]$) of the alpha elements relative to the solar abundance as defined by Salaris, Chieffi & Straniero (1993) and used by Catelan, Pritzl & Smith (2004). These give, respectively, $M_V = 0.55$ and 0.61 at $[\text{Fe/H}] = -1.5$ for $\underline{f} = 3$ (i.e., [alpha/Fe] = 0.3), but the parabolic

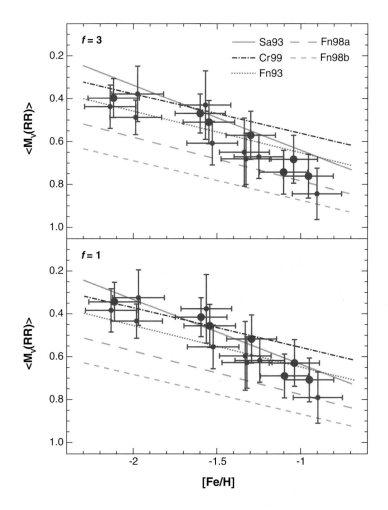

Figure 11
The calibration of $M_V(RR)$ in a number of clusters of different metallicities by Caputo et al. (2000a) using the edges of the theoretical instability strip for the first overtone and the fundamental mode for two different assumptions for the overabundance of the alpha elements as parameterized by f as defined by Salaris, Chieffi & Straniero (1993) and by Catelan, Pritzl & Smith (2004). The linear calibrations from five representative results are shown for comparison. Sa93 is Sandage (1993b); Fn93, Fn98a, and Fn98b are Fernley (1993, 1998a,b); Cr99 is Caretta et al. (2000).

fit of Equation 7 is a better fit, not shown, as

$$M_V(RR) = 1.576 + 1.068([Fe/H]) + 0.242([Fe/H])^2, \qquad (7)$$

valid for [Fe/H] between -1 and -2.2. The slope, b = $1.068 + 0.482$([Fe/H]), varies between 0.59 for [Fe/H] = -1 and 0.10 for [Fe/H] = -2, nicely encompassing all values of the various linear calibrations cited in an earlier section. Equation 7 gives $M_V(RR) = 0.52$ at [Fe/H] = -1.5.

We must mention that the nonlinearity of the M_V([Fe/H]) correlations predicted by the theoretical models cited earlier refers to the ZAHB, i.e., not what is expected to be observed if evolution away from the ZAHB is severe as is suggested, for example, by Lee, Demarque & Zinn (1990). Hence, the nonlinearity of the Caputo et al. calibration of Equation 7 is important because it is based on the observations at a mean evolutionary state of the HB.

Figure 12

Same as **Figure 11** but with the two linear equations listed in the text. A better fit is the parabola of Equation 7 in the text. Diagram is from figure 3 of Caputo et al. (2000a). The parameter f is the measure of the alpha element overabundance (f = antilog [alpha/Fe]) compared to solar. The two Oosterhoff period groups are separated by symbols. The two-line fits are those given by Caputo et al. (2000a).

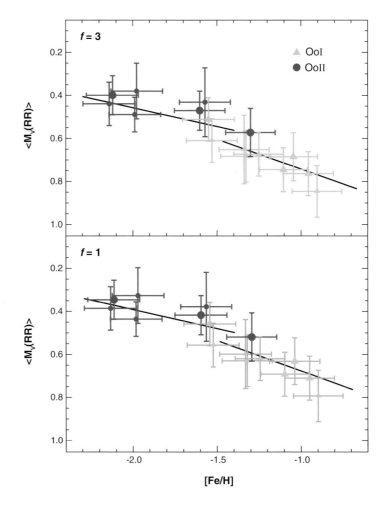

A different approach is by Sandage (2006) using the pulsation equation, upgrading the method used in Sandage (1993b) with improved correlations of the continuum period/metallicity relation, and an improved temperature/metallicity relation at the fundamental blue edge of the instability strip. His nonlinear calibration, zero-pointed from the Clementini et al. (2003) data for RRL for the mean level of the HB in the LMC, is

$$M_V(RR) = 1.109 + 0.600([Fe/H]) + 0.140([Fe/H])^2, \tag{8}$$

valid for [Fe/H] between 0 and -2.0. Equation 8 gives $M_V(RR) = 0.52$ at [Fe/H] = -1.5. The slope, $\underline{b} = 0.60 + 0.28([Fe/H])$, varies from $\underline{b} = 0.6$ to 0.04 between [Fe/H] = 0 and -2.

The nonlinear calibration of McNamara et al. (2004), based on main sequence fittings and on the route through the field SX Phe variables (see below), consists of two linear calibrations below and above [Fe/H] = -1.5 as $M_V(RR) = 0.50$ independent of [Fe/H] for metallicities more metal poor than -1.5, and $M_V(RR) = 1.13 + 0.42([Fe/H])$ for [Fe/H] between -0.5 and -1.5. The calibration is for the mean evolved level of the HB. The assumption by McNamara et al. for a flat M_V-[Fe/H] relation for [Fe/H] more metal poor than -1.5 is consistent with the Caputo et al. data points in **Figure 12**.

Two other nonlinear calibrations from advanced theory of the ZAHB must be mentioned. Catelan, Pritzl & Smith (2004), using oxygen-enhanced opacities, derive

$$M_V = 1.179 + 0.548([Fe/H]) + 0.108([Fe/H])^2, \qquad (9)$$

giving $M_V = 0.60$ at [Fe/H] $= -1.5$ for the ZAHB. This must be made 0.09 mag brighter to account for evolution (Sandage 1993b), giving $M_V = 0.51$ for the average RRL state on the HB. The slope of $\underline{b} = 0.548 + 0.216([Fe/H])$ varies between $\underline{b} = 0.44$ and 0.12 for [Fe/H] between -0.5 and -2.0.

VandenBerg et al. (2000) produced ZAHB models with enhanced oxygen abundance. Their nonlinear ZAHB models average 0.05 mag fainter than those of Catelan et al. (2004), but have nearly the same nonlinear behavior as Equation 9. Their predicted $M_V(RR)$ at [Fe/H] $= -1.5$ for the unevolved ZAHB is 0.65, which when corrected for the average evolution of the HB gives $\langle M_V \rangle = 0.56$ at [Fe/H] $= -1.5$ for the average RRL state.

All of the calibrations discussed here are ~ 0.15 mag brighter than the statistical parallaxes of Layden et al. (1996) and Gould & Popowski (1998) cited earlier, suggesting systematic errors in one or the other of the methods. The main sequence fitting method and the route through SX Phe variables favor the long distance scale, whereas the BBW moving atmospheres methods in the hands of the Carney and the Fernley groups sometimes favor the short distance scale. Reviews of the status up to 2000 are by Gratton (1998), Popowski & Gould (1999), and VandenBerg et al. (2000).

5.3. Main Sequence Fitting

In principal, the cleanest method of HB calibration might seem to be the main sequence fitting method. However, the observational realization that the position of the main sequence is a strong function of metallicity (Sandage & Eggen 1959, Eggen & Sandage 1962), within nearly the same time frame as the theoretical predictions by Stromgren (1952), Reiz (1954), Demarque (1960), and others showing the same thing, greatly complicated the method (Sandage 1970; Sandage & Cacciari 1990).

The method became competitive when the handful of trigonometrically calibrated subdwarf distances available in the 1970s was increased to a useful number with marginally adequate accuracy by the *Hipparcos* parallax satellite. Based on this fundamental advance, a large literature has arisen to solve the intricate problem of correcting the fitting problem to a variable main sequence position due to variations of the chemical composition.

Modern applications of the method are by Reid (1997, 1999), Carretta et al. (2000), Gratton et al. (1997), and VandenBerg et al. (2000), among others, using the *Hipparcos* subdwarf data in concert with the theoretical models of main sequence position together with various applications of the Lutz-Kelker bias (Lutz & Kelker 1973; Hanson 1979; Lutz 1983; Smith 1987).

Much literature exists for the position of the main sequence as a function of metallicity from calculated models. Entrance to this literature can be gained from VandenBerg (1992), Gratton et al. (1997), D'Antona, Caloi & Mazzitelli (1997), Carretta et al. (2000), Straniero, Chieffi, & Limongi (1997) updating Straniero & Chieffi (1991), and VandenBerg et al. (2000) as representative. Extensive references are given to other literature in these citations.

Most of the main sequence fitting calibrations, some of which are discussed above, have as representatives $M_V(ZAHB) = 0.82 + 0.22([Fe/H])$ from Gratton et al. (1997), $M_V(ZAHB) = 0.80 + 0.18 ([Fe/H])$ from Carretta et al. (2000), and $M_V(evolved) = 1.00 + .31([Fe/H])$ by McNamara mentioned earlier. All are consistent with $M_V(evolved) = 0.52$ at $[Fe/H] = -1.5$ to better than 0.05 mag. A summary to 1998 is by Gratton (1998).

A problem with main sequence fitting using *Hipparcos* trigonometric parallaxes of subdwarfs of different metallicities is the application of the Lutz-Kelker bias to the subdwarf ensemble calibration, depending at the 0.1-mag level in a complicated way on the relative parallax error (delta pi/pi). The correction is controversial (see the next section) and can only be avoided at the 0.01-mag level when trigonometric parallaxes of subdwarfs can be known at about five times the accuracy of the *Hipparcos* data (see Sandage & Saha 2002 for a history of the bias and a modern simulation of it).

5.4. The Route Through the SX Phoenicis Stars

The initial discovery of the large class of small amplitude ($A_V < 0.3$ mag), very short period pulsators (periods shorter than 0.25 days to as small as 80 minutes) in the field, that have near main sequence luminosities, was made by Harlan Smith (1955; also unpublished dissertation) and Eggen (1956a,b). Smith called these stars dwarf Cepheids; others called them AI Vel stars (Bessell 1969 in his catalog), or RRL stars (Kukarkin et al. 1969). Eggen (1970, 1979 in his catalog) proposed the term USPC for ultra short period Cepheids, a designation gaining favor in recent years. The class as a whole is known as Delta Scuti stars. A review of what was known by 1967 about these stars that occur in both populations was made by Danziger & Dickens (1967).

A most promising method of RRL calibration in globular clusters is by using the SX Phoenicis USPC, which are the population II subclass of Delta Scuti stars. They are used as templates to step from their well-determined, near main sequence luminosities (from trigonometric parallaxes), to the RRL HB variables that are three magnitudes brighter. The method became possible with the discovery of USPC in globular clusters beginning in the 1990s (Nemec & Mateo 1990a,b; McNamara 1997b for reviews).

Most of the Delta Scuti class are population I, high metal abundance, low space motion stars. However, a small subgroup have high velocity and low metal abundance. They are the population II subtypes, whose prototype examples are SX Phoenicis, CY Aqr, and DY Peg (McNamara & Feltz 1978; Balgin et al. 1973; Breger 1979,

1980, for reviews). This population II subgroup is now universally named SX Phe stars.

An early listing of the known SX Phe stars in globular clusters is by Nemec (1989) and Nemec & Mateo (1990b, their table 2). The position of the dwarf Cepheid instability strip is discussed by McNamara & Powell (1990). A listing complete to 1997 is McNamara (1997b), which also gives a definitive calibration, also of RRL stars.

McNamara (1997b) uses the apparent magnitude difference between the HB and RRL variables in the clusters that contain SX Phe stars. The HB absolute magnitudes are zero-pointed by five SX Phe field stars that have adequate *Hipparcos* trigonometric parallaxes. From these parallaxes he obtained a calibration of the cluster SX Phe stars of

$$M_V(\text{SX Phe}) = -3.725 \log P - 1.933, \tag{10}$$

with a statistical error of 0.05 mag. He later revised that calibration (McNamara 2000) on the basis of a suggestion by Peterson (1999) that a Lutz-Kelker type correction should be applied to the five *Hipparcos* calibrators to give a zero point to Equation 10 of -1.969. However, as mentioned earlier, Lutz-Kelker corrections are controversial, depending crucially on the statistical nature of the sample (Sandage & Saha 2002), which is generally unknown.

Using Equation 10 for each of the USPC in each of the globular clusters in which they occur, the distance modulus of the clusters is determined with high statistical accuracy, from which the absolute magnitude of its RRL variables is determined. In that way McNamara (1997b) derives

$$M_V(\text{evolved HB}) = 0.29[\text{Fe/H}] + 0.90, \tag{11}$$

which gives $M_V(\text{RR}) = 0.52$ at $[\text{Fe/H}] = -1.5$. Equation 11 can be compared with the BBW moving envelope method used by McNamara (1997b), which gave him

$$M_V(\text{BBW, RRL}) = 0.31[\text{Fe/H}] + 0.96 \tag{12}$$

with one method of weighting, and identical with Equation 11 with another. Equation 12 gives $M_V(\text{RR}) = 0.50$ at $[\text{Fe/H}] = -1.5$. Both Equations 11 and 12 refer to the position of the mean evolved HB rather than to the ZAHB.

5.5. Exceptions to the Calibration: The Second Parameter Clusters NGC 6388 and NGC 6441

The results of the preceding four subsections are consistent in the $M_V(\text{RR})$ calibration of the correlation with metallicity that almost certainly is nonlinear, with a stable zero point at $M_V(\text{RR}) = 0.52$ for $[\text{Fe/H}] = -1.5$. However, there are blatant exceptions that warrant caution in assuming that the equations set out in the previous sections are unique and universal.

The exceptions were discovered in the globular clusters NGC 6388 and NGC 6441 (Rich et al. 1997; Pritzl et al. 2000, 2001), which have high metal abundances ($[\text{Fe/H}] \sim -0.5$) yet very long mean periods for their RRL stars, and furthermore, have large star-by-star period shifts relative to the period-amplitude diagrams of "normal" Galactic globulars such as M3 and M15. Furthermore, the morphologies of their HBs are highly abnormal for their metallicities, being nonhorizontal in V and extending across the HR diagram on either side of the RRL instability strip instead of being confined to the red side as in the high metallicity clusters in the Galaxy such as 47 Tuc.

From the abnormal period-amplitude shifts it is certain that the HBs of these clusters are more luminous by about 0.5 mag than the level of the HB in 47 Tuc and about 0.25 mag brighter than for the Oosterhoff II clusters such as M15.

Not only do these clusters violate the metallicity-luminosity equations set out in earlier sections, but they also violate the normal correlations that relate the Fourier component $(\phi)_{31}$ with period and metallicity, discovered by Simon & Clement (1993) and developed by Jurcsik & Kovacs (1996) and Kovacs & Jurcsik (1996, 1997); see also Sandage (2004).

The reasons are not yet understood, but explanations have begun (Sweigart & Catelan 1998; Bono et al. 1997a,b) on several fronts.

NGC 6388 and NGC 6441 may be the extreme examples known so far of the second parameter clusters that to a lesser degree have abnormal period-amplitude correlations and anomalous period shifts. Others with smaller period-amplitude-metallicity anomalies and abnormal HB morphologies include M2, M13, NGC 5986, and NGC 7006. This entire group may not follow the equations of the last sections, with NGC 6388 and NGC 6441 being the most extreme.

It seems likely that we are only near the beginning of an understanding of these and similar aspects of the second parameter problem. Clearly, the next step is to determine the luminosity anomalies of the second parameter clusters at the level of 0.1 mag, and to relate them to the abnormal morphologies of the HBs (Sandage 2006).

5.6. Other Methods and Controversies

A comprehensive review of the status of the RRL absolute magnitude calibration up to 1999 is by Popowski & Gould (1999). In addition to the methods discussed earlier here, they analyze two additional methods, which are (*a*) globular cluster kinematics and (*b*) white dwarf sequence fittings in the HR diagram.

In the cluster kinematic method, used first by Cudworth (1979), the dispersion in radial velocities of individual cluster stars is compared with the dispersion in the proper motions, which require a distance to convert from arcsec per year in the plane of the sky to km s^{-1}. Simple as this appears, it is model-dependent because the three-dimensional velocity distribution of the cluster stars must be known. Any anisotropy must be allowed for. This is usually taken from a fit of Mitchie (1963)-type models (cf. Lupton, Gunn & Griffin 1987) to the cluster light profile, but the method then becomes a combination of observations and theory, giving possible systematic uncertainties.

In the white dwarf (WD) cooling sequence method, the position of the observed WD sequence in a globular cluster is compared with the calibrated sequence of local

WDs whose distances are from trigonometric parallaxes. Entrance to the considerable literature on this method can be gained from Richer et al. (1995, 1997) and Renzini & Fusi Pecci (1988).

The conclusion reached by Popowski & Gould (1999) in their discussion of all seven methods they review, keeping the three methods they consider the most robust (statistical parallax, trigonometric parallax, and internal cluster kinematics), is $\langle M_V(RR)\rangle = 0.71 \pm 0.07$ at [Fe/H] $= -1.6$. This is 0.20 mag fainter than the calibration in Equation 8 here that requires $\langle M_V \rangle = 0.51$ at [Fe/H] $= -1.6$.

A decision between the two calibrations is possible by considering the resulting distance to the LMC based on RRL stars. Popowski & Gould derive $(m-M)^o{}_{LMC} = 18.33 \pm 0.08$ from their calibration. Our Equation 8 here, together with the observation by Clementini et al. (2003) that the mean level of the RRL is $\langle V^0 \rangle = 19.06$ in the LMC, gives $(m-M)^o{}_{LMC} = 18.55$. This agrees to within 0.01 mag with the mean of 13 determinations of the LMC modulus, each of which are independent of the long period Cepheids (Tammann, Sandage & Reindl 2003, their table 6). This result favors Equation 8 for the long distance scale.

6. THE "ABOVE HORIZONTAL BRANCH" POPULATION II (AHB1) VARIABLES IN THE PERIOD RANGE OF 0.8 TO 3 DAYS

In his Mount Wilson survey of the spectra of variable stars in globular clusters, Joy (1949) divided his spectroscopic data into five groups. In addition to the RRL stars, he defined the other four groups on the basis of the light curves as RRL-like variables but with periods of 1–2 days, type II Cepheids (periods of 13 to 19 days) named after the W Vir and RV Tauri field variables (25–90 days), and irregular or semiregular red, longer period variables (60–110 days). These five groups still define the menagerie of Baade's population II variables, both in clusters and in the field (with the addition of the Anomalous Cepheids; see below). Except for the longer period red irregular and semiregular variables that are at the top of the red giant branch, all are pulsators in the Cepheid-RRL instability strip.

The RRL-like variables with periods between 0.8 and 3 days are of particular interest because their existence clarifies the evolutionary state that is the transition from the post-HB to the base of the asymptotic giant branch. These variables provide a test of the models for this very rapid transition phase of evolution. Joy showed that these "short period type II Cepheids" (using an old name) are brighter than the normal RRL stars by up to one mag. They lie above the HB. Field variables of the same type were also known, such as XX Vir (P = 1.3d), SW Tau (P = 1.6d), and, incorrectly, BL Her (P = 1.3d); the latter is not of population II (see below) but is the population I analog.

The first interpretation, subsequently proved to be correct, of the evolutionary state of these stars was by Strom et al. (1970) as post–helium shell burners after the helium core is exhausted. They named such stars AHB, meaning above horizontal branch, because, by then, Joy's bright luminosities of up to one mag brighter than the RRL luminosities (although still in the instability strip) had been confirmed both in the clusters and in the field. Abt & Hardy (1960) derived the higher luminosity for BL Her, as would Wallerstein & Brugel (1979) later for XX Vir. However, Abt & Hardy did not find metal weakening in BL Her, but rather a solar metal abundance, confusing the classification as a population II variable. Because of this, confusion as to population types existed for several years on the status of the "population II BL Her stars" as they were then called (Smith et al. 1978). However, later, Wallerstein & Brugel (1979) found profound metal weakening in XX Vir, and the picture finally became clear.

Earlier, Kwee (1967) had done photometry on a number of field examples of these short period Cepheids (or long period RRL stars) and had described their light curves, from which a start could be made on a morphological separation according to light curve shape between the XX Vir real population II variables and the population I BL Her stars. Their light curves are fundamentally different, permitting separation into separate population classes.

The separation on the basis of light curve shape was also argued, using new observations, by Diethelm (1983), who could define four separate groups among the 1–3 day Cepheids. In a comprehensive study, following the summary paper by Kwee & Diethelm (1984), Diethelm (1990) proposed the division into, and the names of, AHB1 for the XX Vir RRL-like population II brighter than HB stars, AHB2 for the light curves similar to HQ CrA (P = 1.4d) and RT TrA (P = 1.9d) shown in figure 2 of Diethelm (1983), and AHB3 for the population I prototype of BL Her (P = 1.3d) as in figure 3 of Diethelm (1983). The AHB1 are the true population II variables similar in all respects to those in globular clusters. That name is now coming into general use, and is adopted here. As mentioned earlier, the name was invented by Strom et al. (1970). It was first used after that by Kraft (1972). The reader is referred to the atlas of light curves for the four types isolated by Diethelm (1983).

An analysis for luminosities of the data available to 1994 was made by Sandage, Diethelm & Tammann (1994), both for AHB1 field variables and those in globular clusters. Comparison with the post-ZAHB models of Dorman (1992) of evolution toward the AGB was made there.

Dorman's calculated tracks that greatly clarified the evolution status of the AHB1 stars are shown in **Figure 13** for [Fe/H] = −1.66. The blue fundamental edge of the instability strip is drawn as $\log T_e(FBE) = 0.035 M_{bol} + 3.832$ from Iben & Tuggle (1972). The ZAHB is the starting point for the tracks of different mass, marked in solar units along the branch. Tracks of ever decreasing mass cross the blue fundamental edge of the instability strip at ever increasing luminosities. This produces the ABH1 stars.

A detail of the evolution through the strip is shown in **Figure 14** for the Dorman tracks for [Fe/H] = −2.26. Shown are the blue and red edges of the instability strip and tracks for masses from 0.74 to 0.54 threading the strip. Lines of constant period (in days) are shown, based on the pulsation equation of van Albada & Baker (1973).

Comparison of **Figure 14** with similar diagrams that can be made from the Dorman tracks for [Fe/H] = −1.03 and −1.66 shows that the resulting P-L relations (i.e., made by reading the M_{bol} at each period) from the three diagrams are

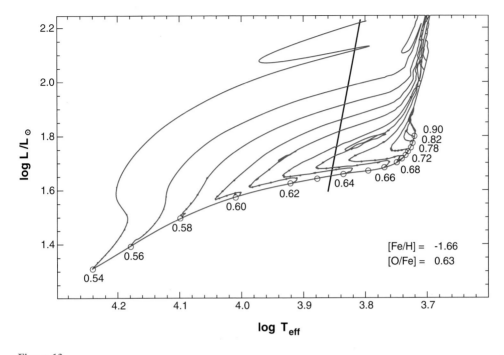

Figure 13
Dorman's (1992) post-zero age horizontal branch (ZAHB) tracks for [Fe/H] = −1.66 for different masses, showing the position of the fundamental blue edge of the instability strip. The mass values are marked for each track as it begins from the ZAHB. Time ticks on the tracks are for every 10^7 years. Diagram is from figure 13a of Dorman (1992) and figure 1 of Sandage, Diethelm & Tammann (1994).

nearly identical. There is no metallicity dependence of M_V at a given period for those luminosities that are brighter than the ZAHB. This is contrary to the strong dependence for the near ZAHB RRL variables one mag fainter that is manifest in a comparison (not shown) of Dorman's other two diagrams for [Fe/H] = −1.03 and −1.66 that are similar to **Figures 13** and **14**. Said differently, Dorman's models give a variation of M_V with [Fe/H] for the ZAHB, which, however, disappears for brighter luminosities in the strip.

An empirical calibration of the P-L relation for the ABH1 population II variables was made by Sandage, Diethelm & Tammann (1994). It was derived by comparison with the RRL (mean evolved state on the HB) luminosities from the Sandage (1993) calibration of $M_V(RR) = 0.94 + 0.30[Fe/H]$. The result in **Figure 15** gives the calibration of

$$M_V = -2.00 \log P - 0.10, \qquad (13)$$

with an rms scatter of 0.29 mag at a given period.

From Dorman's models (and all the subsequent models on the same problem) it can be seen that AHB1 stars cross the instability strip in less than 10^6 years. This is

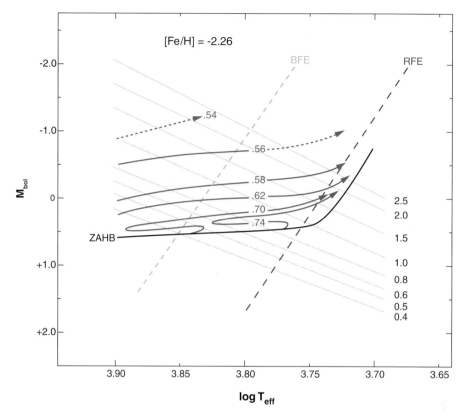

Figure 14
Detail of the evolution tracks through the instability strip as read from a diagram similar to **Figure 13**, but for [Fe/H] = −2.26. The lines of constant period are drawn. The periods are marked in days. The blue and red edges of the strip are drawn for the fundamental mode pulsators. Diagram is from figure 4 of Sandage, Diethelm & Tammann (1994).

100 times faster than the time for crossing the strip for ordinary RRLs on or near the ZAHB. Hence, the AHB1 stars offer the best opportunity to test this prediction for secular period changes due to evolution.

Such period changes at the predicted rate have, in fact, now been observed by Wehlau & Bohlender (1982) for nine AHB1 variables in globular clusters, and by Diethelm (1996) for three field AHB1 variables, confirming the prediction in a most satisfactory way.

7. THE POPULATION II CEPHEIDS

Many summaries of the observational aspects of the population II Cepheids have been made in the past two decades, following the pioneering work by Arp (1955) that was his PhD subject. These later summaries include the reviews and models by

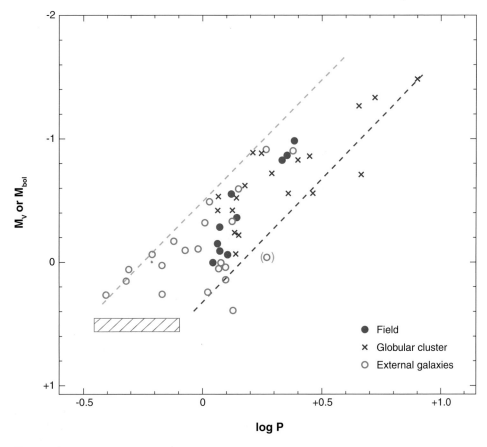

Figure 15
The P-L relation for the above horizontal branch (AHB1) variables in the field, in globular clusters, and in external galaxies known in 1994. The position of the RR Lyrae variables is shown by the hatched rectangle. The fundamental blue and red edge boundaries of the instability strip are estimated from the data. The diagram is from figure 5 of Sandage, Diethelm & Tammann (1994).

Bohm-Vitense et al. (1974), Gingold (1976, 1985), Wallerstein & Cox (1984), Harris (1984, 1985), Nemec, Wehlau & Mendes de Oliveira (1988), Wallerstein (1990, 2002), Whitlock, Feast & Catchpole (1991), Nemec, Nemec & Lutz (1994), McNamara (1995), and the indispensable catalog of Clement et al. (2001). A modern calculation of advanced models of both the population II Cepheids and the anomalous Cepheids (see next section) is by Bono, Caputo & Santolamazza (1997a,b).

A recent census of the population II Cepheids (including the AHB1 stars) in clusters and dwarf spheroidal galaxies is by Pritzl et al. (2003), where the new ABH1 and W Vir Cepheids found in the globular clusters NGC 6388 and NGC 6441 (four in NHC 6388 and six in NGC 6441) are included. (Note that Pritzl et al.

incorrectly call the AHB1 variables in NGC 6388 and NGC 6441 BL Her variables, which, as discussed above, should be named either AHB1 or XX Vir variables. BL Her is the population I analog of the AHB1 population II variables as argued earlier.)

These authors have calibrated the absolute magnitudes of the AHB1, the W Vir, and the RV Tau variables in their list of all such variables in clusters as based on RRL luminosities of $M_V(RR) = 0.17[Fe/H] + 0.82$ from Lee, Demarque & Zinn (1990, their equation 7). This gives $M_V(RR) = 0.57$ at $[Fe/H] = -1.5$, whereas the calibration used by Sandage, Diethelm & Tammann (1994), in **Figure 15** here, uses $M_V(RR) = 0.30 [Fe/H] + 0.94$ from Sandage (1993), which gives $M_V(RR) = 0.49$ at $[Fe/H] = -1.5$. Hence, the calibration by Pritzl et al. (2003, their equation 3), on their scale, is

$$M_V(\text{pop II}) = -1.64 \log P - 0.05, \tag{14}$$

which is 0.08 mag fainter than Equation 13 for the AHB1 stars according to Sandage, Diethelm & Tammann (1994). On the zero point scale of Equation 13, Equation 14 would be $M_V = -1.64 \log P - 0.14$. Therefore, at $\log P = 0.3$ (P = 2 days), the Sandage, Diethelm & Tammann calibration in Equation 13 gives $M_V(2\ d) = -0.56$, whereas the calibration by Pritzl et al. from the rezero pointed Equation 14 is $M_V(2\ d) = -0.63$.

The difference in slope between Equations 13 and 14 is due to the different period range over which each equation has been derived. Equation 14 is preferred for the long period population II Cepheids (P between 13 and 30 days) because this is the range covered by table 8 of Pritzl et al. (2003), far beyond the period range for the AHB1 stars studied by Sandage, Diethelm & Tammann (1994).

It remains only to remark on the difference in slope of the P-L relations for the population II Cepheids, in Equation 14, as -1.64 compared with that of population I Cepheids, as in Equation 4, as -3.09 of Section 3 for the Galactic Cepheids. Hence, the P-L relations of population I and II Cepheids are not parallel, as was assumed when the first distinctions between them were made by Baade and by Swope (circa 1950–1960). This, of course, is due to the different mass-luminosity relations for the type I and II variables. Population II Cepheids are stars in the post-HB phase of evolution on the blue loops (Hoffmeister 1967) on the second giant branch (the AGB) with nearly constant masses of less than 0.8 solar (e.g., Gingold 1976, 1985). In contrast, population I Cepheids are stars with masses that parallel the mass-luminosity relation of their main sequence progenitors that range from about 2 to 30 solar mass, depending on the luminosity.

8. THE ANOMALOUS CEPHEIDS

Up to this point the picture seemed very clean, where the differences in the periods, luminosities, and shapes of the light curves of the population I and II Cepheids, the AHB1 variables, and the RRL stars along the HB of globular clusters were understood as the result of well-known differences in evolution stages and of the relevant stellar masses, which were themselves understood.

Then a complication arose with the discovery of a new class of Cepheid-like variables with periods between 0.8 and 2 days in the dwarf spheroidal galaxies that are companions to the Galaxy (Baade & Swope 1961, Swope 1968, Zinn & Searle 1976 for a review). The shapes of their light curves are distinctly different from either the population I BL Her stars or the population II AHB1 XX Vir stars in the same period range in globular clusters. The light curves of stars of this new class are of small amplitude and are much more symmetrical than the RRL-like asymmetrical rapid rise AHB1 variables of the same period.

At first, no such star was known in globular clusters, then one was discovered in NGC 5466 (Zinn & Dahn 1976). The high velocity, low metal abundance field star XZ Cet (P = 0.82 d) was also discovered with its small amplitude and symmetrical light curve (Dean et al. 1977). The light curves of these stars are so distinctive that they are easily seen to differ from AHB1 stars. Hence such stars can be classified by the morphology of their light curves alone (Diethelm (1983), 1990), but clearly their evolutionary origins are different from the AHB1 stars, understood via **Figures 13** and **14**.

In the dwarf spheroids, these stars, early-on called anomalous Cepheids (AC), are brighter than the HB RRL stars, and are also brighter than the AHB1 stars of the same period. But they are fainter than the classical type I Cepheids (Zinn & Searle 1976, their figure 1). They have been found in every dwarf spheroidal surveyed (Swope 1968; Nemec, Nemec & Lutz 1994 for a comprehensive review), but in only the globular cluster 5466, and two suspected variables in Omega Cen. **Figure 16** here, taken from Nemec, Whelau & Mendes de Oliveira (1988, their figure 13), is a summary of the situation known to 1987 in the B photometric band.

The distinctive place of the ACs in the P-L plane seen in **Figure 16** shows that they must be in a different evolutionary state than either the RRL stars or the population II Cepheids. It was early suggested (Bohm-Vitense et al. 1974, Demarque & Hirshfeld 1975, Norris & Zinn 1975, Zinn & Searle 1976, Zinn & King 1982, Hirshfeld 1980, and undoubtedly others) that they are pulsators with higher mass than either the RRL stars or the AHB1 pulsators with periods of 0.8 to 3 days (the same range as the ACs) and for the blue loop W Vir and RV Tau variables of longer period.

Because such masses are impossible for single stars in the normal evolutionary state of old globular clusters (they have all evolved off the main sequence at this epoch), their high mass must be due to coalescence of a binary star into a single star that then pulsates when it is in the strip, or alternatively, to mass transfer in a contact binary before coalescence, or as a third possibility, to recent star formation. Masses of about 1.5 solar masses are necessary to explain their position in **Figure 16**.

A review of the status of such explanations is by Bono (2003), which, when used together with the earlier definitive review by Nemec, Whelau & Mendes de Oliveira (1988), brings the story up to date. The modern data for the P-L relation discussed by Bono et al. show wide scatter within the boundary lines drawn in **Figure 16**, suggesting a random distribution of mass values and, hence, probably several different evolutionary channels by which the single old stars gain their increased mass.

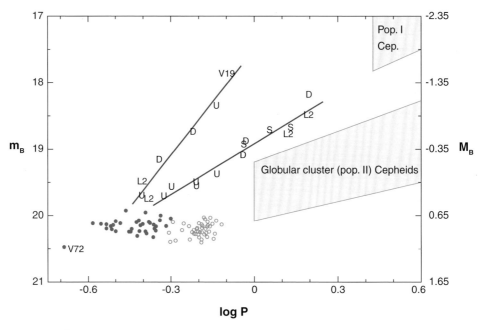

Figure 16

Summary of the B-band P-L relation for anomalous Cepheids (the AC stars) compared with the position of the RRL variables in the Ursa Minor dwarf spheroidal, and compared with the boundary positions of the P-L relations for population I and II variables. The symbols on the lines are for the dwarf spheroidals of U = Ursa Minor, L_2 = Leo II, D = Draco, and S = Sculptor. The mean apparent magnitude of the Ursa Minor RRLs is assumed to be 20.16, with $<M_B(RR)> = 0.8$ to set the zero point in M_B on the right-hand ordinate. The lines through the dwarf spheroidal points are for the fundamental and first overtones. More recent data (Bono et al. 1997, their figure 6) show a wider scatter of the AC stars throughout the region between the two lines, presumably caused by a stochastic distribution of masses. Diagram from Nemec, Whelau & Mendes de Oliveira (1988).

An interesting suggestion is offered by Wallerstein (2002) for the use of the AC to study the origin of at least part of the Galactic halo by accretion of dwarf spheroidals. This is to exploit the absence of AC in globular clusters and their abundance in dwarf spheroidals, and conversely, to make use of the fact that long period type II Cepheids (W Vir and RV Tau) are absent in the dwarf spheroidals but present in globular clusters. Systematic searches for the percentages of each of these types of variables in the distant Galactic halo with the data from surveys such as the current Sloan can be expected to shed light on whether or not an appreciable part of the halo has been formed by such accretions.

But even more basic will be a search for the explanation for the presence and absence of these types of variables in their respective systems. This is perhaps the single most pressing problem remaining for a complete understanding of the astronomy of the population II variables.

FUTURE ISSUES

1. Progress in understanding the reason for the slope and zero point differences in the period-luminosity (P-L) relations between the Galaxy and the LMC can be expected by analysis of the Cepheid data in the SMC, NGC 6822, IC 1613, Wolf-Lundmark-Melotte, and other low-metallicity local galaxies in the same way that the data for the Galaxy have been analyzed (Tammann, Sandage & Reindl 2003; Sandage, Tammann & Reindl 2004). It is yet to be proved or rejected that differences in metallicity are the cause.

2. Clues for or against a metallicity effect on the Cepheid P-L relation can be expected by relating slope and zero point differences to metallicity differences across the disks of galaxies such as M33 and NGC 300 in the presence of metallicity gradients and where the internal absorption is small, in addition to M31 and M101 where the test has already been made.

3. The well-known problem of the origin and evolutionary status of high metal abundance ($-0.5 <$ [Fe/H] $< +0.1$) RR Lyrae (RRL) variables in the field with disk kinematics is still unsolved. Identification of such stars in high metallicity open clusters would provide an opening toward a solution, but no such clusters, nor any other site of origin, has been identified, and the mystery remains. Population I analogues of the AHB1 population II XX Vir exist with the prototype BL Her (0.8 to 3 day) Cepheids, but their evolutionary origin is also a mystery. The two problems are undoubtedly related.

4. Absolute magnitude calibrations are needed of the second parameter effect for RRL stars where the morphology of the horizontal branch (HB) for different metallicity does not follow the pattern for the majority of the Galactic globular clusters. The most blatant cases are NGC 6388 and NGC 6441, where the RRL stars are believed to be 0.25 mag brighter than is predicted by the $M_V(RR)$-[Fe/H] equations given in the text. Hence, although these calibrations are valid for the "normal" RRLs, they are expected to fail at the level of 0.2 mag for second parameter cluster variables. Precision main sequence photometry or any other method accurate at the 0.1-mag level is required to measure $M_V(RR)$ for all second parameter clusters.

5. The HB morphology-metallicity correlations in the dwarf spheroidal companion satellites of the Galaxy, combined with period-amplitude and period-rise-time studies of their RRL stars, can be expected to clarify the morphology-metallicity anomaly present in the dwarf spheroidals. Such studies relate directly to the second parameter problem.

6. The problem remains why Galactic globular clusters contain W Vir population II Cepheids and RV Tauri variables while the dwarf spheroidals do not. And, visa versa, why the dwarf spheroidals contain the anomalous Cepheids (AC; 0.8 to 3 days) that differ from the above horizontal branch (AHB1)

XX Vir variables (the population II dwarf Cepheids), while Galactic globular clusters do not. Whatever the reason, presently it is not understood—although perhaps because the AC stars are young, whereas the AHB1 stars are old (Caputo, private communication)—that difference can be exploited in surveys of variables in the Galactic halo, such as the Sloan, to test if part of the remote halo is formed from accreted dwarf spheriodals. The problem may also be related to the absence of the population II W Vir Cepheids in the high Galactic halo where such stars have always been expected, but whose absence presents a problem; unexpectedly, the field W Vir variables have thick-disk kinematics and thick-disk scale heights in the Galaxy, rather than remote halo values.

ACKNOWLEDGMENTS

Thanks are due to Fillipina Caputo (Rome), to Gissele Clementini, Carla Cacciari, and Flavio Fusi Pecci (Bologna), and to Harold McNamara (Utah) for reading and commenting on an early draft. Special thanks are for John Kormendy for his exemplary editing and his many suggestions for making the text more accessible to nonexperts. It is also a pleasure to thank Roselyn Lowe-Webb, production editor, for her expert copyediting that has improved the text. Thanks are also due to Leona Kershaw who prepared numerous drafts for press, both before and after the editing by Kormendy.

LITERATURE CITED

Abt HA, Hardy RH. 1960. *Ap. J.* 131:155
Adams WA, Joy AH. 1927. *Proc. Natl. Acad. Sci. USA* 13:391
Alcock C. 2000. *Astron. J.* 119:2194
Alcock C, Allsman RA, Alves DR, Axelrod TS, Becker AC, et al. 1998. *Ap. J.* 492:190
Alloin D, Gieren W, eds. 2003. *Stellar Candles for the Extragalactic Distance Scale*. Berlin: Springer-Verlag
Andrievsky SM, Kovtyukh VV, Luck RE, Lépine JRD, Bersier D, et al. 2002. *Astron. Astrophys.* 381:32
Arp HC. 1955. *Astron. J.* 60:317
Baade W. 1926. *Astron. Nachr.* 228:359
Baade W, Swope HH. 1961. *Astron. J.* 66:300
Balgin A, Breger M, Chevalier C, Hauck B, le Contel JM, et al. 1973. *Astron. Astrophys.* 23:221
Baraffe I, Alibert Y. 2001. *Astron. Astrophys.* 371:592
Barnes TG, Evans DS. 1976. *MNRAS* 174:489
Barnes TG, Jeffreys W, Berger J, Mueller PJ, Orr K, Rodriguez R. 2003. *Ap. J.* 592:539
Becker W. 1940. *Z. Astrophys.* 19:289
Bencivenni D, Caputo F, Manteign M, Quarta ML. 1991. *Ap. J.* 380:484
Berdnikov LN, Dambis AK, Voziakova OV. 2000. *Astron. Astrophys. Suppl.* 143:211

Bessell MS. 1969. *Ap. J. Suppl.* 18:195
Bohm-Vitense E, Szhody P, Wallerstein G, Iben I. 1974. *Ap. J.* 194:125
Bono G. 2003. See Alloin & Gieren 2003, p. 85
Bono G, Caputo F, Cassisi S, Castellani V, Marconi M. 1997a. *Ap. J.* 479:279
Bono G, Caputo F, Cassisi S, Incerpi R, Marconi M. 1997b. *Ap. J.* 483:811
Bono G, Caputo F, Marconi M. 1995. *Astron. J.* 110:2365
Bono G, Caputo F, Santolamazza P. 1997a. *Astron. Astrophys.* 317:171
Bono G, Caputo F, Santolamazza P, Cassisi S, Piersimoni A. 1997b. *Astron. J.* 113:2209
Breger M. 1979. *PASP* 91:5
Breger M. 1980. *Ap. J.* 235:153
Cacciari C, Clementini G, eds. 1990. *Confrontation Between Stellar Pulsation and Evolution*, Vol. 11. San Francisco: ASP Conf. Ser.
Cacciari C, Clementini G. 2003. See Alloin & Gieren 2003, p. 105
Caloi V, D'Antona F, Mazzitelli I. 1997. *Astron. Astrophys.* 320:823
Caputo F. 1997. *MNRAS* 284:994
Caputo F, Castellani V, Marconi M, Ripepi V. 2000a. *MNRAS* 316:819
Caputo F, DeRinaldis A, Manteiga M, Pulone L, Quarta ML. 1993. *Astron. Astrophys.* 276:41
Caputo F, Marconi M, Musella I. 2000b. *Astron. Astrophys.* 354:610
Carney BW, Storm J, Jones RV. 1992. *Ap. J.* 386:663
Carretta E, Gratton RG, Clementini G, Fusi Pecci F. 2000. *Ap. J.* 533:215
Cassisi S, Castellani V, Degl'Innocenti S, Salaris M, Weiss A. 1999. *Astron. Astrophys. Suppl.* 134:103
Castellani V, Chieffi S, Pulone L. 1991. *Ap. J. Suppl.* 76:911
Catelan M, Pritzl BJ, Smith HA. 2004. *Ap. J. Suppl.* 155:633
Chen L, Hou JL. 2005. In *The Three-Dimensional Universe with Gaia* (ESA SP-576):159
Chiosi C, Wood P, Bertelli A, Bressan A. 1992. *Ap. J.* 387:320
Clement CM, Muzzin A, Dufton Q, Ponnampalam T, Wang J, et al. 2001. *Astron. J.* 122:2587
Clementini G, Carretta E, Gratton RG, Merighi R, Mould JR, McCarthy JK. 1995. *Astron. J.* 110:2319
Clementini G, Gratton R, Bragalia A, Carretta E, DiFabrizio L, Maio M. 2003. *Astron. J.* 125:1309
Cox JP. 1974. *Rep. Prog. Phys.* 37:563
Cudworth KM. 1979. *Astron. J.* 84:1313
D'Antona F, Caloi V, Mazzitelli I. 1997. *Ap. J.* 477:519
Danziger IJ, Dickens RJ. 1967. *Ap. J.* 149:55
Dean JF, Cousins AWJ, Bywater BA, Warren PR. 1977. *MmRAS* 83:69
Demarque P. 1960. *Ap. J.* 132:366
Demarque P, Hirshfeld AW. 1975. *Ap. J.* 202:346
Demarque P, Zinn R, Lee YW, Yi S. 2000. *Astron. J.* 119:1398
De Santis R, Cassisi S. 1999. *MNRAS* 308:97
Diethelm R. 1983. *Astron. Astrophys.* 124:108

Diethelm R. 1990. *Astron. Astrophys.* 239:186
Diethelm R. 1996. *Astron. Astrophys.* 307:803
Dolphin AE, Saha A, Skillman ED, Tolstoy E, Cole AA, et al. 2001. *Ap. J.* 550:554
Dorman B. 1992. *Ap. J. Suppl.* 81:221
Eggen OJ. 1951. *Ap. J.* 113:367
Eggen OJ. 1956a. *PASP* 68:238
Eggen OJ. 1956b. *PASP* 68:541
Eggen OJ. 1970. *PASP* 82:274
Eggen OJ. 1979. *Ap. J. Suppl.* 41:413
Eggen OJ, Sandage A. 1962. *Ap. J.* 136:735
Feast MF. 1997. *MNRAS* 284:76
Feast MW. 1999. *PASP* 111:775
Feast MW. 2003. See Alloin & Gieren 2003, p. 45
Feast MW, Catchpole RM. 1997. *MNRAS* 286:L1
Feast MW, Walker AR. 1987. *Annu. Rev. Astron. Astrophys.* 25:345
Fernie JD. 1969. *PASP* 81:707
Fernie JD. 1990. *Ap. J. Suppl.* 72:153
Fernie JD. 1994. *Ap. J.* 429:844
Fernie JD, Beattie B, Evans NR, Seager S. 1995. IBVS. 4148. (**http:/ddo.Astro.utoronto.ca/cepheids.html**)
Fernley J. 1993. *Astron. Astrophys.* 268:591
Fernley J. 1994. *Astron. Astrophys.* 284:L16
Fernley J, Carney BW, Skillen I, Cacciari C, Janes K. 1998b. *MNRAS* 293:L61
Fernley JA, Barnes TG, Skillen I, Hawley SL, Hanley SJ, et al. 1998a. *Astron. Astrophys.* 330:515
Ferraro FR, Messineo M, Fusi Pecci F, De Palo MA, Straniero O, et al. 1999. *Astron. J.* 118:1738
Fouque P, Storm J, Gieren W. 2003. See Alloin & Gieren 2003, p. 21
Freedman WL, Madore BF. 1990. *Ap. J.* 365:186
Freedman WL, Madore BF, Gibson BK, Ferrarese L, Kelson DD, et al. 2001. *Ap. J.* 553:47
Fry AM, Carney BW. 1997. *Astron. J.* 113:1073
Fusi Pecci F, Buonanno R, Cacciari C, Corsi CE, Djorgovski G, et al. 1996. *Astron. J.* 112:1461
Gascoigne SCB, Kron GE. 1965. *MNRAS* 130:333
Gautschy A, Saio H. 1995. *Annu. Rev. Astron. Astrophys.* 33:75
Gibson BK, Stetson PB, Freedman WL, Mould JR, Kennicutt RC Jr, et al. 2000. *Ap. J.* 529:723
Gieren W, Storm J, Barnes TG, Fouque P, Pietrzynski G, Kienzle F. 2005. *Ap. J.* 627:224
Gieren WP, Fouque P, Gomez M. 1998. *Ap. J.* 496:17
Gingold RA. 1976. *Ap. J.* 204:116
Gingold RA. 1985. *Mem. Soc. Astron. Ital.* 56:169
Gould A, Popowski P. 1998. *Ap. J.* 508:844
Gratton RG. 1998. *MNRAS* 296:739

Gratton RG, Fusi Pecci F, Carretta E, Clementini G, Corsi CE, Lattanzi M. 1997. *Ap. J.* 491:749
Groenewegen MA, Oudmaijer RD. 2000. *Astron. Astrophys.* 356:849
Grosse E. 1932. *Astron. Nach.* 246:376
Hachenberg O. 1939. *Z. Astrophys.* 18:49
Hanson RB. 1979. *MNRAS* 186:875
Harris HC. 1984. In *Cepheids: Theory and Observations*, ed. BF Madore, p. 232. Cambridge: Cambridge Univ. Press
Harris HC. 1985. *Astron. J.* 90:756
Hirshfeld AW. 1980. *Ap. J.* 241:111
Hoffmeister E. 1967. *Z. Astrophys.* 65:194
Hoyle F, Shanks T, Tanvir NR. 2003. *MNRAS* 345:269
Iben I, Tuggle RS. 1972. *Ap. J.* 178:441
Iben I, Tuggle RS. 1975. *Ap. J.* 197:39
Jones RV, Carney BW, Storm J, Latham DW. 1992. *Ap. J.* 386:646
Joy AH. 1949. *Ap. J.* 110:105
Jurcsik J, Kovacs G. 1996. *Astron. Astrophys.* 312:111
Kanbur SM, Ngeow C. 2002. Poster presented at *Stellar Candles Extragalactic Distance Scale Conf.*, Conception, Chile
Kanbur SM, Ngeow CC. 2004. *MNRAS* 350:962
Kennicutt RC, Stetson PB, Saha A, Kelson D, Rawson DM, et al. 1998. *Ap. J.* 498:181
Kervella P, Bersier D, Mourand D, Nardetto N, Coude du Foresto V. 2004b. *Astron. Astrophys.* 423:327
Kinman TD. 1959. *MNRAS* 119:538
Kovacs G, Jurcsik J. 1996. *Ap. J.* 466:L17
Kovacs G, Jurcsik J. 1997. *Astron. Astrophys.* 322:218
Kovtyukh VV, Wallerstein G, Andrievsky SM. 2005. *PASP* 117:1173
Kraft RP. 1972. In *The Evolution of Population II Stars*, ed. AGD Philip, p. 69. Albany, NY: Dudley Obs.
Kukarkin BV, Kholopov PN, Efremov Yu N, Kukarkina NP, Kurochkin NE, et al. 1969. *General Catalog of Variable Stars*, Vol. 1. Moscow: Akad. Nauka
Kurtz DW, Pollard KP. 2004. *Variable Stars in the Local Group*, Vol. 310. San Francisco: ASP Conf. Ser.
Kwee KK. 1967. *Bull. Astron. Neth.* 19:260
Kwee KK, Diethelm R. 1984. *Astron. Astrophys. Suppl.* 55:77
Landolt A. 1983. *Astron. J.* 88:439
Landolt A. 1992. *Astron. J.* 104:340
Laney CD, Stobie RS. 1986. *MNRAS* 222:449
Layden AC, Hanson RB, Hawley SL, Klemola AR, Hanley CJ. 1996. *Astron. J.* 112:2120
Lee YW, Demarque P, Zinn R. 1990. *Ap. J.* 350:155
Luck RE, Gieren WP, Andrievsky SM, Kovtyukh VV, Fouqué P, et al. 2003. *Astron. Astrophys.* 401:939
Lupton RH, Gunn JE, Griffin RF. 1987. *Astron. J.* 93:1114
Lutz TE. 1983. In *IAU Coll. 76. The Nearby Stars and the Stellar Luminosity Function*, ed. ACD Phillips, AR Upgran, p. 41. Schenectady, NY: Davis

Lutz TE, Kelker DH. 1973. *PASP* 85:573
Macri LM. 2004. See Kurtz & Pollard 2004, p. 33
Makarov VV. 2002. *Astron. J.* 124:3299
McNamara DH. 1995. *Astron. J.* 109:2134
McNamara DH. 1997a. *PASP* 109:857
McNamara DH. 1997b. *PASP* 109:1221
McNamara DH. 1999. *PASP* 111:489
McNamara DH. 2000. In *Delta Scuti and Related Stars*, Vol. 210, ed. M. Breger, MH Montgomery, p. 373. San Francisco: ASP Conf. Ser.
McNamara DH, Feltz KA. 1978. *PASP* 90:275
McNamara DH, Powell JM. 1990. See Cacciari & Clementini 1990, p. 316
McNamara DH, Rose MB, Brown PJ, Ketcheson DI, Maxwell JE, et al. 2004. See Kurtz & Pollard 2004, p. 525
Mitchie RW. 1963. *MNRAS* 125:127
Narayanan VK, Gould A. 1999. *Ap. J.* 523:328
Nemec J, Mateo M. 1990a. In *The Evolution of the Universe of Galaxies*, Vol. 10, ed. RG Kron, p. 134. San Francisco: ASP Conf. Ser.
Nemec J, Mateo M. 1990b. See Cacciari & Clementini 1990, p. 64
Nemec JM. 1989. In *The Use of Pulsating Stars in Fundamental Problems of Astronomy*, ed. EP Schmidt, IAU Symp. 111:215. Cambridge: Cambridge Univ. Press
Nemec JM, Nemec AF, Lutz TE. 1994. *Astron. J.* 108:222
Nemec JM, Whelau A, Mendes de Oliveira C. 1988. *Astron. J.* 96:528
Ngeow CC, Kanbur SM. 2005. *MNRAS* 360:1033
Ngeow CC, Kanbur SM, Nikolaev S, Buonaccorsi J, Cook KH, Welch DL. 2005. *MNRAS* 363:831
Norris J, Zinn R. 1975. *Ap. J.* 202:375
Oosterhoff PT. 1939. *Observatory* 62:104
Oosterhoff PT. 1944. *BAN* 10:55
Peterson JO. 1999. In *Harmonizing Cosmic Distance Scales in a Post-*HIPPARCOS *Era*, Vol. 167, ed. D Egret, A Heck, p. 107. San Francisco: ASP Conf. Ser.
Pietrzynski G, Gieren W, Fouque P, Pont F. 2002. *Astron. J.* 123:789
Pinsonneault MH, Stauffer J, Soderblom DH, King JR, Handon RB. 1998. *Ap. J.* 504:170
Popowski P, Gould A. 1999. In *Post Hipparcos Cosmic Candles*, ed. A Heck, F Caputo. p. 53. Dordrecht: Kluwer
Preston GW. 1959. *Ap. J.* 130:507
Pritzl B, Smith HA, Catelan M, Sweigart AV. 2000. *Ap. J.* 530:L41
Pritzl B, Smith HA, Catelan M, Sweigart AV. 2001. *Astron. J.* 122:2600
Pritzl BJ, Smith HA, Stetson PB, Catelan M, Sweigart AV, et al. 2003. *Astron. J.* 126:1381
Reid IN. 1997. *Astron. J.* 114:161
Reid IN. 1999. *Annu. Rev. Astron. Astrophys.* 37:191
Reiz A. 1954. *Ap. J.* 120:342
Renzini A, Fusi Pecci F. 1988. *Annu. Rev. Astron. Astrophys.* 26:199

Rich RM, Corsi CE, Bellazzini M, Federici L, Cacciari C, Fusi Pecci F. 2001. In *Extragalactic Star Clusters*, IAU Symp. 207, ed. D Geisler, EK Grebel, D Minniti, p. 140. San Francisco: ASP
Rich RM, Sosin C, Djorgovski G, Piotto G, King I, et al. 1997. *Ap. J.* 484:L25
Richer HB, Fahlman GG, Ibata RA, Stetson PB, Bell RA, et al. 1995. *Ap. J.* 451:L17
Richer HB, Fahlman GG, Ibata RA, Pryor C, Bell RA, et al. 1997. *Ap. J.* 484:741
Ritter A. 1879. *Ann. Phys. Chem. Neue. Folge* 8:157
Russell HN. 1927. *Ap. J.* 66:122
Saha A, Thim F, Tammann GA, Reindl B, Sandage A. 2006. *Ap. J. Suppl.* In press. astro-ph/0602572
Saio H, Gautschy A. 1998. *Ap. J.* 498:360
Sakai S, Ferrarese L, Kennicutt RC, Saha A. 2004. *Ap. J.* 608:42
Salaris M, Chieffi A, Straniero O. 1993. *Ap. J.* 414:580
Salaris M, Degl'Innocenti S, Weiss A. 1997. *Ap. J.* 479:665
Sandage A. 1958a. In *Stellar Populations, Specola Vaticana*, ed. D O'Connell. *RicA* 5:41
Sandage A. 1958b. *Ap. J.* 127:515
Sandage A. 1970. *Ap. J.* 162:841
Sandage A. 1972. *Quart. J. RAS* 13:202
Sandage A. 1981a. *Ap. J.* 244:L23
Sandage A. 1981b. *Ap. J.* 248:161
Sandage A. 1982. *Ap. J.* 252:553
Sandage A. 1990a. *Ap. J.* 350:603
Sandage A. 1990b. *Ap. J.* 350:631
Sandage A. 1993a. *Astron. J.* 106:687
Sandage A. 1993b. *Astron. J.* 106:703
Sandage A. 2004. *Astron. J.* 128:858
Sandage A. 2006. *Astron. J.* 131:1750
Sandage A, Bell RA, Tripicco MJ. 1999. *Ap. J.* 522:250
Sandage A, Cacciari C. 1990. *Ap. J.* 350:645
Sandage A, Diethelm R, Tammann GA. 1994. *Astron. Astrophys.* 283:111
Sandage A, Eggen OJ. 1959. *MNRAS* 119:278
Sandage A, Saha A. 2002. *Astron. J.* 123:2047
Sandage A, Tammann GA. 1968. *Ap. J.* 151:531
Sandage A, Tammann GA. 1969. *Ap. J.* 157:683
Sandage A, Tammann GA, Reindl B. 2004. *Astron. Astrophys.* 424:43
Sandage A, Tammann GA, Saha A, Reindl B, Machetto D, Panagia N. 2006. *Ap. J.* In press
Schwarzschild M. 1940. *Cir. Harv. Coll. Obs.* No. 437
Shapley H. 1914. *Ap. J.* 40:448
Shapley H. 1927a. *Cir. Harv. Coll. Obs.* No. 313
Shapley H. 1927b. *Cir. Harv. Coll. Obs.* No. 314
Simon NR, Clement CM. 1993. *Ap. J.* 410:526
Smith H. 1987. *Astron. Astrophys.* 188:233
Smith HA. 1955. *The RR Lyrae Stars*. Cambridge: Cambridge Univ. Press
Smith HA, Lugger JJ, Deming D, Butler D. 1978. *PASP* 90:422

Smith HA, Silbermann NA, Baird SR, Graham JA. 1992. *Astron. J.* 04:1430
Smith HJ. 1955. *Astron. J.* 60:179
Stello D, Nissen PE. 2001. *Astron. Astropyhs.* 374:105
Straniero O, Chieffi A. 1991. *Ap. J. Suppl.* 76:525
Straniero O, Chieffi A, Limongi M. 1997. *Ap. J.* 490:425
Strom SE, Strom KM, Rood RT, Iben I. 1970. *Astron. Astrophys.* 8:243
Stromgren B. 1952. *Astron. J.* 57:65
Sweigart AV, Catelan M. 1998. *Ap. J.* 501:L63
Swope HH. 1968. *Astron. J. Suppl.* 73:204
Tammann GA, Reindl R. 2002. *Astrophys. Space Sci.* 280:165
Tammann GA, Reindl B, Thim F, Saha A, Sandage A. 2002. In *A New Era in Cosmology*, ed. T Shanks, N Metcalfe. 283:258. San Francisco: ASP Conf. Ser.
Tammann GA, Sandage A, Reindl B. 2003. *Astron. Astrophys.* 404:423
Tanvir NR, Hendry MA, Watkins A, Kanbur SM, Berdnikov LN, Ngeow CC. 2005. *MNRAS* 363:749
Thim F, Tammann GA, Saha A, Dolphin A, Sandage A, et al. 2003. *Ap. J.* 590:256
Turner DG, Burke JF. 2002. *Astron. J.* 124:2931
Udalski A, Soszynski I, Szymanski M, Kubiak M, Pietrzynski G, et al. 1999a. *Acta Astron. Sinica* 49:223
Udalski A, Soszynski I, Szymanski M, Kubiak M, Pietrzynski G, et al. 1999b. *Acta Astron. Sinica* 49:437
van Albada TS, Baker N. 1973. *Ap. J.* 185:477
VandenBerg DA. 1992. *Ap. J.* 391:685
VandenBerg DA, Bell RA. 1985. *Ap. J. Suppl.* 58:561
VandenBerg DA, Swenson FJ, Rogers FJ, Iglesias CA, Alexander DR. 2000. *Ap. J.* 532:430
van den Bergh S. 1995. *Ap. J.* 446:39
Walker A, Mack AR. 1988. *Astron. J.* 96:872
Wallerstein G. 1990. See Cacciari & Clementini 1990, p. 56
Wallerstein G. 2002. *PASP* 114:689
Wallerstein GW, Brugel EW. 1979. *Astron. J.* 84:1840
Wallerstein GW, Cox AN. 1984. *PASP* 96:677
Wehlau A, Bohlender D. 1982. *Astron. J.* 87:780
Wesselink A. 1946. *Bull. Astron. Inst. Neth.* 10:91
Whitlock P, Feast M, Catchpole R. 1991. *MNRAS* 248:276
Zinn R, Dahn CC. 1976. *Astron. J.* 81:527
Zinn R, King CR. 1982. *Ap. J.* 262:700
Zinn R, Searle L. 1976. *Ap. J.* 209:734

Stellar Population Diagnostics of Elliptical Galaxy Formation

Alvio Renzini

INAF, Osservatorio Astronomico di Padova, Padua, Italy; email: arenzini@pd.astro.it

Key Words

galaxy evolution, galaxy surveys, stellar populations

Abstract

Major progress has been achieved in recent years in mapping the properties of passively-evolving, early-type galaxies (ETG) from the local universe all the way to redshift ∼2. Here, age and metallicity estimates for local cluster and field ETGs are reviewed as based on color-magnitude, color-σ, and fundamental plane relations, as well as on spectral-line indices diagnostics. The results of applying the same tools at high redshifts are then discussed, and their consistency with the low-redshift results is assessed. Most low- as well as high-redshift ($z \sim 1$) observations consistently indicate (a) a formation redshift $z \gtrsim 3$ for the bulk of stars in cluster ETGs, with their counterparts in low-density environments being on average ∼1–2 Gyr younger, i.e., formed at $z \gtrsim 1.5$–2; (b) the duration of the major star-formation phase anticorrelates with galaxy mass, and the oldest stellar populations are found in the most massive galaxies. With increasing redshift there is evidence for a decrease in the number density of ETGs, especially of the less massive ones, whereas existing data appear to suggest that most of the most-massive ETGs were already fully assembled at $z \sim 1$. Beyond this redshift, the space density of ETGs starts dropping significantly, and as ETGs disappear, a population of massive, strongly clustered, starburst galaxies progressively becomes more and more prominent, which makes them the likely progenitors to ETGs.

1. INTRODUCTION

Following Hubble (1936), we still classify galaxies as ellipticals, spirals, and irregulars (see Sandage 2005, and references therein). This was an eyeball, purely morphological scheme; however, morphology correlates with the stellar population content of these galaxies, with typical ellipticals being redder than the others and showing purely stellar absorption-line spectra with no or very weak nebular emissions. As a consequence, one often refers to early-type galaxies (ETG), even if they are color (or spectral-type) selected rather than morphologically selected. Furthermore, the bulges of spirals of the earlier types show morphological as well as spectral similarities with ellipticals, and one often includes both ellipticals and bulges under the category of galactic spheroids.

Morphologically selected and color- or spectrum-selected samples do not fully overlap. For example, in a recent study of local ($z \sim 0$) ETGs from the Sloan Digital Sky Survey (SDSS) all three criteria were adopted (Bernardi et al. 2006), and the result is reported in **Table 1** (M. Bernardi, private communication).[1] From a sample including $\sim$123,000 galaxies with $14.5 < r_{\rm Petrosian} < 17.77$ and $0.004 < z < 0.08$, ETGs have been selected in turn by each of the three criteria (MOR, COL, SPE), and the resulting numbers of galaxies fulfilling each of them is given on the diagonal of the matrix. Out of the diagonal are the fractions of galaxies that satisfy two of the criteria, as labeled in the corresponding row and column. So, out of the morphologically-selected ETGs, 70% satisfy also the color selection, etc. The correlation between color and morphology selection persists at high redshift, e.g., at $z \sim 0.7$ about 85% of color-selected, red-sequence galaxies are also morphologically early-type, i.e., E/S0/Sa (Bell et al. 2004a), an estimate broadly consistent with the local one, when allowing for the broader morphological criterion.

It is estimated that true ellipticals represent $\sim$22% of the total mass in stars in the local universe, a fraction amounting to $\sim$75% for spheroids (i.e., when including E0s and spiral bulges), whereas disks contribute only $\sim$25% and dwarfs an irrelevant fraction (Fukugita, Hogan, & Peebles 1998). Although earlier estimates gave slightly lower fractions for stars in spheroids versus disks (e.g., Schechter & Dressler 1987; Persic & Salucci 1992), it is now generally accepted that the majority of stars belong to spheroids. **Figure 1** shows separately the mass functions of color-selected ETGs and of blue, star-forming galaxies, also based on the SDSS data (Baldry et al. 2004). Above $\sim 3 \times 10^{10}~M_\odot$ red-sequence galaxies start to increasingly outnumber blue galaxies by a factor that exceeds 10 above $\sim 3 \times 10^{11}~M_\odot$. **Figure 2**, drawn from the same mass functions, shows the contributions to the total stellar mass by red and blue galaxies in the various mass bins, along with the contributions to the total number of galaxies. Then, ETGs represent only 17% of the total number of galaxies in the sample, but contribute $\sim$57% of the total mass. Moreover, $\gtrsim$80% of the stellar mass in ETGs belongs to galaxies more massive than $\sim 3 \times 10^{10}~M_\odot$. Dwarfs are sometimes seen as the "building blocks" of galaxies, but at least in the present universe one cannot build much with them.

---

[1]Here, as well as throughout the whole paper, the definition of the specific criteria adopted by the various authors for the sample selections can be found in the original articles.

Table 1 Morphology- versus color- versus spectrum-selected samples

	MOR	COL	SPE
MOR	37151	70%	81%
COL	58%	44618	87%
SPE	55%	70%	55134

With spheroids holding the major share of stellar mass in galaxies, understanding their evolution—from formation to their present state—is central to the galaxy evolution problem in general. Historically, two main scenarios have confronted each other, the so-called Monolithic Collapse model (Eggen, Lynden-Bell & Sandage 1962; Larson 1974; Arimoto & Yoshii 1987; Bressan, Chiosi & Fagotto 1994), and the Hierarchical Merging model (e.g., Toomre 1977, White & Rees 1978). In the former scenario spheroids form at a very early epoch as a result of a global starburst, and then passively evolve to the present. If the local conditions are keen enough, a spheroid can gradually grow a disk by accreting gas from the environment, hence spheroids precede disks. In the merging model, big spheroids result from the mutual disruption of disks in a merging event, hence disks precede spheroids.

The two scenarios appear to sharply contradict each other, but the contradiction has progressively blurred in recent years. Evidence has accumulated that the bulk of stars in spheroids are old, and most likely formed in major merging events. In the hierarchical merging scenario (the only one rooted in a solid cosmological context)

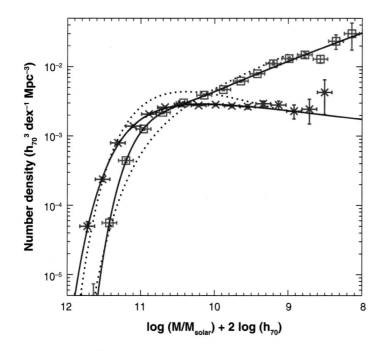

Figure 1

The mass function of local ($z \sim 0$) early-type (*red*) galaxies and late-type (*blue*) galaxies from the Sloan Digital Sky Survey (Baldry et al. 2004). The solid lines represent best-fit Schechter functions. For the blue galaxies the sum of two different Schechter functions is required to provide a good fit. Dotted lines show the mass functions from Bell et al. (2003).

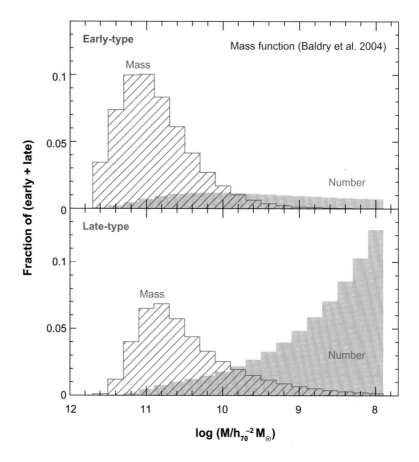

Figure 2

The contributions to the total stellar mass and to the number of galaxies by early-type (*red*) and late-type (*blue*) galaxies in the various mass bins, as derived from the best-fit mass functions of Baldry et al. (2004) shown in **Figure 1**. The relative areas are proportional to the contributions of the early- and late-type galaxies to the total stellar mass and to the number of galaxies. Courtesy of D. Thomas.

successive generations of models have struggled to increase their predicted stellar ages, so as to produce results resembling the opposite scenario. This review will not attempt to trace an history of theoretical efforts to understand the formation and evolution of ellipticals and spheroids. It will rather concentrate on reviewing the accumulating observational evidence coming from the stellar component of these galaxies. Other extremely interesting properties of ellipticals, such as their structural and dynamical properties, their hot gas content, central supermassive black holes, etc., will not be touched in this review, even if they are certainly needed to complete the picture and likely play an important role in the evolution of these galaxies. Also untouched are the internal properties of ETGs, such as color and line-strength gradients indicative of spatial inhomogeneities of the stellar populations across the body of ETGs. A complementary view of ETG formation based on the globular cluster populations of these galaxies is presented by Brodie & Strader (2006).

The "concordance cosmology" ($\Omega_M = 0.3$, $\Omega_\Lambda = 0.7$, $H_0 = 70$) is adopted if not explicitly stated otherwise.

2. SYNTHETIC STELLAR POPULATIONS

The ETGs treated in this review can only be studied in integrated light, hence the interpretation of their photometric and spectroscopic properties needs population synthesis tools. Pioneering unconstrained synthesis using "quadratic programming" (e.g., Faber 1972) was soon abandoned in favor of evolutionary population synthesis, for which foundations were laid down by Beatrice Tinsley in the 1970s (Tinsley & Gunn 1976; Tinsley 1980; Gunn, Stryker & Tinsley 1981). Much progress has been made in the course of the subsequent quarter of a century, especially thanks to the systematic production of fairly complete libraries of stellar evolutionary sequences and stellar spectra.

Several modern population synthesis tools are commonly in use today, including those of Worthey (1994), Buzzoni (1995), Bressan, Chiosi, & Tantalo (1996), Maraston (1998, 2005), Bruzual & Charlot (2003), Fioc & Rocca Volmerange (1997; PEGASE Code), Vázques & Leitherer (2005; Starburst99 Code), Vazdekis et al. (2003), and Gonzĺez Delgado et al. (2005). Although far more reliable than earlier generations of models, even the most recent tools still may suffer from incomplete spectral libraries (especially at high metallicity and for nonsolar abundance ratios), and poorly calibrated mass loss in advanced stages, such as the asymptotic giant branch (AGB). Yet, there is fair agreement among the various models, with the exception of those for ages around ∼1 Gyr, when the contribution by AGB stars is at maximum, and Maraston's models (calibrated on Magellanic Cloud clusters) give appreciably higher near-IR fluxes than the other models.

Only a few "rules of thumb" regarding population synthesis models can be recalled here, which may be useful in guiding the reader through some of the subtleties of their comparison with the observations.

- No evolutionary population synthesis code is perfect. Evolutionary tracks are not perfect and stellar libraries are never really complete. So, any code deficiency will leave its imprint on the results, generating a distortion of the age/metallicity grids used to map plots of one observable versus another. To some extent, such distorsion may lead to spurious correlations/anticorrelations when reading ages and metallicities from overplotted data points.
- Ages derived from best fits to simple stellar populations (SSPs, i.e., single-burst populations) are always luminosity-weighted ages, and in general are more sensitive to the youngest component of the real age distribution. SSP ages should be regarded as lower limits.
- Spectra and colors of SSPs are fairly insensitive to the initial mass function (IMF), because most of the light comes from stars in a narrow mass interval around the mass of stars at the main sequence turnoff.
- The time evolution of the luminosity of a SSP does depend on the IMF, and so does the mass-to-light (M/L) ratio. For example, a now-fashionable IMF that flattens below ∼0.6 $M_\odot$ (e.g., Chabrier 2003) gives M/L ratios a factor of ∼2 lower than a straight Salpeter IMF.
- Stellar ages and metallicities are the main quantities that the analyses of colors and integrated spectra of galaxies are aimed at determining. Yet, for many

observables, age and metallicity are largely degenerate, with a reduced age coupled to an increased metallicity conjuring to leave the spectral energy distribution nearly unchanged. This results primarily from the color (temperature) of the main sequence turnoff, e.g., $(B-V)^{\rm TO}$, (the true clock of SSPs) being almost equally sensitive to age and metallicity changes. Indeed, from stellar isochrones one can derive that $(\partial \log t/\partial [{\rm Fe/H}])^{\rm TO}_{(B-V)} \simeq -0.9-0.35\,[{\rm Fe/H}]$, and a factor of 2 error in estimated metallicity produces a factor of $\sim$2 error in age (Renzini 1992). Red giant branch stars are the major contributors of bolometric luminosity in old stellar populations, and their locus shifts to lower temperatures with both increasing age and metallicity, further contributing to the degeneracy. Thus, from full SSPs, Worthey (1994) estimated that a factor of 3 error in metallicity generates a factor of 2 error in age when using optical colors as age indicators, the so-called 2/3 rule. Several strategies have been devised to circumvent this difficulty and break the age-metallicity degeneracy (see below).

- There are occasionally ambiguities in what is meant by the M/L ratio in the tabulated values. The mass M can be defined either as the mass of gas that went into stars or the mass of the residual population at age t, including the mass in dead remnants (i.e., the original mass diminished by the mass lost by stars in the course of their evolution), or even the mass of the surviving stars, i.e., without including the mass in remnants. Caution should be paid when using tabular values, as different authors may adopt different definitions.

The power of stellar population dignostics stems from the opportunity to age-date the stellar content of galaxies in a fashion that is independent of cosmological parameters. Then, once a cosmology is adopted, ages derived from observations at a lower redshift can be used to predict the properties of the stellar populations of ETGs at a higher one, including their formation redshift. Thus, ages derived for the local elliptical galaxies imply a well-defined color, spectral, and luminosity evolution with redshifts, which all can be subject to direct observational test. The extent to which a consistent picture of ETG formation is emerging from low- and high-redshift observations is the main underlying theme of this review.

3. ELLIPTICAL GALAXIES IN THE LOCAL UNIVERSE

Observations at high redshift are certainly the most direct way to look at the forming galaxies, and a great observational effort is currently being made in this direction. Yet, high-redshift galaxies are very faint, and only few of their global properties can now be measured. Nearby galaxies can instead be studied in far greater detail, and their fossil evidence can provide a view of galaxy formation and evolution that is fully complementary to that given by high-redshift observations. By fossil evidence one refers to those observables that are not related to ongoing, active star formation, and which are instead the result of the integrated past star formation history. At first, studies attempted to estimate ages and metallicities of the dominant stellar populations on a galaxy-by-galaxy basis. But the tools used were still quite rudimentary, being based on largely incomplete libraries of stellar spectra and evolutionary sequences. Hence, through the 1980s progress was relatively slow, and opinions could widely diverge as to whether ellipticals were dominated by old stellar populations—as old as galactic globular clusters—or by intermediate age ones—several billion years younger than globulars (see e.g., O'Connell 1986, Renzini 1986)—with much of the diverging interpretations being a result of the age-metallicity degeneracy. From the beginning of the 1990s, progress has been constantly accelerating, and much of this review concentrates on the developments that took place over the past 15 years.

3.1. Color-Magnitude Relation, Fundamental-Plane, and Line-Indices

3.1.1. The color-magnitude and color-σ relations.
That elliptical galaxies follow a tight color-magnitude (C-M) relation was first recognized by Baum (1959), and in a massive exploration Visvanathan & Sandage (1977) and Sandage & Visvanathan (1978a,b) established the universality of this relation with what continues to be the culmination of ETG studies in the pre-CCD era. The C-M relation looked the same in all nine studied clusters, and much the same in the field as well, though with larger dispersion (at least in part due to larger distance errors). The focus was on the possible use of the C-M relation as a distance indicator; however, Sandage & Visvanathan documented the tightness of the relation and noted that it implies the stellar content of the galaxies to be very uniform. They also estimated that both S0s and ellipticals had to be evolving passively since at least $\sim$1 Gyr ago. **Figure 3** shows

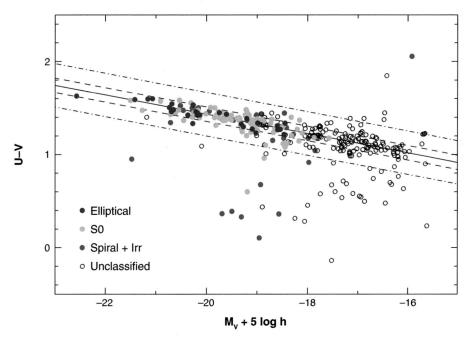

Figure 3

The $(U-V) - M_V$ color-magnitude relation for galaxies that are spectroscopic members of the Coma cluster (Bower et al. 1999).

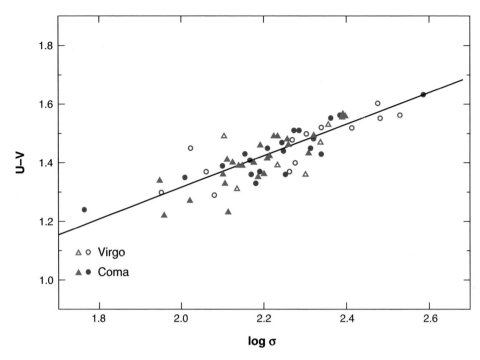

Figure 4

The relation between $(U-V)$ and central velocity dispersion (σ) for early-type galaxies in the Virgo (*open symbols*) and Coma (*filled symbols*) clusters. Red circles represent ellipticals; blue triangles represent S0s. From Bower, Lucey & Ellis (1992).

a modern rendition for the C-M plot for the Coma cluster galaxies, showing how tight it is as well as how closely both S0s and ellipticals follow the same relation, as indeed Sandage & Visvanathan had anticipated.

In a major breakthrough in galaxy dating, Bower, Lucey & Ellis (1992), rather than trying to age-date galaxies one by one, were able to set tight age constraints on all ETGs in Virgo and Coma at once. Noting the remarkable homogeneity of ETGs in these clusters, they estimated the intrinsic color scatter in the color-σ relation (see **Figure 4**) to be $\delta(U-V) \lesssim 0.04$ mag, where σ is the central stellar velocity dispersion of these galaxies. They further argued that, if due entirely to an age dispersion $\delta t = \beta(t_{\rm H} - t_{\rm F})$, such color scatter should be equal to the time scatter in formation epochs, times $\partial(U-V)/\partial t$, i.e.

$$t_{\rm H} - t_{\rm F} \lesssim \frac{0.04}{\beta}\left(\frac{\partial(U-V)}{\partial t}\right)^{-1}, \qquad (1)$$

where $t_{\rm H}$ is the age of the universe at $z=0$, and galaxies are assumed to form before a lookback time $t_{\rm F}$. Bower and colleagues introduced the parameter β, such that $\beta(t_{\rm H} - t_{\rm F})$ is the fraction of the available time during which galaxies actually form. Thus, for $\beta=1$ galaxy formation is uniformly distributed between $t \sim 0$ and $t = t_{\rm H} - t_{\rm F}$, whereas for $\beta < 1$ it is more and more synchronized, i.e., restricted to the fraction

β of time interval $t_H - t_F$. Adopting $\partial(U-V)/\partial t$ from the models of Bruzual (1983), they derived $t_H - t_F < 2$ Gyr for $\beta = 1$ and $t_H - t_F < 8$ Gyr for $\beta = 0.1$, corresponding, respectively, to formation redshifts $z_F \gtrsim 2.8$ and $\gtrsim 1.1$ for their adopted cosmology ($t_H = 15$ Gyr, $q_0 = 0.5$). For the concordance cosmology, the same age constraints imply $z_F \gtrsim 3.3$ and $\gtrsim 0.8$, respectively. A value $\beta = 0.1$ implies an extreme synchronization, with all Virgo and Coma galaxies forming their stars within less than 1 Gyr when the universe had half its present age, which seems rather implausible. Bower and colleagues concluded that ellipticals in clusters formed the bulk of their stars at $z \gtrsim 2$, and later additions should not provide more than ~10% of their present luminosity. Making minimal use of stellar population models, this approach provided, for the first time, a robust demonstration that cluster ellipticals are made of very old stars, with the bulk of them having formed at $z \gtrsim 2$.

As the narrowness of the C-M and color-σ relations sets constraints on the ages of stellar populations in ETGs, their slope can set useful constraints on the amount of merging that may have led to the present-day galaxies. The reason is that merging without star formation increases luminosity and σ, but leaves colors unchanged, thus broadening and flattening the relations. Moreover, merging with star formation makes bluer galaxies, thus broadening and flattening the relations even more. Then, from the constraints set by the slope of the C-M relation, Bower, Kodama, & Terlevich (1998) concluded not only that the bulk of stars in clusters must have formed at high redshift, but also that they cannot have formed in mass units much less than about half their present mass.

3.1.2. The fundamental plane.
Three key observables of elliptical galaxies, namely the effective radius R_e, the central velocity dispersion σ, and the luminosity L (or equivalently the effective surface brightness $I_e = L/2\pi R_e^2$), relate their structural/dynamical status to their stellar content. Elliptical galaxies are not randomly distributed within the 3D space (R_e, σ, I_e), but rather cluster close to a plane, thus known as the fundamental plane (FP), with $R_e \propto \sigma^a I_e^b$ (Dressler et al. 1987, Djorgovski & Davis 1987), where the exponents a and b depend on the specific band used for measuring the luminosity. The projection of the FP over the (R_e, I_e) coordinate plane generates the Kormendy relation (Kormendy 1977), whereas a projection over the (σ, $L = 2\pi R_e^2 I_e$) plane generates the Faber-Jackson relation (Faber & Jackson 1976). At a time when testing the $\Omega_M = 1$ standard cosmology had high priority, the FP was first used to estimate distances, in order to map deviations from the local Hubble flow and construct the gravitational potential on large scales. Its use to infer the properties of the stellar content of galaxies, and set constraints on their formation, came later. Yet, by relating the luminosity to the structural-dynamical parameters of a galaxy, the FP offers a precious tool to gather information on the ages and metallicities of galaxies, at low as well as at high redshifts.

The mere existence of a FP implies that ellipticals (*a*) are well-virialized systems, (*b*) have self-similar (homologous) structures, or their structures (e.g., the shape of the mass distribution) vary in a systematic fashion along the plane, and (*c*) contain stellar populations that must fulfill tight age and metallicity constraints. Here we concentrate on this latter aspect.

To better appreciate the physical implications of the FP, Bender, Burnstein & Faber (1992) introduced an orthogonal coordinate system ($\kappa_1, \kappa_2, \kappa_3$), in which each new variable is a linear combination of $\log \sigma^2$, $\log R_e$, and $\log I_e$. The transformation corresponds to a rotation of the coordinate system such that in the (κ_1, κ_3) projection the FP is seen almost perfectly edge-on. Moreover, if structural homology holds all along the plane, then $\log M/L = \sqrt{3}\kappa_3 +$ const. If σ is virtually unaffected by the dark matter distribution (as currently understood, see Rix et al. 1997), then κ_3 provides a measure of the stellar M/L ratio and $\kappa_1 \propto \log(\sigma^2 R_e) \propto \log M$ a measure of the stellar mass. Bender and colleagues showed that in Virgo and Coma the FP is remarkably "thin," with a 1-σ dispersion perpendicular to the plane of only $\sigma(\kappa_3) \simeq 0.05$, corresponding to a dispersion in the M/L ratio $\lesssim 10\%$ at any position along the plane. Moreover, the FP itself is "tilted," with the M/L ratio apparently increasing by a factor of ~ 3 along the plane, while the mass is increasing by a factor of ~ 100. Note that the tilt does not imply a departure from virialization, but rather a systematic trend of the stellar content with galaxy mass, possibly coupled with a systematic departure from structural homology (e.g., Bender, Burnstein & Faber 1992; Ciotti 1997; Busarello et al. 1997).

The narrowness of the FP, coupled to the relatively large tilt ($\Delta \kappa_3/\sigma(\kappa_3) \simeq 0.35/0.05 = 7$), requires some sort of fine tuning, which is perhaps the most intriguing property of the FP (Renzini & Ciotti 1993). Although unable to identify one specific origin for the FP tilt, Renzini & Ciotti (1993) argued that the small scatter perpendicular to the FP implied a small age dispersion ($\lesssim 15\%$) and high formation redshift, fully consistent with the Bower, Lucey & Ellis (1992) argument based on the narrowness of the C-M and color-σ relations.

The remarkable properties of the FP for the Virgo and Coma clusters were soon shown to be shared by all studied clusters in the local universe. Jørgensen, Franx & Kjærgaard (1996) constructed the FP for 230 ETGs in 10 clusters (including Coma), showing that the FP tilt and scatter are just about the same in all local clusters, thus strengthening the case for the high formation redshift of cluster ETGs being universal. However, Worthey, Trager & Faber (1995) countered that the thinness of the FP, C-M, and color-σ relations could be preserved, even with a large age spread, provided age and metallicity are anticorrelated (with old galaxies being metal poor and young ones being metal rich). This is indeed what Worthey and colleagues reported from their line-indices analysis (see below), indicating a factor of ~ 6 for range in age balanced by a factor of ~ 10 in metallicity (from solar to ~ 10 times solar). If so, then the FP should be thicker in the near infrared, because the compensating effect of metallicity would be much lower at longer wavelength, thus unmasking the full effect of a large age spread (Pahre, Djorgovski, & De Carvalho 1995). But Pahre and colleagues found the scatter of the FP K-band to be the same as in the optical. In addition, its slope implied a sizable variation of $M/L_K \propto M^{0.16}$ along the FP, somewhat flatter than in the optical ($M/L_V \propto M^{0.23}$), still far from the $M/L_K \sim$ constant predicted by Worthey, Trager & Faber (1995).

These conclusions were further documented and reinforced by Pahre, Djorgovski & De Carvalho (1998), Scodeggio et al. (1998), Mobasher et al. (1999), and Pahre, De Carvalho, & Djorgovski (1998), who finally concluded that the origin of the FP tilt defies a simple explanation, but is likely the result of combined age and metallicity trends along the plane (with the most metal-rich galaxies being actually the oldest), plus an unidentified systematic deviation from structural homology. Several possibilities for the homology breaking have been proposed and investigated, such as variation in stellar and/or dark matter content and/or distribution, anisotropy, and rotational support (e.g., Ciotti, Lanzoni & Renzini 1996; Prugniel & Simien 1996; Ciotti & Lanzoni 1997). Recently, Trujillo, Burkert & Bell (2004) argued that for one-fourth the tilt is due to stellar population (i.e., a combination of metallicity and age), and for three-fourths the tilt is due to structural nonhomology in the distribution of the visible matter.

Of special interest is the comparison of the FP in clusters and in the field, because one expects all formation processes to be faster in high-density peaks of the matter distribution. This was tested by Bernardi et al. (2003b, 2006) with a sample of ~40,000 SDSS morphology- and color-selected ETGs spanning a wide range of environmental conditions, from dense cluster cores to very low densities. Bernardi and colleagues found very small, but detectable differences in the FP zero point; the average surface brightness is ~0.08 mag brighter at the lowest density extreme compared to the opposite extreme. As the sample galaxies are distributed in redshift up to $z \sim 0.3$, they used the observed lookback time to empirically determine the time derivative of the surface brightness (hence in a model-independent fashion) and estimated that the 0.08 mag difference in surface brightness implies an age difference of ~1 Gyr, and therefore that galaxies in low-density environments are ~1 Gyr younger compared to those in cluster cores.

3.1.3. The line strength diagnostics. Optical spectra of ETGs present a number of absorption features whose strength must depend on the distributions of stellar ages, metallicities, and abundance ratios, and therefore may give insight over such distributions. To exploit this opportunity, Burstein et al. (1984) introduced a set of indices now known as the Lick/IDS system and started taking measurements for a number of galaxies. The most widely used indices have been the Mg_2 (or Mgb), $\langle Fe \rangle$, and the $H\beta$ indices, measuring, respectively, the strength of $MgH + MgI$ at $\lambda \simeq 5156 - 5197$ Å, the average of two FeI lines at $\lambda \simeq 5248$ and 5315 Å, and of H_β.

A first important result was the discovery that theoretical models based on solar abundance ratios adequately describe the combinations of the values of the $\langle Fe \rangle$ and Mg_2 indices in low-luminosity ETGs, but fail for bright galaxies (Peletier 1989; Gorgas, Efstathiou & Aragón-Salamanca 1990; Faber, Worthey & Gonzales 1992; Davies, Sadler & Peletier 1993; Jørgensen 1997). This implies either that population synthesis models suffered some inadequacy at high metallicity (possibly due to incomplete stellar libraries) or that massive ellipticals were genuinely enriched in magnesium relative to iron, not unlike the halo stars of the Milky Way (e.g., Wheeler, Sneden & Truran 1989). As for the Milky Way halo, such an α-element overabundance may signal a prompt enrichment in heavy elements from Type II supernovae, with the short star-formation timescale having prevented most Type Ia supernovae from contributing their iron while star formation was still active. Yet, a star formation timescale decreasing with increasing mass was contrary to the expectations of galactic wind/monolithic models (e.g., Arimoto & Yoshii 1987), where the star-formation timescale increases with the depth of the potential well (Faber, Worthey & Gonzales 1992). However, as noted by Thomas (1999), the contemporary semi-analytical models did not predict any α-element enhancement at all, no matter whether in low- or high-mass ETGs. Indeed, Thomas, Greggio & Bender (1999) argued that the α-enhancement, if real, was also at variance with a scenario in which massive ellipticals form by merging spirals and required instead that star formation was completed in less than $\sim$1 Gyr. Therefore, assessing whether the α-enhancement was real, and in that case measuring it, had potentially far-reaching implications for the formation of ETGs.

Two limitations had to be overcome in order to reach a credible interpretation of the $\langle Fe \rangle$ − Mgb plots: (*a*) existing synthetic models for the Lick/IDS indices were based on stellar libraries with fixed [α/Fe] (Worthey 1994, Buzzoni 1995), and (*b*) an empirical verification of the reality of the α-enhancement was lacking. In an attempt to overcome the first limitation, Greggio (1997) developed a scaling algorithm that allowed one to use existing models with solar abundance ratios to estimate the Mg overabundance, and she concluded that an enhancement up to [Mg/Fe] $\simeq$ +0.4 was required for the nuclei of the most massive ellipticals (see also Weiss, Peletier & Matteucci 1995). She also concluded that a closed-box model for chemical evolution failed to explain the very high values of the Mg$_2$ index of these galaxies. Indeed, the numerous metal-poor stars predicted by the model would obliterate the Mg$_2$ feature, hence for consistency with the observed Mg$_2$ indices the nuclei of ellipticals had to lack substantial numbers of stars more metal poor than $\sim$0.5 $Z_\odot$. Besides, very old ages ($\gtrsim$10 Gyr) and α-enhancement were jointly required to account for galaxies with strong Mg$_2$. Eventually, Thomas, Maraston & Bender (2003) produced a full set of synthetic models with variable [α/Fe], and Maraston et al. (2003) compared such models to the indices of ETGs and of metal-rich globular clusters of the Galactic bulge, for which the α-enhancement has been demonstrated on a star-by-star basis by high-resolution spectroscopy. The result is displayed in **Figure 5,** showing that indeed the new models indicate for the bulge globulars an enhancement of [α/Fe] $\sim$ +0.3, in agreement with the stellar spectroscopy results, and similar to that indicated for massive ETGs.

Other widely used diagnostic diagrams involved the Hβ index along with $\langle Fe \rangle$ and Mg$_2$ or Mgb. The Balmer lines had been suggested as good age indicators (e.g., O'Connell 1980, Dressler & Gunn 1983), an expectation that was confirmed by the set of synthetic models constructed by Worthey (1994) with the aim of breaking the age-metallicity degeneracy that affects the broad-band colors of galaxies. Worthey's models were applied by Jørgensen (1999) to a sample of 115 ETGs in the Coma cluster, and by Trager et al. (2000) to a sample of 40 ETGs biased toward low-density environments, augmented by 22 ETGs in the Fornax cluster from Kuntschner & Davies (1998), which showed systematically lower Hβ indices. From these samples, and using the Hβ − Mgb and Hβ − $\langle Fe \rangle$ plots from Worthey's models, both Jørgensen and Trager and colleagues concluded that ages ranged from a few to almost 20 Gyr, but age and metallicity were anticorrelated in such a way that the Mg$b - \sigma$, C-M, and FP relations may be kept very tight. Moreover, there was a tendency for ETGs in the

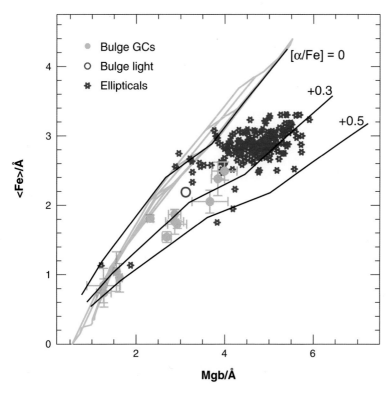

Figure 5

The ⟨Fe⟩ index versus the Mg*b* index for a sample of halo and bulge globular clusters, the bulge integrated light in Baade's Window, and for elliptical galaxies from various sources. Overimposed are synthetic model indices (from Thomas, Maraston & Bender 2003) with solar metallicity ([Z/H] = 0), various α-enhancements as indicated, and an age of 12 Gyr (*black solid lines*). The cyan grid shows a set of simple stellar population models (from Maraston 1998) with solar abundance ratios, metallicities from [Fe/H] = −2.25 to +0.67 (*bottom to top*), and ages from 3 to 15 Gyr (*left to right*). The blue grid offers an example of the so-called age-metallicity degeneracy. From Maraston et al. (2003).

field to appear younger than those in clusters. Nevertheless, Trager and colleagues cautioned that Hβ is most sensitive to even low levels of recent star formation and suggested that the bulk of stars in ETGs may well be old, but a small "frosting" of younger stars drives some galaxies toward areas in the Hβ − Mg*b* and Hβ−⟨Fe⟩ plots with younger SSP ages. Finally, for the origin of the α-enhancement Trager and colleagues favored a tight correlation of the IMF with σ, in the sense of more massive galaxies having a flatter IMF, hence more Type II supernovae. However, with a flatter IMF more massive galaxies would evolve faster in luminosity with increasing redshift, compared to less massive galaxies, an observation which appears to be at variance with the observations (see below).

These conclusions had the merit of promoting further debates. Maraston & Thomas (2000) argued that even a small old, metal-poor component with a blue horizontal branch (like in galactic globulars) would increase the Hβ index, thus making galaxies look significantly younger than they are. Even more embarrassing for the use of the Hβ − Mgb and Hβ − ⟨Fe⟩ plots is that a perverse circulation of the errors automatically generates an anticorrelation of age and metallicity, even where it does not exist. For example, if Hβ is overestimated by observational errors, then age is underestimated, which in turn would reduce Mgb below the observed value unless the younger age is balanced by an artificial increase of metallicity. Trager and colleagues were fully aware of the problem and concluded that only data with very small errors could safely be used. Kuntschner et al. (2001) investigated the effect by means of Monte Carlo simulations, and indeed showed that much of the apparent age-metallicity anticorrelation is a mere result of the tight correlation of their errors. They concluded that only a few outliers among the 72 ETGs in their study are likely to have few-billion-year-old luminosity-weighted ages, and these were typically galaxies in the field or loose groups, whereas a uniformly old age was derived for the vast majority of the studied galaxies. Note also that younger ages are more frequently indicated for S0 galaxies (Kuntschner & Davies 1998). Nevertheless, in a cluster with very tight C-M and FP relations such as Coma, a large age spread at all magnitudes was found for a sample of 247 cluster members (Poggianti et al. 2001), and a sizable age-metallicity anticorrelation was also found for a large sample of SDSS galaxies (Bernardi et al. 2005).

As already alluded to, the main pitfall of the procedure is that the various indices depend on all three population parameters one is seeking to estimate: thus Hβ is primarily sensitive to age, but also to [Fe/H] and [α/Fe]; ⟨Fe⟩ is sensitive to [Fe/H], but also to age and [Mg/Fe]; etc. Thus, the resulting errors in age, [Fe/H], and [Mg/Fe] are all tightly correlated, and one is left with the suspicion that apparent correlations or anticorrelations may be an artifact of the procedure rather than reflecting the real properties of galaxies. In an effort to circumvent these difficulties, Thomas et al. (2005) renounced trusting the results galaxy by galaxy. They rather looked at patterns in the various index-index plots and compared them to mock galaxy samples generated via Monte Carlo simulations that fully incorporated the circulation of the errors. The real result was not a set of ages and metallicities assigned to individual galaxies, but rather age and metallicity trends with velocity dispersion, mass, and environments. Having analyzed a sample of 124 ETGs in high- and low-density environments, Thomas and colleagues reached the following conclusions: (a) all three parameters—age, metallicity, and [α/Fe]—correlate strongly with σ and, on average, follow the relations:

$$\log t/\text{Gyr} = 0.46(0.17) + 0.238(0.32) \log \sigma, \qquad (2)$$

$$[Z/H] = -1.06(-1.03) + 0.55(0.57) \log \sigma, \qquad (3)$$

$$[\alpha/\text{Fe}] = -0.42(-0.42) + 0.28(0.28) \log \sigma, \qquad (4)$$

where quantities in brackets/not in brackets refer to low-density/high-density environments, respectively. (b) For ETGs less massive than $\sim 10^{10}\, M_\odot$ there is evidence for the presence of intermediate-age stellar populations with near-solar Mg/Fe. Instead, massive galaxies ($\gtrsim 10^{11}\, M_\odot$) appear dominated by old stellar populations, whereas at

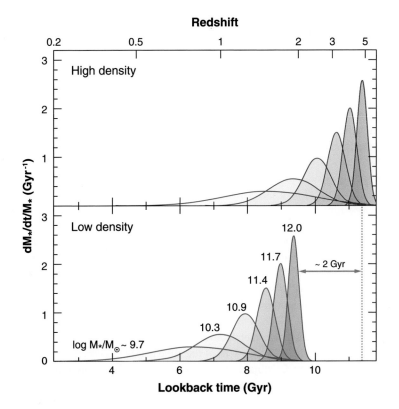

Figure 6
The scenario proposed by Thomas et al. (2005) for the average star formation history of early-type galaxies of different masses, from 5×10^9 $M_\odot$ up to 10^{12} $M_\odot$, corresponding to $\sigma \simeq 100$ to ~ 320 km s^{-1}, for the highest and lowest environmental densities, respectively, in the upper and lower panel.

intermediate masses the strength of Hβ requires either some intermediate age component or a blue horizontal branch (HB) contribution. (c) By and large this picture applies to both clusters and field ETGs, with cluster galaxies having experienced the bulk of their star formation between $z \sim 5$ and 2, and this activity appears to have been delayed by ~ 2 Gyr in the lowest-density environments, i.e., between $z \sim 2$ and ~ 1. **Figure 6** qualitatively summarizes this scenario, in which the duration of star-formation activity decreases with increasing mass (as required by the [Mg/Fe] trend with σ), and extends to younger ages for decreasing mass (as forced by the H$\beta - \sigma$ relation). Note that the smooth star-formation histories in this figure should be regarded as probability distributions, rather than as the actual history of individual galaxies, where star formation may indeed take place in a series of bursts. Qualitatively similar conclusions were reached by Nelan et al. (2005) from a study of ~ 4000 red-sequence galaxies in ~ 90 clusters as part of the National Optical Astronomy Observatory Fundamental Plane Survey. Assuming the most massive galaxies ($\sigma \sim 400$ km s^{-1}.) to be 13 Gyr old, they derived an age of only 5.5 Gyr for less massive galaxies ($\sigma \sim 100$ km s^{-1}). Note that the age-σ scaling of Thomas and colleagues would have given a much older age (~ 9.5 Gyr). Taken together, Equations 2 and 3 imply a trend of M/L_v by a factor of ~ 1.8 along the FP (from $\sigma = 100$ to 350 km s^{-1}), thus accounting for almost two-thirds of the FP tilt.

As extensively discussed by Thomas et al. (2005), one residual concern comes from the possibility that part of the Hβ strength may be due to blue HB stars. Besides a blue HB contribution by low-metallicity stars (especially in less massive galaxies), blue HB stars may also be produced by old, metal-rich populations, and appear to be responsible for the UV upturn in the spectrum of local ETGs (Brown et al. 2000, Greggio & Renzini 1990). In the Thomas et al. sample, some S0 outliers with strong Hβ and strong metal lines would require very young ages and extremely high metallicity (up to $\sim$10 times solar), and may better be accounted for by an old, metal-rich population with a well-developed blue HB.

The $Mg_2 - \sigma$ relation has also been used to quantify environmental differences in the stellar population content. The cluster/field difference turns out to be small, with $\Delta Mg_2 \sim 0.007$ mag, corresponding to $\sim$1 Gyr difference—field galaxies being younger—within a sample including $\sim$900 ETGs (Bernardi et al. 1998), although no statistically significant environmental dependence of both Mg_2 and Hβ was detected within a sample of $\sim$9,000 ETGs from the SDSS (Bernardi et al. 2003a). Still from SDSS, coadding thousands of ETG spectra in various luminosity and environment bins, Eisenstein et al. (2003) detect clear trends with the environment thanks to the resulting exquisite S/N, but the differences are very small, and Eisenstein and colleagues refrain from interpreting them in terms of age/metallicity differences.

These results from the analysis of the Lick/IDS indices, including large trends of age with σ, or even large age-metallicity anticorrelations, have yet to be proven consistent with the FP and C-M relations of the same galaxies as established specifically for the studied clusters. Feeding the values of the indices, the synthetic models return ages, metallicities, and α-enhancements. But along with them the same models also give the colors and the stellar M/L ratio of each galaxy in the various bands, hence allowing one to construct implied FP and C-M relations. It would be reassuring for the soundness of the whole procedure if such relations were found to be consistent with the observed ones. To our knowledge, this sanity check has not been attempted yet. The mentioned trends and correlations, if real, would also have profound implications for the evolution of the FP and C-M relations with redshift, an opportunity that will be exploited below.

3.2. Ellipticals versus Spiral Bulges

The bulges of spiral galaxies are distinguished in "true bulges," typically hosted by S0-Sb galaxies, and "pseudobulges" usually (but not exclusively) in later-type galaxies (Kormendy & Kennicutt 2004). True (classical) bulges have long been known to be similar to ellipticals of comparable luminosity, in both structure, line strengths, and colors (e.g., Bender, Burnstein & Faber 1992; Jablonka, Martin, & Arimoto 1996; Renzini 1999; and references therein). Peletier et al. (1999) were able to quantify this using *Hubble Space Telescope* (HST) WFPC2 (Wide Field Planetary Camera 2) and NICMOS (Near Infrared Camera and Multi-Object Spectrometer) observations, and concluded that most (true) bulges in their sample of 20 spirals (including only 3 galaxies later than Sb) had optical and optical-IR colors similar to those of Coma ellipticals. Hence, like in Coma ellipticals their stellar populations formed

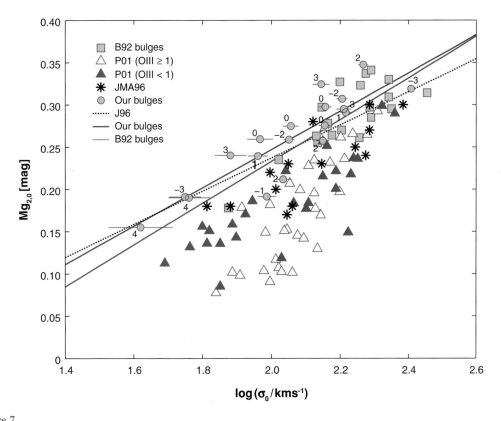

Figure 7
The $Mg_2 - \sigma$ relation for the spiral bulges studied by Falcon-Barroso, Peletier & Balcells (2002), labeled "our bulges" in the insert, are compared to bulges in other samples (B92 = Bender, Burnstein & Faber 1992; P01 = Prugniel, Maubon & Simien 2001, JMA96 = Jablonka et al. 1996). The solid lines are the best fits to the corresponding data, whereas the dashed line shows the average relation for cluster ellipticals from Jørgensen, Franx & Kjærgaard 1996.

at $z \gtrsim 3$, even if most of the galaxies in their sample are in small groups or in the field. More recently, Falcon-Barroso, Peletier & Balcells (2002) measured the central velocity dispersion for the same sample observed by Peletier and colleagues, and constructed the FP for these bulges, showing that they tightly follow the same FP relation as cluster ellipticals, and therefore had to form their stars at nearly the same epoch. The similarity of true bulges and ellipticals includes the tendency of less massive objects to have experienced recent star formation, as indicated by their location in the $Mg_2 - \sigma$ diagram in **Figure 7**. These similarities between true (classical) bulges and ellipticals suggest a similar origin, possibly in merger-driven starbursts at high redshifts. Pseudobulges, instead, are more likely to have originated via secular evolution of disks driven by bars and other deviations from axial symmetry, as extensively discussed and documented by Kormendy & Kennicutt (2004). Several of the objects in the Prugniel and colleagues sample in **Figure 7** are likely to belong the the pseudobulge group. Thomas & Davies (2006) argue that the same scenario depicted in **Figure 6** for ETGs applies to bulges as well, the main difference being that bulges are on average less massive, hence on average younger than ETGs.

Looking near to us, HST and ground-based photometry of individual stars in the Galactic bulge have shown that they are older than at least 10 Gyr, with no detectable trace of an intermediate-age component (Ortolani et al. 1995, Kuijken & Rich 2002, Zoccali et al. 2003). HST/NICMOS photometry of stars in the bulge of M31 has also shown that their H-band luminosity function is virtually identical to that of the Galactic bulge, and by inference should have nearly identical ages (Stephens et al. 2003). These two bulges belong to spirals in a rather small group, and yet appear to have formed their stars at an epoch corresponding to $z \gtrsim 2$, not unlike most ellipticals.

3.3. Summary of the Low-Redshift (Fossil) Evidence

The main observational constraints on the epoch of formation of the stellar populations of ETGs in the near universe can be summarized as follows:

- The C-M, color-σ, and FP relations for ETGs in clusters indicate that the bulk of stars in these galaxies formed at $z \gtrsim 2 - 3$.
- The same relations for the field ETGs suggest that star formation in low-density environments was delayed by $\sim 1 - 2$ Gyr.
- The more massive galaxies appear to be enhanced in Mg relative to iron, which indicates that the duration of the star-formation phase decreases with increasing galaxy mass, having been shorter than ~ 1 Gyr in the most massive galaxies.
- Interpretations of the Lick/IDS indices remains partly controversial, with either an age-metallicity anticorrelation or an increase of both age and metallicity with increasing σ.

These trends are qualitatively illustrated in **Figure 6**, showing that the higher the final mass of the system, the sooner star formation starts and more promptly subsides, in an apparently "antihierarchical" fashion. A trend in which the stellar population age and metallicity are tightly correlated to the depth of the potential well (as measured by σ) argues for star formation, metal enrichment, supernova feedback, merging, and violent relaxation having been all concomitant processes rather than having taken place sequentially.

The fossil evidence illustrated so far is in qualitative agreement with complementary evidence at low as well as high redshift, now relative to star-forming galaxies as opposed to quiescent ones. At low z, Gavazzi (1993) and Gavazzi, Pierini & Boselli (1996) showed that in local (disk) galaxies the specific star-formation rate anticorrelates with galaxy mass, a trend that can well be extended to include fully quiescent ellipticals. On this basis, Gavazzi and collaborators emphasized that mass is the primary parameter controlling the star-formation history of galaxies, with a sharp transition at $L_\mathrm{H} \simeq 2 \times 10^{10} L_\odot$ (corresponding to $\sim 2 \times 10^{10} M_\odot$) between late-type, star-forming galaxies and mostly passive, early-type galaxies (Scodeggio et al. 2002). This transition mass has then been precisely located at $\sim 3 \times 10^{10}\ M_\odot$ with the thorough analysis of the SDSS database (Kauffmann et al. 2003). In parallel, high-redshift observations have shown that the near-IR luminosity (i.e., mass) of galaxies undergoing rapid star formation has declined monotonically from $z \sim 1$ to the present, a trend for which Cowie et al. (1996) coined the term down-sizing. This is becoming a new paradigm for galaxy formation, as the anticorrelation of the specific star-formation rate with mass is now recognized to persist well beyond $z \sim 2$ (e.g., Juneau et al. 2005, Feulner et al. 2006).

4. ELLIPTICAL GALAXIES AT HIGH REDSHIFT

Perhaps the best way of breaking the age-metallicity degeneracy is by looking back in time, studying galaxies at higher and higher redshifts. In the 1990s, this was attempted first with 4m-class telescopes and later, with impressive success, with 8–10m-class telescopes and HST. Studies first focused on cluster ellipticals, and their extension to field galaxies followed with some delay. Thus, the evolution with redshift of various galaxy properties were thoroughly investigated, such as the C-M and Kormendy relations, the luminosity and mass functions, and the FP. Various ongoing surveys are designed to map the evolution with redshift and local environment of all these properties, along with the number density of these galaxies.

4.1. Cluster Ellipticals up to $z \sim 1$

4.1.1. The color-magnitude relation. With the identification of clusters at higher and higher redshifts, from the mid-1990s it became possible to construct their C-M relation, hence to directly assess the rate of evolution of cluster ETGs. Pioneering studies showed a clearly recognizable red sequence in high-redshift clusters, and gave hints that the color evolution up to $z \sim 1$ was broadly consistent with pure passive evolution of the galaxies formed at high redshift (Dressler & Gunn 1990, Aragón-Salamanca et al. 1993, Rakos & Schombert 1995). Subsequent studies fully confirmed these early hints and provided accurate estimates for the formation redshift of the bulk of stars in cluster ellipticals. Thus, replicating the Bower, Lucey & Ellis (1992) procedure for a sample of morphologically-selected ETGs in clusters at $z \sim 0.5$, Ellis et al. (1997) were able to conclude that most of the star formation in ellipticals in dense clusters was completed 5–6 Gyr earlier than the cosmic time at which they are observed, i.e., at $z \gtrsim 3$. Extending these studies to clusters up to $z \sim 0.9$, Stanford, Eisenhardt & Dickinson (1998) showed that pure passive evolution continues all the way to such higher redshift, while the dispersion of the C-M relation remains as small as it is in Virgo and Coma (see **Figure 8**). Thus, Stanford and colleagues concluded that cluster ellipticals formed the bulk of their stars at $z \gtrsim 3$, with the small color dispersion arguing for highly synchronized star-formation histories among galaxies within each cluster, and from one cluster to another. These conclusions were reinforced by several other investigators, e.g., Gladders et al. (1998), Kodama et al. (1998), Nelson et al. (2001), De Lucia et al. (2004), and by van Dokkum et al. (2000), who also cautioned about the "progenitor bias" (see below).

Figure 8

The color evolution of early-type galaxies in clusters out to $z \simeq 0.9$ (Stanford, Eisenhardt & Dickinson 1998; Dickinson 1997). The "blue" band is tuned for each cluster to approximately sample the rest frame U-band, whereas the K band is always in the observed frame. Top panel: the redshift evolution of the zero point of the "blue"–K color-magnitude (C-M) relation relative to the Coma cluster. A purely passive evolution model is also shown. Middle panel: the intrinsic color scatter, having removed the mean slope of the C-M relation in each cluster and the contribution of photometric errors. The intrinsic scatter of Coma galaxies is shown for reference. Bottom panel: the redshift evolution of the slope of the ("blue"–K) – K C-M relation relative to the slope of the corresponding relation for galaxies in Coma.

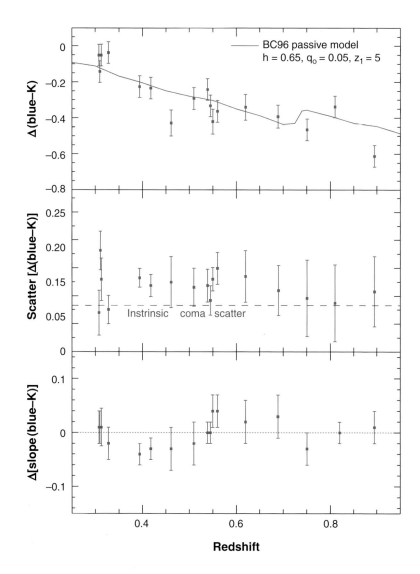

The evolution of the C-M relation was then traced beyond $z = 1$ thanks to the discovery of higher redshift clusters, primarily by the Rosat Deep Cluster Survey (RDCS; Rosati et al. 1998). Using deep HST/ACS (Advanced Camera for Surveys) i- and z-band images, Blakeslee et al. (2003) found a tight red sequence for morphologically selected ETGs in a $z = 1.24$ cluster, implying typical ages of ~ 3 Gyr, and formation redshift $z_F \gtrsim 3$. This was further confirmed by its infrared C-M relation (Lidman et al. 2004). However, some clusters in the range $0.78 < z < 1.27$ appear to have a larger color scatter than others, again with ellipticals in those with tight C-M relation having virtually completed their star formation at $z \gtrsim 3$ (Holden et al. 2004).

Presently, the highest redshift clusters known to be dominated by old, massive ETGs are at $z \simeq 1.4$ (Stanford et al. 2005, Mullis et al. 2005).

The redshift evolution of the color of the red sequence in clusters proved to be a very powerful tool in disentangling ambiguities that are difficult to eliminate based only on $z \sim 0$ observations. From the color evolution of the red sequence in two Abell clusters at $z \sim 0.2$ and ~ 0.4, Kodama & Arimoto (1997) were able to break the age-metallicity degeneracy plaguing most of the global observables of local ellipticals. In principle, because colors depend both on age and metallicity, the slope of the C-M relation could equally well be reproduced with either age or metallicity increasing with increasing luminosity (or σ). However, if age were the dominant effect, then the C-M relation would steepen with lookback time (redshift), as the color of the young galaxies would get more rapidly bluer compared with that of the old galaxies. Instead, the slope of the relation remains nearly the same (see **Figure 8**). Actually, the Kodama & Arimoto argument can be applied also to the color dispersion within a cluster, demonstrating that the tightness of the C-M (and FP) relation in low-z clusters cannot be due to a conspiracy of age and metallicity being anticorrelated (as advocated e.g., by Worthey, Trager & Faber 1995). If so, the color dispersion would rapidly increase with redshift, contrary to what is seen in clusters up to $z \sim 1$ (see **Figure 8**).

4.1.2. The luminosity function. A cross check of the high-formation redshift of ETGs can be provided by looking at their luminosity in distant clusters. If ETGs evolve passively, following a pure luminosity evolution (PLE), then their luminosities should increase with increasing redshift by an amount that depends on the formation redshift and on the slope of the IMF.

Initial attempts at detecting the expected brightening of the characteristic luminosity (M^*) of the luminosity function (LF) were inconclusive, as Barger et al. (1998) failed to detect any appreciable change between clusters at $z = 0.31$ and $z = 0.56$, possibly owing to the small redshift baseline. On the other hand, comparing the LF of $z \sim 0$ clusters to the LF of a sample of 8 clusters at $0.40 < z < 0.48$, Barrientos & Lilly (2003) found a brightening of the characteristic luminosity M^* consistent with passive evolution and high-formation redshift, also in agreement with the $(U - V)$ color evolution of the red sequence. In a major cluster survey, De Propris et al. (1999) explored the evolution of the observed K-band LF in 38 clusters with $0.1 < z < 1$, and compared the results to the Coma LF. With this much larger redshift baseline, De Propris et al. found the trend of K^* with redshift to be consistent with passive evolution and $z_F \gtrsim 2$. They pointed out the agreement with the results based on the color evolution of the red sequence galaxies, but emphasized that this behavior of the LF implies that "not only their stellar population formed at high redshift, but that the assembly of the galaxies themselves was largely complete by $z \sim 1$." Kodama & Bower (2003) and Toft, Soucail & Hjorth (2003) came to the same conclusions by studying the K-band LF of (respectively, two and one) clusters at $z \sim 1$. Breaking the $z = 1$ barrier, Toft et al. (2004) constructed a very deep K-band LF of a rich RDCS cluster at $z = 1.237$, and concluded that the most massive ellipticals that dominate the top end of the LF were already in place in this cluster. They compared the cluster

K-band LF (corresponding to the rest-frame z-band LF) to the z-band LF of local clusters (Popesso et al. 2005) and derived a brightening by ~ 1.4 mag in the rest-frame z-band characteristic magnitude, indeed as expected from PLE. Toft and colleagues also found a substantial deficit of fainter ETGs, which could be seen as a manifestation of the down-sizing effect in a high-redshift cluster, a hint of which was also noticed in other clusters at $z \sim 0.8$ (De Lucia et al. 2004).

However, a more complete study of 3 rich clusters at $1.1 \lesssim z \lesssim 1.3$, including the $z = 1.237$ cluster studied by Toft et al. (2004), did not produce evidence of a down-sizing effect, down to at least 4 magnitudes below K^* (Strazzullo et al. 2006). This study confirmed the brightening of M_z^* and M_K^* by ~ 1.3 mag, consistent with passive evolution of a population that formed at $z \gtrsim 2$, and showed that the massive galaxies were already fully assembled at $z \sim 1.2$, at least in the central regions of the 3 clusters.

4.1.3. The effective radius-surface brightness relation.

An alternative way of detecting the expected brightening of old stellar populations at high redshift is by tracing the evolution of the Kormendy relation, which became relatively easy only after the full image quality of HST was restored. Thus, from HST data, the ETGs in a cluster at $z = 0.41$ were found to be brighter by $\Delta M_K = 0.36 \pm 0.14$ mag or by 0.64 ± 0.3 mag (Barrientos, Schade, & Lopez-Cruz 1996) with respect to local galaxies, consistent (within such large errors) with passive evolution of an old, single-burst stellar population. Schade et al. (1996), using excellent CFHT (Canada-France-Hawaii Telescope) imaging data for 3 clusters at $z = 0.23, 0.43$, and 0.55, detected a progressive brightening in galaxy luminosity at a fixed effective radius that once more was estimated to be consistent with passive evolution and formation at high redshift. No differential evolution with respect to ETGs in the cluster surrounding fields was detected.

Turning to HST data, a systematic brightening in the Kormendy relation, again consistent with passive evolution and high formation redshift, was found by several other groups, eventually reaching redshifts of ~ 1 (see Schade, Barrientos, & Lopez-Cruz 1997, Barger et al. 1999, Ziegler et al. 1999, Holden et al. 2005a, Pasquali et al. 2006).

4.1.4. The fundamental plane.

Besides the high spatial resolution, constructing the FP of high redshift cluster (and field) galaxies requires moderately high-resolution spectroscopy (to get σ), hence a telescope with a large collective area. With one exception, for a few years this was monopolized by the Keck Telescope, and FP studies of high-z ellipticals first flourished at this observatory. In a crescendo toward higher and higher redshifts, the FP was constructed for clusters at $z = 0.39$ (van Dokkum & Franx 1996), $z = 0.58$ (Kelson et al. 1997), $z = 0.83$ (van Dokkum et al. 1998, Wuyts et al. 2004), and finally at $z = 1.25$ and 1.27 (Holden et al. 2005b, van Dokkum & Stanford 2003). The early exception was the heroic study of two clusters at $z = 0.375$ using the 4m-class telescopes at ESO (NTT) and Calar Alto (Bender, Saglia & Ziegler 1996; Bender, Ziegler & Bruzual 1996; Bender et al. 1998).

The redshift evolution of the FP depends on a variety of factors. For passive evolution the FP shifts by amounts that depend on a combination of IMF slope, formation redshift, and cosmological parameters. A systematic trend of the IMF slope with galaxy mass would cause the FP to rotate with increasing redshift (Renzini & Ciotti 1993), as it would do for a similar trend in galaxy age. An age dispersion (Δt) would cause the scatter perpendicular to the FP to increase with redshift, as, for fixed Δt, $\Delta t/t$ increases for increasing redshift, i.e., decreasing galaxy age (t). Clearly, the behavior of the FP with redshift can give a wealth of precious information on the formation of ellipticals, their stellar populations, and to some extent on cosmology also (Bender et al. 1998).

All the quoted FP studies of high-z clusters conclude that the FP actually shifts nearly parallel to itself by an amount that increases with redshift and is consistent with the passive evolution of stellar populations that formed at high redshifts. The FP shifts imply a decrease of the rest-frame M/L ratio $\Delta \log M/L_B \simeq -0.46z$ (van Dokkum & Stanford 2003), but—as emphasized above— the formation redshift one can derive from it depends on both cosmology and the IMF. **Figure 9** illustrates the dependence on the adopted cosmology. Note that for the "old standard" cosmology ($\Omega_M = 1$) galaxies would be older than the universe and therefore the observed FP evolution can effectively rule out this option (see also Bender et al. 1998).

Figure 10 shows the effect of the IMF slope and formation redshift on the expected evolution of the M/L_B ratio. The redshift range $z = 0$ to 1.5 only probes the IMF between ~ 1 and $\sim 1.4\ M_\odot$, which correspond to the masses at the main sequence turnoff M_{TO} of oldest populations at $z = 0$ (age ~ 13 Gyr, $M_{TO} \simeq M_\odot$) and of those at $z = 1.5$ (age ~ 4.5 Gyr and $M_{TO} \simeq 1.4\ M_\odot$). Together, **Figures 9** and **10**

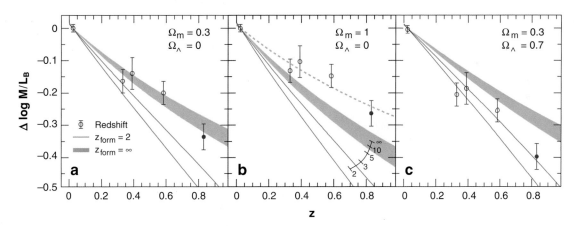

Figure 9

The data points show the redshift evolution of the M/L_B ratio of cluster elliptical galaxies as inferred from the shifts of the fundamental plane. The blue lines refer to the evolution of the M/L_B ratio for stellar populations with a Salpeter initial mass function (IMF) ($s = 2.35$) and formation redshifts as indicated in the left—and middle—panels. The comparison is made for three different cosmologies. The dotted orange line in the middle panel shows a model with $z_F = \infty$ and a steep IMF ($s = 3.35$). From van Dokkum et al. 1998.

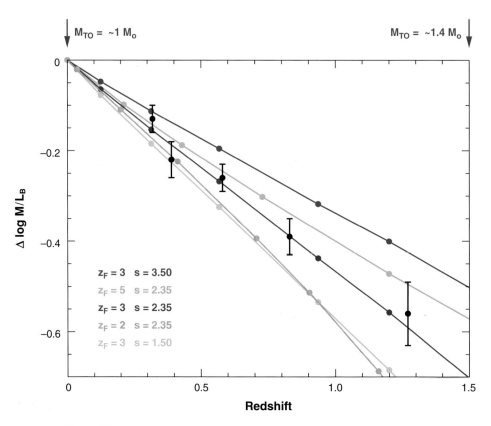

Figure 10

The evolution with redshift of the M_*/L_B ratio of simple stellar populations of solar metallicity and various initial mass function slopes ($dN \propto M^{-s} dM$) and formation redshifts, as indicated. The curves are normalized to their value at $z = 0$. Concordance cosmology ($\Omega_m = 0.3$, $\Omega_\Lambda = 0.7$, $H_0 = 70$) is adopted. The data points (from van Dokkum & Stanford 2003) refer to the shifts in the fundamental plane locations for clusters at various redshifts. Note that for such high formation redshifts the stellar mass at the main sequence turnoff is $\sim 1.4\,M_\odot$ at $z = 1.5$ and $\sim 1.0\,M_\odot$ at $z = 0$, as indicated by the arrows. Adapted from Renzini (2005).

illustrate the formation redshift/IMF/cosmology degeneracies in the FP diagnostics. However, the cosmological parameters are now fixed by other observational evidence, whereas the formation redshift as determined by the color evolution of the cluster red sequence is independent of the IMF slope. In summary, in the frame of the current concordance cosmology, when combining the color and FP evolution of cluster ellipticals one can conclude that the best evidence indicates a formation redshift $z_F \sim 3$ and a Salpeter IMF slope in the pertinent stellar mass range ($1 < M < 1.4\,M_\odot$).

The fact that the cluster FP does not appreciably rotate with increasing redshift is documented down to $\sigma \sim 100\,{\rm km\,s^{-1}}$ for clusters out to $z \sim 0.3$–0.4 (lookback time ~ 4 Gyr) (Kelson et al. 2000, van Dokkum & Franx 1996) and down to $\sigma \sim 150\,{\rm km\,s^{-1}}$ out to $z \sim 0.8$ (lookback time ~ 7 Gyr) (Wuyts et al. 2004). If the large age trends with

σ derived in some of the studies using the Lick/IDS indices were real (see Section 3.1), this would result in a very large rotation of the FP in these clusters. For example, if age were to increase along the FP from 5.5 Gyr to 13 Gyr (at $z = 0$, and for $\sigma = 100$ and 320 km s^{-1}, respectively; see Nelan et al. 2005), then at a lookback time of 4 Gyr ($z \sim 0.4$) the younger population would have brightened by $\Delta M_B \sim 1.33$ mag, and the older one only by ~ 0.46 mag (using models from Maraston 2005), which results in a FP rotation of ~ 0.9 mag in surface brightness. Alternatively, the much shallower age–σ relation derived by Thomas et al. (2005) implies an age increase from ~ 9.5 Gyr to ~ 11.5 Gyr for σ increasing from 180 to 350 km s^{-1}, which implies a rotation of the FP by ~ 0.36 mag in surface brightness by $z = 0.8$, which is still consistent with the hint that in fact there may be a small FP rotation in a cluster at this redshift (Wuyts et al. 2004).

The scatter about the FP of clusters also remains virtually unchanged with increasing redshift; however, some of the claimed age-metallicity anticorrelations derived from the Lick/IDS indices would result in a dramatic increase of the scatter with redshift, causing the FP itself to rapidly blur away.

4.1.5. The evolution of the line indices. The intermediate resolution spectra used for constructing the FP of distant cluster ETGs were also used to measure age-sensitive line indices that can provide further constraints on the formation epoch. Thus, Bender, Ziegler & Bruzual (1996) and Ziegler & Bender (1997) measured the Mgb index of 16 ETGs in their two clusters at $z = 0.375$; they found that the Mg$_2$ – σ relation was shifted toward lower values of the index. From such differences in Mgb index, Ziegler & Bender inferred that the age of the $z = 0.375$ galaxies is about two-thirds of that of ETGs in Coma and Virgo. Therefore, $t(z = 0) - t_F = 1.5 \times [t(z = 0.375) - t_F]$, where t is the cosmic time and t_F the cosmic time when the local and distant cluster ETGs formed (which is assumed to be the same). Adopting the $t - z$ relation for the concordance cosmology, one derives $t_F \sim 1.7$ Gyr, corresponding to $z_F \gtrsim 3$.

From the strength of the Balmer absorption lines (H_δ and H_γ) as age indicators, Kelson et al. (2001) used data for several clusters up to $z = 0.83$ and were able to set a lower limit to the formation redshift $z_F \gtrsim 2.5$, consistent with the above result from the Mgb index.

In summary, the study of the stellar populations in ETGs belonging to distant clusters up to $z \sim 1.3$ have unambiguously shown that these objects have evolved passively from at least $z \sim 2$–3. This came from the color, line strength, and luminosity evolution. Moreover, the brightest cluster members at $z \sim 1$–1.3 and the characteristic luminosity of the LF appear to be brighter than their local counterpart by an amount that is fully consistent with PLE expectations, indicating that these galaxies were already fully assembled at this high redshift. This may not have been the case for less massive galaxies, as their counts may be affected by incompleteness.

4.1.6. Caveats. Although it is well established that ETGs in distant clusters are progenitors to their local analogs and formed at high redshift, some caveats are nevertheless in order. First, as frequently emphasized, the evidence summarized above only proves that at least some cluster galaxies evolved passively from $z \gtrsim 1$ to the present, but other local ETGs may have $z \sim 1$ progenitors that would not qualify as ETGs at that redshift, either morphologically or photometrically. This "progenitor bias" (e.g., van Dokkum & Franx 1996) would therefore prevent us from identifying all the $z \sim 1$ progenitors of local cluster ETGs, some of which may well be still star forming. Second, the slope of the FP is progressively less accurately constrained in higher and higher redshift clusters, because the central velocity dispersion has been measured only for very few cluster members (generally the brightest ones). Third, it is always worth recalling that all luminosity-weighted ages tend to be biased toward lower values by even minor late episodes of star formation. Last, stellar population dating alone only shows when stars were formed, not when the galaxy itself was assembled and reached its observed mass.

4.2. Field versus Cluster Ellipticals up to $z \sim 1$

In the local universe field ellipticals show small, yet detectable differences compared to their cluster counterpart, being possibly ~1 Gyr younger, on average. This Δt difference, if real, should magnify in relative terms and become more readily apparent when moving to high redshift ($\Delta t/t$ is increasing). Using the color evolution of the red sequence and the shift of the FP with redshift, progress in investigating high-z ETGs in the field has been dramatic in recent years, along with the cluster versus field comparison.

Schade et al. (1999) selected ETGs by morphology from the Canada-France Redshift Survey (CFRS; Lilly et al. 1995a,b) and LDSS (Low Dispersion Survey Spectrograph) redshift survey (Ellis et al. 1996), and constructed the rest-frame $(U - V)$ C-M relation for field ETGs in the $0.2 < z < 1.0$ range. They found that the C-M relation becomes progressively bluer with redshift, with $\Delta(U - V) \simeq -0.68 \pm 0.11$ at $z = 0.92$ with respect to the relation in Coma, accompanied by a brightening by ~1 mag in the rest-frame B band, as derived from the Kormendy relation. To be consistent with the color evolution, this brightening should have been much larger than observed if the color evolution were due entirely to the passive evolution of stellar populations formed at high z. Thus, Schade and colleagues reconciled color and luminosity evolution by invoking a residual amount of star formation (adding only ~2.5% of the stellar mass from $z = 1$ to the present), yet enough to produce the observed fast color evolution. Support for such an interpretation comes from about one-third of the galaxies exhibiting weak [O II] emission, which indicates that low-level star formation is indeed fairly widespread. The rate of luminosity evolution was found to be identical to that of cluster ellipticals at the same redshifts, hence no major environmental effect was detected besides the mentioned low level of star formation and a color dispersion slightly broader than in clusters at the same redshift.

With COMBO-17, the major imaging survey project undertaken with the ESO/MPG 2.2m telescope, Wolf et al. (2003) secured deep optical imaging in 17 broad and intermediate bands over a total 0.78 square degree area, from which Bell et al. (2004b) derived photometric redshifts accurate to within $\delta z \sim 0.03$. The bimodality of the C-M relation, so evident at $z \sim 0$ (e.g., Baldry et al. 2004), clearly persists all the way to $z \sim 1.1$ in the COMBO-17 data, and this allowed Bell and colleagues to isolate $\sim$5,000 "red sequence" ETGs down to $R < 24$. As mentioned above, $\sim$85% of such color-selected galaxies appear also to be morphologically early-type on the ACS images of the GEMS (Galaxy Evolution from Morphologies and SED) survey (Rix et al. 2004, Bell et al. 2004a). The rest-frame $(U - V)$ color of ETGs in the COMBO-17 survey evolves by a much smaller amount than that reported by Schade et al. (1999) for the morphologically-selected ETGs, i.e., by only $\sim$0.4 mag between $z = 0$ and 1, as expected for an old stellar population that formed at high redshift ($z_F \gtrsim 2$). This color evolution is also in agreement with the $\sim$1.3 mag brightening of the characteristic luminosity M_B^* in the Schechter fit to the observed LF. Thus, when comparing only the color and M_B^* evolution, the field ETGs in the COMBO-17 sample seem to evolve in much the same fashion as their cluster counterparts. Using COMBO-17 data and GEMS HST/ACS imaging, McIntosh et al. (2005) studied a sample of 728 morphology- and color-selected ETGs, finding that up to $z \sim 1$ the Kormendy relation evolves in a manner that is consistent with the pure passive evolution of ancient stellar populations.

From deep Subaru/Suprime-Cam imaging over a 1.2 deg^2 field [covered by the Subaru-XMM Deep Survey (SXDS)], Kodama et al. (2004) selected ETGs for having the $R - z$ and $i - z$ colors in a narrow range as expected for passively evolving galaxies in the range $0.9 \lesssim z \lesssim 1.1$, and the sampled population included both field and cluster ETGs. They found a deficit of red galaxies in the C-M sequence $\sim$2 mag fainter than the characteristic magnitude (corresponding to stellar masses below $\sim 10^{10} M_\odot$). Less massive galaxies appear to be still actively star-forming, whereas above $\sim 8 \times 10^{10} M_\odot$ galaxies are predominantly passively evolving. This was interpreted as evidence for down-sizing in galaxy formation (à la Cowie et al. 1996), with massive galaxies having experienced most of their star formation at early times and being passive by $z \sim 1$, and many among the less massive galaxies experience extended star-formation histories.

In a comprehensive study also based on deep, wide-field imaging with the Suprime-Cam at the Subaru Telescope, Tanaka et al. (2005) obtained photometric redshifts based on four or five optical bands and constructed the C-M relations for two clusters at $z = 0.55$ and 0.83, and their extended environment. The results are shown in **Figure 11**, where the various plots allow one to visually explore trends with redshift for a given environment, or with an environment for a given redshift. The red sequence appears already in place in the "field" in the highest redshift sample, but no clear color bimodality is apparent. (Note that COMBO-17 does find bimodality at this redshift, possibly due to its photometric redshifts based on many more bands being more accurate.) The color bimodality is instead clearly recognizable in the $z = 0.55$ sample. At higher environmental densities (labeled "group") the C-M relation of the red sequence is clearly recognizable already in the $z = 0.83$ sample, and even more so in the "cluster" sample. Tanaka and colleagues argue that the bright and the faint ends of the red sequence are populated at a different pace in all three environments; the more massive red galaxies are assembled first, i.e., the C-M relation grows from the bright end to the faint end in all three environments (not in the opposite way, as one may naively expect in a hierarchical cosmology), which is interpreted in terms of down-sizing. Note also that the faint end appears to be well in place in "clusters"

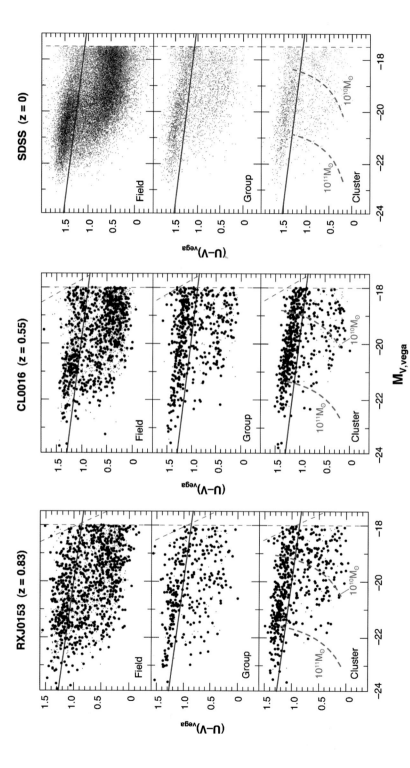

Figure 11

The C-M relations of two high redshift clusters and their surrounding fields at progressive lower density (labeled "group" and "field") are compared with their local counterparts from the Sloan Digital Sky Survey. The red line marks the adopted separation between the red sequence galaxies, and the blue, star-forming galaxies. In the cluster panels the blue dashed lines show the approximate location of galaxies with stellar mass of 10^{10} (*right*) and 10^{11} $M_\odot$ (*left*). From Tanaka et al. (2005).

at $z = 0$ while virtually still lacking in the field (see also Popesso et al. 2005). Using HST/WFPC2 imaging over a $\sim$30 arcmin^2 field including the same $z = 0.83$ cluster, Koo et al. (2005) were able to measure the rest-frame $(U - B)$ colors of the sole bulge component of 92 galaxies with $M_B < -19.5$, part in the cluster itself, part in the surrounding field. Their very red color does not show any environmental dependence, suggesting similarly old ages and high formation redshifts.

The public delivery of the Hubble Deep Field data (HDF; Williams et al. 1996) spurred several studies of field ETGs. Thus, Fasano et al. (1998) applied the Kormendy relation to a sample of morphologically selected ETGs in HDF-North and estimated an increase of the surface brightness at a fixed effective radius that was consistent with a high formation redshift ($z_F \sim 5$), according to the galaxy models by Bressan, Chiosi & Fagotto (1994) and Tantalo et al. (1996).

Although the LF and the C-M and Kormendy relations had already given useful indications on the analogies and differences between cluster and field ETGs, major progress came with the study of the differential evolution (field versus cluster) of the FP with redshift. In early studies, no field/cluster difference had clearly emerged at $z = 0.3$ (Treu et al. 1999), $z \sim 0.4$ (Treu et al. 2001), $z = 0.55$ (van Dokkum et al 2001), and $z = 0.66$ (Treu et al. 2002). But already at these modest redshifts there were hints that the brightest, most massive ETGs in the field closely follow the FP evolution of their cluster counterparts, while less massive ETGs (especially S0s) appear to evolve slightly faster, and hence look younger. This was more accurately quantified for morphologically-selected ETGs up to $z \sim 1$ in the HDF-North by van Dokkum & Ellis (2003), showing a field versus cluster difference $\Delta \ln M/L_B = -0.14 \pm 0.13$ in the FP. This implies that field ETGs are on average younger by only 16% $\pm$ 15% at $\langle z \rangle = 0.88$. Van Dokkum & Ellis also inferred that the bulk of stellar mass in the observed ETGs must have formed at $z \gtrsim 2$ even in the field, and allowed only for minor star formation at lower redshifts. Then, moving to the wider GOODS (Great Observatories Origins Deep Survey) -South field (Giavalisco et al. 2004a), van der Wel et al. (2004, 2005a) constructed the FP for a total of 33 color and morphology-selected ETGs at $0.60 < z < 1.15$ using intermediate resolution spectra taken at the ESO Very Large Telescope (VLT). They also found the most massive galaxies ($M_* > 2 \times 10^{11} M_\odot$) to behave much like their cluster analogs at the same redshifts, while less massive galaxies appeared to be substantially younger. Moreover, all these studies noted the higher proportion of weak [OII] emitters among the field ETGs ($\sim$20%) compared to their cluster counterparts, as well as the higher proportion of galaxies with strong Balmer lines (the K + A Galaxies).

The main limitations of all these early studies were in the small number of objects observed at each redshift, which must go a long way toward accounting for occasional discrepancies in the results. In a major effort to overcome this limitation, Treu et al. (2005a,b) obtained high-resolution spectra at the Keck telescope for 163 morphologically-selected ETGs in the GOODS-North field, which were distributed over the redshift range $0.2 < z < 1.2$. The main results of this study are displayed in **Figure 12**, showing that the most massive ellipticals in the field do not differ appreciably from their cluster analogs in having luminosity-weighted ages implying $z_F \gtrsim 3$. However, the lower the mass the larger the dispersion in the M/L_B ratio,

Figure 12
The offset $\Delta \log M/L_B$ from the fundamental plane of cluster ellipticals at $z \sim 0$ for the early-type galaxies in the GOODS-North field (from Treu et al. 2005b). Different symbols are used for early-type (E+S0) galaxies and bulges in late-type (Sa+Sb) galaxies, as well as for the various stellar mass ranges as indicated. The dotted lines are labeled by various formation redshifts.

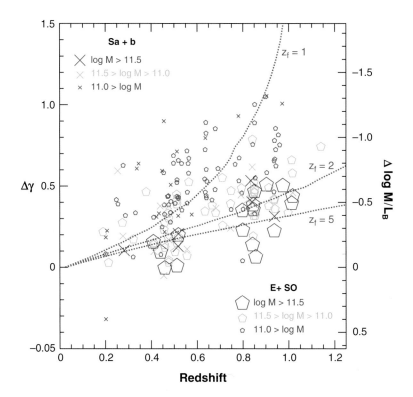

with a definite trend toward lower values with decreasing mass, implying lower and lower formation redshift. This demonstrates that completion of star formation in field galaxies proceeds from the most massive to the less massive ones, as is indeed expected from the down-sizing effect (Cowie et al. 1996, Kodama et al. 2004) and is consistent with the scenario shown in **Figure 6**. This systematic trend in the M/L ratio with galaxy mass results in a "rotation" of the FP with increasing redshift, as less massive galaxies evolve faster in luminosity compared to the more massive, older ones (but one should beware of possible Malmquist bias). This result confirms early hints for a modest rotation of the FP of field ETGs with redshift, as also does a study of the FP of 15 ETGs at $0.9 \lesssim z \lesssim 1.3$ by di Serego Alighieri et al. (2005), a sample drawn from the K20 survey (Cimatti et al. 2002b). **Figure 13** shows the FP for the combined Treu and colleagues and di Serego Alighieri and colleagues samples of ETGs with $\langle z \rangle = 1.1$, where the rotation with respect to the Coma FP is apparent. Note that a similar FP rotation in two clusters at $z = 0.83$ and 0.89 has been recently unambiguously detected by Jørgensen et al. (2006), having extended the σ measurements below ~ 100 km s^{-1}. Somewhat at variance with the FP studies reported above was the Deep Groth Strip Survey result (Gebhardt et al. 2003), in which no difference in the slope was found up to $z \sim 1$ compared to the local FP (hence no down-sizing), which was coupled to a much faster luminosity evolution compared to all other results. Treu et al. (2005b) discuss the possible origins of the discrepancy and attribute

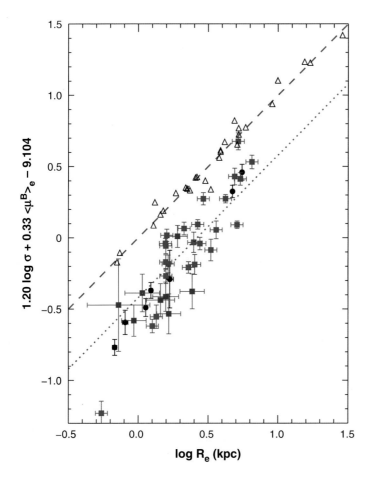

Figure 13
An edge-on view of the fundamental plane for field early-type galaxies at $z \sim 1.1$ from di Serego Alighieri et al. (2005, *red squares* and *black circles*) and Treu et al. (2005b, *blue squares*). Open triangles refer to the Coma ellipticals from Jørgensen, Franx & Kjærgaard (1995) and the dashed line is a best fit to the data. The dotted line is shifted parallel to the dashed line by an amount in surface brightness corresponding to the observed shift of the fundamental plane of galaxy clusters (i.e., $\Delta M/L_B = -0.46z$, van Dokkum & Stanford 2003). The effective surface brightness in the B band (μ_e^B) is in magnitudes per arcsec^2.

it to a combination of selection bias, small number statistics, and relatively low S/N spectra.

From population synthesis models one expects the rate of evolution of the M/L ratio to be slower at longer wavelengths compared to the B band, because it is less affected by the main sequence turnoff moving to cooler temperatures with increasing age. This expectation was qualitatively confirmed by van der Wel et al. (2005b), who found $\Delta \ln(M/L_B) = -(1.46 \pm 0.09)z$ and $\Delta \ln(M/L_K) = -(1.18 \pm 0.1)z$, which appears to be in agreement with the prediction of some models (Maraston 2005), but not of others (Bruzual & Charlot 2003), possibly owing to the different treatment of the AGB contribution.

Instead of measuring central velocity dispersions directly, Kochanek et al. (2000) and Rusin & Kochanek (2005) estimated them from the lens geometry for a sample of (field) lensing ETGs at $0.2 \lesssim z \lesssim 1$. The resulting FP shifts appear to be similar to those of cluster ETGs, although with more scatter, and indicate $z_F \gtrsim 1.5$ for the bulk of the stars in the lensing galaxies.

In summary, like at low redshifts, also at $z \sim 1$ there appear to be small detectable differences between ETGs in high- and low-density regions, but such differences are more evident for faint/low-mass galaxies than for the bright ones.

4.3. Ellipticals Beyond $z \sim 1.3$

Up to $z \sim 1.3$ the strongest features in the optical spectrum of ETGs are the CaII H&K lines and the 4000 Å break. But at higher redshifts these features first become contaminated by OH atmospheric lines, and then move to the near-IR, out of reach of CCD detectors. The lack of efficient near-IR multi-object spectrographs (even in just the J band) has greatly delayed the mapping of the ETG population beyond $z \sim 1.3$. Thus, for almost a decade the most distant spectroscopically confirmed old spheroid was an object at $z = 1.55$ selected for being a radiogalaxy (Dunlop et al. 1996, Spinrad et al. 1997). The spectral features that made the identification possible included a set of FeII, MgII, and MgI lines in the rest-frame near-UV, in the range of $\sim$2580–2850 Å, which is typical of F-type stars. The UV Fe-Mg feature offers at once both the opportunity to measure the redshift and to age-date the galaxy, because it appears only in populations that have been passively evolving since at least a few 10^8 years. It has been also used to age-date local ETGs (e.g., Buson et al. 2004). Thus, using this feature, Spinrad and colleagues inferred an age of $\sim$3.5 Gyr, implying $z_F > 5$ (even in modern cosmology).

This record for the highest redshift ETG was eventually broken by Glazebrook et al. (2004) and Cimatti et al. (2004), using the same features in the rest-frame UV (see **Figure 14**). They reported the discovery of, respectively, five passively evolving galaxies at $1.57 \lesssim z \lesssim 1.85$, and four other such objects at $1.6 \lesssim z \lesssim 1.9$. All being brighter than $K = 20$, these galaxies are quite massive ($M \gtrsim 10^{11} M_\odot$), and hence would rank among the most massive galaxies even in the local universe. This suggests that they were (almost) fully assembled already at this early epoch and, having been passive since at least $\sim$1.1 Gyr, had to form at redshift $\gtrsim$2.7. The four objects found by Cimatti and colleagues are included in the GOODS-South field, and the GOODS deep HST/ACS imaging showed that two objects are definitely elliptical galaxies while the two others are likely to be S0s.

Although breaking the old redshift record was certainly an exciting result, perhaps far more important was the discovery that the surface density of $z > 1.5$ ETGs is indeed much higher than one would have expected from just the Dunlop and colleagues object. Indeed, this galaxy was selected from a catalog of radiogalaxies covering a major fraction of the whole sky, whereas the nine galaxies in the Cimatti and colleagues and Glazebrook and colleagues samples come from a combined area of only 62 arcmin^2.

Further identifications of very high redshift ETGs used this UV feature: McCarthy et al. (2004) reported the discovery of 20 ETGs with $1.3 \lesssim z \lesssim 2.15$ and $K < 20$ as part of the Gemini Deep Deep Survey (GDDS) (including the 5 galaxies from Glazebrook et al. 2004). Within the $\sim$11 arcmin^2 area of the Hubble Ultra Deep Field (HUDF; S. Beckwith et al., in preparation). Daddi et al. (2005b) identified 7 ETGs with $1.39 \lesssim z \lesssim 2.5$ using their HST/ACS grism spectra. For all these objects the stellar

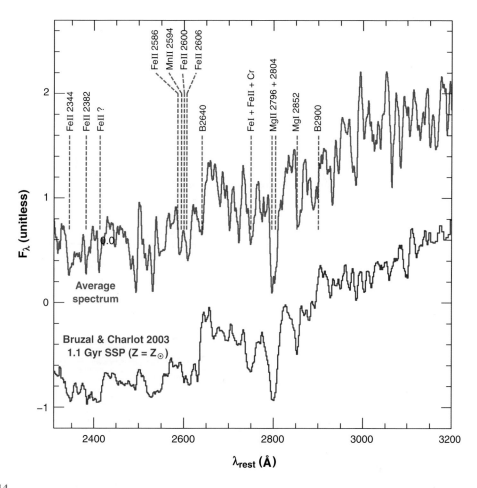

Figure 14

The rest-frame coadded spectrum of the four passively evolving galaxies at $1.6 < z < 1.9$ with the identification of the main spectral features. The spectrum from Bruzual & Charlot (2003) for a 1-Gyr-old SSP (simple stellar population) model of solar metallicity is also shown (*red spectrum*). From Cimatti et al. (2004).

mass derived from the spectral energy distribution (SED; typically extending from the B to the K band) is in excess of $\sim 10^{11} M_\odot$. Over the same field, Yan et al. (2004) identify 17 objects with photometric redshifts between 1.6 and 2.9, whose SED can be best fit by a dominant ~ 2 Gyr old stellar population, superimposed to a low level of ongoing star formation.

Rather than digging deep into small fields, Saracco et al. (2005) searched for bright high-z ETGs over the ~ 160 arcmin^2 field of the MUNICS survey (Drory et al. 2001), and selected objects with $R - K > 5.3$ and $K < 18.5$ for spectroscopic follow up with the low-resolution, near-IR spectrograph on the TNG 3.5m telescope. They identified 7 ETGs at $1.3 \lesssim z \lesssim 1.7$, all with mass well in excess of $10^{11} M_\odot$.

Altogether, these are virtually all the spectroscopically confirmed ETGs at $z > 1.3$ known to date. Given the long integration time needed to get spectroscopic redshifts, it became clear that an effective criterion was indispensable to select high-z ETG candidates. To this end, Daddi et al. (2004) introduced a robust criterion based on the $B − z$ and $z − K$ colors that very effectively selects galaxies at $1.4 \lesssim z \lesssim 2.5$ (the so-called BzKs), and among them separates the star-forming BzKs with $BzK \equiv (z − K)_{AB} − (B − z)_{AB} \geq −0.2$ from the passive ones, with $BzK < −0.2$ and $(z − K)_{AB} > 2.5$. The criterion was tuned using the spectroscopic redshifts from the K20 survey (Cimatti et al. 2002b) and other publicly available data sets. However, Daddi and colleagues showed that synthetic stellar populations of the two kinds (i.e., star-forming and passive) do indeed occupy the corresponding areas in this plot, when redshifted to $1.4 < z < 2.5$. An application of the method to a 320 arcmin² field is shown in **Figure 15** (Kong et al. 2006). In this latter study, it is estimated that the

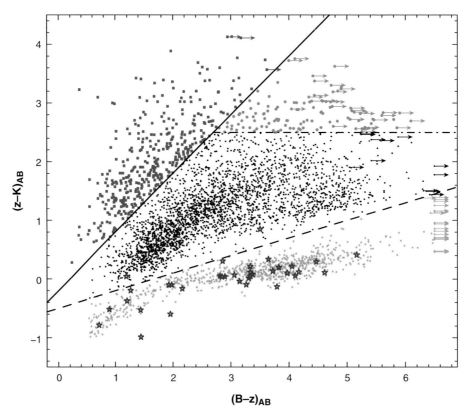

Figure 15

The *BzK* plot introduced by Daddi et al. (2004) is here shown for objects to a limiting magnitude $K_{Vega} = 20$ from a 320 arcmin² field (from Kong et al. 2006). Black dots refer to galaxies at $z \lesssim 1.4$, blue dots refer to star-forming galaxies at $1.4 \lesssim z \lesssim 2.5$, and orange dots refer to passively evolving galaxies at $1.4 \lesssim z \lesssim 2.5$. Green dots are Galactic stars in the same field, while purple stars are local stellar standards.

space density of massive pBzKs (with $K < 20$, stellar mass $\gtrsim 10^{11} M_\odot$, and $\langle z \rangle \simeq 1.7$) is $20 \pm 7\%$ that of $z = 0$ ETGs within the same mass limit. Then there appears to be a sharp drop of passive BzKs beyond $z = 2$, which in part may be due to the available B-band data being not deep enough (Reddy et al. 2005).

Further candidate ETGs with masses up to a few $10^{11} M_\odot$ have been identified at even higher redshifts, such as six objects within the HUDF at a (photometric) redshift >2.8 (Chen & Marzke 2004), where both the redshift and the old age are inferred from the observed break between the J (F110W) and the H (F160W) band being interpreted as the 4000 Å break. One of these objects is undetected in the deep GOODS optical data, but is prominent in the GOODS Spitzer/IRAC 3.5–8 μm images (M. Dickinson et al., in preparation). Thus its SED shows two breaks, one between the z and the J band, and one between the K and the 3.5 μm IRAC band. Identifying them respectively with the Lyman and Balmer breaks, the object would be placed at $z \sim 6.5$, it would be passively evolving with $z_F > 9$, and it would have the uncomfortably large mass of a few $10^{11} M_\odot$ (Mobasher et al. 2005). Lower redshift alternatives give much worse fits to the data, whereas the use of models with strong AGB contribution (Maraston 2005) results in a somewhat less extreme mass and formation redshift.

4.4. Evolution of the Number Density of ETGs to $z \sim 1$ and Beyond

The studies illustrated so far have shown that ETGs exist up to $z \sim 1$, both in clusters and in the field, and are dominated by old stellar populations that formed at $z \gtrsim 2$–3. Moreover, a handful of ETGs has also been identified (over small fields) well beyond $z \sim 1$. Some of these ETGs appear to be as massive as the most massive ETGs in the local universe, demonstrating that at least some very massive ETGs are already fully assembled at $z \gtrsim 1$. However, the expectation is for the number of ETGs to start dropping at some redshift, when indeed entering into the star-formation phase of these galaxies or when they were not fully assembled yet. Therefore, what remained to be mapped by direct observations was the evolution with redshift of the comoving number density of ETGs, and to do so as a function of mass and environment while covering wide enough areas of the sky in order to reduce the bias from cosmic variance. Only in this way could one really overcome the so-called progenitor bias. Because deep and wide surveys require so much telescope time, progress has been slow. Cosmic variance may still be responsible for the apparent discrepancies between galaxy counts from different surveys, but occasionally the interpretation itself of the counts may be prone to ambiguities.

One of the main results of the CFRS was that the number density of red galaxies shows very little evolution over the redshift range $0 < z < 1$ (Lilly et al. 1995b). Following this study, in an attempt to map the number evolution of ETGs all the way to $z \sim 1$, Kauffmann, Charlot & White (1996) extracted 90 color-selected ETGs without [OII] emission from the CFRS redshift catalog. They used a V/V_{max} test and concluded that at $z = 1$ only about one third of bright ETGs had already assembled or had the colors expected for old, pure passively evolving galaxies. However, Im et al.

(1996) identified ∼360 ETGs morphologically selected on archival HST images, and also conducted the $V/V_{\rm max}$ test using photometric redshifts, finding no appreciable number density evolution up to $z \sim 1$ and a brightening consistent with passive evolution. The $V/V_{\rm max}$ test was repeated—again using the CFRS sample—by Totani & Yoshii (1998) who concluded that there was no evolution in the number density up to $z \sim 0.8$. They ascribed the apparent drop at $z > 0.8$ to a color selection bias. No evolution of the space density of morphologically-selected ETGs up to $z \sim 1$ was found by Schade et al. (1999) too, who used the HST imaging of the CFRS and LDSS redshift surveys.

Several attempts at tracing the evolution of the number density of morphologically-selected ETGs to the highest possible redshifts were made using HDF data. In some of these studies very little, if any, change in the space density of ETGs was found up to $z \sim 1$ (e.g., Driver et al. 1998, Franceschini et al. 1998, Im et al. 1999), or up to even higher redshifts when combining the HDF optical data with very deep near-IR data (Benitez et al 1999, Broadhurst & Bouwens 2000). These latter authors emphasized that without deep near-IR data many high-z ETGs are bound to remain undetected, and that spectroscopic incompleteness beyond $z \sim 0.8$ is partly responsible for some of the previous discrepancies. Beyond $z \sim 1$ a drop in the space density of ETGs was detected in several studies including the HDF (Zepf 1997; Franceschini et al. 1998; Barger et al. 1999; Rodighiero, Franceschini & Fasano 2001; Stanford et al. 2004). In particular, Stanford and colleagues applied the $V/V_{\rm max}$ test to a sample of 34 ETGs from the HDF-North that includes deep NICMOS imaging in the H band (F160W) and concluded for a real drop at $z > 1$, but advocated the necessity to explore much wider fields in order to improve the statistics and cope with cosmic variance. Finally, the HDF-South optical data were complemented by ultra-deep JHK imaging at the VLT (the FIRES survey, Labbé et al. 2003, Franx et al. 2003), revealing a population of near-IR galaxies with very red colors ($J - K > 2.3$), called distant red galaxies (DRG), a fraction of which may be ETGs at very high redshifts (see also below). Compared to HDF-North, its Southern equivalent appeared to be much richer in very red galaxies, e.g., including 7 objects with $(V - H)_{\rm AB} > 3$ and $H_{\rm AB} < 25$ while HDF-North has only one. Clearly, exploring much wider fields compared to HDF's ∼5 arcmin² field is imperative in order to make any significant progress.

Passive ETGs formed at very high redshift (e.g., $z > 3$) would indeed have very red colors at $z \gtrsim 1$, and thus they should be found among the so-called extremely red objects (ERO), a class defined for having $R - K > 5$ (or similar color cut), and whose characteristics and relation to ETGs have been thoroughly reviewed by McCarthy (2004). Using a much shallower sample than that from the HDF, but one that covers an area ∼140 times wider than it, Daddi et al. (2000) and Firth et al. (2002) showed that EROs are much more abundant than previously found in smaller fields and are much more strongly clustered than generic galaxies to the same limiting magnitude $K \sim 19$. This made them likely candidates for high-z ETGs, and assuming that ∼70% of EROs are indeed ETGs at $z > 1$, Daddi, Cimatti & Renzini (2000) concluded that most field ellipticals were fully assembled by $z \sim 1$. However, Cimatti et al. (2002a) actually found that out of the 30 EROs with secure redshifts and $K < 19.2$, only

50% are passively evolving objects and these are distributed in the redshift interval $0.8 \lesssim z \lesssim 1.3$, while the other 50% is made by highly-reddened, actively star-forming galaxies. Interestingly, precisely 50% among a sample of 129 EROs with $K < 20.2$ have been detected at 24 μm with Spitzer–MIPS, reinforcing the conclusion that about less than half of EROs are likely to be passive precursors to ETGs (Yan et al. 2004). In fact, the fraction of passive EROs decreases to $\sim$35% on a spectroscopic complete sample to $K = 20$ (Cimatti et al. 2003). Nevertheless, the number density of passive EROs appeared to be broadly consistent with no density evolution of ETGs up to $z \sim 1$ or a modest decrease.

With the COMBO-17 survey Bell et al. (2004b) went a long way toward coping with cosmic variance. With their 5,000 color-selected ETGs up to photometric redshift $z \sim 1.1$, Bell and colleagues were able to construct their rest-frame B-band luminosity functions in nine redshift bins ($0.2 < z < 1.1$), and derived the best-fit Schechter parameters for them using a fixed value of the faint-end slope, $\alpha = -0.6$. They found that the characteristic luminosity M_B^* brightens by $\sim$1.0 mag between $z = 0.25$ and 1.05, consistent with PLE within the errors, and also with the brightening expected from the FP shift ($\Delta \log M/L_B = -0.46\Delta z$), which predicts $\sim$0.9 mag. At the same time, the normalization factor ϕ^* drops by a factor of $\sim$4, but much of the drop is in the highest redshift bins which may be affected by incompleteness. More robust than either ϕ^* or L^* separately is their product $\phi^*L_B^*$, which is proportional to the B-band luminosity density, and this is found to be nearly constant up to $z \sim 0.8$. This is at variance with a PLE scenario that would have predicted an increase by a factor of $\sim$2. Thus, the color of the COMBO-17 red sequence follows nicely the PLE expectation (cf. Section 4.2), but the number density of red-sequence galaxies does not, and Bell and colleagues conclude that the stellar mass in red-sequence galaxies has nearly doubled since $z \sim 1$.

In a major observational effort at the Keck telescope, Faber et al. (2006; DEEP2 project) secured spectroscopic redshifts for $\sim$11,000 galaxies with $R < 24.1$, and also reanalyzed the COMBO-17 data, finding separate best-fit Schechter parameters in various redshift bins up to $z \sim 1.1$. Faber and colleagues emphasize that ϕ^* and M^* are partly degenerate in these fits, for which the faint-end slope was fixed at $\alpha = -0.5$. Thus, between $z = 0.3$ and 1.1, M^* brightens by $\sim$0.47 mag and ϕ^* drops by a factor of $\sim$2.5 for the DEEP2 data, and respectively up by $\sim$0.95 mag and down by a factor of $\sim$4 for the COMBO-17 data. Once more, much of the ϕ^* drops are confined to the last redshift bin, and emphasis is placed on both DEEP2 and COMBO-17 confirming that the B-band luminosity density is nearly constant up to $z \sim 0.8$, along with the implication that the mass density in ETGs has increased, presumably by a factor of $\sim$2 as estimated by Bell et al. (2004b). Extending the analysis from $z = 0.3$ to $z = 0$ (using SDSS data), Faber and colleagues find $\Delta M^* \sim 1.3$ mag and a drop in ϕ^* by a factor of $\sim$4 between $z = 0$ and 1.1, but caution that many of these changes occur between $z \sim 0$ surveys and their first bin (at $z = 0.3$) at one end, and in the last redshift bin at the other end, where the data are said to be the weakest.

Both in COMBO-17 and DEEP2 the shape of the Schechter function for the red-sequence galaxies is assumed constant with redshift. As such, by construction this assumption virtually excludes down-sizing, for which ubiquitous indications have emerged both at low as well as high redshift. Indeed, as alluded to in Kong et al. (2006) and thoroughly documented by Cimatti, Daddi & Renzini (2006), COMBO-17 (and DEEP2) results can also be read in a different way. **Figure 16** shows the evolution of the rest-frame B-band LF from COMBO-17, with the continuous line being the local LF for red-sequence galaxies from Baldry et al. (2004). The local LF has been shifted according to the brightening derived from the empirical FP shift with redshift for cluster ETGs (i.e., by $\Delta M_B = -1.15 \Delta z$, coming from $\Delta(M/L_B) = -0.46z$, taken as the empirical template for passive evolution). From **Figure 16** it is apparent that the brightest part of the LF is fully consistent with purely passive evolution of the most massive galaxies, whereas the fainter part of the LF (below $\sim L^*$) is progressively depopulated with increasing redshift, an effect that only in minor part could be attributed to incompleteness. Therefore, from these data it appears that virtually all the most massive ETGs have already joined the red sequence by $z \sim 1$, whereas less massive galaxies join it later. This is what one would expect from the down-sizing scenario, as exemplified, e.g., in **Figure 6**, as if down-sizing was not limited to stellar ages (stars in massive galaxies are older), but it would work for the assembly itself, with massive galaxies being the first to be assembled to their full size. Being more directly connected to the evolution of dark matter halos, an apparent antihierarchical assembly of galaxies may provide a more fundamental test of the ΛCDM scenario than the mere down-sizing in star formation.

The slow evolution with redshift of the number density of spectrum-selected bright ETGs was also one of the main results of the K20 survey (Pozzetti et al. 2003) and more recently of the VLT VIMOS Deep Survey (VVDS) where the rest-frame B-band LF of ETGs to $I < 24$ is found to be broadly consistent with passive evolution up to $z \sim 1$, with the number density of bright ETGs decreasing by $\sim$40% between $z = 0.3$ and 1.1 (Zucca et al. 2006).

Quite the same scenario, in which the most massive ETGs are already in place at $z \sim 1$ while less massive ones appear later, emerges from the study of the evolution of the stellar mass function for morphologically-selected ETGs in the GOODS fields (Bundy, Ellis & Conselice 2005; Caputi et al. 2006; Franceschini et al. 2006), and especially from the thorough re-analysis by Bundy et al. (2006) of the color-selected ETGs from the DEEP2 survey. No such effect for morphologically-selected ETGs in the GOODS-South field is mentioned in a recent study by Ferreras et al. (2005), who report instead a steep decrease in their number density with redshift. It seems fair to conclude that to fully prove (or disprove) down-sizing in mass assembly, and precisely quantify the effect, one needs to explore the luminosity and mass functions to deeper limits than reached so far, while extending the search to wider areas is needed to overcome cosmic variance. An endeavor of this size requires an unprecedented amount of observing time at virtually all major facilities, both in space and on the ground. The COSMOS project covering 2 square degrees (Scoville 2005) is deliberately targeted to this end, providing public multiwavelength data extending from X rays to radio wavelength that will allow astronomers to map the evolution of galaxies and AGNs in their large-scale structural context, and to derive photometric and spectroscopic redshifts (Lilly 2005) to substantially fainter limits than reached so far. Thus, it will be possible to directly assess the interplay between AGN

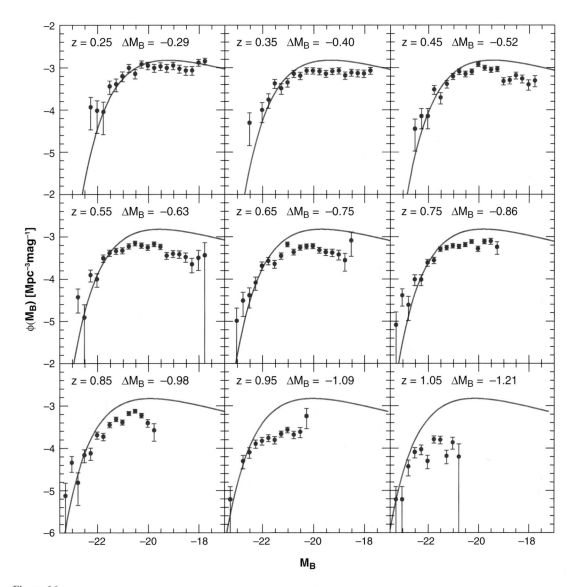

Figure 16

The evolution of the rest-frame B-band luminosity function of early-type (red sequence) galaxies from COMBO-17 (Bell et al. 2004) is compared to the local luminosity function (*solid line*) from Baldry et al. (2004). The local LF has been shifted in magnitude as indicated in each panel, which corresponds to pure passive evolution as empirically derived from the FP shift of cluster of galaxies ($\Delta M/L_B = -0.46z$, van Dokkum & Stanford 2003). Note that there appears to be no number density evolution at the bright end (i.e., for most massive galaxies), whereas at fainter magnitudes there is a substantial decline with redshift of the number density of ETGs. From Cimatti, Daddi & Renzini (2006).

activity, star-formation onset and quenching, merging, mass growth, and morphological differentiation over the largest possible scale, hence promising substantial, perhaps definitive progress in mapping the evolution of ETGs and their progenitors.

It is worth emphasizing that even a low-level of ongoing star formation can make galaxies drop out of our ETG samples, and that theoretical models do not make solid predictions on when star formation is going to cease in a dark matter halo. Therefore, in a broader perspective, what is perhaps more fundamental than the early-type/late-type distinction is the stellar mass of galaxies and the evolution of the galaxy mass function. Several ongoing studies are moving in this direction (e.g., Fontana et al. 2004; Drory et al. 2004, 2005; Bundy, Ellis, & Conselice 2005; Gabasch et al. 2005; Conselice, Blackburne, & Papovich 2005), but extending the discussion to the general mass assembly of galaxies goes beyond the scope of this review.

5. CATCHING ELLIPTICALS IN FORMATION

Immediate precursors for cluster ETGs in the local universe were first identified with a population of blue, star-forming galaxies whose fraction in clusters increases very rapidly with redshift (Butcher & Oemler 1978, 1984, a trend known as the Butcher-Oemler effect), and therefore most such blue, star-forming galaxies had to subside to passive ETGs by $z \sim 0$. Dressler et al. (1997) and Fasano et al. (2000) showed that in clusters up to $z \sim 0.5$ the fraction of true ellipticals is fairly constant, while the fraction of (star-forming) spirals rapidly increases at the expense of the S0s. This was interpreted as evidence that galaxies in clusters that were star-forming spirals at $z \sim 0.5$ have changed their morphology to become passively evolving S0s by $z = 0$, a transformation that may account for much of the Butcher-Oemler effect. In clusters in the same redshift range, a sizable population of ETGs with strong Balmer absorption lines was also identified by Dressler & Gunn (1983), hence called K + A galaxies after the appearance of their spectrum. These ETGs were recognized as poststarburst galaxies, as further documented by Dressler et al. (1999, 2004) and Poggianti et al. (1999).

Thus, the metamorphosis of a fraction of cluster galaxies from star forming to passive is well documented by these studies from $z \sim 0.5$ to $z = 0$. However, these changes seem to affect more spirals and S0s than true ellipticals, and if the bulk of star formation in ETGs took place at very high redshift, then we need to look further out to catch them in formation. From the low-redshift studies we have learned that ETGs are massive, highly clustered, and had to form the bulk of their stars at $z \gtrsim 1.5$–3 (depending on mass and environment) within a short time interval. Indeed, fast star formation is required by the α-overabundance, and by the mere high-formation redshift. At $z \sim 3$ the universe is only ~ 2 Gyr old, hence forming $\sim 10^{11} M_\odot$ of stars in one object between $z = 5$ and $z = 3$ ($\Delta t \sim 1$ Gyr) requires a star-formation rate (SFR) of $\gtrsim 100 M_\odot \mathrm{yr}^{-1}$. Altogether, possible precursors of massive ETGs at low z could be searched among high-z massive, highly clustered, starburst galaxies with very high SFRs (SFR $\gtrsim 100 M_\odot \mathrm{yr}^{-1}$). The rest of this Section is dedicated to mentioning the main results in searching for such objects.

Lyman break galaxies (LBGs) were the first ubiquitous population of galaxies to be identified at $z \sim 3$, and their SFRs often in excess of $\sim 100 \, M_\odot \mathrm{yr}^{-1}$, plus their relatively small size ($R_e \sim 1$–3 kpc), made them natural precursors to local bulges and ETGs (e.g., Giavalisco, Steidel & Macchetto 1996; Steidel et al. 1996; Giavalisco 2002; but see also Giavalisco et al. 1995 for an even earlier attempt at identifying a $z = 3.4$ precursor to ETGs). Recently, Adelberger et al. (2005) show that the 3D correlation length (r_0) of LBGs is such as to match that of local lesser spheroids ($M \lesssim 10^{11} M_\odot$) when the secular increase of r_0 is taken into account. However, Adelberger et al. note that the most massive and most rapidly star forming galaxies at high redshifts are likely to be lost by the Lyman break selection. From **Figure 2** we see that local ETGs with $M > 10^{11} M_\odot$ include almost 50% of the mass in these kinds of galaxies, and $\sim 1/4$ of the total stellar mass at $z = 0$. Therefore, looking at the star-forming precursors of the most massive ETGs refers to a major component of the whole galaxy population. Hence the search was not limited to the LBG technique.

Other obvious candidates are the ultraluminous infrared galaxies (ULIRG, Sanders et al. 1988, Genzel & Cesarsky 2000), detected in mm or sub-mm surveys (e.g., Ivison et al. 1998, Lilly et al. 1999), a class defined for their infrared luminosity (8–1000 μm) exceeding $\sim 10^{12} L_\odot$, and whose typical SFRs ($\gtrsim 200 \, M_\odot \mathrm{yr}^{-1}$) well qualify them for being precursors to massive spheroids, as does their very high (stellar) density ensuring that they would "land" on the fundamental plane as star formation subsides (Kormendy & Sanders 1992, Doyon et al. 1994). ULIRGs as ETG in formation are also advocated by Genzel et al. (2001), who from the resolved kinematics for 12 of them argue that typical ULIRGs are likely precursors to intermediate-mass ETGs rather than to giant ellipticals. However, the internal kinematics of one ULIRG at $z = 2.8$ indicates a mass $\gtrsim 3 \times 10^{11} M_\odot$ (Genzel et al. 2003) making it a likely precursor to a very massive ETG. Blain et al. (2004) note that submillimeter-selected galaxies at $z = 2$–3 appear to be more strongly clustered than LBGs at the same redshifts, which makes them more attractive candidates than LBGs for being the progenitors of the most massive ETGs. Still, their space density falls short by a factor of ~ 10 compared to passive EROs at $z \sim 1$, and they could be the main precursors to EROs only if their duty cycle is very short. Nevertheless, Chapman et al. (2005) argue that the sub-mm galaxies may well be the dominant site of massive star formation at $z = 2$–3, once more making them excellent candidates for being ETG in formation.

The survey and characterization of sub-mm galaxies are currently limited by the modest sensitivity and resolution of existing sub-mm facilities, hence optical/near-IR selections are still the most efficient way of identifying large samples of massive star-forming galaxies at high redshift. For the star-forming *BzK*-selected objects, Daddi et al. (2004) estimate $\langle \mathrm{SFR} \rangle \simeq 200 \, M_\odot \mathrm{yr}^{-1}$, which is typical of ULIRGs, and most of them are clearly mergers on ACS images. Indeed, out of 131 non-AGN (i.e., non X-ray emitter) star-forming sBzKs with $K < 20$ in the GOODS-North field (Dickinson et al. 2005), 82% were individually detected with Spitzer/MIPS at 24 μm (Daddi et al. 2005a). Moreover, by stacking the fluxes of the 131 objects ($\langle z \rangle = 1.9$) from radio to X-rays (i.e., VLA 1.4 GHz, SCUBA 850 and 450 μm, MIPS 24 μm, IRAC 8–3.5 μm, near-IR and optical bands, and Chandra's 0.5–8 keV), Daddi and colleagues showed that the resulting composite SED is an excellent match to that of a template ULIRG with $L_{\rm IR} = 1.7 \times 10^{12} L_\odot$ and $\langle {\rm SFR} \rangle \simeq 250\, M_\odot {\rm yr}^{-1}$, in agreement with the typical SFRs derived from the extinction-corrected UV flux. Two of these BzKs have also been detected at 1.2 mmm with MAMBO, implying a SFR $\sim 1000\, M_\odot {\rm yr}^{-1}$ (Dannerbauer et al. 2006). So, the BzK selection proves to be an excellent way of finding large numbers of ULIRGs at high redshift, whose space density at $z \sim 2$ (~ 1–2×10^{-4} Mpc^{-3}) is about three orders of magnitudes higher than the local density of ULIRGs, and a factor of 2–3 higher than that at $z = 1$. Moreover, the number of star-forming BzKs with $M > 10^{11} M_\odot$ is very close to that of passive BzKs of similar mass, and added together matches the space density of massive ETGs at $z = 0$ (Kong et al. 2006). Hence, it is tantalizing to conclude that as star formation subsides in star-forming BzKs the number of passive ETGs will approach their local density. Finally, worth mentioning is that the majority of samples of $J - K > 2.3$ DRGs in the Extended HDF-South field (Webb et al. 2006) and GOODS-South field (Papovich et al. 2006) have been recently detected at 24 μm with Spitzer/MIPS, indicating that the majority of DRGs are likely to be dusty starburst precursors to ETGs, rather than having already turned into passive ETGs themselves. Moreover, DRGs appear to be distributed over a very broad and nearly flat redshift distribution, from less than 1 to over ~ 3.5 (Reddy et al. 2006).

Moderate redshift precursors to local ETGs are not necessarily star-forming. They may also be less massive ETGs that will merge by $z = 0$, an event now called "dry merging." The merger rate since $z \sim 1.2$ has been estimated by Lin et al. (2004) as part of the DEEP2 survey, concluding that only $\sim 9\%$ of present-day M^* galaxies have undergone a major merger during the corresponding time interval. However, Bell et al. (2006) searched for dry merger candidates over the GEMS field, and based on 7 ETG-ETG pairs estimated that each present-day ETG with $M_v < -20.5$ has undergone 0.5–1 major dry merger since $z \sim 0.7$. This may be at variance with the estimate based on the 3D two-point correlation function of local ETGs in the overwhelming SDSS database. Indeed, each local ETG is found to have less than 1% probability per Gyr of merging with another ETG, hence the dry merging rate appears to be "lower, much lower than the rate at which ETG-hosting DM halos merge with one another" (Hogg 2006), at least for $z \lesssim 0.36$ (Masjedi et al. 2006).

The history of star formation in ETGs has been deduced from the properties of their passively evolving stellar populations in low and high redshifts galaxies, and we know that ETGs and bulges hold at least $\sim 50\%$ of the stellar mass at $z = 0$. Therefore, it is worth addressing here one last issue, even if only in a cursory way: is such an inferred history of star formation consistent with the direct measurements of the star-formation density and stellar mass density at high redshifts? Based on the estimate that the bulk of stars in spheroids formed at $z \gtrsim 3$, it has been suggested that at least $\sim 30\%$ of all stars (and metals) have formed by $z = 3$ (Renzini 1999). This appears to be at least a factor of ~ 3 higher than the direct estimate based on the HDF-North, according to which only 3%–14% of today's stars were in place by $z = 3$ (Dickinson et al. 2003). Data from HDF-South give the higher value 10%–40% (Fontana et al. 2003), most likely as a result of cosmic variance affecting both HDF fields. Based on the ~ 10 times wider field of the K20 survey, $\sim 30\%$ of the stellar mass appears to be in place by $z \sim 2$ (Fontana et al. 2004), but the corrections for incompleteness are large. Drory et al. (2005) find that over the ~ 200 arcmin^2 area of the combined GOODS-South and FORS Deep Field, $\sim 50\%$, $\sim 25\%$, and at least $\sim 15\%$ of the mass in stars is in place, respectively at $z = 1, 2,$ and 3. So, no gross discrepancy has emerged so far between the mass density at $z \sim 3$ as directly measured and as estimated from the fossil evidence at lower redshift. The same holds for the comparison between the star-formation densities as a function of redshift, as inferred from the distribution of stellar ages of ETGs on the one hand, and as directly measured by observations on the other hand (e.g., Madau et al. 1996, Steidel et al. 1999, Giavalisco et al. 2004b). Errors on both sides are still as large as a factor of 2 or 3, but this should rapidly improve thanks to the deep and wide surveys currently under way.

6. EPILOGUE

Almost thirty years ago Toomre (1977) remarked that star formation was observed only in galaxy disks, and further that the final state of merging spirals must be something resembling an elliptical galaxy. Thus, merging spirals to form ellipticals at relatively low redshifts became very popular, especially following the success of CDM theories in accounting for the growth of large-scale structure from tiny initial perturbations.

However, in the intervening three decades an impressive body of evidence on galaxies at low as well as high redshift has accumulated that at least in part contradicts Toomre's assumption. While in the local universe most of the star formation is indeed confined to disks, at $z > 1 - 1.5$ most of it appears to take place in starburst galaxies, such as ULIRGs, whose space density is orders of magnitude higher than in the local universe. Moreover, $\sim 50\%$ of all stars seem to have formed at $z \lesssim 1$ (Dickinson et al. 2003), and to have occurred mostly in disks (Hammer et al. 2005), whereas, if the scenario shown in **Figure 6** is basically correct, then the bulk of star formation in ETGs took place at much higher redshift. At the risk of some simplification, we can say that the era of ETG/spheroid/elliptical formation was largely finished by $z \sim 1$ (if not before), just when the major buildup of disks was beginning (see also Papovich et al. 2005).

The evidence for the stellar populations in ETGs being old, and older in massive galaxies than in less massive ones, has been known for over ten years, along with the evidence for down-sizing and for the anticorrelation of mass and SFR. Theoretical models based on the CDM paradigm have recently incorporated these observational constraints, and have been tuned to successfully reproduce the down-sizing effect in star formation (e.g., De Lucia et al. 2006). In a hierarchical scenario, down-sizing in star formation is indeed natural. Star formation starts first in the highest-density peaks, which in turn are destined to become the most massive galaxies later on. But until recently, models predicted that star formation was continuing all the way to low redshift, as cooling flows were left uncontrasted, thus failing to even produce a red sequence. To get the old and dead massive galaxies we see in nature, such cooling flows (and the accompanying star formation) had to be suppressed in the models, which is now widely accomplished by invoking strong AGN feedback, as first incorporated in ΛCDM simulations by Granato et al. (2001).[2] Yet, the AGN responsibility in switching off star formation remains conjectural at this time, but we became aware that galaxies and supermassive black holes co-evolve, which means we must understand their formation as one and the same problem.

Baryon physics, including star formation, black hole formation and their feedbacks, is highly nonlinear, and it is no surprise if modeling of galaxy evolution relies heavily on many heuristic algorithms, their parameterization, and trials and errors. Dark matter physics, on the contrary, is extremely simple by comparison. Once DM halos are set into motion, there is nothing preventing them from merging with each other under the sole action of gravity, and growing bigger and bigger "galaxies" in an up-sizing process. Thus, the vindication of the ΛCDM paradigm should be found in observations demonstrating that the biggest, most massive galaxies are the first to disappear when going to higher and higher redshifts. This is indeed what has not been seen yet, and actually there may be hints for the contrary.

ACKNOWLEDGMENTS

I thank Ralf Bender, Andrea Cimatti, Emanuele Daddi, Mauro Giavalisco, Laura Greggio, Silvia Pellegrini, and Daniel Thomas for a critical reading of the manuscript and for their valuable suggestions. I am indebted to Mariangela Bernardi, Daniel Thomas, and Sperello di Serego Alighieri for having provided, respectively, **Table 1**, **Figure 2**, and **Figure 13**, specifically for this paper. Finally, I am very grateful to my tutoring editor John Kormendy for his guidance, to Doug Beckner for the final setup of the figures, and to Roselyn Lowe-Webb for her patience in copyediting the manuscript.

LITERATURE CITED

Adelberger KL, Steidel CC, Pettini M, Shapley AE, Reddy NA, et al. 2005. *Ap. J.* 619:697

Aragón-Salamanca A, Ellis RS, Couch WJ, Carter D. 1993 *MNRAS* 262:764

Arimoto N, Yoshii Y. 1987. *Astron. Astrophys.* 173:23

Baldry IK, Galzebrook K, Brinkmann J, Ivezić Z, Lupton LH, et al. 2004. *Ap. J.* 600:681

Barbuy B, Renzini A. 1992. *The Stellar Populations of Galaxies*, IAU Symp. 149. Dordrecht: Kluwer

Barger AJ, Aragón-Salamanca A, Smail I, Ellis, RS, Couch WJ, et al. 1998. *Ap. J.* 501:522

Barger AJ, Cowie LL, Trentham N, Fulton E, Hu EM, et al. 1999. *Astron. J.* 117:102

Barrientos LF, Lilly SJ. 2003. *Ap. J.* 596:129

Barrientos LF, Schade D, Lopez-Cruz O. 1996. *Ap. J.* 460:89

---

[2]See also Ciotti et al. (1991) and Ciotti & Ostriker (1997) for early attempts to suppress cooling flows in ETGs, either with Type Ia supernova feedback alone, or in combination with AGN feedback.

Baum WE. 1959. *Publ. Astron. Soc. Pac.* 71:106
Bell EF, McIntosh DH, Barden M, Wolf C, Caldwell JAR, et al. 2004a. *Ap. J.* 600:L11
Bell EF, McIntosh DH, Katz N, Weinberg MD. 2003. *Ap. J. Suppl.* 149:289
Bell EF, Naab T, McIntosh DH, Somerville RS, Caldwell JAR, et al. 2006. *Ap. J.* 640:241
Bell EF, Wolf C, Meisenheimer K, Rix H-W, Borgh A, et al. 2004b. *Ap. J.* 608:752
Bender R, Burnstein D, Faber SM. 1992. *Ap. J.* 399:462
Bender R, Saglia RP, Ziegler B. 1996. In *The Early Universe with the VLT*, ed. J. Bergeron, p. 105. Berlin: Springer
Bender R, Ziegler B, Bruzual G. 1996. *Ap. J.* 463:L51
Bender R, Saglia RP, Ziegler B, Belloni P, Greggio L, et al. 1998. *Ap. J.* 493:529
Benitez N, Broadhurst T, Bouwens R, Silk J, Rosati P. 1999. *Ap. J.* 515:L65
Bernardi M, Nichol RC, Sheth RK, Miller CJ, Brinkmann J. 2006. *Astron. J.* 131:1288
Bernardi M, Renzini A, da Costa LN, Wegner C, Alonso MV, et al. 1998. *Ap. J.* 508:L143
Bernardi M, Sheth RK, Annis J, Burles S, Finkbeiner DP, et al. 2003a. *Astron. J.* 125:1882
Bernardi M, Sheth RK, Annis J, Burles S, Eisenstein DJ, et al. 2003b. *Astron. J.* 125:1866
Bernardi M, Sheth RK, Nichol RC, Schneider DP, Brinkmann J. 2005. *Astron. J.* 129:61
Blain AW, Chapman SC, Smail I, Ivison R. 2004. *Ap. J.* 611:725
Blakeslee JP, Franx M, Postman M, Rosati P, Holden BP, et al. 2003. *Ap. J.* 596:L143
Bower G, Lucey JR, Ellis RS. 1992. *MNRAS* 254:613
Bower G, Kodama T, Terlevich A. 1998. *MNRAS* 299:1193
Bower G, Terlevich A, Kodama T, Caldwell N. 1999. *ASP Conf. Ser.* 163:211
Bressan A, Chiosi C, Fagotto F. 1994. *Ap. J. Suppl.* 94:63
Bressan A., Chiosi C, Tantalo R. 1996. *Astron. Astrophys.* 311:425
Broadhurst T, Bouwens R. 2000. *Ap. J.* 530:L53
Brodie J, Strader J. 2006. *Annu. Rev. Astron. Astrophys.* 44:193–267
Brown TM, Bowers CW, Kimble RA, Sweigart AV, Ferguson HC. 2000. *Ap. J.* 532:308
Bruzual G. 1983. *Ap. J.* 273:105
Bruzual G, Charlot S. 2003. *MNRAS* 344:1000
Bundy K, Ellis RS, Conselice CJ. 2005. *Ap. J.* 625:621
Bundy K, Ellis RS, Conselice CJ, Taylor JE, Cooper MC, et al. 2006. *Ap. J.* In press (astro-ph/0512465)
Burstein D, Faber SM, Gaskell CM, Krumm N. 1984. *Ap. J.* 287:586
Busarello G, Capaccioli M, Capozziello S, Longo G, Puddu E. 1997. *Astron. Astrophys.* 320:415
Buson LM, Bertola F, Bressan A, Burstein D, Cappellari M. 2004. *Astron. Astrophys.* 423:965
Butcher H, Oemler A. 1978. *Ap. J.* 219:18
Butcher H, Oemler A. 1984. *Ap. J.* 285:426
Buzzoni A. 1995. *Ap. J. Suppl.* 98:69

Caputi KI, McLure RJ, Dunlop JS, Cirasuolo M, Schael AM. 2006. *MNRAS* 366:609
Chabrier G. 2003. *Publ. Astron. Soc. Pac.* 115:736
Chen H-W, Marzke RO. 2004. *Ap. J.* 615:603
Cimatti A, Daddi E, Cassata P, Pignatelli E, Fasano G, et al. 2003. *Astron. Astrophys.* 412:L1
Cimatti A, Daddi E, Mignoli M, Pozzetti L, Renzini A, et al. 2002a. *Astron. Astrophys.* 381:L68
Cimatti A, Daddi E, Renzini A. 2006. *Astron. Astrophys.* In press
Cimatti A, Daddi E, Renzini A, Cassata P, Vanzella E, et al. 2004. *Nature* 430:184
Cimatti A, Mignoli M, Daddi E, Pozzetti L, Fontana A, et al. 2002b. *Astron. Astrophys.* 392:395
Ciotti L. 1997. See da Costa & Renzini 1997, p. 38
Ciotti L, D'Ercole A, Pellegrini S, Renzini A. 1991. *Ap. J.* 376:380
Ciotti L, Lanzoni B 1997. *Astron. Astrophys.* 321:724
Ciotti L, Lanzoni B, Renzini A 1996. *MNRAS* 282:1
Ciotti L, Ostriker JP. 1997. *Ap. J.* 487:L10
Conselice CJ, Blackburne JA, Papovich C. 2005. *Ap. J.* 620:564
Cowie LL, Songaila H, Hu EM, Cohen JG. 1996. *Astron. J.* 112:839
da Costa LN, Renzini A. 1997. *Galaxy Scaling Relations*. Berlin:Springer
Daddi E, Cimatti A, Pozzetti L, Hoeckstra H, Röttgering HJA, et al. 2000. *Astron. Astrophys.* 361:535
Daddi E, Cimatti A, Renzini A. 2000. *Astron. Astrophys.* 362:L45
Daddi E, Cimatti A, Renzini A, Fontana A, Mignoli M, et al. 2004. *Ap. J.* 617:747
Daddi E, Dickinson M, Chary R, Pope A, Morrison G, et al. 2005a. *Ap. J.* 631:L13
Daddi E, Renzini A, Pirzkal N, Cimatti A, Malhotra S, et al. 2005b. *Ap. J.* 626:680
Dannerbauer H, Daddi E, Onodera M, Kong X, Röttgering H, et al. 2006. *Ap. J.* 637:L5
Davies RL, Sadler EM, Peletier RF. 1993. *MNRAS* 262:650
De Lucia G, Poggianti BM, Aragón-Salamanca A, Clowe D, Halliday C, et al. 2004. *Ap. J.* 610:L77
De Lucia G, Springel V, White SDM, Croton D, Kauffmann G. 2006. *MNRAS* 366:499
De Propris R, Stanford SA, Eisenhardt PR, Dickinson M, Elston R. 1999. *Astron. J.* 118:719
Dickinson M. 1997. See da Costa & Renzini 1997, p. 215
Dickinson M, Papovich C, Ferguson HC, Budavari T. 2003. *Ap. J.* 587:25
di Serego Alighieri S, Vernet J, Cimatti A, Lanzoni B, Cassata P, et al. 2005. *Astron. Astrophys.* 442:125
Djorgovski S, Davis M. 1987. *Ap. J.* 313:59
Doyon R, Wells M, Wright GS, Joseph RD, Nadeau D, James PA. 2004. *Ap. J.* 437:L23
Dressler A, Gunn, J. 1983. *Ap. J.* 270:7
Dressler A, Gunn, J. 1990. In *Proc. Edwin Hubble Centennial Symposium, Evolution of the Universe of Galaxies*, ed. RG Kron, p. 200. San Francisco: ASP
Dressler A, Lynden-Bell D, Burnstein D, Davies RL, Faber SM, et al. 1987. *Ap. J.* 313:42

Dressler A, Oemler G, Couch WJ, Smail I, Ellis RS, et al. 1997. *Ap. J.* 490:577
Dressler A, Oemler G, Poggianti B, Smail I, Trager S, et al. 2004. *Ap. J.* 617:867
Dressler A, Smail I, Poggianti B, Butcher HR, Couch WJ, et al. 1999. *Ap. J. Suppl.* 122:51
Driver SP, Fernandez-Soto A, Couch WJ, Odewahn SC, Windorst RA, et al. 1998. *Ap. J.* 496:L93
Drory N, Bender R, Feulner G, Hopp U, Maraston C, et al. 2004. *Ap. J.* 608:742
Drory N, Feulner G, Bender R, Botzler CS, Hopp U, et al. 2001. *MNRAS* 325:550
Drory N, Salvato M, Gabasch A, Bender R, Hopp U, et al. 2005. *Ap. J.* 619:L131
Dunlop J, Peacock J, Spinrad H, Dey A, Jimenez R, et al. 1996. *Nature* 381:581
Eggen OJ, Lynden-Bell D, Sandage AR, 1962. *Ap. J.* 136:748
Eisenstein DJ, Hogg DW, Fukugita M, Nakamura O, Bernardi M, et al. 2003. *Ap. J.* 585:694
Ellis RS, Colless M, Broadhurst T, Heyl J, Glazebrook K. 1996. *MNRAS* 280:235
Ellis RS, Smail I, Dressler A, Couch WJ, Oemler A, et al. 1997. *Ap. J.* 483:582
Faber SM. 1972. *Astron. Astrophys.* 20:361
Faber SM, Jackson R. 1976. *Ap. J.* 204:668
Faber SM, Willmer CNA, Wolf C, Koo DC, Weiner BJ, et al. 2006. *Ap. J.* In press (astro-ph/0506044)
Faber SM, Worthey G, Gonzales JJ. 1992. See Barbuy & Renzini 1992, p. 255
Falcon-Barroso J, Peletier RF, Balcells M. 2002. *MNRAS* 335:741
Fasano G, Cristiani S, Arnouts S, Filippi M. 1998. *Astron. J.* 115:1400
Fasano G, Poggianti B, Couch WJ, Bettoni D, Kjærgaard P, et al. 2000. *Ap. J.* 542:673
Ferreras I, Lisker T, Carollo CM, Lilly S, Mobasher B. 2005. *Ap. J.* 635:243
Feulner G, Gabasch A, Salvato M, Drory N, Hopp U, et al. 2006. *Ap. J.* 633:L9
Fioc M, Rocca Volmerange B. 1997. *Astron. Astrophys.* 326:950
Firth AE, Somerville RS, McMahon RG, Lahav O, Ellis RS, et al. 2002. *MNRAS* 332:617
Fontana A, Donnarumma I, Vanzella E, Giallongo E, Menci N, et al. 2003. *Ap. J.* 594:L9
Fontana A, Pozzetti L, Donnarumma I, Renzini A, Cimatti A, et al. 2004. *Astron. Astrophys.* 424:23
Franceschini A, Rodighiero G, Cassata P, Berta S, Vaccari M, et al. 2006. *Astron. Astrophys.* In press (astro-ph/0601003)
Franceschini A, Silva L, Fasano G, Granato L, Bressan A, et al. 1998. *Ap. J.* 506:600
Franx M, Labbé I, Rudnick G, van Dokkum PG, Daddi E, et al. 2003. *Ap. J.*
Fukugita M, Hogan CJ, Peebles PJE. 1998. *Ap. J.* 503:518
Gabasch A, Hopp U, Feulner G, Bender R, Seitz S, et al. 2006. *Astron. Astrophys.* 448:101
Gavazzi G. 1993. *Ap. J.* 419:469
Gavazzi G, Pierini D, Boselli A. 1996. *Astron. Astrophys.* 312:397
Gebhardt K, Faber SM, Koo DC, Im M, Simard L, et al. 2003. *Ap. J.* 597:239
Genzel R, Baker AJ, Tacconi LJ, Lutz D, Cox P, et al. 2003. *Ap. J.* 584:633
Genzel R, Cesarsky C. 2000. *Annu. Rev. Astron. Astrophys.* 38:761
Genzel R, Tacconi LJ, Rigopoulou D, Lutz D, Tecza M. 2001. *Ap. J.* 563:527

Giavalisco M. 2002. *Annu. Rev. Astron. Astrophys.* 40:579

Giavalisco M, Dickinson M, Ferguson HC, Ravidranath S, Kretchmer C, et al. 2004b. *Ap. J.* 600:L103

Giavalisco M, Ferguson HC, Koekemoer AM, Dickinson M, Alexander DM, et al. 2004a. *Ap. J.* 600:L93

Giavalisco M, Macchetto FD, Madau P, Sparks WB. 1995. *Ap. J.* 441:L13

Giavalisco M, Steidel CC, Macchetto FD. 1996. *Ap. J.* 470:194

Gladders MD, Lopez-Crus O, Tee HKC, Kodama T. 1998. *Ap. J.* 501:577

Glazebrook K, Abraham RG, McCarthy PJ, Savaglio S, Chen H-S, et al. 2004. *Nature* 430:181

González Delgado RM, Cerviño M, Martins LP, Leitherer C, Hauschildt PH. 2005. *MNRAS* 357:945

Gorgas J, Efstathiou G, Aragón-Salamanca A. 1990. *MNRAS* 245:217

Granato GL, Silva L, Monaco P, Panuzzo P, Salucci P, et al. 2001. *MNRAS* 324:757

Greggio L. 1997. *MNRAS* 285:151

Greggio L, Renzini A. 1990. *Ap. J.* 364:35

Gunn JE, Stryker LL, Tinsley BM. 1981. *Ap. J.* 249:48

Hammer F, Flores H, Elbaz D, Zheng XZ, Liang YC, et al. 2005. *Astron. Astrophys.* 430:115

Hogg D. 2006. In *The Fabulous Destiny of Galaxies: Bridging Past and Present*, ed. V Le Brun, A Mazure, S Arnouts, D Burgarella. Paris: Edition Frontieres. In press (astro-ph/0512029)

Holden BP, Blakeslee JP, Postman M, Illingworth GD, Demarco R. 2005a. *Ap. J.* 626:809

Holden BP, Stanford SA, Eisenhardt P, Dickinson M. 2004. *Astron. J.* 127:2484

Holden BP, van der Wel A, Franx M, Illingworth GD, Balkeslee JP, et al. 2005b. *Ap. J.* 620:L83

Hubble EP. 1936. *The Realm of the Nebulae*. New Haven: Yale Univ. Press

Im M, Griffiths RE, Ratnatunga KU, Sarajedini VL. 1996. *Ap. J.* 461:L79

Im M, Griffiths RE, Naim A, Ratnatunga KU, Roche N, et al. 1999. *Ap. J.* 510:82

Ivison RJ, Smail I, Le Borgne J-F, Blain AW, Kneib J-P, et al. 1998. *MNRAS* 298:583

Jablonka P, Martin P, Arimoto N. 1996. *Astron. J.* 112:1415

Jørgensen I. 1997. *MNRAS* 288:161

Jørgensen I. 1999. *MNRAS* 306:607

Jørgensen I, Chiboucas K, Flint K, Bergmann M, Barr J, et al. 2006. *Ap. J.* 639:L9

Jørgensen I, Franx M, Kjærgaard P. 1995. *MNRAS* 276:1341

Jørgensen I, Franx M, Kjærgaard P. 1996. *MNRAS* 280:167

Juneau S, Glazebrook K, Crampton D, McCarthy PJ, Savaglio S, et al. 2005. *Ap. J.* 619:L135

Kauffmann G, Charlot S, White SDM. 1996. *MNRAS* 283:L117

Kauffmann G, Heckman TM, White SDM, Charlot S, Tremonti C, et al. 2003. *MNRAS* 341:54

Kelson DD, Illingworth GD, Franx M, van Dokkum PG. 2001. *Ap. J.* 552:L17

Kelson DD, Illingworth GD, van Dokkum PG, Franx M. 2000. *Ap. J.* 531:184

Kelson DD, van Dokkum PG, Franx M, Illingworth GD, Fabricant D. 1997. *Ap. J.* 478:L13

Kochanek CS, Impey CDE, Lehar J, McLeod BA, Rix H-W, et al. 2000. *Ap. J.* 543:131

Kodama T, Arimoto N. 1997. *Astron. Astrophys.* 320:41

Kodama T, Arimoto N, Barger AJ, Aragón-Salamanca A. 1998. *Astron. Astrophys.* 334:99

Kodama T, Bower R. 2003. *MNRAS* 346:1

Kodama T, Yamada T, Akiyama M, Aoki K, Doi M, et al. 2004. *MNRAS* 350:1005

Kong X, Daddi E, Arimoto N, Renzini A, Broadhurst T, et al. 2006. *Ap. J.* 638:72

Koo DC, Datta S, Willmer CNA, Simard L, Tran K-V, et al. 2005. *Ap. J.* 634:L5

Kormendy J. 1977. *Ap. J.* 218, 333

Kormendy J, Kennicutt RC Jr. 2004. *Annu. Rev. Astron. Astrophys.* 42:603

Kormendy J, Sanders DB. 1992. *Ap. J.* 390:L93

Kuijken K, Rich RM. 2002. *Astron. J.* 124:2054

Kuntschner H, Davies RL. 1998. *MNRAS* 295:L29

Kuntschner H, Lucey RJ, Smith RJ, Hudson MJ, Davies RL 2001. *MNRAS* 323:615

Labbé I, Franx M, Rudnick G, Rix H-W, Moorwood A, et al. 2003. *Astron. J.* 125:1117

Larson RB. 1974. *MNRAS* 166:585

Lidman C, Rosati P, Demarco R, Nonino M, Maineri V. 2004. *Astron. Astrophys.* 416:829

Lilly SJ. 2005. *The Messenger.* 121:42

Lilly SJ, Eales SA, Gear WKP, Hammer F, Le Fèvre O, et al. 1999. *Ap. J.* 518:641

Lilly SJ, Le Fèvre O, Crampton D, Tresse L. 1995a. *Ap. J.* 455:50

Lilly SJ, Tresse L, Hammer F, Crampton D, Le Fèvre O. 1995b. *Ap. J.* 455:108

Lin L, Koo DC, Willmer CNA, Patton DR, Conselice CJ, et al. 2004. *Ap. J.* 617:L9

Madau P, Ferguson HC, Dickinson M, Giavalisco M, Steidel CC, et al. 1996. *MNRAS* 283:1388

Maraston C. 1998. *MNRAS* 300:872

Maraston C. 2005. *MNRAS* 362:799

Maraston C, Greggio L, Renzini A, Ortolani S, Saglia RP, et al. 2003. *Astron. Astrophys.* 400:823

Maraston C, Thomas D. 2000. *Ap. J.* 541:126

Masjedi M, Hogg DW, Cool RJ, Eisenstein DJ, Blanton MR, et al. 2006. *Ap. J.* In press (astro-ph/0512166)

McCarthy PJ. 2004. *Annu. Rev. Astron. Astrophys.* 42:477

McCarthy PJ, Le Borgne D, Crampton D, Chen H-W, Abraham R, et al. 2004. *Ap. J.* 614:L9

McIntosh DH, Bell EF, Rix H-W, Wolf C, Heymans C, et al. 2005. *Ap. J.* 632:191

Mobasher B, Dickinson M, Ferguson HC, Giavalisco M, Wiklind T, et al. 2005. *Ap. J.* 635:832

Mobasher B, Guzmán R, Aragón-Salamanca A, Zepf S. 1999. *MNRAS* 304:225

Mullis CR, Rosati P, Lamer G, Böhringer H, Schwope A, et al. 2005. *Ap. J.* 623:L85

Nelan JE, Smith RJ, Hudson MJ, Wegner GA, Lucey JR, et al. 2005. *Ap. J.* 632:137

Nelson AE, Gonzalez AH, Zaritsky D, Dalcanton JJ. 2001. *Ap. J.* 563:629

Norman CA, Renzini A, Tosi M, eds. 1986. *Stellar Populations.* Cambridge: Cambridge Univ. Press

O'Connell RW. 1980. *Ap. J.* 236:430
O'Connell RW. 1986. See Norman, Renzini & Tosi 1986, p. 167
Ortolani S, Renzini A, Gilmozzi R, Marconi G, Barbuy B, et al. 1995. *Nature* 377:701
Pahre MA, de Carvalho RR, Djorgovski SG. 1998. *Astron. J.* 116:1606
Pahre MA, Djorgovski SG, de Carvalho RR. 1995. *Ap. J.* 453:L17
Pahre MA, Djorgovski SG, de Carvalho RR. 1996. *Ap. J.* 456:L79
Pahre MA, Djorgovski SG, de Carvalho RR. 1998. *Astron. J.* 116:1591
Papovich C, Dickinson M, Giavalisco M, Conselice CJ, Ferguson HC. 2005. *Ap. J.* 631:101
Papovich C, Moustakas LA, Dickinson M, Le Floc'h E, Rieke GH, et al. 2006. *Ap. J.* 640:92
Pasquali A, Ferreras I, Panagia N, Daddi E, Malhotra S, et al. 2006. *Ap. J.* 636:115
Peletier RF. 1989. PhD Thesis, Rijksuniversiteit Groningen
Peletier RF, Balcells M, Davies RL, Andreakis Y, Vazdekis A, et al. 1999. *MNRAS* 310:703
Persic M, Salucci P. 1992. *MNRAS* 258:14 p
Poggianti BM, Bridges TJ, Mobasher B, Carter D, Doi M, et al. 2001. *Ap. J.* 562:689
Poggianti B, Smail I, Dressler A, Couch WJ, Berger AJ, et al. 1999. *Ap. J.* 518:576
Popesso P, Böhringer H, Romaniello M, Voges W. 2005. *Astron. Astrophys.* 433:415
Pozzetti L, Cimatti A, Zamorani G, Daddi E, Menci N, et al. 2003. *Astron. Astrophys.* 402:837
Prugniel Ph, Maubon G, Simien F. 2001. *Astron. Astrophys.* 366:68
Prugniel Ph, Simien F. 1996. *Astron. Astrophys.* 309:749
Rakos KD, Schombert JM. 1995. *Ap. J.* 439:47
Reddy NA, Erb DK, Steidel CC, Shapley AE, Adelberger KL, et al. 2005. *Ap. J.* 633:748
Reddy NA, Steidel CC, Fadda D, Yan L, Pettini M, et al. 2006. *Ap. J.* In press (astro-ph/0602596)
Renzini A. 1986. See Norman, Renzini & Tosi 1986, p. 213
Renzini A. 1992. See Barbuy & Renzini 1992, p. 325
Renzini A. 1999. In *The Formation of Galactic Bulges*, ed. CM Carollo, HC Ferguson, RFG Wyse, p. 9. Cambridge: Cambridge Univ. Press
Renzini A. 2005. In *The Initial Mass Function 50 Years Later*, ed. E Corbelli, F Palla, H Zinnecker, p. 221. Dordrecht: Springer
Renzini A, Ciotti L. 1993. *Ap. J.* 416:L49
Rix H-W, Barden M, Beckwith SVW, Bell EF, Borch A, et al. 2004. *Ap. J. Suppl.* 152:163
Rix H-W, de Zeeuw PT, Cretton N, van der Marel RP, Carollo M. 1997. *Ap. J.* 488:702
Rodighiero G, Franceschini A, Fasano G. 2001. *MNRAS* 324:491
Rosati P, della Ceca R, Norman C, Giacconi R. 1998. *Ap. J.* 492:L21
Rusin D, Kochanek CS. 2005. *Ap. J.* 623:666
Sandage A. 2005. *Annu. Rev. Astron. Astrophys.* 43:581
Sandage A, Visvanathan N. 1978a. *Ap. J.* 223:707
Sandage A, Visvanathan N. 1978b. *Ap. J.* 225:742

Sanders DB, Soifer BT, Elias JH, Madore BF, Matthews K. 1988. *Ap. J.* 325:74
Saracco P, Longhetti M, Severgnini P, della Ceca R, Braito V, et al. 2005. *MNRAS* 357:L40
Schade D, Carlberg RG, Yee HKC, Lopez-Cruz O, Ellingson E. 1996. *Ap. J.* 464:L63
Schade D, Barrientos LF, Lopez-Cruz O. 1997. *Ap. J.* 477:L17
Schade D, Lilly SJ, Crampton D, Ellis RS, Le Fèvre O, et al. 1999. *Ap. J.* 525:31
Schechter PL, Dressler A. 1987. *Astron. J.* 94:563
Scodeggio M, Gavazzi G, Biesole E, Pierini D, Boselli A. 1998. *MNRAS* 301:1001
Scodeggio M, Gavazzi G, Franzetti P, Boselli A, Zibetti S, et al. 2002. *Astron. Astrophys.* 384:812
Scoville N. 2005. In *Multiwavelength Mapping of Galaxy Formation and Evolution*, ed. A Renzini, R Bender, p. 330. Berlin:Springer-Verlag
Spinrad H, Dey A, Stern D, Dunlop J, Peacock J, et al. 1997. *Ap. J.* 484:581
Stanford SA, Dickinson M, Postman M, Ferguson HC, Lucas RA, et al. 2004. *Astron. J.* 127:131
Stanford SA, Eisenhardt PR, Dickinson M. 1998. *Ap. J.* 492:461
Stanford SA, Eisenhardt PR, Brodwin M, Gonzalez AH, Stern D, et al. 2005. *Ap. J.* 634:L129
Steidel CC, Giavalisco M, Pettini M, Dickinson M, Adelberger K. 1996. *Ap. J.* 462:17
Steidel CC, Adelberger KL, Giavalisco M, Dickinson M, Pettini M. 1999. *Ap. J.* 519:1
Stephens AW, Frogel JA, DePoy DL, Freedman W, Gallart C, et al. 2003. *Astron. J.* 125:2473
Strazzullo V, Rosati P, Stanford SA, Lidman C, Nonino M, et al. 2006. *Astron. Astrophys.* 450:909
Tanaka M, Kodama T, Arimoto N, Okamura S, Umetsu K, et al. 2005. *MNRAS* 362:268
Tantalo R, Chiosi C, Bressan A, Fagotto F. 1996. *Astron. Astrophys.* 311:361
Thomas D 1999. *MNRAS* 306:655
Thomas D, Davies RL. 2006. *MNRAS* 366:510
Thomas D, Greggio L, Bender R. 1999. *MNRAS* 302:537
Thomas D, Maraston C, Bender R. 2003. *MNRAS* 339:897
Thomas D, Maraston C, Bender R, Mendez de Oliveira C. 2005. *Ap. J.* 621:673
Tinsley BM. 1980. *Fundam. Cosmic Phys.* 5:287
Tinsley BM, Gunn JE. 1976. *Ap. J.* 203:52
Toft S, Soucail G, Hjorth J. 2003. *MNRAS* 344:337
Toft S, Maineri V, Rosati P, Lidman C, Demarco R, et al. 2004. *Astron. Astrophys.* 422:29
Toomre A. 1977. In *Evolution of Galaxies and Stellar Populations*, ed. BM Tinsley, RB Larson, p. 401. New Haven: Yale University Observatory
Totani T, Yoshii Y. 1998. *Ap. J.* 501:L177
Trager SC, Faber SM, Worthey G, Gonzales JJ. 2000. *Astron. J.* 120:165
Treu T, Ellis RS, Liao TX, van Dokkum PG. 2005a. *Ap. J.* 622:L5
Treu T, Ellis RS, Liao TX, van Dokkum PG, Tozzi P, et al. 2005b. *Ap. J.* 633:174
Treu T, Stiavelli M, Casertano S, Møller P, Bertin G. 1999. *MNRAS* 308:1037

Treu T, Stiavelli M, Bertin G, Casertano S, Møller P. 2001. *MNRAS* 326:237
Treu T, Stiavelli M, Casertano S, Møller P, Bertin G. 2002. *Ap. J.* 564:L13
Trujillo I, Burkert A, Bell EF. 2004. *Ap. J.* 600:L39
van der Wel A, Franx M, van Dokkum PG, Rix H-W. 2004. *Ap. J.* 601:L5
van der Wel A, Franx M, van Dokkum PG, Rix H-W, Illingworth GD, Rosati P. 2005a. *Ap. J.* 631:145
van der Wel A, Franx M, van Dokkum PG, Huang J, Rix H-W, ct al. 2005b. *Ap. J.* 636:L21
van Dokkum PG, Ellis RS. 2003. *Ap. J.* 592:L53
van Dokkum PG, Franx M. 1996. *MNRAS* 281:985
van Dokkum PG, Franx M, Kelson DD, Illingworth GD. 1998. *Ap. J.* 504:L17
van Dokkum PG, Franx M, Kelson DD, Illingworth GD. 2001. *Ap. J.* 553:L39
van Dokkum PG, Franx M, Fabricant D, Illingworth GD, Kelson DD. 2000. *Ap. J.* 541:95
van Dokkum PG, Stanford SA. 2003. *Ap. J.* 585:78
Vazdekis A, Cenarro AJ, Gorgas J, Cardiel N, Peletier RF. 2003. *Astron. Astrophys.* 340:1317
Vázques GA, Leitherer C. 2005. *Ap. J.* 621: 717
Visvanathan N, Sandage A. 1977. *Ap. J.* 216:214
Webb TMA, van Dokkum PG, Egami E, Fazio G, Franx M, et al. 2006. *Ap. J.* 636:L17
Weiss A, Peletier RF, Matteucci F. 1995. *Astron. Astrophys.* 296:73
Wheeler JC, Sneden C, Truran JW Jr. 1989. *Annu. Rev. Astron. Astrophys.* 27:279
White SDM, Rees MJ. 1978. *MNRAS* 183:341
Williams RE, Blacker B, Dickinson M, Dixon WVD, Ferguson HC, et al. 1996. *Astron. J.* 112:1335
Wolf C, Meisenheimer K, Rix H-W, Borch A, Dye S, et al. 2003. *Astron. Astrophys.* 401:73
Worthey G. 1994. *Ap. J. Suppl.* 95:107
Worthey G, Trager SC, Faber SM. 1995. *ASP Conf. Ser.* 86:203
Wuyts S, van Dokkum PG, Kelson DD, Franx M, Illingworth GD. 2004. *Ap. J.* 605:677
Yamada T, Kodama T, Akiyama M, Furusawa H, Iwata I, et al. 2005. *Ap. J.* 634:861
Yan H, Dickinson M, Eisenhardt PRM, Ferguson HC, Grogin NA, et al. 2004. *Ap. J.* 616:63
Yan L, Choi PI, Fadda D, Marleau FR, Soifer BT, et al. 2004. *Ap. J. Suppl.* 154:75
Zepf SE. 1997. *Nature* 390:377
Ziegler BL, Saglia RP, Bender R, Belloni P, Greggio L, et al. 1999. *Astron. Astrophys.* 346:13
Ziegler BL, Bender R. 1997. *MNRAS* 291:527
Zoccali M, Renzini A, Ortolani S, Greggio L, Saviane L, et al. 2003. *Astron. Astrophys.* 399:931
Zucca E, Ilbert O, Bardelli S, Tresse L, Zamorani G, et al. 2006. *Astron. Astrophys.* In press (astro-ph/0506393)

Extragalactic Globular Clusters and Galaxy Formation

Jean P. Brodie and Jay Strader

UCO/Lick Observatory, University of California, Santa Cruz, CA 95064;
email: brodie@ucolick.org, strader@ucolick.org

Key Words

galaxy evolution, globular clusters, star clusters, stellar populations

Abstract

Globular cluster (GC) systems have now been studied in galaxies ranging from dwarfs to giants and spanning the full Hubble sequence of morphological types. Imaging and spectroscopy with the *Hubble Space Telescope* and large ground-based telescopes have together established that most galaxies have bimodal color distributions that reflect two subpopulations of old GCs: metal-poor and metal-rich. The characteristics of both subpopulations are correlated with those of their parent galaxies. We argue that metal-poor GCs formed in low-mass dark matter halos in the early universe and that their properties reflect biased galaxy assembly. The metal-rich GCs were born in the subsequent dissipational buildup of their parent galaxies and their ages and abundances indicate that most massive early-type galaxies formed the bulk of their stars at early times. Detailed studies of both subpopulations offer some of the strongest constraints on hierarchical galaxy formation that can be obtained in the near-field.

1. INTRODUCTION

Globular star clusters (GCs) are among the oldest radiant objects in the universe. With typical masses $\sim 10^4$–10^6 $M_\odot$ (corresponding to luminosities of $\sim M_V = -5$ to -10) and compact sizes (half-light radii of a few parsecs), they are readily observable in external galaxies. The 15 years since the Annual Review by Harris (1991, "Globular Cluster Systems in Galaxies Beyond the Local Group") have seen a revolution in the field of extragalactic GCs. It is becoming increasingly apparent that GCs provide uniquely powerful diagnostics of fundamental parameters in a wide range of astrophysical processes. Observations of GCs are being used to constrain the star formation and assembly histories of galaxies, nucleosynthetic processes governing chemical evolution, the epoch and homogeneity of cosmic reionization, the role of dark matter in the formation of structure in the early universe, and the distribution of dark matter in present-day galaxies. GCs are valuable tools for theoretical and observational astronomy across a wide range of disciplines from cosmology to stellar spectroscopy.

It is not yet widely recognized outside the GC community that recent advances in GC research provide important constraints on galaxy formation that are complementary to in situ studies of galaxies at medium-to-high redshift. The theme of this review is the role of GC systems as tracers of galaxy formation and assembly, and one of our primary aims is to emphasize the current and potential links with results from galaxy surveys at high redshift and interpretations from stellar population synthesis, numerical simulations, and semianalytical modeling. In what follows we will attempt to chronicle the observations that mark recent milestones of achievement and place them in the wider theoretical and observational context. We will focus most closely on work carried out since about 2000. The preceding period is well-covered by the book of Ashman & Zepf (1998) and the Saas-Fee lectures of Harris (2001).[1] Among the significant topics not directly covered are young massive clusters (potential "proto-GCs"), X-ray sources in extragalactic GCs, and ultra-compact dwarf galaxies. Neither do we include a comprehensive discussion of the Galactic GC system.

The fundamental premise in what follows is that GCs are good tracers of the star formation histories of spheroids (early-type galaxies, spiral bulges, and halos), in the sense that major star-forming episodes are typically accompanied by significant GC formation. Low-level star formation (e.g., in quiescent galactic disks) tends to produce few, if any, GCs. Because most of the stellar mass in the local universe is in spheroids ($\sim$75%; Fukugita, Hogan & Peebles 1998), GCs trace the bulk of the star-formation history of the universe. Although the relationships between star formation, GC formation, and GC survival are complex and do not necessarily maintain relative proportions under all conditions, this underlying assumption is supported by a number of lines of argument. Massive star clusters appear to form during all major star-forming events, such as those accompanying galaxy-galaxy interactions (e.g., Schweizer 2001). In these situations, the number of new clusters formed scales with the amount of gas involved in the interaction (e.g., Kissler-Patig, Forbes & Minniti

---

[1]Available online at **http://physwww.mcmaster.ca/∼harris/Publications/saasfee.ps**.

1998). The cluster formation efficiency (the fraction of star formation in clusters) scales with the star-formation rate, at least in spiral galaxies where it can be directly measured at the present epoch (Larsen & Richtler 2000). This may suggest that massive clusters form whenever the star-formation rate is high enough, and that this occurs principally during spheroid formation. Perhaps most importantly, the properties of GCs (especially their metallicities) are correlated with the properties of their host galaxies.

2. COLOR BIMODALITY: GLOBULAR CLUSTER SUBPOPULATIONS

Perhaps the most significant development of the decade in the field of extragalactic GCs was the discovery that the color distributions of GC systems are typically bimodal. Indeed, color bimodality is the basic paradigm of modern GC studies. Nearly every massive galaxy studied to date with sufficiently accurate photometry has been shown to have a bimodal GC color distribution, indicating two subpopulations of GCs. In principle, these color differences can be due to age or metallicity differences or some combination of the two. Due to the well-known degeneracy between age and metallicity (e.g., Worthey 1994), the cause of this bimodality is not readily deduced from optical colors alone. Nonetheless, the significance of the finding was immediately recognized. The presence of bimodality indicates that there have been at least two major star-forming epochs (or mechanisms) in the histories of most—and possibly all—massive galaxies. Subsequent spectroscopic studies (see Section 4) have shown that color bimodality is due principally to a metallicity difference between two old subpopulations.

With our "bimodality-trained" modern eyes, we can see evidence of the phenomenon in the $B - I$ CFHT imaging of NGC 4472 in Couture, Harris & Allwright (1991) and $C - T_1$ CTIO imaging of NGC 5128 by Harris et al. (1992). However, the first groups to propose bimodality (or "multimodality") were Zepf & Ashman (1993) for NGC 4472 and NGC 5128 and Ostrov, Geisler & Forte (1993) for NGC 1399 (in fact, using the Harris and colleagues and Couture and colleagues colors). Observations of the GC systems of galaxies throughout the 1990s provided mounting evidence that bimodality was ubiquitous in massive galaxies. The primary catalyst of this research was the advent of the *Hubble Space Telescope* (HST). The Wide Field and Planetary Camera 2 (WFPC2) provided the spatial resolution and accurate photometry needed to reliably identify GC candidates in galaxies as distant as the Virgo Cluster at 17 Mpc (e.g., Whitmore et al. 1995). At this distance, GCs (with typical half-light radii of 2–3 pc $\sim$0.03–0.04″) are resolvable with the HST and their sizes are measurable with careful modeling of the point spread function. This drove down the contamination from background galaxies and foreground stars to low levels and was a substantial improvement over multiband optical photometry from the ground.

Among the larger and more comprehensive photometric studies using HST/WFPC2 were Gebhardt & Kissler-Patig (1999), Larsen et al. (2001) and Kundu & Whitmore (2001a). Using data from the HST archive, Gebhardt & Kissler-Patig showed that bimodality was a common phenomenon. However, since the imaging was shallow for many of the galaxies in their sample, they failed to find bimodality in ~50% of their 50 galaxies. Taking advantage of deeper data, Larsen et al. (2001) and Kundu & Whitmore (2001a) found statistically significant bimodality in most of their sample galaxies, the majority of which were of early-type. Galaxies that were tentatively identified as unimodal in these studies were later, with improved photometric precision, shown to conform to the bimodality "rule." Indeed, it is important to note that no massive elliptical (E) galaxy has been convincingly shown to lack GC subpopulations. An absence of metal-poor GCs was suggested for both NGC 3311 (Secker et al. 1995) and IC 4051 (Woodworth & Harris 2000), and an absence of metal-rich GCs for NGC 4874 (Harris et al. 2000). However, HST/WFPC2 imaging of NGC 3311 (Brodie, Larsen & Kissler-Patig 2000) revealed a healthy subpopulation of metal-poor GCs. It is now clear that the WFPC2 photometry of the Coma E IC 4051 was not deep enough to securely argue for a uni- or bimodal fit. Finally, the NGC 4874 result was due to a photometric zeropoint error (W. Harris, private communication). The discovery of a massive E that indeed lacked a metal-poor (or metal-rich) subpopulation would be important, but so far no such instances have been confirmed.

The majority of these HST studies were carried out in V- and I-equivalent bands. This choice was largely driven by efficiency considerations [shorter exposure times needed to reach a nominal signal-to-noise (S/N)], despite the fact that other colors, such as $B - I$, offer much better metallicity sensitivity for old stellar populations.

It has been known for some time that the GC system of the Milky Way is also bimodal. The presence of GC subpopulations in the Milky Way was codified by Zinn (1985; see also Armandroff & Zinn 1988) who identified two groups of GCs. "Halo" GCs are metal-poor, nonrotating (as a system), and can be found at large galactocentric radii. "Disk" GCs are metal-rich and form a flattened, rotating population. Later work on the spatial and kinematic properties of the metal-rich GCs by Minniti (1995) and Côté (1999) identified them with the Milky Way bulge rather than its disk (as we shall see below, this association seems to hold for other spirals as well, although see the discussion in Section 3.3). In addition to their sample of early-type galaxies, Larsen et al. (2001) also discussed the GC systems of the Milky Way and NGC 4594 in some detail, pointing out that the locations of the GC color peaks in these spirals were indistinguishable from those of massive early-type galaxies.

The blue (metal-poor) and red (metal-rich) peaks in massive early-type galaxies typically occur at $V - I = 0.95 \pm 0.02$ and 1.18 ± 0.04 (Larsen et al. 2001). These colors correspond to [Fe/H] ~ -1.5 and -0.5 for old GCs (or a bit more metal-rich, depending on the metallicity scale and color-metallicity relation adopted). **Figure 1** shows a histogram of the $V - I$ colors of GCs in the Virgo gE M87, which clearly shows bimodality (Larsen et al. 2001). However, the peak locations are not exactly the same for all galaxies. Before GC bimodality was discovered, van den Bergh (1975) suggested and Brodie & Huchra (1991) confirmed a correlation between the mean color/metallicity of GC systems and the luminosity of their parent galaxies. Brodie & Huchra (1991) also showed that the slope of this relation was very similar to the relation connecting galaxy color and galaxy luminosity, but the GC relation was offset toward lower metallicities by about 0.5 dex. They noted that the similarity in

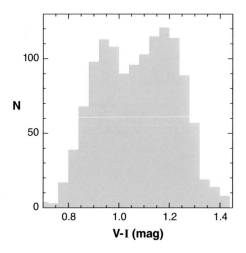

Figure 1

$V − I$ color histogram of globular clusters in the Virgo giant elliptical M87, showing clear bimodality (Larsen et al. 2001; figure from data courtesy of S. Larsen).

slope suggests a close connection between the physical processes responsible for the formation of both GCs and galaxies. Subsequently, a correlation between the color of just the metal-rich GCs and host galaxy luminosity was found by Forbes, Brodie & Grillmair (1997), Larsen et al. (2001), and Forbes & Forte (2001). The slope of this relation was again found to be similar to that of the color-magnitude relation for early-type galaxies ($V − I \propto −0.018 M_V$), suggesting that metal-rich GCs formed along with the bulk of the field stars in their parent galaxies.

With the exception of Larsen et al. (2001), little or no correlation between the color of the metal-poor GCs and host galaxy luminosity was reported in these studies, although Burgarella, Kissler-Patig & Buat (2001) and Lotz, Miller & Ferguson (2004) suggested such a relation might be present, but only for the dwarf galaxies. Larsen and colleagues found a shallow relation for the metal-poor GCs in their sample of 17 massive early-type galaxies, albeit at moderate (3σ) statistical significance. Strader, Brodie & Forbes (2004b) compiled and reanalyzed high-quality data from the literature and found a significant ($>5\sigma$) correlation for metal-poor GCs, extending from massive Es to dwarfs over ~ 10 magnitudes in galaxy luminosity. The relation is indeed relatively shallow ($V − I \propto −0.009\, M_V$, or $Z \sim L^{0.15}$), making it difficult to detect, especially in heterogeneous data sets. This same slope was confirmed by J. Strader, J.P. Brodie, L. Spitler & M.A. Beasley (submitted) and Peng et al. (2006a) for early-type galaxies in Virgo. **Figure 2** shows [Fe/H] versus M_B for both subpopulations; the GC peaks are taken from Strader, Brodie & Forbes (2004b) and J. Strader, J.P. Brodie, L. Spitler & M.A. Beasley (submitted) and have been converted from $V − I$ and $g − z$ using the relations of Barmby et al. (2000) and Peng et al. (2006a), respectively. These data, together with ancillary information about the GC systems, are compiled in **Table 1**. The true scatter at fixed M_B is unclear, because the observational errors vary among galaxies, and there may be an additional component due to small differences between

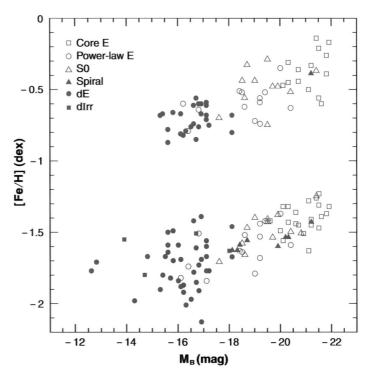

Figure 2

Peak globular cluster (GC) metallicity versus galaxy luminosity (M_B) for metal-poor and metal-rich GCs in a range of galaxies. The points are from Strader et al. (2004b) and J. Strader, J.P. Brodie, L. Spitler & M.A. Beasley (submitted) and have been converted from $V - I$ and $g - z$ to [Fe/H] using the relations of Barmby et al. (2000) and Peng et al. (2006a), respectively. Galaxy types are indicated in the figure key; classifications are in **Table 1**. Linear relations exist for both subpopulations down to the limit of available data.

the $V - I$ and $g - z$ color-metallicity relations. The cutoff in the metal-rich relation at $M_B \sim -15.5$ primarily reflects the magnitude limit of the sample; it may continue to fainter magnitudes, although many such galaxies have only metal-poor GCs. The remarkable inference to be drawn from **Figure 2** is that the peak metallicities of both subpopulations are determined primarily by galaxy luminosity (or mass) across the entire spectrum of galaxy types.

Using their new color-metallicity transformation between $g - z$ and [Fe/H], Peng and colleagues found $Z \sim L^{0.25}$ for metal-rich GCs, which is also consistent with the previous estimates of the slope already noted. The color-metallicity relation appears to be quite nonlinear, as discussed below. Thus, even though the slopes of the metal-poor and metal-rich relations are significantly different in the GC color–galaxy luminosity plane, they are similar in the GC metallicity–galaxy luminosity plane (Peng et al. 2006a; see **Figure 2**). In Section 11 we discuss the constraints on galaxy formation implicit in these relations.

The Advanced Camera for Surveys (ACS) on HST has significantly advanced our understanding of the color distributions of GC systems, offering a wider field of view and improved photometric accuracy compared to WFPC2. Three large studies of Es utilizing HST/ACS have recently been published. As mentioned above, Peng et al. (2006a) and J. Strader, J.P. Brodie, L. Spitler & M.A. Beasley (submitted) studied the GC systems of early-type galaxies (ranging from dwarf to giant) using g and z data taken as part of the ACS Virgo Cluster Survey (Côté et al. 2004). Peng and colleagues investigated all 100 (E and S0) galaxies, while Strader and colleagues focused solely on the Es. Harris et al. (2006) used BI ACS photometry to analyze GCs in eight "BCGs", galaxies that are among the brightest in their respective groups or clusters.

These studies resulted in several new discoveries. First, a correlation was found between color and luminosity for individual metal-poor GCs in some giant Es (the "blue tilt"; see **Figure 3**). This is the first detection of a mass-metallicity relation for GCs. The blue tilt was found by Strader and colleagues in the Virgo giant Es (gEs) M87 and NGC 4649, and by Harris and colleagues in their sample, although the interpretations of the findings differ. The mass-metallicity relation for individual metal-poor GCs may argue for self-enrichment. Strader and colleagues speculated that these metal-poor GCs were able to self-enrich because they once possessed dark-matter halos that were subsequently stripped (see also discussion in Section 12). The M87 data are well-fit by a relation equivalent to $Z \propto M^{0.48}$ over the magnitude range $20 < z < 23.2$, where the turnover of the GC luminosity function (GCLF) is at $z \sim 23$. Harris and colleagues found a similar relation ($Z \propto M^{0.55}$) but suggested that the trend was only present at bright luminosities ($M_I \lesssim -9.5$ to -10, corresponding to $z \lesssim 22$ in the Strader and colleagues Virgo dataset). The color-magnitude diagrams (CMDs) in Strader and colleagues' research on M87 and NGC 4649 appear consistent with a continuation of the correlation to magnitudes fainter than $z = 22$, but do not strongly distinguish between the two interpretations. We do know that the blue tilt phenomenon is not confined to galaxies in high density environments or even just to E galaxies. It has recently been reported for NGC 4594 (L. Spitler, J.P. Brodie, S.S. Larsen, J. Strader, D.A. Forbes & M.A. Beasley, submitted), a luminous Sa galaxy that lies in a loose group. Curiously, the Virgo gE NGC 4472 (also studied by Strader and colleagues) shows no evidence for the blue tilt. If this lack of a tilt is confirmed with better data, it will be a strong constraint on any potential "universal" model for explaining the phenomenon in massive galaxies. The Milky Way itself does not show evidence for the tilt, but this could be due to the small number of metal-poor GCs ($\sim$100) compared to massive galaxies or to the inhomogeneity of metallicities and integrated photometry in current catalogs.

Harris and colleagues suggested that the tilted metal-poor GC relation caused the metal-poor and metal-rich peaks to merge at the brightest GC luminosities, turning a bimodal distribution into a nominally unimodal one. By contrast, Strader and colleagues argued that at these high luminosities there is a separate population of objects with larger-than-average sizes and a range of colors, spanning the metal-poor to metal-rich subpopulations. Indeed, Harris and colleagues find that $\sim$20–30 of the brightest objects in the nearest galaxy in their sample, NGC 1407 (at $\sim$21 Mpc), appear to be extended with respect to normal GCs. The size measurements suggest

Table 1 Properties of GC color distributions

Name	Galaxy Type[a]	Environment[b]	M_B[c] (mag)	MP color[d] (mag)	MR color[d] (mag)	MP [Fe/H][e] (dex)	MR [Fe/H][e] (dex)	Color[f]	S_N[g]	S_N Refs[h]
NGC 4472	core E	Virgo C	−21.9	0.951	1.411	−1.32	−0.17	$g-z$	3.6 ± 0.6	1
NGC 1399	core E	Fornax C	−21.8	0.952	1.185	−1.37	−0.39	$V-I$	5.1 ± 1.2	2
NGC 3309	core(?) E	Hydra C	−21.6	0.947	1.134	−1.39	−0.60	$V-I$	...	...
NGC 4486	core E	Virgo C	−21.5	0.953	1.390	−1.31	−0.21	$g-z$	14.1 ± 1.5	3
NGC 3311	core(?) E	Hydra C	−21.5	0.929	...	−1.47	...	$V-I$	...	...
NGC 4406	core E	Virgo C	−21.5	0.986	1.145	−1.23	−0.56	$V-I$	3.5 ± 0.5	1
NGC 4649	core E	Virgo C	−21.4	0.964	1.424	−1.26	−0.14	$g-z$	4.1 ± 1.0	4
NGC 524	S0	N524 G	−21.4	0.980	1.189	−1.25	−0.37	$V-I$	...	...
NGC 4374	core E	Virgo C	−21.2	0.927	1.322	−1.45	−0.33	$g-z$	1.6 ± 0.3	5
NGC 5322	core(?) E	N5322 G	−21.2	0.942	...	−1.41	...	$V-I$	...	...
NGC 4594	S0/Sa	N4594 G	−21.2	0.939	1.184	−1.43	−0.39	$V-I$	2.1 ± 0.3	1
NGC 4365	core E	Virgo C	−21.1	0.891	1.232	−1.63	−0.50	$g-z$	...	...
NGC 7619	core(?) E	Pegasus C	−21.1	0.973	...	−1.28	...	$V-I$	...	...
NGC 7562	core(?) E	Pegasus C	−20.9	0.920	...	−1.51	...	$V-I$	...	...
NGC 2768	S0	N2768 G	−20.8	0.919	...	−1.51	...	$V-I$	...	...
NGC 4621	transition E	Virgo C	−20.7	0.927	1.305	−1.45	−0.36	$g-z$	...	...
NGC 5813	core E	N5846 G	−20.6	0.935	...	−1.44	...	$V-I$	5.7 ± 1.8	6
IC 1459	core E	I1459 G	−20.6	0.955	...	−1.36	...	$V-I$	...	...
NGC 3115	S0	N3115 G	−20.4	0.922	1.153	−1.50	−0.52	$V-I$	...	...
NGC 4494	power E	N4565 G	−20.4	0.901	1.128	−1.59	−0.63	$V-I$	...	...
NGC 4552	core E	Virgo C	−20.3	0.951	1.334	−1.32	−0.31	$g-z$	...	...
NGC 253	Sc	Sculptor G	−20.3	0.912	...	−1.54	...	$V-I$	...	...
NGC 1404	core(?) E	Fornax C	−20.3	0.938	1.170	−1.43	−0.45	$V-I$	2 ± 0.5	7
M31	Sb	Local G	−20.2	0.912	...	−1.54	...	$V-I$	1.3	8
NGC 3379	core E	Leo I G	−20.1	0.964	1.167	−1.32	−0.47	$V-I$	1.2 ± 0.3	1
NGC 4278	core E	N4631 G	−20.1	0.908	...	−1.56	...	$V-I$	...	...
NGC 4473	power E	Virgo C	−20.0	0.942	1.310	−1.37	−0.35	$g-z$	...	...
NGC 3608	core E	N3607 G	−20.0	0.923	...	−1.49	...	$V-I$	...	...
NGC 1400	S0	Eridanus G	−19.9	0.951	1.164	−1.38	−0.48	$V-I$	...	...

Galaxy	Type	Environment	M_V			[Fe/H]		color		ref
Milky Way	Sbc	Local G	−19.9	0.898	...	−1.60	...	$V − I$	0.7	8
NGC 1023	S0	N1023 G	−19.7	0.912	1.164	−1.54	−0.48	$V − I$	...	...
NGC 4291	core E	N4291 G	−19.6	0.940	...	−1.42	...	$V − I$	...	...
NGC 3384	S0	Leo I G	−19.5	0.942	1.208	−1.41	−0.29	$V − I$	...	...
NGC 3607	S0	N3607 G	−19.5	0.939	1.099	−1.43	−0.75	$V − I$	...	...
NGC 1427	power E	Fornax C	−19.4	0.940	1.153	−1.42	−0.52	$V − I$	3.4 ± 0.6	9
NGC 4478	power E	Virgo C	−19.2	0.882	1.195	−1.68	−0.56	$g − z$	...	...
NGC 4434	power E	Virgo C	−19.2	0.911	1.179	−1.53	−0.59	$g − z$	...	...
NGC 3377	power E	Leo I G	−19.2	0.936	1.103	−1.44	−0.74	$V − I$	...	...
NGC 4564	S0	Virgo C	−19.0	0.935	1.263	−1.40	−0.44	$g − z$	...	...
NGC 4387	power E	Virgo C	−19.0	0.859	1.112	−1.79	−0.72	$g − z$	...	...
NGC 4660	S0	Virgo C	−18.7	0.923	1.320	−1.47	−0.33	$g − z$	...	...
NGC 247	Sd	Sculptor G	−18.7	0.908	...	−1.56	...	$V − I$	...	...
NGC 4733	power E	Virgo C	−18.6	0.918	1.131	−1.52	−0.62	$V − I$	...	...
NGC 4550	S0	Virgo C	−18.6	0.883	1.145	−1.66	−0.56	$V − I$	...	...
NGC 4489	S0	Virgo C	−18.5	0.900	1.260	−1.58	−0.44	$g − z$	...	...
NGC 4551	power E	Virgo C	−18.5	0.890	1.219	−1.64	−0.52	$g − z$	...	...
M33	Scd	Local G	−18.4	0.900	...	−1.59	...	$V − I$	...	...
NGC 4458	power E	Virgo C	−18.4	0.892	1.223	−1.63	−0.51	$g − z$	...	...
NGC 55	Sm	Sculptor G	−18.3	0.892	...	−1.63	...	$V − I$	...	...
IC 3468	dE	Virgo C	−18.1	0.925	1.130	−1.46	−0.68	$g − z$	1.1	10
NGC 300	Sd	Sculptor G	−18.1	0.892	...	−1.63	...	$V − I$	...	...
NGC 4482	dE	Virgo C	−18.1	0.884	1.065	−1.67	−0.80	$g − z$	1.6	10
LMC	dIrr	Local G	−18.0	0.890	...	−1.63	...	$V − I$	0.8	8
NGC 3599	S0	Leo I G	−17.6	0.872	1.112	−1.71	−0.70	$V − I$	...	...
IC 3019	dE	Virgo C	−17.2	0.864	1.093	−1.77	−0.75	$g − z$	1.8	10
IC 3381	dE	Virgo C	−17.1	0.897	1.168	−1.60	−0.61	$g − z$	5.2	10
IC 3328	dE	Virgo C	−17.1	0.905	1.114	−1.56	−0.71	$g − z$	0.9	10
NGC 4318	dE	Virgo C	−17.1	0.884	1.182	−1.67	−0.59	$g − z$	0.7	10

(Continued)

Table 1 (Continued)

Name	Galaxy Type[a]	Environment[b]	$M_B{}^c$ (mag)	MP color[d] (mag)	MR color[d] (mag)	MP [Fe/H][e] (dex)	MR [Fe/H][e] (dex)	Color[f]	$S_N{}^g$	S_N Refs[h]
IC 809	dE	Virgo C	−17.1	0.864	1.129	−1.77	−0.68	$g-z$	3.3	10
IC 3653	power E	Virgo C	−17.1	0.851	N	−1.84	N	$g-z$	0.6	10
IC 3652	dE	Virgo C	−16.9	0.879	1.173	−1.69	−0.60	$g-z$	3.7	10
VCC 543	dE	Virgo C	−16.9	0.794	N	−2.13	N	$g-z$	0.4	10
IC 3470	dE	Virgo C	−16.9	0.938	1.138	−1.39	−0.67	$g-z$	5.0	10
NGC 4486b	power E	Virgo C	−16.8	0.920	1.126	−1.51	−0.64	$V-I$	...	...
IC 3501	dE	Virgo C	−16.8	0.872	1.177	−1.73	−0.60	$g-z$	5.3	10
IC 3442	dE	Virgo C	−16.8	0.837	1.089	−1.91	−0.76	$g-z$	0.8	10
VCC 437	dE	Virgo C	−16.7	0.894	1.042	−1.61	−0.85	$g-z$	1.1	10
IC 3735	dE	Virgo C	−16.7	0.848	1.198	−1.85	−0.56	$g-z$	1.6	10
SMC	dIrr	Local G	−16.7	0.919	...	−1.51	...	$V-I$	1.2	8
IC 3032	dE	Virgo C	−16.6	0.861	1.169	−1.78	−0.61	$g-z$	0.9	10
VCC 200	dE	Virgo C	−16.6	0.931	1.101	−1.42	−0.74	$g-z$	0.5	10
IC 3487	dE	Virgo C	−16.5	0.824	1.086	−1.97	−0.76	$g-z$	1.2	10
IC 3509	power E	Virgo C	−16.4	0.869	1.071	−1.74	−0.79	$g-z$	7.3	10
VCC 1895	dE	Virgo C	−16.3	0.818	1.071	−2.01	−0.79	$g-z$	1.6	10
IC 3647	dE	Virgo C	−16.2	0.845	N	−1.87	N	$g-z$	1.0	10
IC 3383	dE	Virgo C	−16.2	0.834	1.056	−1.92	−0.82	$g-z$	4.6	10
VCC 1627	power E	Virgo C	−16.2	0.755	N	−2.33	N	$g-z$	0.5	10
IC 3693	power E	Virgo C	−16.2	0.779	1.177	−2.21	−0.60	$g-z$	1.0	10
IC 3101	dE	Virgo C	−16.1	0.842	1.134	−1.88	−0.67	$g-z$	4.9	10
IC 798	power E	Virgo C	−16.1	0.854	N	−1.82	N	$g-z$	4.6	10
IC 3779	dE	Virgo C	−16.1	0.880	1.060	−1.69	−0.81	$g-z$	2.0	10
IC 3635	dE	Virgo C	−16.0	0.898	N	−1.59	N	$g-z$	5.0	10
VCC 1993	dE	Virgo C	−16.0	0.851	N	−1.84	N	$g-z$	0.3	10
IC 3461	dE	Virgo C	−15.8	0.919	1.142	−1.49	−0.66	$g-z$	12.1	10

Galaxy	Type	Env	M_B	Color	[Fe/H]	MR?	Color index	S_N	Ref
VCC 1886	dE	Virgo C	−15.8	0.877		N	−1.70	$g-z$	1.5	10
IC 3602	dE	Virgo C	−15.7	0.855		N	−1.82	$g-z$	1.1	10
NGC 205	dE	Local G	−15.6	0.922		N	−1.50	$V-I$	3	8
VCC 1539	dE	Virgo C	−15.6	0.898	1.039	−0.87	−1.59	$g-z$	9.5	10
VCC 1185	dE	Virgo C	−15.6	0.890	1.077	−0.78	−1.64	$g-z$	6.3	10
IC 3633	dE	Virgo C	−15.5	0.884		N	−1.67	$g-z$	3.3	10
IC 3490	dE	Virgo C	−15.4	0.858	1.136	−0.67	−1.80	$g-z$	7.3	10
VCC 1661	dE	Virgo C	−15.3	0.838	1.130	−0.68	−1.90	$g-z$	2.3	10
NGC 185	dE	Local G	−14.8	0.882		N	−1.67	$V-I$	4.6	8
NGC 6822	dIrr	Local G	−14.7	0.850		…	−1.80	$V-I$	1.2	11
NGC 147	dE	Local G	−14.3	0.807		N	−1.98	$V-I$	3.6	8
WLM	dIrr	Local G	−13.9	0.910		N	−1.55	$V-I$	1.7	8
Sagittarius	dSph	Local G	−12.8	0.871		…	−1.71	$V-I$	18.1	11
Fornax	dSph	Local G	−12.6	0.858		N	−1.77	$V-I$	28.8	8

[a] Power/Core E: Ellipticals with power-law or cored central surface brightness distributions. NGC 4621 is a transition between the two groups. The galaxies with (?) are not formally classified—the division has been made between core/power-law Es at $M_B = −20$ (Faber et al. 1997, J. Kormendy, D.B. Fisher, M.E. Cornell & R. Bender, in preparation). Classifications are from J. Kormendy, D.B. Fisher, M.E. Cornell & R. Bender (in preparation, and private communication) for Virgo galaxies, Faber et al. (1997) for many early-type galaxies, and NED for the remainder. The term dE (dwarf elliptical) is often used for all galaxies that are not star forming with $M_B > −18$; here we use it only for galaxies with faint central surface brightness and Sersic n ~ 1 (exponential) profiles.

[b] Local galaxy environment. C: cluster, G: group.

[c] References for M_B can be found in Strader et al. (2004b) and J. Strader, J.P. Brodie, L. Spitler & M.A. Beasley (submitted).

[d] Peak metal-poor (MP) and metal-rich (MR) colors derived from mixture modeling of the GC color distributions. "N" indicates that study of the galaxy has found no metal-rich GCs. Galaxies with "…" either have metal-rich GCs without a determined color peak (typically massive galaxies), or ambiguous evidence for such a subpopulation.

[e] [Fe/H] values converted from the listed colors using the relations of Barmby et al. (2000) and Peng et al. (2006a) for $V-I$ and $g-z$, respectively.

[f] The color listed in columns 5 and 6. The $V-I$ colors are from Strader et al. (2004b); the $g-z$ colors are from J. Strader, J.P. Brodie, L. Spitler & M.A. Beasley (submitted).

[g] V-band specific frequency (S_N). Values are only listed if from a study with sufficient spatial coverage and photometric depth for an accurate estimate of the total GC population. The values from J. Strader, J.P. Brodie, L. Spitler & M.A. Beasley (submitted) have been converted from the B-band S_N by dividing by 2.1; this assumes $B-V = 0.8$.

[h] Literature sources for S_N estimates. 1: Rhode & Zepf (2004) 2: Dirsch et al. (2003). 3: Harris, Harris & McLaughlin (1998). 4: Forbes, Strader & Brodie (2004). 5: Gomez & Richtler (2004). 6: Hopp, Wagner & Richtler (1995). 7: Forbes et al. (1998). 8: Forbes et al. (2000). 10: J. Strader, J.P. Brodie, L. Spitler & M.A. Beasley (submitted). 11: This work; the S_N estimates have been increased from Forbes et al. (2000) to include new GCs in NGC 6822 and Sgr (see text).

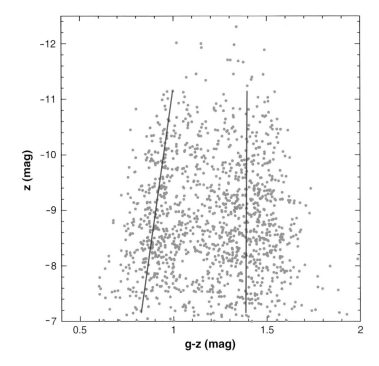

Figure 3

z versus $\sim g - z$ color-magnitude diagram for M87 globular clusters (GCs) (*gray circles*) from J. Strader, J.P. Brodie, L. Spitler & M.A. Beasley (submitted). A correlation between color and luminosity for the bright metal-poor GCs is apparent (the "blue tilt"). The blue and red lines are fitted linear relations.

that there is something qualitatively different about (at least) a subset of the brightest GCs, which has also been recognized in NGC 1399 (Dirsch et al. 2003) and NGC 4636 (Dirsch, Schuberth & Richtler 2005). The sizes and luminosities of the bright intermediate-color objects in these galaxies suggest a relation to the "ultra-compact dwarf" galaxies (UCDs) discovered in both the Fornax and Virgo clusters (e.g., Phillips et al. 2001).

The second significant finding was that the color dispersion of the metal-rich GCs is nearly twice as large as that of the metal-poor GCs. Peng and colleagues, Strader and colleagues, and Harris and colleagues reported essentially the same dispersions in the color distributions of both subpopulations. However, Peng and colleagues and Harris and colleagues adopted different color-metallicity relations, and these led to divergent conclusions about the metallicity distributions of these populations. Harris and colleagues fit a linear relation between $B - I$ and metallicity using Galactic GCs. This has the advantage of being independent of stellar population models but the disadvantage of being yoked to the metallicity distribution of Galactic clusters. There are no low-reddening Galactic GCs with [Fe/H] > -0.5, so the empirical relation is unconstrained at these metallicities, and the data are poorly fit by a linear relation in the very metal-poor regime. Peng and colleagues used a piecewise linear relation broken at $g - z = 1.05$ or [Fe/H] ~ -0.8. This utilized GCs in the Galaxy, M87, and NGC 4472 with both $g - z$ colors and spectroscopic metallicities (these are still ultimately tied to the Galactic GC [Fe/H] scale). The qualitative effect was to "flatten" the relation at low metallicities compared to a linear fit—so small color changes correspond to large metallicity changes—and to "steepen" it at high metallicities.

Consequently, Peng and colleagues found the metal-poor GCs to have a larger metallicity dispersion than the metal-rich GCs: 68% half-width [Fe/H] intervals of ∼0.6 dex and 0.3 dex for the metal-poor and metal-rich GCs in massive galaxies, respectively (these were derived from a nonparametric analysis and thus are not exactly equivalent to a 1σ dispersion for a normal distribution). By contrast, Harris and colleagues deduced mean 1σ dispersions of ∼0.3 dex and 0.4 dex, and noted that the metal-poor and metal-rich subpopulations in the Galaxy have $\sigma = 0.34$ dex and $\sigma = 0.16$ dex. The relative widths of the two Galaxy subpopulations are more consistent with the Peng and colleagues results, though smaller in an absolute sense. However, even if the metallicity dispersion in the metal-poor GCs is larger, the absolute metallicity spread is much smaller. The implications of these differences for the enrichment histories of the two subpopulations remain to be seen.

It seems clear from the new Galactic GC data in Peng et al. (2006a) that a single linear fit is not optimal, but the exact form of the relation at metallicity extremes is poorly constrained. Clearly, identifying the correct form of the $g - z$ to [Fe/H] conversion (and, indeed, conversions for other colors) is essential, because the metallicity spreads in individual subpopulations have important implications for the formation and assembly histories of GC systems. The different color dispersions for the subpopulations also indicate that it is necessary to use heteroscedastic (unequal variance) fits in mixture modeling. Homoscedastic fits will give systematic errors in the peak values, but may be the best option for systems with few GCs. Another important effect of nonlinear color-metallicity relations is that bimodal color distributions can be enhanced or even created from metallicity distributions that are not strongly bimodal (Richtler 2006; Yoon, Yi & Lee 2006).

The last principal finding in these new studies, reported in both Peng et al. (2006a) and J. Strader, J.P. Brodie, L. Spitler & M.A. Beasley (submitted), is that many dEs have metal-rich GC subpopulations. Their colors are consistent with an extrapolation of the GC color–galaxy luminosity relation for massive galaxies, though the fraction of metal-rich GCs tends to be smaller in dEs than in massive Es. Peng and colleagues found that the median fraction of metal-rich GCs is ∼0.15−0.2 in dEs and rises steeply toward more luminous galaxies. Some of this increase is due to the different radial distributions of the two subpopulations; the HST/ACS data preferentially sample the more centrally-concentrated metal-rich GCs. Global fractions of metal-rich GCs in massive Es are likely to be closer to 0.3–0.4 (Rhode & Zepf 2004). The changing fraction of metal-rich GCs with galaxy luminosity, combined with the correlations of GC colors with galaxy luminosity for both subpopulations, fully explains the classic correlation between the mean color/metallicity of a GC system and parent galaxy luminosity (Brodie & Huchra 1991). The overall slope actually measured depends on how the sample is defined. If HST data are used, the metal-rich GCs will be overrepresented and the steep slope of the metal-rich GC relation will dominate the overall relation.

These new results suggest that considerable undiscovered detail may still be hidden in GC color distributions. Intensive use should be made of HST/ACS while it is still operational. It is likely to be the best instrument available for the next ∼15 years for studying the optical color distributions of GCs.

2.1. Scenarios for Bimodality

After bimodality was observed to be a common phenomenon, several scenarios were presented to explain it. In this section, we briefly describe the leading scenarios (see also the review by West et al. 2004), but leave detailed discussion until Section 11. The major merger model of Ashman & Zepf (1992) has the distinction of being the only model to predict bimodality before it was observed. This model evolved from early work suggesting that E galaxies formed in gas-rich major mergers of disk galaxies (Toomre & Toomre 1972; Toomre 1977; Schweizer 1987). Burstein (1987) and Schweizer (1987) suggested that new GCs might be formed in large quantities during the merger process. Ashman & Zepf (1992) and Zepf & Ashman (1993) developed this idea into a predictive model in which the metal-poor GCs are donated by the progenitor spirals and the metal-rich GCs are formed in the gas-rich merger. This model gained enormous support when new HST observations of merging galaxies found large numbers of young massive star clusters (YMCs). The most famous example of this is the Antennae (Whitmore & Schweizer 1995), but several other cases were discovered in the early to mid-1990s (e.g., NGC 1275, Holtzman et al. 1992; NGC 7252, Miller et al. 1997). The interpretation of these YMCs as "proto-GCs" was widely adopted. Determining the extent to which these YMCs have properties consistent with "normal" old GCs is still an active area of research; see Section 8 below.

Several problems with the major merger model were pointed out in Forbes, Brodie & Grillmair (1997), who showed that, when examined in detail, the number and color distributions of GCs in massive Es appeared to be inconsistent with the merger model predictions (see Section 11.1.1). They instead suggested that bimodality could arise as a consequence of a multiphase dissipational collapse. In their scenario, the metal-poor GCs were formed in gaseous fragments during the earliest phases of galaxy formation. GC formation was then truncated at high redshift and resumed after a dormant period of a few Gyr. During this second phase the metal-rich GCs and the bulk of the galaxy field stars were formed. Forbes and colleagues discussed this truncation process in terms of feedback, with gas being expelled from the early cluster-forming clumps. Such gas would later cool and recollapse into the more fully-formed galactic gravitational potential until the local conditions were conducive to renewed star formation. Subsequently, Santos (2003) suggested cosmic reionization as the mechanism for truncating metal-poor GC formation. Although some details differ, similar "in situ" models for GC formation were also presented by Harris & Pudritz (1994) and Harris, Harris & Poole (1999).

In the accretion scenario of Côté, Marzke & West (1998), the metal-rich GCs were formed in situ in a massive seed galaxy, whereas the metal-poor GCs were acquired in the dissipationless accretion of neighboring lower-mass galaxies (see also Hilker 1998; Hilker, Infante & Richtler 1999; and earlier work by Muzzio 1987 and references therein). This works in principle because of the long-known relation between the mean metallicity of the GC system and the mass of the host galaxy (van den Bergh 1975; Brodie & Huchra 1991; Forbes & Forte 2001). Stripping of GCs without the accompanying galaxy light is also a possibility in dense clusters. Côté and colleagues (Côté, Marzke & West 1998; Côté et al. 2000; Côté, West & Marzke 2002) explored the accretion model in detail using Monte Carlo simulations. They showed that the bimodality observed in massive galaxies can be reproduced provided that (*a*) each galaxy has an intrinsic "zero-age" population of GCs, whose metallicity increases with the galaxy's mass, and (*b*) the primordial galactic mass function for low-mass galaxies is a rather steep power law (with $\alpha \sim -2$). The slope is consistent with the halo mass functions predicted by standard ΛCDM models, but much steeper than that actually observed for present-day low-mass galaxies (but see Section 11.2).

This triad of scenarios—major merger, in situ/multiphase, and accretion—were the ones most frequently discussed to explain bimodality throughout the past decade. In a very real sense, the distinctions between these models could be blurred by placing the merger or accretion events at high redshift and allowing for significant gas in the components, so to some extent the debate was semantic. Indeed, Harris (2003) noted that these scenarios could be reclassified in terms of the amount of gas involved. Nevertheless, we have presented these scenarios here to give some context for the discussion of observations that follows.

3. GLOBAL PROPERTIES

3.1. Specific Frequency

Harris & van den Bergh (1981) introduced specific frequency ($S_N = N_{GC} \times 10^{0.4(M_V+15)}$) as a measure of the richness of a GC system normalized to host galaxy luminosity. This statistic has been widely used in toy models to assess the feasibility of galaxy formation mechanisms, e.g., the formation of gEs from disk-disk mergers. N_{GC} was originally calculated by doubling the number of GCs brighter than the turnover of the GCLF. This definition makes both practical and physical sense. The faint end of the GCLF is usually poorly defined (suffering both contamination and incompleteness), and $\sim$90% of the mass of the GC system resides in the bright half. This approach is implicit when fitting Gaussian (or t_5) functions to observed GCLFs, because observations rarely sample the faintest clusters.

S_N comparisons among galaxies are only valid if all galaxies have the same stellar mass-to-light (M/L) ratios. For this reason, Zepf & Ashman (1993) introduced the quantity T—the number of GCs per $10^9 M_\odot$ of galaxy stellar mass. Because M/L is not generally known in detail for a particular galaxy, it is usually applied as a scaling factor that is different for each galaxy type, e.g., stellar $M/L_V = 10$ for Es and 5 for Sbc galaxies like the Galaxy. In this section we quote observational results in terms of S_N, but convert to T for comparisons among galaxies. An even better approach would be to directly estimate stellar masses for individual galaxies. Olsen et al. (2004) discuss the use of K-band magnitudes for this purpose.

Soon after the merger model was proposed, van den Bergh (1982) argued that elliptical galaxies could not arise from the merger of spiral galaxies, because spirals have systematically lower S_N values than ellipticals. Schweizer (1987) and Ashman & Zepf (1992) suggested that this problem could be solved if new GCs were formed in the merging process. Observations of large numbers of YMCs in recent mergers seemed to support this solution. However, the S_N will only increase through the merger if the GC formation efficiency is higher than it was when the existing GC system was formed. Studies of YMCs in nearby spirals (Larsen & Richtler 2000) suggest that galaxies with larger star-formation rates have more of their light in young clusters, but it seems unlikely that local mergers are more efficient at forming GCs than in the gas-rich, violent early universe (see Harris 2001).

In recent years there have been few wide-field imaging studies of the sort needed to accurately estimate S_N. Nonetheless, a trend has emerged that suggests reconsideration of some earlier results. Newer S_N values tend to be lower than older ones. This revision stems, for the most part, from improved photometry that is now deep enough to properly define the GCLF turnover and reject contaminants. Imaging studies that cover a wide field of view are also important, because they avoid uncertain extrapolations of spatial profiles from the inner regions of galaxies. In some cases, like the Fornax gE NGC 1399, the luminosity of the galaxy was underestimated. When corrected, the S_N value for this galaxy was revised downward by a factor of two, from $S_N \sim 12$ to 5–6 (Ostrov, Forte & Geisler 1998; Dirsch et al. 2003).

It has been argued for some time that the evolutionary histories of central cluster galaxies are different than other Es of similar mass. Somehow this special status has resulted in high S_N (or T). In addition, galaxies in high-density environments tend to have higher S_N than those in groups. McLaughlin (1999) argued that high S_N values in central galaxies arise because bound hot gas has been ignored for these galaxies, and that they do not reflect an increased GC formation efficiency with respect to other galaxies (see Blakeslee, Tonry & Metzger 1997 for additional arguments in favor of the hypothesis that high S_N is due to large galactic M/L ratios). The properties of the E galaxy NGC 4636 may also be consistent with this view. Despite its relatively low-density environment, it has $S_N \sim 6$, a value typical of central cluster galaxies (Dirsch, Schuberth & Richtler 2005). However, the galaxy has a dark-matter halo (traced by a halo of hot X-ray gas) that is unusually massive for its luminosity.

The classic Harris (1991) diagram of S_N versus luminosity implied a monotonic increase of S_N with galaxy mass for high-mass galaxies; that diagram also showed an increase toward low luminosities for dwarf galaxies. Both of these trends are less apparent in newer data. The situation for dwarf galaxies is discussed in more detail in Section 10, though it is worth noting here that it is difficult to make robust estimates of the number of GCs in low-mass galaxies outside of the Local Group. In our view, the run of S_N with galaxy mass and environment remains uncertain. Additional discussion may be found in McLaughlin (1999) and the reviews by Elmegreen (1999) and Harris (2001).

3.1.1. Subpopulations. Many of the problems with direct S_N comparisons can be circumvented by considering the metal-poor and metal-rich subpopulations separately. In particular, recent mergers should only affect the metal-rich peak; independent of the details of new star formation, the S_N of metal-poor GCs will not increase. Rhode,

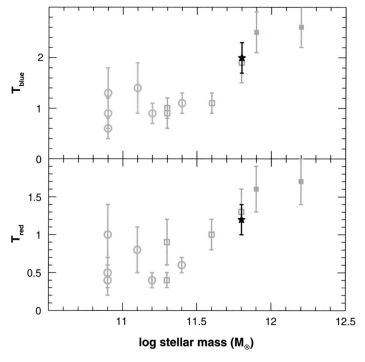

Figure 4

T_{blue} and T_{red} (*top* and *bottom* panels, respectively) versus galaxy mass for a range of spirals and ellipticals (Es) (Rhode, Zepf & Santos 2005). Filled orange squares are cluster Es, open orange squares are field/group Es and S0s, open green circles are field/group spirals, and the black star is the Sa/S0 galaxy NGC 4594. There is a general trend of increasing T_{blue} and T_{red} with galaxy mass (data courtesy of K. Rhode).

Zepf & Santos (2005) have exploited this fact by studying T for metal-poor GCs in 13 massive nearby galaxies, nearly equally split between early- and late-type. In **Figure 4** we show both T_{blue} and T_{red} versus stellar galaxy mass. The T_{blue} values are taken from their paper, and the T_{red} data were kindly provided by K. Rhode (unpublished data). Rhode and colleagues found an overall correlation between T_{blue} and galaxy mass. The spirals are all consistent with $T_{blue} \sim 1$, whereas cluster Es lie higher at $T_{blue} \sim 2$–2.5. NGC 4594 also has $T_{blue} \sim 2$ (note in their classification NGC 4594 is an S0, not an Sa). Other field/group Es, including NGC 5128, NGC 1052, and NGC 3379, have values similar to those of the spirals. Because M/L_V increases with galaxy mass (with a relation as steep as $\propto L^{0.10}$; e.g., Zepf & Silk 1996), it is reasonable to expect at least a weak trend of T with galaxy mass. However, Rhode, Zepf & Santos (2005) argue that this can account for only about one-third of the observed trend. Because no global GC color studies of M87 have been published, this galaxy was not included in the Rhode and colleagues study. Its S_N value is about three times larger than the Virgo gE NGC 4472 (Harris, Harris & McLaughlin 1998). If both galaxies have similar total fractions of metal-poor GCs, the T_{blue} value for M87 would be ~ 8, though this value would be much lower if the mass of its hot gas halo were included along with the stellar mass (McLaughlin 1999).

Based solely on the metal-poor GCs, these comparisons seem to rule out the formation of cluster gEs (and some massive Es in lower-density environments) by major mergers of disk galaxies. However, the relative roles of galaxy mass and environment are still unclear. Spirals in clusters like Virgo and Fornax and more massive field/small group Es must still be studied in detail. The metal-poor GC subpopulations of some lower-mass Es in low-density environments are still consistent with merger formation. The biggest caveat to these interpretations is the effect of biasing—that structure formation is not self-similar. Present-day spirals are mostly located in low-density environments like loose groups and the outskirts of clusters. High-T_{blue} disk galaxies may have been common in the central regions of proto-galaxy clusters at high redshift but have merged themselves out of existence (or could, in some cases, have been converted to S0s) by the present day. Rhode, Zepf & Santos (2005) argue that the observed trend of high T_{blue} for cluster galaxies might be expected in hierarchical structure formation, because halos in high-density environments will collapse and form metal-poor GCs first (see also West 1993). If GC formation is then truncated (by reionization, for example), such halos will have a larger number of metal-poor GCs than similar mass halos in lower density environments. See Section 11 for additional discussion of biasing.

The T_{red} data show a similar correlation with galaxy mass, although with a smaller dynamic range. Note that the T_{red} values for the spirals (∼0.5–1) are normalized to the total galaxy mass in stars. Most have bulge-to-total ratios of <0.3, so if T_{red} had been normalized just to the spheroidal component, it would be substantially higher than plotted. These data appear consistent with the hypothesis of near-constant formation efficiency for metal-rich GCs in both spirals and field Es with respect to spheroidal stellar mass (Forbes, Brodie & Larsen 2001; see also Kissler-Patig et al. 1997). The massive cluster Es have both higher T_{red} and T_{blue}. Again, however, because of the sample under study (with few field Es and no cluster spirals), it is unclear whether environment or mass is the predominant influence. This distinction is moot for the most massive Es because they are found almost exclusively in clusters, but is still relevant for typical Es.

Estimating the spread in T_{red} at a given galaxy mass should help constrain the star-formation histories of early-type galaxies. The stellar mass of an E might have been built entirely through violent, gas-rich mergers (with metal-rich GC formation), or, alternatively, many of the stars could instead have formed quiescently in mature spiral disks (with little metal-rich GC formation). T_{red} for the latter E should be significantly lower than for the former E.

3.2. Radial and Azimuthal Distributions

Most of the existing information on the global spatial distributions of GCs dates from older studies that could not separate GCs into subpopulations. The projected radial distributions are often fitted with power laws over a restricted range in radius, and it is clear that more luminous galaxies have shallower radial distributions (Harris

1986; see the compilation in Ashman & Zepf 1998). Considering the GC system as a whole, typical projected power-law indices range from ~ -2 to -2.5 for some low-luminosity Es (this is also a good fit to the Galactic GC system; Harris 2001) to ~ -1.5 or a bit lower for the most massive gEs. However, it should be kept in mind that power laws provide a poor fit over the entire radial range. Most GC radial distributions have cores, and gradually become steeper in their outer parts. King models capture some of this behavior. In nearly all cases, the GCs have a more extended spatial distribution than the galaxy field stars.

There have been a few wide field imaging programs that considered GC subpopulations separately. In their study of the Virgo gE NGC 4472, Geisler, Lee & Kim (1996) were the first to show clearly that color gradients in GC systems are driven solely by the different radial distributions of the subpopulations. The metal-poor and metal-rich GCs themselves show no radial color gradients. Rhode & Zepf (2001) found a color gradient for the total GC system in NGC 4472 interior to $\sim 8'$, but no gradient when the full radial extent of the GC system (22', i.e., $\sim$110 kpc) was considered. Dirsch et al. (2003) studied the GC system of the Fornax gE NGC 1399 out to $\sim 25'$ ($\sim$135 kpc); this work was extended to larger radii by Bassino et al. (2006). The radial distributions of the two subpopulations are shown in **Figure 5**, along with the profile of the galaxy itself. The metal-poor and metal-rich subpopulations have power-law slopes of ~ -1.6 and ~ -1.9, respectively. These differences persist to large radii, and lead to an overall color gradient in the GC system. The radial

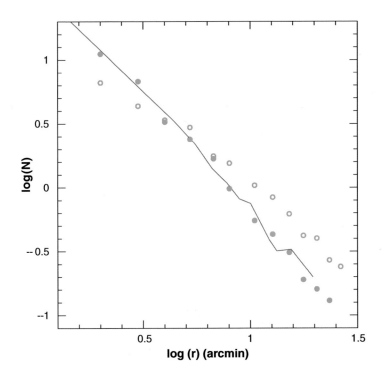

Figure 5

Radial surface density distribution of metal-poor (*open circles*) and metal-rich (*filled circles*) globular clusters (GCs) in the Fornax giant elliptical NGC 1399. The solid line is the scaled galaxy light profile in the R-band. The metal-rich GCs are more centrally concentrated, and closely follow the underlying galaxy light. The radial distribution of the metal-poor GCs is flatter; they dominate the GC system at large radii (Bassino et al. 2006).

distribution of the metal-rich GCs is a close match to that of the galaxy light, and suggests that they formed contemporaneously.

Another notable finding is the rather abrupt truncation of GC systems at a large galactocentric radius. Rhode & Zepf (2001) found that the surface density of GCs in NGC 4472 falls off faster than a de Vaucoleurs or power-law fit at $\sim20'$ (~100 kpc). A similar drop-off occurs at $11'$ in NGC 4406 (Rhode & Zepf 2004), and at $9'$ in NGC 4636 (Dirsch, Schuberth & Richtler 2005). This may be evidence of truncation by the tidal field of the cluster.

Dissipationless mergers (whether disk-disk or E-E) tend to flatten the radial slopes of existing GC systems and create central cores (Bekki & Forbes 2006). These cores are observed and have sizes of a few kiloparsecs (Forbes et al. 1996). These results apply to pre-existing GCs, which certainly include metal-poor GCs, as well as metal-rich GCs that existed before the merger. Thus, the trends predicted by the Bekki & Forbes simulation appear qualitatively consistent with observations. The observations are also consistent with more extensive merger histories for more massive galaxies; these gradually produce larger cores and flatter GC radial distributions. A desirable extension to this work would be to place these simulations in a cosmological framework in which merging occurs according to a full N-body merger tree. Also, as data become available, comparisons between observations and simulations could be restricted to metal-poor GCs. With this approach, any new metal-rich GCs that might have formed in the merger are irrelevant.

Ashman & Zepf (1998) noted that little was known about the two-dimensional spatial distributions of GC systems. Scant progress has been made in the intervening years. Existing data are consistent with the hypothesis that both subpopulations have ellipticities and position angles similar to those of the spheroids of their parent galaxies (e.g., NGC 1427, Forte et al. 2001; NGC 1399, Dirsch et al. 2003; NGC 4374, Gómez & Richtler 2004; NGC 4636, Dirsch, Schuberth & Richtler 2005). This also holds for the dEs in the Local Group (Minniti, Meylan & Kissler-Patig 1996). It may be that in some galaxies (e.g., the E4 NGC 1052; Forbes, Georgakakis & Brodie 2001) the metal-rich GCs follow the galaxy ellipticity more closely than the metal-poor GCs, but whether this is a common phenomenon is unknown. Naïvely, if the metal-rich GCs formed along with the bulk of the galaxy field stars, they should closely trace the galaxy light. The spatial distribution of the metal-poor GCs will depend in detail upon the assembly history of the galaxy.

3.3. Variations with Galaxy Morphology

3.3.1. Spirals. Our views on the GC systems of spiral galaxies are heavily shaped by the properties of the Milky Way (and, to a lesser degree, M31). This is discussed in detail in Section 5. A principal result from the Galaxy is that the metal-poor and metal-rich GCs are primarily associated with the halo and bulge, respectively. Forbes et al. (2001) introduced the idea that the bulge GCs in spirals are analogous to the "normal" metal-rich GCs in early-type galaxies and that in both spirals and field Es the metal-rich GCs have $S_N \sim 1$ when normalized solely to the bulge luminosity. The constancy of bulge S_N appeared consistent with observations of a rather small set of galaxies (Milky Way, M31, M33, NGC 4594) but needed further testing.

Goudfrooij et al. (2003) provided such a test with an HST/WFPC2 imaging study of seven edge-on spirals, ranging across the Hubble sequence (the previously-studied NGC 4594 was part of the sample). Edge-on galaxies were chosen to minimize the effects of dust and background inhomogeneities on GC detection. Corrections for spatial coverage were carried out by comparison to the Galactic GC system; this does not appear to introduce systematic errors (see Goudfrooij et al. 2003 for additional discussion). Kissler-Patig et al. (1999) previously found that one of these galaxies, NGC 4565, has a total number of GCs similar to the Galaxy. The small WFPC2 field of view and the low S_N (or T) values of spirals as compared to Es resulted in the detection of only tens of GCs in some of the Goudfrooij and colleagues galaxies. Although bimodality was not obvious in all of the color distributions, each galaxy had GCs with a range of colors, consistent with multiple subpopulations. Using a color cut to divide the samples into metal-poor and metal-rich GCs, Goudfrooij and colleagues found that all of the galaxies in their study had (*a*) a subpopulation of metal-poor GCs with $S_N \sim 0.5$–0.6 (normalized to total galaxy light), and (*b*) constant bulge S_N, with the exception of the rather low-luminosity Sa galaxy NGC 7814, which appeared to have few GCs of any color. NGC 7814 was later the target of a ground-based study by Rhode & Zepf (2003) with the WIYN telescope. They found a significant metal-rich GC subpopulation with a bulge S_N squarely in the middle of the range found for the other galaxies, which showed that the single WFPC2 pointing on the sparse GC system of NGC 7814 gave an incomplete picture of the galaxy. In this case even small radial leverage was important: Rhode & Zepf found that the surface density of GCs dropped to zero at just 12 kpc (3′) projected.

Chandar, Whitmore, & Lee (2004) studied GCs in five nearby spirals. Their galaxies (e.g., M51, M81) tended to be closer and generally better-studied than those in the Goudfrooij and colleagues sample, but they were also at less favorable inclination angles. As a result, their GC candidate samples were more prone to contamination and more affected by (sometimes unknown amounts of) reddening. This was partially mitigated by their wide wavelength coverage, including (in some cases) U-band imaging, to help constrain the reddening of individual GCs. Chandar and colleagues found evidence for bimodal GC color distributions in M81 and M101. Interestingly, M51, which had rather deep imaging, showed no evidence for metal-rich GCs. They found that NGC 6946 and M101 had subpopulations of clusters with sizes similar to GCs but extending to fainter magnitudes; the typical log-normal GCLF was not seen. In NGC 6946 the imaging was quite shallow and these faint objects were found to be blue, suggesting that they might be contaminants. However, in M101 the imaging was deeper and the faint objects had red colors, consistent with old stellar populations. This may be evidence that some spirals possess clusters unlike those typical of Es, and these clusters may be related to the faint red objects seen in some S0s (see below). Chandar and colleagues also compiled data from the literature on the GC systems of spirals, and argued that S_N/T depends on Hubble type, but not on galaxy mass. This is consistent with the findings of Goudfrooij et al. (2003).

In principle, the S_N value should depend on whether a particular bulge formed "classically," with intense star formation, or through secular processes in quiescent gas disks (e.g., Kormendy & Kennicutt 2004). In the former case we might expect a bulge S_N similar to that found for Es, but in the latter case the star formation is likely to be sufficiently slow and extended that few or no GCs are formed along with the bulge stars. This may result in a rather low bulge S_N compared to Es of similar mass. The implication is that all the bulges of galaxies in the Goudfrooij and colleagues sample formed predominantly through the classical route. The generally old ages of the metal-rich GCs in spirals (see Section 4) is evidence that the majority of bulge star formation, by whatever mechanism, happened at relatively early epochs.

Kormendy & Kennicutt (2004) argue that a large fraction of spiral bulges are built by secular evolution, and that these "pseudobulges" are especially common in late-type spirals. The diagnostics for pseudobulges are many, but include cold kinematics, surface brightness profiles with low Sersic indices, and, in some cases, young stellar populations. It is worth emphasizing that many of these pseudobulges could be composite bulges with young to intermediate-age stars superposed on an old classical bulge. In this case, the bulge S_N could serve as a diagnostic of the degree to which a given bulge can have been built by classical or secular processes. The Milky Way itself could be an example of such a bulge. Its bulge is dominated by an old stellar population, but has a rather low velocity dispersion for its mass. The kinematics of the metal-rich GCs could be consistent with association with either a bulge or a bar (Côté 1999).

3.3.2. S0s. The leading theory for the formation of most S0s involves their transformation from spirals as groups and clusters virialize (e.g., Dressler et al. 1997). This can occur in a variety of ways, including ram pressure stripping and minor mergers that disrupt the disk sufficiently to halt star formation. In this context, it may be more appropriate to compare the GC systems of S0s to those of spirals, rather than make the traditional comparison with Es. Nonetheless, the GC systems of S0s appear to be quite similar to those of Es when compared at fixed mass. Kundu & Whitmore (2001b) studied a variety of S0s with WFPC2 snapshot imaging, and in many galaxies found broad color distributions consistent with multiple subpopulations. Peng et al. (2006b) used deeper imaging from the ACS Virgo Cluster Survey and found color bimodality in nearly all of the massive S0s in their sample. These S0s fall right on the GC color–galaxy luminosity relations of the Es. If indeed S0s descend from spirals, this is yet another piece of evidence that massive galaxies of all types along the Hubble sequence have very similar GC color distributions, and hence are likely to have experienced similar violent formation processes at some point in their history.

One interesting finding so far confined to S0s was the serendipitous discovery of a new class of star cluster, now known informally as the Faint Fuzzies (FFs). These objects were first detected in the nearby (10 Mpc) S0 NGC 1023. Along with a normal, bimodal system of compact GCs, this galaxy hosts an additional population of faint ($M_V > -7$) extended ($R_{eff} \sim 7$–15 pc) star clusters. In deep HST/WFPC2 images, these objects are confined to an annular distribution closely corresponding to the galaxy's isophotes (Larsen & Brodie 2000). Spectroscopic follow-up with Keck/LRIS showed that the FFs are metal-rich ([Fe/H] ~ -0.5), old (>8 Gyr), and rotating in the disk of the galaxy (Brodie & Larsen 2002). With old ages, inferred masses of $\lesssim 10^5 M_\odot$, and sizes about five times larger than a typical globular or open cluster, these objects occupy a distinct region of age–size–mass parameter space for star clusters. As a population, they have no known analogs in the Milky Way or elsewhere in the Local Group. Similar objects have been found in the S0 galaxy NGC 3384 (Brodie & Larsen 2002) and NGC 5195, a barred S0 interacting with M51 (Lee, Chandar & Whitmore 2005; Hwang & Lee 2006). Peng et al. (2006b) found FFs in $\sim 25\%$ of the S0s in the ACS Virgo Cluster Survey. Due to biases in sample selection, however, this fraction is probably not yet well-constrained. The FFs have relatively low surface brightness, and their properties may be consistent with $M \propto R^2$ (unlike GCs, which show no $M - R$ relation). In some cases, their colors are redder than those of the metal-rich GCs in the same galaxy, which may suggest higher metallicities.

Brodie, Burkert, & Larsen (2004) and Burkert, Brodie & Larsen (2005) showed that the properties of the FFs in NGC 1023 were consistent with having formed in a rotating ring-like structure and explored their origin. Numerical simulations suggest that objects with the sizes and masses of FFs can form inside giant molecular clouds, provided star formation occurs only when a density threshold is exceeded. Such special star-forming conditions may be present during specific galaxy-galaxy interactions, in which one galaxy passes close to the center of a disk galaxy, precipitating a ring of star formation. They speculated that the FFs might then be signposts for the transformation of spiral galaxies into lenticulars via such interactions. Alternatively, such conditions might also occur in the inner resonance rings associated with the bars at the centers of disk galaxies. In this case, the old ages of the FFs would suggest that barred disks must have been present at early times.

4. SPECTROSCOPY

The detailed properties of individual extragalactic GCs—ages, metallicities, and chemical abundances—are important constraints on theories of GC and galaxy formation. Emphasis has been placed on accurate age-dating of GCs. To the extent that Es formed in recent gas-rich major mergers, this should be reflected in young age measurements for their metal-rich GCs, while both in situ and accretion scenarios predict old ages for both subpopulations.

In principle, integrated-light spectroscopy of individual GCs offers much stronger constraints on ages and elemental abundances than does broadband photometry. In practice, even low-resolution spectroscopy is challenging. At the distance of Virgo (~ 17 Mpc), GCs are already faint, with the turnover of the GCLF occurring at $V \sim 23.5$. Except within the Milky Way and M31, it has so far been possible to assemble only small samples of high-quality GC data, and these have required significant commitments of 8-m to 10-m telescope time. Historically, spectroscopic samples have been not only small but also biased to the brightest and/or reddest objects in a given galaxy, forcing a reliance on photometric studies for global conclusions about metallicity distributions. As discussed below, this combination of photometry and small-sample spectroscopy has been used to establish that the ubiquitous color bimodality is due primarily to metallicity differences between the subpopulations, without the need to invoke age differences. There are also biases in spatial sampling: the fields of view of the spectrographs used in most studies are too small to sample the outermost GCs, and the GCs in the central regions are generally undersampled due to the minimum length of slitlets and the high central concentration of most GC systems.

There is a fundamental difference between metallicity and age studies in terms of the conclusions that can be derived from small spectroscopic samples. Age substructure can be identified in biased samples of metal-rich GCs, because younger GCs (with a standard GC mass function) will be brighter than their older counterparts. Thus magnitude-limited samples set an upper limit on the proportion of younger GCs in the system. By contrast, to properly sample the metallicity distribution, it is necessary to obtain spectra over a color range representative of the entire system. There is no technical reason why large-sample spectroscopic studies with the new generation of highly-multiplexing spectrographs like Keck/DEIMOS and VLT/VIMOS are not feasible, and such work may proliferate in the near future. In **Figure 6**, we show representative spectra of three GCs in M31: an old metal-poor GC, an old metal-rich GC, and an intermediate-metallicity, ~2 Gyr GC (Beasley et al. 2005).

Most spectroscopic work on extragalactic GCs has utilized Lick/IDS indices (Burstein et al. 1984; Worthey et al. 1994; Trager et al. 1998). These were developed to measure absorption features from ~4000–6400 Å in the spectra of early-type

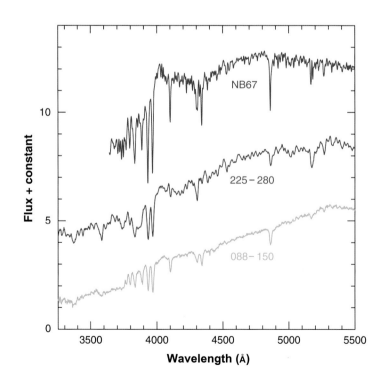

Figure 6

Representative fluxed M31 globular cluster (GC) spectra from Keck/LRIS: from bottom to top, an old metal-poor GC (088–150), an old metal-rich GC (225–280), and an intermediate-age, intermediate-metallicity GC (NB67). The spectrum of NB67 is from Beasley et al. (2005), and has been smoothed with a 5-pixel boxcar. The former two spectra are unpublished data of the authors. NB67 has a metallicity ~ −1 and an age of ~2 Gyr; its combination of strong Balmer and metal lines distinguishes it from the old GCs.

galaxies. Such galaxies typically have large velocity dispersions; thus the low resolution of the Lick system (∼8–11 Å) and wide index bandpasses (tens of Å) required by the IDS (Robinson & Wampler 1972) were not severe impositions. Indices (essentially equivalent widths) are defined in terms of a central bandpass that contains the feature of interest and flanking bandpasses that set the local pseudocontinuum. Observations of distant systems are compared to simple stellar population (SSP) evolutionary models using a set of fitting functions (e.g., Worthey 1994) that predict Lick indices as a function of stellar parameters (e.g., T_{eff}, log g, [Fe/H]). Indices measured on modern spectra must be "corrected" to the Lick/IDS system through observations of standard stars.

This system is not optimal for GCs, which have low velocity dispersions and metallicities compared to galaxies. Although the typical S/N of extragalactic GC spectra has not justified the use of narrower indices until very recently, the opportunity now exists to improve the placement of index and pseudocontinuum bandpasses and to define new indices in the feature-rich wavelength region below 4000 Å. A number of groups are currently defining new index systems on the basis of higher resolution (∼1–3Å) stellar libraries (e.g., the 2.3 Å MILES library; P. Sánchez-Blázquez, R.F. Peletier, J. Jiménez-Vicente, N. Cardiel, J. Falcón-Barroso, et al., submitted), as well as pursuing the direct fitting of models to observed spectra, avoiding indices entirely (e.g., M.J. Wolf, N. Drory, K. Gebhardt & G.J. Hill, submitted).

In the following subsections we discuss observational results on metallicities, ages, and individual elemental abundances of extragalactic GCs.

4.1. Metallicities and Ages

Before proceeding, it is worth noting that the term metallicity is not precisely defined in the context of GCs. Some estimates [e.g., composite principal components analysis (PCA) metallicities; Strader & Brodie 2004] are tied to Galactic GC metallicities on the Zinn & West (1984) scale. This scale measures a nonlinear combination of metals and is unlikely to reflect either [Fe/H] or "true" [Z/H]. The Carretta & Gratton (1997) and the Kraft & Ivans (2003) Fe scales have yet to be extended to the metal-rich regime and calibrations are limited to the metallicity distribution of the Milky Way GC system. This barely touches solar and is poorly matched to the metal-rich regime of GCs in gEs, which extends to solar (and perhaps beyond). The metal-rich GCs in the Galaxy are also few in number and generally highly reddened, as they are concentrated in the bulge. The alternative to direct calibration is to derive metallicities solely from models. A myriad of issues remain with this approach, including uncertain isochrones, meager stellar libraries, corrections to the Lick system, and adjustments for nonsolar abundance ratios (see discussion in Section 4.2). Some authors have adopted the more agnostic [m/H] (with "m" for metallicity) to describe their values; others calculate a "corrected" [Fe/H] from [m/H] or [Z/H] by subtracting an assumed or derived [α/Fe] (e.g., Tantalo, Chiosi & Bressan 1998). Direct comparisons of metallicities derived from different methods can easily be uncertain at the ∼0.2–0.3 dex level.

One method to estimate metallicities for GCs is to simultaneously use multiple line indices; a weighted combination of six indices was used to derive metallicities for individual GCs in the Milky Way and M87 (Brodie 1981; Brodie & Hanes 1986). Brodie & Huchra (1990, 1991) applied a refined version of this approach to GCs in a variety of galaxies. Two other studies derived metallicity calibrations using principal components analysis of indices of Galactic GCs: Gregg (1994) used a large set of narrow indices, and Strader & Brodie (2004) used Galactic GC spectra from Schiavon et al. (2004) to derive a metallicity estimator that is a linear combination of 11 Lick indices. The PCA methods provide accurate metallicities on the Zinn & West (1984) scale in the range $-1.7 \lesssim$ [Fe/H] $\lesssim 0.0$. The other principal method is to simultaneously derive metallicity and age through comparisons to SSP models (see discussion below).

The bulk of spectroscopic studies of extragalactic GCs have been conducted with two instruments, Keck/LRIS and VLT/FORS. Ages, metallicities, and [α/Fe] ratios have been estimated predominantly by measuring Lick indices.

In a Keck/LRIS study, Kissler-Patig et al. (1998) inferred that the majority of GCs in the Fornax gE NGC 1399 were old, but that a small percentage might be as young as a few Gyr. However, the robustness of these conclusions was hampered by the low S/N of the spectra. In another Keck/LRIS program, Cohen, Blakeslee & Ryzhov (1998) studied a sample of ~150 GCs in M87. Despite large errors on the absorption line indices measured in many of their spectra, the size of the sample allowed a good statistical comparison with spectra of Milky Way GCs. The M87 and Milky Way GCs were found to populate similar areas in Mgb versus Hβ and Mgb versus Fe5270 diagrams, suggesting that the M87 GCs have ages and [Mg/Fe] ratios comparable to those of the Galactic GCs. These ages and [α/Fe] ratios are known, from detailed analysis of individual cluster stars, to be >10 Gyr and ~ +0.2–0.3, respectively (e.g., Carney 2001). Beasley et al. (2000) found old ages for coadded (on the basis of $C - T_1$ color) WHT spectra of GCs in NGC 4472. Kuntschner et al. (2002) presented a study of the S0 NGC 3115 showing that GCs in both the metal-poor and metal-rich subpopulations are old.

Subsequent studies undertaken with Keck/LRIS include: Ellipticals—NGC 1399 (Forbes et al. 2001); NGC 4472 (Cohen, Blakeslee & Côté 2003); NGC 4365 (Larsen et al. 2003; Brodie et al. 2005); NGC 3610 (Strader et al. 2003; Strader, Brodie & Forbes 2004a); NGC 1052 (Pierce et al. 2005); NGC 1407 (A.J. Cenarro, J.P. Brodie, J. Strader & M.A. Beasley, in preparation); S0s—NGC 1023 (Brodie & Larsen 2002); NGC 524 (Beasley et al. 2004); Spirals—M81 (Schroder et al. 2002); NGC 4594 (Larsen et al. 2002a). In all cases it was concluded that at most a small fraction of the observed samples of GCs were young or of intermediate age ($\lesssim$5–6 Gyr). In the case studies of the intermediate-age merger remnant E NGC 3610, Strader et al. (2003b) and Strader, Brodie & Forbes (2004a) found that only two out of ten prime candidate intermediate-age GCs had ages consistent with formation in the merger. **Figure 7** shows a typical index-index plot used to derive metallicities and ages for GCs, in this case, from NGC 1407 (A.J. Cenarro, J.P. Brodie, J. Strader & M.A. Beasley, in preparation). Here the GCs studied appear to have ages consistent with Galactic GCs, but extend to higher metallicities.

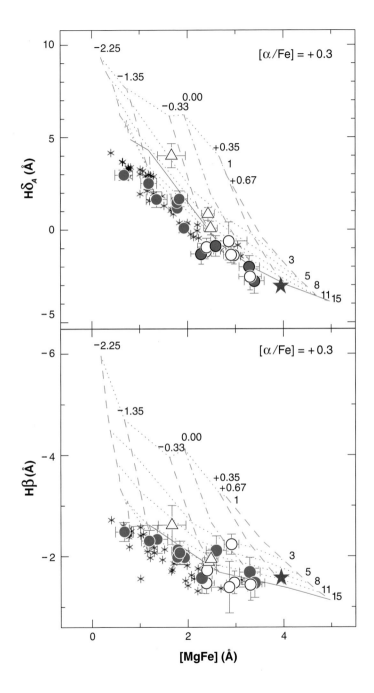

Figure 7
Hδ_A and Hβ versus [MgFe]′ age-metallicity index-index plots for globular clusters (GCs) in NGC 1407 (*circles* and *triangles*; A. J. Cenarro, J. P. Brodie, J. Strader & M. A. Beasley, in preparation) and Galactic GCs (stars, from Schiavon et al. 2004). The blue filled circles are "normal" GCs, the open circles have anomalously high [Mg/Fe] and [C/Fe], and the open triangles are GCs with enhanced Balmer lines (probably due to blue horizontal branches, but younger ages are also a possibility). The large red star is a central $r_e/8$ aperture for NGC 1407 itself. The overplotted grids are Thomas, Maraston, & Korn (2004) models with [α/Fe] = +0.3 and the ages and metallicities indicated. The model grid lines cross at old ages and low metallicities, making exact age estimates impossible.

In an effort to use more of the information available in a spectrum than is contained in a traditional index-index plot, Proctor, Forbes & Beasley (2004; see also Proctor & Sansom 2002) developed a χ^2 minimization routine that simultaneously fit all of the Lick indices to SSP models, exploiting the different sensitivities of each of the indices to age, metallicity, and [α/Fe]. Metallicities derived in this manner are complementary to ones derived using the PCA methods described above, and, for GCs in the range over which the PCA metallicities are calibrated, the agreement is excellent (e.g., A.J. Cenarro, J.P. Brodie, J. Strader & M.A. Beasley, in preparation). Multi-index approaches offer the advantage of minimizing random and systematic problems in any individual index, such as may arise in low S/N data. Moreover, where the GC background is an emission line region (from either sky or the host galaxy), or when individual element abundance anomalies or extreme horizontal branch morphologies may be present in the GC, such precautions are essential (see below).

The best of the Keck data were compiled in a meta-analysis by Strader et al. (2005). This work synthesized more than a decade's worth of effort on the GCs systems of a heterogeneous sample of galaxies, spanning the range from dwarfs to gEs (all but NGC 4594 were of early-type). A typical galaxy had only 10–20 spectra of sufficiently high S/N to be included in the analysis. Lick indices measured from these spectra were combined, and ages were derived through direct differential comparison to indices of Galactic GCs, in part to avoid SSP model uncertainties. This study showed that both the metal-poor and metal-rich GC subpopulations had mean ages as old as (or older than) Galactic GCs. In **Figure 8**, from Strader et al. (2005), PCA metallicity is plotted against a composite Balmer line index of Hβ and Hδ_F (Hγ was excluded because the strong G band lies in its blue pseudocontinuum bandpass). The implication from their sample of galaxies is that GC formation, and by extrapolation, the bulk of star formation in spheroids, took place at early times ($z > 2$). As discussed by Renzini (2006, in this volume), fossil evidence from the local universe and in situ studies at high redshift lead to similar conclusions: Most of the stars in massive Es formed at $z \sim 2$–5. Of course, studies of larger samples of GCs in a wide range of galaxies are needed to further test this result.

This conclusion is still consistent with the existence of young early-type galaxies in the local universe (Trager et al. 2000), because galaxy age estimates are luminosity-weighted. A small "frosting" of recent star formation can easily result in the measurement of young spectroscopic ages for galaxies whose stars are predominately old, and there is an inverse relationship between the age and required burst strength. In addition, radial age gradients in early-type galaxies are often measured. Younger ages are derived from spectra extracted from small (1/8 of the effective radius; r_e) apertures than from larger ($r_e/2$) ones. In any case, a small amount of star formation is expected to produce a correspondingly small subpopulation of young GCs, which could easily have escaped detection in existing samples. High quality spectra of >100 GCs per galaxy, now obtainable with highly multiplexing instruments such as Keck/DEIMOS and VLT/VIMOS, would be necessary to probe these low-level star-forming events.

Puzia et al. (2005; see also Puzia 2003) presented a spectroscopic analysis of a more homogeneous sample of galaxies, comprising 5 Es and 2 S0s, all with $-19.2 > M_B > -21.2$, and mostly in small groups. The low end of this luminosity range is near L^*,

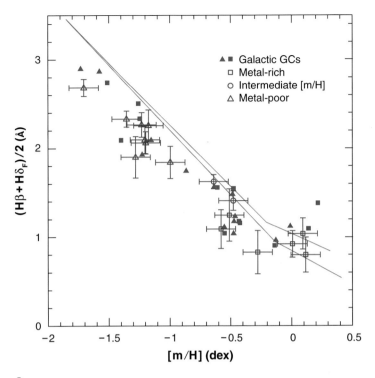

Figure 8
Combined Balmer index versus metallicity for extragalactic globular cluster (GC) subpopulations and Galactic GCs (Strader et al. 2005). The extragalactic subpopulations are plotted as open red triangles (metal-poor), open red circles (intermediate-metallicity), and open red squares (metal-rich), with the subpopulations defined from broadband photometry of the GC systems. Individual Galactic GCs are plotted as filled blue triangles (data from Gregg 1994) and filled blue squares (data from Puzia et al. 2002a). At all metallicities, the extragalactic subpopulations appear coeval with (or older than) the comparison Galactic GCs. This suggests mean ages >10–13 Gyr for the extragalactic GCs. 14 Gyr model lines from Thomas, Maraston & Korn (2004) are superposed ([Z/H] = −2.25 to 0, with [α/Fe] = 0 on bottom and +0.3 on top). These include a blue horizontal branch below [Z/H] = −1.35. At fixed metallicity, older ages lie at weaker Balmer line strength. The data are not fully calibrated to the models so cannot be directly compared, and the offset between the data and the model lines is not significant.

so their sample, like that of Strader et al. (2005), was dominated by early-type galaxies brighter than the knee of the galaxy LF. They included a detailed discussion of the advantages of the different Balmer lines in terms of dynamic range, age-metallicity degeneracy, and ability to correct to the Lick system. Using a combined Balmer line index and the α-insensitive metallicity proxy [MgFe]′ (Thomas, Maraston, & Bender 2003), Puzia and colleagues found 4 GCs (out of their total of 17 high-quality spectra) with index measurements consistent with ages less than 9 Gyr (their **Figure 7**). Although this sample is not sufficient to address the proportion of young GCs that might be present in any individual galaxy, the presence of some younger GCs in these typically group galaxies would be consistent with "downsizing," whereby lower-mass galaxies in lower-density environments tend to have younger mean ages (Cowie et al. 1996). There remains disagreement on the proportion of young GCs present in galaxies of moderate luminosity.

The evidence that most GCs in massive galaxies are quite old is relevant to the distinction between "core" and "power-law" E galaxies seen in the local universe. These designations stem from the surface brightness profiles in the very inner parts of the galaxy. The core galaxies tend to be bright ($M_V \lesssim -21.5$), to have boxy isophotes, and not to rotate; the power-law galaxies are fainter, have disky isophotes, and are at least partially rotationally supported (Kormendy & Bender 1996; Faber et al. 1997). The working hypothesis for the creation of central cores is scouring by merging binary black holes. J. Kormendy, D.B. Fisher, M.E. Cornell & R. Bender (in preparation) have unified this scenario by demonstrating that power-law galaxies have excess light in their central parts, above an extrapolation of a best-fit Sersic profile in the outer parts. Core galaxies have no such excess light. They argue that the distinction between core and power-law galaxies represents Es whose last major merger was dry or wet, respectively (i.e., without or with star formation). In power-law galaxies, black hole scouring may still have occurred, but it is swamped by the light from the stars in the center that formed in the merger. The old GC ages inferred from observations tentatively indicate that, even in power-law galaxies, the wet merger that formed the E either occurred long ago, or produced only a relatively small amount ($\lesssim$10–20%) of the galaxy's current stellar mass. However, much larger spectroscopic samples of GCs in power-law galaxies are needed to better constrain the fraction of mass produced in wet mergers.

4.1.1. Uncertainties. An important uncertainty in age determinations is the level of contribution from hot stars to the integrated spectra of GCs. These can "artificially" enhance the Balmer line strengths beyond the values set by the main sequence turnoff and can lead to an underestimation of the GC age. Blue stragglers and blue horizontal branch stars are the primary offenders—hotter stars, including extreme horizontal branch (HB), AGB-manque, and post-AGB stars have little flux in the optical region where ages are usually derived. At least in the Milky Way, blue stragglers have quite a small effect on a GC's optical integrated spectrum (e.g., Schiavon et al. 2002). It is worth noting that this may not necessarily be the case in extragalactic GCs that have high binary fractions or are very compact.

This leaves blue HB stars as the main systematic source of uncertainty in GC age estimates. HB morphologies are primarily set by GC metallicity, with Balmer line strengths peaking at [Fe/H] ~ -1.3 and decreasing toward lower metallicities as the stars become hotter and H is ionized. The classic "second-parameter problem"—that GCs of a given metallicity can have quite different HB morphologies—is currently unresolved, despite intensive research efforts over several decades. It has frequently been suggested that age is the second parameter (e.g., Chaboyer, Demarque & Sarajedini 1996). However, this does not appear to be true in the few cases where direct turnoff age comparisons can be made by normalizing the main sequence luminosity functions (e.g, Stetson, Vandenbergh, & Bolte 1996). Although age differences certainly will produce changes in HB morphology, it is dangerous to use the HB as an age indicator. As an example, several metal-poor GCs in M33 were studied with HST/WPFC2 by Sarajedini et al. (2000) and found to have unusually red HBs for their metallicities. Sarajedini and colleagues suggested intermediate ages might be the cause. However, Larsen et al. (2002b) derived dynamical masses for these GCs and instead found that they had M/L ratios typical of old GCs. At present there is no known a priori predictor of GC HB morphology. Consequently, various HB diagnostics have been suggested. These include (a) the Ca H + K line strength inversion; enhanced H ϵ from hot stars can increase the Ca H line equivalent width compared to the Ca K line (Rose 1985), (b) near-ultraviolet ($NUV - V$) colors, and (c) a tendency for progressively younger ages to be inferred from increasingly higher-order Balmer lines because of an increasing contribution from putative blue HB stars (Schiavon et al. 2004). The effect of BHB stars on integrated Balmer line strength can be substantial: Puzia et al. (2005) estimated that $H\beta$ may be increased by up to 0.4 Å (based on their analysis of Galactic GCs), while Maraston (2005) inferred an even greater effect from her SSP models at high metallicities (see also Thomas, Maraston & Korn 2004).

Aside from the classic second-parameter effect, there is some evidence that the HB morphologies of GCs in other galaxies can be quite different from those in the Milky Way. For example, in a HST/WFPC2 study by Rich et al. (2005), a sample of 12 M31 GCs appeared offset from the Milky Way HB–metallicity relation. One potential explanation is that these M31 GCs are $\sim$1–2 Gyr younger than their Galactic counterparts. More startling is the finding by Sohn et al. (2006), from an HST/STIS FUV imaging study of bright M87 GCs, that all the GCs have much bluer $FUV - V$ colors than Galactic GCs of the same metallicity. Increasing the age of the M87 GCs relative to Galactic GCs is one potential solution, but it would require implausibly large age differences of 2–4 Gyr to bring the two samples into agreement. Fuel consumption arguments, and the fact that spectroscopic ages of these GCs are old (Cohen, Blakeslee & Ryshov 1998) suggest that the hot stars in M87 GCs are a redistribution of stars blueward of a normal BHB, and not simply an extension of it. A caveat to this finding is the UV-flux limited nature of their sample, which tends to select the most extreme GCs.

A night of 8–10-m telescope time can generate spectra of GCs at the distance of Virgo ($\sim$17 Mpc) from which $H\beta$ line index strengths can be measured to an accuracy of $\sim$0.1 Å for GCs with $V \sim 21.5$ (for fainter GCs, the errors are correspondingly larger). As **Figure 7** illustrates, this translates into an error of $\sim$1–2 Gyr at old ages. Because the isochrones are more widely separated at younger ages, the ability to discriminate fine age differences improves with decreasing age. In these SSP model grids, BHB effects actually cause the isochrones to cross at low metallicities and old ages, setting a fundamental upper limit on our age estimates of $\sim$10 Gyr at these metallicities. It is important to remember that the actual ages of the majority of GCs in the universe could well be older than this.

4.2. [α/Fe]

It is widely recognized that measurements of the [α/Fe] ratio of a stellar population can provide valuable insight into its star-formation history, particularly by constraining the timing and duration of a starburst. Most α-elements (e.g., Mg, Ti, Ca, Si) and one third of the Fe (for the solar mixture) are generally thought to form in Type II supernovae. Because these come from the explosions of massive stars, they closely trace the star-formation rate, and begin to ignite within tens of Myr of the onset of a starburst. The remainder of the Fe is produced in Type Ia supernovae, which typically occur over Gyr timescales. Enhanced [α/Fe] ratios would thus indicate rapid star formation, occurring before there was time for significant enrichment of the interstellar medium by Type Ia supernovae. This is generally the case in early-type galaxies (e.g., Worthey, Faber & Gonzalez 1992; Matteucci 1994; Trager et al. 2000; Thomas et al. 2005). A complication is that the timescale over which fresh ejecta become incorporated into interstellar gas is not necessarily known. Other nucleosynthetic sites, operating on a variety of timescales, also contribute to the cosmic mix (e.g., AGB stars produce a significant amount of s-process and light elements).

The naïve expectation for metal-poor GCs in external galaxies is that they will have supersolar [α/Fe], because they formed early in the universe, before there was substantial metal enrichment. The situation for metal-rich GCs is less clear, as it will depend upon the exact formation mechanism. If they formed along with most of the field stars in massive Es, they should share the [α/Fe] $\sim$ +0.2–0.4 of their parent galaxies. They could perhaps reach even higher values if they formed preferentially early in the starburst. Metal-rich GCs formed at lower redshift from approximately solar metallicity gas in major disk-disk mergers might be expected to have [α/Fe] closer to 0, because it is more difficult to enhance [α/Fe] when starting from a high metallicity (although metallicity gradients in spiral disks must also be considered). The results from extragalactic GCs do not generally agree with these expectations, though much of this may be due to the observational and technical difficulties in accurately estimating [α/Fe], as discussed below.

In theory, it should be very easy to determine [α/Fe] from low-resolution GC spectra. Mg is the element of choice because of the strong Mgb triplet and MgH bandhead in the optical. Trager et al. (2000) described a method for estimating [Mg/Fe] using models by Worthey (1994) and corrections from Trippico & Bell (1995). However, the widespread estimation of this quantity in extragalactic GCs did not occur until Thomas, Maraston & Bender (2003) published the first models to explicitly include the effects of α-enhancement.

In practice, the interpretation of the observations using these models has been complicated. The GCs in the metal-rich subpopulations in a variety of massive galaxies have been found to have [α/Fe] ranging from 0 to $\sim$ +0.3 (e.g., Kuntschner et al. 2002; Beasley et al. 2005) or higher (Puzia et al. 2005 find $\sim$ +0.45 but with large scatter). By contrast, the metal-poor GCs frequently appear to have [α/Fe] $\sim$ 0 or even lower (e.g., Olsen et al. 2004; Pierce et al. 2005), although the convergence of SSP model lines at low metallicities increases the errors on these determinations of [α/Fe]. If [α/Fe] for metal-poor GCs were indeed this low, it would be quite surprising. In the simple view of nucleosynthesis outlined above, supersolar [α/Fe] results from star formation over "short" (<1 Gyr) timescales, whereas solar or subsolar [α/Fe] indicates very extended star-formation histories. Thus we expect the metal-poor GCs to have high [α/Fe].

Galactic GCs from both subpopulations have [α/Fe] $\sim$ +0.3 (e.g., Carney 1996) or perhaps a little lower for some metal-rich bulge GCs. These values are quite closely reproduced by the popular Thomas, Maraston & Bender (2003) SSP models for the Galactic GC data of Puzia et al. (2002a), although the results from applications to extragalactic GCs are mixed (as described above). Current stellar libraries and the Worthey (1994) fitting functions utilize few stars with metallicities in the range of metal-poor GCs, and SSP models may be more uncertain in this regime.

R. Schiavon (submitted) gave a good summary of the issues involved in creating nonsolar [α/Fe] models. These include (*a*) the importance of employing the proper isochrones and luminosity functions, and (*b*) the need to correct the index predictions of the fitting functions. Item (*b*) is usually carried out differentially, using stellar models for stars in representative parts of the CMD. A preliminary attack on this problem by Trippico & Bell (1995), using just three stars, produced broadly similar results to the more thorough treatment of Korn, Maraston & Thomas (2005). Korn and colleagues also discussed in detail the elemental sensitivities of the individual Lick indices.

A potentially confusing difference among models is the treatment of [α/Fe] versus [Fe/H]. Thomas, Maraston & Bender (2003) calculated their models at fixed total metallicity ([Z/H]) and produced supersolar [α/Fe] by lowering [Fe/H], because the dominant component of Z is the α-element O. By contrast, R. Schiavon (submitted) calculated models at fixed [Fe/H]. Schiavon's approach has the benefit of a direct link to the measurable quantity [Fe/H].

These SSP uncertainties, when combined with the wide Lick index bandpasses, which always admit contributions from other elements in addition to the targeted feature, make most current estimates of [α/Fe] untrustworthy. We suggest intensive study of nearby GC systems (e.g., M31), where potentially more accurate results from high-resolution integrated spectroscopy of GCs can be directly compared to low-resolution spectra.

4.3. Abundance Anomalies

In addition to the α-elements, GCs show a variety of abundance anomalies with respect to the solar mixture. This is most clear in the Galaxy, where detailed study of individual stars in GCs is possible (see Gratton, Sneden & Carretta 2004). In extragalactic GCs, the most obvious anomaly is CN-enhancement. This was first reported in M31 GCs by Burstein et al. (1984), and, at least in this galaxy, appears to be due to an excess of N above even the levels in Galactic GCs (see Section 6.1), which themselves are enhanced in N over the solar mixture. Trager (2004) has argued that a similar CN anomaly (presumably due to N) is present in many GC systems, including NGC 3115 (Kuntschner et al. 2002), the old GCs in NGC 3610 (Strader et al. 2003b), and the Fornax dSph (Strader et al. 2003a). It is also present in GCs in the gE NGC 1407 (A.J. Cenarro, J.P. Brodie, J. Strader & M.A. Beasley, in preparation). We must consider very high N abundances to be a generic feature of the early chemical evolution of GC systems (and perhaps their host galaxies). Interestingly, of the two confirmed metal-rich intermediate-age GCs in NGC 3610, one shows the CN anomaly, and one does not.

5. NEAR-IR IMAGING

A new tool for studying GCs emerged during the past few years with the advent of large-format infrared detectors. Kissler-Patig (2000) pointed out that it should be possible to use near-IR (NIR) photometry of GCs to break the age-metallicity degeneracy inherent in optical colors. The $V - K$ color of an old GC is a measure of the temperature of its red giant branch. This temperature is strongly dependent on metallicity but has little sensitivity to age. The prospect of large NIR imaging surveys is exciting because it offers the potential to accurately measure individual ages and mean metallicities for large numbers of GCs in galaxies out to the distance of the Virgo cluster and beyond.

Kissler-Patig and collaborators, primarily using VLT/ISAAC, have taken the lead in such NIR studies. Puzia et al. (2002b) reported a comparison in VIK between the GC systems of the group S0 NGC 3115 and the Virgo E NGC 4365. The GC system of NGC 3115 appeared bimodal in color-color space, whereas the color distribution of GCs in NGC 4365 looked largely unimodal, with a formal peak at supersolar metallicity and intermediate age (∼2–5 Gyr). However, the presence of a large subpopulation of intermediate-age GCs was confusing, as the central light of the galaxy itself is uniformly old (Davies et al. 2001). How could a large number of GCs form without an accompanying field star component? This puzzling result precipitated a flurry of activity on the GC system of NGC 4365. Spectroscopy of a subset of the candidate intermediate-age GCs seemed to confirm their young ages, while demonstrating that other young candidates (selected from their optical/NIR colors) were in fact old (Larsen et al. 2003).

However, follow-up spectroscopy with somewhat higher S/N found no evidence for any intermediate-age clusters. Moreover, when considered in combination with new HST/ACS imaging, these data seemed to point instead to three subpopulations of old GCs, with a very centrally-concentrated, intermediate-metallicity subpopulation filling in the gap between the normal metal-poor and metal-rich GCs (Brodie et al. 2005). A similar intermediate-metallicity subpopulation has been discovered by M.A. Beasley, E.W. Peng, T. Bridges, W.E. Harris, G.L.H. Harris, et al. (in preparation) in the gE NGC 5128. This interpretation is consistent with widerfield K-band photometry obtained by Larsen, Brodie & Strader (2005). These researchers discussed all of the available data and the details of the error estimates, and showed evidence for systematic errors in optical-NIR SSP models. Nonetheless, based on a comparison of HST/NICMOS H-band photometry of 70 GCs in NGC 4365 and 11 GCs in NGC 1399 with SSP models, Kundu et al. (2005) concluded that a large number of intermediate-age (2–8 Gyr) GCs with metallicities up to [Fe/H] = +0.4 are present in NGC 4365. Clearly, the last word has not yet been spoken on this intriguing galaxy.

Hempel et al. (2003) studied the early-type galaxies NGC 5846 and NGC 7192 with NIR photometry, reporting that the former galaxy hosts a large population of intermediate-age GCs. Follow-up U-band photometry in NGC 5846 and NGC 4365 (Hempel & Kissler-Patig 2004) was argued to support the earlier results, although the number of GCs detected was low. Follow-up spectroscopy does not appear to support the presence of a large subpopulation of intermediate-age GCs in NGC 5846 (Puzia et al. 2005). The fraction of GCs with ages formally <8 Gyr in the observed sample of luminous GCs is 15–20% (T. Puzia, private communication).

Goudfrooij et al. (2001a) presented the best evidence to date of a significant population of intermediate-age GCs in any postmerger galaxy. They used a combination of HST and ground-based optical and NIR photometry of GCs in NGC 1316, a merger remnant E in the Fornax cluster. They found metal-rich GCs that were substantially brighter than normal GCs at that distance and deduced photometric ages of $\sim$3 Gyr for these bright red objects. These photometric ages were confirmed by spectroscopy for a small number of the brightest clusters (Goudfrooij et al. 2001b). However, even in this galaxy, the strongest pieces of evidence for an intermediate-age subpopulation are (*a*) the presence of unusually luminous GCs and (*b*) the power-law (rather than log-normal) GC luminosity function of the metal-rich GCs (see also Goudfrooij et al. 2004).

Improvements in NIR SSP models will brighten the prospects for future NIR studies and help this field to reach its full potential. In addition, the launch of the *James Webb Space Telescope* (JWST) in the next decade will allow the collection of high-quality K-band photometry for many GC systems.

6. GLOBULAR CLUSTER–FIELD STAR CONNECTIONS

Other than the Milky Way, the GC systems of only two massive galaxies have been observed in detail, in the sense that a significant number of their GCs have been observed with high S/N spectroscopy and we can study their field stars directly. M31, our sister spiral galaxy in the Local Group, and NGC 5128, one of the nearest massive E galaxies, are both sufficiently well-studied that they can elucidate the role of GCs in tracing star formation in their parent galaxies.

6.1. M31

Though the total mass of M31 is similar to that of the Galaxy (Evans & Wilkinson 2000), it has a GC system about three times as large, with $\sim$400–450 GCs (van den Bergh 1999; Barmby & Huchra 2001). A comprehensive catalog of photometric and spectroscopic data for M31 GCs was assembled by Barmby et al. (2000) and used to show that, broadly speaking, the M31 GC system is quite similar to that of the Galaxy; M31 has an extended halo subpopulation and a more centrally-concentrated rotating bulge/disk subpopulation.

The kinematics of the M31 GC system may be complex. Perrett et al. (2002) found that the metal-rich GCs have a velocity dispersion similar to the bulge ($\sim$150 km/s) and a rotational velocity of 160 km/s. Surprisingly, they also found a large rotational velocity for the metal-poor GCs, ~130 km/s. By contrast, the metal-poor GCs in the Galaxy show little evidence for rotation. Visually, the spatial distribution of the Perrett and colleagues metal-poor GC sample appears to be the superposition of a centrally-concentrated spherical population and a relatively thin disk component. Indeed, Morrison et al. (2004) used the Perrett and colleagues velocities to argue that M31 possesses a rapidly rotating "thin disk" of old metal-poor GCs. Beasley et al. (2004) presented high-S/N spectra of a small number of these putative old metal-poor GCs and showed that they have spectra more similar to young (<1 Gyr), approximately solar metallicity objects. Their luminosities suggest stellar masses comparable to massive Galactic open clusters or low-mass GCs. The presence of young GCs has also been claimed by Burstein et al. (2004) and Puzia, Perrett & Bridges (2005) (both based in part on data from Beasley et al. 2004), and by Fusi Pecci et al. (2005). Cohen, Matthews & Cameron (2005) used Keck NIR/AO imaging to show that four of these very young clusters are actually "asterisms"—either chance groupings of bright stars or a few brighter stars superimposed upon sparse open clusters. A more detailed examination of archival HST imaging indicates that perhaps half of the candidate young clusters are real; previous HST/WFPC2 had confirmed the existence of four such clusters (Williams & Hodge 2001). A high-resolution imaging survey of the disk will be needed to assess the true number of young GCs. In any case, if these interloping objects (asterisms and/or YMCs) are removed from the metal-poor GC candidate list, much of the rotational signature observed by Perrett and colleagues would be erased. The apparent lack of an old metal-poor thin disk of GCs is important, because the existence of this disk would rule out any significant merger having taken place in the last ~10 Gyr. As discussed below, such a merger is the working theory for the existence of intermediate-age stars in M31's halo.

Due to selection effects, including reddening in the inner disk and the paucity of GC candidates at large radii, the overall fraction of metal-poor to metal-rich GCs is still poorly constrained in M31. From their kinematic and metallicity study, Perrett et al. (2002) found that only ~25% of a sample of ~300 GCs belong to the metal-rich (bulge) peak, compared to one third in the Galaxy. If, as discussed above, many of the candidate metal-poor GCs in M31 are interlopers, then the ratio of metal-poor to metal-rich GCs in M31 could be quite consistent with that found for the Galaxy (2:1). There are several ongoing studies of the M31 GC system with MMT/Hectospec that should provide a clearer picture of this situation.

Burstein et al. (1984) found evidence for Balmer line anomalies in M31 GCs compared to GCs in the Milky Way. Their preferred explanation was that the M31 GCs might be younger than their Milky Way counterparts. Brodie & Huchra (1991) and Beasley et al. (2004) compared integrated-light spectra of Milky Way and M31 GCs at the same metallicity and found no evidence for enhanced Balmer lines in M31 GCs. Beasley et al. (2005) extended these results by fitting the Lick indices measured from their sample GCs to stellar population models. They employed the multi-index χ^2 minimization method developed by Proctor, Forbes, & Beasley (2004) to derive ages and metallicities for these clusters. In addition to the normal old metal-poor and metal-rich GC subpopulations, they found a group of six GCs with intermediate metallicities and ages of 3–6 Gyr. A larger number of GCs with similar ages were also reported by Puzia et al. (2005). This group of GCs forms a coherent chemical and kinematic group of objects in M31, consistent with the accretion of an SMC-type galaxy within the last several Gyr (in the Galaxy, there are several comparable GCs associated with the Sgr and CMaj dSphs; Section 10). Ashman & Bird (1993) had previously analyzed M31 GC kinematics and found evidence for about seven distinct groups of GCs. The clustering analysis of Perrett et al. (2003) found about eight such groups, but Monte Carlo simulations indicated that statistically half or more of these should be chance groupings. In an addition to the rapidly growing zoo of stellar clusters, Huxor et al. (2005) have identified three metal-poor objects in the halo of M31 that have luminosities typical of GCs, but sizes from 25–35 pc. They suggest that these could be stripped nuclei of dwarfs, analogous to ω Cen in the Galaxy (Majewski et al. 2000).

Brown et al. (2004) obtained extremely deep HST/ACS photometry of the M31 GC 379-312 ([Fe/H] ~ -0.6) and through isochrone fitting derived a formal age of $10^{+2.5}_{-1}$ Gyr, perhaps ~ 1 Gyr younger than Galactic metal-rich GCs. Rich et al. (2005) carried out photometry deeper than the HB for a sample of 12 M31 GCs. They found the classic second parameter effect, but with an offset that might be explained if the GCs were ~ 1–2 Gyr younger than their Galactic counterparts. These results hint at an interesting relative age difference between the two GC systems, but hard evidence will be difficult to come by. It would require a UV/optical space telescope with an aperture larger than HST to derive turnoff ages for significant numbers of M31 GCs.

Burstein et al. (1984) also discovered CN enhancements in M31 GCs (with respect to Galactic GCs). This result was later confirmed by Brodie & Huchra (1991), who pointed out that CH at 4300 Å did not appear to be anomalous, and argued that the observed CN enhancement might be due to an overabundance of N. Beasley et al. (2004) did not see the enhancement clearly in the Lick CN_2 index, which measures the weaker 4215 Å bandhead, but it is obvious in a different index sensitive to the 3883 Å transition. That N is the culprit in CN variations was confirmed directly by near-UV spectra of Galactic and M31 GCs around the NH band at 3360 Å (Burstein et al. 2004). Observations in this region of the spectrum have been limited by the low fluxes of GCs and the low UV efficiencies of telescopes, instruments, and detectors. As a result, the SSP modeling effort has proceeded slowly, and quantitative [N/Fe] values do not yet exist. Thomas, Maraston & Bender (2003) found that [N/Fe] $\sim +0.8$ was required to match CN in some Galactic GCs; in M31 GCs it must be even higher. Because NH (in the UV) is dominated by light from stars around the main sequence turnoff, the N enhancements are probably primordial in origin and cannot be attributed to noncanonical mixing processes on the red giant branch.

M31's stellar halo, studied in fields from ~ 10–30 projected kpc, appears predominately metal-rich ([m/H] ~ -0.5), a full 1 dex higher than the Milky Way (e.g., Mould & Kristian 1986; Durrell, Harris & Pritchet 1994). In one M31 halo field, Brown et al. (2003) found that $\sim 30\%$ of the stars were of intermediate-age, again unlike the Milky Way. Given the large disk and bulge of M31, it has been suggested that these results may be explained by the contamination of putative halo samples by disk/bulge stars (Worthey et al. 2005). However, this interpretation does not appear to be consistent with recent kinematic studies of a subset of the intermediate-age stars (Rich et al., in preparation). In any case, M31 also appears to have a metal-poor projected r^{-2} stellar halo, like the Galaxy (Kalirai et al. 2006; S.C. Chapman, R. Ibata, G.F. Lewis, A.M.N. Ferguson, M. Irwin, et al., submitted; P. Guhathakurta, J.C. Ostheimer, K.M. Gilbert, R.M. Rich, S.R. Majewski, J.S. Kalirai, D.B. Reitzel & R.J. Patterson, submitted). Numerical simulations of M31 indicate that a relatively minor accretion event (involving an LMC-sized galaxy) would suffice to redistribute the observed number of metal-rich, intermediate-aged stars from the disk into the halo (Font et al. 2006).

Before discussing how field star properties relate to GCs, it is helpful to define the quantity T^n, which is the Zepf & Ashman (1993) T parameter normalized to the mass of a particular stellar population (versus the original definition, which normalizes a single GC subpopulation to the total stellar mass in the galaxy). For spheroids with few (or no) metal-poor stars, T^n_{red} is the same as T_{red}, with modulo uncertainties in the adopted M/L.

The Galaxy has ~100 metal-poor GCs and a halo mass of ~$10^9 M_\odot$ (Carney, Latham, & Laird 1990), resulting in a metal-poor $T^n_{blue} \sim 100$. By contrast, the corresponding metal-rich value is $T^n_{red} \sim 5$ (there are only ~50 metal-rich GCs and bulge mass ~$10^{10} M_\odot$; Kent 1992). Thus the metal-poor GCs formed with an efficiency twenty times higher than the metal-rich GCs with respect to field stars.

In the Galaxy, the association of the metal-poor GCs with halo stars is clear; they have similar peak metallicities, spatial distributions, and kinematics (though the metallicity dispersion of the field stars is much larger than that of the GCs). The situation for the metal-rich GCs is somewhat less well-defined but, in general, their spatial distribution and kinematics are more similar to the bulge than to the thick disk (see Section 7.2). However, the metallicity distribution of the metal-rich GCs does not match that of the bulge. Fulbright, McWilliam & Rich (2006) used Keck/HIRES spectra to recalibrate the metallicity distribution of red clump stars in Baade's Window (Sadler, Rich & Terndrup 1996). They found a peak at [Fe/H] ~ -0.15, roughly 0.3–0.4 dex higher than the peak of the metal-rich GCs. Baade's Window is located beyond one bulge effective radius, so if the bulge has a negative metallicity gradient the mean offset could be even larger.

The metallicity of the M31 bulge is similar to the bulge of the Galaxy (Sarajedini & Jablonka 2005). The GC color distributions in the Milky Way and M31 show no significant differences, suggesting a similar metallicity offset in M31 between the metal-rich GCs and the bulge (as borne out by spectroscopy of individual GCs in M31). The metallicity of the metal-poor component of the stellar halo in M31 is not well-constrained, so no direct comparison can yet be made with the GCs.

6.2. NGC 5128

As one of the nearest (albeit disturbed) giant ellipticals (~4 Mpc), NGC 5128 holds special significance for stellar population studies. W. and G. Harris and coworkers have taken the lead in studying the relationship between GCs and field stars in this galaxy.

Their HST/WFPC2 photometry of red giants in fields from ~8 to 31 kpc (projected radius) give metallicity distributions that peak at [m/H] ~ −0.2 to ~ −0.4, with a slight negative metallicity gradient with galactocentric radius (Durrell, Harris & Pritchet 2001; Harris & Harris 2002). Such metallicities are consistent with those of the metal-rich GCs, modulo uncertainties in the relative zeropoints of the metallicity scales. However, the presence of a radial metallicity gradient in the field stars complicates any comparison with the flat metallicity distribution of the GCs. At what radius should the two be compared? The field star metallicity distribution is reasonably well-fit with an accreting box model like the one needed to solve the Galactic G-dwarf problem (Lynden-Bell 1975). Rejkuba et al. (2005) used an HST/ACS outer halo (38 kpc) with an aim to derive a metallicity distribution for NGC 5128 field stars very similar to that of Harris and colleagues, except that they found a larger number of metal-rich stars (>solar). These stars were missed by Harris and colleagues because of large bolometric corrections in $V - I$ at high metallicities. An age analysis of the Rejkuba and colleagues field, based primarily on the locations of the AGB bump and the red clump, yielded a mean age of ~8 Gyr. However, the data are also consistent with a two-component model, comprising an older base population plus a small percentage of 5–7 Gyr stars. Based on the number of AGB stars (Soria et al. 1996) and Mira variables (Rejkuba et al. 2003), a figure of ~10% has been suggested for the fraction of the total stellar mass in intermediate-age stars. This is consistent with the 7% fraction of intermediate-age GCs in the galaxy (see below).

NGC 5128 possesses only a relatively minor tail of metal-poor "halo" stars, even in the outermost pointing, but (as in M31) it is unclear whether this offers a constraint on the presence or absence of a Milky Way-like metal-poor projected ~r^{-2} halo. Harris & Harris (2002) inferred from the paucity of metal-poor stars that T_{blue}^n must be much larger than T_{red}^n. They argued for a scenario in which the metal-poor GCs formed at the beginning of starbursts in small potential wells. These could be efficiently evacuated by supernova feedback before many accompanying metal-poor stars had a chance to form. In principle, other truncation mechanisms (including reionization) could produce a similar result (see Sections 11 and 12).

There are good theoretical reasons to believe that GCs might form in the initial stages of major star-forming events. This may be when the gas pressure is highest, favoring the creation of massive compact clusters (e.g., Elmegreen & Efremov 1997; Ashman & Zepf 2001). By contrast, the metal-rich GCs are expected to form during a dissipational starburst in a more fully-assembled potential well. Here feedback is less effective, and many field stars form. The scenario in which GCs form in the initial stages of a starburst could also qualitatively reproduce a metallicity offset between metal-rich GCs and field stars. Such an offset is observed between the zeropoints of the metallicity–galaxy luminosity relations for metal-rich GCs and field stars (Forbes, Brodie & Grillmair 1997). However, a difficulty with any such comparison is the existence of radial metallicity gradients in galaxies, because metal-rich GCs show no such gradient. If both metal-poor GC and field star formation are simultaneously truncated at high redshift, their metallicity distributions may show no offset, as is the case in the Galaxy.

GCs in NGC 5128 appear to fall on the structural fundamental plane defined by Galactic and M31 GCs (Harris et al. 2002), although the mean cluster ellipticity in the current (small) sample is higher than that of Galactic GCs and more similar to the distribution in the old GCs of the LMC. Harris and colleagues also identified six massive GCs that had evidence of extratidal light. This could suggest that these objects are stripped nuclei of dEs, or simply that they are normal GCs, but that King models provide a suboptimal fit to their surface brightness profiles. Wide-field photometry in the Washington system by Harris, Harris & Geisler (2004), although complicated by the proximity of NGC 5128 and its low galactic latitude, revealed a total population of ∼1000 GCs ($S_N = 1.4$), evenly divided between metal-poor and metal-rich GCs. Peng et al. (2004a) presented a large photometric and spectroscopic survey of more than 200 GCs in NGC 5128. Some 20–25 of the GCs had high enough S/N to determine spectroscopic ages and metallicities, and they found a wide range of ages for the metal-rich GCs, with a mean age of ∼5 Gyr. Through direct comparison with integrated spectra of Galactic GCs from Cohen, Blakeslee & Ryshov (1998), Peng and colleagues found some evidence for [Mg/Fe] > 0, although perhaps not as high as in the Galaxy.

M.A. Beasley, E.W. Peng, T. Bridges, W.E. Harris, G.L.H. Harris, et al. (in preparation) estimated ages and metallicities for ∼200 GCs in NGC 5128 from deep 2dF spectroscopy (with some overlap with the Peng and colleagues sample). Using the multiline fitting method of Proctor, Forbes & Beasley (2004), they deduced that the bulk of the GCs in this galaxy can be assigned to one of three old subpopulations: the usual metal-poor and metal-rich subpopulations, plus an intermediate-metallicity subpopulation reminiscent of one discovered in the Virgo gE NGC 4365 (Brodie et al. 2005). Evidence for trimodality is also presented in Woodley, Harris & Harris (2005). In **Figure 9**, histograms of the GC $B - I$ and metallicity distributions show the trimodality in the metallicity distribution. As is readily apparent, the nonlinear color–metallicity relationship combines the metal-poor and intermediate-metallicity peaks into a single peak in the $B - I$ histogram. The presence of old intermediate-metallicity GCs is unusual among massive Es. If these GCs have an accompanying subpopulation of field stars, the field star metallicity distribution could be a combination (albeit not necessarily well-mixed) of intermediate-metallicity and metal-rich stars. However, this would be difficult to detect in present data. In addition to the old subpopulations, about 7% of the GCs in the Beasley and colleagues sample appear to be metal-rich with intermediate ages (2–8 Gyr), contrary to previous results. Candidate intermediate-age GCs were specially targeted, so this fraction is probably an upper limit.

The kinematics of the GCs are similar to those observed in the Galaxy: significant rotation is seen in the metal-rich GCs (as defined in Peng, Ford & Freeman 2004), but only weak rotation is apparent in the metal-poor GCs. This result is contrary to that expected to result from a disk-disk merger. In a merger remnant, the inner parts of the galaxy should display very little rotation because angular momentum is transferred to the outer regions. Of course, the remnant properties will depend on the details of the merger, and a statistical sample of Es that might plausibly have formed in recent major mergers does not yet exist. See Section 7 for additional discussion of GC kinematics.

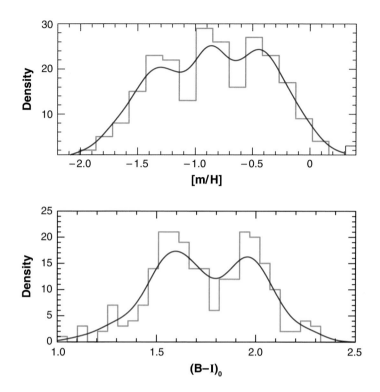

Figure 9
Smoothed kernel density histograms in $V - I$ and [m/H] for globular clusters (GCs) in NGC 5128. The metallicity histogram shows evidence for three distinct subpopulations of GCs, and a comparison of the two histograms indicates the clear nonlinearity of the color–metallicity relation (M.A. Beasley, E.W. Peng, T. Bridges, W.E. Harris, G.L.H. Harris, D.A. Forbes & G. Mackie, in preparation).

6.3. Other Galaxies

An alternative approach to studying the relationship between GCs and field stars has been taken by Forte, Faifer, & Geisler (2005), who assume that the field star subpopulations follow the radial and color distributions of the GC subpopulations and use galaxy surface photometry to decompose the integrated light into metal-poor and metal-rich constituents. For NGC 1399, they find $T_{blue}^n \sim 25$ and $T_{red}^n \sim 4$. This metal-rich value is virtually identical to that of the Galaxy and, as before, the metal-poor value is much larger than the metal-rich one. Although this approach is potentially a powerful tool for studying field stars in galaxies with unresolved stellar populations, substantial future work will be required to establish the veracity of the key underlying assumption: that the radial and metallicity distributions of GCs and their associated field stars are the same.

7. KINEMATICS

7.1. Ellipticals

Because GCs are more extended than the stellar light of galaxies, they are useful kinematic tracers of the dark-matter halo at large galactocentric radii. GC kinematics also encode detailed signatures of the assembly histories of galaxies; the implications of such observations are just beginning to be understood. Large kinematic samples of GCs (>50) have been observed in only four gEs: NGC 1399, M87, NGC 4472, and NGC 5128 (discussed above). Small samples have been studied in several other galaxies, but the numbers are too few to allow the subpopulations to be effectively separated. Aside from noting that there appear to be signatures of strong rotation in the S0s NGC 3115 (Kuntschner et al. 2002) and NGC 524 (Beasley et al. 2004), we will confine our detailed discussion to these four well-studied galaxies.

The largest sample of GC velocities at present exists for NGC 1399, the central gE in the Fornax cluster. Richtler et al. (2004) have analyzed kinematics for 468 GCs at projected radii of 2–9′ (~11–49 kpc). The velocity dispersions for the metal-poor and metal-rich GCs are ~290 km/s and ~255 km/s, respectively, and do not appear to vary with galactocentric radius. They report a slight tangential bias for the metal-poor GCs, but both subpopulations are generally consistent with isotropic orbits. The larger dispersion for the metal-poor GCs is consistent with their extended spatial distribution. Neither subpopulation shows significant rotation, although a weak signature is observed in metal-poor GCs beyond 6′.

It has been suggested from simulations (Bekki et al. 2003) that NGC 1399 might have stripped the outermost GCs from the nearby E NGC 1404, leaving it with an anomalously low S_N (NGC 1399 has a high S_N of ~5–6, typical of cluster Es). Bassino, Richtler & Dirsch (2006) have found that three other Fornax Es—NGC 1374, NGC 1379, and NGC 1387—all have S_N lower than typical of cluster Es, and also suggest that they may have suffered GC loss to NGC 1399. New GC velocities for NGC 1399, out to ~80 kpc, show an asymmetric velocity distribution of the metal-poor GCs at these large radii (Schuberth et al. 2004). This could be interpreted as a signature of GCs that have been stripped from nearby Es but have not yet reached equilibrium in the NGC 1399 potential.

NGC 4472 has been studied by Zepf et al. (2000) and Côté et al. (2003), with ~140 and ~260 velocities, respectively, measured over projected radii of ~1.5–8′ (~7–40 kpc). The velocity dispersions of the metal-poor and metal-rich GCs are ~340 km/s and ~265 km/s, respectively. The metal-poor GCs rotate around the galaxy's minor axis with a velocity of ~90–100 km/s; the metal-rich GCs show weak evidence for counter-rotation around the same axis. There is a hint that the outermost (beyond ~25 kpc) metal-poor GCs may rotate about the major axis, but this conclusion relies on few GCs. The orbits of both subpopulations are consistent with isotropy to the radial limit of the data.

The kinematics of GCs in M87 are rather different from those in either NGC 4472 or NGC 1399. From a sample of ~280 velocities, Côté et al. (2001) find that both the metal-poor and metal-rich GCs rotate around the minor axis at ~160–170 km/s, with respective velocity dispersions of ~365 and ~395 km/s, when averaged over the whole system. The metal-poor GCs actually appear to rotate around the major axis within ~16 kpc but switch to minor axis rotation only at larger radii. Côté and colleagues suggest that this change is due to the increasing gravitational dominance of the cluster potential beyond ~18 kpc. The GC system as a whole appears quite isotropic, but the data are consistent with a tangential bias for the metal-poor GCs and an opposing radial bias for the metal-rich GCs; these biases are poorly constrained in the present sample.

The results for four well-studied gEs (including NGC 5128, discussed in the previous section) present a heterogeneous picture. Although no two galaxies seem to be kinematically similar thus far, we will briefly discuss the extent to which the accumulated information can constrain galaxy formation. K. Bekki and collaborators have used numerical simulations to study the properties of GCs in merging galaxies. Bekki et al. (2002) found that, in disk-disk major mergers, newly formed metal-rich GCs are centrally concentrated and extended along the major axis of the remnant. Angular momentum transferred to the metal-poor GCs causes them to rotate and extend their spatial distribution. These results are strongly dependent on the numerical details of the simulations and could change if, for example, a different prescription for star formation was used. Simulations of dissipationless major mergers for galaxies with both a disk and bulge extended these results (Bekki et al. 2005). However, the initial conditions (velocity dispersion and anisotropy) of the GC system strongly affect the outcome of such simulations. Pre-existing metal-poor and metal-rich GCs acquire significant amounts of rotation beyond ~10 kpc, regardless of the orbital properties of the merging galaxies. Mergers with larger mass ratios leave relatively spherical GC distributions with less rotation. With such a generic prediction of rotation, it is unclear how galaxies like NGC 1399, which shows no rotation in either subpopulation, could have been created. It may be that the merging histories of gEs leave a variety of complex rotational signatures and that much of the observed differences are due to natural variations in remnant properties and projection effects.

It is instructive to consider the results of cosmological simulations of the assembly of dark-matter and stellar halos of massive galaxies. Even though the GCs are more extended than the stellar light of the galaxy, they are still more centrally concentrated than the dark matter, so they are expected to have a lower velocity dispersion than the dark matter at fixed radii (if the anisotropy of the dark matter and GCs are similar). While providing some constraints, present data sets are still too small to fully determine the anisotropy ($\beta = 1 - v_\theta^2/v_r^2$) of the GC system (e.g., Wu & Tremaine 2006). A wide range of numerical simulations suggest that a general relation holds between β and galactocentric radius for both halo tracers and the dark matter: $\beta \sim 0$ (isotropy) in the inner parts, with radial anisotropy increasing outward to $\beta \sim 0.5$ at the half-mass radius of the tracer population (Hansen & Moore 2006; Dekel et al. 2005; Diemand, Madau & Moore 2005; Abadi, Navarro & Steinmetz 2006). Very large samples of GCs (~500–1000) will be needed to test these predictions.

Studies of lower-mass Es are also important, and several groups are pursuing hybrid approaches, using both GCs and planetary nebulae (PNe) to constrain the potential. Romanowsky et al. (2003) suggested, on the basis of PNe kinematics in three "normal" Es (NGC 821, NGC 3379, and NGC 4494), that these galaxies lacked dark matter. A more conventional interpretation is that the observed PNe originate in stars ejected during mergers and are on highly radial orbits (Dekel et al. 2005). GCs (especially those in the metal-poor subpopulation) are expected to have lower orbital anisotropy than PNe, owing to both their extended spatial distribution and the increased probability of destruction for GCs on radial orbits. Two recent studies of GC kinematics in NGC 3379 suggest that the galaxy possesses a "normal" Λ CDM dark halo and are more consistent with the Dekel and colleagues interpretation (Pierce et al. 2006; Bergond et al. 2006). Thus, GCs may be the best tracers of the mass distribution of Es at large radii.

7.2. Disk Galaxies

The best recent review of the kinematics of Galactic GCs is found in Harris (2001). The metal-rich GCs rotate at $\sim$90–150 km/s, and the rotation rises out to a radius of $\sim$8 kpc, beyond which there are few metal-rich GCs. This increasing rotational velocity may represent a transition from bulge to thick disk, though the maximum velocity of $\sim$150 km/s is still less than expected for a typical thick disk in a massive spiral. There is little net rotation over the metal-poor subpopulation as a whole. Moreover, no individual radial bin of metal-poor GCs rotates, but somewhat surprisingly, a strong signature of prograde rotation (140 km/s) is seen in the most metal-poor GCs ([Fe/H] < −1.85). This effect is dominated by very metal-poor GCs in the inner halo. The degree of rotational support is similar in this very metal-poor group ($v/\sigma \sim 1.2$) to the total subpopulation of metal-rich GCs ($\sim$1.3).

M31 has been discussed above: the metal-rich GCs rotate with $v/\sigma \sim 1.1$, and the kinematic state of the metal-poor GCs remains unclear. If the identification of the disk objects as young clusters (no matter what the mass) is secure, then the old metal-poor subpopulation is probably pressure-supported.

The situation in M33 is also uncertain, because the ages of many HST-confirmed star clusters are unknown. Chandar et al. (2002) found that the old metal-poor GCs have a velocity dispersion of $\sim$80 km/s and are not rotating. They suggested that there is a small population of old inner GCs with disk-like kinematics, but an adequate test of this must await a significant increase in the kinematic sample.

Olsen et al. (2004) have suggested that GCs in several Sculptor Group spirals have kinematics consistent with the rotating HI gas disks in these galaxies (in NGC 253, the kinematics are consistent with asymmetric drift from an initial cold rotating disk). Candidates were selected from outside the bright optical disk, so contamination from open clusters in the disk is unlikely to have occurred. Few GCs were observed in each galaxy, and the low systemic velocities of the galaxies inevitably biased the GC selection; those with low velocities cannot be efficiently distinguished from foreground stars. Nonetheless, even with this bias, the line-of-sight velocity dispersions of the GCs are low: from $\sim$35–75 km/s out to $\sim$10 kpc or so. Given the mixed evidence for old disk GCs presented thus far, this preliminary result is worth following up.

The disk galaxies present a somewhat cleaner kinematic picture than the Es, though the number of galaxies studied is still small. If indeed the metal-poor GCs in M31 turn out to be pressure-supported, then this appears to be a common feature of disk galaxies. In the two galaxies with bulges (M31 and the Galaxy) the metal-rich GCs have substantial rotational support. At the very least, these results can serve as valuable starting conditions for simulations of major mergers.

8. LUMINOSITY FUNCTIONS

Excellent reviews of the GCLF, including its use as a distance indicator, are given in Harris (2001) and Richtler (2003). The NED D database (2006 release) gives a comprehensive compilation of GCLF distances. Here we summarize some basic facts and discuss the evolution of the GCLF.

In many massive galaxies studied to date, the GCLF can be well-fit by a Gaussian or t_5 distribution; typical parameters for the normal distribution in Es are peak $M_V \sim -7.4$, with $\sigma \sim 1.4$. In spirals, the peak is similar, but $\sigma \sim 1.2$ may be a more accurate value for the dispersion. The peak luminosity is a convolution of the peaks for the two GC subpopulations; line-blanketing effects (especially in bluer bands) cause the metal-poor GC peak to be slightly brighter than the metal-rich peak, so the exact location of the peak depends on the color distribution of the GC system (Ashman, Conti & Zepf 1995). The standard peak or turnover of this distribution corresponds to a mass of $\sim 2 \times 10^5 M_\odot$. In absolute luminosity or mass space, this distribution is a broken power law with a bright-end slope of ~ -1.8 and a relatively flat faint-end slope of ~ -0.2 (McLaughlin 1994). In some massive galaxies there appears to be a departure below a power law for the most massive GCs (e.g., Burkert & Smith 2000). The expectation of significant dynamical evolution at the faint end of the GCLF has led to suggestions that only the bright half of the GC mass function (GCMF) represents the initial mass spectrum of the GC system. The faint half would then represent the end product of a Hubble time's worth of dynamical destruction. Recent observational and theoretical work on this problem offers some supporting evidence, but also raises some interesting questions.

In the Galaxy, the total sample of GCs is relatively small, especially when divided into subpopulations. This limits our ability to draw general conclusions from this best-studied galaxy. Harris (2001) provided a good overview of the current situation. Within the 8 kpc radius that contains both metal-poor and metal-rich GCs, the subpopulations have quite similar LFs. Considering only the metal-poor GCs, the turnover of the GCLF becomes brighter by nearly ~ 0.5 mag out to 8–9 kpc, then decreases to its initial value. The inner radial trend is the opposite of naïve expectations for dynamical evolution, which should preferentially disrupt low-mass GCs as shocks accelerate mass loss through two-body relaxation. There are too few metal-rich GCs to study the radial behavior of the GCLF in detail. The GCs in the far outer halo (beyond ~ 50 kpc or so) are all quite faint ($M_V < -6$), except for the anomalous GC NGC 2419. This cluster has $M_V = -9.6$ and may be the stripped nucleus of a dwarf galaxy (van den Bergh & Mackey 2004). Of course, this level of detail is not expected to be visible in external galaxies, whose GC systems are generally seen in projection about an unknown axis. This is an important limitation in interpreting the results of the theoretical simulations discussed below.

Fall & Zhang (2001) presented a semianalytic study of GC system evolution in a Milky Way-like galaxy. The mass-loss rate by two-body relaxation is taken to depend only on the details of the cluster's orbit, and not on the mass or concentration of the GC. This appears to be consistent with results from more detailed N-body simulations (e.g., Baumgardt & Makino 2003). Fall & Zhang found that a wide range of initial

GCMFs, including single and broken power laws, eventually evolved to turnover masses similar to those observed. They found that the evolution of the turnover mass was rapid in the first Gyr or two, but that subsequent evolution was slow. The turnover is expected to change substantially with galactocentric radius, due to the preferential destruction of low-mass GCs toward the galaxy center. This can only be avoided if the outer GCs display strong radial anisotropy. However, the assumptions made by Fall & Zhang need to be considered when applying their results to real data. For instance, they used a static spherical potential, whereas a live and/or nonradial potential could increase phase mixing and erase some radial signatures of evolution.

Vesperini (2000; 2001) modeled the evolution of GCLFs of two initial forms, log-normal and power-law functions. The evolution of individual GCs was determined using analytic formulae derived from the simulations of Vesperini & Heggie (1997). GCLFs that are initially log-normal provide a much better fit to the observed data. Dynamical evolution can indeed carve away the low-mass end of a power-law GCLF to produce a log-normal function, but the resulting turnover masses are generally small ($\lesssim 10^5 M_\odot$) and vary significantly with galactocentric radius and from galaxy to galaxy. All three predictions for the evolution of an initial power-law function are inconsistent with observations. However, a GCLF that is initially log-normal suffers little evolution in shape or turnover, and that only in the first few Gyr. If constant initial S_N is assumed, GC destruction leads to a S_N–galaxy luminosity relation of the form $S_N \propto L^{0.67}$. Vesperini et al. (2003) explored M87 in detail and (reminiscent of Fall & Zhang 2001) found that the observed log-normal GCLF, which has a constant turnover with radius (see also Harris, Harris & McLaughlin 1998), could only evolve from a power-law initial GCLF if there was (unobserved) strong radial anisotropy in the GC kinematics. Of course, the details of the simulations are quite important. For example, Vesperini & Zepf (2003) showed that a power-law GCLF and a concentration–mass relation for individual GCs can result in a log-normal GCLF with little radial variation in turnover mass. This is because low-concentration GCs are preferentially destroyed at all radii. This is consistent with the study of Smith & Burkert (2002), who found that the slope of the low-mass part of the GCLF in the Galaxy depends upon GC concentration. Another relevant piece of evidence is the similarity in the GCLF turnover among spirals, Es, and even dEs (see Section 10.1). This suggests either initial GCLFs close to log-normal (such that little dynamical evolution occurs), or that the dominant destruction processes are not specific to particular galaxy types—for example, disk shocking in spirals.

The sole galaxy with a significant subpopulation of intermediate-age GCs and evidence for dynamical evolution is the merger remnant NGC 1316. HST/ACS observations by Goudfrooij et al. (2004) have clearly demonstrated dynamical evolution of its GC system. When they divided the red ($1.03 \leq V - I \leq 1.40$) GCs into two equal radial bins, they found that the outer (beyond ~9.4 kpc) GCs have the power-law LF seen for many systems of YMCs in merging and starbursting galaxies. By contrast, the inner GCs show a LF turnover characteristic of old GC systems. As Goudfrooij and colleagues argued, these observations would appear to provide the conclusive link between YMCs in mergers and old GCs in present-day Es.

However, the difficulty with this interpretation is that the observed location of the metal-rich peak of the inner GCs in NGC 1316 is at $M_V \sim -6$. If the metal-rich GCs were formed in the merger, are ~ 3 Gyr old, and have solar metallicity (consistent with spectroscopic results; Goudfrooij et al. 2001b), Maraston (2005) models with either a Salpeter or Kroupa IMF predict that age-fading to 12 Gyr will result in a peak at $M_V \sim -4.6$. Old metal-rich GCs are observed to have a GCLF turnover at $M_V \sim -7.2$ (e.g., Larsen et al. 2001). Thus ~ 2.6 mag of additional evolution of the GCLF turnover (~ 1.2 mag of dynamical evolution and ~ 1.4 mag of age fading) would be required to turn the new metal-rich GC subpopulation of NGC 1316 into that of a normal E. The 1.2 mag of dynamical evolution needed is far beyond that predicted by even the most "optimistic" models for GC destruction. For example, Fall & Zhang (2001) models for GC system evolution, assuming a power-law initial GCMF, predict <0.2 mag of evolution in the GCLF peak from 3 to 12 Gyr using the same Maraston (2005) models. This is consistent with the finding of Whitmore et al. (2002) that little evolution in the GCLF peak is expected after the first 1.5–2 Gyr, as age-fading balances GC destruction. The implication is that in a Hubble time, the metal-rich GC system of NGC 1316 may not look like that of a normal E galaxy. Fall & Zhang (2001) models, designed to study GC destruction in a Milky Way-like galaxy, may even have limited applicability to NGC 1316. Other models (e.g., those of Vesperini discussed above) predict less evolution, so the expected difference between the suitably evolved young GCs in NGC 1316 and the old metal-rich GCs in local Es could be even greater. A caveat here is that the observed GCLF peak could be the convolution of a brighter, more evolved GC population in the innermost regions with a less evolved population in the outer parts of the bin. Such a convolution would tend to lessen the difference between the observed and expected turnover luminosity.

Thus, while there is some evidence for the paradigm of the evolution of a power-law LF to a log-normal LF, there are still unresolved issues. The installation of WFC3 on a potential future HST servicing mission would allow much more efficient and accurate age-dating of YMCs through its wide-field U-band imaging capability. This would also make it possible to investigate the mass function as a function of age in more detail, and hence directly address questions of dynamical evolution.

9. SIZES

The present-day appearance of a GC is a complicated convolution of the initial conditions of its formation with subsequent internal and external dynamical effects. Of the three scale parameters—core radius (r_c), half-light radius (r_h –"size"), and tidal radius (r_t)—in a King model (only two of which are independent), only r_h is relatively unaffected by dynamical evolution (Spitzer 1987; Meylan & Heggie 1997) and can serve as a probe of GC formation conditions.

A correlation between r_h and galactocentric radius (R) for Galactic GCs was discovered by van den Bergh, Morbey & Pazder (1991). This could not be explained by dynamical evolution of the GC system, because diffuse inner GCs might be expected to be destroyed, but compact distant GCs should (if they existed) have remained intact. Thus, this result represents strong evidence for some degree of in situ formation of Galactic GCs.

HST/WFPC2 imaging of NGC 3115 and M87 revealed that metal-poor GCs are ∼20% larger than metal-rich GCs (Kundu & Whitmore 1998, Kundu et al. 1999). This was confirmed statistically in many early-type galaxies by Larsen et al. (2001) and Kundu & Whitmore (2001a). Explanations offered for this result include

1. It represents an intrinsic formation difference, e.g., the metal-rich GCs formed in a higher-pressure environment.
2. It is a result of projection effects. Because the metal-rich GC spatial distribution is more centrally concentrated than that of the metal-poor GCs, within some given projected radius the metal-rich GCs will tend to lie at smaller R. If there is a strong correlation between size and R (as in the Galaxy), the metal-rich GCs will appear smaller on average than the metal-poor GCs (Larsen & Brodie 2003). This model predicts that size differences will be largest in the inner parts of galaxies and disappear in the outer regions.
3. It is a natural outcome of assuming metal-poor and metal-rich GCs have the same half-mass radii. Because the brightest stars in metal-rich GCs are more massive than in metal-poor GCs, mass segregation leads to a more compact distribution and a smaller half-light radius (Jordán 2004). In this model there should be little change in the relative sizes with galactocentric distance.

Option 1 is not testable at present, so it should be left as a fallback only if the other possibilities can be eliminated. Regarding 2, Larsen & Brodie (2003) showed that the r_h–R correlation in the Galaxy could explain all of the observed size differences between the metal-poor and metal-rich GCs. However, in order to explain the ∼20% size difference in external galaxies, steep r_h–R relations would be required, and the radial distribution of the GCs would need to have a central core like a King profile (this appears to be consistent with observations). Model 3 has a number of critical assumptions upon which its conclusions depend, including identical GC ages and initial mass functions. Small changes in either of these parameters (e.g., an age difference of ∼2 Gyr between the metal-poor and metal-rich GCs) could erase most of the expected size difference. Jordán (2004) also used equilibrium King-Michie models to represent the GCs; full N-body modeling is a desirable next step. In **Figure 10**, taken from Jordán (2004), we show the sizes of GCs in M87 together with a best-fit model of type 3.

Several recent observational results have provided important new constraints. Jordán et al. (2005) studied GC sizes in 67 early-type galaxies with a wide range of luminosity from the ACS Virgo Cluster Survey (Côté et al. 2004). For bright metal-poor GCs they found a significant but rather shallow relationship between r_h and projected R (normalized to the effective radius of the galaxy). In log space the value of the slope is 0.07, compared to ∼0.30 for a similar sample of Galactic GCs. They did not list the fits for individual galaxies, and there are clearly variations, but the bulk of the galaxies do not appear to have r_h–R relations as steep as observed in the Galaxy, so projection effects on the GC subpopulation sizes should be small, and option 2 is not favored. They also found that GCs in bluer/fainter host

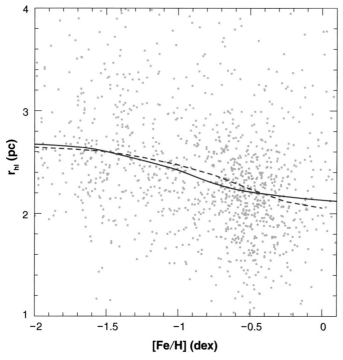

Figure 10

Half-light radii of globular clusters (GCs) in M87 versus [Fe/H] (derived from $g - z$ color). The overplotted solid and dashed lines are a running robust mean and a model fit (based upon King models; Jordán 2004) normalized to the metal-poor GCs. As is typical in massive galaxies, the metal-rich GCs are smaller than the metal-poor GCs (figure courtesy of A. Jordán).

galaxies tend to be slightly larger. Despite these variations, r_h is still relatively constant among galaxies. Thus, with their observed correlations between r_h and galaxy properties, they were able to calibrate r_h as a distance indicator, as suggested by Kundu & Whitmore (2001a).

Wide-field HST/ACS data for NGC 4594 (covering ~6′ × 10′) do not appear consistent with option 3, however. L. Spitler, J.P. Brodie, S.S. Larsen, J. Strader, D.A. Forbes & M.A. Beasley (submitted) found that the ratio of metal-poor to metal-rich GC sizes declines steeply and steadily from ~1.25 in the center to ~1 at the edge of the complete observations. Thus it appears that, at least in NGC 4594, projection effects account for most of the observed size differences.

It seems clear from these results that both projection and segregation mechanisms can play a role in determining the sizes observed for extragalactic GCs. Because each is sensitive to the physical conditions of the GC system, galaxies need to be studied on an individual basis to determine which effects are important. Particularly valuable would be high-resolution, wide-field imaging of GC systems, like the HST/ACS mosaic of NGC 4594 discussed above.

10. DWARF GALAXIES

We have long known that many dwarf galaxies have GC systems (e.g., Fornax dSph; Shapley 1939). In the Galaxy formation model of Searle & Zinn (1978) and in many subsequent studies of GC formation, it was envisioned that metal-poor GCs form in protogalactic dwarf-sized clumps (e.g., Harris & Pudritz 1994; Forbes, Brodie & Grillmair 1997; Côté, Marzke & West 1998; Beasley et al. 2002). The dwarf satellites around massive galaxies like the Milky Way can then be interpreted as the remnants of a large initial population of such objects, most of which merged into the forming protogalaxy. If this process happened at high redshift, most of the fragments could still be gaseous and thus have formed stars and/or contributed gas as they merged. At lower redshift the process could be primarily dissipationless, as envisioned by Côté and colleagues and seen in action through the present-day accretion of the Sgr dSph. What can GCs tell us about this process?

The faint end of the galaxy LF is uniquely accessible in the Local Group. Forbes et al. (2000) provided a good census of GCs in Local Group dwarfs, which has changed only marginally since that time. The candidate GC in the dIrr Aquarius is apparently a yellow supergiant (D. Forbes, private communication), so the lowest-luminosity Local Group galaxies with confirmed GCs are the Fornax ($M_V = -13.1$) and Sgr ($M_V = -13.9$) dSphs, each of which has at least five GCs. The recently discovered CMaj dSph (Martin et al. 2004) appears to have at least four GCs whose properties are distinct from the bulk of the Galactic GC system (Forbes, Strader & Brodie 2004). At least two of the GCs in each of Sgr and CMaj are of intermediate age and metallicity. The LMC has a subpopulation of old metal-poor GCs and a famous "age gap" between the old GCs and a subpopulation of intermediate-age GCs (~ 3 Gyr). It also hosts a number of younger clusters, some of which, like GCs, might be massive enough to survive a Hubble time (e.g., Searle, Wilkinson & Bagnuolo 1980; van den Bergh 1994). The SMC has only one old GC but a more continuous distribution of massive clusters to younger ages (Mighell, Sarajedini & French 1998). Together these results suggest star-formation histories that were at least moderately bursty (e.g., Layden & Sarajedini 2000). Lower-mass Galactic dwarfs (e.g., Leo I; $M_V \sim -12$) do not have GCs. Less is known about the GC systems of similar-mass M31 dwarfs. Grebel, Dolphin & Guhakathurta (2000) suggest a candidate GC in And I ($M_V \sim -12$), but this GC could be a contaminant from M31. Outside the Local Group, Sharina, Sil'chenko & Burenkov (2003) spectroscopically confirmed a GC in the M81 dSph DDO 78, which has a mass intermediate between Fornax and And I. Karachentsev et al. (2000) identified candidate GCs in a number of other M81 dwarfs, but these have not yet been confirmed.

As discussed in Strader et al. (2005), these observations put important constraints on the minimum mass of halos within which metal-poor GCs could form. Fornax and Sgr have total masses of $>10^8 M_\odot$ (Walcher et al. 2003; Law, Johnston & Majewski 2005) and the total mass estimate for And I, assuming it has a similar M/L ratio, is a few $\times 10^7 M_\odot$. This suggests that, at least in a relatively low-density group environment, GCs formed in halos with minimum masses of $\sim 10^7$–$10^8 M_\odot$. Whether GCs typically form in groups of GCs or alone is unknown. Fornax and Sgr each have several GCs, and the dIrr NGC 6822 ($M_V \sim -15.2$) has up to three old GCs, though two of these are located far from the main body of the galaxy (Cohen & Blakeslee 1998; Hwang et al. 2005). The dIrr WLM ($M_V \sim -14.5$) has only one old GC, which is metal-poor (Hodge et al. 1999). A caveat to these arguments is that in some models (e.g., Kravtsov, Gnedin & Klypin 2004), present-day dwarf satellites have undergone significant stripping of dark matter and may have been much more massive initially ($>10^9$–$10^{10} M_\odot$). However, detailed comparisons between observed velocity dispersion profiles and numerical simulations suggest little mass loss due to tidal stripping for well-studied Galactic dSphs (Read et al. 2006). It is important to realize that differences in baryonic mass loss (e.g., due to stellar feedback; Dekel & Silk 1986) may modify the amount of stellar mass in galaxies of similar halo mass. GC kinematics (see Section 10.3) offer one of the best routes to directly determine the total masses of dwarfs outside the Local Group.

10.1. Specific Frequencies and Luminosity Functions

GC systems were discovered around 11 Virgo dwarf ellipticals[2] (dEs) using ground-based (CFHT) imaging (Durrell et al. 1996). The S_N of these galaxies is relatively high, $\sim$3–8, and the GC systems are very centrally concentrated: most GCs are within $<30''$ (2 kpc) at the distance of Virgo, whereas a typical dE has a half-light radius $\lesssim 1$ kpc. J. Strader, J.P. Brodie, L. Spitler & M.A. Beasley (submitted) obtained radial distributions that were consistent with these earlier results, except for a few bright dEs ($M_V \lesssim -18$) where the outermost GCs were found at 7–9 kpc. This is near the limit of the radial coverage of HST/ACS at the distance of Virgo, so it is possible that GCs may be found at even larger radii.

Miller et al. (1998) used HST/WFPC2 snapshot imaging to explore the specific frequencies of a large sample of dEs in the Virgo and Fornax clusters, including galaxies with luminosities as faint as $M_B \sim -13$. They found a dichotomy between nucleated (dE,N) and nonnucleated (dE,noN) galaxies. dE,noN galaxies appeared to have low S_N values ($\sim$3), independent of galaxy luminosity; dE,N galaxies had higher S_N and showed an inverse correlation between S_N and luminosity. It has been suggested that dE galaxies may have originated as dIrrs or low-mass disk galaxies (e.g., Moore, Lake & Katz 1998; see discussion below). Miller and colleagues argued that few dE galaxies could have formed in this manner, as the S_N values even for dE,noN galaxies are larger than expected from age-fading such hosts.

J. Strader, J.P. Brodie, L. Spitler & M.A. Beasley (submitted) revisited these findings in an HST/ACS study of Virgo Es that included 37 dEs. Thirty-two of these have structural parameters consistent with "true" dEs; the other five appear to be faint

---

[2]In deference to common usage, here we utilize the term "dwarf elliptical" for early-type galaxies with low luminosities ($M_B \lesssim -18$). The structural parameters of many of these galaxies differ from those of "classical" Es (e.g., Kormendy 1985, 1987) and may well suggest a different formation history. They are sometimes called spheroidal (Sph) galaxies. A small number of galaxies in this luminosity range have structural parameters consistent with power-law Es (see Section 4).

power-law Es (J. Kormendy, D.B. Fisher, M.E. Cornell & R. Bender, in preparation, and private communication). The ACS images offered superior areal coverage and depth compared to those used by Miller and colleagues, but the Strader and colleagues sample spanned a smaller luminosity range: $-15 \lesssim M_B \lesssim -18$. It was not possible to investigate the differences between dE,N and dE,noN galaxies, because many of the galaxies previously classified as dE,noNs either have faint nuclei or are power-law Es. There may be few true dE,noNs with $M_B \lesssim -15$ in Virgo. Strader and colleagues found no strong correlation between galaxy luminosity and S_N for either dEs or faint power-law Es. The faintest galaxies might have larger S_N, but the effect is not strong. The lack of a $S_N - L$ trend could be due to the more restricted luminosity range of galaxies studied by Strader and colleagues compared to Miller and colleagues.

Interestingly, a bimodal distribution of S_N values was discernible in their sample. As shown in **Figure 11**, more than half of the galaxies were found to have $S_N \sim 1$, whereas the S_N values of the remainder ranged from 3 to 10, with a median at ~ 5. This difference spans the observed luminosity range and does not correlate with either the presence of a nucleus or the color distribution of the GCs. A natural interpretation of

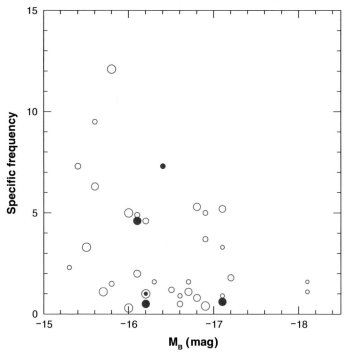

Figure 11

Specific frequency (S_N) of dwarf ellipticals (*open blue circles*) and faint power-law ellipticals (*filled red circles*) versus parent galaxy M_B. The size of the points is proportional to the fraction of blue globular clusters. There is only a weak trend of increasing S_N with decreasing M_B, and no substantial difference between the two galaxy classes. However, there is some evidence for a bimodal distribution of S_N (J. Strader, J.P. Brodie, L. Spitler & M.A. Beasley, submitted).

the S_N differences is that they reflect multiple formation channels for dEs in Virgo. Mechanisms for forming dEs include "harassment," the cumulative effect of many high-speed galaxy encounters (Moore et al. 1996), stripping or age-fading of low-mass disks (Kormendy 1985), or processes similar to those responsible for the formation of more massive Es (this may be most applicable to the faint power-law Es). It is possible that the high S_N galaxies (the Fornax dSph with $S_N \sim 29$ is an extreme example) simply represent those in which stellar feedback during the first major starburst was very effective (e.g., Dekel & Silk 1986). However, if this is the case, a signature should be apparent in the GC color distributions. In particular, the presence of metal-rich GCs (see below) and a continuous (rather than bimodal) distribution of S_N might be expected. Certainly, feedback will be increasingly important with decreasing galaxy mass, and would provide a simple explanation for a relation between S_N and luminosity, should one be confirmed. Photometric, structural, and kinematic studies of these same dEs will be needed to discriminate among the many possible explanations.

These same studies have also afforded the ability to study the GCLF in dEs. The power-law slope (~ -1.8) measured by Durrell et al. (1996) for the massive end of the GCLF in their Virgo dEs is the same as that found in normal Es. However, they measured a GCLF turnover ($M_V \sim -7.0$) that is fainter by ~ 0.4–0.5 mag than that typical for massive galaxies. This result may have been influenced by the difficulty of rejecting contaminants in ground-based data. J. Strader, J.P. Brodie, L. Spitler & M.A. Beasley (submitted) constructed a composite of the 37 dEs in their HST sample, using the outer parts of the images to correct for background contamination. In contrast to Durrell and colleagues, they found that the dE GCLF peak occurs at the same value as in the massive gEs in their sample (M87, NGC 4472, NGC 4649). This comparison was made in z, where there is little dependence of cluster M/L on metallicity over the relevant range, so the differences in GC color distributions between gEs and dEs should not affect the GCLF comparisons.

That the GCLF peaks for the dwarfs and the giants match so well in the Strader and colleagues study is perhaps puzzling. The theoretical expectation is that in low-mass galaxies dynamical friction will act to deplete the GC system, preferentially destroying massive GCs. Such GCs will spiral into the center in less than a Hubble time, forming or contributing to a nucleus. Lotz et al. (2001) performed a semianalytic study of this phenomenon and found that dynamical friction is expected to produce more luminous nuclei than observed. This point, together with the similarity of the GCLF turnovers, implies that dynamical friction has not had a substantial effect on the GC systems of dEs. Lotz et al. (2001) suggested several explanations for the lack of observable consequences of dynamical friction, including extended dark-matter halos around dEs, or tidal torquing of GCs (this latter explanation was also proposed for the Fornax dSph by Oh, Lin & Richer 2000). Goerdt et al. (2006) have used numerical simulations to show that the dynamical friction timescale in Fornax is longer than a Hubble time if its dark-matter halo has a core (instead of the cusp generically predicted in galaxy formation in Λ CDM). Kinematic studies of GCs in dEs, such as those of Beasley et al. (2006), are also beginning to build a better understanding of the halo potentials of dwarf galaxies.

So far we have included little discussion of dIrrs. Their GC systems are quite difficult to study. Indeed, the contrast in our understanding of dEs and dIrrs is analogous to the information gap between Es and spirals. The relatively small GC systems of dIrrs, their ongoing star formation, and the resulting inhomogeneity of the background are serious observational challenges. A HST/WFPC2 study of 11 Virgo and Fornax dIrrs by Seth et al. (2004) found typical S_N values of ∼2, but uncovered two galaxies with much higher S_N values. Stellar M/L values are low in typical dIrrs, which suggests that the S_N values will become much higher after nominal age-fading of the dIrr, as might be expected in transformation to a dE. However, many of the detected objects are unlikely to be old GCs. In the Local Group, the Magellanic Clouds and NGC 6822 each have a population of massive intermediate-age GCs (in addition to a small number of old GCs) that reflect the extended star-formation histories of these galaxies. The same may be true of cluster dIrrs. Spectroscopy will probably be needed to determine the present fraction of intermediate-age GCs, which is crucial for isolating the differences in the formation histories of the various classes of dwarf galaxies. The combination of optical and NIR photometry, once the SSP models have been sufficiently calibrated for old GCs, offers a promising future tool for identifying younger clusters.

10.2. Color Distributions

The classical view of dwarf galaxy GCs is that they are uniformly metal-poor. This is supported by a "combined" metallicity distribution of old GCs in Local Group dwarfs (Minniti, Meylan & Kissler-Patig 1996), which peaks at [Fe/H] ∼ −1.8, 0.3 dex more metal-poor than that of the Galactic metal-poor GC subpopulation.

We now know that reality is more complicated. As noted above, the Sgr dSph (and perhaps the CMa dSph) have two GCs of intermediate metallicity and age. An HST/WFPC2 study of the color distributions of dEs in Virgo and Fornax (Lotz, Miller & Ferguson 2004) revealed a rather wide spread in color, consistent with the presence of metal-rich GCs. However, it was not possible to distinguish subpopulations in their small GC samples.

Sharina, Puzia & Makarov (2005) have published a study of GCs in a large sample of dSphs and dIrrs using a heterogeneous set of HST/WFPC2 images. The metal-poor GC peak appears to be ∼0.1 redder in $V - I$ than the peak found by Lotz, Miller, & Ferguson (2004) for their sample of dEs. There is no ready explanation for this finding, which is inconsistent with previous work. It may be that a photometric zero-point offset is to blame, as suggested by Sharina and colleagues themselves. The GC color distributions of both galaxy types have a tail to redder values. This may reflect a small subpopulation of metal-rich GCs, or it could be due to contamination by background objects. There may also be a few metal-rich GCs in the dIrrs studied by Seth et al. (2004).

The existence of metal-rich GCs in dEs has been shown conclusively by Peng et al. (2006a) and J. Strader, J.P. Brodie, L. Spitler & M.A. Beasley (submitted). A large fraction of Virgo dEs, down to quite faint magnitudes (M_B ∼ −15), were found to have bimodal color distributions, analogous to those observed in massive Es. The slope of the metal-rich GC color–galaxy luminosity relation is not well-constrained at these low luminosities owing to the small number of GCs associated with each galaxy. The new data points are consistent with either: (a) a linear extrapolation to lower magnitudes and bluer colors from the region of massive galaxies, or (b) a slight flattening at the low-mass end of the relation. It is reasonable then to ask whether these metal-rich GCs (or at least a subset) could have intermediate ages, like the two younger GCs discovered in the Sgr dSph. Beasley et al. (2006) obtained high-quality spectra for three metal-rich GCs in the Virgo dE VCC 1087. Their old ages are consistent with those of the metal-poor GCs within the errors. Although spectra of similar quality for a large sample of dEs clearly would be desirable, the results to date suggest that there is no obvious dichotomy in the color distributions of dEs and massive Es.

Using data from Peng et al. (2006a), Forbes (2005) has pointed out a possible link between GC bimodality in dEs and the galaxy color bimodality observed in large surveys (e.g., Bell et al. 2004) below a critical mass of $\sim 3 \times 10^{10} M_\odot$. Above this mass, nearly all galaxies in the Virgo Cluster Survey have bimodal GC systems; below this mass, an increasing fraction of galaxies have unimodal color distributions. One interpretation of this phenomenon is that the critical mass represents a transition from "cold," smooth accretion of gas into halos below the critical mass to "hot" accretion of gas that shocks at the virial radius and is unable to form stars (e.g., Dekel & Birnboim 2006).

10.3. Kinematics

Kinematic studies of GCs in dwarfs are challenging, principally because of the small GC systems and lack of luminous GCs. Puzia et al. (2000) found that the velocity dispersion of seven GCs in the luminous dE NGC 3115 DW1 suggested a relatively high $M/L_V \sim 22 \pm 13$. This could suggest the presence of dark matter, or that its parent S0 NGC 3115 is stripping the outermost GCs.

In the Virgo dE VCC 1087, the GCs rotate at $\sim$100 km/s around the minor axis (Beasley et al. 2006). The sample of 12 GCs is dominated by metal-poor GCs, although it includes 3 GCs whose colors and spectroscopic metallicities are consistent with a metal-rich subpopulation (such subpopulations appear to be common in Virgo dEs; see Section 2 and 10.2). Its GC system has the largest rotational support of any galaxy studied to date, with $v/\sigma \sim 3.6$, typical of a disk. This makes VCC 1087 a prime candidate for a dE that evolved from a disky dIrr. We note in passing that, although the LMC is often considered to have a rotating disk population of old metal-poor GCs (e.g., Schommer et al. 1992), van den Bergh (2004) has argued that the current data do not strongly discriminate between disk and halo kinematics for these GCs.

11. GLOBULAR CLUSTER FORMATION

11.1. Classical Scenarios

In Section 2.1 we described the three principal scenarios that have been suggested as explanations for GC bimodality: major disk-disk mergers, in situ formation through multiphase dissipational collapse, and dissipationless accretion. How do these models account for the other observed properties of GC systems? Here we discuss the arguments made in the literature for and against these scenarios, as well as additional constraints from newer data described in this review.

11.1.1. Major mergers. As noted in Section 2.1, the observation of young massive star clusters in many merger remnants throughout the 1990s gave a significant boost to the major disk-disk merger model for GC bimodality (Ashman & Zepf 1992). Although some of these objects definitely have masses and sizes that should allow them to evolve into old GCs (e.g., Maraston et al. 2004; Larsen, Brodie & Hunter 2004), others may have abnormal IMFs that preclude their long-term survival (e.g., McCrady, Gilbert & Graham 2003; Smith & Gallagher 2001; Brodie et al. 1998), though important uncertainties in dynamical mass estimates due to mass segregation remain (McCrady, Graham & Vacca 2005). In a broader context, despite the fact that YMCs and GCs are remarkably similar in many respects, it remains unclear whether (after a Hubble time of evolution) young GC systems will have properties consistent with those of old GC systems in local galaxies. The issue here is that observations of intermediate-age GCs may be at odds with the expected signatures of a simple dynamical evolution scenario (see Section 8).

Even before color bimodality had been observed, several researchers used GCs to constrain the feasibility of the disk-disk merger picture for forming Es. Harris & van den Bergh (1981) noted that typical Es had more populous GC systems than spirals. Massive disk galaxies have $S_N \sim 1$; Es have $S_N \sim 2$–5 depending on environment, with even higher values for brightest cluster galaxies (BCGs) like M87. This is often termed the "S_N problem." Schweizer (1987) explicitly addressed this concern by suggesting that many new GCs might be formed in the merger. As has been pointed out by multiple researchers, this will only raise the S_N if GCs form with a higher efficiency relative to field stars than they did in the protogalactic era. Because GC formation efficiency appears to increase with star-formation rate, S_N may only increase if the star-formation rate in a present-day merger is higher than it was when the GCs in spirals were originally formed.

Several other problems with the major merger model were pointed out in Forbes, Brodie & Grillmair (1997). For example, they showed that there is a correlation between S_N and the fraction of metal-poor GCs, such that the highest S_N galaxies also have the highest proportion of metal-poor GCs. However, the major merger scenario predicts the opposite behavior: The mechanism to increase the S_N of spirals is the formation of new metal-rich GCs in the merger; this should result in larger metal-rich GC subpopulations in more massive Es. Ashman & Zepf (1998) gave a candid analysis of the then current situation on the merger front and suggested that the gEs that dominated the high S_N end of the Forbes and colleagues relation could have augmented their metal-poor GC population by the accretion of lower-mass galaxies (see below) during their complex formation histories. Moreover, they pointed out that the S_N values in the literature were likely very uncertain because so few galaxies had been scrutinized with high-quality wide-field imaging. As discussed in Section 3, more recent work confirms that S_N values for Es tend to come down with improved observations. In general, however, there are still fewer metal-rich than metal-poor GCs in present-day Es, and this remains in conflict with the major merger prediction. Rhode, Zepf & Santos (2005; see Section 3.1) considered the S_N (or T) values of the individual subpopulations and concluded that massive cluster Es cannot have been formed from mergers of local spirals, although some lower-mass field Es could still have formed in this manner. Harris (2001) reached essentially the same conclusion from an analysis of the required gas content and GC formation efficiencies.

Another constraint on the major merger model can be found in the metal-poor GC metallicity–galaxy luminosity correlation (Strader, Brodie & Forbes 2004b). In the mean, the metal-poor GC subpopulations of spirals have lower metallicities than those of massive Es. This seems to be a strong argument against the major merger scenario. However, this conclusion does not take into account the expected effects of biasing—see Section 11.2. It is notable that even some low-mass dEs have bimodal GC color distributions that follow the same peak relations as massive galaxies (Section 10.2), even though these galaxies have presumably not suffered a major merger. So, even if major disk-disk mergers were a viable route to producing bimodality in some cases, they could not be the sole process in operation. In addition, the ages of metal-rich GCs in Es (see Section 4) imply a formation epoch $z > 2$. This restricts most putative major mergers to higher redshifts.

As discussed in Section 2.1, the Forbes, Brodie & Grillmair (1997) multiphase collapse scenario arose as a response to issues with the merger model. There has been little observational evidence to date against the Forbes and colleagues scenario, but, to a considerable extent, this is because it made few specific predictions of observable quantities. Its usefulness was as a framework within which to consider alternative explanations for GC color bimodality, and it identified aspects of the picture still under consideration, e.g., the need to truncate GC metal-poor formation at high redshift. More recent scenarios described below in Section 11.2 are generally consistent with this broad framework.

We emphasize that the arguments presented here against the major merger scenario apply principally to the formation of massive Es from present-day spirals with relatively small bulges. Current GC observations are consistent with dissipational formation of Es at relatively high redshift ($z > 2$)—including major mergers, as long as the disk progenitors have higher S_N values than spirals in the local universe. Subsequent dissipationless merging could then form the most massive gEs, under the constraint of "biased" merging discussed in Section 11.2.

11.1.2. Dissipationless accretion. The accretion scenario of Côté, Marzke & West (1998) was explicitly designed to be consistent with hierarchical structure formation. It assumes a protogalactic GC metallicity–galaxy mass relation produced through a dissipational process at high redshift. The GC systems of present-day galaxies are envisaged to have formed through subsequent dissipationless merging. Because in this scenario the intrinsic GC metallicities of massive protogalaxies are quite high, such galaxies must accrete large numbers of metal-poor GCs from dwarf galaxies to produce bimodality. Côté, West & Marzke (2002) used Monte Carlo simulations to show that the acquisition of the necessary numbers of low-metallicity GCs required the low-mass end of the the protogalactic mass function to have a very steep slope (~ -2). However, even with such a steep mass function, K.M. Ashman, T.M. Walker & S.E. Zepf (in preparation) found that, when they ran simulations similar to those of Côté and colleagues, color distributions like those observed in massive galaxies occurred in only a small fraction ($\sim 5\%$) of their simulations. Another potential problem is that the accreted dwarfs would be expected to contribute many metal-poor field stars that are not observed, unless the dwarfs are primarily gaseous (e.g., Hilker 1998).

The fact that metal-poor GCs in dwarfs have much lower metallicities than those in massive Es (by 0.5–0.6 dex) would, at first sight, seem to be direct evidence against the accretion scenario. This argument was made in Strader, Brodie & Forbes (2004b). However, this line of reasoning does not account for the effects of biased structure formation, which may be the key to properly understanding the implications of the metal-poor GC metallicity–galaxy mass relation.

As already emphasized, in the light of our current understanding of hierarchical galaxy assembly, all galaxy formation scenarios must be accretion/merger scenarios at some level. The major merger and the accretion models (as published) both provided an important focus for theoretical discussion and observational effort by making fairly explicit predictions against which the observations could be compared. The preponderance of new evidence now suggests that, though elements of each remain viable, the details are pointing us in new directions (see below).

11.2. Hierarchical Merging and Biasing: Recent Scenarios

Beasley et al. (2002) explored GC bimodality in a cosmological context using the semianalytic galaxy formation model of Cole et al. (2000), and this work contained elements of all three classic scenarios. While largely phenomenological, it makes the most specific predictions of any model proposed, and because the scenario is in the context of a full model of cosmological structure formation, it implicitly accounts for many of the issues discussed below (e.g., biasing). Metal-poor GCs were assumed to form in the early universe in gas disks in low-mass dark-matter halos. As in Forbes, Brodie & Grillmair (1997), Beasley and colleagues found it necessary to invoke the truncation of metal-poor GC formation at high redshift (in this case, $z > 5$) in order to produce bimodality. The metal-rich GCs were generally formed during gas-rich mergers. Their predictions for the metal-rich subpopulation included: a correlation between GC metallicity and galaxy luminosity, significant age and metallicity substructure, and decreasing mean ages and metallicities in lower-density environments.

Following suggestions from Santos (2003), Strader et al. (2005) and Rhode, Zepf & Santos et al. (2005) proposed hierarchical scenarios for GC formation intended to account for the metal-poor GC metallicity–galaxy mass relation and the correlation of metal-poor S_N (or T_{blue}) with galaxy mass. Metal-poor GCs are proposed to form in low-mass dark-matter halos at very high redshift, typically $z \sim 10$–15. Halos in high-density environments collapse first. As discussed in Strader and colleagues, this scenario can reproduce the observed correlations with galaxy mass, given reasonable assumptions (including the truncation of metal-poor GC formation at high z, plausibly by reionization). It can also explain other observations, such as the radial distribution of metal-poor GCs (Moore et al. 2006) and possibly the mass–metallicity relation for individual metal-poor GCs (J. Strader, J.P. Brodie, L. Spitler & M.A. Beasley, submitted; Harris et al. 2006). Metal-rich GCs form in the subsequent dissipational merging that forms the host galaxy. When the bulk of this "action" took place is not well constrained, but, for galaxies at $\sim L^*$ and above, most GCs appear to have formed at $z > 2$ (Strader et al. 2005; Puzia et al. 2005). Some additional dissipationless merging for massive Es appears to be required, based on the evolution of the "red sequence" luminosity function of early-type galaxies from $z \sim 1$ to the present (Bell et al. 2004; S. M. Faber, C. N. A. Willmer, C. Wolf, D. C. Koo, B. J. Weiner, et al., submitted) and the dichotomy of core parameters (Section 4.1). The ages of metal-rich GCs do not constrain such dissipationless merging, but in the future the radial distributions and kinematics may offer interesting insights. We call this picture of GC formation the synthesis scenario.

It has been mentioned several times in the preceding sections that biasing is a key factor in understanding structure formation. Could the metal-poor GC subpopulations of massive Es have been built from major mergers of present-day lower-mass disk galaxies or by the accretion of many dwarf galaxies, as both dwarfs and spirals have metal-poor GCs of lower metallicity than those in massive Es?

To simultaneously accommodate (a) the metal-poor GC metallicity–galaxy mass relation and (b) the theoretical and observational evidence that most massive galaxies have undergone some degree of merging/accretion because $z \sim 2$, we must argue that the metal-poor GC relation was different at higher redshift. A present day L^* galaxy cannot have been assembled from present-day sub-L^* galaxies. Instead, the merging must have been biased, in the sense that galaxies with metal-poor GC systems that would lie above the relation connecting GC metallicity and host galaxy mass at $z = 0$ would tend to have merged into more massive galaxies by the present (see **Figure 12** for a schematic diagram of this process). This can be understood as a direct result of hierarchical structure formation: high-σ peaks in the most overdense regions (destined to become, e.g., galaxy clusters) collapse and form metal-poor GCs first. These metal-poor GCs will be more highly enriched than those forming in halos that collapse later, either because they have more time to self-enrich, or because the density of nearby star-forming halos is larger and they could capture more outflowing enriched gas. These first-forming metal-poor GCs will tend to be concentrated toward the center of the overdensity and will quickly agglomerate into larger structures. Similar mass fluctuations in the less-overdense outer regions will tend to be accreted into larger structures more slowly. Some may survive to form more stars and become dwarf satellites of the central galaxy. This picture, at least as it relates to dark-matter halos, is well-understood and accepted. But the important point for GC formation scenarios is that these surviving dwarfs are not representative of the halos that merged to form the central galaxy. The latter collapsed first and may have very different star- (and GC-) formation histories from those that collapsed later.

This process will operate on a variety of scales. For example, the dwarf satellites of the Galaxy have metal-poor GCs with lower metallicities than those of halo GCs in the Galaxy. Moreover, the disk or E galaxies that merged to form gEs like M87 and NGC

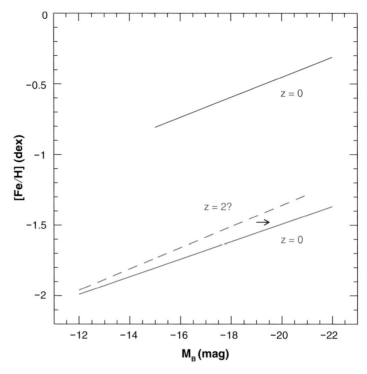

Figure 12
A schematic plot of the evolution of the metal-poor globular cluster metallicity–galaxy luminosity relation due to biased galaxy merging. The solid lines show the $z=0$ relations for both subpopulations; the dashed blue line shows a conceptual metal-poor relation at higher redshift.

4472 must have had metal-poor GCs with metallicities higher than those typical for Es and spirals in the Virgo cluster today. The metal-poor GCs, although they formed at very high redshift, already "knew" to which galaxy they would ultimately belong. The metal-poor relation rules out merger and accretion models, but only in the local universe for structure forming at the present day. Nonetheless, it is consistent with hierarchical galaxy formation, and is a strong end constraint for any galaxy formation

Figure 13
Two snapshots of a high-resolution dark-matter simulation of the formation of a $\sim 10^{12} M_\odot$ galaxy (Moore et al. 2006, Diemand, Madau & Moore 2005). The top panel represents the simulation at $z=12$, and the bottom panel represents the present day. The blue to pink colors indicate dark matter of increasing density, whereas the green regions are those at $z=12$ with virial temperatures $>10^4$ K (such that atomic line cooling is effective, and gas can cool to form stars). At $z=12$, these green regions represent halos with masses 10^8–$10^{10} M_\odot$, and these same green particles are marked in the $z=0$ snapshot. These high-σ peaks collapse in a filamentary structure at high z but are concentrated toward the center of the final galaxy. The boxes can be identified as dwarf satellites of the final galaxy.

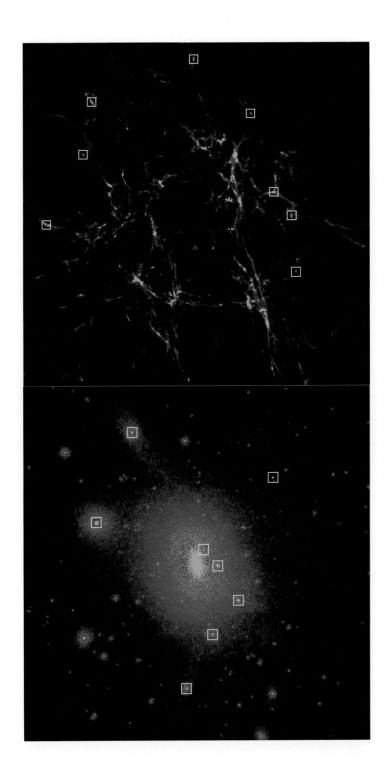

model. To illustrate biasing, **Figure 13** shows two snapshots ($z = 12$ and 0) of a high-resolution dark-matter simulation of the formation of a $10^{12} M_\odot$ galaxy (Diemand, Madau & Moore 2005; Moore et al. 2006). Low-mass, high-σ peaks collapse first in a filamentary structure and end up centrally concentrated in the final galaxy.

12. COSMOLOGICAL FORMATION OF METAL-POOR GLOBULAR CLUSTERS

The preceding sections suggest that the bulk of GCs formed at high redshift. It then follows that they have enormous promise in a cosmological context. In this section we discuss current ideas on the "cosmological" formation of GCs and how this has shaped our overall view of galaxy formation. In what follows, the term cosmological refers to models in which GCs form in low-mass dark-matter halos, before the bulk of their parent galaxy has been assembled.

Soon after CDM cosmology was proposed, Peebles (1984) argued that $10^8 M_\odot$ halos, each hosting several $10^6 M_\odot$ of gas, would be the first to collapse and form stars in the early universe. He suggested GCs as their progeny and noted that the halos of these GCs might be stripped without disrupting the cluster itself. This basic idea has been sustained to the present day, though the mechanism is probably limited to metal-poor GCs. Rosenblatt, Faber & Blumenthal (1988) refined this scenario by suggesting that metal-poor GCs form in 2.8σ halos. This gives a reasonable match to the observed radial distribution of metal-poor Galactic GCs and the mass fraction of metal-poor GCs (with respect to total stellar mass) in a variety of galaxies. The principal problem with this picture—indeed, of GC formation in individual dark-matter halos in general—is that the minimum required baryonic collapse factor (~ 10 or more) would produce GCs with more rotation than that observed, unless they form preferentially in very low-spin halos, or some other mechanism acts to remove angular momentum. A speculative solution might be to form the GC in the core of a larger gas cloud, if the "extraneous" material can be stripped later.

Moore (1996) wrote a short influential paper that described the use of N-body simulations to show that the faint tidal tails observed around some GCs (Grillmair et al. 1995) were inconsistent with the presence of extended dark-matter halos, but consistent with the low M/L ratios observed in the central regions (e.g., Illingworth 1976; Pryor et al. 1989). The observed tidal tails around, e.g., Pal 5 (Odenkirchen et al. 2001) demonstrate that there are at least some present-day GCs that lack dark matter. However, this does not prove that all (or even most) GCs are free of extended dark matter halos, nor does it rule out metal-poor GC formation inside halos that are later stripped away.

This latter idea has been developed in a number of papers by Mashchenko & Sills (2005a,b), who studied the formation and evolution of GCs with individual dark-matter halos. With high-resolution N-body simulations, they found that (depending on the details of the actual collapse) many properties of simulated GCs with halos are similar to those of observed GCs. For example, the central M/L ratios are expected to be quite low. Structure in the halo (e.g., triaxiality, or breaks in the outer parts of the density profile) could easily be manifested as tidal cutoffs, extratidal stars, or eccentric outer contours; such features are not incompatible with the presence of dark matter. When a GC with a dark-matter halo evolves in a tidal field, their simulations indicated that it loses either most [for a Navarro-French-White (NFW) halo] or nearly all (for a Burkert halo) of its dark matter. This finding was confirmed in a very high resolution dark-matter and gas simulation by Saitoh et al. (2006).

Bromm & Clarke (2002) used a simulation with both dark matter and gas to study GC formation at high redshift. As noted by Peebles (1984), $\sim 10^8 M_\odot$ minihalos are expected to collapse out of 3σ fluctuations at $z \sim 15$. At a high fixed gas density threshold Bromm & Clarke created sink particles as "GCs." These GCs initially form inside of halos, but the simultaneous collapse of mass scales results in violent relaxation that erases most of the substructure. The resulting GC mass spectrum is set by that of the dark matter and is a power law with index ~ -1.8. The main problem here is that it is not possible to tell, with the current level of sophistication of their simulations, whether the violent relaxation is real or merely an artifact of insufficient resolution.

Other authors have explored in more detail the triggering mechanism for putative GC formation in dark-matter halos. In the model of Cen (2001), ionization fronts from cosmic reionization shock gas in low-mass halos. The gas is compressed by a factor of ~ 100 and collapses to form metal-poor GCs. To produce the observed numbers of GCs, a large population of low-spin ($\lambda < 0.01$) halos is required. This condition appears to be satisfied when the halo number density is modeled with extended Press-Schecter theory, but whether it would hold in high-resolution cosmological simulations is unknown. This model predicts a power-law GCMF with a slope of ~ -2, similar to that observed at the high-mass end, and has the rather attractive property of predicting no GC mass–radius relation, consistent with the observations. Côté (2002) pointed out that this picture could predict a large number of (unobserved) intergalactic GCs; this objection might be addressed if the ionization fronts are effective above a threshold only met quite close to protogalaxies.

Scannapieco, Weisheit & Harlow (2004) proposed a somewhat similar mechanism, in which gas in minihalos is shock-compressed by galaxy outflows. The momentum of the shock strips the gas from the halo, nicely solving the dark-matter problem. However, this model predicts a mass–radius relation for individual GCs, and the observed lack of such a relation may deal this picture a fatal blow.

Ricotti (2002) suggested that GCs themselves could have reionized the universe. The predicted number of ionizing photons appears to be sufficient, assuming that the escape fraction of such photons is near unity. Such a high escape fraction is qualitatively feasible, given the extended spatial distribution of the GCs with respect to their parent galaxies, but the scenario requires detailed modeling (including radiative transfer) in the proper cosmological setup.

The high-resolution simulation of Kravtsov & Gnedin (2005) offers a glimpse of what should be possible in the future. They performed a gas and dark-matter simulation of the formation of a Milky Way analogue to $z \sim 3$. They were not able to resolve GC formation directly, but assumed GCs formed in the cores of giant molecular clouds (GMCs) when the dynamical time exceeded the cooling time. These GMCs were located in the flattened gas disks of protogalaxies. The resulting mass function

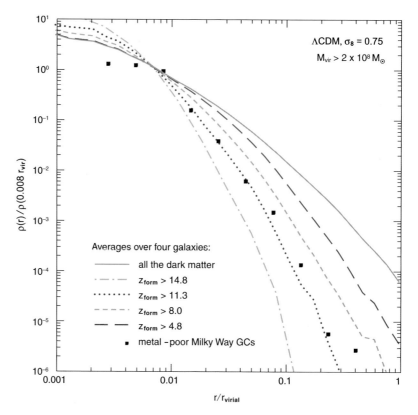

Figure 14

The radial distribution of metal-poor globular clusters in the Galaxy compared to results from numerical simulations of the formation of a Galaxy-like dark-matter halo in a Λ CDM cosmology (Moore et al. 2006). The lines represent the radial distribution of $2 \times 10^8 M_\odot$ mass halos that collapsed before: $z = 14.8$ (*light blue dot-dashed*), $z = 11.3$ (*red dotted*), $z = 8.0$ (*short green dashed*), $z = 4.8$ (*long dark blue dashed*), and the cumulative distribution (*solid*). The higher the redshift of collapse, the rarer the peak. Under the assumption that reionization truncates star formation in such halos, the comparison suggests reionization at $z \sim 10$. This redshift and mass combination corresponds to $\sim 2.5\sigma$ peaks.

appears to be consistent with that of massive GCs, but the metallicity distribution does not: At the end of the simulation, large numbers of (unobserved) [Fe/H] ~ -1 GCs were being formed, even though their simulation included feedback. A desirable future extension of such simulations is to test whether reionization might effectively end metal-poor GC formation where "traditional" stellar feedback cannot.

Moore et al. (2006) have shown that the observed radial distribution of metal-poor GCs in the Galaxy can be reproduced if the GCs are assumed to form in $>2.5\sigma$ peaks of $>2 \times 10^8 M_\odot$ that collapse and form stars before $z \sim 10$ (see **Figure 14**). If the key assumption of truncation by reionization is supported by other evidence, then the surface density distributions of metal-poor GCs could be used to probe reionization in a variety of galaxies, setting limits on the epoch and homogeneity of reionization.

In summary, the cosmological formation of metal-poor GCs is supported by several lines of argument.

1. The ages of metal-poor GCs. The absolute ages of GCs are still poorly known, and must be continually revised in light of advances in a variety of subfields (for example, the recent revision of the $^{14}N(p,\gamma)^{15}O$ reaction rate increases GC ages by 0.7–1 Gyr; Imbriani et al. 2004). If the GC ages are sufficiently close to the age of the universe, cosmological formation becomes a necessity, because low-mass halos as described above are the only existing sites for star formation.
2. The recently discovered correlation between GC metallicity and mass for bright metal-poor GCs in several massive galaxies (J. Strader, J.P. Brodie, L. Spitler & M.A. Beasley (submitted); Harris et al. 2006). Self-enrichment is a potential explanation for this correlation, and it is possible that metals could only be retained in the potential well of a dark-matter halo.
3. The radial distribution of metal-poor GCs. Moore et al. (2006) show that the metal-poor GCs in the Galaxy have a radial distribution consistent with formation in 2.5σ peaks in the dark-matter distribution at $z > 10$.
4. Observations of Local Group dwarfs show that the lowest-mass galaxies with GCs have total masses of $\lesssim 10^8 M_\odot$ (see Section 10), as expected under cosmological formation.

Does this mean that two separate mechanisms are needed to explain metal-poor and metal-rich GC formation? The strongest similarity between the two subpopulations is their mass function; for the less evolved high-mass part of the mass function, these are approximately power laws with indices of ~ -1.8 to -2. Because power-law distributions are a consequence of a variety of physical processes, this similarity does not mandate an identical formation history for both subpopulations. For example, the power-law slope observed in the GC mass function is the same as both that of GMCs in the Galaxy and of low-mass dark-matter halos collapsing at $z \sim 15-20$.

13. FUTURE DIRECTIONS

Those of us actively trying to understand extragalactic globular clusters and their connection to galaxy formation are currently operating in a data-dominated rather than a theory-dominated field. We are accumulating a wealth of observational information that we cannot fully interpret. An urgent need exists for improvements in numerical and semianalytic simulations to help identify GC formation sites and track their spatial, kinematic, chemical, and structural evolution. Models that can resolve the masses and sizes of a typical GC are tantalizingly close to implementation, and their advent will signal a leap in progress toward placing the formation of GCs in its proper cosmological context. With the new generation of "big telescope" wide-field multiplexing spectrographs, such as Keck/DEIMOS, Magellan/IMACS, MMT/Hectospec, and VLT/VIMOS, it is possible to study large samples of GCs in a wide variety of galaxies to carry out detailed tests of these future models, establishing ages, metallicities, and kinematics.

Other important developments will come from the community of SSP modelers. As already mentioned, there remain significant disagreements within this community on the treatment of α-enhancement, HB morphology (the second parameter problem), and underlying stellar synthesis techniques. A new way forward to study extragalactic GCs in detail is the SSP modeling of high-resolution spectra (e.g., Bernstein & McWilliam 2006). This offers the possibility of estimating abundances of light, α-, Fe-peak, and even some strong r- and s-process elements. With an 8–10-m class telescope, this technique can be applied to significant samples of GCs in the Local Group and nearby galaxies, and even to the brightest few GCs in Virgo. This program is in its infancy but could represent a significant leap in our understanding of the detailed formation histories of galaxies. GCs may offer the only route to measuring the abundances of interesting elements that are unobservable in massive galaxies themselves due to their large velocity dispersions. A possibility enabled by multiplexing high-resolution spectroscopy is detailed dynamical modeling of individual Galactic GCs, to determine whether any contain halo dark matter. The discovery of dark matter in GCs would be a "smoking gun" of cosmological GC formation.

With the development of new large-format detectors and CCD mosaics, wide-field optical imaging is poised to address numerous outstanding issues. Obtaining global radial and color distributions for individual GC subpopulations is essential to test scenarios for GC and galaxy formation, especially as models emerge that predict these quantities in detail. Such imaging can also be used to probe the evolution of the GCLF, and provide a definitive test of the scenario in which GCs are formed with a power-law LF that subsequently evolves through various destruction processes to the log-normal LF observed for old GC systems. Because such processes are expected to operate more efficiently at small galactocentric radii, changes in the GCLF of GCs with galactocentric radius would be revealing. Such imaging has the potential to constrain the epoch of reionization from the spatial distribution of metal-poor GCs (as discussed in Section 12).

These ideas cover only a small fraction of the important advances likely to occur in the field over the next decade. The eventual availability of 30-m class telescopes and JWST may allow us to reach the inspiring goal of Renzini (2002): "To directly map the evolution of GCs in galaxies all the way to see them in formation, and eventually stick on the wall a poster with a million-pixel picture of a $z = 5$ galaxy, with all her young GCs around."

ACKNOWLEDGMENTS

We thank many colleagues for reading drafts of this manuscript and for useful discussions, including Keith Ashman, Michael Beasley, Javier Cenarro, Laura Chomiuk, Juerg Diemand, Sandra Faber, Duncan Forbes, Genevieve Graves, Soeren Larsen, John Kormendy, Piero Madau, Joel Primack, Katherine Rhode, Tom Richtler, Brad Whitmore, and Steve Zepf. We also thank Michael Beasley, Andres Jordán, and Tom Richtler for permission to use their figures. Michael Beasley, Soeren Larsen, and Katherine Rhode provided data used to create other figures. Finally, we thank Takayuki Saitoh and Marsha Wolf for providing advance copies of their articles.

Support was provided by NSF Grants AST-0206139 and AST-0507729, an NSF Graduate Research Fellowship, and STScI grant GO-9766.

LITERATURE CITED

Abadi M, Navarro J, Steinmetz M. 2006 In press (astro-ph/0506659)
Armandroff TE, Zinn R. 1988. *Astron. J.* 96:92
Ashman KM, Bird CM. 1993. *Astron. J.* 106:2281
Ashman KM, Conti A, Zepf SE. 1995. *Astron. J.* 110:1164
Ashman KM, Zepf SE. 1992. *Ap. J.* 384:50
Ashman KM, Zepf SE. 1998. *Globular Cluster Systems*. New York: Cambridge Univ.
Ashman KM, Zepf SE. 2001. *Astron. J.* 122:1888
Barmby P, Huchra JP. 2001. *Astron. J.* 122:2458
Barmby P, Huchra JP, Brodie JP, Forbes DA, Schroder LL, Grillmair CJ. 2000. *Astron. J.* 119:727
Bassino LP, Faifer FR, Forte JC, Dirsch B, Richtler T, et al. 2006. *Astron. Astrophys.* 45:789
Bassino LP, Richtler T, Dirsch B. 2006. *MNRAS* 367:156
Baumgardt H, Makino J. 2003. *MNRAS* 340:227
Beasley MA, Baugh CM, Forbes DA, Sharples RM, Frenk CS. 2002. *MNRAS* 333:383
Beasley MA, Brodie JP, Strader J, Forbes DA, Proctor RN, et al. 2004. *Astron. J.* 128:1623
Beasley MA, Brodie JP, Strader J, Forbes DA, Proctor RN, et al. 2005. *Astron. J.* 129:1412
Beasley MA, Strader J, Brodie JP, Cenarro AJ, Geha M. 2006. *Ap. J.* 131:814
Beasley MA, Sharples RM, Bridges TJ, Hanes DA, Zepf SE, et al. 2000. *MNRAS* 318:1249
Bekki K, Beasley MA, Brodie JP, Forbes DA. 2005. *MNRAS* 363:1211
Bekki K, Couch WJ, Drinkwater MJ, Shioya Y. 2003. *MNRAS* 344:399
Bekki K, Forbes DA. 2006. *Astron. Astrophys.* 445:485
Bekki K, Forbes DA, Beasley MA, Couch WJ. 2002. *MNRAS* 335:1176
Bekki K, Forbes DA, Beasley MA, Couch WJ. 2003. *MNRAS* 344:1334
Bell EF, Wolf C, Meisenheimer K, Rix H, Borch A, et al. 2004. *Ap. J.* 608:752
Bergond G, Zepf SE, Romanowsky AJ, Sharples RM, Rhode KL. 2006. *Astron. Astrophys.* 448:155
Bernstein R, McWilliam A. 2006. Resolved stellar populations. In *ASP Conf. Ser.*, ed. D Valls-Gabaud, M Chavez. San Francisco: ASP. In press
Blakeslee JP, Tonry JL, Metzger MR. 1997. *Astron. J.* 114:482
Brodie JP. 1981. *Integrated spectrophotometric properties of globular clusters*. PhD thesis. Cambridge Univ. 94 pp.
Brodie JP, Burkert A, Larsen SS. 2004. In *ASP Conf. Ser. 322: The Formation and Evolution of Massive Young Star Clusters*, ed. H Lamers, L Smith, A Nota, p. 139. San Francisco: ASP
Brodie JP, Hanes DA. 1986. *Ap. J.* 300:258
Brodie JP, Huchra JP. 1990. *Ap. J.* 362:503

Brodie JP, Huchra JP. 1991. *Ap. J.* 379:157
Brodie JP, Larsen SS. 2002. *Astron. J.* 124:1410
Brodie JP, Larsen SS, Kissler-Patig M. 2000. *Ap. J. Lett.* 543:19
Brodie JP, Schroder LL, Huchra JP, Phillips AC, Kissler-Patig M, Forbes DA. 1998. *Astron. J.* 116:691
Brodie JP, Strader J, Denicol G, Beasley MA, Cenarro AJ, et al. 2005. *Astron. J.* 129:2643
Bromm V, Clarke CJ. 2002. *Ap. J. Lett.* 566:L1
Brown TM, Ferguson HC, Smith E, Kimble RA, Sweigart AV, et al. 2003. *Ap. J. Lett.* 592:17
Brown TM, Ferguson HC, Smith E, Kimble RA, Sweigart AV, et al. 2004. *Ap. J. Lett.* 613:125
Burgarella D, Kissler-Patig M, Buat V. 2001. *Astron. J.* 121:2647
Burkert A, Brodie J, Larsen S. 2005. *Ap. J.* 628:231
Burkert A, Smith GH. 2000. *Ap. J.* 542:95
Burstein D. 1987. See Faber 1987, p. 47
Burstein D, Faber SM, Gaskell CM, Krumm N. 1984. *Ap. J.* 287:586
Burstein D, Li Y, Freeman KC, Norris JE, Bessell MS, et al. 2004. *Ap. J.* 614:158
Carney B. 2001. *Star Clusters: Saas-Fee Advanced Course*, 28:1. New York: Springer-Verlag
Carney BW. 1996. *Publ. Astron. Soc. Pac.* 108:900
Carney BW, Latham DW, Laird JB. 1990. *Astron. J.* 99:572
Carretta E, Gratton RG. 1997. *Astron. Astrophys. Suppl.* 121:95
Cen R. 2001. *Ap. J.* 560:592
Chaboyer B, Demarque P, Sarajedini A. 1996. *Ap. J.* 459:558
Chandar R, Bianchi L, Ford HC, Sarajedini A. 2002. *Ap. J.* 564:712
Chandar R, Whitmore B, Lee MG. 2004. *Ap. J.* 611:220
Cohen JG, Blakeslee JP. 1998. *Astron. J.* 115:2356
Cohen JG, Blakeslee JP, Côté P. 2003. *Ap. J.* 592:866
Cohen JG, Blakeslee JP, Ryzhov A. 1998. *Ap. J.* 496:808
Cohen JG, Matthews K, Cameron PB. 2005. *Ap. J. Lett.* 634:45
Cole S, Lacey CG, Baugh CM, Frenk CS. 2000. *MNRAS* 319:168
Côté P. 1999. *Astron. J.* 118:406
Côté P. 2002. *New Horizons in Globular Cluster Astronomy. ASP Conf. Ser.*, ed. G Piotto, G Meylan, G Djorgovski, M Riello, p. 457. San Francisco: ASP
Côté P, Blakeslee JP, Ferrarese L, Jordán A, Mei S, et al. 2004. *Ap. J. Suppl.* 153:223
Côté P, Marzke RO, West MJ. 1998. *Ap. J.* 501:554
Côté P, Marzke RO, West MJ, Minniti D. 2000. *Ap. J.* 533:869
Côté P, McLaughlin DE, Cohen JG, Blakeslee JP. 2003. *Ap. J.* 591:850
Côté P, McLaughlin DE, Hanes DA, Bridges TJ, Geisler D, et al. 2001. *Ap. J.* 559:828
Côté P, West MJ, Marzke RO. 2002. *Ap. J.* 567:853
Couture J, Harris WE, Allwright JWB. 1991. *Ap. J.* 372:97
Cowie LL, Songaila A, Hu EM, Cohen JG. 1996. *Astron. J.* 112:839
Davies RL, Kuntschner H, Emsellem E, Bacon R, Bureau M, et al. 2001. *Ap. J. Lett.* 548:33 *Ap. J. Lett.* 548:33

Dekel A, Birnboim Y. 2006. *MNRAS* 368:2

Dekel A, Silk J. 1986. *Ap. J.* 303:39

Dekel A, Stoehr F, Mamon GA, Cox TJ, Novak GS, Primack JR. 2005. *Nature* 437:707

Diemand J, Madau P, Moore B. 2005. *MNRAS* 364:367

Dirsch B, Richtler T, Geisler D, Forte JC, Bassino LP, Gieren WP. 2003. *Astron. J.* 125:1908

Dirsch B, Schuberth Y, Richtler T. 2005. *Astron. Astrophys.* 433:43

Dressler A, Oemler A Jr, Couch WJ, Smail I, Ellis RS, et al. 1997. *Ap. J.* 490:577

Durrell PR, Harris WE, Pritchet CJ. 1994. *Astron. J.* 108:2114

Durrell PR, Harris WE, Pritchet CJ. 2001. *Astron. J.* 121:2557

Durrell PR, McLaughlin DE, Harris WE, Hanes DA. 1996. *Ap. J.* 463:543

Elmegreen BG. 1999. *ApSS* 269:469

Elmegreen BG, Efremov YN. 1997. *Ap. J.* 480, 235

Evans NW, Wilkinson MI. 2000. *MNRAS* 316:929

Faber SM, ed. 1987. *Nearly Normal Galaxies: From the Planck Time to the Present. Proc. 8th Santa Cruz Summer Workshop Astron. Astrophys*, p. 47. New York: Springer-Verlag

Faber SM, Tremaine S, Ajhar EA, Byun Y-I, Dressler A, et al. 1997. *Astron. J.* 114:1771

Fall SM, Zhang Q. 2001. *Ap. J.* 561:751

Font AS, Johnston KV, Guhathakurta P, Majewski SR, Rich RM. 2006. *Ap. J.* 131:1436

Forbes DA. 2005 *Ap. J. Lett.* 635:137

Forbes DA, Beasley MA, Brodie JP, Kissler-Patig M. 2001. *Ap. J. Lett.* 563:143

Forbes DA, Brodie JP, Grillmair CJ. 1997. *Astron. J.* 113:1652

Forbes DA, Brodie JP, Larsen SS. 2001. *Ap. J. Lett.* 556:83

Forbes DA, Forte JC. 2001. *MNRAS* 322:257

Forbes DA, Franx M, Illingworth GD, Carollo CM. 1996. *Ap. J.* 467:126

Forbes DA, Georgakakis AE, Brodie JP. 2001. *MNRAS* 325:1431

Forbes DA, Grillmair CJ, Williger GM, Elson RAW, Brodie JP. 1998. *MNRAS* 293:325

Forbes DA, Masters KL, Minniti D, Barmby P. 2000. *Astron. Astrophys.* 358:471

Forbes DA, Strader J, Brodie JP. 2004. *Astron. J.* 127:3394

Forte JC, Faifer F, Geisler D. 2005. *MNRAS* 357:56

Forte JC, Geisler D, Ostrov PG, Piatti AE, Gieren W. 2001. *Astron. J.* 121:1992

Fukugita M, Hogan CJ, Peebles PJE. 1998. *Ap. J.* 503:518

Fulbright JP, McWilliam A, Rich RM. 2006. *Ap. J.* 636:821

Fusi Pecci F, Bellazzini M, Buzzoni A, De Simone E, Federici L, Galleti S. 2005. *Astron. J.* 130:554

Gebhardt K, Kissler-Patig M. 1999. *Astron. J.* 118:1526

Geisler D, Lee MG, Kim E. 1996. *Astron. J.* 111:1529

Grillmair CJ, Freeman KC, Irwin M, Quinn RJ. 1995 *Astron. J.* 109:2553

Goerdt T, Moore B, Read JI, Stadel J, Zemp M. 2006. *MNRAS* 368:1073

Gomez M, Richtler T. 2004. *Astron. Astrophys.* 415:499

Goudfrooij P, Alonso MV, Maraston C, Minniti D. 2001a. *MNRAS* 328:237

Goudfrooij P, Gilmore D, Whitmore BC, Schweizer F. 2004. *Ap. J. Lett.* 613:121

Goudfrooij P, Mack J, Kissler-Patig M, Meylan G, Minniti D. 2001b. *MNRAS* 322:643

Goudfrooij P, Strader J, Brenneman L, Kissler-Patig M, Minniti D, Edwin Huizinga J. 2003. *MNRAS* 343:665

Gratton R, Sneden C, Carretta E. 2004. *Annu. Rev. Astron. Astrophys.* 42:385

Grebel EK, Dolphin AE, Guhathakurta P. 2000. *Astron. Ges. Abstr. Ser.* 17:79 (Abstr.)

Gregg MD. 1994. *Astron. J.* 108:2164

Hansen SH, Moore B. 2006. *New Astron. Rev.* 11:333

Harris GLH, Geisler D, Harris HC, Hesser JE. 1992. *Astron. J.* 104:613

Harris GLH, Harris WE, Geisler D. 2004. *Astron. J.* 128:723

Harris GLH, Harris WE, Poole GB. 1999. *Astron. J.* 117:855

Harris WE. 1986. *Astron. J.* 91:822

Harris WE. 1991. *Annu. Rev. Astron. Astrophys.* 29:543

Harris WE. 2001. *Star Clusters: Saas-Fee Advanced Course*, 28:223. New York: Springer-Verlag

Harris WE. 2003. *Extragalactic Globular Cluster Systems*, ed. M Kissler-Patig, *ESO Astron. Symp.*, p. 317. Berlin: Springer-Verlag

Harris WE, Harris GLH. 2002. *Astron. J.* 123:3108

Harris WE, Harris GLH, Holland ST, McLaughlin DE. 2002. *Astron. J.* 124:1435

Harris WE, Harris GLH, McLaughlin DE. 1998. *Astron. J.* 115:1801

Harris WE, Kavelaars JJ, Hanes DA, Hesser JE, Pritchet CJ. 2000. *Ap. J.* 533:137

Harris WE, Pudritz RE. 1994. *Ap. J.* 429:177

Harris WE, van den Bergh S. 1981. *Astron. J.* 86:1627

Harris WE, Whitmore BC, Karakla D, Okoń W, Baum WA, et al. 2006. *Ap. J.* 636:90

Hempel M, Hilker M, Kissler-Patig M, Puzia TH, Minniti D, Goudfrooij P. 2003. *Astron. Astrophys.* 405:487

Hempel M, Kissler-Patig M. 2004. *Astron. Astrophys.* 428:459

Hilker M. 1998. *The center of the Fornax cluster: dwarf galaxies, cD halo, and globular clusters*. PhD thesis. Sternwarte Bonn. 86 pp.

Hilker M, Infante L, Richtler T. 1999. *Astron. Astrophys. Suppl.* 138:55

Hodge PW, Dolphin AE, Smith TR, Mateo M. 1999. *Ap. J.* 521:577

Holtzman JA, Faber SM, Shaya EJ, Lauer TR, Groth J, et al. 1992. *Astron. J.* 103:691

Hopp U, Wagner SJ, Richtler T. 1995. *Astron. Astrophys.* 296:633

Huxor AP, Tanvir NR, Irwin MJ, Ibata R, Collett JL, et al. 2005. *MNRAS* 360:1007

Hwang N, Lee MG. 2006. *Ap. J. Lett.* 638:79

Hwang N, Lee MG, Lee JC, Park W, Park HS, et al. 2005. *IAU Colloq. 198: Near-Fields Cosmology with Dwarf Elliptical Galaxies*, p. 257

Illingworth G. 1976. *Ap. J.* 204:73

Imbriani G, Costantini H, Formicola A, Bemmerer D, Bonetti R, et al. 2004. *Astron. Astrophys.* 420:625

Jordán A. 2004. *Ap. J. Lett.* 613:117

Jordán A, Côté P, Blakeslee JP, Ferrarese L, McLaughlin DE, et al. 2005. *Ap. J.* 634:1002

Kalirai JS, Gilbert KM, Guhathakurta P, Majewski SR, Ostheimer JC, et al. 2006. *Ap. J.* In press

Karachentsev ID, Karachentseva VE, Dolphin AE, Geisler D, Grebel EK, et al. 2000. *Astron. Astrophys.* 363:117

Karachentsev ID, Sharina ME, Grebel EK, Dolphin AE, Geisler D, et al. 2000. *Ap. J.* 542:128

Kent SM. 1992. *Ap. J.* 387:181

Kissler-Patig M. 2000. In *Reviews in Modern Astronomy*, Vol. 13, *New Astrophysical Horizons*, ed. RE Schielicke, p. 13. Hamburg, Ger.: Astron. Ges.

Kissler-Patig M, Ashman KM, Zepf SE, Freeman KC. 1999. *Astron. J.* 118:197

Kissler-Patig M, Brodie JP, Schroder LL, Forbes DA, Grillmair CJ, Huchra JP. 1998. *Astron. J.* 115:105

Kissler-Patig M, Forbes DA, Minniti D. 1998. *MNRAS* 298:1123

Kissler-Patig M, Kohle S, Hilker M, Richtler T, Infante L, Quintana H. 1997. *Astron. Astrophys.* 319:470

Kormendy J. 1985. *Ap. J.* 295:73

Kormendy J. 1987. See Faber 1987, p. 163

Kormendy J, Bender R. 1996. *Ap. J.* 464:119

Kormendy J, Kennicutt RC Jr. 2004. *Annu. Rev. Astron. Astrophys.* 42:603

Korn AJ, Maraston C, Thomas D. 2005. *Astron. Astrophys.* 438:685

Kraft RP, Ivans II. 2003. *Publ. Astron. Soc. Pac.* 115:804

Kravtsov AV, Gnedin OY. 2005. *Ap. J.* 623:650

Kravtsov AV, Gnedin OY, Klypin AA. 2004. *Ap. J.* 609:482

Kundu A, Whitmore BC. 1998. *Astron. J.* 116:2841

Kundu A, Whitmore BC. 2001a. *Astron. J.* 121:2950

Kundu A, Whitmore BC. 2001b. *Astron. J.* 122:1251

Kundu A, Whitmore BC, Sparks WB, Macchetto FD, Zepf SE, Ashman KM. 1999. *Ap. J.* 513:733

Kundu A, Zepf SE, Hempel M, Morton D, Ashman KM, et al. 2005. *Ap. J. Lett.* 634:41

Kuntschner H, Ziegler BL, Sharples RM, Worthey G, Fricke KJ. 2002. *Astron. Astrophys.* 395:761

Larsen SS, Brodie JP. 2000. *Astron. J.* 120:2938

Larsen SS, Brodie JP. 2003. *Ap. J.* 593:340

Larsen SS, Brodie JP, Beasley MA, Forbes DA. 2002a. *Astron. J.* 124:828

Larsen SS, Brodie JP, Beasley MA, Forbes DA, Kissler-Patig M, et al. 2003. *Ap. J.* 585:767

Larsen SS, Brodie JP, Hunter DA. 2004. *Astron. J.* 128:2295

Larsen SS, Brodie JP, Huchra JP, Forbes DA, Grillmair CJ. 2001. *Astron. J.* 121:2974

Larsen SS, Brodie JP, Sarajedini A, Huchra JP. 2002b. *Astron. J.* 124:2615

Larsen SS, Brodie JP, Strader J. 2005. *Astron. Astrophys.* 443:413

Larsen SS, Richtler T. 2000. *Astron. Astrophys.* 354:836

Law DR, Johnston KV, Majewski SR. 2005. *Ap. J.* 619:807

Layden AC, Sarajedini A. 2000. *Astron. J.* 119:1760

Lee MG, Chandar R, Whitmore BC. 2005. *Astron. J.* 130:2128

Lotz JM, Miller BW, Ferguson HC. 2004. *Ap. J.* 613:262

Lotz JM, Telford R, Ferguson HC, Miller BW, Stiavelli M, Mack J. 2001. *Ap. J.* 552:572

Lynden-Bell D. 1975. *Vistas Astron.* 19:299
Majewski SR, Patterson RJ, Dinescu DI, Johnson WV, Ostheimer JC, et al. 2000. *Proc. 35th Liege Int. Astrophys. Colloq.: The Galactic Halo: From Globular Cluster to Field Stars*, ed. A Noels, P Magain, D Caro, E Jehin, G Parmentier, AA Thoul, p. 619. Liege, Belg.: Inst. Astrophys. Geophys.
Maraston C. 2005. *MNRAS* 362:799
Maraston C, Bastian N, Saglia RP, Kissler-Patig M, Schweizer F, Goudfrooij P. 2004. *Astron. Astrophys.* 416:467
Martin NF, Ibata RA, Bellazzini M, Irwin MJ, Lewis GF, Dehnen W. 2004. *MNRAS* 348:12
Mashchenko S, Sills A. 2005a. *Ap. J.* 619:243
Mashchenko S, Sills A. 2005b. *Ap. J.* 619:258
Matteucci F. 1994. *Astron. Astrophys.* 288:57
McCrady N, Gilbert AM, Graham JR. 2003. *Ap. J.* 596:240
McCrady N, Graham JR, Vacca WD. 2005. *Ap. J.* 621:278
McLaughlin DE. 1994. *Publ. Astron. Soc. Pac.* 106:47
McLaughlin DE. 1999. *Astron. J.* 117:2398
Meylan G, Heggie DC. 1997. *Astron. Astrophys. Rev.* 8:1
Mighell KJ, Sarajedini A, French RS. 1998. *Astron. J.* 116:2395
Miller BW, Lotz JM, Ferguson HC, Stiavelli M, Whitmore BC. 1998. *Ap. J. Lett.* 508:133
Miller BW, Whitmore BC, Schweizer F, Fall SM. 1997. *Astron. J.* 114:2381
Minniti D. 1995. *Astron. J.* 109:1663
Minniti D, Meylan G, Kissler-Patig M. 1996. *Astron. Astrophys.* 312:49
Moore B. 1996. *Ap. J. Lett.* 461:13
Moore B, Diemand J, Madau P, Zemp M, Stadel J. 2006. *MNRAS* 368:536
Moore B, Katz N, Lake G, Dressler A, Oemler A Jr. 1996. *Nature* 379:613
Moore B, Lake G, Katz N. 1998. *Ap. J.* 495:139
Morrison HL, Harding P, Perrett K, Hurley-Keller D. 2004. *Ap. J.* 603:87
Mould J, Kristian J. 1986. *Ap. J.* 305:591
Muzzio JC. 1987. *Publ. Astron. Soc. Pac.* 99:245
Odenkirchen M, Grebel EK, Rochosi CM, Dehnen W, Ibata R, et al. 2001. *Ap. J.* 548:L165
Olsen KAG, Miller BW, Suntzeff NB, Schommer RA, Bright J. 2004. *Astron. J.* 127:2674
Oh KS, Lin DNC, Richer HB. 2000. *Ap. J.* 531:727
Ostrov P, Geisler D, Forte JC. 1993. *Astron. J.* 105:1762
Ostrov PG, Forte JC, Geisler D. 1998. *Astron. J.* 116:2854
Peebles PJE. 1984. *Ap. J.* 277:470
Peng EW, Jordán A, Côté P, Blakeslee JP, Ferrarese L, et al. 2006a. *Ap. J.* 639:95
Peng EW, Côté P, Jordán A, Blakeslee JP, Ferrarese L, et al. 2006b. *Ap. J.* 639:838
Peng EW, Ford HC, Freeman KC. 2004. *Ap. J.* 602:705
Perrett KM, Bridges TJ, Hanes DA, Irwin MJ, Brodie JP, et al. 2002. *Astron. J.* 123:249
Perrett KM, Stiff DA, Hanes DA, Bridges TJ. 2003. *Ap. J.* 589:790

Phillipps S, Drinkwater MJ, Gregg MD, Jones JB. 2001. *Ap. J.* 560:201
Pierce M, Beasley MA, Forbes DA, Bridges T, Gebhardt K, et al. 2006. *MNRAS* 366:1253
Pierce M, Brodie JP, Forbes DA, Beasley MA, Proctor R, Strader J. 2005. *MNRAS* 358:419
Proctor RN, Forbes DA, Beasley MA. 2004. *MNRAS* 355:1327
Proctor RN, Sansom AE. 2002. *MNRAS* 333:517
Pryor C, McClure RD, Fletcher JM, Hesser JE. 1989. *Astron. J.* 98:596
Puzia TH. 2003. *Extragalactic globular cluster systems*. PhD thesis. Ludwig-Maximilians Univ. 262 pp.
Puzia TH, Kissler-Patig M, Brodie JP, Schroder LL. 2000. *Astron. J.* 120:777
Puzia TH, Kissler-Patig M, Thomas D, Maraston C, Saglia RP, et al. 2005. *Astron. Astrophys.* 439:997
Puzia TH, Perrett KM, Bridges TJ. 2005. *Astron. Astrophys.* 434:909
Puzia TH, Saglia RP, Kissler-Patig M, Maraston C, Greggio L, et al. 2002a. *Astron. Astrophys.* 395:45
Puzia TH, Zepf SE, Kissler-Patig M, Hilker M, Minniti D, Goudfrooij P. 2002b. *Astron. Astrophys.* 391:453
Read JI, Wilkinson MI, Evans NW, Gilmore G, Kleyna JT. 2006. *MNRAS* 367:387
Rejkuba M, Greggio L, Harris WE, Harris GLH, Peng EW. 2005. *Ap. J.* 631:262
Rejkuba M, Minniti D, Silva DR, Bedding TR. 2003. *Astron. Astrophys.* 411:351
Renzini A. 2002. *New Horizons in Globular Cluster Astronomy*. ASP Conf. Ser., ed. G Piotto, G Meylan, G Djorgovski, M Riello, p. 603. San Francisco: ASP
Rhode KL, Zepf SE. 2001. *Astron. J.* 121:210
Rhode KL, Zepf SE. 2003. *Astron. J.* 126:2307
Rhode KL, Zepf SE. 2004. *Astron. J.* 127:302
Rhode KL, Zepf SE, Santos MR. 2005. *Ap. J. Lett.* 630:21
Rich RM, Corsi CE, Cacciari C, Federici L, Fusi Pecci F, et al. 2005. *Astron. J.* 129:2670
Richtler T. 2003. *LNP Vol. 635: Stellar Candles for the Extragalactic Distance Scale* 635:281
Richtler T. 2006. *Bull. Astron. Soc. India*. In press
Richtler T, Dirsch B, Gebhardt K, Geisler D, Hilker M, et al. 2004. *Astron. J.* 127:2094
Ricotti M. 2002. *MNRAS* 336:L33
Romanowsky AJ, Douglas NG, Arnaboldi M, Kuijken K, Merrifield MR, et al. 2003. *Science* 301:1696
Robinson LB, Wampler EJ. 1972. *Publ. Astron. Soc. Pac.* 84:161
Rose JA. 1985. *Astron. J.* 90:1927
Rosenblatt EI, Faber SM, Blumenthal GR. 1988. *Ap. J.* 330:191
Sadler EM, Rich RM, Terndrup DM. 1996. *Astron. J.* 112:171
Saitoh TR, Koda J, Okamoto T, Wada K, Habe A. 2006. *Ap. J.* 640:22
Santos MR. 2003. In *Extragalactic Globular Cluster Systems, Proc. ESO Workshop, Garching, Ger.*, ed. M Kissler-Patig, p. 348. Berlin: Springer-Verlag
Sarajedini A, Geisler D, Schommer R, Harding P. 2000. *Astron. J.* 120:2437

Sarajedini A, Jablonka P. 2005. *Astron. J.* 130:1627
Scannapieco E, Weisheit J, Harlow F. 2004. *Ap. J.* 615:29
Schiavon RP, Faber SM, Castilho BV, Rose JA. 2002. *Ap. J.* 580:850
Schiavon RP, Rose JA, Courteau S, MacArthur LA. 2004. *Ap. J. Lett.* 608:33
Schommer RA, Olszewski EW, Suntzeff NB, Harris HC. 1992. *Astron. J.* 103:447
Schroder LL, Brodie JP, Kissler-Patig M, Huchra JP, Phillips AC. 2002. *Astron. J.* 123:2473
Schuberth Y, Richtler T, Dirsch B, Hilker M, Larsen SS. 2004. *Astronomische Nachtrichten* 325:62
Schweizer F. 1987. *Philos. Trans. R. Soc. London Ser. A* 358:2063
Searle L, Wilkinson A, Bagnuolo WG. 1980. *Ap. J.* 239:803
Searle L, Zinn R. 1978. *Ap. J.* 225:357
Secker J, Geisler D, McLaughlin DE, Harris WE. 1995. *Astron. J.* 109:1019
Seth A, Olsen K, Miller B, Lotz J, Telford R. 2004. *Astron. J.* 127:798
Sharina ME, Puzia TH, Makarov DI. 2005. *Astron. Astrophys.* 442:85
Sharina ME, Sil'chenko OK, Burenkov AN. 2003. *Astron. Astrophys.* 397:831
Smith GH, Burkert A. 2002. *Ap. J. Lett.* 578:51
Smith LJ, Gallagher JS. 2001. *MNRAS* 326:1027
Sohn ST, O'Connell RW, Kundu A, Landsman WB, Burstein D, et al. 2006. *Astron. J.* 131:866
Soria R, Mould JR, Watson AM, Gallagher JS III, Ballester GE, et al. 1996. *Ap. J.* 465:79
Spitzer L. 1987. *Dynamical Evolution of Globular Clusters*. Princeton, NJ: Princeton Univ. 196 pp.
Stetson PB, Vandenberg DA, Bolte M. 1996. *Publ. Astron. Soc. Pac.* 108:560
Strader J, Brodie JP. 2004. *Astron. J.* 128:1671
Strader J, Brodie JP, Cenarro AJ, Beasley MA, Forbes DA. 2005. *Astron. J.* 130:1315
Strader J, Brodie JP, Forbes DA. 2004a. *Astron. J.* 127:295
Strader J, Brodie JP, Forbes DA. 2004b. *Astron. J.* 127:3431
Strader J, Brodie JP, Forbes DA, Beasley MA, Huchra JP. 2003a. *Astron. J.* 125:1291
Strader J, Brodie JP, Schweizer F, Larsen SS, Seitzer P. 2003b. *Astron. J.* 125:626
Tantalo R, Chiosi C, Bressan A. 1998. *Astron. Astrophys.* 333:419
Thomas D, Maraston C, Bender R. 2003. *MNRAS* 339:897
Thomas D, Maraston C, Bender R, de Oliveira CM. 2005. *Ap. J.* 621:673
Thomas D, Maraston C, Korn A. 2004. *MNRAS* 351:L19
Toomre A. 1977. In *Evolution of Galaxies and Stellar Populations*, ed. BM Tinsley, RB Larson, p. 40. *Proc. Conf. Yale Univ.* New Haven: Yale Univ. Obs.
Toomre A, Toomre J. 1972. *Ap. J.* 178:623
Trager SC. 2004. *Carnegie Obs. Astrophys. Ser.* Vol. 4: *Origin and Evolution of the Elements*, ed. A McWilliam, M Rauch, p. 391. Cambridge: Cambridge Univ. Press
Trager SC, Faber SM, Worthey G, González JJ. 2000. *Astron. J.* 120:165
Trager SC, Worthey G, Faber SM, Burstein D, González JJ. 1998. *Ap. J. Suppl.* 116:1
Tripicco MJ, Bell RA. 1995. *Astron. J.* 110:3035
van den Bergh S. 1975. *Annu. Rev. Astron. Astrophys.* 13:217
van den Bergh S. 1982. *Publ. Astron. Soc. Pac.* 94:459

van den Bergh S. 1994. *The Local Group*, ed. RC Smith, J Storm. Garching: Eur. South. Obs.
van den Bergh S. 1999. *Astron. Astrophys. Rev.* 9:273
van den Bergh S. 2004. *Astron. J.* 127:897
van den Bergh S, Mackey AD. 2004. *MNRAS* 354:713
van den Bergh S, Morbey C, Pazder J. 1991. *Ap. J.* 375:594
Vesperini E. 2000. *MNRAS* 318:841
Vesperini E. 2001. *MNRAS* 322:247
Vesperini E, Heggie DC. 1997. *MNRAS* 289:898
Vesperini E, Zepf SE. 2003. *Ap. J. Lett.* 587:97
Vesperini E, Zepf SE, Kundu A, Ashman KM. 2003. *Ap. J.* 593:760
Walcher CJ, Fried JW, Burkert A, Klessen RS. 2003. *Astron. Astrophys.* 406:847
West MJ. 1993. *MNRAS* 265:755
West MJ, Côté P, Marzke RO, Jordán A. 2004. *Nature.* 427:31
Whitmore BC, Schweizer F. 1995. *Astron. J.* 109:960
Whitmore BC, Schweizer F, Kundu A, Miller BW. 2002. *Astron. J.* 124:147
Whitmore BC, Sparks WB, Lucas RA, Macchett FD, Biretta JA. 1995. *Ap. J. Lett.* 454:73
Williams BF, Hodge PW. 2001. *Ap. J.* 548:190
Woodley KA, Harris WE, Harris GLH. 2005. *Astron. J.* 129:2654
Woodworth SC, Harris WE. 2000. *Astron. J.* 119:2699
Worthey G. 1994. *Ap. J. Suppl.* 95:107
Worthey G, España A, MacArthur LA, Courteau S. 2005. *Ap. J.* 631:820
Worthey G, Faber SM, Gonzá lez JJ. 1992. *Ap. J.* 398:69
Worthey G, Faber SM, Gonzá lez JJ, Burstein D. 1994. *Ap. J. Suppl.* 94:687
Wu X, Tremaine S. 2006. *Ap. J.* 643:210
Yoon SJ, Yi SK, Lee YW. 2006. *Science* 311:1129
Zepf SE, Ashman KM. 1993. *MNRAS* 264:611
Zepf SE, Beasley MA, Bridges TJ, Hanes DA, Sharples RM, et al. 2000. *Astron. J.* 120:2928
Zepf SE, Silk J. 1996. *Ap. J.* 466:114
Zinn R. 1985. *Ap. J.* 293:424
Zinn R, West MJ. 1984. *Ap. J. Suppl.* 55:45

First Fruits of the *Spitzer Space Telescope*: Galactic and Solar System Studies[*]

Michael Werner,[1] Giovanni Fazio,[2] George Rieke,[3] Thomas L. Roellig,[4] and Dan M. Watson[5]

[1]Astronomy and Physics Directorate, Jet Propulsion Laboratory, California Institute of Technology, Pasadena, California 91109; email: mwerner@sirtfweb.jpl.nasa.gov

[2]Harvard-Smithsonian Center for Astrophysics, Cambridge, Massachusetts 02138; email: gfazio@cfa.harvard.edu

[3]Steward Observatory, University of Arizona, Tucson, Arizona 85721; email: grieke@as.arizona.edu

[4]Astrophysics Branch, NASA/Ames Research Center, Moffett Field, California 94035; email: thomas.l.roellig@nasa.gov

[5]Department of Physics and Astronomy, University of Rochester, Rochester, New York 14627; email: dmw@pas.rochester.edu

First published online as a Review in Advance on June 5, 2006

The *Annual Review of Astrophysics* is online at astro.annualreviews.org

doi: 10.1146/annurev.astro.44.051905.092544

Copyright © 2006 by Annual Reviews. All rights reserved

0066-4146/06/0922-0269$20.00

[*]The U.S. Government has the right to retain a nonexclusive, royalty-free license in and to any copyright covering this paper.

Key Words

brown dwarfs, circumstellar disks, exoplanets, infrared astronomy, planetary system formation, space technology, star formation, stellar evolution

Abstract

The *Spitzer Space Telescope*, launched in August 2003, is the infrared member of NASA's Great Observatory family. *Spitzer* combines the intrinsic sensitivity of a cryogenic telescope in space with the imaging and spectroscopic power of modern infrared detector arrays. This review covers early results from *Spitzer* that have produced major advances in our understanding of our own solar system and phenomena within the Galaxy. *Spitzer* has made the first detection of light from extrasolar planets, characterized planet-forming and planetary debris disks around solar-type stars, showed that substellar objects with masses smaller than 10 M_{Jup} form through the same processes as do solar-mass stars, and studied in detail the composition of cometary ejecta in our Solar System. *Spitzer*'s major technical advances will pave the way for yet more powerful future instruments. *Spitzer* should operate with full capabilities well into 2009, enabling several additional cycles of discovery and follow-up.

1. INTRODUCTION

KBO: Kuiper Belt object

ISO: Infrared Space Observatory

The launch of the National Aeronautics and Space Administration's (NASA) *Spitzer Space Telescope* (*Spitzer*) in 2003 provided the scientific community with the most powerful tool yet available for astronomical explorations between 3.6 and 160 μm. As the infrared member of NASA's family of Great Observatories, *Spitzer* has been used very successfully in multispectral studies with its companion observatories, the *Chandra X-Ray Observatory* and the *Hubble Space Telescope* (HST). *Spitzer* has also observed objects currently accessible only in the infrared, most notably detecting radiation from extrasolar planets for the first time. This review presents a selection of early results from *Spitzer* that have advanced our understanding of the solar system and phenomena within the Galaxy. We include studies of protoplanetary and planetary debris disks where *Spitzer* results—often in combination with those from other perspectives—appear to paint a fairly complete preliminary picture. In other areas, such as observations of extrasolar planets, brown dwarf spectroscopy, and Kuiper Belt object (KBO) studies, we highlight particularly provocative or exciting results from *Spitzer*. Some topics have been largely omitted, so that this review does not fully encompass the scope of *Spitzer* results already in hand; most notably, readers interested in interstellar matter should consult individual *Spitzer* papers and refer to van Dishoeck (2004) and Cesarsky & Salama (2005) for a comprehensive summary of results from the Infrared Space Observatory (ISO). This review includes primarily results submitted for publication as of January 1, 2006. Early *Spitzer* results on extragalactic science are reported by Armus (2006) and Chary, Sheth & Teplitz (2006). The history of *Spitzer* is described by Werner (2006) and Rieke (2006).

2. OBSERVATORY OVERVIEW

Detailed technical descriptions of *Spitzer* and its three focal plane instruments are provided in the September 2004 special issue of the *Astrophysical Journal Supplement* (Werner et al. 2004 and following papers) and also in the proceedings of the International Society for Optical Engineering (Roellig et al. 2004b and following papers). We thus limit this technical description to those features of *Spitzer* most directly related to its scientific performance. We also highlight the innovations demonstrated by *Spitzer* that might be most applicable to future missions.

2.1. Orbit

Spitzer utilizes an Earth-trailing heliocentric orbit. As seen from Earth, *Spitzer* recedes at about 0.1 AU per year. For *Spitzer*, the Earth-trailing orbit has several major advantages over near-Earth orbits. The principal advantage is the distance from Earth and its heat; this facilitates the extensive use of radiative cooling, which makes *Spitzer*'s cryo-thermal design extremely efficient. The orbit also permits excellent sky viewing and observing efficiency. *Spitzer* is constrained to point no closer than 80° toward and no further than 120° from the Sun, but even with these constraints 35% of the sky is visible at any time, and the entire sky is visible every six months. Surprisingly, the solar orbit has enabled *Spitzer* to carry out several unique science investigations. These include: (*a*) conducting in situ studies of the Earth-trailing density enhancement in the zodiacal cloud, which *Spitzer* passes directly through, and (*b*) obtaining constraints on the nature of objects responsible for background star microlensing toward the Magellanic Clouds. *Spitzer* operates autonomously in this orbit; once or twice per day the observatory reorients itself so that a fixed X-band antenna mounted on the bottom of the spacecraft points to the Earth and downlinks 12 to 24 hours of stored data via the Deep Space Network (DSN).

MIPS: Multiband Imaging Photometer for *Spitzer*

IRS: Infrared Spectrograph

IRAC: Infrared Array Camera

IRAS: Infrared Astronomy Satellite

2.2. Architecture

The overall configuration of the flight system is shown in **Figure 1**. The spacecraft and solar panel were provided by Lockheed Martin. The telescope, cryostat, and associated shields and shells make up the Cryogenic Telescope Assembly (CTA), built by Ball Aerospace, who also built the Multiband Imaging Photometer for *Spitzer* (MIPS) and Infrared Spectrograph (IRS) instruments. The third instrument, the Infrared Array Camera (IRAC), was built at NASA's Goddard Space Flight Center.

2.2.1. Cryo-thermal system.
Most of the CTA was at room temperature at launch; only the science instrument cold assemblies and the superfluid helium vessel were cold within the cryostat vacuum shell. This allowed a much smaller vacuum pressure vessel and a smaller observatory mass than the cold launch architecture used in the Infrared Astronomy Satellite (IRAS) and ISO missions. A combination of passive radiative cooling and helium boil-off vapor cools the components of the CTA after launch, using a concept developed by F. Low and described in Lysek et al. (1995). Radiative cooling works for *Spitzer* because the solar orbit allows the spacecraft always to be oriented with the solar array pointed toward the Sun, while the Earth is so distant that its heat input is negligible. The system of reflective and emitting shells and shields, which is always shadowed by the solar array, rejects almost all the heat that leaks inward while radiating the small amount not rejected into the cold of deep space. As a result, the outer shell of the CTA achieves a temperature of 34 to 34.5 K solely by radiative cooling.

With such a cold outer shell, a small amount of helium vapor suffices to maintain the telescope at its operating temperature, which can be as low as 5.5 K. The performance of the cryo-thermal system on-orbit has exceeded expectation. Repeated measurements of the system's helium volume suggest that *Spitzer*'s cryogenic lifetime will exceed five years, extending well into 2009.

Following cryogen depletion, the telescope will warm up, but it will still be colder than the outer shell; current estimates suggest that the temperature will be <30 K. At this temperature, both the instrumental background and the detector dark current should be low enough for *Spitzer* to continue natural background-limited operations in the shortest wavelength IRAC bands (3.6 and 4.5 μm). Additional information on the measured on-orbit thermal performance of the *Spitzer* CTA can be found in Finley, Hopkins & Schweickart (2004).

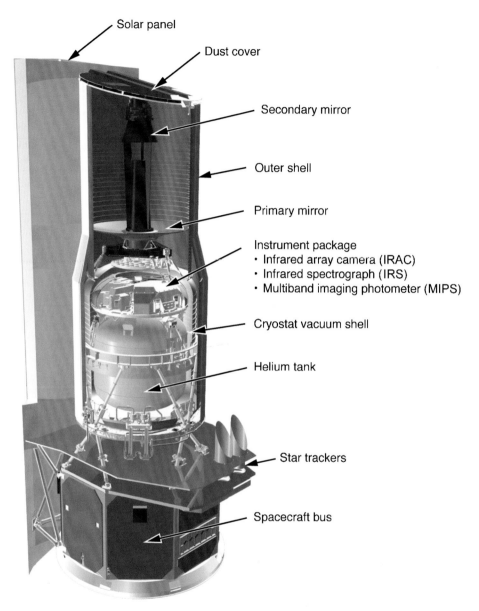

Figure 1

The *Spitzer Space Telescope*. The outer shell forms the boundary of the Cryogenic Telescope Assembly (CTA), which incorporates the telescope, the cryostat, the helium tank, and the three instruments, the Multiband Imaging Photometer, the Infrared Spectrograph, and the Infrared Array Camera. Principal Investigator-led teams provided the instruments, while Ball Aerospace provided the remainder of the CTA. Lockheed Martin provided the solar panel and spacecraft bus. The observatory is approximately 4.5 m tall and 2.1 m in diameter; the mass at launch was 861 kg. The dust cover atop the CTA was jettisoned 5 days after launch. (Image courtesy of Ball Aerospace.)

2.2.2. Optics.
Spitzer has an f/1.6 primary mirror with a diameter of 85 cm and a Ritchey-Chretien Cassegrain optical design with system f-ratio f/12. The telescope optics and metering structure are constructed entirely of beryllium so that changes in both the telescope prescription and its alignment with the focal plane are minimized as the telescope cools on-orbit. A focus mechanism was used to adjust the axial position of the secondary mirror to optimize the focus as the telescope reached its equilibrium temperature. The resulting image quality is excellent, and the *Spitzer* telescope provides diffraction-limited performance at all wavelengths greater than 5.5 μm. The image diameter (FWHM) at 5.5 μm is ∼1.3 arcsec. The three instruments are confocal to within their depths of focus.

SSC: *Spitzer* Science Center

2.2.3. Pointing and reaction control.
Spitzer's pointing and control system includes redundant gyroscopes and star trackers for sensing the pointing position of the telescope, and reaction wheels for moving the telescope. Visible light sensors sharing the cold focal plane determine the telescope's line of sight relative to that of the warm star trackers. The thermal design of *Spitzer* and the fixed position of the Sun relative to the spacecraft in the solar orbit give *Spitzer* a very high degree of thermal stability. Thus the warm-to-cold alignment varies by less than 0.5 arcsec over timescales of days. The star tracker has proven to be very accurate, with a noise-equivalent angle of approximately 0.11 arcsec using an average of 37 tracked stars. The excellent alignment stability and star-tracker accuracy allow *Spitzer* to point the telescope boresight to within <0.45 arcsec (1-σ radial uncertainty) of the desired position. *Spitzer* achieves excellent pointing stability of ∼0.03 arcsec (1-σ) for times up to 600 seconds.

2.2.4. Instruments.
Spitzer's three instruments occupy the multiple instrument chamber behind the primary mirror. They share a common focal plane, with their fields of view defined by pickoff mirrors. The instruments achieve great scientific power with uncomplicated design through the use of state-of-the-art infrared detector arrays in formats as large as 256×256 pixels. For broadband imaging and low spectral resolution spectroscopy, *Spitzer* has achieved sensitivities close to or at the levels established by the natural astrophysical backgrounds—principally the zodiacal light—encountered in Earth orbit. The only moving part in the instrument payload is a scan mirror in the MIPS.

Together, the three instruments provide imaging and photometry in eight spectral bands between 3.6 and 160 μm and spectroscopy and spectrophotometry between 5.2 and 95 μm. **Table 1** summarizes the characteristics and performance of the instruments. Compared to previous space infrared missions, most notably ISO, *Spitzer* brings a factor of ∼10–100 times improvement in limiting point source sensitivity over most of its wavelength band. In addition, the arrays in use on *Spitzer* provide 100–1000 times more pixels than previously available, leading to major increases in efficiency for both imaging and spectroscopy.

The *Spitzer* Science Center (SSC) at Caltech provides instrument handbooks and detailed performance and operability information on its Web site at http://ssc.spitzer.caltech.edu. The SSC, responsible for *Spitzer* science operations, serves as the main interface between *Spitzer* and its scientific user community. The

Table 1 *Spitzer* instrumentation summary

λ (μm)	Array type	λ/Δλ	F.O.V.	Pixel size (arcsec)	Sensitivity[a] (5σ in 500 sec)
IRAC: Infrared Array Camera—P.I. Giovanni Fazio, Smithsonian Astrophysical Observatory					
3.6	InSb	4.7	5.2′ × 5.2′	1.22	1.3 μJy
4.5	InSb	4.4	5.2′ × 5.2′	1.21	2.7 μJy
5.8	Si:As (IBC)	4.0	5.2′ × 5.2′	1.22	18 μJy
8.0	Si:As (IBC)	2.7	5.2′ × 5.2′	1.22	22 μJy
MIPS: Multiband Imaging Photometer for *Spitzer*—P.I.-George Rieke, University of Arizona					
24	Si:As (IBC)	4	5.4′ × 5.4′	2.5	110 μJy
70 wide	Ge:Ga	3.5	5.25′ × 2.6′	9.8	7.2 mJy
70 fine	Ge:Ga	3.5	2.6′ × 1.3′	5.0	14.4 mJy
55–95	Ge:Ga	14–24	0.32′ × 3.8′	9.8	200 mJy
160	Ge:Ga (stressed)	4	0.53′ × 5.3′	16	24 mJy
IRS: Infrared Spectrograph—P.I. Jim Houck, Cornell University					
5.2–14.5	Si:As (IBC)	60–127	3.6″ × 57″	1.8	400 μJy
13–18.5 (peakup imaging)	Si:As (IBC)	3	1′ × 1.2′	1.8	75 μJy
9.9–19.6	Si:As (IBC)	600	4.7″ × 11.3″	2.4	1.5×10^{-18} W m^{-2}
14–38	Si:Sb (IBC)	57–126	10.6″ × 168″	5.1	1.7 mJy
18.7–37.2	Si:Sb (IBC)	600	11.1″ × 22.3″	4.5	3×10^{-18} W m^{-2}

[a]Sensitivity numbers are indicative of *Spitzer* performance in low-background sky. Confusion is not included. Detailed performance estimates should be based on tools available at http://ssc.spitzer.caltech.edu/obs

Jet Propulsion Laboratory, California Institute of Technology, managed *Spitzer* development for NASA and continues as the overall managing center while carrying out mission operations with the assistance of Lockheed Martin, Denver.

3. STELLAR EVOLUTION

Stars comprise most of the visible matter in the universe, making understanding star formation and evolution essential to our exploration of other fundamental questions in the field of astrophysics. In this section, we report *Spitzer*'s studies of key milestones in the stellar life cycle.

Stars form deep within dense clouds of molecular gas and dust, hidden from view at optical wavelengths but accessible in the infrared. As a young star evolves and sheds the dusty cloak of its birth, a remnant of the stellar accretion disk often remains. If this disk is dense and massive enough to support the formation of planets, it is termed a "protoplanetary" disk. In well-studied nearby regions, these disks have masses ∼0.001–0.01 $M_\odot$ and sizes ∼100 AU. As the protoplanetary disk dissipates through continued evolution and, perhaps, the formation of planets, a planetary debris disk, formed by the inevitable pulverization of the macroscopic bodies formed in the protoplanetary disk, arises to take its place. Throughout this evolution the disks are inviting targets for study in the infrared because the dust they contain is heated by the star, while the gas, when present, may be warmed by accretion, viscosity, or gas-grain collisions. The study of these protoplanetary disks remains a particularly active area of *Spitzer* research, and the initial results presented in Section 3.1 will be greatly augmented in the next several years.

YSO: young stellar object

A star leaving the main sequence passes through a series of stages that require exploration in the infrared. Copious mass loss cloaks the star in an envelope of dust that characterizes the post-main sequence asymptotic giant branch and supergiant phases. From here, lower-mass stars evolve into planetary nebulae consisting of a hot and compact white dwarf primary exciting a diffuse envelope of ejecta that radiates copiously in the infrared. More massive stars may explode as Type II supernovae, initially characterized by similar infrared-bright ejecta envelopes. At later stages, supernovae excite infrared radiation from pre-existing circumstellar and interstellar material swept up by the expanding blast wave.

Objects below the $\sim 0.08\ M_\odot$ limit required to ignite and sustain nuclear burning become brown dwarfs. These objects span almost two decades in mass from that of Jupiter ($0.0015\ M_\odot$, or $1\ M_{Jup}$) to $\sim 0.08\ M_\odot$. They may have temperatures ~ 3000 K when formed but cool rapidly and spend most of their lives at much lower temperatures, again accessible to study primarily in the infrared.

Our current understanding of star and planet formation, including but also going far beyond *Spitzer*'s contributions, is summarized by Reipurth, Jewitt & Keil (2006).

Many of the results presented in this section and throughout this review are the initial fruits of *Spitzer*'s Legacy Science Programs. Readers should refer to the sidebar to gain some appreciation of the scope of these programs, which will release large amounts of data—both from *Spitzer* and from ancillary sources—prior to the end of 2006. Substantial data sets are also being produced by the larger Guaranteed Time Observer (GTO) programs and by Legacy-scale programs proposed in response to the first calls for General Observer (GO) proposals.

3.1. Formation of Stars, Substellar Objects, and Protoplanetary Disks

Lada & Wilking (1984) showed that it was useful to sort young stellar objects (YSOs) of roughly solar mass and below by the spectral energy distribution (SED) of their near- and mid-infrared excesses. In their classification system, YSO classes I, II, and III correspond to the degree to which the central object is dust-embedded. Class I corresponds to especially deeply embedded objects with heavily veiled spectra and, usually, associated outflows and Herbig-Haro objects; Class II corresponds to classical T Tauri stars with ongoing accretion and substantial circumstellar disks; and Class III corresponds simply to reddened photospheres. Adams, Lada & Shu (1987) showed that the three classes probably comprise an evolutionary sequence, initially dominated in appearance by an infalling envelope and a dense, optically-thick disk, both of which gradually dissipate. Efforts to observe stages of protostellar evolution earlier than Class I identified a rarer class of object—dubbed Class 0 YSOs—with even

THE *SPITZER* LEGACY SCIENCE PROGRAM

The original *Spitzer* Legacy Science Program enabled major science observing projects early in the *Spitzer* mission, creating a substantial and coherent database of archived observations. The six legacy projects used a total of 3160 hours of *Spitzer* observing time and released all their data to the public archive, accompanied by post-pipeline data products and analysis tools for use by the scientific community. Three of these projects focus on galactic studies, and results from these projects are presented throughout this article:

The GLIMPSE project (P.I.-E. Churchwell, Wisconsin) surveyed the galactic plane between latitude $\pm 1°$ and longitude 10–65° on both sides of the galactic center. The survey used all four IRAC bands and provides data on galactic structure, high-mass star formation, and supernova remnants and infrared dark clouds throughout the inner galaxy.

The c2d project (P.I.-N. Evans, Texas) includes complete IRAC and MIPS maps of five large molecular clouds—each covering about 4 square degrees— maps of ~135 smaller cores, and photometry and spectroscopy of about 200 stars, emphasizing objects younger than 10 million years. c2d is designed to trace the early stages of stellar evolution from starless cores to the formation of planet-forming disks.

The FEPS project (P.I.-M. Meyer, Arizona) has obtained photometry and spectroscopy of more than 300 main-sequence stars aged 3 Myr to 3 Gyr to study the evolution of dust and gas in postaccretion disk systems.

SED: spectral energy distribution

redder SEDs, as well as resolved, spheroidal shapes at submillimeter wavelengths, and extremely faint central objects (e.g., André, Ward-Thompson & Barsony 1993, Myers & Ladd 1993). **Figure 2** shows *Spitzer*-obtained spectra of typical examples of the YSO classes to illustrate the decrease in dustiness along the sequence from Class 0 to Class III. Lada (2005) reviews this overall scenario.

Spitzer observations are producing complete studies of the spatial and temporal evolution of large samples of stars and protoplanetary systems through all of these stages; the *Spitzer* samples include objects with masses below the hydrogen-burning limit that might properly be considered young substellar objects. We also present early *Spitzer* results on the formation of higher mass stars, both in the Milky Way and in the nearby Large Magellanic Clouds (LMC); *Spitzer*'s sensitivity permits detailed study of individual massive protostars at the ~55 kpc distance of the LMC.

3.1.1. Young clusters. We have long known that most stars form in clusters. However, mid-infrared observations are needed to establish the evolutionary state of the stars. *Spitzer*'s MIPS and IRAC arrays, with their 5 × 5 arcmin fields of view and high sensitivities, quickly survey large areas at mid-infrared wavelengths, detecting large numbers of YSOs, determining their SEDs, and mapping their spatial distribution.

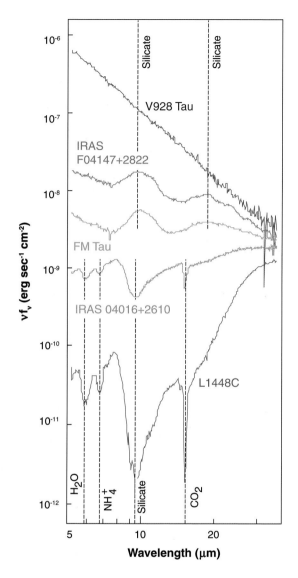

Figure 2

Spitzer Infrared Spectrograph spectra of exemplars of the four classes of young stellar objects: the Class 0 object L1448C (Najita et al., in preparation); the Class I object IRAS 04016+2610 (Watson et al. 2004); two Class II objects with substantially different small-dust composition, FM Tau and IRAS F04147+2822 (E. Furlan, L. Hartmann, N. Calvet, P. D'Alessio, R. Franco-Hernandez, et al., submitted; D.M. Watson, J.D. Leisenring, E. Furlan, C.J. Bohac, B. Sargent, et al., submitted); and the binary Class III object V928 Tau (E. Furlan, L. Hartmann, N. Calvet, P. D'Alessio, R. Franco-Hernandez, et al., submitted). Identifications are shown for the most prominent spectral features, which are due to ices and minerals. The spectra of FM Tau, IRAS F04147, and V928 Tau have been offset by factors of 50, 200, and 10^4, respectively.

IRAC observations show that the IRAC color-color diagram is a very powerful diagnostic tool for rapidly identifying and classifying young stars (Allen et al. 2004, Megeath et al. 2004). **Figure 3**, based on data from observations of four young stellar clusters, illustrates this. Investigators are beginning to use *Spitzer*'s ability to sort and classify YSOs to explore such questions as the timescales of the various stages of, and the propagation or distribution of, star formation through a cloud. For example, in **Figure 3** the number of objects with colors between those of Class II and Class III is relatively small. This may indicate that the transition between Classes II and III is short compared to the duration of Class II (Gutermuth et al. 2006), but the overlap

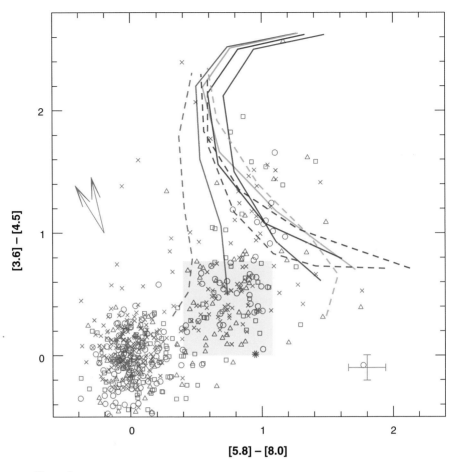

Figure 3

Model colors and measured Infrared Array Camera (IRAC) colors for four young stellar clusters—S140 (*squares*), S171 (*circles*), NGC 7129 (*triangles*), and Cep C (*crosses*) from Allen et al. (2004). Representative error bars are shown. The data fall into three main groups: a group at (0,0) that contains background and foreground stars and Class III objects with no infrared excess in the IRAC bands, a group that occupies the Class II region (within the *blue box*), and a group that lies along the Class I theoretical locus with luminosity $L > 1$ $L_\odot$. The identification of the Class I and II sources is based on the agreement of the colors with the distribution of model colors (Kenyon, Calvet & Hartmann 1993; Calvet et al. 1994; D'Alessio et al. 1998, 1999, 2001, 2005b; D'Alessio, Calvet & Hartmann 2001). Extinction vectors are shown for $A_v = 30$ mag, using the two extremes of the six vectors calculated by Megeath et al. (2004). (Reproduced by permission of the Am. Astron. Soc.)

of Class I and II suggests that the timescale for the final dissipation or settling of the envelope is comparable to the duration of the Class I phase.

Hartmann et al. (2005b) confirm the IRAC color-color classification scheme by comparing it to the independently known properties of a well-studied sample of YSOs in Taurus. They also observe that IRAC colors are in good agreement with recent improved disk models, and in general agreement with the models for protostellar envelopes. Among the systems with disks, they observe a strong correlation between IRAC excess emission and signatures of accretion as inferred from emission line profiles or UV excess. Young et al. (2005) report MIPS observations of the Chameleon II molecular cloud and make a first effort to extend this classification into the MIPS bands. They show a [24] versus [24]–[70] color magnitude plot and identify ∼16 objects brighter than fifth magnitude (∼50 mJy) at 24 μm as lying within the cloud. Within this group, the known Class 0 and Class I sources are redder than the Class II objects. However, all sources generally lie blueward of the current models by Young & Evans (2005) and by Whitney et al. (2003).

Megeath et al. (2004) report on the spatial distribution of the YSOs in the four clusters used to validate IRAC's photometric classification. The clusters are similar in distance but span a range of far-infrared luminosities, molecular gas masses, and cluster membership. The results suggest that a significant number of stars in each region form outside the dense stellar cluster identified from ground-based observations, which typically has a diameter of ∼1 pc. Megeath et al. (2004) also find that the distribution of sources differs from cluster to cluster, but overall the Class I objects appear to be more spatially localized than the Class II objects, as might be expected from the evolutionary sequence proposed above. NGC 7129 illustrates the occurrence of star formation outside of a cluster core. Gutermuth et al. (2004) augment IRAC data on NGC 7129 with ground-based near-infrared data and with MIPS data on the cluster from Muzerolle et al. (2004), finding that half of the YSOs are located outside the 0.5 pc cluster core and that star formation is continuing in this halo. They also determine that approximately half of the stars in the cluster core have disks. Additionally, Muzerolle et al. (2004) identify in NGC 7129 several objects with photospheric colors out to 8 μm but excesses at 24 μm; these objects must have central clearings and are probably further examples of the transition disks discussed in detail below (see **Figure 8**).

Teixeira et al.'s (2006) MIPS/IRAC observations (**Figure 4**) of the massive young cluster NGC 2264 identify a set of dense filaments delineated in the submillimeter continuum, which are shown by MIPS images to be dotted with bright 24-μm sources with quasi-uniform separation. The majority of these are Class I objects. The preferred spacing between these 24-μm sources, ∼20 arcsec or ∼0.08 pc at the distance of NGC 2264, is remarkably close to the Jeans length for these filaments. Teixeira et al. (2006) suggest that the filaments may have formed through turbulent motions of the cloud and subsequently thermalized and fragmented into star-forming cores. Young et al. (2006), combining *Spitzer* and Magellan data, resolve the submillimeter source IRAS 12 S1 in NGC 2264—previously thought to be a single Class 0 object—into a dense cluster of embedded, low-mass YSOs. This cluster's estimated dynamical lifetime of only a few times ten thousand years highlights its youth.

3.1.2. Large area surveys. S.T. Megeath, K.M. Flaherty, J. Hora, R. Gutermuth, L.E. Allen, et al. (submitted) have mapped the Orion A and B molecular clouds, covering an area of 5.6 square degrees, with sensitivity to detect objects below the hydrogen burning limit at an age of 1 Myr. Initial results show that approximately half

Figure 4

Image of the Spokes cluster in NGC 2264. Color coding: blue = IRAC 3.6 μm; green = IRAC 8.0 μm; red = MIPS 24 μm. The region shown is about 5 × 10 arcmin in extent. North is up, and east is to the left. The image shows unusual linear alignments of the brightest 24-μm sources; Teixeira et al. (2006) show that they preferentially have a separation of 20″, close to the Jeans length estimated for the parent cloud of the cluster given the adopted distance of 800 pc. (From Teixeira et al. 2006, reproduced by permission of the Am. Astron. Soc.)

of the young stars and protostars identified in this survey are found in dense clusters surrounding the two regions of recent massive star formation, NGC 2024 and the Orion Nebula, whereas the other half are found in lower density environments such as L1641. This suggests that the rate and/or density of star formation may be enhanced by the presence of OB stars in the molecular cloud.

As part of the c2d Legacy program (see sidebar), Harvey et al. (2006) analyzed a 0.89-square-degree IRAC map of the Serpens dark cloud and identified more than 240 YSO candidates. They discovered a particularly rich area of star and substellar object formation about a degree southwest of the well-studied Serpens Core. These observations also suggest that a population of infrared-excess sources exists in Serpens at least down to luminosities ~ 0.001 $L_\odot$. Determining the nature of these low luminosity objects will require deeper imaging and spectroscopy.

3.1.3. The youngest objects. Jørgensen et al. (2005) observed the Class 0 protostar 16293-2422 with the IRS, obtaining the first detection of this archetypal object shortward of 60 μm. Their detailed modeling of the SED of this object suggests a central cavity of radius ~ 600 AU, which they note is comparable to the centrifugal radius of the envelope. *Spitzer* observations of L1014, a dense core previously thought to be starless, showed the presence of an embedded infrared point source that Young et al. (2004a) classified as a Class 0 object. Its luminosity of at most a few tenths $L_\odot$ makes this among the lowest luminosity protostars known and suggests the formation of a brown dwarf. Bourke et al. (2005) eliminated the possibility that this might be a background object. They used the Submillimeter Array to discover an associated compact molecular outflow, which is among the smallest known—only ∼500 AU in size and $<10^{-4}$ $M_\odot$ in mass. Finally, Huard et al. (2006) obtained very deep near-infrared images that showed a scattered-light nebula like those typically associated with protostars. T.L. Huard et al. (in preparation) report that about 20% of cores formerly thought to be starless contain objects of very low luminosity that may be similar to the object in L1014.

Spitzer studies of the mid-infrared spectra of Class 0 and I YSOs reveal several strong absorption features of ices, silicate features that display both emission and absorption, and emission lines of abundant low-excitation ions, atoms, and molecules. Relative intensities suggest that the emission lines arise in fairly low density gas (100–1000 cm^{-3}), probably indicating an origin in the bipolar outflows and shocks associated with these objects (E. Furlan, M. McClure, N. Calvet, L. Hartmann, P. D'Alessio, et al., submitted) rather than in gas associated with the envelopes or disks (cf. **Figure 5**). The 10 and 20 μm silicate features exhibit a combination of emission and absorption that is consistent with the flattening of the envelope and disk, along with a range of inclinations with respect to the line of sight (Watson et al. 2004). Their profiles resemble in detail the absorptions produced by interstellar dust grains; the silicates must be mostly amorphous, though faint hints of crystalline silicate absorption have been noted (Ciardi et al. 2005).

The ice absorption features seen in Class I and 0 YSOs by *Spitzer* (Boogert et al. 2004, Noriega-Crespo et al. 2004; **Figures 2** and **5**) also appear in ISO mid-infrared spectra of more massive deeply embedded objects (see van Dishoeck 2004). Most prominent are the features due to the ices of water ($\lambda = 6.0$ and 11 μm), HCOOH ($\lambda = 7.3$ μm), methane (7.8 μm), and carbon dioxide ($\lambda = 15.2$ μm), and a still unidentified feature at 6.8 μm usually associated with NH_4^+. The *Spitzer* targets include much fainter and lower-mass YSOs than previously studied, and the ice features in these objects tend to be stronger relative to the silicate features than in the more massive objects (**Figure 6**). Studies of the Class I YSOs in Taurus (Watson et al. 2004; see also E. Furlan, M. McClure, N. Calvet, L. Hartmann, P. D'Alessio, et al., submitted) suggest that the ice absorption arises primarily in the YSO envelopes rather than in the disks. Comparison of these feature strengths to models indicates that most of the ice-feature absorption originates in the region of the envelope outside the centrifugal radius, as its strength is much less dependent upon inclination than that of the silicate feature. Thus, these observations suggest the accumulation of ices in the solid components of protoplanetary systems, seen before they settle to the disk to begin incorporation into planetesimals. Pontoppidan et al. (2005) sound a cautionary note for the interpretation of ice absorption spectra in YSOs with a detailed analysis of the spectrum of the source CRBR 2422.8–3423 showing that much of the absorption arises in the dense foreground Ophiuchus cloud. Fortunately, the Taurus Class I YSOs shown in **Figure 6** are not observed through such a foreground cloud.

3.1.4. Brown dwarf formation. Brown dwarfs with ages of only a few million years have not cooled appreciably and form a smooth continuum in luminosity and color with the slightly more massive late M stars. It had been thought that, below a certain mass, brown dwarfs might form as companions to more massive stars rather than

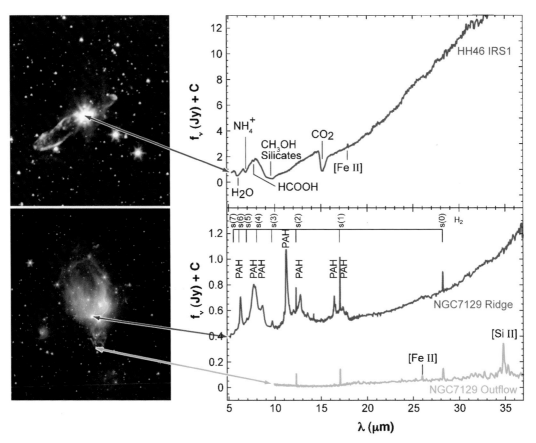

Figure 5

Outflows, nebulae, and embedded sources. Lower left: Infrared Array Camera (IRAC) image of the NGC 7129 region, ∼10′ on a side. The distance to NGC 7129 is ∼1.25 kpc. The NGC 7129 image is rotated so that the vertical direction is southwest on the sky; the HH46–47 image is rotated so that south is up and east is to the right. Color coding (for both *left* images): blue = 3.6 μm, green = 4.5 μm, and red = 8.0 μm. Lower right: Infrared Spectrograph (IRS) spectra obtained over ∼20″ × 20″ regions at the indicated positions of the NGC 7129 outflow (*green*) and the reflection nebulosity (*red*) (Morris et al. 2004). The outflow spectrum stops at the IRS high-resolution limit of 10 μm; *Infrared Space Observatory* (ISO) spectra (e.g., Rosenthal, Bertoldi & Drapatz 2000) show that H_2 and CO emission lines from outflows cluster in IRAC band 2 (4–5 μm), giving the outflow its green color in this image. The reflection nebulosity looks red because of strong PAH emission at 8 μm. Upper left: IRAC image of the HH46–47 system, ∼6′ on a side. Upper right: Spectrum of the bright central source embedded between the lobes in HH46–47 (Noriega-Crespo et al. 2004). The deep absorption features arise in the torus that confines the outflow. See van Dishoeck (2004) for similar results from ISO. (Right image reproduced by permission of the Am. Astron. Soc. Top left image credits: NASA/JPL-Caltech/A. Noriega-Crespo [SSC/Caltech], Digital Sky Survey. Bottom left image credit: NASA/JPL-Caltech/T. Megeath [Harvard-Smithsonian Cent. for Astrophys.])

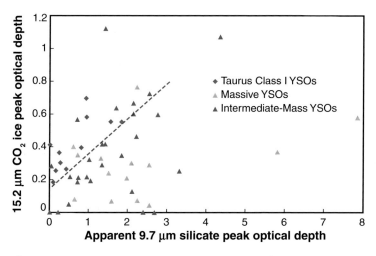

Figure 6

Peak optical depth in the 15.2-μm feature of CO_2 ice, plotted as a function of the apparent (i.e., excess of absorption over emission) peak optical depth of the 10-μm silicate feature, for low-mass Class I objects in Taurus (F. Markwick-Kemper, in preparation; see also Watson et al. 2004) observed by *Spitzer*, and toward intermediate- and high-mass YSOs observed by the *Infrared Space Observatory* (Alexander et al. 2003, Gibb et al. 2004).

through the gravitational collapse sequence outlined above. However, as summarized by Luhman (2006), extensive studies from *Spitzer* indicate that the frequency of occurrence and spectral properties of disks around young brown dwarfs with masses $\lesssim 10\ M_{Jup}$ do not differ from those of disks around young stars. Thus, there is as yet no evidence that brown dwarfs form differently than stars.

For example, Luhman et al. (2005c) suggest from *Spitzer* data that for both the IC348 and Chameleon I star-forming clusters the fraction of substellar objects in the mass range $<0.08\ M_\odot$ with evidence for disks—40% to 50% in each case—does not differ significantly from the fraction of protostars with masses $0.1\ M_\odot$ to $0.7\ M_\odot$ showing evidence for disks. The statistics on substellar objects result only from IRAC observations and would not include possible transition disks (**Figure 8**) such as Muzerolle et al. (2006) identified using MIPS observations in IC348. Thus, these results may be treated as lower limits on the fraction of both stellar and substellar objects having disks. Hartmann et al. (2005b) report similar results demonstrating the prevalence of disks around brown dwarfs in their survey of Taurus.

Luhman et al. (2005a,b) identified young, low-mass brown dwarfs with evidence for disks, most notably a brown dwarf in Chameleon with an estimated mass of $8\ M_{Jup}$. (The mass estimate, based on comparison with brown dwarf models, is uncertain by about a factor of two.) Even taking into account the uncertainties, the masses of these disk-bearing young brown dwarfs fall squarely into the range of the planetary companions identified around nearby stars by radial velocity measurements. The SEDs of the disks around these low-mass brown dwarfs closely resemble those of the disks around true protostars. Thus the raw materials for planet formation exist around free-floating planetary-mass bodies, and the extrasolar planets closest to Earth may orbit brown dwarfs.

3.1.5. High-mass star formation. As part of the GLIMPSE Legacy program (see sidebar) Churchwell et al. (2004) and Whitney et al. (2004a) used *Spitzer* observations of the giant H II region RCW 49 to identify hundreds of stars forming within the clouds surrounding RCW 49. Some of the YSOs are massive B stars and are therefore very young, suggesting their formation was triggered by stellar winds and shocks generated by the older (2–3 Myr) massive central cluster.

Allen et al. (2005) surveyed W5/AFGL 4029, S255, and S235, to gain a better understanding of the processes involved in high-mass star formation. This group investigated bright-rimmed molecular clouds, where an edge-on molecular cloud surface is externally illuminated by nearby young massive stars (see **Figure 7**). Using the color-color diagnostics described earlier, they determined the spatial distribution and surface densities of young stars (classes I and II) and noted that the Class I objects cluster tightly on the edge of the molecular cloud, very close to the bright rim that delineates the interface between the molecular cloud and the adjacent H II region, whereas the Class II objects are more widely dispersed. These data again support the evolutionary sequence underlying the YSO classification and suggest that either the propagation of the bright rim into the cloud or the effects of previous stellar generations triggers ongoing star formation in a self-sustaining fashion. Reach et al. (2004) noted similar phenomenology in their observations of the optically dark globule IC 1396A, which has a bright rim illuminated by a nearby O star.

The Midcourse Space Experiment (MSX) satellite discovered a class of infrared dark clouds that appear in silhouette at 8 μm against the bright mid-infrared emission of the galactic plane, indicative of high extinction (Egan et al. 1998). Rathborne et al. (2005) present a MIPS 24-μm image that reveals three embedded protostars in G34.4 + 0.2, a $\sim$5-arcmin-long filamentary cloud of this type. Each protostar has a luminosity between 10,000 and 30,000 $L_\odot$ and a projected main sequence mass $\sim$10 $M_\odot$. The authors suggest that the infrared dark clouds play an important role in massive star formation throughout the Galaxy. Beuther, Sridharan & Saito (2005) studied the dark cloud IRDC 18223-3. The results indicate an extremely young massive protostellar object hidden at the center of the core that causes hints of outflow activity both in molecular line emission and in IRAC imaging.

Spitzer's ability to study high-mass star formation in the LMC and other nearby galaxies seen from a favorable external perspective provides an excellent complement to its studies of similar phenomena within the Milky Way. The individual objects identified in LMC studies (e.g., Chu et al. 2005, Jones et al. 2005) have luminosities of $\sim 10^3$ to 10^4 $L_\odot$ and inferred masses ranging upward from $\sim$10 $M_\odot$. Van Loon et al. (2005) present the mid-infrared spectrum of a $\sim$20 $M_\odot$ YSO in the LMC, finding a different pattern of ice abundances from that seen in our Galaxy.

3.1.6. Outbursts and outflows. McNeil's Nebula, illuminated by V1647 Ori, rose in brightness by about four magnitudes at I-band over a period of a few months (Briceño et al. 2004). Serendipitous MIPS and IRAC observations of V1647 Ori postoutburst,

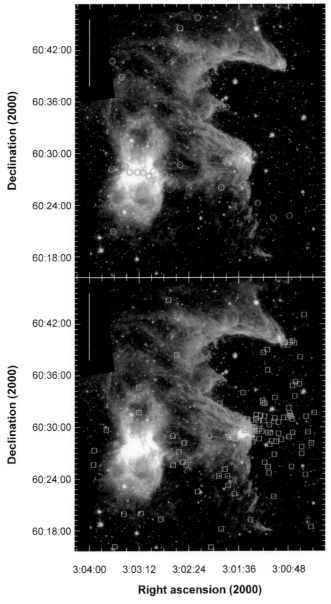

Figure 7

Infrared Array Camera (IRAC) images of the GL4029 region, which lies at a distance of about 2 kpc. A bright star just to the right of the region imaged sculpted the complex structure shown out of a molecular cloud. In the top image, circles show the positions of Class I objects identified by their IRAC colors; similarly, squares in the bottom image show the positions of Class II objects, which are much more widely distributed. Color coding by IRAC Band: blue = 3.6 μm, green = 4.5 μm, orange = 5.8 μm, and red = 8.0 μm. (Image credit: NASA/JPL-Caltech/L. Allen [Harvard-Smithsonian Cent. for Astrophys.])

PAH: polycyclic aromatic hydrocarbon

combined with data from previous surveys, provided a rare opportunity to study eruptive events in YSOs (Muzerolle et al. 2005). These observations show that the event is most likely due to a sudden increase in the accretion luminosity of the source from ∼3 to ∼44 $L_\odot$. They modeled the source as a Class I object with star, disk, and envelope and suggested that outbursts such as this may contribute to the clearing of Class I envelopes.

Spitzer can study the energetic outflows that accompany pre-main sequence evolution. **Figure 5** shows how the different IRAC bands distinguish polycyclic aromatic hydrocarbon (PAH)-dominated reflection nebulae from knots and filaments excited by outflows in NGC 7129 (Gutermuth et al. 2004, Morris et al. 2004). The dramatic images of the HH46-47 outflow presented by Noriega-Crespo et al. (2004) show two lobes largely unseen at visible wavelengths (**Figure 5**) and apparently dominated by shock-heated emission. By contrast, the spectrum of the exciting star, seen through an edge-on disk, shows deep ice absorption features.

3.2. Evolution of Protoplanetary Disks

For a YSO of solar luminosity, the excess emission detectable by the *Spitzer* instruments probes the dust distribution between a few tenths and a few tens of an astronomical unit from the star, with the inner material dominating the short wavelength emission. Thus, the spectrum or SED of the infrared radiation serves neatly as a proxy for the radial structure of the disk, and the absence of an excess at the shortest wavelengths signals exhaustion or consumption of the solid material from which terrestrial planets can form.

Hartmann (2005) reviews our understanding of disk evolution as it stood prior to the availability of *Spitzer* data. Protoplanetary disks appear to be very common around stars at an age of 1 Myr. These disks are rarer at 10 Myr, and no optically thick disks have been found around solar-type stars older than 30 Myr (e.g., Sicilia-Aguilar et al. 2005 and references therein). To understand how disks evolve over this 30-million-year period, *Spitzer* observations were designed to investigate disk evolution in clusters ranging in age from less than 1 Myr to upward of 100 million years. By exploring the structure of disks in varying age ranges, *Spitzer* can determine the epoch of terrestrial planet formation and study the transition of protoplanetary disks into optically thin remnants and finally into planetary debris disks.

3.2.1. Timescales for disk evolution in clusters.
Combining IRAC data with ground-based near-infrared fluxes, Gutermuth et al. (2004) determined that 54 ± 14% of stars in the core of the 1-Myr-old cluster NGC 7129 exhibit disks that increase the stellar brightness by ∼50% or more over the expected photospheric level at 4.5 μm. Like the other studies described in this section, the Gutermuth et al. (2004) survey reaches stellar masses below the hydrogen burning limit. Lada et al. (2006) used IRAC and MIPS to examine the ∼300 members of IC 348 to investigate the frequency and nature of the circumstellar disk population in the cluster. IC 348 has an intermediate age (2–3 Myr) and lies close enough to permit a robust disk census at the peak of the stellar initial mass function. The fraction of disk-bearing stars is

44 ± 7%; however, only 31 ± 5% of these stars are surrounded by robust, optically thick disks. These measurements indicate that in the 2–3 Myr since the cluster formed, 70% of the stars have lost most or all of their primordial circumstellar disks. The disk fraction peaks for stars of K6–M2 spectral types, with masses similar to the Sun. Thus, planet formation appears most favorable around the solar mass stars in this cluster. Evidence for disk evolution among young brown dwarfs was presented by Muzerolle et al. (2006), who combined the MIPS 24-μm data with the IRAC data in IC 348 for spectral types later than M6. Two objects among the six investigated show evidence for inner disk clearing. Their model for one of the objects suggests a clearing scale of ∼0.5–0.9 AU; one plausible explanation is that a planet formed very rapidly close to the central object.

f: fractional luminosity

Sicilia-Aguilar et al. (2005) examined two clusters (Tr 37 and NGC 7160) in the Cepheus OB2 region, aged 4 and 10 Myr, respectively, both about 900 pc from the Earth. These clusters exhibited significantly different disk fractions (48% and 4%, respectively). The younger cluster shows evidence of significant disk evolution: *Spitzer* observations (3.6–24 μm; Sicilia-Aguilar et al. 2006) showed that about 10% of the objects with disks in Tr 37 have photospheric fluxes at wavelengths <4.5 μm and excesses at longer wavelengths, indicating an optically thin inner disk. Similarly, Megeath et al. (2005) show that the ∼5-Myr-old eta Cha association shows a larger fraction of disks evolving into the optically thin stage than has been found in younger clusters. Young et al. (2004b) used IRAC and MIPS to observe the cluster NGC 2547, which is ∼30 Myr old. They derived a 3.6-μm emitting disk fraction of <7% but found that ∼25% of the stars showed an excess at 24 μm, suggestive of a cool disk with a central hole. It therefore appears well-established that the warm inner disks seen by IRAC dissipate on timescales less than ∼10 Myr, whereas 24-μm excesses persist significantly longer.

The nearby TW Hydrae association, with an age of 8–10 Myr, provides interesting insights into disk evolution (Uchida et al. 2004, Low et al. 2005). Four of the 24 stars studied in this grouping show 24-μm excesses at a factor of ∼100 over the expected photospheres—two of these are accreting T Tauri stars and two are debris systems with unusually high values of fractional luminosity, f, defined as a debris system's ratio of debris disk luminosity to star luminosity. Only one of the remaining 20 stars shows any evidence for a 24-μm excess, and that excess is no more than a factor of two. This sharp division between stars with very large excesses and stars with no excesses at all—particularly in such a young population—suggests that the transitions between these states occur very quickly. Zuckerman & Song (2004) review the use of nearby clusters, such as the TW Hydrae association, for the study of disk evolution.

An exception to the rule that dust disks around newborn stars disappear in a few million years was the discovery by *Spitzer* of a dusty disk orbiting St 34, an accreting M star binary with an estimated age of ∼25 Myr (Hartmann et al. 2005a). They suggest that the dust persists because dynamical effects of the binary star have inhibited planet formation, which otherwise would have dissipated the orbiting material.

3.2.2. Spectroscopic studies of disk evolution. The possible contribution of PAH emission and silicate emission or absorption in the IRAC channels introduces uncertainty in the interpretation of colors in young clusters. This is alleviated by extracting the dust features and the underlying continuum separately from IRS spectra. Furlan et al. (2005a) recently constructed the first such "emission-feature free" mid-infrared color-color diagram for a complete sample of Class II objects in the Taurus cloud. The match between the data and models is poor unless dust grains are heavily depleted by settling to the disk midplane; it appears that in most of these Class II objects, small dust grains are depleted from the upper reaches of the disks by factors of 100–1000. Settling of dust to the midplane of a protoplanetary disk due to gas drag forces may initiate the planetary formation process by stimulating grain-grain collisions and the coalescence of increasingly massive solid particles (Lada 2005).

The coalescence process also leaves an imprint on the shape and crystallinity of the 10 and 20 µm silicate emission features, as originally shown by ISO for Herbig Ae stars (van Boekel et al. 2005 and references therein). *Spitzer* spectroscopy has identified and studied in detail this coalescence in solar mass YSOs (Forrest et al. 2004, Kessler-Silacci et al. 2006). In addition, Apai et al. (2005) report IRS studies of young brown dwarfs that demonstrate both grain growth and the settling of grains to the midplane in the disks around these substellar objects.

Disks with central clearings are visible above the stellar photosphere only at longer wavelengths than are radially continuous disks. IRS spectra can exploit this phenomenon in detail to study the structure of the disks and the range of mechanisms responsible for the central clearings in these transition disks. A spectroscopic survey of 150 members of Taurus-Auriga (Forrest et al. 2004; E. Furlan, L. Hartmann, N. Calvet, P. D'Alessio, R. Franco-Hernandez, et al., submitted) has so far yielded four confirmed transition disks. The spectra (**Figure 8**) show clearly that the inner edges of the disks are very sharp, accurately modeled in each case as a single-temperature blackbody wall with a thin atmosphere, and that the central clearings are practically devoid of small dust grains (Calvet et al. 2005).

As enumerated by D'Alessio et al. (2005a), radiative, stellar-wind, or gas-drag mechanisms fail to produce central clearings in dusty disks within the ages of the Taurus stars. However, a companion, formed after both the disk and central star, could produce a clearing through its orbital resonances that prevents the exterior disk from moving further in but allows the inner material to accrete onto the star. Simulations by Quillen et al. (2004) and Varniere et al. (2006a,b) show that central clearings such as that in the CoKu Tau/4 disk (**Figure 8**) can be produced by companions of the Saturn-Jupiter class in less than 10^5 years. Establishing that massive planetary companions produce these gaps would challenge theories of planet formation because the ages of the stars in these systems (1–2 Myr) lie in the no-man's-land between the timescales for giant-planet formation in the two leading models. To form a Jovian planet by core-accretion processes (e.g., Pollack et al. 1996) takes upward of 10 Myr. To form one by growth of gravitational instabilities in a gaseous disk takes only on the order of 1000 years (e.g., Boss 2005); however, the disks may be gravitationally unstable only in the earliest evolutionary phases, and a planet formed too early may migrate inward to the star within 1 Myr.

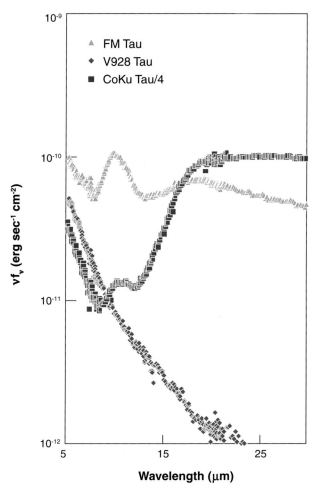

Figure 8

Infrared Spectrograph spectra of three Class II/III young stellar objects in Taurus. CoKu Tau/4 shows a flux deficit in the 5–15 μm region, indicative of a central clearing in its circumstellar disk (Forrest et al. 2004, revised and reproduced by permission of the Am. Astron. Soc.). CoKu Tau/4 is an example of a transition disk, as discussed in Section 3.2.2.

Spitzer spectra can further constrain disk evolution by studying the gas content of disks of various ages. Lahuis et al. (2006) report the discovery of gas phase absorption due to C_2H_2, HCN, and CO_2 in the low-mass Class I YSO IRS 46 in Ophiuchus. The high rotational temperatures suggest that this material is located in the terrestrial planet zone of the disk. This is the only YSO in a sample of 100 that shows these features. This may reflect mainly the need for both favorable geometry and a large intrinsic line width to make the lines visible in absorption, as many of these ∼1 Myr old objects should retain substantial amounts of their primordial gas.

In the case of HD105, with an age of ∼30 Myr, Hollenbach et al. (2005) obtain upper limits on the intensity of several infrared emission lines and show that less than 1 M_{Jup} of gas exists between 1 and 40 AU, so that giant-planet formation cannot occur in this disk at present; however, there may be enough residual gas in the disk to influence the dynamics of the dust.

3.2.3. Crystalline silicates among small dust grains in Class II young stellar objects.
Evolution of silicate dust grain composition in protoplanetary disks has been studied for more than a decade, via ground-based (e.g., Knacke et al. 1993, Honda et al. 2003, Kessler-Silacci et al. 2005) and ISO spectroscopy (van Dishoeck 2004). *Spitzer* spectra of Class II YSOs in this domain show evidence of silicate compositions ranging from one indistinguishable from that of amorphous interstellar grains to one heavily dominated by crystalline minerals such as pyroxenes, olivines, and silica. In **Figure 2** the spectra of FM Tau and IRAS F04147+2822 serve as examples of the extremes. The latter shows the same crystalline silicate emission features attributable to forsterite (crystalline Mg_2SiO_4) that appear in the spectra shown in **Figure 9**. ISO spectra of more massive luminous Herbig Ae stars show a similar variety (van Boekel et al. 2005).

Among Class II YSOs, *Spitzer* spectra indicate that the crystalline mass fraction among the small dust grains varies from very small upper limits below those placed on interstellar grains (<2%; Kemper, Vriend & Tielens 2004), to nearly 100% (Forrest et al. 2004, Uchida et al. 2004, Kessler-Silacci et al. 2006, Sargent et al. 2006; also D.M. Watson, J.D. Leisenring, E. Furlan, C.J. Bohac, B. Sargent, W.J. Forrest, et al., submitted). Not even the lowest-mass objects are immune to the large range of thermal processing implied, and several brown dwarfs and brown-dwarf candidates have displayed rich crystalline silicate spectra (Apai et al. 2005, Furlan et al. 2005a, Sargent et al. 2006; also B. Merin, et al., in preparation; see **Figure 9**). The quantification of the degree of crystallinity can be confounded by the simultaneous evolution of the particle size distribution. In several cases, detailed modeling suggests that the *Spitzer* spectra indicate substantial fractions of large (>1 μm), porous dust grains (Kessler-Silacci et al. 2006, Sargent et al. 2006). There are even indications of differences in the composition of amorphous grain material among the disks of low-mass Class II objects (Sargent et al. 2006). These strong indications of grain processing correlate with the observation of dust grain growth and settling to the disk midplane (Furlan et al. 2005b; E. Furlan, L. Hartmann, N. Calvet, P. D'Alessio, R. Franco-Hernandez, et al., submitted), which in turn implies the potential in these objects for rocky planetesimal growth and planet formation. As shown in **Figure 9**, *Spitzer* spectra of small bodies in the solar system show crystalline silicate emission very similar to that seen in protoplanetary and debris disks.

The origin of the crystalline silicates remains a mystery (see Molster & Kemper 2005). E. Furlan, L. Hartmann, N. Calvet, P. D'Alessio, R. Franco-Hernandez, et al. (submitted), Kessler-Silacci et al. (2006), and D.M. Watson, J.D. Leisenring, E. Furlan, C.J. Bohac, B. Sargent, et al. (submitted) have studied the emission spectra of large samples of Class II objects in order to understand how the amorphous interstellar silicates convert to crystalline form. They find no significant

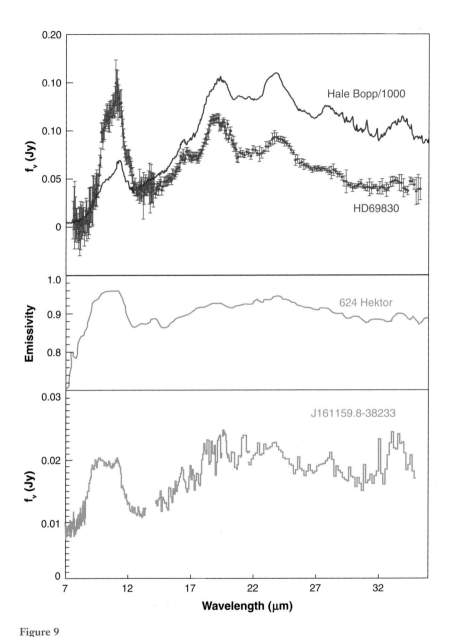

Figure 9

Infrared Spectrograph low-resolution spectra of the 2-Gyr-old K star HD69830 (*top*; Beichman et al. 2005a); the Trojan asteroid 624 Hektor (emissivity spectrum, *middle*; J.P. Emery, D.P. Cruikshank & J. Van Cleve, submitted); and the candidate brown dwarf J161159.8-38233 in Lupus (*bottom*; B. Merin, et al., in preparation) are compared with the spectrum of comet Hale Bopp (*top*; Crovisier et al. 1996). All sources show the characteristic emission features of Mg-rich crystalline silicates. (*Top spectra* reproduced by permission of the Am. Astron. Soc. *Middle spectra* reprinted from *Icarus* with permission from Elsevier.)

correlation between crystallinity and any property of the disks or central stars, including stellar mass, effective temperature, and accretion rate, indicating no systematic influence of radiation. There is also no correlation with disk mass or disk temperature, nor with the ratio of disk mass to stellar mass, which discounts a connection with disk-based heating mechanisms such as spiral shocks. A correlation of crystalline mass fraction with X-ray properties (flaring, in particular) remains to be explored, as does whether the structural evolution of the disks, driven by the formation and migration of planets, produces the lack of correlation between crystallinity and stellar or disk properties[1] (D.M. Watson, J.D. Leisenring, E. Furlan, C.J. Bohac, B. Sargent, et al., submitted).

3.3. Post-Main-Sequence Stellar Evolution

Post-main-sequence stellar evolution virtually always involves mass loss and dust condensation. The resulting phenomena that can be studied with *Spitzer* range from strong infrared continua associated with dust, through interesting chemistry on dust grains and in large molecules, to emission lines from both molecular and atomic transitions in the gas. Studies of galactic and extragalactic supernovae and their remnants, discussed below, illustrate the variety of analyses already under way with *Spitzer* data. Here, we touch very briefly on other *Spitzer* explorations of post-main-sequence topics. Blommaert et al. (2005) discuss ISO's extensive work on this subject (see other articles in Volume 119 of *Space Science Reviews* as well).

Spitzer has unique power to identify post-main-sequence stars for further study. For example, Soker & Subag (2005) predict that there should be a large population of undiscovered planetary nebulae in the plane of the Milky Way, and Cohen et al. (2005) demonstrate the power of the GLIMPSE survey to reveal them. *Spitzer* enables studies of post-main-sequence stars in a variety of extragalactic environments, such as the circumnuclear region in M31 (see images in Gordon et al. 2006). Characterizing post-main-sequence evolution in nearby low-metallicity galaxies such as the Magellanic Clouds will allow sampling of an earlier epoch in our own Galaxy. Examples include the study of known planetary nebulae in the Magellanic Clouds (J. Bernard-Salas, J.R. Houck, P.W. Morris, G.C. Sloan, S.R. Pottasch, D.J. Barry, in preparation), the search for LMC RCrB stars by Kraemer et al. (2005), and the IRS spectral atlas of the LMC compiled by C. Buchanan, J. Kastner, B. Forrest, S. Raghvendra, M. Egan, D. Watson, et al., (in preparation).

In addition to its role in discovering and classifying post-main-sequence stars, *Spitzer* is revealing interesting aspects of their evolution and the chemistry in their mass-loss shells. Hora et al. (2004) examine IRAC images of planetary nebulae that effectively locate hot dust and trace large structures of molecular gas identified through molecular hydrogen emission within some of the IRAC photometric bands. Su et al.

---

[1]Note that high spatial resolution spectra with large ground-based telescopes have found spatial variations of the dust spectrum and crystallinity with position across large disks around A-type stars (Weinberger, Becklin & Zuckerman 2003; van Boekel et al. 2004).

(2004) study the planetary nebula NGC 2346 in the MIPS bands, finding interesting changes in morphology with increasing wavelength. These include a circumstellar ring that becomes optically thin at 70 μm and, most significantly, a shell of cold dust from previous mass-loss episodes detected at 160 μm. Although such cold dust halos might be expected around many planetaries, so far their detection has proven elusive (W. Latter, private communication).

IRS spectra are discovering unexpected aspects of the mineralogy and chemistry in circumstellar material. For example, Markwick-Kemper, Green & Peeters (2005) report possible oxygen-rich material in the carbon-rich outflow of the Red Rectangle; conversely, Jura et al. (2006) report carbon-rich material (PAHs) orbiting the oxygen-rich red giant HD233517 and suggest that the material may have been synthesized in situ in a long-lived disk created when the star engulfed a companion. Sloan et al. (2006) report unexpected structure in the silicate emission from the LMC Mira variable HV 2310, partly attributable to crystalline grains. In addition, *Spitzer* observations of mid-infrared fine structure and recombination lines have been used to diagnose ionization conditions, wind properties, and abundances in objects ranging from planetary nebulae (J. Bernard-Salas, J.R. Houck, P.W. Morris, G.C. Sloan, S.R. Pottasch, D.J. Barry, in preparation) to dramatically mass-losing stars such as Sakurai's Object (Evans et al. 2006) and Wolf-Rayet stars (Morris, Crowther & Houck 2004; Crowther, Morris & Smith 2006).

3.4. Supernovae

Infrared observations of supernovae and supernova remnants can provide important insight into the origin and dissemination of heavy elements, both by detecting fine-structure emission lines from these materials and by enabling the study of dust created in and heated by the supernovae. The infrared spectrum of a supernova remnant also provides samples of pre-existing circumstellar and interstellar matter and can be diagnostic of the interaction of the supernova with its environment. Dwek & Arendt (1992) report that IRAS detected ∼30% of the known galactic supernova remnants, providing evidence for collisional (as opposed to radiative) heating of the remnant dust and for grain processing in some remnants. All of the detected remnants radiate more power in the infrared than in the X ray. More recently, ISO has studied the Kepler, Tycho, Cas A, and Crab Nebula remnants (Douvion, Lagage & Cesarsky 1999, Douvion et al. 2001), finding a range of properties, as expected from the known variety of supernovae. In Cas A, Arendt, Dwek & Moseley (1999) find a rich infrared emission-line spectrum and a broad 22-μm emission feature, indicative of a class of silicate minerals different from that typically associated with interstellar dust grains.

The explosion of SN1987A in the LMC provided a unique opportunity for detailed infrared observations of the earliest stages of a supernova (Wooden et al. 1993, Wooden 1995). A rich emission spectrum including contributions from Fe, Ni, Co, Ar, CO, and SiO persisted for more than 400 days, and the shift of the bolometric luminosity into the infrared following the onset of dust formation was recorded ∼530 days after the explosion. Bouchet et al. (2004) detected the faint infrared glow of the warm dust in the remnant more than 16 years after the event and present mid-infrared images of the previously ejected circumstellar ring now being shock-heated by the supernova ejecta.

Many of the broad lines in the SN1987A spectrum had brightness $\gtrsim$10 Jy even a year after the explosion. The IRS has measured sources fainter than $\sim$1 mJy; thus *Spitzer* can study in detail supernovae in galaxies as distant as $\sim$10 Mpc. Barlow et al. (2005), Kotak et al. (2005), and Stanimirovic et al. (2005) have already reported *Spitzer* results on extragalactic supernovae. For supernova remnants within our galaxy, *Spitzer*'s large-scale maps have already yielded surprising results, and *Spitzer* can obtain detailed spatially resolved spectra of both gas phase and dust emission.

3.4.1. Imaging observations. Within a supernova remnant, a small amount of mass condenses into small, silicate-rich grains and other refractory materials such as aluminum oxide. During the first few years after the supernova explosion, this dust may be heated by the radioactive ejecta of new heavy elements from the supernova (Bouchet et al. 2004 and references therein). Later, the emission—with contributions from both freshly synthesized and pre-existing grains—arises from very small grains stochastically heated by the energetic electrons also responsible for the thermal X-ray emission (Dwek & Arendt 1992). Hines et al. (2004) report *Spitzer* images of the young supernova remnant Cas A, which they discuss in terms of this process. They estimate that a total dust mass of only $\sim 3 \times 10^{-3}$ $M_\odot$ accounts for the mid-infrared emission from the remnant. They also call attention to a region at the edge of Cas A where a shock wave produced by the expanding supernova remnant excites infrared radiation from the dust in an adjacent molecular cloud.

A far larger mass of cold dust—about 3 $M_\odot$—had been reported from submillimeter observations of Cas A (Dunne et al. 2003). However, Krause et al. (2004) use a combination of *Spitzer* data and millimeter-wave molecular line observations to argue that this material is foreground interstellar gas and dust, a result confirmed by Wilson & Batria (2005). Stanimirovic et al. (2005) report observations of 1E 0102.2-7219, a $\sim$1000-year-old supernova remnant in the Small Magellanic Cloud similar to Cas A. They estimate $\sim 8 \times 10^{-4}$ $M_\odot$ of hot dust is visible in this object and point out that this value is $\sim$100 times below recent theoretical predictions; *Spitzer* observations of young remnants should explore this discrepancy futher.

Lee (2005) and Reach et al. (2006) search the GLIMPSE data for supernova remnant detections. Each group detects about 20% of the supernova remnants known in the radio band within the GLIMPSE survey region. Reach et al. (2006) show that a broad variety of emission mechanisms dominate the various IRAC bands.

3.4.2. Infrared echoes. Supernovae can also emit through infrared echoes, when nearby dust is heated by the outgoing energy flash from the explosion (e.g., Douvion et al. 2001). A Type II progenitor loses material in a $\sim$1500-km s^{-1} wind at rates of 10^{-6} $M_\odot$ per year. This wind will sweep up any surrounding material to generate a wind-blown bubble. Van Marle, Langer & García-Segura (2004) calculate that by the time a 25-$M_\odot$ star becomes a supernova it can have blown a bubble of $\sim$35-pc radius into a surrounding interstellar medium of density 3×10^{-23} g cm^{-3}. Thus, for up to $\sim$100 years after the explosion, we can expect an infrared echo generated

Figure 10

Two images of the Cas A supernova remnant and its environs at 24 μm, taken one year apart, superposed to show the propagation of a light echo (Krause et al. 2005). The blue image dates from November 2003, whereas the orange image was taken in December 2004. The strip shown is approximately 20 arcmin in length, oriented with east down and north to the left. [Image credit: NASA/JPL-Caltech/O. Krause (Steward Observatory)]

in material that the star has lost into this bubble, as in SN 2002hh in NGC 6946. Barlow et al. (2005) model the *Spitzer* observations of this event and show that they can be explained as emission by 0.1–0.15 $M_\odot$ of dust (and hence some 10 $M_\odot$ of dust and gas) ejected by the progenitor prior to the explosion.

An infrared echo can also arise as the flash propagates through the surrounding interstellar medium (ISM), including any stellar ejecta that have penetrated into the ISM, as found for Cas A by Krause et al. (2005). In this case, *Spitzer* traces the progress of the flash through the ISM, resulting in the illusion of speed-of-light motions as the expanding flash heats various cloudlets. The pattern of apparent motions close to the remnant (**Figure 10**) suggests a more recent event than the main explosion, possibly a giant flare from the central neutron star occurring around 1952. This hypothesis will be tested by tracking these motions throughout the *Spitzer* mission.

3.4.3. Infrared spectra. T. L. Roellig, T. Onaka & K.-W. Chan (in preparation) present complete *Spitzer* 5.2–36-μm spectra of a bright knot in the Kepler remnant showing emission lines of Ni, Ar, S, Ne, and Fe superposed on an underlying continuum. The emission line spectrum is consistent with J-shock heating of material with interstellar abundances (see Hollenbach & McKee 1989). The continuum can be approximated by emission from collisionally heated dust grains, but a detailed fit will require more sophisticated models.

Kotak et al. (2005) measured the Ni II line at 6.62 μm in SN 2004dj in NGC 2403 at a distance of 3.1 Mpc. The measured line intensity implies a stable nickel mass of at least 2.2×10^{-4} $M_\odot$ and suggests that the progenitor to the supernova was a red supergiant with initial mass ~10–15 $M_\odot$. Kotak et al. (2005) also review the advantages of mid-infrared fine-structure lines for determining heavy element abundances in supernovae: (*a*) the far fewer line transitions in the infrared result in less blending; (*b*) the infrared lines have greatly reduced sensitivity to extinction; and (*c*) the infrared line strengths are insensitive to temperature.

3.5. Brown Dwarfs

The initial detections of brown dwarfs occurred in the 1990s. Since then, systematic surveys reviewed by Kirkpatrick (2005) have shown that brown dwarfs compare in number with all other stars in the neighborhood of the Sun. Brown dwarfs are divided into three main subclasses, L, T, and Y; L-class objects (1400–2600 K) are below the hydrogen-burning limit with no methane, T-class objects (600–1400 K) are below the hydrogen-burning limit with methane, and the as-yet undiscovered Y-class objects (150–600 K) would be cool enough to have water clouds. As all brown dwarfs steadily cool during their lives, there is no equivalent to a main sequence, and an object's spectral type and mass do not correspond uniquely. Even Jupiter, with mass ~0.0015 $M_\odot$, has a substantial residual internal heat source and falls into the brown dwarf category by some determinations.

Spitzer can study brown dwarfs in detail, both photometrically and spectroscopically. Searching for and studying brown dwarfs in young clusters has proven particularly fruitful for *Spitzer* investigations, as discussed in Section 3.1.4. The discussion here focuses on *Spitzer*'s studies of older, field brown dwarfs.

3.5.1. Spectroscopy of field brown dwarfs. IRS can obtain mid-infrared spectra of field brown dwarfs for comparison with model atmospheres of very low-mass objects. Roellig et al. (2004a) observed an M star, an L dwarf, and a T dwarf binary system, and report excellent agreement between their observed spectra and the most recent atmospheric models. Roellig et al. (2004a) also report the first detection of ammonia in the atmosphere of a brown dwarf, as well as the first detection of the predicted methane feature located at 7.8 µm. Currently available data permit the construction of a mid-infrared MLT spectral sequence (Cushing et al. 2006). **Figure 11** shows objects in this sequence ranging from early M dwarfs to cool T dwarfs. As predicted by the models, the spectra become progressively more complicated as the effective temperature falls. The region around the transition between the L and T spectral types is particularly interesting, because the models predict that atmospheric clouds of silicates and iron should become directly visible in the mid-infrared in these objects. At higher effective temperatures the clouds do not form, whereas at lower effective temperatures the clouds are physically located below the photosphere. In that location, they are not directly observable, although their influence on the atmospheric energy transport can affect the observable photosphere above. The *Spitzer* data clearly demonstrate the effects of cloud opacity in the L–T transition regime, as evidenced by a flattening in the atmospheric gas-phase absorption features, due to the much broader solid state features from the dust clouds.

These data support detailed comparison of spectra and models. **Figure 12** shows an IRS spectrum of Gl570D, a T8 dwarf with an effective temperature of 800 K. The strong methane, water, and ammonia absorption bands are immediately visible

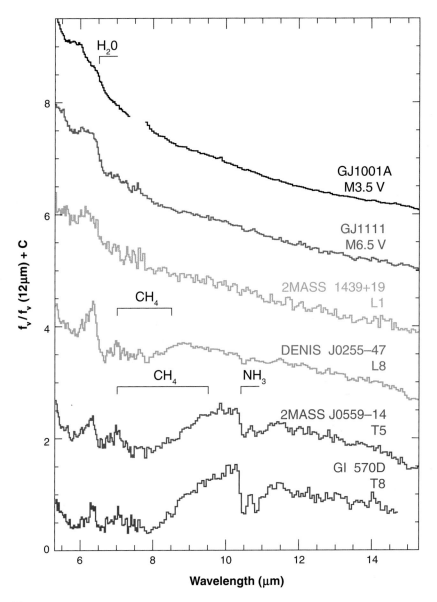

Figure 11

A selection of M, L, and T brown dwarf spectra taken with the *Spitzer* Infrared Spectrograph. The effective temperatures of these objects range from approximately 3400 K for the M3.5V object down to 800 K for the T8 dwarf. The molecular species responsible for the obvious features are indicated. The smaller-scale spectral features are also due to water, methane, and ammonia, with their relative importance depending on the temperature. Because there are no observed temperature inversions in the atmospheres of these objects, all of the features are seen in absorption (Cushing et al. 2006).

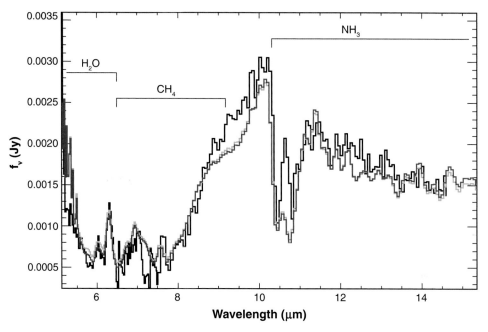

Figure 12
The observed spectrum of the T8 dwarf Gl570D, plotted in black together with two model spectra. The blue and red lines indicate models with and without clouds, respectively. Although the cloud model predicts that the clouds will lie below the photosphere in such a cool object (resulting in near-perfect agreement between the red and blue lines), the observed spectrum is brighter than expected in the 9.8-μm region near the silicate feature (Cushing et al. 2006). The models compared to the data are taken from Marley et al. (2002, 2003) and Saumon et al. (2003).

in this spectrum. Two model atmospheres are overlaid on these data, one with and one without clouds. The agreement is reasonable except for the region between the ammonia bands near 10.6 μm. The similarity of the model spectra with and without clouds shows that for these very low temperature objects, any clouds are located too far below the photosphere to greatly affect the mid-infrared spectrum. More careful analysis of the spectra shows that the spectral features are indicative of nonequilibrium chemistry in the brown dwarf's atmosphere. The IRS should be capable of studying the atmospheres of even cooler Y-dwarfs when or if they are discovered.

3.5.2. White dwarf–brown dwarf binary systems.
Farihi, Zuckerman & Becklin (2005) used *Spitzer* to identify an unresolved brown dwarf companion to the white dwarf GD 1400. Because the brown dwarf dominates the infrared radiation from the system, it should be possible to obtain spectra of such companions to compare with those of the field brown dwarfs and perhaps to illuminate the evolution of the parent binary system.

4. PLANETARY SYSTEM FORMATION AND EVOLUTION

During and long after the dissipation of a protoplanetary disk, collisions of asteroids or evaporation of comets around a main sequence star inject dust into its circumstellar environment. We refer to the tenuous disk that forms and regenerates through this process as a planetary debris disk. With rare exceptions, these systems are too diffuse and faint for detection, let alone detailed study, by warm telescopes. The diagnostic capabilities of IRS augment the photometric capabilities of IRAC and MIPS to take us far beyond the simple identification and characterization of such systems, enabling detailed studies of disk composition and structure.

The infrared is the prime wavelength band for direct detection of extrasolar planets because the planet/star flux ratio, though small, should be considerably larger than in the visible spectrum. The sensitivity and stability of *Spitzer*'s instruments have allowed full exploitation of this advantage, leading to the direct detection of radiation from several extrasolar planets with orbital geometries particularly favorable for this purpose.

Spitzer studies of both extrasolar planets and debris disks can be correlated directly to phenomena in our own solar system. The extrasolar planets appear similar in mass and size to Jupiter; the debris disks are exactly analogous to solar system dust structures such as the zodiacal cloud; and the objects that feed the debris disks have local counterparts in the comets, asteroids, KBOs, and other small bodies within the solar system. *Spitzer* has begun systematic studies of these solar system phenomena.

4.1. Evolution of Debris Disks

Debris disks are produced from larger bodies and respond gravitationally to the planets whose space they share, so their properties can reveal the character of extrasolar planetary systems. Overall, the dust density and rate of dust production decline with time, as radiation pressure or radiation drag remove the small particles from the system and the parent planetesimals are no longer replenished. In the current-day solar system, the level of debris generation is very low. The asteroid belt has a fractional luminosity $f \leq 10^{-7}$. It was therefore a surprise when the IRAS mission demonstrated that a number of nearby stars have excess emission with $f \geq 10^{-4}$ (Aumann et al. 1984). A second surprise was that these systems were in orbits with radii of roughly 100 AU. This discovery, made prior to our finding the first KBOs, exemplifies how debris disks can reveal aspects of other planetary systems relevant to understanding the solar system.

Less than 10^{-2} Earth masses of dust are required to produce $f \sim 10^{-4}$. Although the dust is much colder than the star, the surface area of the finely divided small particles is large enough that their far-infrared radiation dominates that of the star by factors of 50 to 100 at 60 μm and 100 μm for the prominent disks discovered by IRAS [see Backman & Paresce (1993) for a discussion of these and other debris disk fundamentals].

4.1.1. *Spitzer*'s role. IRAS and ISO show that debris disks with $f \gtrsim 10^{-5}$ occur around perhaps 15% of nearby main sequence stars of spectral types K through A. However, the formal uncertainties in this estimate are large. These missions also established a rough timescale of a few hundred million years for the decay of the systems around young stars (Habing et al. 2001, Spangler et al. 2001, Decin et al. 2003), but found that some much older stars also have large excesses (Decin et al. 2003). A general summary of debris disk studies after the completion of these two missions can be found in Zuckerman (2001) and Caroff et al. (2004). HST has imaged scattered light from a number of circumstellar disks, but with the exception of the one around Fomalhaut (Kalas, Graham & Clampin 2005), successful imaging has been achieved only for very young systems. In the submillimeter spectral range, a few of the nearest debris disks have been imaged (Wyatt et al. 2003 and references therein) and a few others have been measured photometrically. In most cases these imaging experiments have established that the dust in these systems has a disk- or ring-like distribution, as is assumed throughout this discussion.

Spitzer's contributions derive largely from its ability to search a virtually unlimited number of stars for debris-derived infrared excess emission. At 24 μm, *Spitzer* can detect as little as 10% above the bare photosphere, limited by our ability to estimate the intrinsic photospheric emission. At 70 μm, *Spitzer* can measure levels 20–30% above the photosphere. **Table 2** summarizes many of the debris disk surveys carried out with *Spitzer* to date. With spectroscopy, such searches can be expanded both to constrain the location of the debris and to probe its mineralogical content. *Spitzer*'s relatively high angular resolution allows imaging of prominent, nearby debris systems as well as improved discrimination against spurious debris disks caused by stellar heating of nearby interstellar material or a background source in the large IRAS beam (e.g., Kalas et al. 2002).

4.1.2. Spectral properties. The broad SEDs of the great majority of debris disks are remarkably similar, with shapes indicating that the material is at a temperature of about 70 K and therefore 10–100 AU from the star. Hence, these regions appear to be analogous to the Kuiper Belt. For old stars of roughly solar type (late F to early K), *Spitzer* spectra generally show no detectable debris emission at wavelengths short of 33 μm (Kim et al. 2005, Beichman et al. 2006), indicating that these rings of debris usually terminate at their inner edges with little material within the zone where the terrestrial planets lie in the solar system.

A-type stars behave differently; a detectable excess at 24 μm almost always accompanies a large excess at 70 μm (K.Y.L. Su et al., in preparation). However, the two well-imaged A-star debris systems, Fomalhaut and Vega, show Keplerian systems of debris at radius ∼100 AU in the submillimeter range (and also in the far-infrared for Fomalhaut). Thus, the 24-μm emission arises from tenuous material inside the Kuiper Belt analog region. That is, despite the differences in SED, the A stars may not host qualitatively different structures from the debris around cooler stars.

Only a small number of extreme systems have been found that depart substantially from this overall pattern. For example, *Spitzer* has identified only a few cases of

24-μm-only excesses, which might suggest asteroid dust, as appears to be the case for Zeta Lep (cf. Chen & Jura 2001). A particularly interesting example, the ~2-Gyr-old K star HD69830, has a 24-μm excess due almost entirely to a large population of small crystalline grains (Beichman et al. 2005a) within a few astronomical units of the star, but without a substantial reservoir of cooler material. The grains are known to be less than a few micrometers in size because of the strong emission features due to crystalline silicates in the 10-μm region (**Figure 9**). Models of HD69830 demonstrate that these small grains have very short lifetimes before their destruction or expulsion from the system. Their presence in such numbers indicates their recent generation by a transient phenomenon such as a supercomet or a collision in a densely populated asteroid belt (Beichman et al. 2005a). Similar conclusions may apply to other systems of this type (Chen et al. 2005a,b; Song et al. 2005; Hines et al. 2006); even among this group HD69830 may be unique with its >2 Gyr stellar age.

Uzpen et al. (2005) report a number of debris disks in the GLIMPSE database on the basis of excess emission in the 8-μm IRAC band. However, this sample may be contaminated owing to the distance of the GLIMPSE sources and the strong interstellar PAH emission at 8 μm. IRS spectroscopy can test the true nature of the excesses.

4.1.3. Direct imaging. *Spitzer* has imaged well four debris disks as of this writing: Fomalhaut, Vega, eps Eridani, and beta Pic (however, the last is too young to be considered a mature debris system). The imaging tests the assumptions made in analyzing the SEDs of the many spatially unresolved systems explored by *Spitzer*; thus, the marked variety of behavior even within this small sample challenges any interpretation critically dependent on a particular assumption about the disk geometry. Fomalhaut does indeed behave as expected, with a circumstellar ring of radius ~110 AU prominent at 70 μm (**Figure 13**) and through the submillimeter, but filled in at 24 μm by particles being dragged into the star. Asymmetries in the visible image may reflect the gravitational influence of a planet orbiting interior to the ring (Kalas, Graham & Clampin 2005); at 70 μm the ring shows asymmetry consistent with that seen in the visible. Beta Pic is large in extent (K.Y.L. Su, personal communication), as indicated also by previous optical and mid-infrared ground-based imaging, consistent with arguments that grains are being ejected by photon pressure. Telesco et al. (2005) suggest from ground-based observations that a recent catastrophic planetesimal collision generated a prominent cloud of tiny grains within this disk. Given the youth of the star (10–20 Myr; Lanz, Heap & Hubeny 1995), this behavior may provide important insight into the embryo planet phase in the evolution of terrestrial planets.

Spitzer images, combined with those at other wavelengths, highlight the importance of processes that segregate particles by size, producing images that vary in appearance with wavelength. Eps Eridani shows a ~60-AU radius ring in the submillimeter range, but the ring is already filled at 70 μm, and the 70-μm emission extends exterior to the submillimeter ring as well (D. Backman, et al., in preparation). The Vega images are yet more striking (**Figure 13**). Debris can be traced to a radius of nearly 1000 AU. Su et al. (2005) fit the characteristics of the system with small grains

Table 2 Published results from *Spitzer* debris disk surveys carried out mainly at 24 and 70 μm

Reference[a]	Sample	24-μm-only excesses	70-μm-only excesses	24- & 70-μm excesses	Range of f	Comments
Beichman et al. 2005	26 FGK stars with RV companions, older than ~1 Gyr	none	6 (0.23)	none	$>1.2 \times 10^{-4}$ to $>1 \times 10^{-5}$	
Bryden et al. 2006	69 FGK stars with median age ~4 Gyr	1 (0.01)	7 (0.10)	none	$>6 \times 10^{-6}$ to $>6.8 \times 10^{-5}$	
Chen et al. 2005a	40 F and G stars in Sco-Cen OB group. Ages 5 to 20 Myr	7 (0.17)	none	7 (0.17)	4×10^{-5} to 3.00×10^{-3}	No photospheres detected at 70 μm
Chen et al. 2005b	39 A through M dwarfs. Ages 12 to 600 Myr	1 (0.026)	4 (0.10)	2 (0.05)	2×10^{-6} to 7.70×10^{-4}	
T.N. Gautier III, G.H. Rieke, J. Stansberry, G.C. Bryden, K.R. Stapelfeldt, et al. (in preparation)	30 local M stars not thought to be young	none	none	none	NA	See text
Gorlova et al. 2004	63 stars in M47, age ~100 Myr, types A through M	11 (0.17)	NA	NA	NA	Mix of objects includes several late-type debris disk candidates
Kim et al. 2005	~35 solar type stars from the FEPS sample	1 (0.03)	5 (0.14)	none	2.3×10^{-4} to $<3 \times 10^{-5}$	70-μm-only detections aged ~0.7–3 Gyr. IRS data used as well

Study	Sample					Comments
Low et al. 2005	24 stars in TW Hydra association, age ~8–10 Myr, types A through M	none	1 (0.04)	5 (0.21)	0.27 to 8.60×10^{-4}	Includes objects still in accretion phase. Some 160 μm detections reported as well
Rieke et al. 2005	266 A stars aged <10 to 800 Myr	72 (0.27)	70-μm data on this sample in preparation by K.Y.L. Su et al. (see text)			Max 24 μm excess declines as 150 Myr^{-1}
Stauffer et al. 2005	20 G dwarfs in the Pleiades, age ~100 Myr	3 (0.15)	none	none	1.6×10^{-4} or less	No photospheres detected at 70 μm. IRS data used as well

[a] Use this table to point to papers and programs of particular interest to you. In general, the results of each study are too complex to be summarized completely in tabular form. Studies summarized here are primarily based on MIPS photometry at 24 and 70 μm. Beichman et al. (2006) show that Low Resolution IRS Spectra can pick up disks at levels of f or in wavelength regions difficult to achieve with photometry alone and use the technique to set limits on asteroidal emission from stars with radial velocity companions.

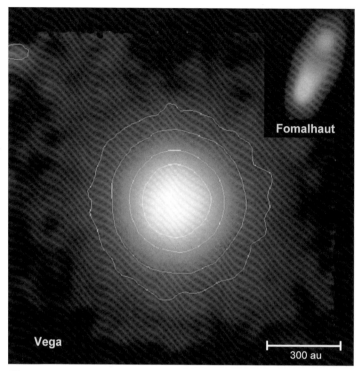

Figure 13

The 70 µm Multiband Imaging Photometer for *Spitzer* (MIPS) image of the debris disk around Fomalhaut (Stapelfeldt et al. 2004) is shown on the same spatial scale as that around Vega (Su et al. 2005) to illustrate the great variety in appearance of debris systems with similar spectral energy distributions. A logarithmic display is used for Vega to highlight the great extent of the dust orbiting this star; by contrast, the smaller Fomalhaut disk has a sharp outer edge. (Reproduced by permission of the Am. Astron. Soc.)

(~10-µm radius) that are being ejected by photon pressure. The fits indicate that the grains originate in a Keplerian ring of objects detected in the submillimeter range at a radius of about 90 AU, where a large collision may have taken place on the order of a million years ago, setting up the collisional cascade responsible for the small grains. As with HD69830, the high loss rate for these grains makes it implausible that the Vega system has always had its current appearance.

4.1.4. Evolution. The new *Spitzer* results emphasize the role of individual planetesimal collisions in producing large populations of grains that can dominate the radiometric properties of these systems for a few million years. These events complicate comparisons with most of the existing models, which assume a smooth and continuous evolution (the models of Kenyon & Bromley 2004, 2005 and Grogan, Dermott & Durda 2001 are exceptions). They also make it difficult to determine to what extent the variety in observed disk properties reflects fundamental differences in the disks.

Spitzer confirms previous indications that strong excesses, suggesting active terrestrial planet building (and destruction), occur commonly around stars less than 100 Myr in age (Gorlova et al. 2004; Chen et al. 2005a,b; Rieke et al. 2005). A-type stars are attractive targets for tracking disk evolution because a significant 70-μm excess is almost always accompanied by a detectable excess at 24 μm (K.Y.L. Su et al., in preparation). Because of the high angular resolution of the MIPS 24-μm band, its reduced susceptibility to cirrus confusion emission, and its high sensitivity to stellar photospheres, A-star excesses ∼10% above the photospheric output can be detected at 24 μm to a distance of ∼500 pc, putting the nearest young stellar clusters within range. Rieke et al. (2005) report a preliminary study based on this capability, presenting a compilation of 266 A stars, some of which were observed by IRAS and ISO. In total, 27% of the stars have 24-μm excesses. Even in the youngest age range (<25 Myr), ∼50% of the stars show no excesses, while some young stars have extremely large excesses. As shown in **Figure 14**, the envelope of the maximum excess decays with age, t, roughly as 150 Myr·t^{-1}.

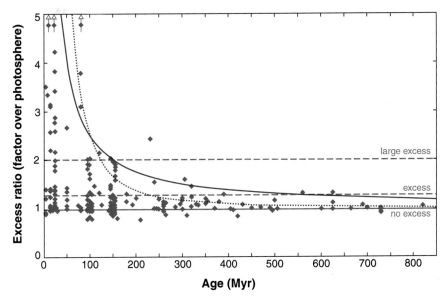

Figure 14

24-μm excess versus age for a large sample of A stars (Rieke et al. 2005). Excess emission is indicated as the ratio of the measured flux density to that expected from the stellar photosphere alone: a value of 1 represents no excess (*solid gray line*). Additional dashed gray horizontal lines show the threshold for detection of an excess (1.25) and the one for a large excess (2). The upward-pointing arrows are, from left to right, HR 4796A, βPic, and HD 21362. The thin solid line is an inverse time dependence, whereas the thin dashed line is inverse time squared. Age uncertainties are roughly a factor of 1.5 below 200 Myr (where ages are almost entirely from cluster and moving group membership) and roughly a factor of 1.5–2 above 200 Myr (where many ages are assigned by placing the stars on the Hertzsprung-Russell diagram). (Reproduced by permission of the Am. Astron. Soc.)

Again, these results highlight the stochastic evolution of debris disks, as stars more than 1 Gyr in age would hardly ever have significant excesses according to the 150 Myr·t^{-1} enveloping behavior, yet many of them do. In the combined results from Kim et al. (2005) and Bryden et al. (2006), 13 ± 4% of old, solar-like (late F to early K) stars in the solar neighborhood have substantial excess emission at 70 μm, although virtually none have excesses at 24 μm. The limiting f for the F-to-K stars is $\sim 10^{-5}$.

In evaluating these debris-disk differences among stellar types, we need to distinguish between effects due to age and those due to stellar mass. A typical solar-type debris system circles a star with a main sequence lifetime exceeding that of an A star. Observations of solar-type stars in the Pleiades, at 120 Myr, indicate a significant incidence ($\sim$15%) of 24-μm excesses (Stauffer et al. 2005; N. Gorlova, private communication). It appears to take 100–200 Myr for debris to clear from the terrestrial planet zone in a typical planetary system, independent of stellar mass between solar and A-type. Jura (1990), using IRAS data, showed that excesses are rare in G giants, former A stars that have left the main sequence. This suggests that the high rate of A-star detection arises in part because of their youth and the slow clearing of debris systems in general.

4.1.5. Dependence on other factors. Detection of debris to a given level of f becomes more difficult as the stellar luminosity and temperature decrease along the main sequence. The stellar luminosity is directly proportional to T^4 (where T is its temperature), whereas the Rayleigh-Jeans photospheric flux density above which the infrared excess must be detected is directly proportional to T. Thus, for a given contrast limit above the photosphere, the limiting f will scale as T^{-3}.

Debris disk excesses in M stars have been elusive, because of this increase in detection difficulty with decreasing stellar luminosity and because of the intrinsic faintness of these targets. The only M stars known to have excesses at 25 μm and beyond are T Tauri stars and young objects such as AU Mic (Song et al. 2002; Kalas, Liu & Matthew 2004). T.N. Gautier III, G.H. Rieke, J. Stansberry, G.C. Bryden, K.R. Stapelfeldt, et al. (in preparation) have studied 30 nearby early M stars not expected to be particularly young and find no definite excesses at either 24 or 70 μm. They also show that the average excess over the photosphere for these stars at 70 μm is at least a factor of $\sim$6 lower than the average excess level for the F-, G-, and K-star sample of Bryden et al. (2006); by the argument given above, however, the M stars may still have f comparable to that of their more massive counterparts.

Bryden et al. (2006) show the incidence of debris disks to be roughly independent of stellar metallicity, in dramatic contrast to the rapid increase in radial velocity planet detections with increasing metallicity (Fischer & Valenti 2005). Beichman et al. (2005b) report 70-μm excesses around 6 of 26 FGK stars with known radial velocity planetary companions, a rate of 23%. Supplementing this result with additional ongoing GTO and GO observations (G. Bryden, personal communication) yields a net of 7 excesses in 37 stars with known planets, or a rate of 19 ± 7%. Considering the small sample, this value does not differ significantly from the detection rate of 13 ± 4% for similar, single stars (Kim et al. 2005, Bryden et al. 2006). J.A. Stansberry & D.E. Trilling (personal communication) also find a similar detection rate for 70-μm excesses around binary stars. The incidence of detectable excesses thus appears surprisingly independent of metallicity or the presence of companions.

Spitzer has identified debris-disk-like structures in a surprising variety of environments. The white dwarf Giclas 29–38 shows an infrared excess most naturally attributed to a circumstellar dust cloud or debris disk (Chary, Zuckerman & Becklin 1999). Reach et al. (2006) have obtained *Spitzer* observations of this system—including the first infrared spectra—that strongly support this identification. The spectrum shows a strong emission feature at 10 μm, well modeled as emission from small carbonaceous and silicate particles, which would be located just a few solar radii from the star. A possible explanation is that comets and/or asteroids survived the post-main-sequence evolution of the main sequence progenitor (estimated to have been a ~ 3-$M_\odot$ A star) and have now been disrupted to produce the cloud of particles radiating at 10 μm. At the other end of the mass spectrum, Kastner et al. (2006) identify debris-disk analogs around two highly luminous B[e] stars in the LMC. They suggest that the grains may arise in Kuiper Belts around these stars while pointing out that such structures might have difficulty surviving the luminous main sequence lifetime of the O-star progenitors of the B[e] stars. Finally, Wang, Chakrabarty & Kaplan (2006) report *Spitzer* detection of excess infrared emission from an X-ray pulsar identified as a neutron star. They attribute the infrared radiation to a "fall-back disk" produced around the neutron star by ejecta from the original supernova explosion.

4.1.6. Comparison with the Solar System. From the crater record on the Moon and other arguments, we know that Earth grew up in a dangerous neighborhood. The heavy bombardment of the Earth and Moon ended about 700 Myr after the formation of the Sun (see Gomes et al. 2005 and references therein). Throughout the period of bombardment, there would have been a continuous production of debris, probably with spikes in grain production whenever a particularly large collision occurred. Strom et al. (2005) attribute the period of heavy bombardment to dynamical instabilities triggered by the orbital migration of the giant planets. This raises the intriguing possibility that the sporadic outbursts producing the stochastic behavior of the planetary debris disks described above may trace the dynamical evolution of their host planetary systems.

The Kuiper Belt is the region within the Solar System most analogous to the debris disks around nearby stars. As summarized in Kim et al. (2005), Backman, Dasgupta & Stencel (1995) used COBE data to place an upper limit of $f = 10^{-6}$ on the debris system associated with the Kuiper Belt. Models by Moro-Martín & Malhotra (2003) suggest that the actual level of emission should only be a factor of a few below this limit. Thus, the debris disk emission of the solar system probably falls about two orders of magnitude fainter than the faintest systems being detected with *Spitzer*. However, the observations are consistent with the detected systems defining the high end of a statistical distribution of debris disks including, near the average level, the one in the solar system (Kim et al. 2005, Bryden et al. 2006). That is, the detected systems appear to have a strong family resemblance to the debris in our own system.

Spitzer's studies therefore have direct relevance to understanding the conditions under which potentially habitable planets form and evolve around hundreds of nearby stars. The events and processes occurring in these systems must have analogs to those that shaped the solar system. This connection was fragmentary and incomplete prior to the launch of *Spitzer*, but it is now firmly established.

4.2. External Planet Occultation Results

The first confirmed extrasolar planet orbiting a solar-type star—in the 51 Pegasi system (Mayor & Queloz 1995)—was also the first example of a "hot Jupiter" or "roaster," that is, a planet with a mass comparable to that of Jupiter that orbits within ~0.05 AU of its parent star. The origins of these planets and their orbits remain a puzzle; however, their temperatures (close to 1000 K) and the amount of infrared radiation they produce should make them detectable by *Spitzer* out to distances of 200 pc.

To do this, Charbonneau et al. (2005) and Deming et al. (2005) exploited the favorable case where the orbital inclination permits mutual eclipses of the star and the planet. The disappearance of the planet behind the star during the secondary eclipse leads to a small drop in the infrared radiation from the star-planet system. As shown in **Figure 15**, Deming et al. (2005) used MIPS to detect the secondary eclipse in the HD209458 system with an amplitude at 24 μm of $0.26 \pm 0.046\%$, whereas Charbonneau et al. (2005) used IRAC to detect the eclipse in the TrES-1 system with amplitude $0.066 \pm 0.013\%$ at 4.6 μm and $0.225 \pm 0.036\%$ at 8 μm. Such measurements provide unique constraints on the temperature, Bond albedo, and perhaps the composition of the planets, which in turn provide clues to how they formed. In addition, the *Spitzer* data determine the time of the secondary eclipse precisely, which constrains the eccentricity of the orbit. In the two cases studied to date, the temperature of the planets was found to be ~1100 K; for TrES-1, the Bond albedo was found to be 0.31 ± 0.14. In each case, the timing of the secondary eclipse suggests that the orbital eccentricity of the planet is too low for tidal dissipation to be a significant heat source, ruling out tidal heating as an explanation for the unexpectedly large radius of the planet orbiting HD209458.

We can anticipate further results of this type from *Spitzer*, including studies of additional transiting planets. However, the measurements of TrES-1 and HD209458 have already led to interesting speculations about these planets. As one example, Burrows, Hubeny & Sudarsky (2005) show that agreement with the *Spitzer* results appears better for a model in which the incident starlight is reradiated primarily by the day side—as seen in the eclipse measurements—rather than uniformly by the whole disk. On the other hand, Barman, Hauschildt & Allard (2005) report that significant day-to-night energy redistribution is required to reproduce the observations. Resolution of this issue, which may be possible with additional *Spitzer* data, will clarify the nature of the mass motions in the extrasolar planets' atmospheres that would be required to redistribute the incident energy. Looking at the *Spitzer* data from a different perspective, Fortney et al. (2005) suggest that the very red 4.5–8 μm color of TrES-1 results from a metallicity enhancement by a factor of 3–5 in the planet

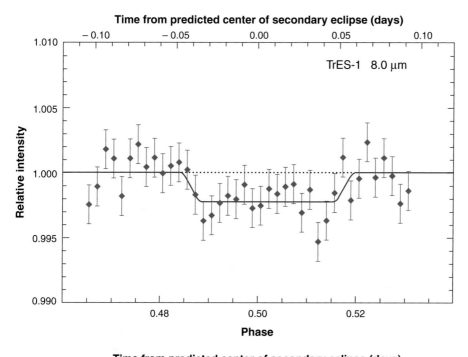

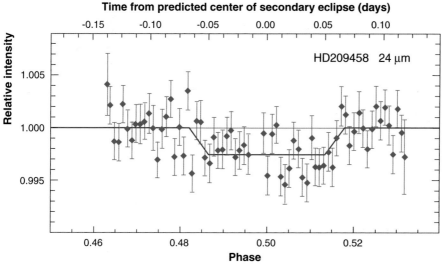

Figure 15

Charbonneau et al. (2005) and Deming et al. (2005) detected light from secondary eclipses of, respectively, the TrES-1 (*top*) and HD209458 (*bottom*) systems. The drop in relative intensity in each plot coincides with the disappearance of the planet behind the star. Solid red lines show best-fit secondary-eclipse models (Charbonneau et al. 2005, revised and reproduced by permission of the Am. Astron. Soc. Deming et al. 2005 adapted by permission from Macmillan Publishers Ltd: *Nature*, copyright 2006).

over the parent star; this may speak to the mode of formation of the planet. Other issues that *Spitzer* observations will illuminate include the chemical and molecular composition of the planets' atmospheres and, perhaps, the presence or absence of clouds or dust within them.

4.3. Solar System Studies

Spitzer's capabilities for the study of small bodies in the solar system—asteroids, comets, planetary satellites, and KBOs—take on new significance with the flood of data being returned on extrasolar planets and planetary systems. Combining *Spitzer* studies of the solar system with observations of extrasolar planetary systems opens the exciting possibility of comparative planetology studies, in which we can apply our understanding of our own solar system to analysis of the extrasolar systems. At the same time, the full range of extrasolar systems accessible to *Spitzer*, together with the favorable external and complete views we have of them, may improve our understanding of our home planetary system.

4.3.1. Asteroids. The ecliptic plane survey portion of the First Look Survey (Meadows et al. 2004) has made measurements over 0.13 square degree areas at 8 and 24 µm at a solar elongation of 115° and ecliptic latitudes of 0° and 5° that are sensitive to main belt objects ∼0.5 km in diameter. In the ecliptic plane, the results translate to an asteroid number density of 154 ± 37 per square degree brighter than 0.1 mJy (5-σ) at 8 µm. This number is consistent with the low side of the predictions of several previous asteroid population models. Interestingly, the *Spitzer* counts appear to fall less steeply with ecliptic latitude than predicted by the models, suggesting that the subkilometer asteroids studied for the first time by *Spitzer* at these wavelengths may be more broadly distributed in ecliptic latitude than are the larger objects.

The Trojan asteroids, gravitationally trapped in Jupiter's stable Lagrange points, may be among the most primitive objects in the solar system. J.P. Emery, D.P. Cruikshank & J. Van Cleve (submitted) have used the IRS low-resolution modules from 5.2 to 37 µm to obtain the first thermal infrared spectra of Trojan asteroids. The spectra show broad spectral features suggesting that fine-grained silicates—both crystalline and amorphous—cover the surfaces of these objects. These results support the suggestion that the red spectral reflectivity of these objects is due to silicates rather than to organic material.

The spectrum of the asteroidal emission—in excess of the predictions of the standard thermal model—is strikingly similar to that of comet Hale-Bopp and of the debris disk orbiting the nearby solar-type star HD69830 (**Figure 9**). *Spitzer* observations of comet Schwassmann-Wachmann 1 (SW-1), which has a semimajor axis of ∼6 AU, show similar features (Stansberry et al. 2004). Together, the results suggest that the middle part of the solar system, the transition region between rocky and icy objects, may not contain an abundance of organic materials.

4.3.2. Comets. *Spitzer*'s ability to measure thermal emission from comets at large heliocentric distances permits determination of the temperatures, sizes, and albedos of cometary nuclei. Stansberry et al. (2004) found SW-1 to have a nuclear radius of

27 km and an albedo of 0.025. Lisse et al. (2005) studied comet Tempel 1 when it was about 3.7 AU from the Sun. These measurements determined that the nucleus has semimajor and semiminor axes of $\sim 7.2 \pm 0.9$ km and 2.3 ± 0.3 km, respectively, and that the albedo is 0.04 ± 0.01. The very low albedo for these cometary nuclei, which is consistent with other determinations (Lamy et al. 2004), contrasts sharply with the high albedos determined by *Spitzer* for a number of large KBOs and trans-Neptunian objects (see Section 4.3.3).

Spitzer joined many ground- and space-based telescopes in observing the Deep Impact encounter with comet Tempel 1 on July 4, 2005. Before impact, a smooth ambient coma surrounded the nucleus; afterwards, ejecta from the comet flowed out at velocities ~ 200 ms^{-1} for 40–50 hours. C.M. Lisse, J. VanCleve, A.C. Adams, M.F. A'Hearn, Y.R. Fernández, et al. (submitted) use IRS spectra to identify an intriguing variety of materials within the ejecta from the impact, including clay, crystalline silicates, and carbonates as well as polycyclic aromatic hydrocarbons (see **Figure 16**). The structure in the spectrum of the ejecta contrasts sharply with the smooth spectrum of the cometary nucleus prior to the encounter (Lisse et al. 2005) and is similar to, but with even stronger features than, the spectra of comet Hale-Bopp and the debris disk around HD 69830 (**Figure 9**). The *Spitzer* results show that Tempel 1 contains materials characteristic of a wide range of temperatures and water content, suggesting that the comet somehow managed to agglomerate from material that formed throughout the solar system.

Cometary mass loss within our own solar system connects with the studies of planetary debris disks, providing a local and dramatic example of the processes that maintain the extrasolar systems. *Spitzer* can connect these processes by studying not only the relatively bright comae and dust tails of active comets but also the debris trails found by IRAS to fill cometary orbits as a result of the accumulation of dust lost over many orbital periods (Reach et al. 2005, Gehrz et al. 2006). These extended, low-surface-brightness trails are not readily seen with warm telescopes.

4.3.3. Kuiper Belt and trans-Neptunian objects. Intensive studies during the past decade (see Luu & Jewitt 2002) have identified hundreds of KBOs in the region beyond Neptune, which was previously known to be inhabited only by Pluto and its satellite, Charon. These objects have been discovered through visible surveys, which provide orbital parameters and have identified dynamical families. However, the visible observations provide limited information about the physical properties of individual KBOs, because of the degeneracy between the albedo and the size of an object seen only in reflected light. *Spitzer*'s ability to measure thermal emission from KBOs breaks this degeneracy and allows determination of the albedos and radii of individual KBOs. For the KBO 2002 AW197, Cruikshank et al. (2005) use MIPS photometry to determine an albedo of 0.17 ± 0.03 and a diameter of 700 ± 50 km, larger than all but one main belt asteroid. They also suggest that the surface of 2002 AW197 has low but nonzero thermal inertia, indicating either a very porous surface or the presence of a material with low intrinsic thermal inertia, such as amorphous H_2O ice. Additional information is available in the not infrequent case of a binary KBO. For the binary KBO 1999 TC36, Stansberry et al. (2006) determine a system albedo of

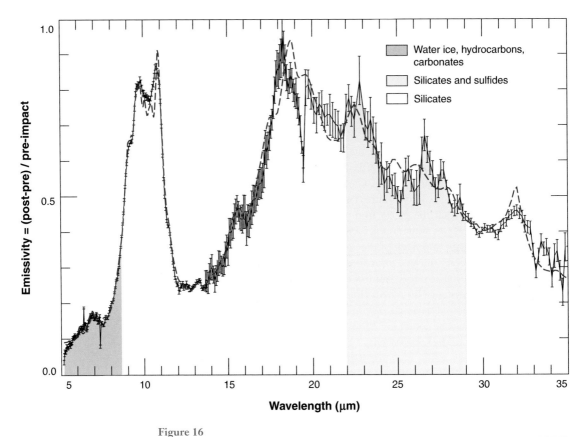

Figure 16
The data points show the emissivity spectrum of the ejecta from comet Tempel 1 as measured 45 minutes after the Deep Impact encounter (C.M. Lisse, J. VanCleve, A.C. Adams, M.F. A'Hearn, Y.R. Fernández, et al., submitted). The red line is the best fit model. The color coding indicates the dominant materials in each wavelength interval.

0.08 and an effective diameter of 405 km from MIPS photometry. The average density for the two components, using the system mass determined from HST observations of the binary orbit and considering the uncertainties in the albedo and the diameter, lies between 0.3 and 0.8 g cm^{-3}. A higher and perhaps more plausible density would be indicated if the primary were itself a binary.

M. Brown and colleagues recently announced the discovery of three unusually large and bright KBOs (Brown, Trujillo & Rabinowitz 2005; M.E. Brown, et al. in preparation). All three have been studied with *Spitzer*. *Spitzer* detected the two smaller objects, 2003 EL61 and 2005 FY9; they appear to have albedos ∼0.7 and radii ∼900 km (about 75% the size of Pluto). The larger object, 2003 UB313, is more distant and was not detected by *Spitzer*. If its albedo is comparable with that of the other two objects, which would be consistent with *Spitzer*'s upper limit on its flux, it is about 20% larger than Pluto and would qualify for consideration as the tenth planet. All three of these large objects occupy high-inclination orbits and belong to the scattered Kuiper Belt population.

5. CONCLUSIONS

The results from *Spitzer* presented above permit the following conclusions to be stated with confidence:

1. Protoplanetary disks around solar-type stars evolve quickly, suggesting that terrestrial planet formation is well under way within less than 10 Myr. The disks dissipate from the inside out.
2. Substellar objects with mass as low as ~ 10 M_{Jup} form by the same gravitational collapse process that gives rise to more massive stars. Their protoplanetary disks appear similar in structure and composition to those orbiting young solar-mass stars.
3. Although supernova remnants harbor only small masses of dust, large signals are detected from infrared echoes as the pulse from the supernova expansion heats surrounding circumstellar and interstellar matter.
4. Current models provide reasonably accurate predictions of the thermal infrared spectra of old, cool brown dwarfs.
5. Planetary debris disks evolve stochastically. Statistically, these systems show a gradual decline in dust content with time that can be punctuated even at late stages by events that inject large amounts of new material into the circumstellar environment.
6. The character and frequency of planetary debris disks do not vary dramatically with stellar type, metallicity, or the presence or absence of planets.
7. Kuiper Belt objects can have a wide range of albedos, from the very low values suspected prior to *Spitzer* to ~ 0.1–0.2 for moderate sized ones and ~ 0.7 for very large ones. *Spitzer* results indicate that KBOs may also have very low densities (<1 g cm^{-3}).

Looking ahead to a retrospective *Annual Reviews* chapter that might appear several years after the end of the *Spitzer* mission, we can anticipate that at least the following advances would be reported.

Spitzer's ability to survey large areas and to study numerous objects in great detail—spectroscopically and photometrically—will lead to great advances in our understanding of star and planetary system formation. *Spitzer* will pin down the timescales for disk evolution, characterize the relationship between disk structure and dust composition, and clarify the importance of starless cores and infrared dark clouds in low-mass and high-mass star formation, respectively. On larger spatial scales, *Spitzer* surveys should lead to the identification of the mechanisms by which star formation propagates through large molecular complexes and permit study of the role of radiation, winds, and molecular outflows in triggering or inhibiting star formation.

Spitzer studies of regions of star formation may identify substellar objects with masses as low as ~ 1 M_{Jup} forming in isolation by gravitational collapse. At the same time, *Spitzer* spectra of field brown dwarfs should permit detailed searches for small deviations from the models that may signal processes such as cloud formation and atmospheric circulation.

Spitzer will have studied at least several handfuls of transiting planets at a level adequate to permit detailed tests of theoretical predictions and comparisons with the giant planets in the solar system. These studies should include not only photometric measurements but also the first thermal infrared spectra of extrasolar planets. In addition, *Spitzer* spectroscopy of material in protoplanetary and planetary debris disks, as well as of comets and asteroids in the solar system, will continue to permit comparisons between the composition of extrasolar and local planetary materials. These and other examples, including *Spitzer*'s continuing studies of KBOs, will assure continued traffic across the interface between planetary science and extrasolar planet studies, to the great benefit of both disciplines.

Finally, detailed *Spitzer* studies of many phenomena, including star formation, post-main-sequence stellar evolution, and supernovae, will extend far beyond the boundaries of our own Galaxy and well into the local universe. Just as our studies of extrasolar planetary systems illuminate phenomena within our own solar system, exploring these familiar objects in the varying environments of nearby galaxies should provide important insights into the fundamental nature of the underlying astrophysical processes.

ACKNOWLEDGMENTS

The authors acknowledge the contributions of many people across the project who worked with us to bring *Spitzer* to fruition. We thank L. Allen, M.J. Barlow, M. Brown, D. Charbonneau, C. Lada, K. Luhman, R. Kirshner, T. Megeath, T. Soifer, K. Stapelfeldt, J. Stauffer and, particularly, E. van Dishoeck for helpful discussions and comments on the manuscript and many colleagues for making their *Spitzer* results available prior to publication. G. Fazio acknowledges NASA support through contracts 1062296 and 1256790 issued by JPL/Caltech. G. Rieke acknowledges NASA support through contract 1255094 issued by JPL/Caltech. T. Roellig acknowledges NASA support through WBS# 420579.04.03.02.02. D. Watson acknowledges NASA support through contract 960803 with Cornell University issued by JPL/Caltech, and through Cornell subcontract 31219-5714 to the University of Rochester. M. Werner thanks Charles Alcock and the Center for Astrophysics, and Rosemary and Robert Putnam, for hospitality during the preparation of the paper. He also thanks Mary Young, Jim Jackson and, in particular, Courtney Young for editorial assistance. This research was carried out in part at the Jet Propulsion Laboratory, California Institute of Technology, under a contract with the National Aeronautics and Space Administration.

LITERATURE CITED

Alexander RD, Casali MM, André P, Persi P, Eiroa C. 2003. *Astron. Astrophys.* 401:613–24

Allen LE, Calvet N, D'Alessio P, Merin B, Hartmann L, et al. 2004. *Ap. J. Suppl.* 154(1):363–66

Allen LE, Hora JL, Megeath ST, Deutsch LK, Fazio GG, et al. 2005. *IAU Symp.* 227:352–57

André P, Ward-Thompson D, Barsony M. 1993. *Ap. J.* 406(1):122–41

Apai D, Pascucci I, Bouwman J, Natta A, Henning T, Dullemond CP. 2005. *Science* 310(5749):834–36

Arendt RG, Dwek E, Moseley SH. 1999. *Ap. J.* 521(1):234–45

Armus L, ed. 2006. ***ASP Conf. Ser.*** **In press**

Aumann HH, Beichman CA, Gillett FC, de Jong T, Houck JR, et al. 1984. *Ap. J.* 278:L23–27

Backman DE, Dasgupta A, Stencel RE. 1995. *Ap. J.* 450:L35–38

Backman DE, Paresce F. 1993. In *Protostars and Planets III*, ed. EH Levy, JI Lunine, pp. 1253–1304. Tucson: Univ. Ariz. Press

Barlow MJ, Sugerman BEK, Fabbri J, Meixner M, Fisher RS, et al. 2005. *Ap. J.* 627(2):L113–16

Barman TS, Hauschildt PH, Allard F. 2005. *Ap. J.* 632(2):1132–39

Beichman CA, Bryden G, Gautier TN, Stapelfeldt KR, Werner MW, et al. 2005a. *Ap. J.* 626(2):1061–69

Beichman CA, Bryden G, Rieke GH, Stansberry JA, Trilling DE, et al. 2005b. *Ap. J.* 622(2):1160–70

Beichman CA, Tanner A, Bryden G, Stapelfeldt KR, Werner MW, et al. 2006. *Ap. J.* 639(2):1166–76

Beuther H, Sridharan TK, Saito M. 2005. *Ap. J.* 634(2):L185–88

Blommaert JADL, Cami J, Szczerba R, Barlow MJ. 2005. *Space Sci. Rev.* 119(1–4):215–43

Boogert ACA, Pontoppidan KM, Lahuis F, Jørgensen JK, Augereau JC, et al. 2004. *Ap. J. Suppl.* 154(1):359–62

Boss AP. 2005. *Ap. J.* 629(1):535–48

Bouchet P, De Buizer JM, Suntzeff NB, Danziger IJ, Hayward TL, et al. 2004. *Ap. J.* 611(1):394–98

Bourke TL, Crapsi A, Myers PC, Evans NJ II, Wilner DJ, et al. 2005. *Ap. J.* 633(2):L129–32

Briceño C, Vivas AK, Hernández J, Calvet N, Hartmann L, et al. 2004. *Ap. J.* 606(2):L123–26

Brown ME, Trujillo CA, Rabinowitz DL. 2005. *Ap. J.* 635(1):L97–100

Bryden G, Beichman CA, Trilling DE, Rieke GH, Holmes EK, et al. 2006. *Ap. J.* 636(2):1098–113

Burrows A, Hubeny I, Sudarsky D. 2005. *Ap. J.* 625(2):L135–38

Calvet N, D'Alessio P, Watson DM, Franco-Hernández R, Furlan E, et al. 2005. *Ap. J.* 630(2):L185–88

Calvet N, Hartmann L, Kenyon SJ, Whitney BA. 1994. *Ap. J.* 434(1):330–40

Caroff L, Moon LJ, Backman D, Praton E, eds. 2004. *ASP Conf. Ser.* 324

Cesarsky CJ, Salama A, eds. 2005. ***Space Sci. Rev.*** **119(1–4)**

Charbonneau D, Allen LE, Megeath ST, Torres G, Alonso R, et al. 2005. *Ap. J.* 626(1):523–29

Chary R, Sheth K, Teplitz H. 2006. *ASP Conf. Ser.* In press

> Includes numerous papers and abstracts reporting early *Spitzer* results.

> Topical reviews of *Infrared Space Observatory* results in key science areas.

Chary R, Zuckerman B, Becklin EE. 1999. In *The Universe as Seen by ISO*, ed. P Cox, MF Kessler, ESA-SP 427, p. 289

Chen CH, Jura M. 2001. *Ap. J.* 560(2):L171–74

Chen CH, Jura M, Gordon KD, Blaylock M. 2005a. *Ap. J.* 623(1):493–501

Chen CH, Patten BM, Werner MW, Dowell CD, Stapelfeldt KR, et al. 2005b. *Ap. J.* 634(2):1372–84

Chu YH, Gruendl RA, Chen CHR, Whitney BA, Gordon KD, et al. 2005. *Ap. J.* 634(2):L189–92

Churchwell E, Whitney BA, Babler BL, Indebetouw R, Meade MR, et al. 2004. *Ap. J. Suppl.* 154(1):322–27

Ciardi DR, Telesco CM, Packham C, Gómez Martin C, Radomski JT, et al. 2005. *Ap. J.* 629(2):897–902

Cohen M, Green AJ, Roberts MSE, Meade MR, Babler B, et al. 2005. *Ap. J.* 627(1):446–53

Crovisier J, Brooke TY, Hanner MS, Keller HU, Lamy P, et al. 1996. *Astron. Astrophys.* 315:L385–88

Crowther PA, Morris PW, Smith JD. 2006. *Ap. J.* 636(2):1033–44

Cruikshank DP, Stansberry JA, Emery JP, Fernández YR, Werner MW, et al. 2005. *Ap. J.* 624(1):L53–56

Cushing MC, Roellig TL, Van Cleve JE, Sloan GC, Wilson JC, et al. 2006. *Ap. J.* In press

D'Alessio P, Calvet N, Hartmann L. 2001. *Ap. J.* 553(1):321–34

D'Alessio P, Calvet N, Hartmann L, Lizano S, Cantó J. 1999. *Ap. J.* 527(2):893–909

D'Alessio P, Canto J, Calvet N, Lizano S. 1998. *Ap. J.* 500(1):411–28

D'Alessio P, Hartmann L, Calvet N, Franco-Hernández R, Forrest WJ, et al. 2005a. *Ap. J.* 621(1):461–72

D'Alessio P, Merín B, Calvet N, Hartmann L, Montesinos B. 2005b. *Rev. Mex. Astron. Astrofis.* 41:61–67

Decin G, Dominik C, Waters LBFM, Waelkens C. 2003. *Ap. J.* 598(1):636–44

Deming D, Seager S, Richardson LJ, Harrington J. 2005. *Nature* 434(7034):740–43

Douvion T, Lagage PO, Cesarsky CJ. 1999. *Astron. Astrophys.* 352:L111–15

Douvion T, Lagage PO, Cesarsky CJ, Dwek E. 2001. *Astron. Astrophys.* 373:281–91

Dunne L, Eales S, Ivison R, Morgan H, Edmunds M. 2003. *Nature* 424(6946):285–87

Dwek E, Arendt RG. 1992. *Annu. Rev. Astron. Astrophys.* 30:11–50

Egan MP, Shipman RF, Price SD, Carey SJ, Clark FO, Cohen M. 1998. *Ap. J.* 494(2):L199–202

Evans A, Geballe TR, Tyne VH, Pollacco D, Eyres SPS, Smalley B. 2006. *MNRAS* In press

Farihi J, Zuckerman B, Becklin EE. 2005. *Astron. J.* 130(5):2237–40

Finley PT, Hopkins RA, Schweickart RB. 2004. *Proc. SPIE* 5487:26–37

Fischer DA, Valenti J. 2005. *Ap. J.* 622(2):1102–17

Forrest WJ, Sargent B, Furlan E, D'Alessio P, Calvet N, et al. 2004. *Ap. J. Suppl.* 154(1):443–47

Fortney JJ, Marley MS, Lodders K, Saumon D, Freedman R. 2005. *Ap. J.* 627(1):L69–72

> **Basic physics relevant to dust in supernova remnants.**

Furlan E, Calvet N, D'Alessio P, Hartmann L, Forrest WJ, et al. 2005a. *Ap. J.* 621(2):L129–32

Furlan E, Calvet N, D'Alessio P, Hartmann L, Forrest WJ, et al. 2005b. *Ap. J.* 628(1):L65–68

Gehrz RD, Reach WT, Woodward Ce, Kelley MS. 2006. *Adv. Space Res.* In press

Gibb EL, Whittet DCB, Boogert ACA, Tielens AGGM. 2004. *Ap. J. Suppl.* 151(1):35–73

Gomes R, Levison HF, Tsiganis K, Morbidelli A. 2005. *Nature* 435(7041):466–69

Gordon K, Bailin J, Engelbracht CW, Rieke GH, Misselt KA, et al. 2006. *Ap. J. Lett.* 638(2):L87–92

Gorlova N, Padgett DL, Rieke GH, Muzerolle J, Morrison JE, et al. 2004. *Ap. J. Suppl.* 154(1):448–52

Grogan K, Dermott SF, Durda DD. 2001. *Icarus.* 152(2):251–67

Gutermuth RA, Megeath ST, Pipher JL, Allen T, Williams JP, et al. 2006. In *Star Formation in the Era of Three Great Observatories*, Cambridge, MA. In press

Gutermuth RA, Megeath ST, Muzerolle J, Allen LE, Pipher JL, et al. 2004. *Ap. J. Suppl.* 154(1):374–78

Habing HJ, Dominik C, Jourdain de Muizon M, Laureijs RJ, Kessler MF, et al. 2001. *Astron. Astrophys.* 365:545–61

Hartmann L. 2005. *ASP Conf. Ser.* 341:131–44

Hartmann L, Calvet N, Watson DM, D'Alessio P, Furlan E, et al. 2005a. *Ap. J.* 628(2):L147–50

Hartmann L, Megeath ST, Allen L, Luhman K, Calvet N, et al. 2005b. *Ap. J.* 629(2):881–96

Harvey PM, Chapman N, Lai S-P, Evans NJ II, Allen L, et al. 2006. *Ap. J.* In press

Hines DC, Rieke GH, Gordon KD, Rho J, Misselt KA, et al. 2004. *Ap. J. Suppl.* 154(1):290–95

Hines DC, Backman DE, Bouwman J, Hillenbrand LA, Carpenter JM, et al. 2006. *Ap. J.* 638(2):1070–79

Hollenbach D, McKee CF. 1989. *Ap. J.* 342:306–36

Hollenbach D, Gorti U, Meyer M, Kim JS, Morris P, et al. 2005. *Ap. J.* 631(2):1180–90

Honda M, Kataza H, Okamoto YK, Miyata T, Yamashita T, et al. 2003. *Ap. J.* 585(1):L59–63

Hora JL, Latter WB, Allen LE, Marengo M, Deutsch LK, Pipher JL. 2004. *Ap. J. Suppl.* 154(1):296–301

Huard TL, Myers PC, Murphy DC, Crews LJ, Lada CJ, et al. 2006. *Ap. J.* 640(1):391–401

Jones TJ, Woodward CE, Boyer ML, Gehrz RD, Polomski E. 2005. *Ap. J.* 620(2):731–43

Jørgensen JK, Lahuis F, Schöier FL van Dishoeck EF, Blake GA, et al. 2005. *Ap. J.* 631(1):L77–80

Jura M. 1990. *Ap. J.* 365:317–20

Jura M, Bohac CJ, Sargent B, Forrest WJ, Green J, et al. 2006. *Ap. J.* 637(1):L45–48

Kalas P, Graham JR, Beckwith SVW, Jewitt DC, Lloyd JP. 2002. *Ap. J.* 567(2):999–1012

Reviews the status of disk evolution prior to the deluge of *Spitzer* data.

Kalas P, Graham JR, Clampin M. 2005. *Nature* 435(7045):1067–70
Kalas P, Liu MC, Matthews BC. 2004. *Science* 303(5666):1990–92
Kastner JH, Buchanan CL, Sargent B, Forrest WJ. 2006. *Ap. J.* 638(1):L29–32
Kemper F, Vriend WJ, Tielens AGGM. 2004. *Ap. J.* 609(2):826–37
Kenyon SJ, Calvet N, Hartmann L. 1993. *Ap. J.* 414(2):676–94
Kenyon SJ, Bromley BC. 2004. *Ap. J.* 602(2):L133–36
Kenyon SJ, Bromley BC. 2005. *Astron. J.* 130(1):269–79
Kessler-Silacci JE, Augereau JC, Dullemond CP, Geers V, Lahuis F, et al. 2006. *Ap. J.* 639(1):275–91
Kessler-Silacci JE, Hillenbrand LA, Blake GA, Meyer MR. 2005. *Ap. J.* 622(1):404–29
Kim JS, Hines DC, Backman DE, Hillenbrand LA, Meyer MR, et al. 2005. *Ap. J.* 632(1):659–69

> Up-to-date review on brown dwarfs by a leading researcher in the field.

Kirkpatrick JD. 2005. *Annu. Rev. Astron. Astrophys.* 43:195–245

Knacke RF, Fajardo-Acosta SB, Telesco CM, Hackwell JA, Lynch DK, Russell RW. 1993. *Ap. J.* 418:440
Kotak R, Meikle P, van Dyk SD, Höflich PA, Mattila S. 2005. *Ap. J.* 628(2):L123–26
Kraemer KE, Sloan GC, Wood PR, Price SD, Egan MP. 2005. *Ap. J.* 631(2):L147–50
Krause O, Birkmann SM, Rieke GH, Lemke D, Klaas U, et al. 2004. *Nature* 432(7017):596–98
Krause O, Rieke GH, Birkmann SM, Le Floc'h E, Gordon KD, et al. 2005. *Science* 308(5728):1604–06

> Summary of star-formation observations and theory pre-*Spitzer*.

Lada CJ. 2005. *Prog. Theor. Phys. Suppl.* 158:1–23

Lada CJ, Muench AA, Luhman KL, Allen L, Hartmann L. 2006. *Astron. J.* 131(3):1574–607
Lada CJ, Wilking BA. 1984. *Ap. J.* 287:610–21
Lahuis F, van Dishoeck EF, Boogert ACA, Pontoppidan KM, Blake GA, et al. 2006. *Ap. J.* 636(2):L145–48
Lamy P, Toth I, Fernandez YR, Weaver HA. 2004. In *Comets II*, ed. HU Keller, HA Weaver, pp. 223–64. Tucson: Univ. Ariz. Press
Lanz T, Heap SR, Hubeny I. 1995. *Ap. J.* 447(2):L41–44
Lee HG. 2005. *J. Korean Astron. Soc.* 38(4):385–414
Lisse CM, A'Hearn MF, Groussin O, Fernández YR, Belton MJS, et al. 2005. *Ap. J.* 625(2):L139–42
Low FJ, Smith PS, Werner M, Chen C, Krause V, et al. 2005. *Ap. J.* 631(2):1170–79
Luhman KL, Adame L, D'Alessio P, Calvet N, Hartmann L, et al. 2005a. *Ap. J.* 635(1):L93–96
Luhman KL, D'Alessio P, Calvet N, Allen LE, Hartmann L, et al. 2005b. *Ap. J.* 620(1):L51–54
Luhman KL, Lada CJ, Hartmann L, Muench AA, Megeath ST, et al. 2005c. *Ap. J.* 631(1):L69–72
Luhman KL. 2006. In *Protostars and Planets V.* Tucson: Univ. Ariz. Press. In press

> Summary of the rapidly evolving field of Kuiper Belt objects.

Luu JX, Jewitt DC. 2002. *Annu. Rev. Astron. Astrophys.* 40:63–101

Lysek MJ, Israelsson UE, Garcia RD, Luchik TS. 1995. In *Advances in Cryogenic Engineering*, ed. P. Kittel New York: Plenum 41:1143
Markwick-Kemper F, Green JD, Peeters E. 2005. *Ap. J.* 628(2):L119–22

Marley MS, Seager S, Saumon D, Lodders K, Ackerman AS, et al. 2002. *Ap. J.* 568(1):335–42

Marley MS, Ackerman AS, Burgasser AJ, Saumon D, Lodders K, Freedman R. 2003. *IAU Symp.* 211:333

Mayor M, Queloz D. 1995. *Nature* 378(6555):355–59

Meadows VS, Bhattacharya B, Reach WT, Grillmair C, Noriega-Crespo A, et al. 2004. *Ap. J. Suppl.* 154(1):469–74

Megeath ST, Allen LE, Gutermuth RA, Pipher JL, Myers PC, et al. 2004. *Ap. J. Suppl.* 154(1):367–73

Megeath ST, Hartmann L, Luhman KL, Fazio GG. 2005. *Ap. J.* 634(1):L113–16

Molster F, Kemper C. 2005. *Space Sci. Rev.* 119(1–4):3–28

Moro-Martín A, Malhotra R. 2003. *Astron. J.* 125(4):2255–65

Morris PW, Crowther PA, Houck JR. 2004. *Ap. J. Suppl.* 154(1):413–17

Morris PW, Noriega-Crespo A, Marleau FR, Teplitz HI, Uchida KI, Armus L. 2004. *Ap. J. Suppl.* 154(1):339–45

Muzerolle J, Adame L, D'Alessio P, Calvet N, Luhman KL, et al. 2006. *Ap. J.* In press

Muzerolle J, Megeath ST, Flaherty KM, Gordon KD, Rieke GH, et al. 2005. *Ap. J.* 620(2):L107–10

Muzerolle J, Megeath ST, Gutermuth RA, Allen LE, Pipher JL, et al. 2004. *Ap. J. Suppl.* 154(1):379–84

Myers PC, Ladd EF. 1993. *Ap. J.* 413(1):L47–50

Noriega-Crespo A, Morris P, Marleau FR, Carey S, Boogert A, et al. 2004. *Ap. J. Suppl.* 154(1):352–58

Pollack JB, Hubickyj O, Bodenheimer P, Lissauer JJ, Podolak M, Greenzweig Y. 1996. *Icarus.* 124(1):62–85

Pontoppidan KM, Dullemond CP, van Dishoeck EF, Blake GA, Boogert ACA, et al. 2005. *Ap. J.* 622(1):463–81

Quillen AC, Blackman EG, Frank A, Varnière P. 2004. *Ap. J.* 612(2):L137–40

Rathborne JM, Jackson JM, Chambers ET, Simon R, Shipman R, Frieswijk W. 2005. *Ap. J.* 630(2):L181–84

Reach WT, Kuchner MJ, von Hippel T, Burrows A, Mullallly F, et al. 2005. *Ap. J.* 635(2):L161–64

Reach WT, Rho J, Tappe A, Pannutti TG, Brogan CL, et al. 2006. *Astron. J.* 131(3):1479–500

Reach WT, Rho J, Young E, Muzerolle J, Fajardo-Acosta S, et al. 2004. *Ap. J. Suppl.* 154(1):385–90

Reipurth B, Jewitt D, Keil K, ed. 2006. *Protostars and Planets V*. Tucson: Univ. Ariz. Press. In press

Rieke G. 2006. *The Last of the Great Observatories: Spitzer and the Era of Faster, Better, Cheaper at NASA*. Tucson: Univ. Ariz. Press. In press

Rieke GH, Su KYL, Stansberry JA, Trilling D, Bryden G, et al. 2005. *Ap. J.* 620(2):1010–26

Roellig TL, Van Cleve JE, Sloan GC, Wilson JC, Saumon D, et al. 2004a. *Ap. J. Suppl.* 154(1):418–21

Proceedings of a major conference touching on many of *Spitzer*'s science themes.

Comprehensive history of the *Spitzer Space Telescope*.

Roellig TL, Werner MW, Gallagher DB, Irace WR, Fazio GG, et al. 2004b. *Proc. SPIE.* 5487:38–49

Rosenthal D, Bertoldi F, Drapatz S. 2000. *Astron. Astrophys.* 356:705–23

Sargent B, Forrest WJ, D'Alessio P, Najita J, Li A., et al. 2006. *Ap. J.* In press

Saumon D, Marley MS, Lodders K, Freedman RS. 2003. *IAU Symp.* 211:345

Sicilia-Aguilar A, Hartmann LW, Hernández J, Briceño C, Calvet N. 2005. *Astron. J.* 130(1):188–209

Sicilia-Aguilar A, Hartmann L, Calvet N, Megeath ST, Muzerolle J, et al. 2006. *Ap. J.* 638(2):897–919

Sloan GC, Devost D, Bernard-Salas J, Wood PR, Houck JR. 2006. *Ap. J.* 638(1):472–77

Soker N, Subag E. 2005. *Astron. J.* 130(6):2717–24

Song I, Weinberger AJ, Becklin EE, Zuckerman B, Chen C. 2002. *Astron. J.* 124(1):514–18

Song I, Zuckerman B, Weinberger AJ, Becklin EE. 2005. *Nature* 436(7049):363–65

Spangler C, Sargent AI, Silverstone MD, Becklin EE, Zuckerman B. 2001. *Ap. J.* 555(2):932–44

Stanimirovic S, Bolatto AD, Sandstrom K, Leroy AK, Simon JD, et al. 2005. *Ap. J.* 632(2):L103–06

Stansberry JA, Grundy WG, Emery JP, Margot JL, Cruikshank DP, et al. 2006. *Ap. J.* In press

Stansberry JA, Van Cleve J, Reach WT, Cruikshank DP, Emery JP, et al. 2004. *Ap. J. Suppl.* 154(1):463–68

Stapelfeldt KR, Holmes EK, Chen C, Rieke GH, Su KYL, et al. 2004. *Ap. J. Suppl.* 154(1):458–62

Stauffer JR, Rebull LM, Carpenter J, Hillenbrand L, Backman D, et al. 2005. *Astron. J.* 130(4):1834–44

Strom RG, Malhotra R, Ito T, Yoshida F, Kring DA. 2005. *Science* 309(5742):1847–50

Su KYL, Kelly DM, Latter WB, Misselt KA, Frank A, et al. 2004. *Ap. J. Suppl.* 154(1):302–08

Su KYL, Rieke GH, Misselt KA, Stansberry JA, Moro-Martín A, et al. 2005. *Ap. J.* 628(1):487–500

Teixeira PS, Lada CJ, Young ET, Marengo M, Muench A, et al. 2006. *Ap. J.* 636(1):L45–48

Telesco CM, Fisher RS, Wyatt MC, Dermott SF, Kehoe TJJ, et al. 2005. *Nature* 433(7022):133–36

Uchida KI, Calvet N, Hartmann L, Kemper F, Forrest WJ, et al. 2004. *Ap. J. Suppl.* 154(1):439–42

Uzpen B, Kobulnicky HA, Olsen KAG, Clemens DP, Laurance TL, et al. 2005. *Ap. J.* 629(1):512–25

van Boekel R, Min M, Leinert C, Waters LBFM, Richichi A, et al. 2004. *Nature* 432(7016):479–82

van Boekel R, Min M, Waters LBFM, de Koter A, Dominik C, et al. 2005. *Astron. Astrophys.* 437(1):189–208 van Dishoeck EF. 2004. *Annu. Rev. Astron. Astrophys.* **42(1):119–67**

> Summarizes outstanding work from ISO on spectroscopy of interstellar and circumstellar matter and includes wavelength regions not covered by *Spitzer* and higher spectral resolution observations.

van Loon JT, Oliveira JM, Wood PR, Zijlstra AA, Sloan GC, et al. 2005. *MNRAS* 364(1):L71–75

van Marle AJ, Langer N, García-Segura G. 2004. *Rev. Mex. Astron. Astrofis.* 22:136–39

Varniere P, Bjorkman JE, Frank A, Quillen AC, Carciofi AC, et al. 2006a. *Ap. J.* 637(2):L125–28

Varniere P, Blackman EG, Frank A, Quillen AC. 2006b. *Ap. J.* 640(2):1110–14

Wang Z, Chakrabarty D, Kaplan DL. 2006. *Nature* 440:772–75

Watson DM, Kemper F, Calvet N, Keller LD, Furlan E, et al. 2004. *Ap. J. Suppl.* 154(1):391–95

Weinberger AJ, Becklin EE, Zuckerman B. 2003. *Ap. J.* 584(1):L33–37

Werner MW. 2006. The *Spitzer space telescope*: new views of the cosmos. In *ASP Conf. Ser.* **In press**

> A short and personal history of the *Spitzer space telescope*.

Werner MW, Roellig TL, Low FJ, Rieke GH, Rieke M, et al. 2004. *Ap. J. Suppl.* 154(1):1–9

Whitney BA, Indebetouw R, Babler BL, Meade MR, Watson C, et al. 2004a. *Ap. J. Suppl.* 154(1):315–21

Whitney BA, Wood K, Bjorkman JE, Cohen M. 2003. *Ap. J.* 598(2):1079–99

Wilson TL, Batria W. 2005. *Astron. Astrophys.* 430:561–66

Wooden DH. 1995. *ASP Conf. Ser.* 73:405–12

Wooden DH, Rank DM, Bregman JD, Witteborn FC, Tielens AGGM, et al. 1993. *Ap. J. Suppl.* 88(2):477–507

Wyatt MC, Holland WS, Greaves JS, Dent WRF. 2003. *Earth, Moon, and Planets.* 92(1):423–34

Young CH, Evans NJ II. 2005. *Ap. J.* 627(1):293–309

Young CH, Jørgensen JK, Shirley YL, Kauffmann J, Huard T, et al. 2004a. *Ap. J. Suppl.* 154(1):396–401

Young ET, Lada CJ, Teixeira P, Muzerolle J, Muench A, et al. 2004b. *Ap. J. Suppl.* 154(1):428–32

Young ET, Teixeira PS, Lada CJ, Muzerolle J, Persson SE, et al. 2006. *Ap. J.* 642:972–78

Young KE, Harvey PM, Brooke TY, Chapman N, Kauffmann J, et al. 2005. *Ap. J.* 628(1):283–97

> Pre-*Spitzer* review of debris disks post-ISO and -IRAS.

Zuckerman B. 2001. *Annu. Rev. Astron. Astrophys.* **39:549–80**

Zuckerman B, Song I. 2004. *Annu. Rev. Astron. Astrophys.* **42(1):685–721**

> Discusses nearby groups of young stars that are promising candidates for study with *Spitzer*.

Populations of X-Ray Sources in Galaxies

G. Fabbiano

Harvard-Smithsonian Center for Astrophysics, Cambridge, Massachusetts 02138;
email: gfabbiano@cfa.harvard.edu

Key Words

galaxies, globular clusters, stellar populations, supersoft X-ray sources, ultraluminous X-ray sources (ULXs), X-ray binaries, X-ray sources

Abstract

Today's sensitive X-ray observations allow the study of populations of X-ray binaries in galaxies as distant as 20–30 Mpc. Photometric diagrams and luminosity functions applied to these populations provide a direct probe of the evolved binary component of different stellar populations. The study of the X-ray populations of E and S0 galaxies has revamped the debate on the formation and evolution of low-mass X-ray binaries (LMXBs) and on the role of globular clusters in these processes. Though overall stellar mass drives the amount of X-ray binaries in old stellar populations, the amount of sources in star-forming galaxies is related to the star-formation rate. Short-lived, luminous, high-mass X-ray binaries (HMXBs) dominate these young X-ray populations. The most luminous sources in these systems are the debated ultraluminous X-ray sources (ULXs). Observations of the deep X-ray sky, and comparison with deep optical surveys, are providing the first evidence of the X-ray evolution of galaxies.

1. *CHANDRA*: A NEW PARADIGM

This review comes almost two decades after the 1989 *Annual Review* article on the X-ray emission from galaxies (Fabbiano 1989), and a few words on the evolution of this field are in order. In 1989, the *Einstein Observatory* (Giacconi et al. 1979), the first imaging X-ray telescope, opened up the systematic study of the X-ray emission of normal galaxies. The *Einstein* images, in the $\sim$0.3–4 keV range, with resolutions of $\sim$5″ and $\sim$45″ (see the *Einstein Catalog and Atlas of Galaxies*, Fabbiano, Kim & Trinchieri 1992) showed extended and complex X-ray emission, and gave the first clear detection of individual luminous X-ray sources in nearby spiral galaxies, other than the Milky Way. The first ultraluminous (nonnuclear) X-ray sources (ULXs) were discovered with *Einstein*, and the suggestion was advanced that these sources may host >100 $M_\odot$ black holes, a topic still intensely debated. Hot diffuse halos were discovered in elliptical galaxies and used as a means of estimating the mass of the dark matter associated with these galaxies, but their ubiquity and properties were hotly debated. Super-winds from active star-forming galaxies (e.g., M82), an important component of the ecology of the universe, were first discovered with *Einstein*. All these topics are discussed in the 1989 review.

The subsequent X-ray observatories *ROSAT* (Truemper 1983) and *ASCA* (Tanaka, Inoue, Holt 1994) expanded our knowledge of the X-ray properties of galaxies (see, e.g., a review summary in Fabbiano & Kessler 2001), but did not produce the revolutionary leap originated by the first *Einstein* observations. The angular resolution of these missions was comparable (*ROSAT*) or inferior (*ASCA*, with 2 arcmin resolution) to that of *Einstein*, but the *ROSAT* spectral band (extending down to $\sim$0.1 keV) and lower background provided a better view in some cases of the cooler X-ray components (halos and hot outflows), while the wide spectral band ($\sim$0.5–10 keV) and better spectral resolution of the *ASCA* CCD detectors allowed both the detection of emission lines in these hot plasmas and the spectral decomposition of integrated emission components (e.g., Matsushita et al. 1994). Overall, however, many of the questions raised by the *Einstein* discoveries remained (see Fabbiano & Kessler 2001).

It is only with *Chandra*'s subarcsecond angular resolution (Weisskopf et al. 2000), combined with photometric capabilities commensurable with those of *ASCA*, that the study of normal galaxies in X rays has taken a second revolutionary leap. With *Chandra*, populations of individual X-ray sources, with luminosities comparable to those of Galactic X-ray binaries, can be detected at the distance of the Virgo Cluster and beyond; the emission of these sources can be separated from the diffuse emission of hot interstellar gases, both spatially and spectrally; detailed measures of the metal abundance of these gaseous components can be attempted (e.g., Martin, Kobulnicky & Heckman 2002; Soria & Wu 2002; Fabbiano et al. 2004a; Baldi et al. 2006a,b); and quiescent supermassive nuclear black holes can be studied (e.g., Fabbiano et al. 2004b, Pellegrini 2005, Soria et al. 2006).

Here, I will concentrate only on one aspect of the emission of normal galaxies, the study of their populations of discrete X-ray sources, with emphasis on compact accreting binary systems (**Table 1**). I avoid detailed discussions of the properties of

Table 1 Accretion-powered XRBs in galaxies (L_X more than a few 10^{36} erg s^{-1})

LMXB	Low Mass X-ray Binaries
	Neutron Star (NS) or Black Hole (BH) + later than type A star
	Time-variable: orbital periods, flares, bursts
	Spectral/luminosity states in BH XRBs
	On average soft spectra with $kT \sim 5$–10 keV
	Associated with old stellar populations
	Found in the stellar field (bulges) and in Globular Clusters
	Generally believed to be long-lived: lifetimes $\sim 10^{8-9}$ years
	Exception is model of Bildsten & Deloye (2004), Section 3.6
	Discussed in Section 3
HMXB	High Mass X-ray Binaries
	NS or BH + OB star
	Time-variable luminosities and spectra
	Orbital periods, outburst, rapid flaring, pulsations
	On average harder spectra than LMXBs, but BH binaries may have similarly soft spectra
	Associated with young stellar populations (e.g., spiral arms)
	Short-lived: lifetimes $\sim 10^{6-7}$ years
	Discussed in Section 4
ULX	Ultraluminous X-ray sources of debated nature
	$L_X > 10^{39}$ ergs s^{-1} (> Eddington luminosity of NS or ~ 5 $M_\odot$ BH)
	Proposed as intermediate mass BH candidates (> 100 $M_\odot$)
	Tend to be found in active star-forming environments
	Discussed in Section 6
SSS	Super-Soft Sources (black body $kT \sim 15$–80 eV)
	Nuclear burning White Dwarf binaries
	Discussed in Section 5
QSS	Quasi-Soft Sources discovered with *Chandra*
	$kT \sim 100$–300 eV
	Some exhibit a hard spectral tail
	Nature is still debated
	Discussed in Section 5, Section 6

individual nearby galaxies, which are covered in the earlier review by Fabbiano & White (2006, based on publications up to 2003). Also, I do not discuss the properties of the hot interstellar medium (ISM) and of low-level nuclear emission, which were all included in the 1989 review. The field has expanded enough since then that these topics deserve separate reviews. Most of the work discussed in this review is the result of the study of high-resolution *Chandra* images. Whenever relevant (for the most nearby galaxies, and the spectral and time-variability study of ULXs), I will also discuss observations with *XMM-Newton* (the European Space Agency X-ray telescope, which has an effective area ~ 3 times larger than *Chandra*, but significantly coarser angular resolution, $\sim 15''$).

This review proceeds as follows: Section 2 is a short discussion of the observational and analysis approaches opened by the availability of high resolution, sensitive X-ray data; Section 3 reviews the results on the old X-ray binary population found in early-type galaxies and spiral bulges; Section 4 addresses the work on the younger X-ray source population of spiral and irregular galaxies; Section 5 and Section 6 discuss two classes of rare X-ray sources, to the understanding of which recent observations of many galaxies have contributed significantly: supersoft sources (SSSs) and ULXs; Section 7 concludes this review with a short discussion of the properties of the galaxies observed in deep X-ray surveys. Section 3, Section 4, and Section 6 are the most substantial. They all start with brief historical introductions, summarize the observational evidence that identifies the X-ray sources with X-ray binaries (XRBs), discuss the X-ray luminosity functions as a means to compare and characterize the XRB populations, address the constraints deriving from the association of the X-ray sources with stellar or other features, and conclude with reviews of the theoretical work and interpretations. Throughout this review, I try to give the reader a feeling for the evolving state of the field by highlighting the different points of view and unresolved questions.

2. POPULATION STUDIES IN X-RAYS

It is well known that the Milky Way hosts both old and young luminous X-ray source populations, reflecting its general stellar makeup. In the luminosity range detectable in most external galaxies with typical *Chandra* observations ($>10^{37}$ erg s^{-1}), these Galactic populations are dominated by XRBs, and include both low-mass X-ray binaries (LMXBs) and high-mass X-ray binaries (HMXBs) (**Table 1**). A few young supernova remnants (SNRs) may also be expected. At lower luminosities, reachable with *Chandra* in Local Group galaxies, Galactic sources include accreting white dwarfs and more evolved SNRs (see, e.g., the review by Watson 1990 for a census of Galactic X-ray sources; Grimm, Gilfanov & Sunyaev 2002 for a study of the X-ray luminosity functions of the Galactic X-ray source populations; White, Nagase & Parmar 1995 for a review of the properties of Galactic X-ray binaries; Verbunt & van den Heuvel 1995 for a review on the formation and evolution of XRBs; Fender & Belloni 2004 on the spectral states of black-hole binaries). **Figure 1** shows the cumulative X-ray luminosity functions (XLFs) of LMXBs and HMXBs in the Galaxy (Grimm, Gilfanov & Sunyaev 2002). Note the high luminosity cut off of the LMXB XLF and the power-law distribution of the HMXB XLF; these basic characteristics are echoed in the XRB populations of external galaxies (Section 3.3 and Section 4.2).

Figure 2 shows two typical observations of galaxies with *Chandra*: the spiral M83 (Soria & Wu 2003) and the elliptical NGC4697 (Sarazin, Irwin & Bregman 2000), both observed with the ACIS CCD detector. The images are color coded to indicate the energy of the detected photons (*red*, 0.3–1 keV; *green*, 1–2 keV; and *blue*, 2–8 keV). Populations of point-like sources are easily detected above a generally cooler diffuse emission from the hot ISM. Note that luminous X-ray sources are relatively sparse by comparison with the underlying stellar population, and can be detected individually

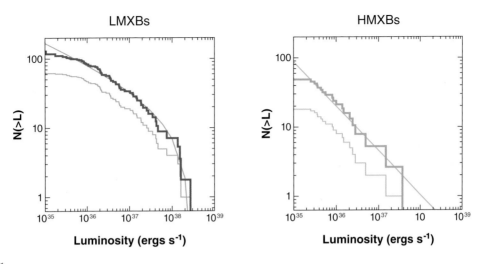

Figure 1

Cumulative X-ray luminosity functions (XLFs) of Galactic low-mass X-ray binaries (LMXBs, *left*) and high-mass X-ray binaries (HMXBs, *right*), from figure 12 of Grimm, Gilfanov & Sunyaev (2002). A mass of $2.5 M_\odot$ for the companion star was used as a boundary between LMXBs and HMXBs. The thin and thick histograms are the apparent and volume-corrected distributions, respectively. The lines are the best fits to the volume-corrected distributions: a power law with cumulative slope -0.64 ± 0.15 for the HMXBs and a power law (slope, -0.26 ± 0.08) truncated at $\sim 2.7 \times 10^{38}$ erg s^{-1} for the LMXBs. Note that a similar HMXB power law is also found in the X-ray binaries populations of star-forming galaxies (Section 4.2); the LMXB power law is flatter than those found in the populations of E and S0 galaxies (≤ -1, for $L_X > 2 \times 10^{37}$ erg s^{-1}, Section 3.3), and it may reflect the wider luminosity range covered by the Milky Way observations and a complex XLF shape.

with the *Chandra* subarcsecond resolution (excluding the crowded circumnuclear regions).

The X-ray CCD detectors (present both in *Chandra* and *XMM-Newton*) provide us with a data-hypercube of the observed area of the sky, where each individually detected photon is tagged with a two-dimensional position, energy, and time of arrival. Therefore, for each detected source, we can measure flux (and luminosity), spectral (or photometric) parameters, and time variability. For the most intense sources, it is also possible to study the time variability of spectra if the galaxy has been observed at different epochs (which is still rare in the available data set; see, e.g., Fabbiano et al. 2003a,b). To analyze this wealth of data two approaches have been taken: (*a*) a photometric approach, consisting of X-ray color-color diagrams and color-luminosity diagrams, and (*b*) X-ray luminosity functions.

2.1. X-Ray Photometry

The use of X-ray colors to classify X-ray sources is not new. For example, White & Marshall (1984) used this approach to classify Galactic XRBs, and Kim, Fabbiano & Trinchieri (1992) used *Einstein* X-ray colors to study the integrated X-ray emission of galaxies. Given the lack of standard X-ray photometry to date, different definitions

Figure 2

Chandra ACIS images of M83 (*left*, box is 8.57 × 8.86 arcmin) and NGC4697 (*right*, box is 8.64 × 8.88 arcmin). See text for details. Both images are from the Web page **http://chandra.harvard.edu/photo/category/galaxies.html**; credit NASA/CXC).

of X-ray colors have been used in different works; in the absence of instrument corrections, these colors can only be used for comparing data obtained in the same observational set up. Colors, however, have the advantage of providing a spectral classification tool when only a limited number of photons are detected from a given source, which is certainly the case for most X-ray population studies in galaxies.

Compared with the traditional X-ray data analysis approach of deriving spectral parameters via model fitting, color-color diagrams provide a relatively assumption-free comparison tool. Early *Chandra*-based examples of this approach can be found in Zezas et al. (2002a,b) and Prestwich et al. (2003). Color diagrams are used frequently to classify the discrete source populations of galaxies (see Sections 3 and 4); although useful, it is important to remember that some ambiguity in the outcome is unavoidable. Both the choice of spectral boundaries and the sensitivity of a given telescope-detector combination are important (see, for example, the identification of supersoft and quasi-soft sources, Pietsch et al. 2005, Di Stefano et al. 2004; Section 5). Moreover, both spectrum and flux of XRBs may vary in time, so that the classification of a given source may change when repeated observations become available.

2.2. X-Ray Luminosity Functions

Luminosity functions are a well-known tool of observational astrophysics. XLFs have been used to characterize different XRB populations in the Milky Way (e.g., Grimm, Gilfanov & Sunyaev 2002; see **Figure 1**), but these studies have always required a model of the spatial distribution of the sources, and of the intervening absorption, in order to estimate their luminosities; these corrections are inherently sources of uncertainty. External low-inclination galaxies, instead, provide clean source samples all at the same distance. Moreover, the detection of X-ray source populations in a wide range of different galaxies allows us to explore global population differences that may be connected with the age and or metallicity of the parent stellar populations. XLFs establish the observational basis of X-ray population synthesis (Belczynski et al. 2004). The first early attempts to construct and compare XLFs of X-ray source populations in external galaxies include the comparisons of the XLFs of M31 and of the disk of M81 with *Einstein* data (Fabbiano 1988; see also Fabbiano 1995), concluding that in M81 there is a relative surplus of very luminous sources, and the conclusion of a flat XLF in M101, connected with massive accreting young binaries (Trinchieri, Fabbiano & Romaine 1990). These early results are in general agreement with the trends suggested by the XLFs of LMXB and HMXB in the Milky Way (**Figure 1**) and with the results discussed in this review.

In principle XLFs are simple to construct, but care must be taken to apply corrections for observational biases and statistical effects. These include the incomplete detection of low-luminosity sources that may cause flattening of the XLF at the low-luminosity end; the artificial "brightening" of threshold sources because of statistical fluctuations (Eddington bias; Eddington 1913); the varying amount of diffuse emission around the source from a hot ISM (e.g., Zezas & Fabbiano 2002), which affects the detection threshold; and source confusion in crowded regions especially near the galaxy centers. In the case of *Chandra* the detection efficiency is also affected by the radial dependence of the degradation of the mirror resolution off-axis (see Kim & Fabbiano 2003, 2004; Gilfanov 2004). These low-luminosity biases have not been treated consistently in the literature, giving rise in some cases to potentially spurious results (Section 3.3). For galaxies extending over large angular sizes, the effect of background active galactic nuclei (AGNs) and stellar interlopers in the XLF must also be considered (e.g., Finoguenov & Jones 2002, Gilfanov 2004, Grimm et al. 2005).

At the high luminosity end, the paucity of very luminous X-ray sources in galaxies makes uncertain the parameterization of the XLF of individual populations; this problem has been approached by coadding "consistent" samples of X-ray sources (Kim & Fabbiano 2004). This same effect is responsible for uncertainties in the measurement of the total X-ray luminosity of a galaxy from the relatively small number of X-ray sources detected in short or insensitive observations (Gilfanov, Grimm & Sunyaev 2004b).

Compact X-ray sources are notorious for their variability and this variability could in principle also affect the XLF, which is typically derived from a snapshot of a given galaxy. However, repeated *Chandra* observations in the cases of NGC5128 (Kraft et al. 2001), M33 (Grimm et al. 2005) and the Antennae galaxies (Zezas et al. 2004) empirically demonstrate that the XLF is remarkably steady against individual source variability.

3. OLD X-RAY BINARY POPULATIONS

At variance with most previous reviews of X-ray observations of galaxies, which tend to concentrate first on nearby well-studied spiral and irregular galaxies, I will begin by discussing the X-ray populations of old stellar systems: E and S0 galaxies. By comparison with spirals, these galaxies present fairly homogeneous stellar populations, and therefore one can assume that their XRB populations are also more uniform, providing a "cleaner" baseline for population studies.

I begin this Section with a historical note (Section 3.1), followed by a summary of the detection of ubiquitous discrete X-ray source populations in spheroids and their spectral and variability properties (Section 3.2), which point to LMXB populations. I then address the characterization of these populations by means of XLFs (Section 3.3), and give an overview of the association of these sources with globular clusters (GCs) and of the properties of GC sources in comparison with field sources (Section 3.4). Finally, I summarize the discussion generated by these results for the dependence of LMXB formation in GCs on the metallicity and dynamical properties of the cluster (Section 3.5), and address the current debate on the formation and evolution of the entire LMXB populations, including both formation in GCs and evolution of field native binary systems (Section 3.6). I conclude this Section with a short discussion of results that may suggest evolution of the X-ray populations of some E and S0 galaxies (Section 3.7).

3.1. Low-Mass X-Ray Binaries in Early-Type Galaxies: There They Are—Past and Present

In the 1989 review (Fabbiano 1989) I argued that LMXBs should be present in E and S0s and might even dominate the X-ray emission of some of these galaxies. This was a controversial issue at the time, because LMXBs could not be detected individually, and their presence was supported only by statistical considerations (e.g., Trinchieri & Fabbiano 1985). Although the spectral signature of LMXBs was eventually detected (Kim, Fabbiano & Trinchieri 1992; Fabbiano, Kim & Trinchieri 1994; Matsushita et al. 1994), uncontroversial detection of samples of these sources in all early-type galaxies has become possible only with the subarcsecond resolution of *Chandra* (such a population was first reported in NGC4697, where 80 sources were detected by Sarazin, Irwin & Bregman 2000; see **Figure 2**).

A statistical analysis of a large sample of early-type galaxies observed with *Chandra* is still to come, but the results so far confirm the early conclusion (see Fabbiano 1989; Kim, Fabbiano & Trinchieri 1992; Eskridge, Fabbiano & Kim 1995a,b) that LMXBs account for a very large fraction of the X-ray emission of some early-type galaxies (those formerly known as "X-ray faint," i.e., devoid of large hot gaseous halos): for example, in NGC4697 (Sarazin, Irwin & Bregman 2000) and NGC1316 (Kim & Fabbiano 2003) the fraction of detected counts attributable to the hot ISM is $\sim$23% and $\sim$50%, respectively. In both cases, given the harder spectrum of LMXBs, these sources dominate the integrated luminosity in the 0.3–8 keV range. In NGC1316 the integrated LMXB emission, including nondetected LMXBs with luminosities below threshold, could be as high as 4×10^{40} erg s^{-1}. Sivakoff, Sarazin & Irwin (2003) reach similar conclusions for NGC4365 and NGC4382.

Although this review focuses on the X-ray binary populations, I cannot help remarking that the *Chandra* results demonstrate unequivocally that ignoring the contribution of the hidden emission of LMXBs was a source of error in past analyses. In

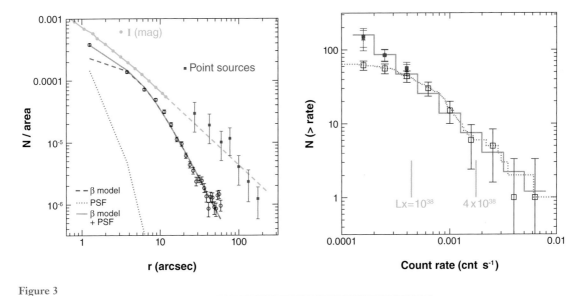

Figure 3

Left: Radial profile of the low-mass X-ray binary distribution (*red dots*) in NGC1316 compared with the profiles of the optical light (*green*) and diffuse hot interstellar medium emission (*black data points* and best-fit model); Right: X-ray luminosity function (XLF) before (*squares* are the binned data and the *dotted line* gives the unbinned XLF) and after completeness correction (*filled dots*; the *solid line* gives the best-fit power-law model, binned to resemble the data). These figures are figures 6 and 10 of Kim & Fabbiano (2003).

particular, estimates of galaxy dynamical mass were affected, as discussed in the 1989 review (see also Trinchieri, Fabbiano & Canizares 1986). NGC1316 (Kim & Fabbiano 2003) provides a very clear illustration of this point. In this galaxy the LMXBs are distributed like the optical light and dominate the emission at large radii. Instead, the hot ISM follows a steeper profile (see **Figure 3**, *left*), with temperature possibly decreasing at larger radii, suggestive of galactic winds. Use of lower-resolution *Einstein* data, with the assumption that the entire detected emission originated from a hot ISM in hydrostatic equilibrium, resulted in a large mass estimate for this galaxy (2.0×10^{12} $M_\odot$; Forman, Jones & Tucker 1985). This result is not sustained by the present data; because the gaseous halo is less extended than assumed in the *Einstein* paper, its temperature is lower (because the *Einstein* spectrum was clearly contaminated by the harder LMXB emission), and the halo may not be in hydrostatic equilibrium.

Estimates of the metal abundance of the hot ISM also must be reconsidered. In NGC1316, spectral analysis of the integrated X-ray emission obtained with *ASCA* suggested extremely subsolar values (0.1 solar, Iyomoto et al. 1998). This extremely low metallicity is typical of *ASCA* results for E and S0 galaxies, and cannot be reconciled with the predictions of stellar evolution (e.g., Arimoto et al. 1997). Spectral analysis of the NGC1316 *Chandra* data (Kim & Fabbiano 2003), after subtraction of the detected LMXBs, and taking into account the unresolved LMXB component, allows larger metallicities of the hot ISM (up to 1.3 $Z_\odot$), more in keeping with the expected values.

3.2. Source Spectra and Variability

Populations of several tens to hundreds of sources have been detected in E and S0 galaxies with *Chandra* (see review by Fabbiano & White 2006), and their number is growing as more galaxies are observed and the depth of the observations increase. With the exception of a few SSSs reported in some galaxies (see Irwin, Athey & Bregman 2003; Humphrey & Buote 2004), the X-ray colors and spectra of these sources are consistent with those expected of LMXBs, and consistent with those of the LMXBs of M31 (Blanton, Sarazin & Irwin 2001; Sarazin, Irwin & Bregman 2001; Finoguenov & Jones 2002; Irwin, Athey & Bregman 2003; Kim & Fabbiano 2003; Sivakoff, Sarazin & Irwin 2003; Humphrey & Buote 2004; Jordan et al. 2004; Kim & Fabbiano 2004; Randall, Sarazin & Irwin 2004; Trudolyubov & Priedhorsky 2004; David et al. 2005).

The most extensive spectral study to date is that of Irwin, Athey & Bregman (2003), who studied 15 nearby early-type galaxies observed with *Chandra*. They found that the average spectrum of sources fainter than 10^{39} erg s^{-1} is remarkably consistent from galaxy to galaxy, irrespective of the distance of the sources from the center of the galaxy. These spectra can be fitted with either power laws with photon index $\Gamma = 1.56 \pm 0.02$ (90%) or with bremsstrahlung emission with kT = 7.3 ± 0.3 keV. Sources with luminosities in the $(1–2) \times 10^{39}$ erg s^{-1} range instead have softer spectra, with power law $\Gamma \sim 2$, consistent with the high-soft emission of black-hole binaries (with masses of up to $15 M_\odot$ expected for these luminosities, based on the Eddington limit). Within the errors, these results are consistent with those reported in other studies, although sources in different luminosity ranges are usually not studied separately in these works. Jordan et al. (2004) confirm the luminosity dependence of the average source spectrum in M87; their color-color diagram suggests a spectral softening for sources more luminous than 5×10^{38} erg s^{-1}.

Relatively little is known about the time variability of these sources, because repeated monitoring of the same galaxy is not generally available. Type I X-ray bursts have been detected in some GC sources in M31, identifying these sources as neutron star LMXBs (Pietsch & Haberl 2005). Time variable sources and at least five transients (dimming by a factor of at least 10) have been detected in NGC5128, with two *Chandra* observations (Kraft et al. 2001). Variable sources are also detected with two observations of NGC1399, taken two years apart (Loewenstein, Angelini & Mushotzky 2005). Sivakoff, Sarazin & Irwin (2003) report time variability in a few sources in NGC4365 and NGC4382 within 40ks *Chandra* observations; Humphrey & Buote (2004) report two variable sources in NGC1332. Sivakoff, Sarazin & Jordan (2005) report short-timescale X-ray flares from 3 out of 157 sources detected in NGC4697; two of these flares occur in GC sources and are reminiscent of the superbursts found in Galactic neutron star binaries, while the third could originate from a black-hole binary. Maccarone (2005) suggests that these flares may be periodic events resulting from periastron accretion of eccentric binaries in dense globular clusters. The spectral characteristics of the point sources detected in E and S0 galaxies, their luminosities, and their variability, show that these sources are compact accreting X-ray binaries.

3.3. X-Ray Luminosity Functions of Low-Mass X-Ray Binary Populations

The luminosities of individual sources range from the detection threshold (typically a few 10^{37} erg s^{-1}, depending on the distance of the galaxy and the observing time) up to $\sim 2 \times 10^{39}$ erg s^{-1}. XLFs have been derived in most *Chandra* studies of early-type galaxies, and modeled to characterize their functional shape (power-law slopes, eventual breaks) and normalization. In the following I first review the work on the shape of the XLF and then discuss the drivers of the normalization (i.e., the total LMXB content of a galaxy).

The high luminosity ($L_X >$ a few 10^{37} erg s^{-1}) shape of the XLF has been parameterized with models consisting of power laws or broken power laws. The overall shape (in a single power-law approximation in the range of $\sim 7 \times 10^{37}$ to a few 10^{39} erg s^{-1}) is fairly steep, i.e., with a relative dearth of high luminosity sources, when compared with the XLFs of star-forming galaxies (Section 4.2; see also Kilgard et al. 2002, Colbert et al. 2004, Fabbiano & White 2006), but the details of these shapes and the related presence of breaks have been a matter of some controversy.

Two breaks have been reported in the XLFs of E and S0 galaxies: the first is a break at ~ 2–5×10^{38} erg s^{-1}, near the Eddington limit of an accreting neutron star, first reported by Sarazin, Irwin & Bregman (2000) in NGC4697, which may be related to the transition in the XLF between neutron stars and black-hole binaries (Blanton, Sarazin & Irwin 2001 in NGC1553; Finoguenov & Jones 2002 in M84; Kundu, Maccarone & Zepf 2002 in NGC4472; Jordan et al. 2004 in M87; Gilfanov 2004, Kim & Fabbiano 2004, and also Di Stefano et al. 2003 for the XLF of the Sa Sombrero galaxy, NGC4594); the second is a high luminosity break at $\sim 10^{39}$ erg s^{-1}, first reported by in NGC720 by Jeltema et al. (2003; see also Sivakoff et al. 2003, Jordan et al. 2004). Both breaks are somewhat controversial, because the interpretation of the observed XLFs is crucially dependent on a proper completeness correction (see Section 2.2).

Kim & Fabbiano (2003, 2004) show that incompleteness effects are particularly relevant for the detection of the Eddington break, because the typical exposure times of the data and the distances of the target galaxies in most cases conspire to produce a spurious break at just this value (see **Figure 3**, *right*, for an example). Interestingly, no break was required in the case of NGC5128 (Kraft et al. 2001), where the proximity of this galaxy rules out incompleteness near the neutron star Eddington luminosity. An apparent Eddington break that disappears after correction for completeness is also found by Humphrey & Buote (2004) for the XLF of NGC1332. Similarly, Eddington breaks are absent in NGC4365 and NGC4382 (Sivakoff, Sarazin & Irwin 2003), whereas a high luminosity cut-off at 0.9–3.1 $\times 10^{39}$ erg s^{-1} could be allowed; these researchers also consider the effect of incompleteness in their results.

Other recent papers, however, do not discuss, or do not apply, completeness corrections to the XLFs, so their conclusions on the presence of Eddington breaks need to be confirmed. Randall, Sarazin & Irwin (2004) report a break at $\sim 5 \times 10^{38}$ erg s^{-1} in NGC4649, with large uncertainties, but do not discuss the derivation of the XLF. Jordan et al. (2004) derive and fit the XLF of M87, and compare it with their own fit of those of NGC4697 and M49 (NGC4472), using the data from Sarazin, Irwin & Bregman (2001) and Kundu, Maccarone & Zepf (2002) respectively. However, completeness corrections are not applied, although the low-luminosity data are not fitted. Jordan et al. (2004) report breaks at 2–3×10^{38} erg s^{-1} in all cases, or a good fit with a single power law truncated at 10^{39} erg s^{-1}. Note that these results are not consistent with those of Kim & Fabbiano (2004) where the corrected XLFs of NGC4647 and NGC4472 are well fitted with single unbroken power laws.

Kim & Fabbiano (2004) derive corrected luminosity functions for a sample of 14 E and S0 galaxies, including some with previously reported breaks, and find that all the individual corrected XLFs are well fitted with single power laws with similar differential slopes (-1.8 to -2.2; cumulative slopes are -0.8 to -1.2) in the observed luminosity range. None of these fits require an Eddington break. However, a break may be hidden by the poor statistics in each case. The statistical consistency of the individual power laws justifies coadding the data to obtain a high significance composite XLF of early-type galaxies (**Figure 4** *left*). This composite XLF is not consistent with a single power law, suggesting a break at $(5 \pm 1.6) \times 10^{38}$ erg s^{-1}. The best-fit differential slope is -1.8 ± 0.2 in the few 10^{37} to 5×10^{38} erg s^{-1} luminosity range for the coadded XLF; at higher luminosity, above the break, the differential slope is steeper (-2.8 ± 0.6). These results are confirmed by the independent work of Gilfanov (2004), who analyzes four early-type galaxies, included in the Kim & Fabbiano (2004) sample (**Figure 4** *right*); however, Gilfanov's differential slope for the high luminosity portion of the XLF is somewhat steeper (-3.9 to -7.3). Both the

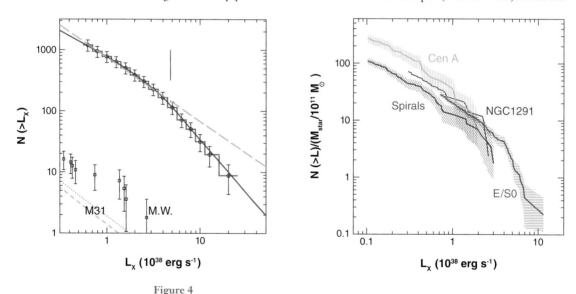

Figure 4

Left: Cumulative X-ray luminosity function (XLF) of 14 E and S0 galaxies (figure 7 of Kim & Fabbiano 2004), with the single power-law best fit (*dashed line*), and the broken power-law model (*solid line*); the M31 and Milky Way low-mass X-ray binary (LMXB) XLFs are sketched in the left lower corner. Right: Cumulative LMXB XLFs from figure 5 of Gilfanov (2004). Note the similarity of the XLFs and the break at $\sim 5 \times 10^{38}$ erg s^{-1} in the E/S0 XLF.

Kim & Fabbiano and Gilfanov analyses are consistent with a cut-off of the XLF of LMXBs at a few 10^{39} erg s^{-1}. A more recent paper (Xu et al. 2005) is in agreement with the above conclusions, reporting a consistent Eddington break in the corrected XLF of NGC4552; on the basis of a simulation, this paper concludes that the break may or may not be detected in any individual galaxy XLF, given the relatively small number of sources present in each case.

The $(5 \pm 1.6) \times 10^{38}$ erg s^{-1} break is at somewhat higher luminosity than would be expected from the Eddington luminosity of normal neutron star binaries. It may be consistent with the luminosity of the most massive neutron stars (3.2 ± 1 M$_\odot$; see Ivanova & Kalogera 2006), He-enriched neutron star binaries (1.9 ± 0.6 M$_\odot$; see Ivanova & Kalogera 2006), or low-mass black-hole binaries. This break may be caused by the presence of both neutron star and black-hole binary populations in early-type galaxies; it may also be the consequence of a true high luminosity break in the XLF (e.g., Sivakoff, Sarazin & Irwin 2003). Whatever the cause, the shape of the XLF points to a dearth of very luminous sources in E and S0 galaxies. Note that a high luminosity cut-off is also present in the XLF of Galactic LMXBs (**Figure 1**).

With the exception of NGC5128 (Cen A), for which the XLF has been measured down to $\sim 2 \times 10^{36}$ erg s^{-1} (Kraft et al. 2001, Voss & Gilfanov 2006), the available *Chandra* data does not allow the detection of LMXBs in E and S0 galaxies with luminosities below the mid or high 10^{37} erg s^{-1} range. By including Cen A and the LMXB (bulge) population of nearby spirals (Milky Way; see **Figure 1**, M31, M81) in his study, Gilfanov (2004) suggests that the XLF flattens below 10^{37} erg s^{-1}. A recent reanalysis of the Cen A data confirms this result (Voss & Gilfanov 2006). Direct deep observations of "normal" early-type galaxies are needed to see if this suggestion is generally valid; a legacy *Chandra* program will provide the necessary data for NGC3379 and NGC4278 by the end of 2007. These future studies may show complex behavior in the low-luminosity XLFs. For example, in M31 a radially dependent XLF break has been reported in the bulge, which could be related to an increasingly older population at the inner radii (Kong et al. 2002). Also, the GC XLF of M31 has a distinctive break at $2-5 \times 10^{37}$ erg s^{-1} (Kong et al. 2003, Trudolyubov & Priedhorsky 2004). The discovery of a similar break in the E and S0 XLFs may argue for a GC-LMXB connection in these galaxies. The "outburst peak luminosity—orbital period" correlation (King & Ritter 1998) predicts a break at this luminosity if a large fraction of the sources are short-period neutron star systems. This is intriguing, because the formation of ultracompact LMXBs is favored in Milky Way GCs (Bildsten & Deloye 2004; see also Section 3.6).

The normalization of the XLF is related to the number of LMXBs in a given galaxy. X-ray-optical/near-IR correlations in bulge-dominated spirals observed with *Einstein* (Fabbiano, Gioia & Trinchieri 1988; Fabbiano & Shapley 2002) had suggested a connection between the number of LMXBs and the overall stellar content of a galaxy. This connection has now been demonstrated to hold for the LMXB populations of E and S0 galaxies (Gilfanov 2004; Kim & Fabbiano 2004). That stellar mass is the main regulator of the number of LMXBs in a galaxy is not surprising, considering that LMXBs are long-lived systems, but there may be other effects. White, Sarazin & Kulkarni (2002) suggested a link with GC specific frequency (the number of GC per unit light in a galaxy) using low-resolution *ASCA* data. Kim & Fabbiano (2004; see also Humphrey & Buote 2004 for general agreement with this correlation in the case of NGC1332) find a correlation between K-band luminosity (which is proportional to stellar mass) and integrated LMXB luminosity, but also note that this correlation has more scatter than would be expected in terms of measurement errors. This scatter appears correlated with the GC-specific frequency, confirming a role of GCs in LMXB evolution.

3.4. Association of Low-Mass X-Ray Binaries with Globular Clusters: The Facts

In virtually all E and S0 galaxies with good coverage of GCs, both from the ground and better from *Hubble*, a fraction of the LMXBs is found in GCs (see earlier reviews by Verbunt & Lewin 2006, Fabbiano & White 2006). Sarazin, Irwin & Bregman (2000) first reported this association in NGC4697 and speculated on a leading role of GCs in LMXB formation, revisiting the original suggestion of Grindlay (1984) for the evolution of bulge sources in the Milky Way. Below, I summarize the observational results on the association of LMXBs with GCs from the large body of papers available in the literature. In Sections 3.5 and 3.6, I will discuss the implications of these results.

3.4.1. Statistics. It appears that in general $\sim$4–5% of the GCs in a given galaxy are likely to be associated with a LMXB (e.g., NGC1399—Angelini, Loewenstein & Mushotzky 2001; NGC4472—Kundu, Maccarone & Zepf 2002; NGC1553, NGC4365, NGC4649, NGC4697—Sarazin et al. 2003; NGC1339—Humphrey & Buote 2004; M87—Jordan et al. 2004, Kim et al. 2006). Not surprisingly, as first noticed by Maccarone, Kundu & Zepf (2003), the number of LMXBs associated with GCs varies, depending on the GC-specific frequency of the galaxy, which is also a function of the morphological type. Sarazin et al. (2003) point to this dependence on the galaxy Hubble type, with the fraction of LMXBs associated with GCs increasing from spiral bulges (MW, M31) $\sim$10–20%, to S0s $\sim$20% (NGC1553, Blanton, Sarazin & Irwin 2001; see also NGC5128, where 30% of the LMXBs are associated with GCs, Minniti et al. 2004), E $\sim$50% (NGC4697—Sarazin, Irwin & Bregman 2000; NGC4365—Sivakoff, Sarazin & Irwin 2003; NGC4649—Randall, Sarazin & Irwin 2004; see also NGC4552, with 40% of sources in GCs—Xu et al. 2005), cD $\sim$ 70% (in NGC1399—Angelini, Loewenstein & Mushotzky 2001; see also M87, where 62% of the sources are associated with GCs—Jordan et al. 2004).

3.4.2. Dependence on low-mass X-ray binary and globular cluster luminosity. In NGC1399 (Angelini, Loewenstein & Mushotzky 2001) the most luminous LMXBs are associated with GCs. No significant LMXB luminosity dependence of the LMXB-GC association is instead seen in NGC4472 (Kundu, Maccarone & Zepf 2002) or in the four galaxies studied by Sarazin et al. (2003); if anything, a weak trend is present in the opposite sense. The reverse is, however, consistently observed: more luminous GCs are more likely to host a LMXB (Angelini, Loewenstein & Mushotzky 2001;

Kundu, Maccarone & Zepf 2002; Sarazin et al. 2003; Minniti et al. 2004; Xu et al. 2005; Kim et al. 2006); this trend is also observed in M31 (Trudolyubov & Priedhorsky 2004). Kundu, Maccarone & Zepf (2002) suggest that this effect is just a consequence of the larger number of stars in optically luminous GCs. Sarazin et al. (2003) estimate that the probability per optical luminosity of LMXBs to be found in a GC is $\sim 2.0 \times 10^{-7}$ LMXBs per $L_{\odot,I}$ for $L_X \geq 3 \times 10^{37}$ erg s^{-1}.

This probability is consistent with past estimates based on the Milky Way and is a few hundred times larger than the probability of LMXBs occurring in the field per unit integrated stellar light in a galaxy, in agreement with the conclusion that dynamical interactions in GCs favor LMXB formation (Clark 1975).

3.4.3. Dependence on globular cluster color.
The probability that a GC hosts a LMXB is not a function of the GC luminosity alone. GC color is also an important variable, as first reported by Angelini, Loewenstein & Mushotzky (2001) in NGC1399 and Kundu, Maccarone & Zepf (2002, see also Maccarone, Kundu & Zepf 2003) in NGC4472, and confirmed by subsequent studies (e.g., Sarazin et al. 2003, Jordan et al. 2004, Minniti et al. 2004, Xu et al. 2005, Kim et al. 2006). In particular, the GC populations in these galaxies tend to be bi-modal in color (e.g., Zepf & Ashman 1993), and LMXBs preferentially are found in red, younger and/or metal-rich clusters (V-I >1.1), rather than in blue, older and/or metal-poor ones. The association of LMXBs with high metallicity GCs was observed in the Galaxy and M31 (Bellazzini et al. 1995, Trudolyubov & Priedhorsky 2004). In NGC4472, red GCs are three times more likely to host a LMXB than blue ones (Kundu, Maccarone & Zepf 2002). Similarly, in M87, which has a very rich LMXB population, the fraction of red GCs hosting a LMXB is 5.1% ± 0.7% versus 1.7% ± 0.5% for blue GCs (Jordan et al. 2004), also a factor of three discrepancy. In a sample of six ellipticals yielding 285 LMXB-GC associations (Kim et al. 2006), the mean probability for a LMXB-GC association is 5.2%, the probability of a blue GC to host a LMXB is $\sim$2% for all galaxies except NGC1399 (where it is 5.8%), while that of LMXB-red GC association is generally larger, but varies from one galaxy to another (2.7% to 13%).

3.4.4. X-ray colors.
Maccarone, Kundu & Zepf (2003) reported that in NGC4472 LMXBs associated with blue GCs have harder "stacked" X-ray spectra than those in red GCs. However, this result is not confirmed by the analysis of the much larger sample of sources assembled by Kim et al. (2006), where no significant differences are found in the X-ray colors of LMXBs associated with either red or blue GCs. Also, no significant differences are found in the X-ray colors of LMXBs in the field or in GCs (Sarazin et al. 2003; Kim et al. 2006).

3.4.5. Spatial distributions of field and globular cluster low-mass X-ray binaries.
To obtain additional constraints on LMXB formation and evolution, the radial distributions of the LMXBs have been compared with those of the GCs and of the field stellar light. Some of these comparisons have used the entire sample of detected LMXBs, irrespective of GC counterpart; others have also investigated differences between the LMXBs associated with GCs and those in the field.

Investigating the overall LMXB distribution in NGC4472, Kundu, Maccarone & Zepf (2002) suggest that it follows more closely the distribution of the GCs than the stellar light (which differ, with the GC one being more extended) and infer an evolutionary connection of all LMXBs with GCs (see Section 3.6). Other authors instead conclude that overall the LMXB distribution and the stellar light trace each other in E and S0 galaxies (NGC1316, Kim & Fabbiano 2003, shown in **Figure 3** *left*; NGC1332, Humphrey & Buote 2004). As for the XLFs, incompleteness may affect these comparisons and account for some of the discrepant reports: sources may be missed in the crowded inner parts of a galaxy, resulting in an apparently more extended distribution than the real one (see Kim & Fabbiano 2003, Gilfanov 2004).

Comparisons of the radial distributions of field and GC X-ray sources do not reveal any measurable differences (Sarazin et al. 2003; Jordan et al. 2004; Kim et al. 2006). A first comparison of these LMXB distributions with those of the stellar light and GCs was attempted in M87, but was inconclusive, given the statistical uncertainties (Jordan et al. 2004). With their significantly larger LMXB and GC samples, Kim et al. (2006) instead find that the LMXB radial profiles, regardless of association with either red or blue GCs, are closer to the more centrally peaked field stellar surface brightness distribution, than to the overall flatter GC distributions (**Figure 5**). The implication of this result for the GC sources is that the probability of a GC being associated with a LMXB increases at smaller galactocentric radii.

3.4.6. X-ray luminosity functions of field and globular cluster low-mass X-ray binaries.
No significant differences have been found in the XLFs of LMXBs in the field and in GCs (Kundu, Maccarone & Zepf 2002; Jordan et al. 2004). The coadded XLFs of field and GC LMXBs in six ellipticals (Kim et al. 2006) are also consistent within the errors, with a similar percentage of high luminosity sources with $L_X > 10^{39}$ erg s^{-1}.

This similarity of field and GC XLFs does not extend, however, to the X-ray populations of the Sombrero galaxy (Di Stefano et al. 2003) and M31 (from a comparison of the XLFs of bulge and GC sources; Trudoyubov & Priedhorsky 2004). In both cases, the GC XLFs show a more pronounced high luminosity break than the field (bulge) XLFs. In M31 the XLF of GC sources is relatively more prominent at the higher luminosities than that of field LMXBs; in the Sombrero galaxy, GC sources dominate the emission in the $1-4 \times 10^{38}$ erg s^{-1} range, but there is a high luminosity tail in the field XLF, which, however, could be due to contamination from a younger binary system belonging to the disk of this galaxy (see Di Stefano et al. 2003).

3.5. Metallicity and Dynamical Effects in Globular Cluster Low-Mass X-Ray Binary Formation

The preferential association of LMXBs with red clusters could be either an age or a metallicity effect. A correlation between the number density of binaries and the metallicity of GCs was first suggested by Grindlay (1987), who ascribed this effect to a flatter IMF in higher metallicity GCs, resulting in a larger number of neutron stars and thus LMXBs. Kundu et al. (2003) argue that metallicity is the main driver, based on the

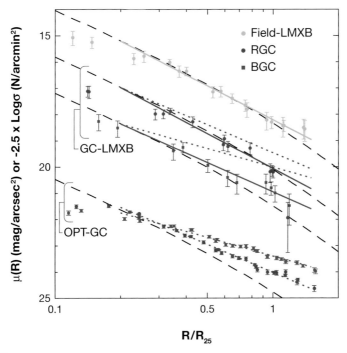

Figure 5

Radial distributions of low-mass X-ray binaries (LMXBs) in the field (*green*), in red GCs (*red*) and in blue GCs (*blue*), compared with the best-fit GC distributions (*red* and *blue* at the bottom of the figure), plotted versus the radius normalized to R_{25} for a sample of six galaxies. The flattening of the distributions at small radii is likely to be an incompleteness effect. The dotted blue and red lines are the best-fit models of the red and blue GC distributions. The solid lines show the best fits of the LMXB distributions. The black dashed lines represent the stellar light distribution (Kim et al. 2006).

absence of any correlations of LMXB association with different age GC populations in NGC4365. Maccarone, Kundu & Zepf (2004) propose irradiation-induced winds in metal-poor stars to speed up evolution and account for the observed smaller numbers of LMXBs in blue GCs. These winds, however, would cause absorption and thus harder X-ray spectra. Although these authors tentatively reported this spectral effect in NGC4472, studies of a larger sample of sources do not confirm this conclusion (see Section 3.4.4).

Jordan et al. (2004) revisit the IMF-metallicity effect, because the resulting increase in the number of neutron stars agrees with their conclusion that the probability that a GC contains a LMXB is driven by the dynamical properties of the cluster. Based on their study of M87, these researchers propose that the probability p_X for a given GC to generate a LMXB has the form $p_X \sim \Gamma \rho_0^{-0.42 \pm 0.11} (Z/Z_\odot)^{0.33 \pm 0.1}$, where Γ is a parameter related to the tidal capture and binary-neutron star exchange rate and ρ_0 is the central density of the cluster. This conclusion agrees with three-dimensional hydrodynamical calculations of the dynamical formation of ultracompact binaries in

GCs, from red giant and neutron star progenitors (Ivanova et al. 2005). Kim et al. (2006) also invoke dynamical effects to explain the increasing probability of LMXB-GC association at smaller galactocentric radii. They suggest that the GCs nearer to the galaxy centers are likely to have more compact cores and higher central densities to survive tidal disruption, compared with the GCs at the outskirts, characteristics that would also increase the chance of dynamical LMXB formation.

3.6. Constraints on the Formation and Evolution of Low-Mass X-Ray Binaries: Field Binaries or Globular Cluster Sources?

The formation processes of LMXBs have been debated since these sources were discovered in the Milky Way (see Giacconi 1974). LMXBs may result from the evolution of a primordial binary system, if the binary is not disrupted when the more massive star undergoes collapse and a supernova event, or may be formed by capture of a companion by a compact remnant in GCs (see Grindlay 1984, reviews by Verbunt 1993, Verbunt & van den Heuvel 1995). The same scenarios are now being debated for the LMXB populations of E and S0 galaxies. If GCs are the principal (or sole) birthplaces, formation kicks or evaporation of the parent cluster have been suggested as an explanation for the existence of field LMXBs in these galaxies (see e.g., Kundu, Maccarone & Zepf 2002).

The correlation of the total LMXB luminosity in a galaxy with the GC specific frequency (White, Sarazin & Kulkarni 2002; Kim & Fabbiano 2004; see Section 3.3) suggests that GCs are important in the formation of LMXBs. White, Sarazin & Kulkarni (2002) proposed formation in GCs as the universal LMXB formation mechanism in early-type galaxies. Other authors have supported this hypothesis, because of the similarity of field and GC LMXB properties (see Section 3.4; e.g., Maccarone, Kundu & Zepf 2003). However, this conclusion is by no means certain or shared by all. Beside uncertainties in the correlations (Kim & Fabbiano 2004), the relationship between the fraction of LMXBs found in GCs and the GC specific frequency (see Section 3.4.1) is consistent with the simple relationship expected if field LMXBs originate in the field while GC LMXBs originate in GCs (Juett 2005; Irwin 2005). This picture would predict different spatial distributions of field and GC LMXBs, an effect not seen so far, although, as Juett (2005) notes, the prevalence of LMXBs in red (more centrally concentrated) GCs and the effect of supernova kicks in the distribution of binaries may make the two distributions less distinguishable.

Piro & Bildsten (2002) and Bildsten & Deloye (2004) compare the observational results with theoretical predictions for the evolution of field and GC binaries. Piro & Bildsten remark that the large X-ray luminosities of the LMXBs detected in early-type galaxies ($>10^{37}$ and up to 10^{39} erg s^{-1}) imply large accretion rates ($>10^{-9}$ M$_\odot$ yr^{-1}). In an old stellar population these sources are likely to be fairly detached binaries that accumulate large accretion disks over time, and undergo transient X-ray events when accretion is triggered by disk instabilities. These transients would have recurrence times greater than 100 years and outbursts of 1–100 years duration. In this picture field binaries should be transient, a prediction that is supported by the detection of transients in the NGC5128 LMXB population (Kraft et al. 2001) and by the discovery of a population of quiescent X-ray binaries in the Sculptor dwarf spheroidal galaxy (Maccarone et al. 2005). Piro & Bildsten also point out that GC sources tend to have shorter orbital periods and would be persistent sources, reducing the fraction of transients in the LMXB population. Interestingly, Trudolyubov & Priedhosky (2004) report only one recurrent transient in their study of GC sources in M31, although 80% of these sources show some variability; however, they also find six persistent sources in the 10^{38} erg s^{-1} luminosity range.

Bildsten & Deloye (2004) instead look at ultracompact binaries formed in GCs to explain the bulk of the LMXBs detected in E and S0 galaxies. A motivation for this work is the large probability of finding LMXBs in GCs (per unit optical light, see Section 3.4.2), which makes formation in GCs more efficient than in the field. Ultracompact binaries would be composed of an evolved low-mass donor star (a white dwarf), filling its Roche lobe, in a 5–10 minute orbit around a neutron star or a black hole. The entire observable life of such a system is $\sim 10^7$ years, much shorter than the age of the galaxies and the GCs, therefore their total number would be indicative of their birth rate. From this consideration Bildsten & Deloye derive a XLF with a functional slope in excellent agreement with the measurements of Kim & Fabbiano (2004) and Gilfanov (2004). Bildsten & Deloye also predict a break at $L_X \sim 10^{37}$ erg s^{-1} in the XLF, which would correspond to the luminosity below which such a system would be a transient. As discussed in Section 3.3, there is some evidence of a low-luminosity break in the composite XLF of Gilfanov (2004), which, however, includes data from spiral bulges as well.

Confirmation of this break in a number of E and S0 populations by itself would not be proof of the Bildsten & Deloye scenario, because the break may occur from the evolution of field binaries. For example, a flattening of the XLF at the lower luminosities is found in the population synthesis of Pfahl, Rappaport & Podsiadlowski (2003, their figure 3), if irradiation of the donor star from the X-ray emission of the compact companion is considered in the model. More recently, Postnov & Kuranov (2005) have proposed that the mean shape of the XLF of Gilfanov (2004) can be explained by accretion on neutron star from Roche lobe overflow driven by gravitational wave emission, below $\sim 2 \times 10^{37}$ erg s^{-1}, and by magnetic stellar winds at higher luminosities. Optical identification of X-ray sources with GCs and an estimate of the transient fraction at different luminosities would help to discriminate among possible scenarios; planned deep time-monitoring *Chandra* observations may provide the observational constraints.

The nature of the most luminous sources in E and S0 galaxies (those with L_X above the 5×10^{38} erg s^{-1} break, Kim & Fabbiano 2004) is the subject of a recent paper by Ivanova & Kalogera (2006). These researchers point out that only a small fraction of these luminous sources are associated with GCs (at least in M87, see Jordan et al. 2004) and that they are too luminous to be explained easily with accreting neutron star systems that may form in GCs (Kalogera, King & Rasio 2004). With the assumption that these sources are accreting black-hole binaries, these authors explore their nature from the point of view of the evolution of field native binaries. In this picture most donor stars would be of low enough mass (<1–1.5 $M_\odot$ given the age of the stellar populations in question) that the binary would be a transient (see Piro & Bildsten 2002) and therefore populate the XLF only when in outburst emitting at the Eddington luminosity; this would happen from main-sequence, red-giant, and white-dwarf donors. In this case the XLF is a footprint of the black-hole mass spectrum in these stellar populations, which is an important ingredient for linking the massive star progenitors with the resulting black hole. Ivanova & Kalogera derive a differential slope of ~ -2.5 for the black-hole mass spectrum, and an upper black-hole mass cut-off at ~ 20 $M_\odot$, to be consistent with the observed cumulative XLF of Kim & Fabbiano (2004) and Gilfanov (2004). Depending on the magnetic breaking prescription adopted, either red-giant donors or main-sequence donors would dominate the source population. A word of caution is in order here, because the similar shape of GC and field LMXB XLFs (Kim et al. 2006, see Section 3.4.6) suggests that high-luminosity black-hole sources may also be found in GCs, at odds with theoretical discussions (e.g., Kalogera, King & Rasio 2004).

3.7. Young Early–Type Galaxies and Rejuvenation

There have been some puzzling and somewhat controversial results suggesting that the stellar populations of some early-type galaxies may not be uniformly old, as implied by their optical characteristics, but may hide a small fraction of younger stars, which give rise to luminous and easily detectable X-ray binaries. Rejuvenation (e.g., by a merger event or close encounter with a dwarf galaxy) has been suggested to explain the presence of very luminous and asymmetrically distributed X-ray source populations in some galaxies [NGC720—Jeltema et al. 2003; NGC4261 (shown in **Figure 6**) and NGC4697—Zezas et al. 2003]. Sivakoff, Sarazin & Carlin (2004) report an exceptionally luminous population of 21 sources with $L_X > 2 \times 10^{39}$ erg s^{-1} (in the ULX regime, see Section 6) in the X-ray bright elliptical NGC1600, which is twice the number of sources that would be expected from background AGNs and suggests an XLF slightly flatter than in most ellipticals. In all these cases, however, both cosmic variance affecting the background AGN density and distance uncertainties may play a role. Moreover, Giordano et al. (2005) report the identification of the NGC4261 sources with GCs, undermining the suggestion that they may be linked to a rejuvenation event.

The behavior opposite the one just discussed is reported in an X-ray and optical study of the nearby lenticular galaxy NGC5102 (Kraft et al. 2005). In this galaxy, where the stellar population is young ($<3 \times 10^9$ years old), and where there is evidence of two recent bursts of star formation, a definite lack of X-ray sources is observed. NGC5102 has also a very low specific frequency of GC (~ 0.4). Kraft et al. speculate that the lack of LMXBs may be related either to insufficient time for the evolution of a field binary and/or to the lack of GCs.

4. YOUNG XRB POPULATIONS

The association of luminous X-ray sources (HMXBs, and the less luminous SNRs) with the young stellar population has been known since the dawn of X-ray astronomy (see Giacconi 1974). The presence of X-ray source populations in Local Group

Figure 6

The left panel shows a *Chandra* image of NGC4261; note the distribution of the luminous point sources, which clearly do not follow the optical light shown in the right panel; Both images are from **http://chandra.harvard.edu/photo/category/galaxies.html**; credit NASA/CXC; Zezas et al. (2003).

and nearby spiral and irregular galaxies was clearly demonstrated by the early *Einstein* observations (see the 1989 review, Fabbiano 1989; and the *Einstein Catalog and Atlas of Galaxies*, Fabbiano, Kim & Trinchieri 1992). Luminous HMXBs are expected to dominate the emission of star-forming galaxies (Helfand & Moran 2001). These sources, resulting from the evolution of a massive binary system where the more massive star has undergone a supernova event, are short-lived ($\sim 10^{6-7}$ years) and constitute a marker of recent star formation: their number is likely to be related to the galaxy star-formation rate (SFR). This X-ray population–SFR connection was first suggested as a result of the analysis of the sample of normal galaxies observed with *Einstein*, where a strong correlation was found between global X-ray and FIR emission of late-type star-forming galaxies (Fabbiano & Trinchieri 1985; Fabbiano, Gioia & Trinchieri 1988; David, Forman & Jones 1991; Shapley, Fabbiano & Eskridge 2001; Fabbiano & Shapley 2002), and has been confirmed by analyses of *ROSAT* observations (Read & Ponman 2001; Lou & Bian 2005).

Though HMXBs are likely to dominate the X-ray emission of galaxies where star formation is most violent, they are also expected to be found in more normal spirals (see the Milky Way population; Section 2), albeit in smaller numbers and mixed with more aged X-ray populations. In bulge-dominated spirals, HMXBs may constitute only a small fraction of the X-ray-emitting population; for example, witness the strong correlation between X-ray and H-band luminosity found in these systems

(Shapley, Fabbiano & Eskridge 2001; Fabbiano & Shapley 2002). The study of HMXB populations is then less straightforward than that of LMXBs, because in many cases HMXBs must be culled from the complex X-ray source populations of spiral galaxies.

Below, I first discuss the detection and characterization of populations of X-ray sources in spiral and irregular galaxies with X-ray imaging and photometry (Section 4.1), and then review the work on the XLFs of these star-forming populations (Section 4.2).

4.1. X-Ray Source Populations of Spiral and Irregular Galaxies

Reflecting the complex stellar populations of these galaxies, *Chandra* and *XMM-Newton* observations are discovering complex X-ray source populations. Typically, in each observed galaxy, from a few tens to well over hundreds of sources have been detected. Time variability and spectral analysis have been carried out for the most luminous sources, confirming that these sources are accreting binaries; the results are reminiscent of the spectral and temporal-spectral behavior of Galactic XRBs, including soft and hard spectral states [e.g., M31 (NGC224)—Trudolyubov, Borodzin & Priedhorsky 2001; Trudolyubov et al. 2002a; Kaaret 2002; Kong et al. 2002; Williams et al. 2004; Trudolyubov & Priedhorsky 2004; Pietsch, Freyberg & Haberl 2005; M33 (NGC598)—Grimm et al. 2005; Pietsch et al. 2004; NGC1068—Smith & Wilson 2003; NGC1637—Immler et al. 2003; NGC2403—Schlegel & Pannuti 2003; M81 (NGC3031)—Swartz et al. 2003; M108 (NGC3556)—Wang, Chavez & Irwin 2003; NGC4449—Miyawaki et al. 2004; M104 (NGC4594; Sombrero)—Di Stefano et al. 2003; M51 (NGC5194/95)—Terashima & Wilson 2004; M83 (NGC5236)—Soria & Wu 2002; M101 (NGC5457)—Pence et al. 2001; Jenkins et al. 2004, 2005].

Most detected sources, however, are too faint for detailed analysis; their position relative to the optical image of the galaxy (e.g., bulge, arms, disk, GCs), their X-ray colors, and in some cases optical counterparts have been used to aid in classification. Typically, as demonstrated by Prestwich et al. (2003) who applied this method to five galaxies (**Figure 7**), color-color diagrams can discriminate between harder XRB candidates (with relatively harder neutron star HMXBs and softer LMXBs; rare black-hole HMXBs would also belong to this "softer" locus), softer SNR candidates, and very soft sources (SSSs, with emission below 1 keV).

Similarly, XRBs, SNRs, and SSSs are found with *XMM-Newton* X-ray colors in IC342, where most sources are near or on the spiral arms, associating them with the young stellar population (Kong 2003). In M33, the Local Group Scd galaxy with a predominantly young stellar population, *Chandra* and *XMM-Newton* colors, luminosities, and optical counterparts indicate a prevalence of (more luminous) HMXBs and a population of (fainter) SNRs (Pietsch et al. 2004; Grimm et al. 2005). In M83 (**Figure 2**, *left*), the X-ray source population can be divided into three groups, based on their spatial, color, and luminosity distributions (Soria & Wu 2003): fainter SSSs, soft sources (with no detected emission above 2 keV), and more luminous and harder XRBs. The positions of the soft sources are strongly correlated with current star-formation regions, as indicated by Hα emission in the spiral arms and the starburst nucleus, strongly suggesting that they may be SNRs.

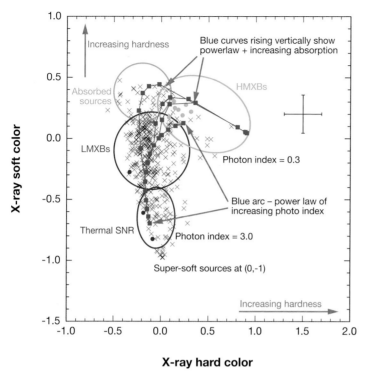

Figure 7

Chandra color-color diagram (figure 4 of Prestwich et al. 2003). Energy bands chosen for this diagram are S = 0.3–1 KeV, M = 1–2 KeV, and H = 2–8 KeV. Soft color = (M−S)/(S + M + H), hard color = (H−M)/(S + M + H).

Chandra X-ray colors (or hardness ratios) were also used to study the X-ray source populations of M100 (NGC4321; Kaaret 2001), M101 (Jenkins et al. 2005), NGC1637 (Immler et al. 2003), NGC4449 (Summers et al. 2004), NGC5494 (the Sombrero galaxy, a Sa, with a predominantly older stellar population; Di Stefano et al. 2003), and the star-forming merging pair NGC4038/9, the Antennae galaxy (Fabbiano, Zezas & Murray 2001; Fabbiano et al. 2004a), where spectral and flux variability is revealed by color-color and color-luminosity diagrams (Fabbiano et al. 2003a,b; Zezas et al. 2002a,b, 2005). Colbert et al. (2004) employ *Chandra* color diagrams to classify the X-ray source populations in their survey of 32 nearby galaxies of all morphological types, suggesting that hard accreting X-ray pulsars do not dominate the X-ray populations and favoring softer black-hole binaries.

These spectral, photometric and time-variability studies all point to the prevalence of XRB emission at the higher luminosities in the source population, in agreement with what is known from the X-ray observations of the Milky Way and Local Group galaxies (e.g., Helfand & Moran 2001). Comparison of accurate *Chandra* source positions with the stellar field in three nearby starburst galaxies shows that some of these sources experience formation kicks, displacing them from their parent star cluster (Kaaret et al. 2004), as observed in the Milky Way. The X-ray luminosity functions, which are discussed below, can then be considered as reflecting the XRB contribution (LMXBs, and HMXBs), with relatively little contribution from the SNRs. This point is confirmed by the direct comparison of HMXB and SNR luminosity functions in M33 (Grimm et al. 2005).

4.2. X-Ray Luminosity Functions—the X-Ray Luminosity Function of the Star-Forming Population

The XLF of LMXB populations (at high luminosity, at least) is well defined by the study of early-type galaxies, which have fairly uniform old stellar populations, and where little if any contamination from a young X-ray source population is expected (Section 3.3). The XLFs of late-type galaxies (spirals and irregulars) are instead the sum of the contributions of different X-ray populations, of different age and metallicity. This complexity was clearly demonstrated by the first detailed studies of nearby galaxies, including a comparison of different stellar fields of M31, with *XMM-Newton* and *Chandra*, yielding different XLFs (e.g., Trudolyubov et al. 2002b, Williams et al. 2004, Kong et al. 2003), and the *Chandra* observations of M81. In this Sab galaxy, the XLF derived from disk sources is flatter than that of the bulge (Tennant et al. 2001; shown in **Figure 8**, *left*). In the disk itself, the XLF becomes steeper with increasing distance from the spiral arms. In the arms the XLF is a pure power law with cumulative slope -0.48 ± 0.03 (Swartz et al. 2003), pointing to a larger presence of high luminosity sources in the younger stellar population.

To derive the XLF of HMXB populations, to a first approximation, one must either evaluate the different contributions of older and younger source populations to the XLFs of spiral galaxies, as discussed above in the case of M81, or study galaxies where the star-formation activity is so intense as to produce a predominantly young X-ray source population. Both approaches suggest that the HMXB XLF is overall flatter than that of the LMXBs, with a cumulative power-law slope of -0.6 to -0.4 (to be compared with a cumulative slope of ≤ -1 for LMXBs, e.g., Kim & Fabbiano 2004); in other words, young HMXB populations contain on average a larger fraction of very luminous sources than the old LMXB populations (see the comparisons of Eracleous et al. 2002; Kilgard et al. 2002; Zezas & Fabbiano 2002; Colbert et al. 2004). These comparisons also show that flatter XLF slopes of about -0.4 to -0.5 (cumulative) are found in intensely star-forming galaxies, such as the merging pair NGC4838/9 (the Antennae galaxies) and M82 (Zezas & Fabbiano 2002; Kilgard et al. 2002). In particular, Kilgard et al. (2002) find a correlation of the power-law slope with the 60-μm luminosity of the galaxy (shown in **Figure 8**, *right*), which suggests that such a flat power law may describe the XLF of the very young HMXB population. A comparison of the XLFs of dwarf starburst galaxies with those of spirals (Hartwell et al. 2004) is consistent with the above picture; cumulative XLF slopes for spirals are -1.0 to -1.4, whereas slopes for starbursts are lower, -0.4 to -0.8. The connection of the slope with the SFR is demonstrated by comparisons with the 60/100-μm ratio, 60-μm luminosity, and FIR/B ratio.

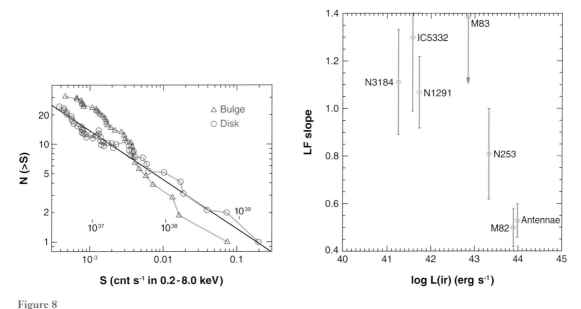

Figure 8

Left: Bulge and disk X-ray luminosity functions (XLFs) of M81 (figure 3 of Tennant et al. 2001). Right: XLF slope versus infrared luminosity for seven galaxies (figure 2 of Kilgard et al. 2002). Younger stellar populations have flatter XLFs.

Grimm, Gilfanov & Sunyaev (2003) took these considerations a significant step further by comparing the XLFs of 10 star-forming galaxies (taken from the literature), observed with *Chandra* and *XMM-Newton*, with the HMXB luminosity functions of the Small Magellanic Cloud and the Milky Way. They suggest that there is a universal XLF of star-forming populations, stretching over four decades in luminosity ($\sim 4 \times 10^{36}$–10^{40} erg s^{-1}), with a simple power law with cumulative slope -0.6. They reach this conclusion by considering that the XLFs of star-forming galaxies are dominated by young and luminous short-lived HMXBs, whose number would be proportional to the SFR per unit stellar mass; when normalized relative to SFR, the XLFs these authors consider in their study collapse into a single -0.6 power law. Postnov (2003) suggests that this empirically observed slope might result from the mass-luminosity and mass-radius relations in wind-accreting high mass binaries. The possibility of a "universal" HMXB XLF is an interesting result, although there are clear variations in the individual luminosity functions used by Grimm, Gilfanov & Sunyaev (2003), with slopes ranging from ~ -0.4 (the Antennae; Zezas & Fabbiano 2002, where the data were corrected for incompleteness at the low luminosities) to -0.8 (M74-NGC628; Soria & Kong 2002).

A number of XLFs of spiral galaxies have cumulative slopes close to the -0.6 slope of Grimm, Gilfanov & Sunyaev (2003) (IC342—Kong 2003; Bauer, Brandt & Lehmer 2003; NGC5253—Summers et al. 2004; NGC4449—Summers et al. 2003; NGC2403—Schlegel & Pannuti 2003; NGC6946—Holt et al. 2003; NGC1068— Smith & Wilson 2003; NGC2146—Inui et al. 2005). However, different and more complex XLFs are also observed, pointing to complexity or evolution of the X-ray source populations. In NGC1637 (Immler et al. 2003), the cumulative XLF is reported to follow a power law of slope -1 for the entire luminosity range covered ($\sim 6 \times 10^{36}$–10^{39} erg s^{-1}); though a possible break of the XLF ($L_X > 1 \times 10^{37}$ erg s^{-1}) is reported, there is no discussion of completeness correction. In NGC2403 (Schlegel & Pannuti 2003), the XLF has cumulative slope -0.6, but this galaxy does not follow the XLF slope; FIR correlation of Kilgard et al. (2002) suggesting it may have stopped forming stars, and we may be observing it after the massive stellar population has evolved, but the HMXBs are still emitting. In NGC6946 (Holt et al. 2003), though the cumulative XLF slope is generally consistent with the Grimm, Gilfanov & Sunyaev (2003) conclusions, differences are seen comparing the XLF of the sources in the spiral arms (slope -0.64) with that of sources within two arcminutes of the starburst central region, which is flatter (-0.5). The XLF of NGC5194 (M51, Terashima & Wilson 2004) follows an unbroken power law with slope -0.9.

In a detailed *Chandra* study of M83, a grand design spiral with a nuclear starburst, Soria & Wu (2003) find that the XLFs of different groups of sources identified by their X-ray colors differ. SSSs (see Section 5), which are found in regions with little or no Hα emission, have steep XLFs, typical of old populations; soft SNR candidates, which tend to be associated with the spiral arms, also have fairly steep XLFs, although they extend to luminosities higher than that of the SSSs; the hard XRB candidates dominate the overall X-ray emission, and therefore the overall XLF. For these sources differences in the XLF are also found, which can be related to stellar age. The XLF of the actively star-forming central region is a power law with cumulative slope -0.7; the XLF of the outer disk has a break at $L_X \sim 8 \times 10^{37}$ erg s^{-1}, follows a power law with slope -0.6 below the break, and gets considerably steeper at higher luminosities (-1.6). This type of broken power law has also been found in the disk of M31 (Williams et al. 2004, Shirey et al. 2001). In M83, a dip is seen in the XLF at $\sim 3 \times 10^{37}$ erg s^{-1}, corresponding to 100–300 detected source counts (well above source detection threshold), for sources in the disk and spiral arms where confusion is not a concern. The XLF rises again (toward lower luminosities) after the $\sim 3 \times 10^{37}$ erg s^{-1} dip, so incompleteness effects are not likely here. Soria & Wu (2003) speculate that this complex XLF (shown in **Figure 9**) may result from an older population of disk sources mixing with a younger (but aging) population of spiral arm sources.

The highest reaches of a star-forming XLF are found in the Cartwheel galaxy (Wolter & Trinchieri 2004), whose detected XRB population is dominated entirely by ULXs. This XLF has a slope consistent with that of Grimm, Gilfanov & Sunyaev (2003) and a large normalization, which suggests a SFR of ~ 20–25 M$_\odot$ yr^{-1}.

Grimm et al. (2005) and Shtykovskiy & Gilfanov (2005) explore the lowest luminosity reaches of the HMXB XLF with the *Chandra* survey of M33 (reaching $\sim 10^{34}$ erg s^{-1}) and with the *XMM-Newton* observations of the Large Magellanic Cloud (reaching $\sim 3 \times 10^{33}$ erg s^{-1}), respectively. In both galaxies, a large number of the detected sources are background AGNs. In M33, the XLF, corrected for interlopers and incompleteness, is consistent with the HMXB XLF of the Milky Way (Grimm,

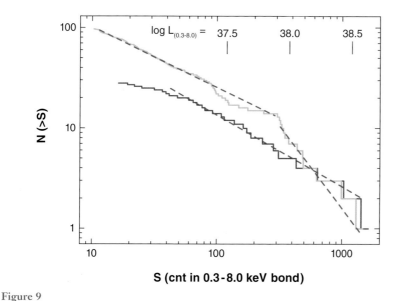

Figure 9

X-ray luminosity functions of inner regions (*red*, d < 60″ from galactic center) and outer disk (*green*) of M83 (figure 11 of Soria & Wu 2003).

Gilfanov & Sunyaev 2002). In the Large Magellanic Cloud, the corrected XLF, with spurious sources removed and rescaled for the SFR, globally fits the HMXB XLF of Grimm, Gilfanov & Sunyaev (2003) at the high luminosities ($\sim 10^{37}$ erg s^{-1}). The dearth of low-luminosity sources in this XLF leads Shtykovskiy & Gilfanov to suggest a propeller effect (i.e., the magnetic field stopping the accretion flow away from the pulsar surface of the pulsar for relatively low accretion rates). Observing regions of intense local star formation (as indicated by $H\alpha$ and FIR maxima), where no HMXBs are found, these researchers also suggest an age effect: these star-forming regions could be too young for HMXBs to have evolved, as HMXBs take on the order of 10 Myr to emerge after the star-formation event. Tyler et al. (2004) advanced a similar suggestion in their comparison of $H\alpha$, mid-IR, and *Chandra* images of 12 nearby spiral galaxies.

In conclusion, the XLFs of sources in a given galaxy reflect the formation, evolution, and physical properties of the X-ray source populations. These differences are evident, for example, in different regions of M81 and M83 (see **Figures 8** and **9**), by comparing elliptical and spiral galaxies and by comparing star-forming galaxies with different SFRs. These differences may be related to the aging of the X-ray source population, which will be gradually depleted of luminous young (and short-lived) sources associated with more massive, faster-evolving donor stars, and also to metallicity effects (Wu 2001; Belczynski et al. 2004). In the future, these X-ray population studies will constitute the baseline against which to compare models of X-ray population synthesis. An early effort toward this end can be found in Belczynski et al. (2004).

5. SUPER-SOFT SOURCES (SSSs) AND QUASI-SOFT SOURCES (QSSs)

SSSs, as a new class of luminous X-ray sources, were discovered with *ROSAT*. These sources— first found in the Miky Way, M31, the Magellanic Clouds and NGC55— are detected only at energies below 1 keV and are characterized by spectra that can be fitted with black-body temperatures ~15–80 eV (see review by Kahabka & van den Heuvel 1997). Their bolometric luminosities are in the 10^{36}–10^{38} erg s^{-1} range, and they are believed to be nuclear-burning white dwarfs (van den Heuvel et al. 1992).

Chandra observations have led to the discovery of populations of very soft sources in several galaxies. These newly discovered populations stretch the spectral definition of SSSs, including both slightly harder sources, typically fitted with black-body temperatures ~100–300 eV and sources with a small extra hard component in addition to a typical SSS spectrum (dubbed QSSs; see Di Stefano & Kong 2003, Di Stefano et al. 2004). A new class of supersoft ULXs has also been found (in M101, Mukai et al. 2003; in the Antennae, Fabbiano et al. 2003b; see Section 6). Time variability has been reported in some cases, supporting the idea that these sources are accretion binaries. In M31, a comparison of *Chandra* and *ROSAT* SSSs establishes a variability timescale of several months (Greiner et al. 2004); in NGC300 a luminous (10^{39} erg s^{-1}) variable SSS is found in *XMM-Newton* data, with a possible 5.4 hr period when in low state (Kong & Di Stefano 2003); the supersoft ULXs in M101 and the Antennae are both highly variable (Mukai et al. 2003, 2005; Kong, Di Stefano & Yuan 2004; Fabbiano et al. 2003b).

These very soft sources are associated with both old and young stellar populations. They are found in the elliptical galaxies NGC1332 (Humphrey & Buote 2004) and NGC4967 (Di Stefano & Kong 2003, 2004), in the Sombrero galaxy (an Sa; Di Stefano et al. 2003), and in a number of spirals (M31—Kahabka & van den Heuvel 1997; Kong et al. 2002; Di Stefano et al. 2004; M81—Swartz et al. 2002; M101—Pence et al. 2001; Di Stefano & Kong 2003, 2004; M83—Di Stefano & Kong 2003, 2004; Soria & Wu 2003; M51—Di Stefano & Kong 2003, 2004; Terashima & Wilson 2004; IC342—Kong 2003; NGC300 with *XMM-Newton*—Kong & Di Stefano 2003; NGC4449—Summers et al. 2003). Very soft sources are found both in the arms of spiral galaxies, suggesting systems of 10^8 years of age or younger (see, e.g., Di Stefano & Kong 2004), and in the halo and bulges, suggesting older counterparts; very soft sources in bulges tend to concentrate preferentially nearer the nuclei (Di Stefano et al. 2003, 2004). A QSS is associated with a GC in the Sombrero galaxy (Di Stefano et al. 2003). Pietsch et al. (2005) report a significant association of SSSs with optical novae in both M31 and M33.

As discussed in several of the above-mentioned papers, these results, and the spectral and luminosity regimes discovered with *Chandra* and *XMM-Newton*, strongly suggest that these very soft sources may constitute a heterogeneous population, including both hot white dwarf systems (SSSs), and black-hole (or neutron star) binaries (QSSs, supersoft ULXs). Pietsch et al. (2005) stress that classic SSSs with black-body temperatures below 50 eV can be cleanly identified only using X-ray colors with boundaries below 1 keV. Contamination by supernova remnants has also been pointed out in a recent study of M31, when using the color choice of most *Chandra* papers (Orio 2006).

6. ULTRALUMINOUS X-RAY SOURCES

The most widely used observational definition of ULXs is that of sources detected in the X-ray observing band-pass with luminosities of at least 10^{39} erg s^{-1}, implying bolometric luminosities clearly in excess of this limit. ULXs (also named intermediate luminosity X-ray objects—IXOs; Colbert & Ptak 2002) were first detected with *Einstein* (Long & Van Speybroeck 1983; see the review by Fabbiano 1989). These sources were dubbed super-Eddington sources, because their luminosity was significantly in excess of the Eddington limit of a neutron star ($\sim 2 \times 10^{38}$ erg s^{-1}), suggesting accreting objects with masses of 100 $M_{\odot}$ or larger. Because these masses exceed those of stellar black holes in binaries (which extend up to $\sim$30 $M_{\odot}$; Belczynski, Sadowski & Rasio 2003), ULXs could then be a new class of astrophysical objects, possibly unconnected with the evolution of the normal stellar population of a galaxy. They could represent the missing link in the black-hole mass distribution, bridging the gap between stellar black holes and the supermassive black holes found in the nuclei of early-type galaxies. These "missing" black holes have been called intermediate mass black holes (IMBH), and could be the remnants of the collapse of primordial stars in the early universe (Madau & Rees 2001; Volonteri, Haardt & Madan 2003; Volonteri & Perna 2005; see the review by Bromm & Larson 2004), or they could be forming in the core collapse of young dense stellar clusters (e.g., Miller & Hamilton 2002).

Conversely, ULXs could represent a particularly high-accretion stage of X-ray binaries, possibly with a stellar black-hole accretor (King et al. 2001), or even be powered by relativistic jets, as in Doppler-beamed microquasars (Koerding, Falke & Markoff 2002). Other models have also been advanced (very young SNRs, e.g., Fabian & Terlevich 1996; young Crab-like pulsars, Perna & Stella 2004), but cannot explain the bulk of the ULXs, given their spectral and time variability characteristics that point to accretion systems (see below).

Given these exciting and diverse possibilities it is not surprising that ULXs have generated a large amount of both observational and theoretical work. An in-depth discussion of all this work is beyond the scope of the present review. Among the recent reviews on ULXs presenting different points of view are those of Fabbiano (2004), Miller & Colbert (2004), Mushotzky (2004), and Fabbiano & White (2006). Two recent short articles in *Nature* and *Science* (McCrady 2004 and Fabbiano 2005) are also useful examples of different perspectives on this subject: McCrady argues for the IMBH interpretation of ULXs, whereas Fabbiano instead concludes that although a few very luminous ULXs are strong candidates for IMBHs, the majority may be just sources at the upper luminosity end of the normal XRB population.

Here, I will discuss the main points of the current debate on ULXs, as they pertain to the discourse on X-ray populations, quoting only recent and representative work. In Section 6.1 I discuss the association of ULXs with star formation, in Section 6.2 their spectral and time variability, suggesting the presence of accreting binaries, and in Section 6.3 identification with optical and radio objects. In Section 6.4 I give a summary of the current theoretical debate on the nature of ULXs.

6.1. Association of ULXs with Active Star-Forming Stellar Populations

From a population point of view it is useful to see where we find ULXs. The heightened recent interest in ULXs has spurred a number of studies that have sought to take a systematic view of these sources. These include both works using the *Chandra* data archive and those revisiting the *ROSAT* data and the literature. From a minisurvey of 13 galaxies observed with *Chandra*, including both ellipticals and spirals, Humphrey et al. (2003) suggested a star-formation connection on the basis of a strong correlation of the number of ULXs per galaxy with the 60-μm emission and a lack of correlation with galaxy mass. Swartz et al. (2004) published spectra, variability, and positions for 154 ULXs in 82 galaxies from the *Chandra* ACIS archive, confirming their association with young stellar populations, especially those of merging and colliding galaxies. This conclusion is in agreement with that of Grimm, Gilfanov & Sunyaev (2003), based on a comparison of XLFs of star-forming galaxies (see Section 4.2). The strong connection of ULXs with star formation is also demonstrated by the analysis of a catalog of 106 ULXs derived from the *ROSAT* HRI observations of 313 galaxies (Liu & Bregman 2005). Liu & Mirabel (2005) instead compile a catalog of 229 ULXs from the literature, together with optical, IR, and radio counterparts, when available; they observe that the most luminous ULXs (those with $L_X > 10^{40}$ erg s^{-1}), which are the most promising candidates for IMBHs, can be found in either intensely star-forming galaxies or in the halo of ellipticals (the latter, however, are likely to be background QSOs, see below). The association of ULXs with high SFR galaxies is exemplified by the discovery of 14 of these sources in the Antennae galaxies, the prototype galaxy merger (**Figure 10**).

As discussed in Section 3.3, the XLFs of E and S0 galaxies are rather steep, i.e., the number of very luminous sources in these LMXB populations is relatively small, especially in comparison with star-forming galaxies; however, sources with luminosities in excess of 10^{39} erg s^{-1} exist (see an earlier discussion of this topic in Fabbiano & White 2005; see also Section 3.7).

Several authors have considered the statistical association of ULXs with early-type galaxies (E and S0s, old stellar populations). Swartz et al. (2004) find that the number of ULXs in early-type galaxies scales with galaxy mass and can be explained with the high luminosity end of the XLF (see Gilfanov 2004 and discussion in Section 3.3). They also point out that ULX detections in early-type galaxies are significantly contaminated by background AGNs, in agreement with the statistical works of Ptak & Colbert (2004) and Colbert & Ptak (2002), which are based on *ROSAT* HRI (5″ resolution) observations of galaxies. Irwin, Bregman & Athey (2004) find that sources in the 1×10^{39} erg s^{-1} –2×10^{39} erg s^{-1} luminosity range are likely to belong to the associated galaxies and have spectra consistent with those of Galactic black-hole binaries (Irwin, Athey & Bregman 2003). The sample of sources more luminous than 2×10^{39} erg s^{-1} (if placed at the distance of the associated galaxy) is

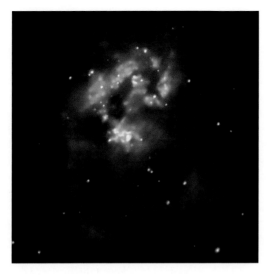

Figure 10

Left: *Chandra* ACIS image of the Antennae from two years of monitoring (4.8′ side box; from the web page http://chandra.harvard.edu/photo/category/galaxies.html; credit NASA/CXC; Fabbiano et al. 2004a). The regions of most intense emission, where most of the X-ray sources are clustered, correspond to regions of intense star formation. Note that the region of most intense star formation is obscured in X ray (*blue*). Right: *Spitzer* IR (3.6–8-μm in *red*) and optical composite image from the web page http://www.spitzer.caltech.edu/Media/releases/ssc2004-14/ssc2004-14a.shtml (Credit: NASA/JPL-Caltech/Z. Wang).

instead consistent with the expected number and spatial distribution of background AGNs.

This growing body of results demonstrates that ULXs are associated with the star-forming population. The presence of ULXs in early-type galaxies has been debated, but there is no strong statistical evidence for the existence of a population of sources with $L_X > 2 \times 10^{39}$ erg s^{-1} in these galaxies. In the following I will only discuss ULXs in star-forming galaxies.

6.2. Spectra and Time Variability from *Chandra* and *XMM-Newton*

Chandra and *XMM-Newton* work has confirmed that ULXs are compact accreting sources, building on the more limited observations of nearby ULXs with *ASCA* (Makishima et al. 2000, Kubota et al. 2001). Flux-color transitions have been observed in a number of ULXs, suggesting the presence of an accretion disk (in the Antennae—Fabbiano, Zezas & Murray 2001; Fabbiano et al. 2003a,b, 2004a; Zezas et al. 2006; M101—Jenkins et al. 2004; NGC7714—Soria & Motch 2004; M33—LaParola et al. 2003; Dubus, Charles & Long 2004; Foschini et al. 2004; Ho II X-1—Dewangan et al. 2004; and a sample of 5 ULXs in different galaxies monitored with *Chandra*—Roberts et al. 2004). Some of these spectra and colors are consistent with or reminiscent of those of black-hole binaries (see above references and Colbert et al. 2004, Liu et al. 2005). A recent spectral survey with *XMM-Newton* finds different spectral types, suggesting either spectral variability or a complex source population (Feng & Kaaret 2005).

Shorter-term time variability is also consistent with the presence of X-ray binaries and accretion disks. In particular a ULX in NGC253 has recently been shown to be a recurrent transient (Bauer & Pietsch 2005). Moreover, features in the power density spectra have been used to constrain the mass of the accreting black hole (Strohmayer & Mushotzky 2003; Soria et al. 2004). In the very luminous M82 ULX ($L_x > 10^{40}$ erg s^{-1}, $L_{bol} \sim 10^{41}$ erg s^{-1}), which is the most compelling IMBH candidate, Strohmayer & Mushotzky (2003) detect a 55mHz QPO, also confirmed by Fiorito & Titarchuck (2004).

The most statistically significant spectra are those obtained with *XMM-Newton* in nearby very bright ULXs where confusion with unresolved emission in the detection area is not severe. In several cases, a composite power-law plus very soft accretion disk is the simplest model that optimizes the fit to the observed spectra (there are exceptions, e.g., for the 10^{41} erg s^{-1} ULX in NGC2276, where a multicolored disk model is preferred, Davis & Mushotzky 2004). A very soft component was first reported by Kaaret et al. (2003) for the ULX NGC5408 X-1, and soon after by Miller et al. (2003) for the ULX NGC1313 X-2, with temperatures of $\sim$110 and 150 eV, respectively. These soft components would be consistent with the emission of an accretion disk surrounding an IMBH of nearly 1000 $M_\odot$ (but see a more recent estimate of 100 $M_\odot$ for NGC1313 X-2, Zampieri et al. 2004). Similar soft components were found in other ULXs (Miller, Fabian & Miller 2004a; Miller et al. 2004; Jenkins et al. 2005; Roberts et al. 2005).

Unfortunately, these results are not the smoking gun that one may have hoped for to conclusively demonstrate the presence of IMBHs in ULXs. Two other models have been proposed that fit the data equally well, but are consistent with normal stellar black-hole masses. One is the slim disk model (e.g., Watarai et al. 2005, Ebisawa et al. 2004, advanced to explain the emission of an accretion disk in a high accretion mode; see Foschini et al. 2005, Roberts et al. 2005). The second model is a physical Comptonized disk model (Kubota, Makishima & Done 2004). Although both models are significantly more complex than the power-law + soft-component model, nature can easily be wicked, and the models are physically motivated. The controversy is raging, given the tantalizing possibility of proving the discovery of IMBHs (see Fabian, Ross & Miller 2004; Miller, Fabian & Miller 2004b; Wang et al. 2004; Goad et al. 2006).

The recurrent variable very soft ULX in M101 provides an excellent case study to illustrate the difficulty of reaching a firm conclusion on the presence of an IMBH. Given their very soft spectra, SSSs and QSSs in the ULX luminosity range are IMBH candidates (in Sombrero—Di Stefano et al. 2003; M101—Mukai et al. 2003, 2005; Kong, Di Stefano & Yuan 2004; the Antennae—Fabbiano et al. 2003b). These sources are too luminous to be explained in terms of hot white dwarfs, unless the emission is beamed, which is unlikely (e.g., Fabbiano et al. 2003b).

The expanding black-hole photosphere of a stellar black hole was first suggested to explain the M101 very soft ULX (Mukai et al. 2003), but the subsequent detection of a hard power-law component and low/hard–high/soft spectral variability pointed to a Comptonized accretion disk in a black-hole binary (Kong, Di Stefano & Yuan 2004; Mukai et al. 2005); but what kind of black hole?

Based on the *XMM-Newton* spectrum, which can be fitted with an absorbed blackbody, and implies outburst luminosities in the 10^{41} erg s^{-1} range, Kong, Di Stefano & Yuan (2004) advanced the IMBH candidacy. Mukai et al. (2005), instead, argue for a 20–40 M$_\odot$ stellar black-hole counterpart. Their main point is that the high L$_X$ derived in the previous study results from the adoption of an emission model with a considerable amount of line-of-sight absorption; the colors of the optical counterpart, instead, are consistent with very little absorption; moreover, if the obscuring material were close to the black hole it would be most likely ionized (warm absorber). Adopting an accretion disk plus emission line model, Mukai et al. (2005) obtain luminosities in the 10^{39} erg s^{-1} range. They also use the variability power density spectrum of the source to constrain the emission state, and with the luminosity, the mass of the black hole. This is another example where a considerable amount of ambiguity exists in the choice of the X-ray spectral model, and X-ray spectra alone may not give the conclusive answer. The luminous optical counterpart makes this source an obvious candidate for future studies aimed at obtaining the mass function of the system.

6.3. Counterparts at Other Wavelengths

As shown by the example at the end of Section 6.2, identification of ULXs may be crucial for understanding their nature. Three main classes of counterparts have been discussed in the literature: stellar counterparts, ionized or molecular nebulae, and radio sources. Stellar counterparts tend to have very blue colors, suggesting early-type stars, although the colors could also arise from the optical emission of the accretion disk (Kaaret, Ward & Zezas 2004; Liu, Bregman & Seitzer 2004; Soria et al. 2005; Zampieri et al. 2004; Rappaport, Podsiadlowski & Pfahl 2005; see also Fabbiano & White 2006 for earlier references). These counterparts would point to the high accretion rate model of ULXs, if they were indeed early-type stars (e.g., King et al. 2001; Rappaport, Podsiadlowski & Pfahl 2005). However, even ignoring the uncertainty on the nature of the optical counterpart, these results cannot firmly constrain the nature of the compact object. A recent paper by Copperwheat et al. (2005) proposes a model, including irradiation by X-rays of both the accretion disk and the companion star, which when supplemented by variability data and IR photometry, could be used to constrain both the nature of the companion star and the mass of the accreting black hole.

Nebular counterparts suggest isotropic emission in some cases, and therefore a truly large L$_X$, thus arguing against a substantial amount of beaming and pointing to fairly massive black holes (Roberts et al. 2003; Pakull & Mirioni 2003; Kaaret, Ward & Zezas 2004); radio counterparts have alternately been found consistent with either beamed sources or IMBHs (Kaaret et al. 2003; Neff, Ulvestad & Campion 2003; Miller, Mushotzy & Neff 2005; Koerding, Colbert & Falke 2005). Optical variability studies of the stellar counterparts are needed to firmly measure the mass of the system. The new generation of large-area, high-resolution, optical telescopes are likely to solve the nature of these ULXs.

In more distant systems, like the Antennae (D ~ 19 Mpc—Zezas et al. 2002a,b, 2005) or the Cartwheel galaxies (D ~ 122 Mpc—Gao et al. 2003, King 2004, Wolter & Trinchieri 2004), where spectacular populations of ULXs are detected, individual stellar counterparts cannot be detected. However, comparison with the optical emission field also provides very interesting results. In the Antennae, ULXs tend not to coincide with young star clusters, suggesting that either the system has been subject to a supernova formation kick to eject it from its birthplace (thus implying a normal HMXB with a stellar mass black hole; Zezas et al. 2002b; Sepinsky, Kalogera & Belczynski 2005), or that the parent cluster has evaporated, in the core collapse model of IMBH formation (e.g., Zwart et al. 2004). However, a recent paper suggests that some of these displacements may be reduced with better astrometric corrections (Clark et al. 2005). In the Cartwheel, the ULXs are associated with the most recent expanding star-formation ring, setting strong constraints to the IMBH hypothesis and favoring the high accretion HMXB scenario (King 2004). It must be said, however, that given the distance of this galaxy, and the lack of time monitoring, it cannot be excluded that the ULXs may represent clumps of unresolved sources.

Higher red-shift galaxy and QSO counterparts to ULXs have also been found in some cases (Masetti et al. 2003; Arp, Gutierrez & Lopez-Corredoira 2004; Gutierrez & Lopez-Correidora 2005; Burbidge et al. 2004; Galianni et al. 2005; Clark et al. 2005; see also H. Arp & E.M. Burbidge, submitted; Ghosh et al. 2005). Although at this point these identifications are still few and consistent (within small number statistics) with chance coincidences with background AGN, some of the above authors (H. Arp, E.M. Burbidge, G. Burbidge and collaborators) have raised the hypothesis of a physical connection between the QSO and the parent galaxy; clearly this possibility cannot be extended to the entire body of ULXs, given the results of other identification campaigns (see above).

6.4. Models of ULX Formation and Evolution, and Paths for Future Work

I will summarize here some of the more recent theoretical work on ULXs, and refer the reader to the reviews cited earlier for details on earlier work. As I've already noted, the two principal lines of thought are: (*a*) most ULXs are IMBHs; (*b*) most ULXs are luminous X-ray sources of "normal" stellar origin and the IMBH explanation should be sought only for the ULXs with $L_{bol} > 10^{41}$ erg s^{-1} (the M82 ULX—Kaaret et al. 2001, Matsumoto et al. 2001; the NGC2276 ULX—Davis & Mushotzky 2004; the most luminous ULX in the Cartwheel galaxy—Gao et al. 2003, Wolter & Trinchieri 2004; and the variable ULX in NGC7714—Soria & Motch 2004; Smith, Struck & Nowak 2005).

The stellar evolution camp was originally stimulated by the abundance of ULXs in star-forming galaxies (King et al. 2001; King 2004) and by the apparently universal shape of the XLF of the star-forming population (Grimm, Gilfanov & Sunyaev 2003; see Section 4.2). The variability and spectra of these systems (see Section 6.2) point to accretion binaries. In this paradigm, the problem is to explain the observed luminosities. Both relativistic (Koerding, Falke & Markoff 2002) and nonrelativistic beaming (King et al. 2001), and super-Eddington accretion disks (Begelman 2002; both spectra and observed variability patterns can be explained, M. Belgelman, private communication) have been suggested as a way to explain the source luminosities inferred from the observations. With the exception of relativistic beaming, these mechanisms can account for a factor of 10 enhancement of the luminosity above the Eddington value. If black-hole masses of a few tens solar masses exist (Belczynski, Sadowski & Rasio 2004), most or all the ULXs could be explained this way. For example, Rappaport, Podsiadlowski & Pfahl (2005) have combined binary evolution models and binary population synthesis, finding that for donors with $M \geq 10\, M_\odot$, accretion binaries can explain the ULXs, with modest violation of the Eddington limit.

The IMBH camp has generated a larger volume of papers. IMBHs may be remnants of collapse in the early universe (e.g., Heger & Woosley 2002; Islam, Taylor & Silk 2004; Van der Marel 2004), or may result from the collapse of dense stellar clusters (e.g., Gurkan, Freitag & Rasio 2004; Zwart et al. 2004.). In the cosmological remnant options, one would expect IMBHs to be particularly abundant in the more massive elliptical galaxies, contrary to the observed association with star-forming galaxies (Zezas & Fabbiano 2002). However, IMBHs would not be visible unless they are fueled, and fuel is more readily available in star-forming galaxies, in the form of dense molecular clouds (Schneider et al. 2002; Krolik 2004). Accretion from a binary companion is an efficient way of fueling an IMBH, and consequently a number of papers have explored the formation of such binaries via tidal capture in GCs. In this picture, the ULX may not be still associated with the parent cluster because of cluster evaporation (Hopman, Zwart & Alexander 2004; Li 2004; Zwart, Dewi & Maccarone 2004; Zwart et al. 2004). A twist to the cosmological hypothesis is given by the suggestion that the very luminous ULXs, with $L_{bol} > 10^{41}$ erg s^{-1} such as the M82 ULX, may be the nuclei of satellite galaxies, switching on in the presence of abundant fuel (King & Dehnen 2005).

Some of this work has resulted in predictions that can be directly compared with the data, and complement the tests based on the study of the optical and multiwavelength counterparts discussed in Section 6.3. In particular, the slope and normalization of the high-luminosity XLFs of star-forming galaxies have been reproduced in both IMBH (Islam, Taylor & Silk 2004; Krolik 2004) and jet models (Koerding, Colbert & Falke 2004). Zezas & Fabbiano (2002) discuss the effect of either a beamed population of ULXs or a population of IMBHs in the context of the XLF of the Antennae. Gilfanov, Grimm & Sunyaev (2004b) predict a change of slope in the L_X-SFR relation, where L_X is the total X-ray luminosity of a galaxy, if a new population of IMBHs is present at the higher luminosities (see Section 7).

Other properties have also been investigated, including the X-ray spectral distribution (Section 6.2), the presence of radio emission from IMBHs and the comparison of radio and X-ray properties with those of AGN and stellar black-hole Galactic binaries (e.g., Merloni, Heinz & Di Matteo 2003), and time variability-based tests. The latter include studying the QPO frequency (which may be a function of backhole mass, Abramowicz et al. 2004), the observation of long-term transient behavior (expected from IMBH binaries, whereas thermal-timescale mass transfer onto stellar black holes would produce stable disks; Kalogera et al. 2004) and the detection of eclipses (expected more frequently in stellar black-hole binaries than in IMBHs, Pooley & Rappaport 2005). These time variability tests require long-term monitoring of ULXs and future larger X-ray telescopes.

7. X-RAY EMISSION AND GALAXY EVOLUTION

Chandra observations of galaxies at high redshift ($z > 0.1$), either from identification of deep survey sources or from stacking analysis of distant galaxy fields, have been reviewed recently in the literature (see Fabbiano & White 2006; Brandt & Hasinger 2005) and will not be discussed in detail here. In summary, the emission from normal galaxies becomes an increasingly greater component of the X-ray emission at the deepest X-ray counts (Bauer et al. 2004; Ranalli, Comastri & Setti 2005); moreover, the hard X-ray emission is a direct diagnostic of star formation, as demonstrated by the good FIR-X-ray correlations and by the work on the XLFs of star-forming galactic populations discussed earlier in this review (Fabbiano & Shapley 2002; Grimm, Gilfanov & Sunyaev 2003; Ranalli, Comastri & Setti 2003; Colbert et al. 2004; Gilfanov, Grimm & Sunyaev 2004a; Persic et al. 2004). It is clear that the study of the global properties and luminosity functions of galaxies at different redshifts can give information in this area, and this work is beginning to gather momentum (e.g., Georgakakis et al. 2003; Norman et al. 2004; Hornschemeier et al. 2005; Ranalli, Comastri & Setti 2005), given the availability of *XMM-Newton* surveys of the nearby universe and the increasingly deep *Chandra* surveys.

Enhanced star formation early in the life of a galaxy is expected to produce enhancements in its X-ray emission at different epochs, related to the formation and evolution of HMXB and LMXB populations (Ghosh & White 2001). Lehmer et al. (2005) report such an effect in their stacking analysis of Lyman break galaxies in the *HST GOODS* fields covered by deep (1–2 Ms exposure) *Chandra* fields (**Figure 11**). Conversely, if the SFR is independently known, the relation between the integrated luminosity of galaxies and the SFR can be used to measure the maximum luminosity of a HMXB and the presence of a very high luminosity IMBH population not related to stellar sources (Gilfanov, Grimm & Sunyaev 2004b). These authors, based on the XLF-SFR connection (Grimm, Gilfanov & Sunyaev 2003), explore the statistical properties of a population of discrete sources and demonstrate that a break is expected in the relation between the total X-ray luminosity L_X of the galaxies and the SFR, which depends on the high luminosity cut-off of the XRB population. Comparing the local galaxy sample with the *Hubble* Field North galaxies, they suggest a cut-off luminosity $\sim 5 \times 10^{40}$ erg s^{-1} for HMXBs. They also suggest that a population of very luminous IMBHs ($L_X > 10^{40}$ erg s^{-1}) would reveal itself with a steeper L_X-SFR relation at higher luminosities and star-formation regimes.

The *Chandra* observations of nearby galaxies have significantly increased our understanding of X-ray source populations, and will deepen this understanding in the next several years. These results and the tantalizing possibility of studying the X-ray evolution of galaxies show that future very deep, high-resolution X-ray observations

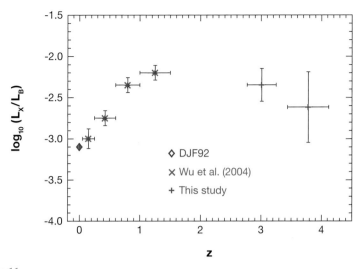

Figure 11

Evolution of the X-ray-to-optical ratio of galaxies with redshift, peaking at z ~ 1.5–3 (from figure 5 of Lehmer et al. 2005).

are essential. Unfortunately, there are no plans for X-ray missions with comparable angular resolution to follow *Chandra*. These types of studies may have to wait for several decades for a large area, subarcsecond resolution X-ray mission (e.g., the *Generation-X* mission, for which a NASA-funded vision study is in progress).

ACKNOWLEDGMENTS

This review has benefited from comments and discussions with several colleagues, including Dong-Woo Kim, Martin Elvis, Andrew King, Mitch Begelman, Andreas Zezas, Albert Kong, Sergey Trudolyubov, Konstantin Postnov, Massimo Persic, Tom Maccarone, Curt Struck, Tim Roberts, Phil Kaaret, Wolfgang Pietsch, Xiang-Dong Li, and John Kormendy. I thank A. Tennant, M. Gilfanov, R. Soria, E. Kim, Z. Wang, A. Prestwhich and B. Lehmer for providing figures. I am indebted to the NASA ADS for making my job of searching the literature so much easier.

LITERATURE CITED

Abramowicz MA, Kluzniak W, McClintock JE, Remillard RA. 2004. *Ap. J.* 609:L63
Angelini L, Loewenstein M, Mushotzky RF. 2001. *Ap. J.* 557: L35
Arimoto N, Matsushita K, Ishimaru Y, Ohashi T, Renzini A. 1997. *Ap. J.* 477:128
Arp H, Gutierrez CM, Lopez-Corredoira M. 2004. *Astron. Astrophys.* 418:877
Baldi A, Raymond JC, Fabbiano G, Zezas A, Roths AH, et al. 2006a. *Ap. J. Suppl.* 162:113
Baldi A, Raymond JC, Fabbiano G, Zezas A, Roths AH, et al. 2006b. *Ap. J.* 636:158

Bauer FE, Alexander DM, Brandt WN, Schneider DP, Treister E, et al. 2004. *Astron. J.* 128:2048

Bauer FE, Brandt WN, Lehmer B. 2003. *Astron. J.* 126:2797

Bauer M, Pietsch W. 2005. *Astron. Astrophys.* 442:925

Begelman MC. 2002. *Ap. J.* 568:L97

Belczynski K, Kalogera V, Zezas A, Fabbiano G. 2004. *Ap. J.* 601:L147

Belczynski K, Sadowski A, Rasio FA. 2004. *Ap. J.* 611:1068

Bellazzini M, Pasquali A, Federici L, Ferraro FR, Fusi Pecci F. 1995. *Ap. J.* 439:687

Bildsten L, Deloye CJ. 2004. *Ap. J.* 607:L119

Blanton EL, Sarazin CL, Irwin JA. 2001. *Ap. J.* 552:106

Brandt WN, Hasinger G. 2005. *Annu. Rev. Astron. Astrophys.* 43:827

Bromm B, Larson RB. 2004. *Annu. Rev. Astron. Astrophys.* 42:79

Burbidge ME, Burbidge G, Arp HC, Zibetti S. 2004. *Ap. J. Suppl.* 153:159

Clark DM, Christopher MH, Eikenberry SS, Brandl BR, Wilson JC, et al. 2005. *Ap. J.* 631:L109

Clark GW. 1975. *Ap. J.* 199:L143

Colbert EJM, Heckman TM, Ptak AF, Strickland DK, Weaver KA. 2004. *Ap. J.* 602:231

Colbert EJM, Ptak AF. 2002. *Ap. J. Suppl.* 143:25

Copperwheat C, Cropper M, Soria R, Wu K. 2005. *MNRAS* 362:79

David LP, Forman W, Jones C. 1991. *Ap. J.* 369:121

David LP, Jones C, Forman W, Murray SS. 2005. *Ap. J.* 635:1053

Davis SD, Mushotzky RF. 2004. *Ap. J.* 604:653

Dewangan GC, Miyaji T, Griffiths RE, Lehmann I. 2004. *Ap. J.* 608:L57

Di Stefano R, Kong AKH. 2003. *Ap. J.* 592:884

Di Stefano R, Kong AKH. 2004. *Ap. J.* 609:710

Di Stefano R, Kong AKH, Greiner J, Primini FA, Garcia MR, et al. 2004. *Ap. J.* 610:247

Di Stefano R, Kong AKH, VanDalfsen AL, Harris WE, Murray SS, Delain KM. 2003. *Ap. J.* 599:1067

Dubus G, Charles PA, Long KS. 2004. *Astron. Astrophys.* 425:95

Ebisawa K, Zycki PT, Kubota A, Mizuno T, Watarai K. 2004. *Prog. Theor. Phys.* 155:67

Eddington AS. 1913. *MNRAS* 73:359

Eracleous M, Shields JC, Chartas G, Moran EC. 2002. *Ap. J.* 565:108

Eskridge PB, Fabbiano G, Kim DW. 1995a. *Ap. J. Suppl.* 97:141

Eskridge PB, Fabbiano G, Kim DW. 1995b. *Ap. J.* 442:523

Fabbiano G. 1988. *Ap. J.* 325:544

Fabbiano G. 1989. *Annu. Rev. Astron. Astrophys.* 27:87

Fabbiano G. 1995. See Lewin et al. 1995, p. 390

Fabbiano G. 2004. *Rev. Mex. Astron. Astrofis. (Ser. Conf.)* 20:46

Fabbiano G. 2005. *Science* 307:533

Fabbiano G, Baldi A, King AR, Ponman TJ, Raymond J, et al. 2004a. *Ap. J.* 605:L21

Fabbiano G, Baldi A, Pellegrini S, Siemiginowska A, Elvis M, et al. 2004b. *Ap. J.* 616:730

Fabbiano G, Gioia IM, Trinchieri G. 1988. *Ap. J.* 324:749

Fabbiano G, Kessler MF. 2001. In *The Century of Space Science*, ed. JAM Bleeker, J Geiss, MCE Huber, I:561. Dordrecht: Kluwer Acad.

Fabbiano G, Kim DW, Trinchieri G. 1992. *Ap. J. Suppl.* 80:531

Fabbiano G, Kim DW, Trinchieri G. 1994. *Ap. J.* 429:94

Fabbiano G, King AR, Zezas A, Ponman TJ, Rots A, Schweizer F. 2003b. *Ap. J.* 591:843

Fabbiano G, Shapley A. 2002. *Ap. J.* 565:908

Fabbiano G, Trinchieri G. 1985. *Ap. J.* 296:430

Fabbiano G, White NE. 2006. See Lewin & van der Klis 2006, p. 475

Fabbiano G, Zezas A, King AR, Ponman TJ, Rots A, Schweizer F. 2003a. *Ap. J.* 584:L5

Fabbiano G, Zezas A, Murray SS. 2001. *Ap. J.* 554:1035

Fabian AC, Ross RR, Miller JM. 2004. *MNRAS* 355:359

Fabian AC, Terlevich R. 1996. *MNRAS* 280:L5

Fender R, Belloni T. 2004. *Annu. Rev. Astron. Astrophys.* 42:317

Feng H, Kaaret P. 2005. *Ap. J.* 633:1052

Finoguenov A, Jones C. 2002. *Ap. J.* 574:754

Fiorito R, Titarchuk L. 2004. *Ap. J.* 614:L113

Forman W, Jones C, Tucker W. 1985. *Ap. J.* 293:102

Foschini L, Ebisawa K, Kawaguchi T, Cappelluti N, Grandi P, et al. 2005. *Advances in Space Research*, Special Issue *Proc. 35th COSPAR, Paris, Fr., 18–25 July 2004.* In press

Foschini L, Rodriguez J, Fuchs Y, Ho LC, Dadina M, et al. 2004. *Astron. Astrophys.* 416:529

Galianni P, Burbidge EM, Arp H, Junkkarinen V, Burbidge G, Zibetti S. 2005. *Ap. J.* 620:88

Gao Y, Wang QD, Appleton PN, Lucas RA. 2003. *Ap. J.* 596:L171

Georgakakis A, Georgantopoulos I, Stewart GC, Shanks T, Boyle BJ. 2003. *MNRAS* 344:161

Ghosh KK, Swartz DA, Tennant AF, Wu K, Saripalli L. 2005. *Ap. J.* 623:815

Ghosh P, White NE. 2001. *Ap. J.* 559:L97

Giacconi R. 1974. In *X-ray Astronomy*, ed. R Giacconi, H Gursky, p. 155. Dordrecht: Reidel

Giacconi R, Branduardi G, Briel U, Epstein A, Fabricant D, et al. 1979. *Ap. J.* 230:540

Gilfanov M. 2004. *MNRAS* 349:146

Gilfanov M, Grimm HJ, Sunyaev R. 2004a. *MNRAS* 347:L57

Gilfanov M, Grimm HJ, Sunyaev R. 2004b. *MNRAS* 351:1365

Giordano L, Cortese L, Trinchieri G, Wolter A, Coppi M, et al. 2005. *Ap. J.* 634:272

Goad MR, Roberts TP, Reeves JN, Uttley P. 2006. *MNRAS* 365:191

Greiner J, Di Stefano R, Kong A, Primini F. 2004. *Ap. J.* 610:261

Grimm HJ, Gilfanov M, Sunyaev R. 2002. *Astron. Astrophys.* 391:923

Grimm HJ, Gilfanov M, Sunyaev R. 2003. *MNRAS* 339:793

Grimm HJ, McDowell J, Zezas A, Kim DW, Fabbiano G. 2005. *Ap. J. Suppl.* 161:271

Grindlay JE. 1984. *Adv. Space Res.* 3:19

Grindlay JE. 1987. In *IAU Symp. 125, Origin and Evolution of Neutron Stars*, ed. D Helfand, J Huang, p. 173. Dordrecht: Reidel
Gurkan MA, Freitag M, Rasio FA. 2004. *Ap. J.* 604:632
Gutierrez CM, Lopez-Corredoira M. 2005. *Ap. J.* 622:L89
Hartwell JM, Stevens IR, Strickland DK, Heckman TM, Summers LK. 2004. *MNRAS* 348:406
Heger A, Woosley SE. 2002. *Ap. J.* 567:532
Helfand D, Moran EC. 2001. *Ap. J.* 554:27
Holt SS, Schlegel EM, Hwang U, Petre R. 2003. *Ap. J.* 588:792
Hopman C, Zwart SFP, Alexander T. 2004. *Ap. J.* 604:L101
Hornschemeier AE, Heckman TM, Ptak AF, Tremonti CA, Colbert EJM. 2005. *Astron. J.* 129:86
Humphrey PJ, Buote DA. 2004. *Ap. J.* 612:848
Humphrey PJ, Fabbiano G, Elvis M, Church MJ, Balucinska-Church M. 2003. *MNRAS* 344:134
Immler S, Wang QD, Douglas CL, Schlegel EM. 2003. *Ap. J.* 595:727
Inui I, Matsumoto H, Tsuru T, Koyama K, Matsushita S, et al. 2005. *Publ. Astron. Soc. Jpn.* 57:135
Irwin JA. 2005. *Ap. J.* 631:511
Irwin JA, Athey AE, Bregman JN. 2003. *Ap. J.* 587:356
Irwin JA, Bregman JN, Athey AE. 2004. *Ap. J.* 601:L143
Islam RR, Taylor JE, Silk J. 2004. *MNRAS* 354:443
Ivanova N, Kalogera V. 2006. *Ap. J.* 636:985
Ivanova N, Rasio FA, Lombardi JC Jr, Dooley KL, Proulx ZF. 2005. *Ap. J.* 621:L109
Iyomoto N, Makishima K, Tashiro M, Inoue S, Kaneda H, et al. 1998. *Ap. J.* 503:L31
Jeltema TE, Canizares CR, Buote DA, Garmire GP. 2003. *Ap. J.* 585:756
Jenkins LP, Roberts TP, Warwick RS, Kilgard RE, Ward MJ. 2004. *MNRAS* 349:404
Jenkins LP, Roberts TP, Warwick RS, Kilgard RE, Ward MJ. 2005. *MNRAS* 357:401
Jordan A, Cote P, Ferrarese L, Blakeslee JP, Mei S, et al. 2004. *Ap. J.* 613:279
Juett AM. 2005. *Ap. J.* 621:L25
Kaaret P. 2001. *Ap. J.* 560:715
Kaaret P. 2002. *Ap. J.* 578:114
Kaaret P, Alonso-Herrero A, Gallagher JS, Fabbiano G, Zezas A, Rieke MJ. 2004. *MNRAS* 348:L28
Kaaret P, Corbel S, Preswich AH, Zezas A. 2003. *Science* 299:365
Kaaret P, Prestwich AH, Zezas A, Murray SS, Kim DW, et al. 2001. *MNRAS* 321:L29
Kaaret P, Ward MJ, Zezas A. 2004. *MNRAS* 351:L83
Kahabka P, van den Heuvel EPJ. 1997. *Annu. Rev. Astron. Astrophys.* 35:69
Kalogera V, Henninger M, Ivanova N, King AR. 2004. *Ap. J.* 603:L41
Kalogera V, King AR, Rasio FA. 2004. *Ap. J.* 601:L171
Kilgard RE, Kaaret P, Krauss MI, Prestwich AH, Raley MT, Zezas A. 2002. *Ap. J.* 573:138
Kim DW, Fabbiano G. 2003. *Ap. J.* 586:826
Kim DW, Fabbiano G. 2004. *Ap. J.* 611:846
Kim DW, Fabbiano G, Trinchieri G. 1992. *Ap. J.* 393:134

Kim E, Kim DW, Fabbiano G, Lee MG, Park HS, et al. 2006. *Ap. J.* In press
King AR. 2004. *MNRAS* 347:L18
King AR, Davies MB, Ward MJ, Fabbiano G, Elvis M. 2001. *Ap. J.* 552:L109
King AR, Dehnen W. 2005. *MNRAS* 357:275
King AR, Ritter H. 1998. *MNRAS* 293:L42
Koerding E, Colbert E, Falke H. 2004. *Prog. Theor. Phys.* 155:365
Koerding E, Colbert E, Falke H. 2005. *Astron. Astrophys.* 436:427
Koerding E, Falke H, Markoff S. 2002. *Astron. Astrophys.* 382:L13
Kong AKH. 2003. *MNRAS* 346:265
Kong AKH, Di Stefano R. 2003. *Ap. J.* 590:L13
Kong AKH, Di Stefano R, Garcia MR, Greiner J. 2003. *Ap. J.* 585:298
Kong AKH, Di Stefano R, Yuan F. 2004. *Ap. J.* 617:L49
Kong AKH, Garcia MR, Primini FA, Murray SS, Di Stefano R, McClintock JE. 2002. *Ap. J.* 577:738
Kraft RP, Kregenow JM, Forman WR, Jones C, Murray SS. 2001. *Ap. J.* 560:675
Kraft RP, Nolan LA, Ponman TJ, Jones C, Raychaudhury S. 2005. *Ap. J.* 625:785
Krolik JH. 2004. *Ap. J.* 615:383
Kubota A, Makishima K, Done C. 2004. *Prog. Theor. Phys. Suppl.* 155:19
Kubota A, Mizuno T, Makishima K, Fukazawa Y, Kotoku J. 2001. *Ap. J.* 547:L119
Kundu A, Maccarone TJ, Zepf SE. 2002. *Ap. J.* 574:L5
Kundu A, Maccarone TJ, Zepf SE, Puzia TH. 2003. *Ap. J.* 589:L81
La Parola V, Damiani F, Fabbiano G, Peres G. 2003. *Ap. J.* 583:758
Lehmer BD, Brandt WN, Alexander DM, Bauer FE, Conselice CJ, et al. 2005. *Astron. J.* 129:1
Lewin WHG, van der Klis M, eds. 2006. *Compact Stellar X-ray Sources in Normal Galaxies.* Cambridge, UK: Cambridge Univ. Press
Lewin WHG, van Paradijs J, van den Heuvel EPJ, eds. 1995. *X-ray Binaries.* Cambridge, UK: Cambridge Univ. Press
Li XD. 2004. *Ap. J.* 616:L119
Liu JF, Bregman JN. 2005. *Ap. J. S.* 157:59
Liu JF, Bregman JN, Lloyd-Davies E, Irwin J, Espaillat C, Seitzer P. 2005. *Ap. J.* 621:L17
Liu JF, Bregman JN, Seitzer P. 2004. *Ap. J.* 602:249
Liu QZ, Mirabel IF. 2005. *Astron. Astrophys.* 429:1125
Loewenstein M, Angelini L, Mushotzky RF. 2005. *Chin. J. Astron. Astrophys.* 5(Suppl.):159
Long KS, Van Speybroeck LP. 1983. In *Accretion Driven X-ray Sources*, ed. W Lewin, EPJ van den Heuvel, p. 117. Cambridge, UK: Cambridge Univ. Press
Lou YQ, Bian FY. 2005. *MNRAS* 358:1231
Maccarone TJ. 2005. *MNRAS* 364:971
Maccarone TJ, Kundu A, Zepf SE. 2003. *Ap. J.* 586:814
Maccarone TJ, Kundu A, Zepf SE. 2004. *Ap. J.* 606:430
Maccarone TJ, Kundu A, Zepf SE, Piro AL, Bildsten L. 2005. *MNRAS* 364:L61
Madau P, Rees MJ. 2001. *Ap. J.* 551:L27
Makishima K, Kubota A, Mizuno T, Ohnishi T, Tashiro M, et al. 2000. *Ap. J.* 535:632

Martin CL, Kobulnicky HA, Heckman TM. 2002. *Ap. J.* 574:663

Masetti N, Foschini L, Ho LC, Dadina M, Di Cocco G, et al. 2003. *Astron. Astrophys.* 406:L27

Matsumoto H, Tsuru TG, Koyama K, Awaki H, Canizares CR, et al. 2001. *Ap. J.* 547:L25

Matsushita K, Makishima K, Awaki H, Canizares CR, Fabian AC, et al. 1994. *Ap. J.* 436:L41

McCrady N. 2004. *Nature* 428:704

Merloni A, Heinz S, Di Matteo T. 2003. *MNRAS* 345:1057

Miller JM, Fabbiano G, Miller MC, Fabian AC. 2003. *Ap. J.* 585:L37

Miller JM, Fabian AC, Miller MC. 2004a. *Ap. J.* 607:931

Miller JM, Fabian AC, Miller MC. 2004b. *Ap. J.* 614:L117

Miller JM, Zezas A, Fabbiano G, Schweizer F. 2004. *Ap. J.* 609:728

Miller MC, Colbert EJM. 2004. *Int. J. Mod. Phys. D* 13:1

Miller MC, Hamilton DP. 2002. *MNRAS* 330:232

Miller N, Mushotzky RF, Neff SG. 2005. *Ap. J.* 623:L109

Minniti D, Rejkuba M, Funes JG, Akiyama S. 2004. *Ap. J.* 600:716

Miyawaki R, Sugiko M, Kokubun M, Makishima K. 2004. *Publ. Astron. Soc. Jpn.* 56:591

Mukai K, Pence WD, Snowden SL, Kuntz KD. 2003. *Ap. J.* 582:184

Mukai K, Still M, Corbet RHD, Kuntz KD, Barnard R. 2005. *Ap. J.* 634:1085

Mushotzky R. 2004. *Prog. Theor. Phys. Suppl.* 155:27

Neff SG, Ulvestad JS, Campion SD. 2003. *Ap. J.* 599:1043

Norman C, Ptak A, Hornschmeier A, Hasinger G, Bergeron J, et al. 2004. *Ap. J.* 607:721

Orio M. 2006. *Ap. J.* In press

Pakull MW, Mirioni L. 2003. *Rev. Mex. Astron. Astrofis. (Ser. Conf.)* 15:197

Pellegrini S. 2005. *Ap. J.* 624:155

Pence WD, Snowden SL, Mukai K, Kuntz KD. 2001. *Ap. J.* 561:189

Perna R, Stella L. 2004. *Ap. J.* 615:222

Persic M, Rephaeli Y, Braito V, Cappi M, Della Ceca R, et al. 2004. *Astron. Astrophys.* 419:849

Pfahl E, Rappaport S, Podsiadlowski P. 2003. *Ap. J.* 597:1036

Pietsch W, Fliri J, Freyberg MJ, Greiner J, Haberl F, et al. 2005. *Astron. Astrophys.* 442:879

Pietsch W, Freyberg M, Haberl F. 2005. *Astron. Astrophys.* 434:483

Pietsch W, Haberl F. 2005. *Astron. Astrophys.* 430:L45

Pietsch W, Misanovic Z, Haberl F, Hatzidimitriou D, Ehle M, Trinchieri G. 2004. *Astron. Astrophys.* 426:11

Piro AL, Bildsten L. 2002. *Ap. J.* 571:L103

Pooley D, Rappaport S. 2005. *Ap. J.* 634:L85

Postnov KA. 2003. *Astron. Lett.* 29:372

Postnov KA, Kuranov AG. 2005. *Astron. Lett.* 31:7

Prestwich AH, Irwin JA, Kilgard RE, Krauss MI, Zezas A, et al. 2003. *Ap. J.* 595:719

Ptak A, Colbert E. 2004. *Ap. J.* 606:291

Ranalli P, Comastri A, Setti G. 2003. *Astron. Astrophys.* 399:39
Ranalli P, Comastri A, Setti G. 2005. *Astron. Astrophys.* 440:23
Randall SW, Sarazin CL, Irwin JA. 2004. *Ap. J.* 600:729
Rappaport SA, Podsiadlowski Ph, Pfahl E. 2005. *MNRAS* 356:401
Read AM, Ponman TJ. 2001. *MNRAS* 328:127
Roberts TP, Goad MR, Ward MJ, Warwick RS. 2003. *MNRAS* 342:709
Roberts TP, Warwick RS, Ward MJ, Goad MR. 2004. *MNRAS* 349:1193
Roberts TP, Warwick RS, Ward MJ, Goad MR, Jenkins LP. 2005. *MNRAS* 357:1363
Sarazin CL, Irwin JA, Bregman JN. 2000. *Ap. J.* 544:L101
Sarazin CL, Irwin JA, Bregman JN. 2001. *Ap. J.* 556:533
Sarazin CL, Kundu A, Irwin JA, Sivakoff GR, Blanton EL, Randall SW. 2003. *Ap. J.* 595:743
Schlegel EM, Pannuti TG. 2003. *Astron. J.* 125:3025
Schneider R, Ferrara A, Natarajan P, Omukai K. 2002. *Ap. J.* 571:30
Sepinsky J, Kalogera V, Belczynski K. 2005. *Ap. J.* 621:L37
Shapley A, Fabbiano G, Eskridge PB. 2001. *Ap. J. Suppl.* 137:139
Shirey R, Soria R, Borozdin K, Osborne JP, Tiengo A, et al. 2001. *Astron. Astrophys.* 365:L195
Shtykovskiy P, Gilfanov M. 2005. *Astron. Astrophys.* 431:597
Sivakoff GR, Sarazin CL, Carlin JL. 2004. *Ap. J.* 617:262
Sivakoff GR, Sarazin CL, Irwin JA. 2003. *Ap. J.* 599:218
Sivakoff GR, Sarazin CL, Jordan A. 2005. *Ap. J.* 624:L17
Smith BJ, Struck C, Nowak MA. 2005. *Astron. J.* 129:1350
Smith DA, Wilson AS. 2003. *Ap. J.* 591:138
Soria R, Cropper M, Pakull M, Mushotzky R, Wu K. 2005. *MNRAS* 356:12
Soria R, Fabbiano G, Graham AW, Baldi A, Elvis M, et al. 2006. *Ap. J.* 640:126
Soria R, Kong A. 2002. *Ap. J.* 572:L33
Soria R, Motch C. 2004. *Astron. Astrophys.* 422:915
Soria R, Motch C, Read AM, Stevens IR. 2004. *Astron. Astrophys.* 423:955
Soria R, Wu K. 2002. *Astron. Astrophys.* 384:99
Soria R, Wu K. 2003. *Astron. Astrophys.* 410:53
Strohmayer TE, Mushotzky RE. 2003. *Ap. J.* 586:L61
Summers LK, Stevens IR, Strickland DK, Heckman TM. 2003. *MNRAS* 342:690
Summers LK, Stevens IR, Strickland DK, Heckman TM. 2004. *MNRAS* 351:1
Swartz DA, Ghosh KK, McCollough ML, Pannuti TG, Tennant AF, Wu K. 2003. *Ap. J. Suppl.* 144:213
Swartz DA, Ghosh KK, Suleimanov V, Tennant AF, Wu K. 2002. *Ap. J.* 574:382
Swartz DA, Ghosh KK, Tennant AF, Wu K. 2004. *Ap. J. Suppl.* 154:519
Tanaka Y, Inoue H, Holt SS. 1994. *Publ. Astron. Soc. Jpn.* 46:L37
Tennant AF, Wu K, Ghosh KK, Kolodziejczak JJ, Swartz DA. 2001. *Ap. J.* 549:L43
Terashima Y, Wilson AS. 2004. *Ap. J.* 601:735
Trinchieri G, Fabbiano G. 1985. *Ap. J.* 296:447
Trinchieri G, Fabbiano G, Canizares CR. 1986. *Ap. J.* 310:637
Trinchieri G, Fabbiano G, Romaine S. 1990. *Ap. J.* 356:110
Trudolyubov S, Priedhorsky W. 2004. *Ap. J.* 616:821

Trudolyubov SP, Borozdin KN, Priedhorsky WC. 2001. *Ap. J.* 563:L119

Trudolyubov SP, Borozdin KN, Priedhorsky WC, Osborne JP, Watson MG, et al. 2002a. *Ap. J.* 581:L27

Trudolyubov SP, Borozdin KN, Priedhorsky WC, Osborne JP, Watson MG, et al. 2002b. *Ap. J.* 571:L17

Truemper J. 1982. *Adv. Space Res.* 2(4):241

Tyler K, Quillen AC, LaPage A, Rieke GH. 2004. *Ap. J.* 610:213

van den Heuvel EPJ, Bhattacharya D, Nomoto K, Rappaport SA. 1992. *Astron. Astrophys.* 262:97

Van der Marel RP. 2004. In *Coevolution of Black Holes and Galaxies*, ed. LC Ho, p. 37. Cambridge, UK: Cambridge Univ. Press

Verbunt F. 1993. *Annu. Rev. Astron. Astrophys.* 31:93

Verbunt F, Lewin WHG. 2006. See Lewin & van der Klis 2006, p. 341

Verbunt F, van den Heuvel EPJ. 1995. See Lewin et al. 1995, p. 457

Volonteri M, Haardt F, Madau P. 2003. *Ap. J.* 582:559

Volonteri M, Perna R. 2005. *MNRAS* 358:913

Voss R, Gilfanov M. 2006. *Astron. Astrophys.* 447:71

Wang QD, Chaves T, Irwin JA. 2003. *Ap. J.* 598:969

Wang QD, Yao Y, Fukui W, Zhang SN, Williams R. 2004. *Ap. J.* 609:113

Watarai K, Ohsuga K, Takahashi R, Fukue J. 2005. *Publ. Astron. Soc. Jpn.* 57:513

Watson MG. 1990. In *Windows on Galaxies*, ed, G Fabbiano, JS Gallagher, A Renzini, p. 17. Dordrecht: Kluwer

Weisskopf MC, Tananbaum HD, Van Speybroeck LP, O'Dell SL. 2000. *Proc. SPIE* 4012:2

White NE, Marshall FE. 1984. *Ap. J.* 281:354

White NE, Nagase F, Parmar AN. 1995. See Lewin et al. 1995, p. 1

White RE III, Sarazin CL, Kulkarni SR. 2002. *Ap. J.* 571:L23

Williams BF, Garcia MR, Kong AKH, Primini FA, King AR, et al. 2004. *Ap. J.* 609:735

Wolter A, Trinchieri G. 2004. *Astron. Astrophys.* 426:787

Wu K. 2001. *Publ. Astron. Soc. Aust.* 18:443

Xu Y, Xu H, Zhang Z, Kundu A, Wang Y, Wu XP. 2005. *Ap. J.* 631:809

Zampieri L, Mucciarelli P, Falomo R, Kaaret P, Di Stefano R, et al. 2004. *Ap. J.* 603:523

Zepf SE, Ashman KM. 1993. *MNRAS* 264:611

Zezas A, Fabbiano G. 2002. *Ap. J.* 577:726

Zezas A, Fabbiano G, Baldi A, King AR, Ponman TJ, et al. 2004. *Rev. Mex. Astron. Astrofis. (Ser. Conf.)* 20:53

Zezas A, Fabbiano G, Baldi A, Schweizer F, King AR, et al. 2006. *Ap. J. Suppl.* In press

Zezas A, Fabbiano G, Rots AH, Murray SS. 2002a. *Ap. J. Suppl.* 142:239

Zezas A, Fabbiano G, Rots AH, Murray SS. 2002b. *Ap. J.* 577:710

Zezas A, Hernquist L, Fabbiano G, Miller J. 2003. *Ap. J.* 599:L73

Zwart SFP, Baumgardt H, Hut P, Makino J, McMillan SLW. 2004. *Nature* 428:724

Zwart SFP, Dewi J, Maccarone T. 2004. *MNRAS* 355:413

Diffuse Atomic and Molecular Clouds

Theodore P. Snow[1] and Benjamin J. McCall[2]

[1]Center for Astrophysics and Space Astronomy, University of Colorado, Boulder, Colorado 80309; email: theodore.snow@colorado.edu

[2]Departments of Chemistry and Astronomy, University of Illinois at Urbana-Champaign, Urbana, Illinois 61801; email: bjmccall@uiuc.edu

Key Words

interstellar medium, interstellar molecules, spectroscopy

Abstract

Diffuse interstellar clouds have long been thought to be relatively devoid of molecules, because of their low densities and high radiation fields. However, in the past ten years or so, a plethora of polyatomic molecules have been observed in diffuse clouds, via their rotational, vibrational, and electronic transitions. In this review, we propose a new systematic classification method for the different types of interstellar clouds: diffuse atomic, diffuse molecular, translucent, and dense. We review the observations of molecules (both diatomic and polyatomic) in diffuse clouds and discuss how molecules can be utilized as indicators of the physical and chemical conditions within these clouds. We review the progress made in the modeling of the chemistry in these clouds, and the (significant) challenges that remain in this endeavor. We also review the evidence for the existence of very large molecules in diffuse clouds, and discuss a few specific clouds of particular interest.

1. INTRODUCTION

Given that the density of the diffuse interstellar medium is, on average, far lower than those of even the best laboratory vacuums, it was surprising to astronomers in the late 1930s to discover that molecules exist in space (see comments by Eddington 1926 and by Dalgarno 2000). Yet, beginning in the 1930s when high-resolution optical spectroscopy was pioneered at the Mount Wilson Observatory and at the Dominion Astrophysical Observatory, sharp absorption lines seen in the spectra of several distant stars were soon recognized as being caused by the diatomic molecules CN, CH, and CH^+ (Dunham 1937; Swings & Rosenfeld 1937; McKellar 1940; Douglas & Herzberg 1941; Adams 1941; for an overview of early work on interstellar molecules, see Feldman 2001). Today astronomers are aware of the existence of several additional diatomic molecules and a number of polyatomics in the diffuse interstellar medium (ISM)—and a host of as-yet unidentified very large species such as the carriers of the unidentified infrared bands (UIBs) and the carriers of the diffuse interstellar bands (DIBs), a large collection of optical absorption features that have been noted in interstellar spectra since the 1920s. Despite initial expectations to the contrary, the diffuse ISM is chemically rich.

To persist in the harsh environment of interstellar space, molecules must form at rates sufficient to counterbalance their destruction. Molecular destruction occurs primarily by direct photodissociation or predissociation, which is a two-step process in which the molecule is first photoexcited to an unstable state and then dissociates. The formation of small molecules is generally thought to occur via gas-phase two-body (usually ion-neutral) reactions, although grain surface reactions can also be important. It is also possible that some complex species are formed in red supergiant stellar outflows or through the destruction of solid dust grains that are shattered by shocks.

Interstellar molecules can be identified through their electronic, vibrational, and rotational spectra. Typically, electronic transitions of simple molecules arise in the ultraviolet (UV) or visible portion of the spectrum; vibrational bands lie at infrared (IR) wavelengths; and rotational lines are seen at radio wavelengths. Hence the study of interstellar molecules necessarily involves a wide range of observational techniques and instruments. Diffuse clouds are unique in that the same line of sight can, in principle, be observed by all of these techniques, providing a far broader understanding of the physical and chemical state of diffuse clouds than is possible for other cloud types.

In addition to gas-phase molecules, solid dust grains (which contain about 1% of the mass in diffuse clouds) play a very significant role in controlling the physics and chemistry. Some molecules are formed on dust grain surfaces, and in addition the absorption and scattering of starlight by dust controls the attenuation of starlight passing through interstellar clouds. Radiation (especially in the UV) plays a very strong role in governing the physical and chemical state of interstellar molecules; hence dust extinction is a crucial parameter in modeling the chemistry of diffuse clouds. For recent reviews of dust in the ISM, see Draine (2003), Whittet (2003), or Witt, Clayton & Draine (2004), each of which provides comprehensive discussions of dust observations, properties, and models.

2. THE CLASSIFICATION OF DIFFUSE CLOUDS

2.1. General Comments

Early models (e.g., McKee & Ostriker 1977) classified the ISM into three phases: the Cold Neutral Medium (CNM), often referred to as clouds; the Warm Ionized Medium or Warm Neutral Medium (WIM or WNM), which is sometimes considered the boundary layers of the CNM; and the Hot Ionized Medium (HIM), which is sometimes referred to as the intercloud medium or the coronal gas. These phases are thought to be in approximate pressure equilibrium with one another (see Savage & Sembach 1996 or Cox 2005 for a general description of the physical conditions and phases in the galactic ISM).

The CNM itself appears to contain a variety of cloud types, spanning a wide range of physical and chemical conditions. The densest clouds that are most protected from UV radiation from stars are variously referred to as dense clouds, dark clouds, or molecular clouds. The most tenuous clouds, fully exposed to starlight, are usually called diffuse clouds. Clouds that fall in between these two extremes are often referred to as translucent clouds. Unfortunately, the application of these categories has not been uniformly consistent in the literature; here we propose a new systematic classification for cloud types. Our proposed classifications are summarized in **Table 1**, and illustrated using a chemical model in **Figure 1**.

We wish to emphasize that the ISM is inherently complex in its structure, and though theorists and observers prefer to think of isolated, homogenous clouds, most real sightlines probably consist of a mixture of different types of clouds. In some cases, a sightline may consist of a concatenation of discrete clouds, whereas in other cases the gas may have an "onion-like" structure, with dense cloud material in the center, surrounded by translucent gas, which is in turn surrounded by more diffuse gas.

Because of this complexity, our classification of "cloud types" is intended to reflect the local conditions in a parcel of gas, rather than the overall properties of a larger structure. In particular, one must keep in mind that the definitions do not refer to line-of-sight properties. Although line-of-sight properties are the most easily observed ones, observational technology (especially the development of ultra-high resolution spectrographs) is now making it possible for astronomers to better estimate local properties (or, at least, the average properties of individual parcels of gas).

Table 1 Classification of Interstellar Cloud Types

	Diffuse Atomic	Diffuse Molecular	Translucent	Dense Molecular
Defining Characteristic	$f^n_{H_2} < 0.1$	$f^n_{H_2} > 0.1$ $f^n_{C^+} > 0.5$	$f^n_{C^+} < 0.5$ $f^n_{CO} < 0.9$	$f^n_{CO} > 0.9$
A_V (min.)	0	~0.2	~1–2	~5–10
Typ. n_H (cm^{-3})	10–100	100–500	500–5000?	$>10^4$
Typ. T (K)	30–100	30–100	15–50?	10–50
Observational Techniques	UV/Vis H I 21-cm	UV/Vis IR abs mm abs	Vis (UV?) IR abs mm abs/em	IR abs mm em

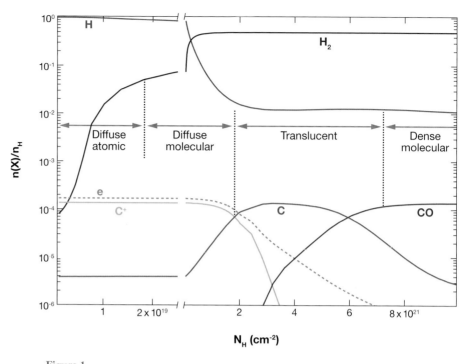

Figure 1

Results from photodissociation region model [with $n_H = 100$ cm^{-3} and $\chi_{UV} = 1$] from Neufeld et al. (2005), illustrating the revised definitions of cloud types.

Our proposed classification of cloud types may be easily reconciled with the working definition that infrared astronomers sometimes use to designate "diffuse" versus "dense" sightlines. For an infrared astronomer, a line of sight showing no evidence of ice coatings on grains is considered "diffuse" even though the total visual extinction may be ten or more magnitudes. For instance, several lines of sight in Cygnus, such as the very well-studied clouds toward Cygnus OB2 12, have large total extinctions but are considered diffuse because they show no ices. In our view these sightlines may be interpreted simply as extended aggregations of diffuse atomic and/or diffuse molecular clouds as defined below.

2.2. Definitions of Key Quantities

There are a number of important quantities that are useful in describing lines of sight, and for classifying the nature of parcels of gas. To make our notation clear, we collect the definitions of these quantities here.

We define the local number density (in cm^{-3}) of a certain species X to be n(X). The directly observable quantity is not the number density, but the column density (essentially the integral of the number density along the line of sight), which we denote N(X). For a given atom Y, we define the total number density of its nuclei to be n_Y; for example, $n_H = n(H) + 2n(H_2)$. For the case of carbon, most nuclei are in the form of C^+, C, or CO, so we can approximately write $n_C \approx n(C^+) + n(C) + n(CO)$. Similar expressions can be written in terms of column densities, e.g., $N_H = N(H) + 2N(H_2)$. The fraction of an atom's nuclei that are in a particular form is usually denoted as f_X (for example, the fraction of hydrogen in the form of H_2 is customarily denoted f_{H_2}). However, this notation is ambiguous because it can refer either to the local conditions or to a line-of-sight average. We define f^n to be the local fraction, in terms of number densities, and f^N to be the line-of-sight fraction, in terms of column densities. Thus, $f^n_{H_2} = 2n(H_2)/n_H$ and $f^N_{H_2} = 2N(H_2)/N_H$. Similarly, $f^n_{CO} = n(CO)/n_C$, etc.

Because hydrogen nuclei are by far the most abundant, the total amount of material along a line of sight can be approximated by N_H. Bohlin, Savage & Drake (1978) showed that there is a good correlation between N_H and the reddening E_{B-V}: In sightlines with $E_{B-V} < 0.5$, $N_H = 5.8 \times 10^{21}$ cm^{-2} mag$^{-1} \times E_{B-V}$. This relation was later extended up to $E_{B-V} \sim 1$ by Rachford et al. (2002). Although the quantity E_{B-V} is easier to determine observationally (by simply comparing the apparent color of the background star to the expected color based on its spectral type), the total visual extinction A_V is more commonly used as a stand-in for N_H. For the remainder of this review, we assume that $R_V \equiv A_V/E_{B-V} \sim 3$, so that $N_H \sim 1.7 \times 10^{21}$ cm^{-2} mag$^{-1} \times A_V$, and consider only A_V.

2.3. Diffuse Atomic Clouds

Diffuse atomic clouds represent the regime in the ISM that is fully exposed to the interstellar radiation field, and consequently nearly all molecules are quickly destroyed by photodissociation. Hydrogen is mainly in neutral atomic form, and atoms with ionization potentials less than that of hydrogen (most notably carbon) are almost fully ionized, providing abundant electrons. The paucity of molecules implies that very little chemistry occurs in this gas. Mysteriously, these clouds seem to be where most of the diffuse interstellar band carriers (see Section 6.2) thrive.

Many sightlines with low extinction seem to pass exclusively through diffuse atomic gas. Such sightlines typically have N_H less than about 5×10^{20} cm^{-2}, and are sufficiently optically thin to be observable by means of visible and UV absorption-line measurements. Diffuse atomic clouds typically have a fairly low density ($\sim$10–100 cm^{-3}), and temperatures of 30–100 K. We again wish to emphasize that the defining characteristic of these clouds is the low molecular fraction ($f^n_{H_2} < 0.1$), not the overall line-of-sight properties or the density. However, in sightlines that only cross diffuse atomic clouds, $f^N_{H_2}$ will also be <0.1.

2.4. Diffuse Molecular Clouds

Diffuse molecular clouds represent the regime where the interstellar radiation field is sufficiently attenuated, at least at the individual wavelengths that dissociate H_2, that the local fraction of hydrogen in molecular form, $f^n_{H_2}$ becomes substantial (>0.1). However, enough interstellar radiation is still present to photoionize any atomic carbon, or to photodissociate CO, such that carbon is predominantly still in the form of C^+ ($f^n_{C^+} > 0.5$).

In steady state, diffuse molecular clouds must necessarily be surrounded by diffuse atomic gas, in order to provide the shielding of radiation. This means that most sightlines that cross a diffuse molecular cloud will also cross diffuse atomic gas. [One exception is the sightline toward HD 62542; see Section 7.2.1.] Therefore, one has to be careful in interpreting line-of-sight quantities such as $f^N_{H_2}$, which will always be lower than the $f^n_{H_2}$ within the diffuse molecular cloud. For example, even a cloud that is fully molecular inside ($f^n_{H_2} = 1$) is likely to be along a line of sight with $f^N_{H_2}$ considerably less than unity.

The presence of abundant H_2 in diffuse molecular clouds permits chemistry to begin in earnest. Molecules are observed in these clouds in absorption in the UV/visible (e.g., CO, CH, CN, C_2, C_3), in the infrared (CO, H_3^+), and at millimeter wavelengths (e.g., HCO^+, OH, C_2H). These clouds typically have densities on the order of 100–500 cm^{-3}, and temperatures that range from 30–100 K.

Diffuse molecular clouds can be observed in sightlines with a wide range of visual extinction, or total hydrogen column density. The lower limit of N_H where molecular hydrogen begins to shield itself from interstellar radiation is a few times 10^{20} cm^{-2}, or $A_V \sim 0.2$.

2.5. Translucent Clouds

With sufficient protection from interstellar radiation, carbon begins to transition from ionized atomic form into neutral atomic (C) or molecular (CO) form. To call attention to this transition, van Dishoeck & Black (1989) defined such material as "translucent clouds." The chemistry in this regime is qualitatively different than in the diffuse molecular clouds, both because of the decreasing electron fraction and because of the abundance of the highly reactive C atoms. Although the original definition was cast in terms of isolated cloud structures with a certain range of A_V, van Dishoeck (1998) expanded the definition to be more general, in order to encompass gas on the periphery of dense molecular clouds, for example.

In many ways, the translucent cloud regime is the least well understood of all the cloud types. This is partly because of a relative lack of observational data, but also because theoretical models do not all agree on the chemical behavior in this transition region. In some models, there is a zone where the abundance of C exceeds that of C^+ and CO; in others the peak abundance of C falls below that of C^+ and CO. To cope with this uncertainty, we propose a working definition of translucent cloud material as gas with $f^n_{C^+} < 0.5$ and $f^n_{CO} < 0.9$. This definition reflects the fact that C^+ is no longer the dominant form of carbon as it converts to neutral or molecular form, but also excludes the dense molecular clouds, where carbon is almost exclusively CO. We stress that this working definition may need to be changed as better observational and/or theoretical understanding of these clouds is achieved.

Translucent cloud material must, in steady state, be surrounded by diffuse molecular cloud material, and is not expected to be present for sightlines with A_V less than about unity, because of the insufficient shielding of radiation. Thus, sightlines with $A_V > 1$ are candidates for hosting translucent clouds. However, a high value of A_V alone does not imply the presence of translucent material, as it could represent simply a pile-up of diffuse molecular clouds along the line of sight (e.g., Cygnus OB2 12). Once again, it is the local parameters that define this cloud type. Such high extinction sightlines are still observable with optical absorption line techniques, and can be observed at millimeter wavelengths in both absorption and emission. In the UV, increasing extinction makes observations very difficult. As discussed in Section 3.1 below, it has not yet been possible to observe (at UV wavelengths) sightlines that are clearly dominated by translucent clouds.

A series of papers by Gredel, van Dishoeck, Black and others have sought to study translucent clouds through high-resolution optical studies of CH and CN together with mm-wave observations of CO (Gredel et al. 1992, 1994; van Dishoeck et al. 1991). This work has led to at least two cases where translucent clouds seem to be present (HD 169454, Jannuzi et al. 1988; and HD 210121, Gredel et al. 1992).

2.6. Dense Molecular Clouds

With increasing extinction, carbon becomes almost completely molecular ($f^n_{CO} \sim 1$), defining the regime of dense molecular clouds. The chemistry is again qualitatively different, as the electron abundance is very low (cosmic-ray ionization being the dominant source) and the reactive C is replaced by the very stable CO. This regime is entered only in sightlines with $A_V > 5–10$; again, not all such sightlines will contain dense cloud material, and if dense cloud material is present it is likely to be surrounded by translucent material.

These clouds are typically self-gravitating, and are most often observed by IR absorption and mm-wave emission methods. Their densities are typically at least 10^4 cm^{-3}, and their kinetic temperatures are typically on the order of 10–50 K in the quiescent regions. There is a very rich literature on dense cloud chemistry, both from observational and theoretical perspectives. In fact, the earliest chemical models were focused on these environments (e.g., the seminal papers by Solomon & Klemperer 1972, Herbst & Klemperer 1973, and Watson 1973). Most of the more than 130 currently known interstellar molecules were found through observations of microwave rotational transitions in such clouds, starting with the discovery of OH by Weinreb et al. (1963), followed by a host of other new detections such as CO, NH$_3$, H$_2$O, and H$_2$CO (for an early review of radio-wavelength interstellar molecular observations, see Rank et al. 1971).

Much of the recent work on dense clouds has been directed toward hot cloud cores where star formation is incipient (for a recent paper summarizing the current state of modeling of such clouds, see Smith et al. 2004). We do not further discuss dense clouds in this review.

3. OBSERVATIONS OF MOLECULES IN DIFFUSE CLOUDS

3.1. The Most Abundant Molecule: H$_2$

Molecular hydrogen has always been expected to be abundant in diffuse clouds (e.g., Eddington 1937; Strömgren 1939; Hollenbach et al. 1971) but because it is symmetric and homonuclear, it possesses no electric-dipole allowed vibrational or rotational transitions. Therefore, the only probes of H_2 in the diffuse ISM are the far-UV electronic transitions in the Lyman and Werner bands lying below 1115 Å (Spitzer & Zabriskie 1959).

Observing these transitions was not possible until the advent of space-based far-UV spectrographs. The first success, using rocket-based spectroscopy, was by Carruthers (1970), who detected strong H_2 absorption in the line of sight toward ξ Persei. In 1972, the *Copernicus* orbital observatory was launched, with the primary scientific goal of observing interstellar H_2 absorption (Rogerson et al. 1973). *Copernicus* detected molecular hydrogen in more than 100 sightlines with A_V less than about 1.5. *Copernicus* H_2 surveys can be found in Spitzer et al. (1973) and Spitzer, Cochran & Hirshfeld (1974). An extensive review of H_2 studies based on *Copernicus* observations can be found in Shull & Beckwith (1982). The essential results of the *Copernicus* H_2 surveys were: (*a*) in the observed sightlines, $f^N_{H_2}$ ranged from nearly 0 to about 0.6, with a discontinuous transition from $<10^{-4}$ up to >0.01 at about $N_H \approx 5 \times 10^{20}$ cm^{-2} (Spitzer et al. 1973; Savage et al. 1977); (*b*) significant rotational excitation in the ground electronic and vibrational state was detected everywhere, as discussed in Section 4.2 below.

The *Far Ultraviolet Spectroscopic Explorer* (FUSE) observatory (launched in 1999 and still in operation at this writing; Moos et al. 2000) provides spectra over a broader far-UV wavelength region than *Copernicus*, with similar spectral resolving power and far greater sensitivity, thus allowing the detection and analysis of H_2 toward much dimmer stars (see **Figure 2**). FUSE has allowed the detection and analysis of molecular hydrogen in a wide range of environments not accessible to *Copernicus*, such as high galactic latitude clouds formed in the halo of the Milky Way (Shull et al. 2000; Richter et al. 2001), the Magellanic Clouds (Tumlinson et al. 2002), and more heavily reddened diffuse cloud sightlines (Snow et al. 2000; Rachford et al. 2001, 2002). The general findings of the FUSE H_2 surveys are consistent with those from *Copernicus*, but have extended them into new regimes.

The highest line-of-sight molecular fraction measured with FUSE is $f^N_{H_2} \sim 0.8$, for the line of sight toward X Persei (aka HD 24534), a diffuse cloud with one of the richest molecular contents yet found (Lien 1984; Sheffer et al. 2002b). Rachford et al. (2001, 2002) searched for evidence of translucent clouds in highly reddened sightlines, but failed to identify any line of sight that seems convincingly dominated by a translucent cloud. Rachford et al. concluded that, up to $A_V \approx 3$, most lines of sight consist of a summation of multiple diffuse clouds, each with a low to moderate hydrogen molecular fraction, but none containing purely molecular hydrogen or purely neutral and molecular carbon. The inability of FUSE to detect translucent clouds should not be taken as evidence that such clouds do not exist; rather, it is a reflection of the fact that it cannot observe stars lying behind clouds with large extinction ($A_V > 3$), and that it has relatively low spectral resolution.

Some absorption lines of vibrationally excited H_2 lie in the mid-UV, and have been studied with spectrographs [Goddard High Resolution Spectrograph (GHRS), and Space Telescope Imaging Spectrograph (STIS)] on the *Hubble Space Telescope* (HST). The first detection was toward ζ Oph (Federman et al. 1995). Subsequently, Meyer

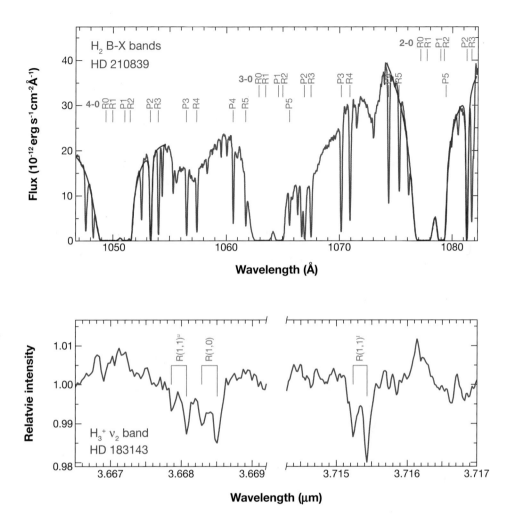

Figure 2

(*Top*) Spectrum of H_2 obtained with *Far Ultraviolet Spectroscopic Explorer* (FUSE) and model fit in red (adapted from Rachford et al. 2002); (*bottom*) spectrum of H_3^+ (adapted from McCall et al. 2002).

et al. (2001) found a rich spectrum of absorption lines from vibrationally excited H_2 toward the star HD 37903, which is the exciting star in the reflection nebula NGC 2023. Meyer et al. also detected vibrationally excited H_2 toward three other stars (HD 37021, HD 37061, and HD 147888). More recently Boissé et al. (2005) have detected absorption from vibrationally excited H_2 toward HD 34078, another star embedded in circumstellar gas, also finding lines arising from rotational levels as high as $J = 11$ of the ground electronic and vibrational state. The mechanisms for populating H_2 in such highly excited states is discussed in Section 4.2 below.

3.2. HD in Diffuse Clouds

The H_2 isotopomer HD was first detected in *Copernicus* spectra by Spitzer et al. (1973) and Morton (1975), and has subsequently been observed in many sightlines by FUSE. The HD lines arising from the lowest-lying rotational levels ($J = 0$ and $J = 1$) are far weaker than their counterparts for H_2, but are in many cases still strong enough to be saturated, requiring a curve-of-growth analysis for column density determinations. To date, HD analyses from FUSE spectra have been published only for a few stars (Ferlet et al. 2000; Lacour et al. 2005) though undoubtedly many more detections reside in the FUSE archives. In addition, Lacour et al. have reanalyzed *Copernicus* spectra for several stars, resulting in a uniform survey of HD abundances in some 17 sightlines.

The ratio of HD to H_2—more specifically, $N(HD)/2N(H_2)$—in the Lacour et al. survey ranges from a few times 10^{-7} to several times 10^{-6}. This is somewhat higher than the values found earlier from *Copernicus* data, but those values were based on two strong lines that were saturated.

It would be useful if the HD/H_2 ratio were a reliable indicator of the atomic ratio D/H, because the latter is an important indicator of the early expansion rate of the universe and therefore valuable in cosmology (e.g., Schramm & Turner 1998; Olive, Steigman & Walker 2000). But for three distinct reasons the comparison of HD to H_2 is difficult to apply to the cosmological problem.

First, in diffuse molecular clouds where H_2 is self-shielding (see Section 5.2.1) the far-UV lines of HD remain optically thin, or at best only marginally saturated, and therefore HD is not self-shielded. This favors H_2 and leads to a reduced HD/H_2 ratio, smaller than the atomic D/H value. Second, the lower mobility of D atoms on grains as compared to H atoms greatly reduces the formation rate of HD as compared to H_2, again acting to reduce the HD/H_2 ratio relative to the H/D ratio. But third, HD has a gas-phase formation channel, through the "chemical fractionation" reaction $H_2 + D^+ \rightarrow HD + H^+$ (Ferlet et al. 2000), which tends to enhance the HD/H_2 ratio and depends on the cosmic-ray ionization rate (Black & Dalgarno 1973).

As noted by Ferlet et al. (2000), if HD column densities reach the level where the molecule becomes self-shielding, then HD becomes the dominant reservoir of deuterium, and the HD/H_2 ratio would reflect the overall D/H value. So far, no observed HD column densities have come close to the level where self-shielding would occur, and in general we conclude that the HD/H_2 ratio is not useful for determining the cosmological D/H ratio, although it may be useful in constraining the rate of cosmic-ray ionization (which produces D^+).

3.3. Heavier Diatomics in Diffuse Clouds

3.3.1. Carbon monoxide.
In addition to electronic transitions in the UV, CO possesses (unlike H_2) dipole-allowed rotational transitions in the mm-wave band and vibrational transitions in the infrared. Because the electronic transitions are the strongest, the most sensitive searches for CO in diffuse clouds have been performed in the UV, but CO has been detected in diffuse clouds by the other methods as well.

In the radio, both emission and absorption lines of CO have been detected in diffuse clouds. Direct comparisons of CO and H_2 can be made if radio CO emission is observed in the same sightlines as far-UV H_2 absorption. Such comparisons are risky due to the differing regions that are sampled, but assignment of the mm-wave emission to the same clouds that produce the H_2 absorption can be made on the basis of velocity matches. Knapp & Jura (1976), Kopp et al. (1996), and Liszt (1997) have made such comparisons, generally finding a CO/H_2 ratio of about 10^{-6}, consistent with expectations based on diffuse cloud gas-phase chemistry models.

A few detections of the CO fundamental vibrational band near 4.6 μm have been reported for diffuse clouds on the basis of ground-based observations (McCall et al. 1998; Shuping et al. 1999). For the most part IR studies of CO gas-phase absorption have been applied to dense clouds and not diffuse regions—but with improved spectral resolving power and sensitivity, further pursuit of the IR gas-phase absorption in diffuse clouds might be profitable (especially given the current lack of any UV instruments capable of observing the electronic bands).

The first UV detection of CO was made toward ζ Ophiuchi, using a rocket-borne spectrograph (Smith & Stecher 1971). Several far-UV bands of CO were subsequently detected by *Copernicus* observers in a number of sightlines (Jenkins et al. 1973; Morton 1975; Snow 1975; Smith, Stecher & Krishna Swamy 1978; Wannier, Penzias & Jenkins 1982).

In principle, UV absorption line measurements should be most useful for comparing CO and H_2, because the exact same pathlength can be probed using both molecules. However, early attempts to derive CO abundances using *Copernicus* were hampered by uncertainties surrounding the oscillator strengths for the transitions, difficulties in resolving the rotational structure within each electronic band, and saturation effects. The oscillator strength issues were largely resolved by the systematic compilations of Morton & Noreau (1994) and Eidelsberg et al. (2004 and references cited therein). Better resolution of rotational structure became possible with GHRS and later STIS on HST, yielding column density determinations impacted only by saturation effects (e.g., Sheffer et al. 1992; Lambert et al. 1994; Wannier et al. 1999; Kaczmarczyk 2000a,b).

The issue of line saturation can be avoided by observing the weak intersystem bands of CO in the UV, which are usually unsaturated even in sightlines with high total column densities. The difficulty in analyzing these transitions, which involve perturbations between electronic states, lies in a lack of accurate laboratory wavelengths and f-values. For some time, results from studies of these intersystem bands appeared to be inconsistent with those based on the allowed electronic transitions, but using a combination of theory and observations, Sheffer et al. (2002a) were able to determine consistent empirical f-values for some seven intersystem bands of CO and apply them to the determination of the CO column density and excitation in the line of sight toward ζ Ophiuchi. In parallel with this work, laboratory studies of the intersystem bands yielded oscillator strengths consistent with the astronomically derived values (Eidelsberg et al. 2004). Very recently, P. Sonnentrucker, D. Welty, J. Thorburn & D. York (submitted) have taken advantage of the revised f-values for the intersystem bands of CO to conduct a survey of several lines of sight where other diatomics such as H_2, C_2, and CN are also measured, finding that CO correlates more strongly with CN than with H_2, and that current models underestimate the amount of CO present in diffuse molecular clouds. The close correspondence between CO and CN has also been discussed by Pan et al. (2005).

Because the HST no longer has any UV spectroscopic capability (although several observations reside in its archives), CO observations are now being conducted in the far-UV using FUSE. The first such study, by Sheffer et al. (2003), reports the detection of 11 Rydberg bands of CO toward the star HD 203374A. Because the FUSE spectral resolving power is insufficient to separate the rotational lines within the bands, Sheffer et al. made use of high-resolution optical data on CH and UV data on the longer-wavelength CO bands (from STIS) to determine the line-of-sight velocity structure and hence the accurate column density of CO. They were then able to calculate empirical f-values for the far-UV lines. Their most significant result was that the far-UV transitions that are not blocked by strong H_2 absorption have higher f-values than previously believed, which means that these transitions contribute more strongly to the predissociation of CO than in standard models, though this is offset by enhanced self-shielding.

The ratio of ^{12}CO to ^{13}CO is an indicator of the isotopic ratio of carbon, of interest for galactic nucleosynthesis studies. In principle, lines from the different isotopic species can be observed, and the isotopic ratio determined, from observations in the UV, IR, or radio. However, the determination of the isotopic ratio is complicated by selective photodissociation (when ^{12}CO self-shields, ^{13}CO is more easily destroyed), saturation corrections (^{12}CO lines have larger optical depth), and chemical fractionation (which enhances ^{13}CO; Sheffer et al. 2002b). Federman et al. (2003) discuss these processes in more detail, and present UV observations with the HST/GHRS that suggest that the value of $^{12}CO/^{13}CO$ toward two sightlines in Ophiuchus is $\sim$120, higher than the terrestrial value of 90.

The carbon monoxide oxygen isotopic ratios ($C^{16}O/C^{17}O$ and $C^{16}O/C^{18}O$) have also been derived from UV observations, and are larger than the corresponding overall isotope ratios in local interstellar gas (Sheffer, Lambert & Federman 2002b). Sheffer et al. explain this as being caused by the favored photodestruction of $C^{17}O$ and $C^{18}O$ relative to $C^{16}O$, which is more self-shielded.

3.3.2. Other diatomics seen at UV/optical wavelengths. The first three interstellar molecules ever detected, CH, CH^+, and CN (see Section 1), were diatomics, but detection of additional diatomics has generally been a challenge, as most species fall well below H_2, CO, CH, CH^+, and CN in abundance.

After extensive early searches and initial detections made only with some difficulty (Souza & Lutz 1977, 1978; Snow 1978), diatomic carbon is now routinely observed with high-S/N high-resolution spectrographs on large telescopes (Hobbs, Black & van Dishoeck 1983; van Dishoeck & de Zeeuw 1984; van Dishoeck & Black 1986a; Lambert, Sheffer & Federman 1995; Thorburn et al. 2003). The Phillips band system of C_2, which has several bands in the far red and near-IR, has proven most readily observable. The first detection, by Souza & Lutz (1977), was of the (1,0) band near 1.014 μm toward the highly reddened distant B star Cygnus OB2 12 (see Section 2.1), while most subsequent work has been focused on the (3,0) and especially the (2,0) bands (at 7720 and 8765 Å, respectively). The fundamental (0,0) band of C_2 near 1.2 μm has not, to our knowledge, been detected.

Because C_2 is a homonuclear diatomic, there are no allowed dipole transitions between rotational levels of the ground state, and many rotational levels are populated. Rotational lines arising from levels up to J = 12 or higher are typically observed. As discussed in Section 4.4, the rotational excitation of C_2 is a useful indicator of the local radiation field intensity and gas density, as these levels are populated by both radiative and collisional pumping, as in the case of H_2 (van Dishoeck & Black 1982).

The first molecule detected through radio observations, the OH radical, also has electronic transitions at 1222 Å and 3078 Å. The far-UV line was detected by Snow (1976a) using the *Copernicus* satellite, whereas the near-UV line was first detected from the ground, with great effort due to atmospheric ozone interference (Crutcher & Watson 1976a; Chaffee & Lutz 1977; Felenbok & Roueff 1996; Federman, Weber & Lambert 1996). The near-UV OH features are more easily observed from space, though the transition lies near the long-wavelength cut-off for instruments such as the *International Ultraviolet Explorer* (IUE) and the HST—and was not covered at all by *Copernicus* or FUSE. Though the OH λ3078 feature was readily detectable with the GHRS and the STIS, only a few observations have been reported (e.g., Snow et al. 1994). Microwave absorption due to OH in diffuse sightlines was observed by several researchers, as summarized by Crutcher (1979) and Liszt & Lucas (1996). Based on both the optical and the radio observations, the column densities of OH, at the level of 10^{-8} of the total hydrogen, are generally consistent with gas-phase chemistry models for diffuse clouds (e.g., Black & Dalgarno 1977, van Dishoeck & Black 1986b).

Another hydride, NH, has an electronic transition at 3358 Å, which is accessible from the ground, but proved to be very difficult to detect. Gas-phase chemistry models show no rapid reaction sequence leading to NH, so a detection would suggest that some other process, such as grain surface formation, must be responsible. The first intensive search for NH was negative (Crutcher & Watson 1976b), consistent with gas-phase chemistry, but eventually NH was detected (Meyer & Roth 1991) at a level suggesting either that current models are overlooking important gas-phase reactions or that grain surface formation may be important even in diffuse clouds (Crawford & Williams 1997). Recently, D.E. Welty (private communication) has detected NH in an additional three sightlines (HD 62542, HD 73882, and Walker 67).

Another diatomic not expected to be abundant according to gas-phase models is N_2, which has electronic transitions only in the far-UV. An early search based on *Copernicus* data proved negative (Lutz, Owen & Snow 1979), but recently Knauth et al. (2004) have claimed a detection toward HD 124314 and have reported an additional detection toward 20 Aquilae (D. Knauth; private communication), based on FUSE spectra in both cases. Because of blending with interstellar H_2 lines, blending with a prominent stellar photospheric line, and the possibility of contamination by N_2 in the upper atmosphere, it is very difficult to conclude unambiguously that interstellar N_2 has been detected. But, as discussed by Snow (2004), Knauth et al. have addressed these issues and make a convincing case for detection. The inferred column density of N_2 in the sightlines where the detections are claimed is higher than expected from gas-phase models for diffuse clouds, but lower than expected for dense molecular clouds, suggesting that dust grain surface reactions may be responsible. In view of the observational complexities, clearly it is important to search for N_2 in additional lines of sight, and Knauth and colleagues are doing so, again using FUSE.

Jura (1974), followed by Dalgarno et al. (1974), realized that ionized chlorine can undergo rapid reactions with H_2 to form HCl^+, which then leads to H_2Cl^+ and HCl (through electron-ion recombination). Jura predicted that HCl might be detectable through an electronic transition at 1291 Å, but this line was not detected in a subsequent search using *Copernicus* data (Jura & York 1978). The idea lay dormant for some time, until Federman et al. (1995), using HST/GHRS spectra, detected HCl absorption toward ζ Oph. The detected column density was consistent with the predictions of models.

3.3.3. Diatomics seen in the millimeter-wave.
Many molecules, diatomics and larger, have dipole-allowed transitions in the mm-wave spectral region, suggesting that radio observations of molecular emission and absorption from diffuse clouds, in combination with UV, optical, and IR absorption measures, could provide detailed information on the chemistry of these clouds. Radio observations have the significant advantage that velocity components are resolved and that many more species are accessible than at optical and UV wavelengths. In practice there are difficulties, as diffuse clouds have lower molecular column densities than the dense clouds typically observed at radio frequencies, and mm-wave emission can arise from an extended region rather than solely from the pencil beam where the observed absorption lines arise. Important efforts to combine radio and optical data on specific diffuse clouds have been carried out by Crutcher (1985) and Gredel et al. (1994).

Such difficulties are eliminated when performing absorption measurements at millimeter waves; however, there are relatively few suitable continuum sources in our galaxy. Liszt & Lucas (2002, and references cited therein) have pioneered the use of extragalactic radio emitters, in particular AGN/blazars, as continuum sources against which to measure mm-wave absorption lines. A major advantage of using extragalactic background sources is that they randomly sample the ISM, as opposed to OB stars, which are often directly associated with concentrations of interstellar matter.

The result of their work is that a significant heritage of radio absorption-line detections has been compiled for diffuse clouds. Among the diatomics detected in diffuse clouds are CO (both emission and absorption; Liszt & Lucas 1998), CH (emission; Liszt & Lucas 2002), CN (absorption; Liszt & Lucas 2001), and OH (emission and absorption; Liszt & Lucas 1996, Liszt 1997). Other diatomics found in absorption include the sulfur-bearing species CS, and SO (Lucas & Liszt 2002) as well as SiO (Lucas & Liszt 2000a).

3.4. Triatomic and Polyatomic Molecules in Diffuse Clouds

Given that gas-phase chemistry can form abundant diatomic molecules in diffuse clouds, a few polyatomics are expected as well, according to models (e.g., van Dishoeck & Black 1986b). Polyatomics in diffuse clouds took much longer to find than diatomics, but a combination of optical absorption measurements, IR detections of vibrational transitions, and millimeter-wave radio absorption and emission observations has recently revealed a surprisingly rich chemistry in these clouds.

3.4.1. H_3^+ in diffuse clouds.
The most fundamental polyatomic molecule is the ion H_3^+, whose presence was predicted originally by Martin, McDaniel & Meeks (1961) in regions where H_2 is partially ionized, formed through the reaction

$$H_2^+ + H_2 \rightarrow H_3^+ + H.$$

Once formed, H_3^+ can donate its extra proton to just about any atom or molecule, thereby initiating a network of ion-molecule reactions that is thought to be responsible for the formation of many interstellar molecules. For a detailed review of the chemistry of, and the search for, interstellar H_3^+, see Geballe (2000).

Initially it was thought that H_3^+ would be important only in dense clouds, where H_2 is dominant and ionization is provided by cosmic rays, and so the first searches were aimed at dark molecular clouds with embedded protostars that could be used as infrared continuum sources. Because H_3^+ has no tightly-bound excited electronic states, there are no detectable spectral features in the visible or UV; and because H_3^+ has no permanent dipole moment, it has no allowed rotational transitions either. The only spectroscopic probe of this molecule, therefore, is a set of ro-vibrational lines between 3.5 and 4.0 μm. Meaningful searches for these transitions required accurate frequencies, which were determined experimentally by Oka (1980), and f-values, which were computed by J. K. G. Watson (private communication).

Geballe & Oka (1996) finally detected H_3^+ toward two deeply embedded protostars (W33A and GL2136), and this momentous discovery was followed by several additional detections in dense clouds (McCall et al. 1999, Brittain et al. 2004). Though significant quantities of H_3^+ were not expected in diffuse clouds, this species was soon detected in several diffuse sightlines (McCall et al. 1998, 2002, Geballe et al. 1999; see **Figure 2**). One of the first detections of H_3^+ in a classical diffuse cloud was made recently (McCall et al. 2003) toward the star ζ Persei, which has visual extinction $A_v \approx 1$ mag.

The discovery of H_3^+ in diffuse lines of sight was quite surprising, and the abundance of H_3^+ in diffuse clouds remains a challenge to understand (Section 5.4.2).

3.4.2. Triatomics observed via optical absorption.
Triatomic carbon, C_3, was long expected to exist in diffuse clouds, but proved to be elusive for a time. The strong electronic band of C_3, near 4050 Å, appears in many cometary emission spectra, and given its high f-value appeared to be the best choice for interstellar absorption searches. After some failed searches (Clegg & Lambert 1982; Snow, Seab & Joseph 1988), Haffner & Meyer (1995) reported a tentative detection in the line of sight toward HD 147889 in the ρ Ophiuchi cloud. Subsequently the first clear-cut detection was made by Maier et al. (2001), who found C_3 absorption in three diffuse clouds (ζ Ophiuchi, 20 Aquilae, and ζ Persei). More recently, some 15 detections of interstellar C_3 have been reported by Oka et al. (2003), using the Apache Point Observatory

(APO) 3.5-m telescope. The resolving power of the APO spectra was insufficient for resolving the individual rotational lines, but then Ádámkovics, Blake, & McCall (2003) obtained much higher-resolution spectra of C_3 in 10 sightlines and were able to extract the rotational excitation (see Section 4.4).

H_2O and CO_2 have been sought through UV absorption in diffuse clouds without success. Snow (1975) and Snow & Smith (1981) used *Copernicus* to search for water vapor in the line of sight toward ζ Ophiuchi, with no detections—though the latter search revealed a tentative feature at the right wavelength that might have represented a marginal detection. However, Spaans et al. (1998) conducted an intensive search for the 1240 Å transition of H_2O toward HD 154368, placing a limit on the column density of 9×10^{12} cm^{-2}, or about 10^{-8} times the total hydrogen column density, which is more stringent than the *Copernicus* limit for ζ Oph.

In the case of CO_2, upper limits for electronic transitions in the far UV were reported by Snow (1980) based on *Copernicus* spectra, though the reported limit was not sensitive enough to strongly challenge gas-phase chemistry models, which do not predict much CO_2 in any case.

Tentative detections of CH_2 have been reported by Lyu et al. (2001) based on HST/GHRS spectra of HD 154368 and ζ Ophiuchi; if the detections are real, they correspond to abundances of CH_2 in good agreement with predictions of gas-phase chemical models.

3.4.3. Polyatomics seen in millimeter-wave absorption.
As mentioned in Section 3.3.3 above, detections of millimeter-wave absorption are extraordinarily useful because absorption (as opposed to emission) measurements necessarily apply to exactly the same interstellar column that is studied via infrared, optical, and UV absorption-line observations, thus allowing a truly multispectral analysis of the chemistry in a given line of sight. Liszt & Lucas (2002, and references therein) have conducted extensive surveys of millimeter-wave absorption lines using extragalactic continuum sources.

The remarkable result of the millimeter-wave absorption line work is that many polyatomic species, including notably HCO^+, are detected in absorption along diffuse sightlines over a range of galactic latitudes (Liszt & Lucas 1994, 1996, Hogerheijde et al. 1995). Other polyatomic species detected in absorption include HCN (Lucas & Liszt 1994), H_2CO (Liszt & Lucas 1995), C_2H and C_3H_2 (Cox, Guesten & Henkel 1988; Lucas & Liszt 2000b), H_2S and HCS^+ (Lucas & Liszt 2002), NH_3 (Liszt, Lucas, & Pety 2006), and HOC^+ (Liszt, Lucas & Black 2004); several others were sought and not detected (see **Table 3**).

Some of the extragalactic continuum sources observed by Liszt & Lucas are also bright enough at visible wavelengths for optical interstellar line measurements with large telescopes. Tappe (2004) and Tappe & Black (2004; and A. Tappe & J. H. Black, in preparation) have taken advantage of this opportunity to compare the radio and optical data to develop a more complete picture of the observed diffuse clouds and their physical and chemical states, and found a very good correspondence between the optical and millimeter-wave absorption components. This work conclusively demonstrates that these extragalactic sightlines pass through the same types of diffuse clouds

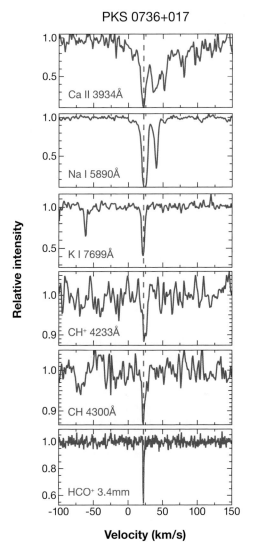

Figure 3
Comparison of optical and millimeter-wave spectra of PKS 0736 + 017, adapted from Tappe (2004). The HCO$^+$ data are from Lucas & Liszt (1996).

that are usually studied in absorption at optical and UV wavelengths against OB stars. **Figure 3** shows the comparison of optical and millimeter-wave absorption lines toward the source PKS 0736 + 017. Taken together with the Lucas & Liszt surveys, it is now clear that the chemistry of polyatomic molecules in diffuse clouds is surprisingly rich. Clearly, much work is left to be done to fully understand these observations; as an initial step, Tappe (2004) has explored the question of whether the observations support a connection between CH$^+$ and the formation of HCO$^+$.

4. MOLECULES AS INDICATORS OF CLOUD PHYSICAL CONDITIONS

In the diffuse ISM, most molecules exist in their ground electronic and vibrational states. The population of the various rotational states is generally controlled by both collisional and radiative processes. Collisions with atoms and other molecules (especially H and H_2) can cause either rotational excitation or de-excitation. For molecules with dipole moments (and hence, allowed rotational transitions), excitation can occur by stimulated absorption in the cosmic microwave background (CMB), and de-excitation can occur by spontaneous emission and by stimulated emission. Molecules without dipole moments can still be excited by radiation, by absorbing photons in an electronic transition. If all of the relevant molecular processes are understood, the observed rotational excitation can provide valuable information about the physical conditions in a cloud, such as the density, kinetic temperature, and the intensity of the radiation field.

4.1. CN and the Cosmic Microwave Background

The CN radical was one of the first interstellar molecules to be detected, and immediately its rotational excitation was noted by McKellar (1940, 1941), who estimated the excitation temperature of CN to be about 2.3 K. Some have cited this as a "pre-discovery" of the CMB (e.g., Feldman 2001), though McKellar apparently did not speculate that the excitation might be caused by a pervasive radiation field.

The rotational excitation of CN remained unexplained until 1965, when the CMB was discovered (Penzias & Wilson 1965, Dicke et al. 1965). It was soon realized that the excess excitation of CN could be explained by radiative equilibrium with the CMB (Field & Hitchcock 1966, Thaddeus & Clauser 1966)—and this was, in fact, the first definitive measurement of the temperature of the CMB (a temperature of about 2.6 K was the consensus interpretation at the time; see the review by Thaddeus 1972). A much more precise determination of the excitation temperature of CN in several sightlines has been presented by Roth & Meyer (1995), who also provide a useful review of previous observations.

Of course more recent measurements from above Earth's atmosphere have provided a more direct measure of the temperature of the CMB (Fixsen & Mather 2002), but it is noteworthy that a simple analysis of the rotational excitation of a diatomic interstellar molecule revealed the essential result well before the more sophisticated direct measurements of the CMB spectrum were possible.

4.2. Rotational Excitation of H_2: A Case of UV Pumping

As a homonuclear molecule with no dipole-allowed vibrational or rotational transitions, molecular hydrogen is a premier indicator of physical conditions in diffuse interstellar clouds. Far-UV observations with *Copernicus* and FUSE show that in most diffuse clouds there is a substantial population of H_2 in rotational levels up to about $J = 6$, and that the relative populations can usually be interpreted in terms of two

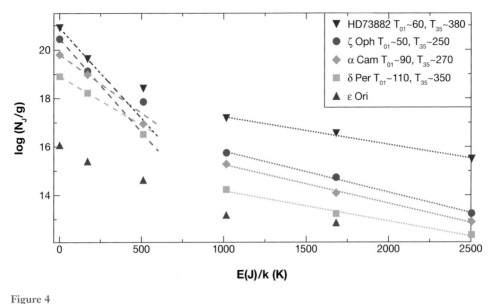

Figure 4

Boltzmann diagram of H_2 in five diffuse cloud sightlines, illustrating the low T_{01} and higher T_{35}. Column densities from Spitzer et al. (1973), Spitzer, Cochran & Hirshfeld (1974), and Snow et al. (2000).

distinct excitation temperatures. As illustrated in **Figure 4**, the relative populations of the low-lying levels ($J = 0$ and $J = 1$, in particular) generally correspond to a temperature of 50 to 150 K with a mean around 80 K, whereas the higher-J levels fit a temperature of a few to several hundred Kelvin (Spitzer et al. 1973; Spitzer & Cochran 1973; Spitzer, Cochran & Hirshfeld 1974; Spitzer & Zweibel 1974; Savage et al. 1977; Rachford et al. 2002; Browning et al. 2003).

The temperature derived from the $J = 0$ and $J = 1$ levels of H_2 (T_{01}) is generally assumed to represent the kinetic temperature (T_{kin}) of the gas, because the lines arising from these states are usually saturated and therefore self-shielding, making radiative pumping out of these states relatively inefficient. However, for T_{01} to be rigorously equal to the T_{kin}, the balance between $J = 0$ and $J = 1$ would have to be dominated by collisional transitions. The real situation is more complex, because $J = 0$ and $J = 1$ belong to different nuclear spin configurations (para and ortho, respectively) and cannot interconvert through ordinary collisions, but only via reactive collisions with H^+ or H_3^+, or by collisions with paramagnetic grains. Although the rate of such reactive collisions may well be sufficient to ensure that $T_{01} = T_{kin}$, the complexity of the situation should not be forgotten.

Black & Dalgarno (1973) proposed that the high-J excitation was the result of radiative pumping, in which absorption of UV photons was followed by a cascade back to excited vibrational and rotational levels of the ground electronic state. This mechanism was mentioned by Spitzer & Cochran (1973), among others, and was later modeled by Spitzer & Zweibel (1974) and Jura (1975a,b; see also Black & Dalgarno

1977; Black, Hartquist & Dalgarno 1978; Federman Glassgold & Kwan 1979). The analysis of the relative populations of the high-J levels can provide information about the radiation field, the density, and perhaps also the distribution of rotational levels populated when H_2 is initially formed. More recently, the excitation of H_2 has been modeled in great detail by many researchers using improved cross sections, and taking into account the effects of vibrational excitation (e.g., Browning, Tumlinson & Shull 2003, and references therein).

The early *Copernicus* results indicated that the high-J lines in most diffuse cloud sightlines are formed in environments with radiation fields much more intense than the average interstellar field. Several *Copernicus*-based studies (references cited above) led to the conclusion that the material where the excited H_2 resides is close to a source of UV radiation, presumably the hot star that is being observed. Velocity studies of the high-J lines seemed to support this notion, because several cases were found where the absorption lines arising from excited levels showed negative velocities, as if they were formed in gas expanding away from the star. The spectral resolving power of *Copernicus* was only marginally sufficient for such a conclusion, however, and higher-resolution studies by Jenkins et al. (1989) and Jenkins & Peimbert (1997) did not show such a trend.

As mentioned in Section 3.1, a few cases have been found where an observable fraction of the H_2 exists in excited vibrational levels of the ground electronic state. The first detection, by Federman et al. (1995) was toward ζ Ophiuchi, and suggested that the UV radiation field pumping the H_2 was not much higher than the average interstellar value (in contrast to the *Copernicus* results with rotationally excited H_2). However, observations of vibrationally excited H_2 by Meyer et al. (2001) toward HD 37903 and by Boissé et al. (2005) toward HD 34078 have demonstrated a close interaction between the molecular gas and the radiation fields of these particular stars.

FUSE data have yet to be fully exploited for studies of the high-J lines in more heavily-reddened lines of sight than those analyzed by Browning et al. (2003). At least one individual case has been well studied, that of HD 185418 (Sonnentrucker et al. 2003), but no general survey of rotational excitation toward reddened stars has yet been completed, though one is under way (B. Rachford, private communication). For HD 185418, Sonnentrucker et al. show that the excited H_2 is widely distributed and not concentrated near the star where the UV radiation field is intense, suggesting instead that the high-J lines in this case may be populated collisionally in a warm neutral gas, an idea previously proposed by Gry et al. (2002) for other sightlines.

4.3. An Intermediate Case: CO

The rather small dipole moment (0.1 Debye) of CO implies that it is less efficient than, say, CN at relaxing rotational excitation through spontaneous emission. Therefore, CO does not reach equilibrium with the CMB like CN. On the other hand, the fact that CO does have a nonzero dipole moment implies that it will not reach equilibrium with the kinetic temperature of the gas via collisions, because spontaneous emission does have some effect. The critical density (the density at which the collisional de-excitation rate equals the spontaneous emission rate) for the $J = 1-0$ transition of CO is around 1000 cm^{-3}, so only at densities higher than this would thermal equilibrium be reached.

Consequently, one would expect the rotational excitation temperature of CO, at the densities of diffuse clouds, to be somewhere in between the CMB temperature (2.7 K) and the kinetic temperature of the gas (~50 K). Observationally, the CO rotational excitation temperature is typically 5 to 15 K (Smith et al. 1978, Lambert et al. 1994, Shuping et al. 1999, Sonnentrucker et al. 2003). The observed rotational excitation can be used to infer the gas density and temperature. Detailed models for CO excitation have been presented by Goldsmith (1972), van Dishoeck & Black (1987), Lyu, Smith & Bruhweiler (1994), and Warin, Benayoun & Viala (1996), among others.

4.4. Excitation Analysis of C_2 and C_3

As homonuclear symmetric molecules, both C_2 and C_3 lack permanent dipole moments, so neither can quickly relax radiatively from excited rotational levels of the ground electronic and vibrational state. Hence the excitation levels of both of these molecules can be interpreted to provide information on both the collisional temperature (from the low-lying rotational populations) and the presence and nature of extrathermal excitation mechanisms (from the higher-J states).

The interpretation of C_2 excitation is very similar to the results of H_2 rotational analysis as described above. One advantage over the H_2-based studies is that the absorption lines of C_2 are almost always optically thin and unsaturated (see **Figure 5**),

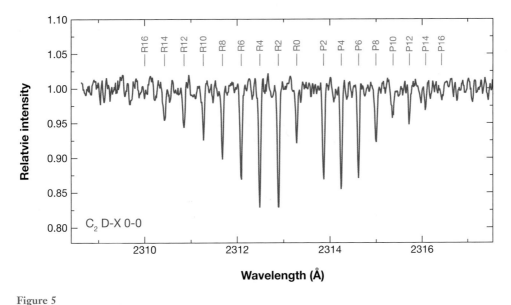

Figure 5

Spectrum of the Mulliken band of C_2 toward X Per (HD 24534), adapted from P. Sonnentrucker, D. Welty, J. Thorburn & D. York (submitted).

which simplifies the derivation of the relative rotational populations (saturation effects hamper the analysis of some H_2 high-J lines, especially in sightlines with significant reddening and large H_2 column densities). Because the energy separations among rotational levels in C_2 are smaller than in H_2, many more levels are observed (up to $J = 14$ in some cases), adding to the precision of the analysis. The C_2 excitation is also sensitive to a different portion of the radiation field, because the pumping occurs mainly through the Phillips band system near 1 µm rather than in the UV (as in H_2). The use of C_2 as an indicator of diffuse cloud physical conditions was modeled in detail by van Dishoeck & Black (1982). One of us (B.J.M.) has used this model to construct a calculator that derives values of both n and T from equivalent widths of C_2 lines; this calculator is available online at dibdata.org. The temperatures derived from C_2 are generally consistent with T_{01} of H_2, within the mutual uncertainties; for example, toward ζ Persei, $T(C_2) = 80 \pm 15$ K and $T_{01} = 58 \pm 8$ K, and toward o Persei, $T(C_2) = 60 \pm 20$ K and $T_{01} = 48 \pm 5$ K.

The spectrum of C_3 is more complicated, and Ádámkovics, Blake & McCall (2003) developed a novel method for obtaining rotational populations from spectra containing closely-spaced lines up to $J = 30$. As in the cases of H_2 and C_2, the low-J levels of C_3 are populated by collisions while the upper states show evidence of extrathermal excitation, which is most likely due to radiative pumping. Very good agreement with the earlier C_2 results was found for the inferred gas kinetic temperature, so the assumption of collisional equilibrium for the low-J states appears sound. But understanding the excitation of the high-J states requires detailed modeling. Roueff et al. (2002) carried out a detailed analysis of C_3 excitation in four lines of sight (including HD 210121; see Section 7.2.3 of this review), but have not yet published the details of their method.

5. CHEMICAL MODELS OF DIFFUSE CLOUDS

5.1. Physical and Chemical Processes

Most of the physical and chemical processes that occur in diffuse clouds are the same ones that operate in dense clouds; the major qualitative difference between these two types of clouds is that photoprocesses play an important role in diffuse cloud chemistry, whereas in dense clouds the large dust extinction suppresses them. The chemistry of diffuse and dense clouds has been reviewed by van Dishoeck (1998).

In terms of the heating and cooling balance of clouds, photons provide the dominant heating force in diffuse clouds via the photoelectric effect on dust, which injects relatively hot electrons into the gas. In terms of the ionization, abundant photons with energies >11.26 eV will photoionize nearly all of the carbon atoms in diffuse clouds, leaving a substantial fractional ionization $n(e)/n_H \sim 10^{-4}$, which is orders of magnitude higher than in dense clouds. Finally, in terms of chemistry, photodissociation substantially reduces the abundance of molecules, for example $H_2 + h\nu \rightarrow H + H$ or $CH + h\nu \rightarrow C + H$.

Aside from photon-induced reactions, much of the chemistry of diffuse clouds can be explained in terms of the gas-phase processes that occur in dense clouds. These processes include ion-molecule reactions (e.g., $CH_2^+ + H_2 \rightarrow CH_3^+ + H$), which typically proceed with a Langevin rate coefficient of $\sim 10^{-9}$ cm^3 s^{-1}; dissociative recombination with electrons (e.g., $H_3^+ + e^- \rightarrow H + H + H$), which typically have faster rate coefficients of $\sim 10^{-7}$ cm^3 s^{-1}; radiative association reactions, which can be very slow (e.g., $C^+ + H_2 \rightarrow CH_2^+ + h\nu$, k $\sim 10^{-16}$ cm^3 s^{-1}) or quite fast (e.g., $C^+ + C_4 \rightarrow C_5^+ + h\nu$, k $\sim 10^{-9}$ cm^3 s^{-1}) depending on the size of the reactants; and charge exchange reactions (e.g., $H^+ + O \rightarrow O^+ + H$, which is endothermic by only 227 K). Neutral-neutral reactions involving radicals (e.g., $CH + O \rightarrow CO + H$) probably also play an important role, but we have less information about their rate coefficients, which have only recently become the subject of sustained experimental efforts. Finally, collisions with cosmic rays (high energy bare nuclei produced in supernovae) are very important in ionizing atoms and molecules with high ionization potentials, especially H, He, and H_2.

The role of interstellar dust grains is perhaps the most uncertain of all. It is clear that H_2 is produced abundantly on grains, by H atoms sticking to grains, tunneling between different binding sites, and eventually combining to form H_2, which is ejected from the grain by its large enthalphy of formation. This reaction is critically important, as gas-phase formation mechanisms for H_2 are incredibly slow. [However, in the early universe before heavy elements and solid dust existed, H_2 must have formed in the gas phase, probably by associative detachment of H with H^-; McDowell 1961, Dalgarno & McCray 1973; see also the very recent summary by Glover et al. 2006].

It is still an open question whether grain surface production in diffuse clouds is efficient only for H_2 (because H atoms tunnel easily and H_2 has such a low freezing point), or whether other molecules can be efficiently synthesized through grain chemistry and returned to the gas phase. van Dishoeck (1998) has suggested that the low observed column density of NH may provide an upper limit on the formation rate of molecules other than H_2 on grains. If other grain-surface reactions prove to be important, this could revolutionize our understanding of diffuse cloud chemistry. One additional grain "reaction" that seems to be very important (Lepp et al. 1988, Liszt 2003, Wolfire et al. 2003) is the neutralization of atomic ions by grains or large molecules (e.g., $H^+ + \text{grain} \rightarrow H + \text{grain}^+$), which proceeds much more efficiently than gas-phase radiative recombination (e.g., $H^+ + e^- \rightarrow H + h\nu$).

5.2. Key Transitions in Diffuse Clouds

To a first approximation, the chemical conditions in a parcel of gas are determined by the state of hydrogen and carbon. If hydrogen is ionized (H^+), as in the hot ionized medium, essentially no chemistry occurs. If hydrogen is neutral and completely atomic (H), as in diffuse atomic clouds, a limited chemistry can occur but the efficient ion-molecule reactions are deprived of their best reagent, H_2. When hydrogen becomes molecular (H_2), as in diffuse molecular, translucent, and dense molecular clouds, chemistry becomes more effective. Within this realm, the state of carbon plays a determining role: if carbon is photoionized (C^+), as in diffuse molecular clouds, there are abundant electrons available to destroy molecular ions through dissociative recombination, and ion-molecule chemistry is partially quenched. As carbon begins to transition into the form of C or CO, in translucent clouds, the presence of the reactive C radical drives a more active chemistry. Once carbon becomes completely molecular (CO), as in dense clouds, electrons become much scarcer, and a rich ion-molecule chemistry can thrive.

5.2.1. The hydrogen transition. Within diffuse clouds, the state of (neutral) hydrogen is determined by the interplay of two processes: the production of H_2 from H atoms on grain surfaces, and the destruction of H_2 following absorption of photons. The rate of H_2 formation can be written (Hollenbach, Werner & Salpeter 1971) as $R_G = (1/2)\gamma \langle v_H \rangle n_g n(H) \langle \sigma_g \rangle$, where γ is the fraction of atoms striking a grain that eventually forms a molecule, $\langle v_H \rangle$ is the average velocity of H atoms, n_g and $n(H)$ are the number density of grains and H atoms, and $\langle \sigma_g \rangle$ is the mean cross-section of a grain. It is typically assumed (without evidence to the contrary) that γ and $\langle \sigma_g \rangle$ are independent of location in a cloud, and that $n_g \propto n_H$. With these assumptions, the H_2 formation rate scales as $T^{1/2} n_H n(H)$. Note that H_2 formation may not be as simple as these assumptions suggest (Biham et al. 2001, Cazaux & Tielens 2004).

The destruction of H_2 by photons in diffuse clouds is not a direct process, because the threshold for direct photodissociation is 14.7 eV. Instead, H_2 absorbs line photons in the Lyman (B-X) bands, inducing transitions from the ground state into various vibrational levels in the $B\,^1\Pi_u^+$ electronic state; these transitions have a certain probability of being followed by radiative decays into unbound dissociative (v > 14) levels of the electronic ground state. The Werner (C-X) bands make a substantially smaller contribution to dissociation (Spitzer 1978). The rate of H_2 destruction then depends only on the number density of H_2 molecules and on the radiation flux in the discrete transitions of the Lyman band. In the inner regions of a cloud, this radiation flux is attenuated by all of the H_2 molecules in the outer regions of the cloud, leading to a "self-shielding" of the H_2.

Models suggest that the transition to self-shielding is a fairly sharp one: Once a critical column density of H_2 has been built up, the fraction of hydrogen in the form of H_2 rises dramatically. This effect is seen in observations of H and H_2 by *Copernicus* (Savage et al. 1977): In sightlines with E(B-V) > 0.08 (corresponding to $N_H > 5 \times 10^{20}$ cm^{-2}), the column density of H_2 always exceeds 10^{19} cm^{-2}, whereas for E(B-V) <0.08, there is a wide range in H_2 column densities, presumably reflecting variations in number density in different sightlines. Models (e.g., van Dishoeck & Black 1986b) predict that even in the central regions of diffuse clouds with n $\sim$ 100 cm^{-3} a substantial fraction ($\sim$1/4) of hydrogen remains in atomic form. In clouds with densities $\sim$1000 cm^{-3}, the atomic hydrogen fraction declines to <1% (van Dishoeck & Black 1988).

5.2.2. The carbon transition. Roughly speaking, the state of carbon in diffuse clouds is determined by processes involving CO. In dense molecular clouds, nearly all carbon is in the form of CO, due to its chemical stability; that this is not the case in diffuse clouds is owing to the effect of photodissociation. CO is formed primarily following the reaction $C^+ + OH \rightarrow CO^+ + H$. When CO^+ is formed, it reacts with H_2 to form HCO^+, which then recombines with an electron to form CO. The

OH necessary for CO formation is formed following the slightly endothermic charge exchange between O and H^+ (formed by cosmic-ray ionization of H), followed by reactions of O^+ with H_2 to form OH^+, H_2O^+, and H_3O^+, followed by dissociative recombination. Other channels that may be important are the reaction of H_3^+ with O, and the neutral-neutral reaction of CH with O.

The major destruction pathway of CO is photodissociation. Much like the case of H_2, there are no pathways for direct photodissociation of CO at wavelengths longer than 912 Å. Unlike H_2, CO is destroyed by discrete absorptions into predissociated electronically excited states (which are quasi-bound with a finite lifetime). There are over 30 predissociated bands lying to the red of 912 Å (van Dishoeck & Black 1988, Viala et al. 1988), making the problem more complicated than in the case of H_2. With increasing depth into a diffuse cloud, the photodissociation rate will be reduced by self-shielding, and also by shielding by coincident lines of H and H_2.

The state of carbon results from the competition between production of CO by ion-neutral (and perhaps neutral-neutral) reactions and the (depth-dependent) destruction of CO by photons. Diffuse atomic and diffuse molecular clouds contain relatively little CO (it is all photodissociated to O and C, which is then photoionized to C^+). In translucent clouds, photoionization becomes less efficient and many models suggest a zone where carbon is predominantly in neutral atomic form (C). With sufficient depth (e.g., leading into a dense cloud) CO will become the dominant form of carbon. Once C or CO becomes dominant, the C^+ abundance drops by orders of magnitude. The chemical impact of this transition is profound, as it also results in a drop in the electron fraction and the removal of dissociative recombination as an important process in destroying molecular ions.

The exact location of the transition from C^+ to CO depends not only on A_V, but also on the number density, the strength and spectrum of the incident radiation field, and other parameters. According to the models of van Dishoeck & Black (1988) and Jannuzi et al. (1988), the cross-over point where $n(C^+) \sim n(CO)$ occurs at $A_V \sim 0.6$, or $N_H \sim 10^{21}$ cm^{-2} for clouds with $n_H \sim 700$ cm^{-3}. The C^+/CO transition is much less sharp than the H/H_2 transition, according to these models. Newer models (e.g., Neufeld et al. 2005) show a transition from C^+ to C around $A_V \sim 1$, and a distinct transition from C to CO around $A_V \sim 3$. Given the sensitivity of this transition to model parameters, these differences are not too surprising. However, the earlier models also used a too-low H_3^+ dissociative recombination rate coefficient, and adopted higher densities to fit the observed properties of specific clouds. Also, the newer models include the effect of C^+ recombination with large molecules (M.G. Wolfire, private communication).

5.3. Chemical Models

To our knowledge, a "complete" model of a diffuse cloud—that is, one where the complete physical and chemical structure is solved self-consistently—has not yet been attempted. Such a model would require treating the equations of hydrostatic equilibrium, thermal balance, radiative transfer, ionization balance, and chemical processes. In the following subsections we briefly discuss the models that have been constructed that treat some subset of these equations. Before doing so, we wish to note that a great number of publications have considered a small subset of the chemistry [e.g., CH and CH$^+$ chemistry; Federman (1982)]; while these have led to a much better understanding of parts of diffuse cloud chemistry, we focus here on more comprehensive numerical models. We also wish to direct the reader's attention to the model of Turner (2000), which considers the chemistry in both diffuse and dense clouds, with a focus on molecules observed at radio wavelengths.

5.3.1. Early models. The first comprehensive model of a diffuse cloud was reported by Glassgold & Langer (1974). In many ways, this model comes close to the "complete" model discussed above, in that it utilizes pressure, thermal, electrical, and chemical balance equations. Like most of the models that would follow, it is a one-dimensional model, and assumes steady state (that the formation rate for each species equals its destruction rate). It does not make an assumption for the total number density, but instead treats it as a free parameter, subject to pressure balance with an assumed external pressure. This model's greatest shortcoming is the very small set of species considered; it only explicitly treats the atoms H, He, C, and Ca (along with their ions) and the molecules H_2, H_2^+, and H_3^+. The same researchers (Glassgold & Langer 1975) later considered the transition of C^+ to CO, but not in the context of their comprehensive model.

A much more complete treatment of the chemistry was given by Black & Dalgarno (1977) and Black, Hartquist & Dalgarno (1978); their models included at least 18 atoms and more than 30 molecules. They modeled a diffuse cloud as consisting of two regions (a "cold core" and a "warm envelope") with constant number density and temperature, and varied these parameters (while maintaining pressure balance) to best fit the observed abundances of atomic and molecular species in clouds toward ζ Ophiuchi and ζ Persei.

5.3.2. van Dishoeck & Black models. The next major advance in diffuse cloud modeling came nearly a decade later, in a paper aptly entitled "Comprehensive Models of Diffuse Interstellar Clouds: Physical Conditions and Molecular Abundances" (van Dishoeck & Black 1986b), in which improved models are compared with the sightlines toward ζ Ophiuchi, χ Ophiuchi, ζ Persei, and o Persei. These models utilized a somewhat larger chemical network (schematic diagrams of the chemistry are given in their figures 4–8), improved the treatment of radiative transfer in the H_2 lines, and also computed the rotational excitation of H_2, C_2, and CO. Furthermore, these models assumed hydrostatic equilibrium, including the gravitational potential of the gas, yielding continuous density gradients through the clouds. One drawback of these models is that thermal balance was not explicitly treated, but instead polytropic equations of state were assumed. This work indeed represents the most comprehensive modeling of diffuse clouds that had been done at the time, and it has not yet been surpassed in the completeness of its treatment by subsequent models. One unfortunate feature of these models is that they adopted a value for the H_3^+ dissociative recombination rate that is about three orders of magnitude too small (this value was in fashion at the time of these models).

Subsequently, these models were extended to denser sightlines with higher visual extinctions, referred to by the researchers as "translucent" clouds. This work began with two models of the HD 169454 sightline (Jannuzi et al. 1988), and was then extended to even higher densities and visual extinctions along with an improved treatment of the photodissociation of CO (van Dishoeck & Black 1988). Direct comparisons to the HD 29647 and HD 147889 clouds were also made (van Dishoeck & Black 1989).

5.3.3. Meudon models. An ongoing program of modeling of diffuse clouds has been conducted in Meudon, beginning with the work of Viala (1986). This model, which assumes a constant number density and temperature, was applied to ζ Ophiuchi (Viala, Roueff & Abgrall 1988) and later to ζ Persei (Heck et al. 1993). Subsequent work has focused on the possible role of turbulence (e.g., Falgarone et al. 1995) or shocks (e.g., Flower & Pineau des Forets 1998) to explain the anomalous abundance of CH^+ (discussed below). The models were also extended by Le Bourlot et al. (1993) to treat dense photodissociation regions (PDRs), using a variety of isochoric, isobaric, and self-gravitating models. More details of the model, including the incorporation of deuterium chemistry, have been given by Le Petit et al. (2002). The "Meudon PDR model" has very recently been applied to the classical diffuse cloud toward ζ Persei (Le Petit et al. 2004), in an effort to understand the large observed column density of H_3^+ in that sightline (McCall et al. 2003), and has also been applied to the case of HD 34078 (Boissé et al. 2005); the application of this model to diffuse clouds is reviewed in Le Petit et al. (2006).

5.3.4. Cloudy model. An emerging development in this field is the recent application of the workhorse spectral synthesis code Cloudy (reviewed in Ferland 2003) to diffuse clouds. Shaw et al. (2006) have performed constant-density simulations to reproduce the observed column densities toward HD 185418, and deduce a slightly higher number density and a substantially higher cosmic-ray ionization rate than a previous study had indicated (Sonnentrucker et al. 2003). It will be exciting to see what impact Cloudy will have as additional sightlines are modeled.

5.4. Challenges to the models

Although the models of diffuse clouds are quite advanced (compared to those used for dense clouds) in terms of their treatment of physical processes (especially radiative processes), they are still in a relatively immature state in terms of the number of chemical species and reactions considered. Furthermore, the models are hampered by our lack of a good physical understanding of clouds, and especially of processes such as turbulence and their coupling to the chemistry. Not surprisingly, there are many mysteries in diffuse cloud chemistry left unsolved by the current generation of models. The eventual solution of these mysteries will likely lead us to an improved understanding of the physical and/or chemical conditions in the diffuse ISM.

5.4.1. CH$^+$. The large observed abundance of CH$^+$ (first identified by Douglas & Herzberg 1941) has been a constant problem for steady-state equilibrium models of the chemistry. The fundamental difficulty is that the reaction C$^+$ + H$_2$ → CH$^+$ + H is endothermic by 4650 K and does not proceed at interstellar temperatures of $\sim$100 K. The only other known formation mechanism is the radiative association reaction C$^+$ + H → CH$^+$ + hν (or C$^+$ + H$_2$ → CH$_2^+$ + hν followed by photodissociation of CH$_2^+$), which has a very small rate coefficient $\sim$10^{-16} cm^3 s^{-1} (Graff et al. 1983). Not only is CH$^+$ expected to be produced very slowly, but it is destroyed rapidly by reaction with H, H$_2$, or electrons. Consequently, models such as van Dishoeck & Black's (1986b) underpredict the CH$^+$ abundance by about two orders of magnitude.

There seems no way out of this quandary, unless some very important formation pathway is being overlooked or unless all of the destruction pathways are slower than expected (for example, due to quantal effects at low temperatures). The most popular solution to this problem has been to invoke some sort of nonequilibrium process that provides enough energy to overcome the endothermicity of the reaction C$^+$ + H$_2$ → CH$^+$ + H.

The challenge is to identify a physical mechanism that can drive this endothermic reaction without "breaking" the rest of the chemistry. For example, a simple invocation of a shock is difficult, as it would also drive other endothermic reactions such as O + H$_2$ → OH + H, thereby overproducing OH (which is well-explained by current models) and it would lead to velocity shifts between CH$^+$ and CH (which are not observed). To avoid these problems, attention has focused on magnetohydrodynamic (MHD) effects, which offer the possibility of accelerating C$^+$ ions with less effect on the temperature of neutrals. Although much work has been done on individual MHD shocks (e.g., Draine & Katz 1986), a more recent intriguing suggestion is that the CH$^+$ production may be due to ubiquitous but intermittent dissipation of MHD turbulence (e.g., Joulain et al. 1998, Falgarone et al. 2005); this suggestion may also help account for the observed rotational excitation of H$_2$.

5.4.2. H$_3^+$. The observed large column densities of H$_3^+$ in diffuse cloud sightlines (McCall et al. 2002) pose another challenge to chemical models. The problem is that H$_3^+$ is produced following cosmic-ray ionization of H$_2$ (an infrequent occurrence), and destroyed efficiently by dissociative recombination with electrons, which are abundant in diffuse clouds. Based on simple steady-state models, the H$_3^+$ abundance should be one to two orders of magnitude lower in diffuse clouds than in dense clouds, but H$_3^+$ column densities are observed to be comparable in the two environments.

There seem to be only three possible solutions to this puzzle: (*a*) an increase in the H$_2$ ionization rate, (*b*) a decrease in the electron recombination rate coefficient, or (*c*) a decrease in the electron abundance. A low recombination rate coefficient seems to be ruled out by recent storage-ring measurements at low temperature (McCall et al. 2004, Kreckel et al. 2005) and supporting theoretical calculations (Kokoouline & Greene 2003). A decrease in electron abundance, at least in one sightline, has been ruled out by the observation of H$_3^+$ toward ζ Persei, where the electron fraction is known from UV measurements of C$^+$ and H$_2$ (McCall et al. 2003). The possibility of a low electron abundance has not been ruled out in the higher extinction sightlines of McCall et al. (2002), but seems unlikely; a search is currently underway for H_3^+ in additional classical diffuse cloud sightlines (B. J. McCall, private communication).

The observations seem to suggest that the enhanced H_3^+ abundance is a by-product of higher ionization rates than normally assumed, either by X-ray and UV photoionization (Black 2000) or by an enhanced cosmic-ray ionization rate (McCall et al. 2002, 2003). MHD shocks have been considered, but are unable to produce H_3^+ (Le Petit, Roueff & Herbst 2004). The suggestion of a higher cosmic-ray ionization rate in diffuse clouds has recently been revived by Liszt (2003); this theory was shown to be marginally compatible with OH chemistry (Le Petit et al. 2004), and has received some theoretical support from Padoan & Scalo (2005). In attempting to reconcile their observations with gas-phase chemistry models, McCall et al. (2003) required an ionization rate some 40 times higher than the canonical value. The repercussions of this enhancement for other molecular species need to be fully analyzed, and very recent models (e.g., Le Petit, Roueff & Herbst 2004) have been only marginally consistent with the observations.

5.4.3. Other challenges. There are a host of other challenges to chemical models. One is explaining the abundance and excitation of C_3 (Ádámkovics, Blake & McCall 2003) and the observed low upper limits to the column densities of C_4 and C_5 (Maier, Walker & Bohlender 2004), which models predict to be more abundant than C_3 at large visual extinctions (Roueff et al. 2002). Models (e.g., van Dishoeck & Black 1986b) also underpredict the column densities of CO. Sufficient production of molecules observed at radio wavelengths, including HCO^+, C_2H, and C_3H_2, poses perhaps an even greater challenge. Finally, the existence of the large polyatomic molecules responsible for the unidentified infrared bands and the diffuse interstellar bands is perhaps the biggest mystery of all (see below).

6. VERY LARGE MOLECULES IN DIFFUSE CLOUDS

One of the most astonishing developments in diffuse cloud research in the past twenty years has been the realization that very large organic molecules may be highly abundant. Here we refer to molecules containing tens or hundreds of atoms, rather than the diatomics and several-atom molecules discussed above—and in fact much more massive than the largest species yet identified in dense molecular clouds.

There are two general lines of evidence for the presence of these large organics in the diffuse ISM: a series of infrared emission features known as the Unidentified Infrared Bands (UIBs); and the long-known but still unidentified diffuse interstellar bands (DIBs) seen in the visible absorption spectra of reddened stars.

6.1. Unidentified Infrared Bands (UIBs)

Gillette, Forrest & Merrill (1973) first noticed a series of unidentified infrared emission features in H II regions, and subsequent observations revealed the same bands in carbon-rich planetary nebulae and some carbon star atmospheres. The strongest UIBs occur at 3.3, 6.2, 7.7, and 11.3 μm, wavelengths indicative of vibrations in sp^2-hybridized hydrocarbons, such as aromatic molecules. Allamandola & Norman (1978) and Allamandola, Greenberg & Norman (1979) suggested that these features might be due to vibrational emission from complex hydrocarbons (in solid state) following UV excitation. Duley & Williams (1981) suggested polycyclic aromatic hydrocarbons (PAHs) as the responsible species, but still in solid form (loosely attached to small interstellar grains). Léger & Puget (1984) suggested that free PAH molecules, i.e., in the gas phase, could be responsible for the observed emission (for a comprehensive review of the PAH hypothesis, see Allamandola, Tielens & Barker 1989).

The general mechanism invoked to explain the UIBs is absorption of a UV photon, followed by relaxation to an excited vibrational state of the ground electronic state, and subsequent emission of IR photons as the molecule transitions to the ground vibrational state. This hypothesis was developed in the work of Sellgren (1984), who showed that only very small grains or large molecules could be stochastically heated sufficiently to produce the observed IR emission following the absorption of a single UV photon. As reviewed by Geballe (1997) and Peeters et al. (2004), UIBs have been observed in a wide range of astrophysical environments where gas and dust is exposed to radiation, including planetary nebulae, H II regions, reflection nebulae, post-AGB stars, young stellar objects, and novae.

The existence of UIB carriers in diffuse molecular clouds also seems probable. There have been some detections of the UIB bands in absorption (Sellgren et al. 1995, Bregman et al. 2000, Chiar et al. 2000, Song et al. 2003), but these have been in dense cloud sightlines or toward the Galactic Center, where both dense and diffuse gas are present. Observations using ISO (Kahanpää et al. 2003) and IRTS (Sakon et al. 2004) have shown the presence of the UIBs in nondescript regions of the galactic plane, suggesting that the UIBs may be prevalent throughout the diffuse ISM. Most researchers in the field agree that the UIBs are caused by some sort of aromatic hydrocarbons, or carbonaceous material that contains aromatic groups. As discussed by Tokunaga (1997), the identification of the UIBs with gas-phase classical PAH molecules (such as coronene) is very popular but is not definitive, and other potential carriers could include very small particles of materials such as quenched carbonaceous composite (QCC), hydrogenated aromatic hydrocarbon (HAC), and coals.

Despite the uncertainties surrounding the exact nature of the carriers of the UIBs, it is clear that they represent an important component of the molecular and/or grain inventory of the ISM in general, and likely also in diffuse molecular clouds.

6.2. Diffuse Interstellar Bands (DIBs)

The first notice given to diffuse unidentified features in the spectra of distant stars came more than 80 years ago, when Heger (1922) reported, on the basis of spectra obtained in 1919, a pair of "stationary" features at 5780 and 5797 Å in spectroscopic binaries. A decade later, Merrill and others took up the pursuit of these features (e.g., Merrill 1934, 1936). The term "diffuse interstellar bands" was coined: "diffuse" because the absorption features are broader than atomic lines and resemble typical molecular bands, and "interstellar" because by then there was general acceptance of the notion that the "stationary" lines were formed in the space between the stars.

Several prominent astronomers of the day considered it likely that the DIBs were formed by molecules in the ISM (e.g., Russell 1935, Eyster 1937, Swings 1937), but in time the emphasis swung to solid-state absorption as the origin of the DIBs (e.g., Herbig 1963, Duley 1968, Huffman 1970, and many others—see review by Snow 1995) as formation mechanisms for maintaining a sufficient population of gas-phase molecules were not known. But in the 1960s and 1970s, when the first radio-wavelength detections of molecules occurred (Weinreb et al. 1963, Rank et al. 1971) and concurrently astronomers realized that rapid molecular formation through ion-neutral reactions was possible, molecules in the diffuse ISM received new respect. The notion that the DIBs are formed by molecules regained favor, and several researchers proposed specific mechanisms for producing the observed features by molecular transitions, including some types of transitions that could occur in relatively small and simple species (e.g., Danks & Lambert 1976, Douglas 1977). Smith, Snow & York (1977) summarized arguments against a solid-state origin of the DIBs and those favoring a molecular carrier. The most powerful of arguments favoring gas-phase carriers are the observed constancy of DIB wavelengths and profiles from sightline to sightline (these should vary if solid-state transitions were responsible; see, for example, Smith, Snow & York 1977; also M. Drosback, T. Snow, J. Thorburn, S. Friedman, L. Hobbs, et al., in preparation) and the existence of fine structure in some DIBs (e.g., Sarre et al. 1995).

As suggested by Herbig (1993, 1995), the majority of the strong DIBs are most abundant in the atomic hydrogen gas and are not closely associated with dense molecular clouds or perhaps even diffuse molecular clouds. These inferences are based on the fact that the correlations of DIBs with atomic hydrogen are much stronger than correlations with H_2, and that the DIBs are weak (relative to reddening) in dense clouds (e.g., Wampler 1966; Adamson, Whittet, & Duley 1991). Thus many of the DIB carriers appear to exist primarily in the diffuse atomic clouds, where hydrogen is atomic and the radiation environment is harsh. However, a subset of the DIBs (called the "C_2 DIBs"; Thorburn et al. 2003) does appear to be abundant in diffuse molecular clouds (Ádámkovics, Blake & McCall 2005). The DIB carriers are likely to be fairly large molecules and the breadth of the DIBs is likely due to unresolved rotational structure and/or intramolecular vibrational relaxation (IVR). The behavior of the DIBs with respect to incident radiation field further suggests that at least some DIB carriers may be ionized molecules (e.g., Sonnentrucker et al. 1997).

Although the exact nature of the molecules responsible for the DIBs is still not known, some clues can be gleaned from observational evidence. It is widely assumed that the DIB carriers are composed of cosmically abundant elements (such as H, C, O, and N), out of concern that molecules containing trace elements may not be abundant enough to account for the large number of DIBs (at least 200; Tuarisig et al. 2000) and their overall strength. It is also widely assumed that DIBs are "organic" in nature, because carbon has a sufficiently rich chemistry to support the wide variety of individual species probably needed to explain the DIBs (for a review of the arguments favoring large organic species as DIB carriers, see Snow 2001, and references cited therein).

The following simple order-of-magnitude argument shows that there is enough carbon available to produce the DIBs. A hypothetical line of sight with $N_H \sim 10^{22}$ cm^{-2} (roughly equivalent to the most reddened stars routinely observed for DIBs) would possess a total equivalent width of DIBs of about 22 Å. Assuming the DIB transitions are unsaturated, the column density of DIB carriers is of order $N = 10^{14}/f$, where f is the typical oscillator strength for the transitions. If each carrier molecule contains m carbon atoms, then the required carbon column density is $N_C = 10^{14}$ m/f. According to interstellar gas abundance studies (e.g., Snow & Witt 1995), the amount of carbon left over after accounting for the observed gas-phase carbon plus the amount of carbon needed to satisfy dust models is at most 10 ppm, or a column density of roughly 10^{17} cm^{-2} toward our hypothetical reddened star. As long as the ratio m/f is less than 10^3, there should be enough carbon available. Because the f-value may approach or even exceed unity for allowed transitions in large molecules, there is plenty of carbon. This argument does not immediately rule out atoms containing less abundant elements, as long as f is large and m is small.

Much of the speculation about which specific molecules form the DIBs is centered on two classes of molecules: PAHs and carbon chains. The speculation about PAHs is motivated by the likely aromatic nature of the UIB carriers and their probable presence in diffuse clouds; in contrast, the speculation about carbon chains is motivated by the fact that many of the molecular species observed in cold dense molecular clouds are carbon chains (though carbon chains represent only a small fraction of the total molecular mass in these clouds—the apparent preponderance of carbon chains is in part a selection effect, because these species have high dipole moments and are therefore more likely to be detectable than other classes of molecules). Constraints on possible carriers in the PAH and carbon chain families have recently been discussed by Ruiterkamp et al. (2005) and Maier, Walker & Bohlender (2004), respectively. However, we should keep in mind the rich diversity of chemistry, and consider the likelihood that we have simply not yet thought of the best candidates for the DIB carriers.

Because many of the molecules proposed as possible carriers have accessible electronic transitions in the UV, useful insight into the nature of the DIB carriers may be gained by extending the study of DIBs to UV wavelengths. However, several obstacles make this difficult: the far-UV extinction rise limits studies to much lower A_V than optical observations; hot stars have intrinsically complex UV spectra; and other atomic and molecular interstellar lines create confusion, especially below 1108 Å where strong H_2 bands begin to dominate the spectra of reddened stars. Two general searches for UV DIBs have been carried out without any significant detections (Snow, York & Resnick 1977, using IUE data; Clayton et al. 2003, on the basis of high-quality HST/STIS spectra). Tripp, Cardelli & Savage (1994) reported a weak feature near 1369 Å in a HST/GHRS spectrum of ζ Oph that they suggested might be a UV DIB, but Watson (2001) later assigned this feature to a predissociative transition of the CH molecule. One of the goals of the science team for the Cosmic Origins Spectrograph is to extend the search for UV DIBs to greater extinction than has been probed so far.

Although observational studies can provide important insights into the nature of the DIB carriers, their rigorous identification will only come through laboratory studies showing precise and specific spectral matches with the observed astronomical features. Such laboratory studies are challenging, because most DIB carrier candidates are not stable under terrestrial conditions and are difficult to produce in large quantities for spectroscopic study. Until recently, most work was done using frozen rare gas matrices (where large number densities could be accumulated), but such "matrix spectra" suffer from wavelength shifts and band broadening, which are difficult to quantify with sufficient precision to unambiguously compare with interstellar (gas-phase) absorption features. The formidable task of measuring gas-phase spectra of carbon chains and PAH ions has been undertaken by a few groups in the past decade, with some success (Tulej et al. 1998, Ruiterkamp et al. 2002, Biennier et al. 2003), but as yet these efforts have yielded no specific identifications of DIB carriers. There have been some near matches (such as in the case of C_7^-; Tulej et al. 1998), but these have not stood up to the standards of high-resolution spectroscopy once higher quality laboratory (Lakin et al. 2000) and astronomical (McCall et al. 2001) spectra became available. In addition to these spectroscopic studies, the chemical viability of various DIB carrier candidates has been tested through laboratory measurements of reaction rates with the most common neutral reaction partners expected in the diffuse ISM, such as atomic and molecular hydrogen, atomic oxygen, and atomic nitrogen (Snow et al. 1995; Le Page et al. 1999a,b; Barckholtz, Snow & Bierbaum 2001).

The problem of the DIBs, which has been called the longest standing mystery in all of spectroscopy, has received much more attention within the past decade or so than previously, when these unidentified features were widely considered to be curiosities but not necessarily significant. The enhanced focus on the DIBs has come about owing to the availability of very high-quality spectra that allow very precise measurements of wavelengths and profiles, the growing recognition that the DIB carriers may represent an enormous reservoir of complex organics in the diffuse ISM, and the involvement of chemists as well as astronomers in the pursuit. In contrast to the UIBs, which are very broad and represent vibrational transitions that generally indicate only the type of chemical bond involved, the DIBs are relatively narrow and represent electronic transitions that tend to be unique for every molecule. These properties of the DIBs provide hope that the coming decades may finally yield the identification of at least some of the DIB carriers. The ultimate identification of the DIB carriers will likely more than double the known inventory of interstellar molecules, and will open a whole new window into the chemistry of diffuse clouds. Additionally, once the DIB carriers are known and their chemistry understood, the DIBs may represent a powerful multidimensional probe of the physical conditions (temperature, density, radiation field, etc.) in diffuse clouds.

6.3. The Formation of Large Molecules in Diffuse Clouds

Gas-phase chemistry models for diffuse clouds (e.g., van Dishoeck & Black 1986b, Le Petit et al. 2004) normally do not address the formation of large species such as PAHs or carbon chains—though some work on modeling their effects on the chemistry and heating of clouds has been done (Lepp & Dalgarno 1988, Lepp et al. 1988).

One possibility is that these large organic species form in the gas phase in dense molecular clouds, and are then cycled back into the diffuse ISM (e.g., Bettens & Herbst 1996). But it is not clear that sufficient quantities of these species to account for the UIBs and the DIBs can be produced in this way. Another possibility is that the large organics are formed in the outflows from carbon-rich giant stars late in their evolution (Frenklach & Feigelson 1989; Latter 1991; Cherchneff, Barker & Tielens 1992; Giard et al. 1994). Under highly constrained conditions (density, velocity law of the outflow, and initial temperature) it may be possible to form the observed quantities in this way, and it is widely accepted that carbon star atmospheres may be a major source of large organics in the general ISM. Yet another closely related possibility is that large molecules in the diffuse ISM are the result of shock-induced destruction of graphitic or carbonaceous grains, some of which may have initially formed in carbon star outflows. For example, graphite essentially consists of stacked layers of fused carbon rings with very weak bonding between layers. When such a grain is shattered in a shock, the fragments will be planar fused carbon ring species, which, if they acquire peripheral hydrogen atoms, would become PAHs.

Any discussion of the origin of the large organics in diffuse clouds is necessarily speculative at this point. But apparently these species are present, and it is of great interest to understand how they came to be.

7. SPECIFIC CLOUDS OF SPECIAL INTEREST

Having reviewed the general properties and the models of diffuse clouds, here we turn our attention to a small number of specific sightlines that are often cited, either as "typical" diffuse clouds, or as possibly dominated by diffuse molecular clouds.

7.1. Prototypical Examples of Molecule-Bearing Diffuse Clouds

Many of the studies of diffuse cloud molecules cited in this review have been based on a small number of widely-studied "typical" sightlines toward certain hot stars. These stars get most of the attention not necessarily because their sightlines are typical, but because of their bright apparent magnitudes, their intrinsically clean spectra, and the evidence for their having significant molecular column densities. Hence these cases are in fact highly selected with a bias toward high molecular abundances.

7.1.1. ζ Ophiuchi (HD 149757; O9Ve, V = 2.58, Av = 1.06). By far the most frequently cited diffuse cloud sightline is that toward ζ Ophiuchi. The total hydrogen column density is 1.4×10^{21} cm^{-2}, with some 56% of the nuclei in molecular form (Morton 1975). Most of the species cited in **Table 2** have been observed in this line of sight, many of them first detections. Given this history, ζ Oph has been a target of intense scrutiny in several surveys of interstellar lines, including the very high-S/N visible-wavelength study of Shulman, Bortolot & Thaddeus (1974), in which several new species were detected; and the classic *Copernicus*-based survey of UV spectral lines carried out by Morton (1974, 1975), which remains a standard in the field.

Table 2 Molecules detected in diffuse molecular clouds

Weight	Species	Method	Target	$N(X)/N_H$	Reference
2	H_2	UV	ζ Oph	0.56	1
3	HD	UV	ζ Oph	4.5 (−7)	2
3	H_3^+	IR	ζ Per	5.1 (−8)	3
13	CH	Optical	ζ Oph	1.5 (−9)	4
13	CH^+	Optical	ζ Oph	2.4 (−8)	5
14	$^{13}CH^+$	Optical	ζ Oph	3.5 (−10)	6
15	NH	Optical	ζ Oph	6.2 (−10)	7
17	OH	UV	ζ Oph	3.3 (−8)	8
24	C_2	Optical	ζ Oph	1.3 (−8)	9
25	C_2H	mm abs.	BL Lac	1.8 (−8)	10
26	CN	Optical	ζ Oph	1.9 (−9)	11
27	HCN	mm abs.	BL Lac	2.6 (−9)	12
27	HNC	mm abs.	BL Lac	4.4 (−10)	12
28	N_2	UV	HD 124314	3.1 (−8)	13
28	CO	UV	X Per	6.4 (−6)	14
29	HCO^+	mm abs.	BL Lac	1.5 (−9)	15
29	HOC^+	mm abs.	BL Lac	2.2 (−11)	15
29	^{13}CO	UV	X Per	8.9 (−8)	16
29	$C^{17}O$	UV	X Per	7.4 (−10):	16
30	$C^{18}O$	UV	X Per	2.1 (−9):	16
30	H_2CO	mm abs.	BL Lac	3.7 (−9)	17
36	C_3	Optical	ζ Oph	1.1 (−9)	18
36	HCl	UV	ζ Oph	1.9 (−10)	19
38	C_3H_2	mm abs.	BL Lac	6.4 (−10)	10
44	CS	mm abs.	BL Lac	1.6 (−9)	20
64	SO_2	mm abs.	BL Lac	≤8.2 (−10)	20

Note: To facilitate comparisons we include, where possible, data on different molecules for the same source. One exception is CO, where there are ζ Oph data but we chose to list X Per instead, to allow a comparison among all the observed isotopomers. For millimeter-wave studies we have listed only BL Lac because an estimated $N_H \sim 1.7 \times 10^{21}$ cm^{-2} was available. Liszt & Lucas have detected, in other lines of sight, the additional species SO, SiO, H_2S, and HCS^+. Other values of N_H are 1.4×10^{21} cm^{-2} (ζ Oph), 1.6×10^{21} cm^{-2} (ζ Per), 3.1×10^{21} cm^{-2} (HD 124314), and 2.2×10^{21} cm^{-2} (X Per).
References: (1) Savage et al. 1997; (2) Lacour et al. 2005; (3) McCall et al. 2003; (4) Danks et al. 1984; (5) Lambert & Danks 1986; (6) Stahl & Wilson 1992; (7) Crawford & Williams 1997; (8) Roueff 1996; (9) Lambert et al. 1995; (10) Lucas & Liszt 2000b; (11) Federman et al. 1984; (12) Liszt & Lucas 2001; (13) Knauth et al. 2004; (14) Sheffer et al. 2002a; (15) Liszt et al. 2004; (16) Sheffer et al. 2002b; (17) Liszt & Lucas 1995; (18) Oka et al. 2003; (19) Federman et al. 1995; (20) Lucas & Liszt 2002.

This sightline has also been the subject of many chemical modeling efforts. For example Black & Dalgarno (1977) developed the first comprehensive diffuse cloud model taking into account not only an extensive network of chemical reactions but also physical processes and the effects of dust extinction. Their best fit to the line of sight consisted of two clouds (whose existence and velocity separation were inferred from Doppler structure in the observed absorption lines): one low temperature, high-density cloud containing most of the mass and nearly all of the molecules; and a secondary cloud of higher temperature and lower density, blue-shifted with respect to the main component, that is more highly ionized and produces the high-J lines of H_2. Later, more comprehensive models of diffuse cloud chemistry often used the ζ Oph sightline for comparison (e.g., van Dishoeck & Black 1986b; see also comments in Heck et al. 1993).

7.1.2. X Persei (HD 24534; O9Ve, V = 6.13, Av = 2.05). As an X-ray source (attributed to a neutron star companion), X Per is the only early-type star bright enough in both UV and X-ray wavelengths to allow high-S/N observations of both the gas absorption lines and the gas-plus-dust X-ray absorption edges, thus allowing a complete analysis of the depletions, the dust content, and the chemical and physical state of the gas in the same line of sight. Recently Cunningham, McCray & Snow (2004) have taken advantage of this to evaluate the distribution of oxygen between gas and dust, yielding useful constraints on dust models.

Earlier, the sightline toward X Per had been recognized as having a very rich molecular content (Lien 1983, 1984), with strong optical CH and CN lines as well as CO emission observed at millimeter-wavelengths and UV absorption bands of CO observed with the IUE. More recently, UV CO bands have been observed with the HST/GHRS and analyzed by Kaczmarczyk (2000a,b), who used a statistical fitting method to several bands to derive a column density of roughly 10^{-5} times the total hydrogen abundance in the line of sight. More recently Sheffer et al. (2002b) found similar results based on STIS spectra of intersystem CO lines. Though X Per is fainter than ζ Oph, it shows more promise as a standard for observing and interpreting diffuse molecular clouds, particularly because of its unique status as an X-ray source.

7.1.3. Other sightlines. Other widely-studied "typical" diffuse cloud sightlines that have been the focus of various molecular studies include ζ Persei and o Persei (Chaffee 1974; Snow 1975, 1976b, 1977; Snow, Lamers & Joseph 1987; Crutcher & Watson 1976a,b; McCall et al. 2003); and several of the bright OB stars in the ρ Ophiuchi cloud (Carrasco, Strom & Strom 1973; Snow & Cohen 1974; Snow & Jenkins 1980; Snow 1983) where the UV dust extinction is low and the molecular fraction small.

7.2. Three Sightlines Dominated by Diffuse Molecular Clouds

Most sightlines are best interpreted as due to a mixture of diffuse atomic and diffuse molecular clouds, but there are a few cases where the diffuse atomic material seems to be relatively insignificant or even absent, thus revealing the properties of diffuse molecular clouds alone. Here we describe three such cases, each of which is characterized by an extremely steep far-UV extinction rise, exceptionally high molecular abundances, and relatively weak strengths of the most common diffuse interstellar bands (which are mostly associated with diffuse atomic gas). A fourth case, that toward HD 29647 behind the Taurus dark cloud (Snow & Seab 1980, Crutcher 1985), is confounded by extreme observational difficulties because of a profusion of photospheric lines in the visible spectrum.

7.2.1. HD 62542 (B5V; V = 8.04; Av = 1.1).

This star lies behind a portion of the Gum nebula, and was found to have a UV extinction curve very similar to that of HD 29647, with a very broad and weak 2175 Å bump and a steep far-UV rise (Cardelli & Savage 1988), which may be attributed to mantle growth over small carbonaceous grains, suppressing the extinction bump (Mathis & Cardelli 1992, Mathis 1994). The infrared extinction curve is very close to the galactic average, in strong contrast to the unusual UV curve (Whittet et al. 1993), suggesting that no significant grain growth has occurred. Visible-wavelength spectroscopic studies by Cardelli et al. (1990) and by Gredel, van Dishoeck & Black (1993) showed that the molecular abundances are very high for the amount of extinction, and that the density is high enough to cause excess excitation of CN above the radiation temperature of the cosmic microwave background. Detailed modeling of the CN rotational excitation (Black & van Dishoeck 1991) yielded a gas kinetic temperature of 50 K and a cloud density in the range from 500 to 1000 cm^{-3}. There is enough UV flux for FUSE observations, resulting in a direct measurement of N(H$_2$) (Rachford et al. 2002), but N(H) cannot be measured owing to the late spectral type of the star (at spectral type B5 and later, stellar Lyman-α absorption totally obscures the interstellar line), so f_{H_2} is unknown. Ádámkovics, Blake & McCall (2005) have used optical surrogates to estimate $f^N_{H_2} \sim 0.8$. High-resolution optical spectra show that the line of sight is dominated by a single component (D. Welty, private communication).

As discussed by Snow et al. (2002), the DIBs in this sightline are extremely weak. Ádámkovics, Blake & McCall (2005) have pressed the issue with higher-resolution and S/N spectra, and now have reported the clear detection of several DIBs, most of them among the so-called "C$_2$ DIBs." They conclude that the line of sight toward HD 62542 is dominated by a single diffuse molecular cloud core, without any accompanying outer diffuse atomic cloud. This is consistent with the speculation by Cardelli et al. (1990) that the diffuse outer portions of the cloud could have been evaporated by the intense UV radiation of the nearby O stars ζ Puppis and γ^2 Velorum.

7.2.2. HD 204827 (B0V; V = 7.94; Av = 3.0).

This sightline was identified by Cardelli, Clayton & Mathis (1989) as having anomalously steep far-UV extinction, and recent measurements have revealed very high molecular abundances. This line of sight is distinguished by having by far the strongest C$_3$ absorption yet observed (Oka et al. 2003). The breadth of the C$_3$ lines suggests the possibility that there are at least two strong molecular components—in addition, Pan et al. (2004) found several velocity components in CN toward this star. Unfortunately, because of its relatively faint apparent magnitude and especially because of its very steep UV extinction rise, HD 204827 is far too dim in the UV for HST or FUSE observations. As a consequence we have no direct information on its hydrogen column density or molecular fraction. Using molecular and DIB proxies, Ádámkovics, Blake & McCall (2005) have estimated a H$_2$ molecular fraction of about 0.6, and suggest that this sightline consists

Table 3 Molecules not detected in diffuse molecular clouds

Weight	Species	Method	Target	$N(X)/N_H$	Reference
14	CH_2	UV	ζ Oph	$\leq 2.4\,(-8)$	1
15	NH^+	UV	ζ Oph	$<8.7\,(-10)\,f^{-1}$	2
18	H_2O	UV	HD 154368	$<2\,(-9)$	3
24	NaH	UV	ζ Oph	$<1.2\,(-9)$	2
25	MgH	UV	ζ Oph	$<3.0\,(-8)$	2
25	MgH^+	UV	ξ Per	$<1.0\,(-10)\,f^{-1}$	2
26	CN^+	UV	ξ Per	$<5.3\,(-10)$	2
28	CO^+	UV	ζ Oph	$<4.1\,(-9)$	2
28	AlH	UV	ζ Oph	$<1.3\,(-10)\,f^{-1}$	2
29	SiH	UV	ζ Oph	$<1.0\,(-9)$	2
29	N_2H^+	mm abs.	various	—[b]	4
30	NO	UV	ξ Per	$<1.2\,(-6)$	2
30	NO^+	UV	ζ Oph	$<5.9\,(-8)$	5
32	O_2	UV	o Per	$<1.8\,(-10)\,f^{-1}$	2
33	SH	UV	o Per	$<7.6\,(-11)\,f^{-1}$	2
37	C_3H	mm abs.	various	—[b]	6
41	CaH	UV	ζ Oph	$<4.9\,(-11)\,f^{-1}$	2
41	CH_3CN	mm abs.	various	—[b]	4
44	CO_2	UV	ζ Oph	$<1.5\,(-9)\,f^{-1}$	2
44	SiO	UV	ζ Oph	$<1.7\,(-10)\,f^{-1}$	2
48	C_4	Optical	ζ Oph	$<2.6\,(-10)$	7
49	C_4H	mm abs.	various	—[b]	6
50	HC_4H^+	Optical	HD 207198	$<1.4\,(-9)$	8
52	$C_4H_4^+$	Optical	—[a]	$<2\,(-10)$	9
60	C_5	Optical	ζ Oph	$<5.3\,(-11)$	7
61	C_5H	Optical	—[a]	$<1\,(-10)$	10
73	C_6H	Optical	HD 207198	$<3\,(-10)$	8
74	HC_6H^+	Optical	HD 210839	$<1.3\,(-10)$	8
75	HC_5N^+	Optical	HD 207198	$<1.7\,(-10)$	8
76	$C_6H_4^+$	Optical	—[a]	$<5\,(-11)$	11
76	NC_4N^+	Optical	HD 210839	$<1\,(-10)$	8
85	C_7H	Optical	—[a]	$<1\,(-10)$	10
86	HC_7H	Optical	—[a]	$<1\,(-9)$	12
97	C_8H	Optical	HD 207198	$<3\,(-10)$	8
98	HC_8H^+	Optical	HD 207198	$<3\,(-10)$	8
100	$C_8H_4^+$	Optical	—[a]	$<6\,(-11)$	9
109	C_9H	Optical	—[a]	$<1\,(-10)$	10
110	HC_9H	Optical	—[a]	$<1\,(-9)$	12
121	$C_{10}H$	Optical	HD 207198	$<5\,(-10)$	8
134	$HC_{11}H$	Optical	—[a]	$<5\,(-9)$	12

(*Continued*)

Table 3 (*Continued*)

Weight	Species	Method	Target	N(X)/N$_H$	Reference
145	C$_{12}$H	Optical	HD 207198	<5 (−10)	8
158	HC$_{13}$H	Optical	—[a]	<5 (−9)	12
720	C$_{60}$	Optical	Cyg OB2 8A	<4.8 (−11)	13

[a]Limit is not from a dedicated search, but from absence of known DIB coincident with laboratory wavelength; N$_H$ is assumed to be 10^{22} cm^{-2}.
[b]No value of N$_H$ is available for these extragalactic sightlines.
References: (1) Lyu et al. 2001; (2) Tabulated in Snow 1979; (3) Spaans et al. 1998; (4) Liszt & Lucas 2001; (5) Federman et al. 1995; (6) Lucas & Liszt 2000b; (7) Maier et al. 2004; (8) Motylewski et al. 2000; (9) Araki et al. 2004; (10) Ding et al. 2002; (11) Araki et al. 2005; (12) Ding et al. 2003; (13) Herbig 2000.

of both a diffuse cloud core (diffuse molecular cloud) and an envelope (diffuse atomic cloud material).

7.2.3. HD 210121 (B3V; V = 7.67; A$_V$ = 1.2). This star lies at high galactic latitude and appears to lie behind the core of a molecular cloud (Désert, Bazell & Boulanger 1988; de Vries & van Dishoeck 1988). Gredel et al. (1992) published a comprehensive analysis of this sightline, including millimeter-wave emission observations and optical absorption line data. Welty & Fowler (1992) carried out another broad study of this line of sight, incorporating high-resolution optical spectra with UV measurements of the extinction curve and of interstellar gas absorption lines (obtained with the IUE). Based on these two analyses, the radiation field impinging on the cloud appears to be less intense than in most sightlines, and the material along the sightline appears to be clumpy in nature. Based on an analysis of C$_3$ absorption, Roueff et al. (2002) inferred that the characteristic density is several times higher than that in more classical diffuse clouds. Larson, Whittet & Hough (1996) examined the extinction and polarization properties of the dust, and concluded that small grains dominated and little grain growth has occurred.

It appears that the line of sight toward HD 210121 is a very good case where a single cloud complex can be studied without interference from foreground material, because of the high galactic latitude of the star and the evidence of a rich chemistry not dominated by radiation.

8. CONCLUDING REMARKS

The past three decades have seen enormous advances in our understanding of the physical and chemical complexity of diffuse interstellar clouds. In this review we have attempted to summarize recent work on diffuse clouds containing a molecular component, which we call diffuse molecular clouds under the nomenclature that we are proposing for categorizing interstellar clouds.

We have summarized the known interstellar molecules in diffuse molecular clouds and have described how the observations can be used to derive information on cloud physical conditions such as density and radiation field intensity. We have found that much of the chemistry of diffuse molecular clouds can be well represented by existing models—though no models that take into account all of the processes known to be important are yet available.

Several outstanding mysteries remain: The origins of some simple species (e.g., CH^+ and HCO^+) are not understood; the high abundance of H_3^+ in diffuse molecular gas is a mystery; the possible role of grain surface reactions in the formation of molecules other than H_2 is as-yet not defined; and perhaps most puzzling are the origin and role of large organic species whose presence is inferred from observations of IR emission features and the optical diffuse interstellar bands (we do not even know whether the same family of molecules is responsible for both phenomena).

We can suggest several areas for future research, some realistic in the near term and others that will take time. First and foremost, existing telescopes and spectrographs can significantly improve the optical and IR spectra of reddened stars to the point where species expected to be abundant will either be detected or their absence will conflict with existing models. The list of species in **Table 3**, which consists of upper limits only, provides a starting point.

Another potentially fruitful area for future work might be to obtain high-sensitivity IR absorption spectra of reddened stars behind diffuse molecular clouds, to achieve overlap between optical (and possibly UV) absorption line measurements and IR absorption lines in the same lines of sight. A wealth of new physical data will become available if this can be done. Like most ISM absorption-line projects on reddened stars, this will require substantial amounts of observing time.

We advocate further intensive searches for millimeter-wave absorption due to polyatomic species, along the line of the Liszt & Lucas surveys, toward additional sources for which optical and UV (and in principle, IR) absorption-line spectra exist or can be obtained.

Finally, we emphasize the need for more sensitive UV spectroscopic instruments to probe denser clouds through absorption line techniques. When/if the Cosmic Origins Spectrograph is installed aboard the HST, it may become feasible to perform sensitive searches for the UV electronic transitions of PAHs and other large organic species.

ACKNOWLEDGMENTS

We are deeply indebted to the following peers who read various versions of this manuscript and contributed helpful comments and suggestions: John Black, Alex Dalgarno, Steve Federman, Roger Ferlet, Tom Geballe, Harvey Liszt, John Maier, John Mathis, Takeshi Oka, Evelyne Roueff, Peter Sarre, Paule Sonnentrucker, Xander Tielens, Ewine van Dishoeck, Dan Welty, Mark Wolfire, and Don York. T. P. S. acknowledges the help of Lynsi Aldridge in preparing the reference list, and B. J. M. thanks Susanna Widicus Weaver for a careful reading of the manuscript. We thank Meredith Drosback, Brian Rachford, Achim Tappe, Dan Welty, and Mark Wolfire for assistance, especially in assembling the data for the illustrations. T. P. S. acknowledges NASA grants NAG5-11487 and NG04GL34G, and NASA contract NAG5-12279 for support of research used in the preparation of this review. B. J. M. acknowledges support from the NSF (CHE 04-49592), NASA (NNG05GE59G), the Camille and Henry Dreyfus Foundation, and the American Chemical Society Petroleum Research Fund.

LITERATURE CITED

Ádámkovics M, Blake GA, McCall BJ. 2003. *Ap. J.* 595:235–46
Ádámkovics M, Blake GA, McCall BJ. 2005. *Ap. J.* 625:857–63
Adams WS. 1941. *Ap. J.* 93:11–23
Adamson AJ, Whittet DCB, Duley WW. 1991. *MNRAS* 252:234–45
Allamandola LJ, Greenberg JM, Norman CA. 1979. *Astron. Astrophys.* 77:66–74
Allamandola LJ, Norman CA. 1978. *Astron. Astrophys.* 66:129–35
Allamandola LJ, Tielens AGGM, Barker JR. 1989. *Ap. J. Suppl.* 71:733–75
Araki M, Cias P, Denisov A, Fulara J, Maier JP. 2004. *Can. J. Chem.* 82:848–53
Araki M, Motylewski T, Kolek P, Maier JP. 2005. *PCCP* 7:2138–41
Barckholtz C, Snow TP, Bierbaum VM. 2001. *Ap. J.* 547:L171–74
Bettens RPA, Herbst E. 1996. *Ap. J.* 468:686–93
Biennier L, Salama F, Allamandola LJ, Scherer JJ. 2003. *J. Chem. Phys.* 118:7863–72
Biham O, Furman I, Pirronello V, Vidali G. 2001. *Ap. J.* 553:595–603
Black JH. 2000. *Philos. Trans. R. Soc. London Ser. A* 358:2515
Black JH, Dalgarno A. 1973. *Ap. J.* 184:L101–4
Black JH, Dalgarno A. 1977. *Ap. J. Suppl.* 34:405–23
Black JH, Hartquist TW, Dalgarno A. 1978. *Ap. J.* 224:448–52
Black JH, van Dishoeck EF. 1991. *Ap. J.* 369:L9–12
Bohlin RC, Savage BD, Drake JF. 1978. *Ap. J.* 224:132–42
Boissé P, Le Petit F, Rollinde E, Roueff E, Pineau des Forêts G, et al. 2005. *Astron. Astrophys.* 429:509–23
Bregman JD, Hayward TL, Sloan GC. 2000. *Ap. J.* 544:L75–78
Brittain SD, Simon T, Kulesa C, Rettig TW. 2004. *Ap. J.* 606:911–16
Browning MK, Tumlinson J, Shull JM. 2003. *Ap. J.* 582:810–22
Cardelli JA, Clayton GC, Mathis JS. 1989. *Ap. J.* 345:245–56
Cardelli JA, Edgar RJ, Savage BD, Suntzeff NB. 1990. *Ap. J.* 362:551–62
Cardelli JA, Savage BD. 1988. *Ap. J.* 325:864–79
Carrasco L, Strom SE, Strom KM. 1973. *Ap. J.* 182:95–109
Carruthers GR. 1970. *Ap. J.* 161:L81–85
Cazaux S, Tielens AGGM. 2004. *Ap. J.* 604:222–37
Chaffee FH. 1974. *Ap. J.* 189:427–40
Chaffee FH, Lutz BL. 1977. *Ap. J.* 213:394–404
Cherchneff I, Barker JR, Tielens AGGM. 1992. *Ap. J.* 401:269–87
Chiar JE, Tielens AGGM, Whittet DCB, Schutte WA, Boogert ACA, et al. 2000. *Ap. J.* 537:749–62
Clayton GC, Gordon KD, Salama F, Allamandola LJ, Martin PG, et al. 2003. *Ap. J.* 592:947–52
Clegg RSS, Lambert DL. 1982. *MNRAS* 201:723–33
Cox DP. 2005. *Annu. Rev. Astron. Astrophys.* 43:337–85

Cox P, Guesten R, Henkel C. 1988. *Astron. Astrophys.* 206:108–16
Crawford IA, Williams DA. 1997. *MNRAS* 291:L53–56
Crutcher RM. 1979. *Ap. J.* 234:881–89
Crutcher RM. 1985. *Ap. J.* 288:604–17
Crutcher RM, Watson WD. 1976a. *Ap. J.* 203:L123–26
Crutcher RM, Watson WD. 1976b. *Ap. J.* 209:778–81
Cunningham NJ, McCray RA, Snow TP. 2004. *Ap. J.* 611:353–59
Dalgarno A. 2000. In *Astrochemistry: From Molecular Clouds to Planetary Systems, IAU Symp. 197*, ed. YC Minh, EF van Dishoeck, pp. 1–12. San Francisco: ASP
Dalgarno A, de Jong T, Oppenheimer M, Black JH. 1974. *Ap. J.* 192:L37–39
Dalgarno A, McCray RA. 1973. *Ap. J.* 181:95–100
Danks AC, Federman SR, Lambert DL. 1984. *Astron. Astrophys.* 130:62–66
Danks AC, Lambert DL. 1976. *MNRAS* 174:571–86
Désert FX, Bazell D, Boulanger F. 1988. *Ap. J.* 334:815–40
de Vries CP, van Dishoeck EF. 1988. *Astron. Astrophys.* 203:L23–26
Dicke RH, Peebles PJE, Roll PG, Wilkinson DT. 1965. *Ap. J.* 142:414–19
Ding H, Pino T, Güthe F, Maier JP. 2002. *J. Chem. Phys.* 117:8362–67
Ding H, Schmidt TW, Pino T, Boguslavskiy AB, Güthe F, Maier JP. 2003. *J. Chem. Phys.* 119:814–19
Douglas AE. 1977. *Nature* 269:130–32
Douglas AE, Herzberg GH. 1941. *Ap. J.* 94:381–81
Draine BT. 2003. *Annu. Rev. Astron. Astrophys.* 41:241–89
Draine BT, Katz N. 1986. *Ap. J.* 310:329–402
Duley WW. 1968. *Nature* 218:153
Duley WW, Williams DA. 1981. *MNRAS* 196:269–74
Dunham T. 1937. *Publ. Astron. Soc. Pac.* 49:26–28
Eddington AS. 1937. *Observatory* 60:99
Eddington AS. 1926. *Bakerian Conf. Proc. R. Soc. London Ser. A* 111:424
Eidelsberg M, Lemaire JL, Fillion JH, Rostas F, Federman SR, Sheffer Y. 2004. *Astron. Astrophys.* 424:355–61
Eyster EH. 1937. *Ap. J.* 86:486–88
Falgarone E, Pineau des Forets G, Roueff E. 1995. *Astron. Astrophys.* 300:870–80
Falgarone E, Verstraete L, Pineau des Forets G, Hily-Blant P. 2005. *Astron. Astrophys.* 433:997–1006
Federman SR. 1982. *Ap. J.* 257:125–34
Federman SR, Cardelli JA, van Dishoeck EF, Lambert DL, Black JH. 1995. *Ap. J.* 445:325–29
Federman SR, Danks AC, Lambert DL. 1984. *Ap. J.* 287:219–27
Federman SR, Glassgold AE, Kwan J. 1979. *Ap. J.* 227:466–73
Federman SR, Weber J, Lambert DL. 1996. *Ap. J.* 463:181–90
Federman SR, Lambert DL, Sheffer Y, Cardelli JA, Andersson B-G, van Dishoeck EF, Zsargó J. 2003. *Ap. J.* 591:986–99
Feldman PA. 2001. *Can. J. Phys.* 79:89–100
Felenbok P, Roueff E. 1996. *Ap. J.* 465:L57–60
Ferland GJ. 2003. *Annu. Rev. Astron. Astrophys.* 41:517–44

Ferlet R, André M, Hébrard G, Lecavelier des Etangs A, Lemoine M, et al. 2000. *Ap. J.* 538:L69–72
Field GB, Hitchcock JL. 1966. *Ap. J.* 146:1–6
Fixsen DJ, Mather JC. 2002. *Ap. J.* 581:817–22
Flower DR, Pineau des Forets G. 1998. *MNRAS* 297:1182–88
Frenklach M, Feigelson ED. 1989. *Ap. J.* 341:372–84
Geballe TR. 1997. In *ASP Conf. Ser.* 122, ed. YJ Pendelton, AGGM Tielens, pp. 119–28. San Francisco: ASP
Geballe TR. 2000. *Philos. Trans. R. Soc. London Ser. A* 358:2503–13
Geballe TR, McCall BJ, Hinkle KH, Oka T. 1999. *Ap. J.* 510:251–57
Geballe TR, Oka T. 1996. *Nature* 384:334–35
Giard M, Lamarre JM, Pajot F, Serra G. 1994. *Astron. Astrophys.* 286:203–10
Gillette FC, Forest WJ, Merrill KM. 1973. *Ap. J.* 183:87–93
Glassgold AE, Langer WD. 1974. *Ap. J.* 193:73–91
Glassgold AE, Langer WD. 1975. *Ap. J.* 197:347–50
Glover SCO, Savin DW, Jappsen AK. 2006. *Ap. J.* 640:553–68
Goldsmith PF. 1972. *Ap. J.* 176:597–610
Graff MM, Moseley JT, Roueff E. 1983. *Ap. J.* 269:796–802
Gredel R, van Dishoeck EF, Black JH. 1993. *Astron. Astrophys.* 269:477–95
Gredel R, van Dishoeck EF, Black JH. 1994. *Astron. Astrophys.* 285:300–21
Gredel R, van Dishoeck EF, de Vries CP, Black JH. 1992. *Astron. Astrophys.* 257:245–26
Gry C, Boulanger F, Nehmé C, Pineau des Fôrets G, Habart E, Falgarone E. 2002. *Astron. Astrophys.* 391:675–80
Haffner L, Meyer DM. 1995. *Ap. J.* 453:450–53
Heck EL, Flower DR, Le Bourlot J, Pineau des Forets G, Roueff E. 1993. *MNRAS* 262:795–99
Heger ML. 1922. *Lick Obs. Bull.* No. 337:141–48
Herbig GH. 1963. *Ap. J.* 137:200–12
Herbig GH. 1993. *Ap. J.* 407:142–56
Herbig GH. 1995. *Annu. Rev. Astron. Astrophys.* 33:19–74
Herbig GH. 2000. *Ap. J.* 542:334–43
Herbst E, Klemperer W. 1973. *Ap. J.* 185:505–34
Hobbs LM, Black JH, van Dishoeck EF. 1983. *Ap. J.* 271:L95–99
Hogerheijde MR, de Geus EJ, Spaans M, von Langevelde HJ, van Dishoeck EF. 1995. *Ap. J.* 441:L93–96
Hollenbach DJ, Werner MW, Salpeter EE. 1971. *Ap. J.* 163:165–80
Huffman DR. 1970. *Ap. J.* 161:1157–60
Jannuzi BT, Black JH, Lada CJ, van Dishoeck EF. 1988. *Ap. J.* 332:995–1008
Jenkins EB, Drake JF, Morton DC, Rogerson JB, Spitzer L, et al. 1973. *Ap. J.* 181:L122–27
Jenkins EB, Lees JF, van Dishoeck EF, Wilcots EM. 1989. *Ap. J.* 343:785–810
Jenkins EB, Peimbert A. 1997. *Ap. J.* 477:265–80
Joulain K, Falgarone E, Pineau des Forêts G, Flower D. 1998. 340:241–56
Jura M. 1974. *Ap. J.* 190:L33

Jura M. 1975a. *Ap. J.* 197:575–80
Jura M. 1975b. *Ap. J.* 197:581–86
Jura M, York DG. 1978. *Astrophysics* 219:861–69
Kaczmarczyk G. 2000a. *MNRAS* 312:794–806
Kaczmarczyk G. 2000b. *MNRAS* 316:875–84
Kahanpää J, Matilla K, Lehtinen K, Leinert C, Lemke D. 2003. *Astron. Astrophys.* 405:999–1012
Knapp GR, Jura M. 1976. *Ap. J.* 209:782–92
Knauth DC, Andersson BG, McCandliss SR, Moos HW. 2004. *Nature* 429:636–38
Kokoouline V, Greene CH. 2003. *Phys. Rev. A* 68:012703
Kopp M, Gerin M, Roueff E, Le Bourlot J. 1996. *Astron. Astrophys.* 305:558–71
Kreckel H, Motsch M, Mikosch J, Glosík J, Plasil R, et al. 2005. *Phys. Rev. Lett.* 95:263201
Lacour S, André MK, Sonnentrucker P, Le Petit F, Welty DE, et al. 2005. *Astron. Astrophys.* 430:967–77
Lakin NM, Pachkov M, Tulej M, Maier JP, Chambaud G, Rosmus P. 2000. *J. Chem. Phys.* 113:9586–92
Lambert DL, Danks AC. 1986. *Ap. J.* 303:401–15
Lambert DL, Sheffer Y, Federman SR. 1995. *Ap. J.* 438:740–49
Lambert DL, Sheffer Y, Gilliland RL, Federman SR. 1994. *Ap. J.* 420:756–71
Larson KA, Whittet DCB, Hough JH. 1996. *Ap. J.* 472:755–59
Latter WB. 1991. *Ap. J.* 377:187–91
Le Bourlot J, Pineau des Forêts G, Roueff E, Flower DR. 1993. *Astron. Astrophys.* 267:233–54
Léger A, Puget J. 1984. *Astron. Astrophys.* 137:L5–8
Le Page V, Bierbaum VM, Keheyan Y, Snow TP. 1999a. *J. Am. Chem. Soc.* 121:9435–46
Le Page V, Keheyan Y, Snow TP, Bierbaum VM. 1999b. *Int. J. Mass. Spectrom.* 185/186/187:949–59
Le Petit F, Roueff E, Herbst E. 2004. *Astron. Astrophys.* 417:993–1002
Le Petit F, Roueff E, Le Bourlot J. 2002. *Astron. Astrophys.* 390:369–81
Le Petit F, Nehmé C, Le Boulot J, Roueff E. 2006. *Ap. J.* In press
Lepp S, Dalgarno A. 1988. *Ap. J.* 324:553–56
Lepp S, Dalgarno A, van Dishoeck EF, Black JH. 1988. *Ap. J.* 329:418–24
Lien DJ. 1983. *The interstellar medium along the line of sight toward X Per.* PhD Diss., Univ. Ill.
Lien DJ. 1984. *Ap. J.* 287:L95–98
Liszt HS. 1997. *Astron. Astrophys.* 322:962–74
Liszt HS. 2003. *Astron. Astrophys.* 398:621–30
Liszt HS, Lucas R. 1994. *Ap. J.* 431:L131–34
Liszt HS, Lucas R. 1995. *Astron. Astrophys.* 299:847–56
Liszt HS, Lucas R. 1996. *Astron. Astrophys.* 314:917–26
Liszt HS, Lucas R. 1998. *Astron. Astrophys.* 339:561–74
Liszt HS, Lucas R. 2001. *Astron. Astrophys.* 370:576–85
Liszt HS, Lucas R. 2002. *Astron. Astrophys.* 391:693–704

Liszt HS, Lucas R, Black JH. 2004. *Astron. Astrophys.* 428:117–20
Liszt HS, Lucas R, Pety J. 2006. *Astron. Astrophys.* 448:253–59
Lucas R, Liszt HS. 1994. *Astron. Astrophys.* 282:L5–8
Lucas R, Liszt HS. 1996. *Astron. Astrophys.* 307:237–52
Lucas R, Liszt HS. 2000a. *Astron. Astrophys.* 355:327–32
Lucas R, Liszt HS. 2000b. *Astron. Astrophys.* 358:1069–76
Lucas R, Liszt HS. 2002. *Astron. Astrophys.* 384:1054–61
Lutz BL, Owen T, Snow TP. 1979. *Ap. J.* 227:159–62
Lyu CH, Smith AM, Bruhwieler FC. 1994. *Ap. J.* 426:254–68
Lyu CH, Smith AM, Bruhwieler FC. 2001. *Ap. J.* 560:865–70
Maier JP, Lakin MN, Walker GAH, Bohlender DA. 2001. *Ap. J.* 553:267–73
Maier JP, Walker GAH, Bohlender DA. 2004. *Ap. J.* 602:286–90
Martin DW, McDaniel EW, Meeks ML. 1961. *Ap. J.* 134:1012–13
Mathis JS. 1994. *Ap. J.* 422:176–86
Mathis JS, Cardelli JA. 1992. *Ap. J.* 398:610–20
McCall BJ, Geballe TR, Hinkle KH, Oka T. 1998. *Science* 279:1910–13
McCall BJ, Geballe TR, Hinkle KH, Oka T. 1999. *Ap. J.* 522:338–48
McCall BJ, Hinkle KH, Geballe TR, Moriarty-Schieven GH, Evans NJ, et al. 2002. *Ap. J.* 567:391–406
McCall BJ, Huneycutt AJ, Saykally RJ, Djuric NJ, Dunn GH, et al. 2004. *Phys. Rev. A* 70:052716
McCall BJ, Huneycutt AJ, Saykally RJ, Geballe TR, Djuric N, et al. 2003. *Nature* 422:500–2
McCall BJ, Thorburn JA, Hobbs LM, Oka T, York DG. 2001. *Ap. J.* 559:L49–53
McDowell MRC. 1961. *Observatory* 81:240–43
McKee CF, Ostriker JP. 1977. *Ap. J.* 218:148–69
McKellar A. 1940. *Publ. Astron. Soc. Pac.* 52:187–92
McKellar A. 1941. *Publ. Astron. Soc. Pac.* 53:233–35
Merrill PW. 1934. *Publ. Astron. Soc. Pac.* 46:206–7
Merrill PW. 1936. *Ap. J.* 83:126–28
Meyer DM, Lauroesch JT, Sofia UJ, Draine BT, Bertoldi F. 2001. *Ap. J.* 553:L59–62
Meyer DM, Roth KC. 1991. *Ap. J.* 376:L49–52
Moos HW, Sembach KR, Friedman SD, Kruk JW, Sonneborn G, et al. 2000. *Ap. J.* 538:L1–6
Morton DC. 1974. *Ap. J.* 193:L35–39
Morton DC. 1975. *Ap. J.* 197:85–115
Morton DC, Noreau L. 1994. *Ap. J. Suppl.* 95:301–43
Motylewski T, Linnartz H, Vaizert O, Maier JP, Galazutdinov GA, et al. 2000. *Ap. J.* 531:312–20
Neufeld DA, Wolfire MG, Schilke P. 2005. *Ap. J.* 628:260–74
Oka T. 1980. *Phys. Rev. Lett.* 45:531–34
Oka T, Thorburn JA, McCall BJ, Friedman SD, Hobbs LM, et al. 2003. *Ap. J.* 582:823–29
Olive KA, Steigman G, Walker TP. 2000. *Phys. Rep.* 333:389–407
Padoan P, Scalo J. 2005. *Ap. J.* 624:L97–100

Pan K, Federman SR, Cunha K, Smith VV, Welty DE. 2004. *Ap. J. Suppl.* 151:313–43
Pan K, Federman SR, Sheffer Y, Andersson BG. 2005. *Ap. J.* 633:986–1004
Peeters E, Allamandola LJ, Hudgins DM, Hony S, Tielens AGGM. 2004. In *Astrophysics of Dust*, ASP Conf. Ser. 309, ed. AN Witt, GC Clayton, BT Draine, p. 141. San Francisco: ASP
Penzias A, Wilson R. 1965. *Ap. J.* 142:419–21
Rachford BL, Brian L, Snow TP, Tumlinson J, Shull J, et al. 2001. *Ap. J.* 555:839–49
Rachford BL, Brian L, Snow TP, Tumlinson J, Shull J, et al. 2002. *Ap. J.* 577:221–44
Rank DM, Townes CH, Welch WJ. 1971. *Science* 174:1083–1101
Richter P, Sembach KR, Wakker B, Savage BD. 2001. *Ap. J.* 562:L181–84
Rogerson JB, Spitzer L, Drake JF, Dressker K, Jenkins EB, et al. 1973. *Ap. J.* 181:L97–100
Roth KC, Meyer DM. 1995. *Ap. J.* 441:129–43
Roueff E. 1996. *MNRAS* 279:L37–40
Roueff E, Felenbok P, Black JH, Gry C. 2002. *Astron. Astrophys.* 384:629–37
Ruiterkamp R. Cox NLJ, Spaans M, Kaper L, Foing BH, et al. 2005. *Astron. Astrophys.* 432:515–29
Ruiterkamp R, Halasinski T, Salama F, Foing BH, Allamandola LJ, et al. 2002. *Astron. Astrophys.* 390:1153–70
Russell HN. 1935. *MNRAS* 95:610–35
Sakon I, Onaka T, Ishihara D, Ootsubo T, Yamamura I, et al. 2004. *Ap. J.* 609:203–19
Sarre PJ, Miles JR, Kerr TH, Hibbins RE, Fossey SJ, et al. 1995. *MNRAS* 277:L41–43
Savage BD, Drake JF, Budich W, Bohlin RC. 1977. *Ap. J.* 216:291–307
Savage BD, Sembach KR. 1996. *Annu. Rev. Astron. Astrophys.* 4:279–330
Schramm DN, Turner MS. 1998. *Rev. Mod. Phys.* 70:303–18
Sellgren K. 1984. *Ap. J.* 277:623–33
Sellgren K, Brooke TY, Smith RG, Geballe TR. 1995. *Ap. J.* 449:L69–72
Shaw G, Ferland GJ, Srianand R, Abel NP. 2006. *Ap. J.* 639:941–50
Sheffer Y, Federman SR, Andersson BG. 2003. *Ap. J.* 597:L29–32
Sheffer Y, Federman SR, Lambert DL. 2002a. *Ap. J.* 572:L95–98
Sheffer Y, Federman SR, Lambert DL, Cardelli JA. 1992. *Ap. J.* 397:382–91
Sheffer Y, Lambert DL, Federman SR. 2002b. *Ap. J.* 574:L171–74
Shull JM, Beckwith W. 1982. *Annu. Rev. Astron. Astrophys.* 20:163–90
Shull JM, Tumlinson J, Jenkins EB, Moos HW, Rachford BL, et al. 2000. *Ap. J.* 538:L73–76
Shulman S, Bortolot VJ, Thaddeus P. 1974. *Ap. J.* 193:97–102
Shuping RS, Snow TP, Crutcher RM, Lutz BL. 1999. *Ap. J.* 520:149–57
Smith AM, Stecher TP. 1971. *Ap. J.* 164:L43–47
Smith AM, Stecher TP, Krishna Swamy KS. 1978. *Ap. J.* 220:138–48
Smith IWM, Herbst E, Chang Q. 2004. *MNRAS* 350:323–30
Smith WH, Snow TP, York DG. 1977. *Ap. J.* 218:124–32
Snow TP. 1975. *Ap. J.* 201:L21–24
Snow TP. 1976a. *Ap. J.* 204:L127–30
Snow TP. 1976b. *Ap. J.* 204:759–74
Snow TP. 1977. *Ap. J.* 216:724–37

Snow TP. 1978. *Ap. J.* 202:L93–96
Snow TP. 1979. *Astrophys. Space Sci.* 66:453–66
Snow TP. 1980. In *Proc. IAU Symp.* 87, ed. BH Andrew, pp. 247–54. Dordrecht: Reidel
Snow TP. 1983. *Ap. J.* 269:L57–59
Snow TP. 1995. In *The Diffuse Interstellar Bands*, ed. AGGM Tielens, TP Snow, pp. 379–93. Dordrecht: Kluwer
Snow TP. 2001. *Spectrochim. Acta Part A* 57:615–26
Snow TP. 2004. *Nature* 429:615–16
Snow TP, Black JH, van Dishoeck EF, Burks G, Crutcher RM. 1994. *Ap. J.* 465:245–63
Snow TP, Cohen JG. 1974. *Ap. J.* 194:313–22
Snow TP, Jenkins EB. 1980. *Ap. J.* 241:161–72
Snow TP, Lamers HJGLM, Joseph CL. 1987. *Ap. J.* 321:952–57
Snow TP, Le Page V, Keheyen Y, Bierbaum VM. 1995. *Nature* 391:259–60
Snow TP, Rachford BL, Tumlinson J, Shull JM, Welty DE, et al. 2000. *Ap. J.* 538:L65–68
Snow TP, Seab CG. 1980. *Ap. J.* 242:L83–86
Snow TP, Seab CG, Joseph CL. 1988. *Ap. J.* 335:185–87
Snow TP, Smith WH. 1981. *Ap. J.* 250:163–65
Snow TP, Welty DE, Thorburn J, Hobbs LM, McCall BJ, et al. 2002. *Ap. J.* 573:670–77
Snow TP, Witt AN. 1995. *Science* 270:1455–60
Snow TP, York DG, Resnick M. 1977. *Publ. Astron. Soc. Pac.* 89:758–64
Solomon PM, Klemperer W. 1972. *Ap. J.* 178:389–422
Song IO, Kerr TH, McCombie J, Sarre PJ. 2003. *MNRAS* 346:L1–5
Sonnentrucker P, Cami J, Ehrenfreund P, Foing BH. 1997. *Astron. Astrophys.* 327:1215–21
Sonnentrucker P, Friedman SD, Welty DE, York DG, Snow TP. 2003. *Ap. J.* 596:350–61
Souza SP, Lutz BL. 1977. *Ap. J.* 213:L129–30
Souza SP, Lutz BL. 1978. *Ap. J.* 216:L49–51
Spaans M, Neufeld D, Lepp S, Melnick GJ, Stauffer J. 1998. *Ap. J.* 503:780–84
Spitzer L. 1978. *Physical Processes in the Interstellar Medium*, p. 124. New York: Wiley
Spitzer L, Cochran WD. 1973. *Ap. J.* 186:L23–27
Spitzer L, Cochran WD, Hirshfeld A. 1974. *Ap. J. Suppl.* 28:373–89
Spitzer L, Drake JF, Morton DC, Jenkins EB, Rogerson JB, et al. 1973. *Ap. J.* 181:L116–19
Spitzer L, Zabriskie FR. 1959. *Publ. Astron. Soc. Pac.* 71:412–20
Spitzer L, Zweibel EG. 1974. *Ap. J.* 191:L127–30
Stahl O, Wilson TL. 1992. *Astron. Astrophys.* 254:327–30
Strömgren B. 1939. *Ap. J.* 89:526–47
Swings P. 1937. *MNRAS* 97:212–15
Swings P, Rosenfeld L. 1937. *Ap. J.* 86:483–86
Tappe A. 2004. *Interstellar absorption across the electromagnetic spectrum*. PhD Diss. Onsala Space Obs.

Tappe A, Black JH. 2004. *Astron. Astrophys.* 423:943–54
Thaddeus P. 1972. *Annu. Rev. Astron. Astrophys.* 10:305–34
Thaddeus P, Clauser JF. 1966. *Phys. Rev. Lett.* 16:819–22
Thorburn JA, Hobbs LM, McCall BJ, Oka T, Welty DE, et al. 2003. *Ap. J.* 584:339–56
Tokunaga AT. 1997. In *ASP Conf. Ser.* 122, ed. YJ Pendelton, AGGM Tielens, pp. 149–60. San Francisco: ASP
Tripp TM, Cardelli JA, Savage BD. 1994. *Astron. J.* 107:645–50
Tuairisg SÓ, Cami J, Foing BH, Sonnentrucker P, Ehrenfreund P. 2000. *Astron. Astrophys. Suppl.* 142:225–38
Tulej M, Kirkwood DA, Pachkov M, Maier JP. 1998. *Ap. J.* 506:L69–73
Tumlinson J, Shull JM, Rachford BL, Browning MK, Snow TP, et al. 2002. *Ap. J.* 566:857–79
Turner BE. 2000. *Ap. J.* 542:837–60
van Dishoeck EF. 1998. In *The Molecular Astrophysics of Stars and Galaxies*, ed. TW Hartquist, DA Williams, p. 53. Oxford: Clarendon
van Dishoeck EF, Black JH. 1982. *Ap. J.* 258:533–47
van Dishoeck EF, Black JH. 1986a. *Ap. J.* 307:332–36
van Dishoeck EF, Black JH. 1986b. *Ap. J. Suppl.* 62:109–45
van Dishoeck EF, Black JH. 1987. In *Physical Processes in Interstellar Clouds*, ed. G Morfill, MS Scholer, p. 241. Dordrecht: Reidel
van Dishoeck EF, Black JH. 1988. *Ap. J.* 334:771–802
van Dishoeck EF, Black JH. 1989. *Ap. J.* 340:273–97
van Dishoeck EF, de Zeeuw T. 1984. *MNRAS* 206:383–406
van Dishoeck EF, Phillips TG, Black JH, Gredel R. 1991. *Ap. J.* 366:141–62
Viala YP. 1986. *Astron. Astrophys.* 64:391–437
Viala YP, Roueff E, Abgrall H. 1988. *Astron. Astrophys.* 190:215–36
Wampler EJ. 1966. *Ap. J.* 144:921–36
Wannier P, Andersson BG, Penprase BE, Federman SR. 1999. *Ap. J.* 510:291–304
Wannier PG, Penzias AA, Jenkins EB. 1982. *Ap. J.* 254:100–7
Warin S, Benayoun JJ, Viala YP. 1996. *Ap. J.* 308:535–64
Watson JKG. 2001. *Ap. J.* 555:472–76
Watson WD. 1973. *Ap. J.* 183:L17–20
Weinreb S, Barrett AH, Meeks ML, Henry JC. 1963. *Nature* 200:829
Welty DE, Fowler JR. 1992. *Ap. J.* 393:193–205
Whittet DCB. 2003. *Dust in the Galactic Environment.* London: Inst. Phys. 2nd ed.
Whittet DCB, Martin PG, Fitzpatrick EL, Massa D. 1993. *Ap. J.* 408:573–78
Witt AN, Clayton GC, Draine BT. 2004. In *Astrophysics of Dust*, ed. AN Witt, GC Clayton, BT Draine. San Francisco: Astron. Soc. Pac.
Wolfire MG, McKee CF, Hollenbach D, Tielens AGGM. 2003. *Ap. J.* 587:278–311

Observational Constraints on Cosmic Reionization

Xiaohui Fan,[1] C.L. Carilli,[2] and B. Keating[3]

[1]Steward Observatory, University of Arizona, Tucson, Arizona 85721; email: fan@sancerre.as.arizona.edu
[2]National Radio Astronomy Observatory, Socorro, New Mexico 87801; email: ccarilli@nrao.edu
[3]Department of Physics, University of California, San Diego, California 92093; email: bkeating@ucsd.edu

Key Words

cosmic reionization, galaxy formation, observational cosmology

Abstract

Observations have set the first constraints on the epoch of reionization (EoR), corresponding to the formation epoch of the first luminous objects. Studies of Gunn-Peterson (GP) absorption indicate a rapid increase in the neutral fraction of the intergalactic medium (IGM) from $x_{HI} < 10^{-4}$ at $z \leq 5.5$, to $x_{HI} > 10^{-3}$, perhaps up to 0.1, at $z \sim 6$, while the large scale polarization of the cosmic microwave background (CMB) implies a significant ionization fraction extending to higher redshifts, $z \sim 11 \pm 3$. These results, as well as observations of galaxy populations, suggest that reionization is a process that begins as early as $z \sim 14$, and ends with the "percolation" phase at $z \sim 6$ to 8. Low luminosity star-forming galaxies are likely the dominant sources of reionizing photons. Future low-frequency radio telescopes will make direct measurements of HI 21-cm emission from the neutral IGM during the EoR, and measurements of secondary CMB temperature anisotropy will provide details of the dynamics of the reionized IGM.

1. INTRODUCTION

The baryonic pregalactic medium (PGM) evolves in three distinct phases. At high redshifts ($z > 1100$) the PGM is hot, fully ionized, and optically thick to Thomson scattering, and hence coupled to the photon field. As the universe expands, the PGM cools, and eventually recombines, leaving a surface of last scattering (the cosmic microwave background, CMB), plus a neutral PGM. This neutral phase lasts from $z = 1100$ to $z \sim 14$. At some point between $z \sim 14$ and 6, hydrogen in the PGM is "reionized," due to UV radiation from the first luminous objects, leaving the fully reionized intergalactic medium (IGM) seen during the "realm of the galaxies" ($6 > z > 0$). The ionized, dense PGM at very high redshift has been well studied through extensive observations of the CMB. Likewise, the reionized, rarified IGM at low redshift has been well characterized through QSO absorption line studies. The middle phase—the existence of a neutral IGM during the so-called dark ages (Rees 1998), and the process of reionization of this medium—is the last directly observable phase of cosmic evolution that remains to be verified and explored. The epoch of reionization (EoR) is crucial in cosmic structure formation studies, because it sets a fundamental benchmark indicating the formation of the first luminous objects, either star-forming galaxies or active galactic nuclei (AGN).

Cosmic reionization has been discussed within the larger context of mostly theoretical reviews on the formation of the first luminous objects (Loeb & Barkana 2001; Barkana & Loeb 2001; Ciardi & Ferrara 2005). The past several years have seen the first observational evidence for a neutral IGM, and the first constraints on the process of reionization. In this review we do not repeat the theory of early structure formation, but focus on these recent observations of reionization and the evolution of the neutral IGM. Such constraints come from optical and near-IR spectroscopy of the most distant QSOs, and study of galaxy populations at the highest redshifts, among others. Detections of the IGM Thomson optical depth come from observations of the large angular scale polarization of the CMB (Kogut et al. 2003, Spergel et al. 2006, Page et al. 2006). These measurements yield a $\geq 3\sigma$ detection of polarized emission from the CMB, and future observations of the small-scale CMB anisotropy will complement the large angular scale polarization by probing the fine details of the reionization epoch. We also review the potential sources of reionization, including star-forming galaxies, AGN, and decaying particles. We conclude with a discussion of low-frequency radio telescopes that are being constructed to detect the neutral IGM directly through the 21-cm line of neutral hydrogen.

We focus our review on the observations of hydrogen reionization. Neutral helium reionization likely happened at a similar epoch to that of neutral hydrogen, due to their similar ionization potential and recombination rate. Reionization of He II happens at a much lower redshift—the He II GP effect is observed in the spectrum of quasars at $z \sim 3$ (e.g., Jakobsen et al. 1994; Anderson et al. 1999; Davidsen, Kriss & Wei 1996; Kriss et al. 2001; Shull et al. 2004; Zheng et al. 2004). Evolution of the IGM temperature (e.g., Ricotti, Gnedin & Shull 2000; Schaye et al. 2000; Theuns et al. 2002a,b) also suggests a rapid change of IGM temperature at $z \sim 3$, consistent with the onset of He II reionization. For a detailed recent review of He II reionization, see Ciardi & Ferrara (2005).

2. A BASIC MODEL OF REIONIZATION

In order to put recent observations in context, we include a brief, mostly qualitative, description of the reionization process. Again, for more extensive reviews of early structure formation, see Loeb & Barkana (2001), Barkana & Loeb (2001), and Ciardi & Ferrara (2005).

Assuming reionization is driven by UV photons from the first luminous sources (stars or AGN), analytic (e.g., Madau, Haardt & Rees 1999; Miralda-Escudé, Haehnelt & Rees 2000; Wyithe & Loeb 2003), and numerical (e.g., Gnedin 2000; Razoumov et al. 2002; Sokasian, Abel & Hernquist 2002; Ciardi, Stoehr & White 2004; Paschos & Norman 2005), calculations suggest that reionization follows a number of phases. The first phase will be relatively slow, with each UV source isolated to its own Strömgren sphere. Eventually, these spheres grow and join, and reionization proceeds much faster owing to the combination of accelerating galaxy formation, plus the much larger mean free path of ionizing photons in the now porous IGM. This stage is known as "overlap," or "percolation." The last phase entails the etching-away of the final dense filaments in the IGM by the intergalactic UV radiation field, leading to a fully ionized IGM (neutral fraction, $x_{HI} \equiv \frac{n_{HI}}{n_H} \sim 10^{-5}$, at $z=0$).

There are a number of important caveats to this most basic model. First, causality, cosmic variance, and source clustering will set an absolute minimum width in redshift for the fast phase of reionization, $\Delta z > 0.15$, and the process itself is likely to be much more extended in time (Wyithe & Loeb 2004a). In particular, Furlanetto & Oh (2005) point out that the overlap phase is a relatively local phenomenon with significant differences along different lines of sight. Second, the recombination time for the mean IGM becomes longer than the Hubble time at $z \geq 8$, allowing for the possibility of complex reionization histories. As shown by a number of semianalytical works (e.g., Cen 2003a,b; Chiu, Fan & Ostriker 2003; Wyithe & Loeb 2003; Haiman & Holder 2003), adding feedback from an early generation of galaxy formation, or adjusting star-formation efficiency at $z > 6$–10, will produce a prolonged, or multiepoch, reionization history. Third, it remains an open question whether reionization proceeds from high to low density regions (inside-out; Ciardi & Madau 2003; Sokasian, Abel & Hernquist 2002; Iliev et al. 2006), or from low to high density regions (outside-in; Gnedin 2000; Miralda-Escudé, Haehnelt & Rees 2000). Fourth, a very different picture arises if we assume reionization by decaying fundamental particles or by penetrating hard X rays from the first luminous sources. In this case, the reionization process is not localized to individual Strömgren spheres, but occurs more uniformly throughout the IGM (Section 8.4). Lastly, we have been necessarily vague in terms of precise redshifts and timescales for the various phases—these are the main concern of this review.

Throughout this review, we adopt a standard concordance cosmology (Spergel et al. 2003, 2006) unless stated otherwise.

3. OBSERVATIONS OF THE GUNN-PETERSON EFFECT AT $z \sim 6$

Gunn & Peterson (1965) first proposed using Lyα resonance absorption in the spectrum of distant quasars as a direct probe to the neutral hydrogen density in the IGM at high redshift (see also Field 1959, Shklovsky 1964, Bahcall & Salpeter 1965, Scheuer 1965). For objects beyond reionization, neutral hydrogen in the IGM creates complete GP absorption troughs in the quasar spectrum blueward of Lyα emission. Observations of the GP effect directly constrain the evolution of neutral hydrogen fraction and the ionization state of the IGM.

3.1. Gunn-Peterson Effect: Basics

The Gunn-Peterson (1965) optical depth to Lyα photons is

$$\tau_{GP} = \frac{\pi e^2}{m_e c} f_\alpha \lambda_\alpha H^{-1}(z) n_{HI}, \qquad (1)$$

where f_α is the oscillator strength of the Lyα transition, $\lambda_\alpha = 1216$ Å, $H(z)$ is the Hubble constant at redshift z, and n_{HI} is the density of neutral hydrogen in the IGM. At high redshifts,

$$\tau_{GP}(z) = 4.9 \times 10^5 \left(\frac{\Omega_m h^2}{0.13}\right)^{-1/2} \left(\frac{\Omega_b h^2}{0.02}\right) \left(\frac{1+z}{7}\right)^{3/2} \left(\frac{n_{HI}}{n_H}\right) \qquad (2)$$

for a uniform IGM. Even a tiny neutral fraction, $x_{HI} \sim 10^{-4}$, gives rise to complete GP absorption. This test is only sensitive at the end of the reionization when the IGM is already mostly ionized, and the absorption saturates for the higher neutral fraction in the earlier stage.

At $z < 5$, the IGM absorption is resolved into individual Lyα forest lines; their number density increases strongly with redshift $[N(z) \propto (1+z)^{2.5}$; see Rauch 1998]. Earlier attempts to study GP absorption concentrated on measuring the amount of flux between individual Lyα forest lines using high-resolution spectroscopy to place limits on the diffuse neutral IGM. Webb et al. (1992) and Giallongo et al. (1994) found $\tau < 0.05$ at $z \sim 4$. Songaila et al. (1999) and Fan et al. (2000) place upper limits of $\tau = 0.1$ and 0.4 based on lack of complete GP troughs in SDSS (Sloan Digital Sky Survey) J0338 + 0021 ($z = 5.00$) and SDSS J1044 − 0125 ($z = 5.74$), respectively. However, at $z > 5$, even with a moderately high resolution spectrum, Lyα forest lines overlap severely, making it impossible to find a truly "line-free" region.

A more accurate picture of the IGM evolution interprets the Lyα forest as a fluctuating GP effect: absorption arises from low-density gas in the IGM that is in approximate thermal equilibrium between photoionization heating by the UV background and adiabatic cooling due to the Hubble expansion, rather than as discrete Lyα forest clouds (Bi 1993; Cen et al. 1994; Zhang, Anninos & Norman 1995; Hernquist et al. 1996). The neutral hydrogen fraction and therefore the GP optical depth depend on the local density of the IGM. By studying the evolution of the average transmitted flux or effective optical depth, one can trace the evolution of the UV ionizing background and neutral fraction of the IGM. At high-redshift, the IGM is highly clumpy (Miralda-Escudé, Haehnelt & Rees 2000), which must be taken into account in order to estimate the IGM ionization from observations.

3.2. Complete Gunn-Peterson Troughs: Phase Transition or Gradual Evolution?

The SDSS (York et al. 2000) provides large samples of luminous quasars over $0 < z < 6.5$. Fan et al. (2000, 2001a,b, 2003, 2004, 2006a) carried out a survey of i-dropout quasars ($z > 5.7$) using the SDSS imaging data, resulting in the discovery of 19 luminous quasars in this redshift regime (**Figure 1**). Other multicolor survey projects (e.g., Mahabal et al. 2005; AGES, Cool et al. 2006; QUEST, Djorgovski, Bogosavljevic & Mahabal 2006; CFHT Legacy survey, Willott et al. 2005) are also searching for quasars at similar redshifts. They provide by far the best probes of IGM evolution toward the end of the EoR.

Songaila (2004) summarized the evolution of transmitted flux over a wide redshift range ($2 < z < 6.3$) using high signal-to-noise, moderate-resolution ($R \geq 5000$) observations of 50 quasars. Strong evolution of Lyα absorption at $z_{abs} > 5$ is evident, and the transmitted flux quickly approaches zero at $z > 5.5$ (**Figure 2**). At $z_{abs} > 6$, complete absorption troughs begin to appear: the GP optical depths are $\gg 1$, indicating a rapid increase in the neutral fraction. However, is this evolution the result of a smooth transition due to gradual thickening of the Lyα forest (e.g., Songaila & Cowie 2002, Songaila 2004)? Or is it a reflection of a more dramatic change in the IGM evolution, marking the end of reionization (e.g., Becker et al. 2001, Djorgovski et al. 2001, Fan et al. 2002)?

Djorgovski et al. (2001) detected an extended dark gap in the spectrum of SDSS J1044 − 0125 ($z = 5.74$), at $z_{abs} = 5.2$–5.6, with $\tau_{GP} > 4.6$. The first clear-cut GP trough was discovered in the spectrum of SDSS J1030 + 0524 ($z = 6.28$; Becker et al. 2001, Fan et al. 2001a,b; see **Figure 4**), which showed complete GP absorption at $5.95 < z_{abs} < 6.15$ in both Lyα and Lyβ transitions. A high S/N spectrum presented in White et al. (2003) placed a stringent limit $\tau_{GP} > 6.3$ in Lyα. However, Songaila (2004) suggested that constraints based on Lyα alone do not deviate from a simple extrapolation from lower redshift by a large factor, consistent with a relatively smooth evolution in the ionization state.

The GP optical depth $\tau \propto f\lambda$, where f and λ are the oscillator strengh and rest-frame wavelength of the transition. For the same neutral density, the GP optical depths of Lyβ and Lyγ are factors of 6.2 and 17.9 smaller than that of Lyα. Therefore, Lyβ can provide more stringent constraints on the IGM ionization state when Lyα absorption saturates. Complete Lyβ absorption implies a significant rise in the neutral fraction toward SDSS J1030 + 0524. Songaila (2004) pointed out that in a clumpy IGM, the simple conversion factor of 6.2 between Lyα and Lyβ optical depths is an upper limit. Using high order Lyman lines requires proper treatment of the clumpy IGM. Oh & Furlanetto (2005) and Fan et al. (2006b) found $\tau^\alpha/\tau^\beta \sim 2$–3 and $\tau^\alpha/\tau^\gamma \sim 4$–6 for a clumpy IGM.

Fan et al. (2006b) measured the evolution of GP optical depths along the line of sight of the nineteen $z > 5.7$ quasars from the SDSS (**Figure 3**). They found that at

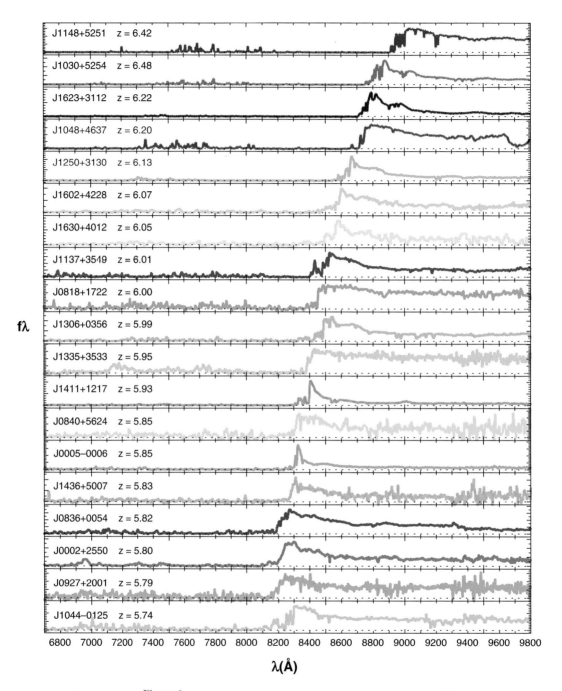

Figure 1

Moderate resolution spectra of nineteen SDSS quasars at $5.74 < z < 6.42$. Adapted from Fan et al. (2006b).

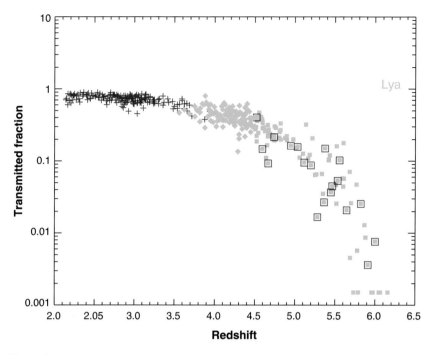

Figure 2

Transmitted flux blueward of Lyα emission as a function of redshift from $z \sim 2$ to 6.3 using Keck/ESI and HIRES data. Large open red squares are points data for possible BAL quasars. Flux was computed in bins of 15 Å. Adapted from Songaila (2004).

$z_{abs} < 5.5$, the optical depth can be best fit as $\tau \propto (1+z)^{4.3}$, whereas at $z_{abs} > 5.5$, the evolution of optical depth accelerates: $\tau \propto (1+z)^{>10}$. There is also a rapid increase in the variation of optical depth along different lines of sight: $\sigma(\tau)/\tau$ increases from $\sim 15\%$ at $z \sim 5$, to $> 30\%$ at $z > 6$, in which τ is averaged over a scale of ~ 60 comoving Mpc. **Figure 4** (White et al. 2003) compares the GP absorption in the two highest redshift quasars at the time of this writing: SDSS J1030 + 0524 ($z = 6.28$) shows complete absorption. On the other hand, SDSS J1148 + 5251 ($z = 6.42$) shows clear transmission in Lyα and Lyβ transitions at the same redshift range, clearly indicating that this line of sight is still highly ionized (Oh & Furlanetto 2005, White et al. 2005). The increased variance in the IGM optical depth at the highest redshifts is further discussed in Songaila (2004), Djorgovski, Bogosavljevic & Mahabal (2006) and Fan et al. (2006b).

Wyithe & Loeb (2004b) and Furlanetto & Oh (2005) pointed out that at the end of reionization, the size of ionized bubbles grows rapidly. They estimate an observed bubble size of tens of Mpc at $z \sim 6$ and a scatter in the observed redshift of overlap along different lines of sight to be $\Delta z \sim 0.15$. Wyithe & Loeb (2005) further showed the GP optical depth variations to be of order unity on a scale of ~ 100 comoving

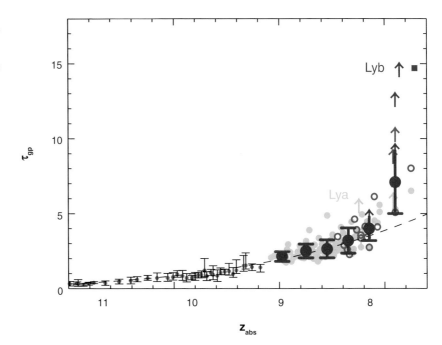

Figure 3

Evolution of optical depth with combined Lyα and Lyβ results. The dash line is for a redshift evolution of $\tau_{GP} \propto (1+z)^{4.3}$. At $z > 5.5$, the best-fit evolution has $\tau_{GP} \propto (1+z)^{>10.9}$, indicating an accelerated evolution. The large open symbols with error bars are the average and standard deviation of optical depth at each redshift. The sample variance also increases rapidly with redshift. Adapted from Fan et al. (2006b).

Mpc. The increased variance on the IGM optical depth supports the interpretation that the IGM is close to the end of reionization.

With the emergence of complete GP troughs at $z > 6$, it becomes increasingly difficult to place stringent limits on the optical depth and neutral fraction of the IGM. Songaila & Cowie (2002) suggested using the distribution of optically thick, dark gaps in the spectrum as an alternative statistic. Fan et al. (2006b) examined the distribution of dark gaps with $\tau > 2.5$ among their sample of SDSS quasars at $z > 5.7$, and showed a dramatic increase in the average length of dark gaps at $z > 6$ (**Figure 5**), similar to the model prediction of Paschos & Norman (2005). Gap statistics provides a powerful new tool to characterize the IGM ionization at the end of reionization (S. Gallerani, R. Choudhury & A. Ferrara, submitted; K. Kohler, N. Gnedin, A.J.S. Hamilton, submitted), and can be sensitive to larger neutral fractions, as they carry higher-order information than optical depth alone. Songaila & Cowie (2002), Pentericci et al. (2002), and Fan et al. (2002, 2006b) also studied using the distribution function of transmited fluxes (e.g., Rauch et al. 1997) or statistics of threshold crossing (e.g., Miralda-Escudé et al. 1996).

3.3. Estimating the Ionization State and the Neutral Fraction

Measurements of GP optical depth can be used to derive the ionization state of the IGM, using parameters such as the mean UV ionizing background, the mean free path of UV photons, and neutral hydrogen fractions. The evolution of GP absorption is usually described in the context of photoionization. Weinberg et al. (1997) present

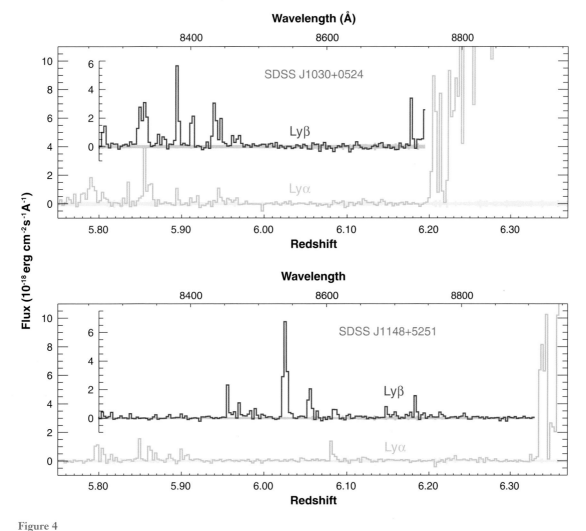

Figure 4

Close-up of the Lyα and Lyβ troughs in the two highest redshift quasars currently known. SDSS J1030 + 0524 ($z = 6.28$) shows complete GP absorption in both Lyα and Lyβ, where SDSS J1148 + 5251 ($z = 6.42$) has clear transmission, suggesting a large line-of-sight variance at the end of reionization. Adapted from White et al. (2003).

the basic formalism to use the IGM optical depth to measure the cosmic baryon density. McDonald et al. (2000) and McDonald & Miralda-Escudé (2001) expanded this work to study the evolution of the ionizing background at $z < 5.2$ by comparing the observed transmitted flux to that of artificial Lyα forest spectra created from cosmological simulations. Although slightly different in technical details, a number of works (Cen & McDonald 2002, Fan et al. 2002, Lidz et al. 2002, Songaila & Cowie 2002, Songaila 2004) followed the same formalism to calculate the evolution of the

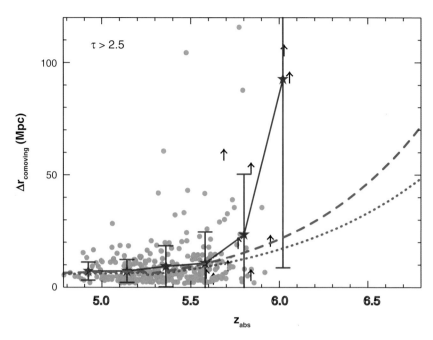

Figure 5

Distributions of dark gaps, which are defined as regions in the spectra where all pixels have an observed optical depth larger than 2.5 for Lyα transition. Upward arrows are gaps immediately blueward of the quasar proximity zone, therefore the length is only a lower limit. Solid lines with error bars are average depth lengths with 1-σ dispersion at each redshift bin. Long dark gaps start to appear at $z \sim 5.6$, with the average gap length increasing rapidly at $z > 6$, marking the end of reionization. It is compared with the simulation of Paschos & Norman (2005), in which dashed and dotted lines are for moderate and high spectral resolutions. The simulation has an overlapping redshift at $z \sim 7$. Adapted from Fan et al. (2006b).

ionizing background at $z > 5$ and reached consistent results: from $z \sim 5$ to $z > 6$ there is an order-of-magnitude decrease in the UV background, and the mean free path of UV photons is shown to be <1 physical Mpc at $z > 6$ (Fan et al. 2006b). This scale is comparable to the clustering scale of star-forming galaxies at these redshifts (e.g., Hu et al. 2005, Kashikawa et al. 2006, Malhotra et al. 2005). High-redshift star-forming galaxies, which likely provide most of the UV photons for reionization, are highly biased and clustered at similar physical scales to the mean free path. The assumption of a uniform UV background is no longer valid. There is a clear indication of a nonuniform UV background at the end of reionization (Fan et al. 2006b). However, current observations at $z > 6$ are based on a handful of quasars. A much larger sample is needed to quantify the variation of the ionization state.

Using the same data, Fan et al. (2002, 2006b), Lidz et al. (2002), and Cen & McDonald (2002) find that at $z > 6$ the volume-averaged neutral fraction of the IGM has increased to $>10^{-3.5}$. The results are displayed in **Figure 6**. It is important to note that this is strictly a lower limit, owing to the large optical depths in the Lyα line.

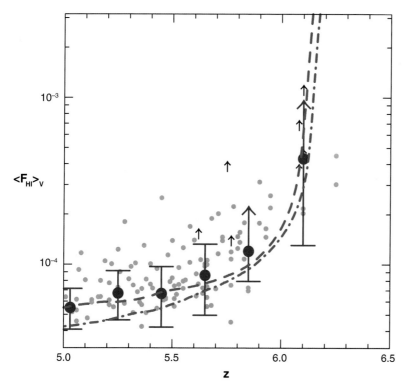

Figure 6

Evolution of volume-averaged neutral hydrogen fraction of the IGM. The solid points with error bars are measurements based on the 19 high-redshift quasars. The dashed and dashed-dotted-dashed lines are volume-averaged results from the A4 and A8 simulations of Gnedin (2004). The neutral fraction inferred from the observations is comparable to the transition from overlap stage to postoverlap stage of reionization in simulations (Gnedin 2000). Adapted from Fan et al. (2006b).

Further, the presence of transmitting pixels and the finite length of dark gaps in the quasar spectrum can also be used to place an independent upper limit of <10–30% on the neutral fraction (Furlanetto, Hernquist & Zaldarriaga 2004; Fan et al. 2006b). At higher neutral fraction, the GP damping wing from the average IGM (see Section 4.4) could wipe out any isolated HII region transmissions.

In summary, the latest work on GP absorption toward the highest redshift QSOs implies a qualitative change in the nature of Lyα absorption at $z \sim 6$, including: (*a*) a sharp rise in the power-law index for the evolution of GP optical depth with redshift, (*b*) a large variation of optical depth between different lines of sight, and (*c*) a dramatic increase in the number of dark gaps in the spectra. The GP results indicate that the IGM is likely between $10^{-3.5}$ and $10^{-0.5}$ neutral at $z \sim 6$. Although saturation in the GP part of the spectrum remains a challenge, the current results are consistent with conditions expected at the end of reionization, during the transition from the percolation, or overlap, stage to the postoverlap stage of reionization, as suggested by numerical simulations (Gnedin 2000, 2004; Ciardi et al. 2001; Razoumov et al. 2002; Paschos & Norman 2005; Gnedin & Fan 2006). Sources at higher redshift, and estimators sensitive to larger neutral fractions, are needed to probe deeper into the reionization era. Wide-field, deep, near-IR surveys using dedicated 4-meter class telescopes, such as UKIDSS (A. Lawrence, S. Warren, O. Almaini, A. Edge, N. Hambly, et al., submitted) and VISTA (McPherson et al. 2004), will likely discover quasars at $z \sim 8$ in the next decade (see also Section 5.2 for GRB observations).

4. COSMIC STRÖMGREN SPHERES AND SURFACES

Luminous quasars produce a highly ionized HII region around them even when the IGM is still mostly neutral otherwise. The presence of (time bounded) cosmic Strömgren spheres (CSS) around the highest redshift SDSS QSOs has been deduced from the observed difference between the redshift of the onset of the GP effect and the systemic redshift of the host galaxy (Walter et al. 2003; White et al. 2003; Wyithe & Loeb 2004a; Wyithe, Loeb & Carilli 2005; Fan et al. 2006b). The physical size of these spheres is typically ~ 5 Mpc at $z > 6$, or an order-of-magnitude larger in volume than the typical spheres expected from clustered galaxy formation (Furlanetto, Zaldarriaga & Hernquist 2004), owing to the extreme luminosity of the QSOs ($\sim 10^{14}$ L$_\odot$). The size of the Strömgren spheres is determined by the UV luminosity of the QSO, the HI density of the IGM, and the age of the QSO. Wyithe & Loeb (2004a) and Wyithe, Loeb & Barnes (2005) use QSO demographics, plus the distribution of CSS sizes, to infer a mean neutral IGM fraction of $x_{HI} \geq 0.1$ at $z \sim 6$.

A complementary analysis considers the "cosmic Strömgren surface" around the $z = 6.28$ QSO SDSS J1030 + 0524. Using high quality spectra (see **Figure 4**), Mesinger & Haiman (2004) show that the apparent size of the CSS around this QSO is smaller using the Lyα rather than the Lyβ line. They attribute this difference to the damping wing of Lyα, and from this infer an IGM neutral fraction of $x_{HI} \geq 0.2$.

However, Oh & Furlanetto (2005) highlight the many significant uncertainties in both these calculations, including: knowledge of the systemic redshift of the source, unknown 3D geometry, QSO age distribution, effect of large scale structure and clustering of ionizing sources around early luminous quasars, and often noisy spectra taken in a difficult wavelength range. In particular, recombination in denser regions can be a significant photon sink, and hence the clumping factor becomes a key parameter. Yu & Lu (2005) estimate that these factors could result in up to an order of magnitude of uncertainty in the derived neutral fraction. Also, there remains debate as to whether these techniques measure a mass- or a volume-averaged neutral fraction. Fan et al. (2006b) show that over $5.7 < z < 6.4$, the average size of the quasar Strömgren spheres decreases by ~ 2.5, consistent with an increase of the neutral fraction by a factor of ~ 15 over this redshift range, assuming the Strömgren radius $R_s \propto F(HI)^{-1/3}$ and other conditions being the same. Measurements of CSS are sensitive to a much larger neutral fraction than using the GP effect, but are also strongly affected by systematics. More examples of $z > 6$ QSOs over a larger range in intrinsic luminosity are required to verify these calculations.

5. OTHER IGM PROBES

In this section, we describe other IGM probes that are either sensitive to a large neutral fraction, or could be extended to higher redshift in the near future.

5.1. Evolution of Metal Absorption Systems

The early star formation that presumably started cosmic reionization also enriched the ISM and IGM with heavy elements. Detailed metal enrichment models in the early universe are reviewed in Ciardi & Ferrara (2005). Songaila (2001, 2005) and Pettini et al. (2003) found that the total CIV abundance Ω(CIV) does not significantly evolve over $1.5 < z < 5.5$. Schaye et al. (2003) found a lower bound of [C/H] ~ -4 even in the underdense regions of the IGM at $z < 4$. Madau, Ferrara & Rees (2001, see also Stiavelli, Fall & Panagia 2004) found that by assuming massive stars ionized the Universe with yield from Population II or III stars, the Universe would have a mean metallicity of $\sim 10^{-3 \sim -4} Z_\odot$ after reionization, comparable to those found in the high-redshift quasar absorption lines. Although these measurements do not directly constrain the reionization history, they suggest that one could detect numerous metal absorption lines even at $z > 6$.

Furlanetto & Loeb (2003) suggest using observations of low-ionization absorption lines such as C II or OI to constrain properties of the stellar population that ionized the Universe at $z > 6$, such as star-formation efficiency and escape fraction from supernova winds. To directly constrain the reionization history, Oh (2002) suggested using OI $\lambda 1302$ absorption as a tracer of the neutral fraction. OI has almost identical ionization potentials to H and should be in tight charge exchange equilibrium with H, while its lower abundance means that it would not saturate even when the Universe was mostly neutral: $\tau_{OI}^{eff} = 10^{-6}(\frac{\langle Z \rangle}{10^{-2} Z_\odot}) \tau_{HI}^{eff}$. OI (and Si II $\lambda 1260$) forests could provide combined constraints on the reionization and metal enrichment histories. Oh (2002) predicted that a handful of OI lines could be detected in the GP trough redshift regions of known $z > 6$ quasars when observed at high resolution. Becker et al. (2006) obtained high-resolution, high S/N spectra of a sample of six quasars at $z > 5$ using Keck/HERES and detected an OI system up to $z = 6.26$. They did not find a dense OI forest, consistent with the high degree of IGM ionization at $z \sim 6$. However, it is puzzling that the line of sight of SDSS J1148 + 5251 ($z = 6.42$), which has the highest ionization fraction at $z > 6$ (Section 3.3), also has the highest density of OI lines, raising the possibility that low metallicity, not high ionization, may be the cause for the lack of OI lines. Detailed modeling of IGM enrichment is needed to interpret high-redshift metal line results.

5.2. Gamma-Ray Bursts

Gamma-ray bursts (GRBs) are the most powerful explosions in the Universe, and could be detected at $z > 10$. At high-redshift, the time dilation means that their afterglow will fade away $(1 + z)$ times slower, aiding rapid spectroscopic follow-up observations (e.g., Ciardi & Loeb 2000) to probe the IGM evolution. GRB afterglow has been detected up to $z = 6.30$ (SWIFT GRB 050904; Kawai et al., 2006; Tagliaferri et al. 2005; J. Haislip, M. Nysewander, D. Reichart, A. Levan, N. Tanvir, submitted; P.A. Price, L.L. Cowie, T. Minezaki, B.P. Schmidt, A. Songaila & Y. Yoshii, submitted).

For a largely neutral IGM, $\tau \sim 10^5$, the damping wing of the GP trough arising from the large GP optical depth of the neutral medium will extend into the red side of the Lyα emission line (Miralda-Escudé 1998). For $z \sim 6$, at ~ 10 Å redward of Lyα of the host galaxy, the optical depth is of order unity for a neutral IGM. However, this GP damping wing test cannot be applied to luminous quasars, due to the proximity effect from the quasar itself, as shown by Madau & Rees (2000) and Cen & Haiman (2000), and discussed in Section 4. Absorption spectra of GRBs, however, are not affected by the proximity effect and can be used to probe the existence of damping wings and measure IGM neutral fractions up to order of unity. However, strong internal absorption from the neutral hydrogen of the ISM (log N(HI) > 21) in the host galaxy appears to be ubiquitous among GRB afterglow spectra (e.g., Chen et al. 2005). Such internal absorption, or gas infall in the host galaxy environment (Barkana & Loeb 2004), will complicate the interpretation of GRB observations. At large distance from line center, the damping wing from a diffused IGM has a profile $\tau \propto \Delta \nu^{-1}$, instead of $\tau \propto \Delta \nu^{-2}$ for a discrete absorber. Totani et al. (2006) fit a Lyα absorption profile of GRB 050904 ($z = 6.30$) with contributions from the internal damped Lyα absorption and the diffuse IGM damping wing simultaneously, and obtain a conservative limit on the IGM neutral fraction $x_{HI} < 0.6$.

5.3. Evolution of the IGM Thermal State

Reionization will photo-heat the IGM to several times 10^4 K. After this episode, the IGM will gradually cool mostly due to the Hubble expansion. Because of its long cooling time, the IGM will retain some of its thermal memory of reionization—earlier reionization leads to a cooler IGM at lower redshifts. Theuns et al. (2002a) used absorption line width measurements to estimate the IGM temperature at $z = 2$–4, and found an average temperature of $\sim 2.5 \times 10^4$ at $z \sim 3.5$. This temperature constrains reionization to be $z_{\rm reion} < 9$. Hui & Haiman (2003) carried out a similar analysis and considered different ionizing sources and different ionization histories. **Figure 7** shows their models with different reionization redshifts and a quasar-like ionizing spectrum. They concluded that for all models where the Universe was ionized at $z > 10$, and remained ionized thereafter, the IGM would have reached an asymptotic temperature too cold compared with observations. The IGM thermal history measurement requires the ionized fraction of the IGM to have of order unity changes at late ($6 < z < 10$) epoch, regardless of whether there is a very early episode of reionization. This independent constraint on the reionization history is consistent with GP measurements, and it points to a rapid ionization transition at low redshift. However, this is a difficult observation, and the interpretation of IGM temperature evolution may be further complicated by IGM heating during He II reionization at lower redshift (e.g., Sokasian, Abel & Hernquist 2002).

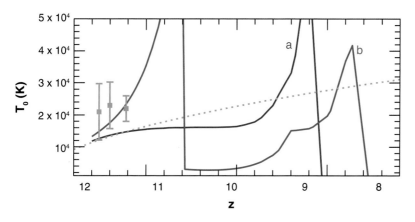

Figure 7

Evolution of IGM temperature for a quasar-like ionizing spectrum, assuming different reionization history. In models where the Universe is reionized early and remained so afterward, the IGM temperature is too cold when comparing to measurements at $z = 2$–4. Model A has a single reionization episode at $z = 14$. Model B also has early reionization but experienced a drop in ionizing flux until $z \sim 7$ at which time it undergoes a second period of reionization. To prevent overcooling, a major heating event that resulted in large changes in the IGM ionization state would have to occur at $z = 6$–10. However, a different theory of He II reionization history may complicate this intepretation. Adapted from Hui & Haiman (2003).

5.4. Luminosity Function and Line Profiles of Lyα Galaxies

Surveys of galaxies with strong Lyα emission lines through narrow-band imaging in selected dark windows of the night sky OH emission forest have proven to be a powerful technique for discovering the highest redshift galaxies (e.g., Hu et al. 2002, 2004; Kodaira et al. 2003; Rhoads et al. 2003; Kurk et al. 2004; Malhotra & Rhoads 2004; Martin & Sawicki 2004; Santos et al. 2004; Tran et al. 2004; Taniguchi et al. 2005). At the time of this writing, more than 100 Lyα galaxy candidates have been found at $z \sim 6.5$, including >30 with spectroscopic confirmations, plus the most distant galaxy with a confirmed spectroscopic redshift, SFJ J132418.3 + 271455 at $z = 6.589$ (Ajiki et al. 2003). Currently, ambitious surveys of Lyα galaxies at even higher redshift, through windows in the near-IR, are underway (e.g., Barton et al. 2004, Horton et al. 2004).

Lyα galaxies represent a significant fraction of star-forming galaxies at high redshift (Bouwens et al. 2006). Properties of Lyα galaxies directly probe the IGM neutral fraction. As described above, in a largely neutral IGM, the GP damping wing extends to the redside of Lyα emission. Without a large Strömgren sphere, the intrinsic Lyα emission will be considerably attenuated. In the simplest picture, one predicts: (*a*) the Lyα galaxy luminosity function will decrease sharply in an increasingly neutral IGM, even if the total star-formation rate in the Universe remains roughly constant, and (*b*) the Lyα profiles will have a stronger red wing and a smaller average equivalent width before the onset of reionization.

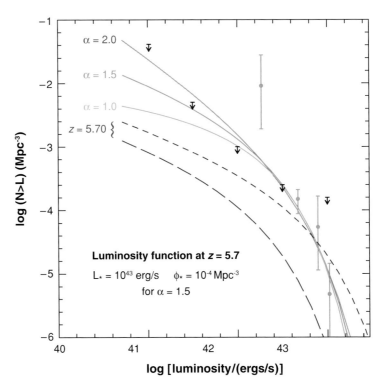

Figure 8

The luminosity function of Lyα galaxies at $z \sim 6.5$ and $z \sim 5.7$, from the calculations in Malhotra & Rhoads (2004). The luminosity function shows little evolution between these redshifts, consistent with a largely ionized IGM at $z \sim 6.5$. Adapted from Malhotra & Rhoads (2004).

Malhotra & Rhoads (2004) and Stern et al. (2005) combined the LALA survey of Lyα galaxies (Rhoads et al. 2003) with other Lyα surveys in the literature to determine the luminosity function of Lyα galaxies at $z = 6.5$ and 5.7. They found no evolution between these two redshift bins, consistent with the IGM being largely ionized by $z \sim 6.5$ (**Figure 8**). Hu et al. (2005) constructed Lyα line profiles at $z = 6.5$ and 5.7 from their surveys using Keck and Subaru, and found a similar lack of evolution. The interpretations of these results, however, require more detailed modeling. Haiman (2002), Santos (2004), and Cen, Haiman & Mesinger (2005) showed that the local HII regions around Lyα galaxies reduce the attenuations of Lyα flux. Gnedin & Prada (2004), Wyithe & Loeb (2005), R. Cen (submitted), and Furlanetto, McQuinn & Hernquist (2005) further demonstrated that the clustering of ionizing sources increases the HII region size and further reduces the attenuation. Haiman & Cen (2005) estimated that the lack of evolution in the Lyα luminosity function is consistent with the neutral fraction $x_{HI} < 0.25$, when no clustering is considered (see also Malhotra & Rhoads 2005). Including large scale clustering, the constraint on the IGM neutral fraction becomes less stringent (Furlanetto, Zaldarriaga & Hernquist 2004). Santos

(2004) showed that the presence of galactic winds may also play a crucial role in determining the observed Lyα fluxes. Haiman & Cen (2005) suggest studies of Lyα profiles as a function of luminosity, compared with low-redshift samples, could provide robust diagnostics to the IGM ionization state. Furlanetto, McQuinn & Hernquist (2005) showed that at large neutral fraction, only galaxies in large ionized bubbles can be detected in Lyα emission, which have larger bias than individual galaxies. With future large area Lyα surveys, it may be possible to measure the size distribution of HII regions during reionization to constrain reionization history.

6. COSMIC MICROWAVE BACKGROUND PROBES OF REIONIZATION

The cosmic microwave background (CMB) is a rich "fossil record" of the early universe. The CMB not only reveals the universe's initial conditions, but also its structure and dynamics from the time of the CMB's origin 400,000 years after the Big Bang to today. A CMB photon arriving in a terrestrial, balloon-borne, or space-based CMB observatory is encoded with the conditions of the universe along its 13.7 billion year journey. In particular, because the CMB is the oldest observable electromagnetic radiation it acts as a backlight, illuminating the transition from bulk-neutrality to complete reionization. Both the temperature anisotropy and polarization of the CMB help reconstruct the details of the EoR. As we will see, the CMB's polarization and temperature anisotropy are completely complementary probes, both in their spatial structure and in their dependence on the conditions of the EoR that they probe.

6.1. The Generation of CMB Polarization and Temperature Anisotropy During the Epoch of Reionization

Thomson scattering produces CMB polarization only when free-electron scatterers are illuminated by an anisotropic photon distribution (Hu & White 1997). Moreover, the photon anisotropy, when decomposed into spherical harmonics $Y_{\ell,m}$ in the rest frame of the electron, must possess a nonzero quadrupole term ($\ell = 2$)—no other term contributes due to the orthogonality of the spherical harmonics. Both free-electrons and anisotropically distributed photons were present during the transition from complete ionization to the EoR. This epoch produced the primary CMB polarization signal, referred to as E-mode or gradient mode owing to its symmetry properties under parity transformations (Kamionkowski, Kosowsky & Stebbins 1997; Zaldarriaga & Seljak 1997). The antisymmetric component of CMB polarization is referred to as B-mode or curl mode and arises in inflationary cosmological models that predict a primordial gravitational wave background. The E-mode polarization anisotropy is produced by the same perturbations that produce temperature anisotropy and peaks at $\simeq 10'$ angular scales (or multipoles $\ell \simeq 1000$) with an amplitude approximately 10% of the CMB temperature anisotropy at large ($>10°$) angular scales.

We begin this section by briefly reviewing how the EoR produces new CMB polarization and temperature anisotropy. The physics of CMB polarization and anisotropy has been treated in numerous sources. We refer the reader to Hu & White 1997;

Kamionkowski, Kosowsky & Stebbins 1997; Zaldarriaga & Seljak 1997; and Dodelson 2003 for analytic and theoretical treatments and Seljak & Zaldarriaga 1996 for numerical calculations. Ultimately, the CMB data, in combination with the 21-cm emission and Lyα absorbtion by neutral HI, will allow for a detailed reconstruction of the physics of the EoR.

6.1.1. Large angular scale CMB polarization and reionization. Large-angular scale CMB polarization as a probe of the ionization history of the universe has been considered in, e.g., Zaldarriaga 1997, Keating et al. 1998, and Kaplinghat et al. 2003. The importance of large angular scale CMB polarization is that, unlike the 21-cm and Lyα HI-absorption measurements discussed elsewhere in this review, the polarization of the CMB is sensitive to ionized hydrogen (H II) as it is generated by Thomson scattering.

Reionization produces free-electrons that Thomson-scatter CMB photons, producing CMB polarization. For angular scales smaller than the horizon at reionization the scattering damps the CMB temperature in direction **n** as $T'(\mathbf{n}) = e^{-\tau} T(\hat{\mathbf{n}})$, where τ is the Thomson optical depth. Reionization, therefore, damps the temperature anisotropy power angular spectrum as $C_\ell^{T'} = e^{-2\tau} C_\ell^T$. The damping of the temperature power spectrum is degenerate with the primordial power spectrum's amplitude, A, for scales smaller than the horizon at last scattering.

Fortunately, a new feature (e.g., Zaldarriaga 1997) in the E-mode power spectrum develops at large angular scales that breaks the degeneracy between A and τ. The degeneracy (for the CMB temperature anisotropy) and its breaking (using CMB polarization) are illustrated in real-space in **Figure 9**. **Figure 10** shows the polarization angular power spectrum associated with these simulations, along with current observational results (discussed further in Section 6.2). Note in **Figure 10** the new large angular scale (low-ℓ) peak, which discriminates between models with and without reionization.

The polarization anisotropy of the CMB reveals the ionization condition of the universe in ways that the temperature anisotropy cannot. Following recombination at $z = 1089 \pm 1$ (Spergel et al. 2003), primordial hydrogen remained neutral until the epoch of reionization. The term last-scattering is only approximately true because we know that the universe is nearly 100% ionized at present ($z = 0$). Therefore some fraction (a few to $\sim$10%) of the photons scattered again following decoupling. If reionization was instantaneous, only the primordial temperature quadrupole at the "last-scattering surface" ($z \simeq 1100$) projected to the redshift of reionization contributes to the CMB polarization at the <1 μK level (at >10° scales). This reionization feature has been detected by WMAP and is interpreted as a detection of $\tau = 0.10 \pm 0.03$ or $\tau = 0.09 \pm 0.03$ when all six cosmological parameters are fit to all of the WMAP data (TT,TE,EE) (Page et al. 2006).

Whereas CMB temperature anisotropy is generated by processes occurring after reionization [e.g., gravitational redshifts due to matter inhomogeneity along the line of sight—the Integrated Sachs-Wolfe (ISW) effect], the large-scale polarization of CMB is not similarly affected. The large-scale CMB polarization is a cleaner and more precise probe of the onset of the reionization epoch than the anisotropy for

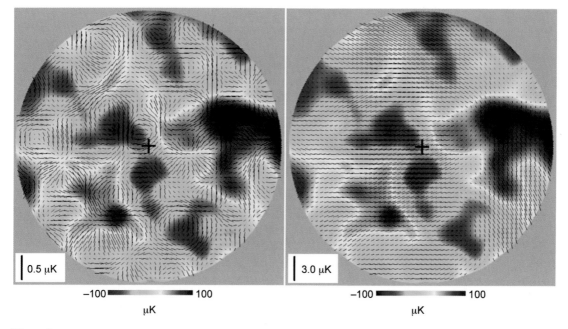

Figure 9

Large angular scale temperature (color scale) and polarization (lines) simulations in real-space. The left panel shows a simulation with no reionization and the right panel shows the effects of instantaneous reionization with $\tau = 0.17$. Both maps and cosmic microwave background observables are smoothed with a 4° beam. Reionization does not affect the large-scale temperature pattern but it produces polarization roughly ten-times larger at large scales—note change in scale for polarization between the two panels. The maximum polarization in the right panel is $\simeq 800$ nK. Figures courtesy of Eric Hivon.

two reasons. The first is that the ratio of the polarized intensity to the total intensity is unchanged when the CMB propagates through an absorbing medium (such as the postreionization intergalactic plasma)—both polarization components are attenuated equally. The second is that both polarization components incur equal gravitational redshifts (blueshifts) when climbing out of (falling into) time-varying gravitational potential wells from last-scattering to today because gravity cannot distinguish between the two linear polarization states. This is equivalent to asserting that CMB polarization does not experience the ISW effect, and so the low-multipole E-mode spectrum is not contaminated by the ISW effect.

High precision observations of the large-scale polarization will reveal more of the EoR's details. The polarization power spectra are sensitive to the duration of reionization. With upcoming observations the dynamics of the EoR can be constrained, complementing 21-cm observations (Section 8), by providing (modest) redshift information, most notably regarding the transition from partial to total reionization by constraining $x_{HI}(z)$, which is difficult using 21-cm data alone. This would be a powerful probe of "double reionization" models (Cen 2003b). A cosmic variance limited

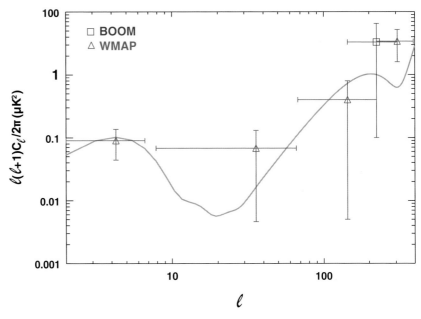

Figure 10

Measurements of the large angular scale gradient-mode polarization power spectrum C_ℓ^E for angular scales greater than $1° \Leftrightarrow \ell < 200$. The solid line is the polarization power spectrum for the Wilkinson Microwave Anisotropy Probe (WMAP) best-fit cosmological model, with $\tau = 0.09$ (Page et al. 2006, Spergel et al. 2006).

experiment, with sufficient control over instrumental systematics and foregrounds, can distinguish the transition from partial to total reionization using details of the polarization power spectra (Holder et al. 2003, Kaplinghat et al. 2003). Kaplinghat and colleagues show that large-scale CMB polarization can constrain the redshift of the transition from partial to total reionization with precision $\sigma_{z_{ri}} = 333\sigma_\tau/\sqrt{z_{ri}}$, where σ_τ is the precision with which the experiment measures the optical depth. Upcoming measurements, such as Planck, will produce percent-level constraints on partial reionization or double reionization scenarios and discriminate between models with identical τ, but very different ionization histories (Kaplinghat et al. 2003).

Reionization also produces a new low-ℓ peak in the B-mode polarization pattern, which is absent in models without reionization. Even a modest ionization fraction increases the observability of the B-mode (curl-mode) polarization. This is significant because gravitational lensing of large-scale structure converts E-mode power into B-mode power (Zaldarriaga & Seljak 1998, Hu & Okamoto 2002). This contaminant increases with increasing multipoles up to $\ell \sim 1000$. With the new, low-ℓ structure, a more stringent limit on the inflationary-generated gravitational wave background can be obtained than would be possible without reionization (Kaplinhat, Knox & Song 2003).

6.1.2. Small angular scale CMB polarization and reionization. Because even the primary CMB polarization is so small, observing effects to second order in τ or distinguishing the details of the Strömgrensphere percolation process is extremely challenging using CMB polarization. Fortunately, the small angular scale CMB anisotropy will provide a rich data set with vital information on the fine details of the reionization epoch.

6.1.3. The effects of reionization on CMB temperature anisotropy. CMB polarization probes the onset of reionization and distinguishes partial from complete reionization, but it cannot probe the details of the ionization percolation process as it is sensitive to the integrated Thomson optical depth to the last scattering surface. In fact, the reionization details that can be provided by large scale polarization are known to be limited by cosmic variance (Hu & Holder 2003) and the smoothing inherent in large-scale polarization measurements. Fortunately, the CMB temperature anisotropy at small angular scales may reveal many of these interesting features (Aghanim et al. 1996; Gruzinov & Hu 1998; Knox, Scoccimarro & Dodelson 1998; Haiman & Knox 1999; Barkana & Loeb 2001; Santos et al. 2003; Zahn et al. 2005). Thus, both the bulk and fine details of reionization will be probed by the complementary measurements of large-scale CMB polarization anisotropy and small-scale temperature anisotropy observations.

In noninstantaneous models of reionization (due to discrete ionizing sources), reionization proceeds in a "patchy" manner where the Strömgren spheres of ionized hydrogen surrounding the sources eventually coalesce to produce the fully-ionized universe observed out to $z \sim 6$. This patchiness induces secondary CMB temperature anisotropy (Aghanim et al. 1996; Gruzinov & Hu 1998; Knox, Scoccimarro & Dodelson 1998; McQuinn et al. 2005). In addition to the damping of the CMB temperature anisotropy power spectrum, there are two interesting secondary temperature anisotropy effects: (a) the kinetic Sunyaev-Zel'dovich (kSZ) effect and (b) the Ostriker-Vishniac (OV) effect.

The kSZ effect is due to the motion of regions of reionized electrons along the line of sight. It is similar to the kinematic SZ effect produced by inverse Compton scattering of the CMB by electrons in moving galaxy clusters (Sunyaev & Zeldovich 1980), but here it refers to the motions of the ionized electrons in the PGM during the EoR. At very small angular scales (<4') the primary CMB anisotropy signal is damped due to photon diffusion out of overdense regions (Silk damping) prior to decoupling. Here the thermal SZ effect dominates over the primary CMB anisotropy, but at the thermal SZ null (218 GHz) the EoR kSZ effect may be the dominant source of CMB temperature anisotropy.

During reionization, electron spatial inhomogeneity can arise in two ways: either by a variable ionization fraction, x_i, or by a constant x_i and a variable baryon density. The latter effect is the OV effect, which is primarily generated by structures in the nonlinear regime, i.e., at low-z. The OV effect is the result of a patchy reionized medium. Much progress has been made due to the development of fast structure formation codes (e.g., smooth particle hydrodynamics, SPH), which trace the reionization mechanism (because they trace the formation of baryonic structures). The reionization efficiency of the baryonic structures is the only free parameter, and many of these approaches, e.g., Zahn et al. (2005), incorporate semianalytic reionization models into their simulations. Using an extended Press-Schechter formalism to model the reionization process Zahn et al. (2005) find that patchy reionization makes a large contribution to the CMB anisotropy at scales of 6' ($\ell \sim 2000$), which exceeds the kSZ effect. In fact, ignoring these secondary EoR effects will lead to significant biases in the estimation of cosmological parameters derived from the primary CMB anisotropy, which further underscores the vital importance of their measurement (Santos et al. 2003, Zahn et al. 2005).

Both the kSZ and the OV effects occur at small scales ($<0.1°$, $\ell > 2000$) and probe both the homogeneity and efficiency of the reionization process. The shapes of the predicted temperature anisotropy power spectra for these secondary effects appear to be robust to changes in the reionization model (Knox 2003, Santos et al. 2003, Zahn et al. 2005), although the peak-position of the secondary power spectra occurs when $x_{HI} \sim 0.5$, which is model-dependent (Furlanetto, McQuinn & Hernquist 2005). In contrast, the overall amplitude of the secondary spectra appears to be significantly model-dependent. However, the amplitude depends only weakly on the EoR and more strongly on the duration of the patchy phase, making it complementary to the CMB polarization measurements of the optical depth (McQuinn et al. 2005). For completeness we close this subsection on secondary effects by mentioning that both the polarized kSZ and OV effects are expected to be nearly negligible and practically impossible to measure (Seshadri & Subramanian 1998, Hu 2000).

6.2. Current CMB Results and Future Prospects

The first detection of reionization phenomena using CMB observations came with NASA's Wilkinson Microwave Anisotropy Probe (WMAP), which reported a nearly 5σ detection of a reionization feature in the large angular scale ($>10°$) temperature-polarization cross-correlation function ($\langle TE \rangle$) with only a single year of data (Kogut et al. 2003). This detection was the first use of CMB polarization to measure the physics of the EoR. It also was the first detection of the Thomson optical depth, τ, rather than the lower limits (provided by Lyα measurements) or upper limits (provided by CMB temperature anisotropy measurements). WMAP's initial detection of τ in 2003 using the TE data and subsequent detection using the E-mode polarization only (Page et al. 2006) also determine the primordial power spectrum amplitude A, $\sigma_A \simeq 2\sigma_\tau$ (Spergel et al. 2003). WMAP's detection illustrates CMB polarization's power to probe vitally important cosmological parameters that are essentially unobservable using temperature anisotropy alone.

WMAP's three-year data set (Page et al. 2006) reports several detections of the E-mode and temperature-polarization cross-correlation power spectrum, especially at low-ℓ (due to WMAP's ability to map most of the sky). CBI, DASI, and BOOMERANG (F. Piacentini, P. Ade, J. Bock, J. Bond, J. Borrill, et al., submitted)

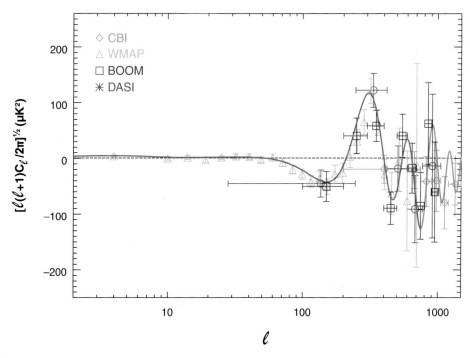

Figure 11

Measurements of the temperature-polarization cross-correlation power spectrum C_ℓ^{TE}. The solid line is the power spectrum for the Wilkinson Microwave Anisotropy Probe (WMAP) best fit cosmological model, with $\tau = 0.09$ (Page et al. 2006, Spergel et al. 2006).

have also detected the $\langle TE \rangle$ cross-correlation spectrum, at smaller angular scales than WMAP.

WMAP's first-year results are relatively robust to the choice of additional data sets and priors used to constrain cosmological parameters (including τ), such as SDSS (Tegmark et al. 2004) and Lyα forest measurements (Seljak et al. 2005).

In **Figures 10** and **11** observational CMB polarization and temperature-polarization cross-correlation power spectra are shown for the angular scales relevant for probing the EoR. The primary effects of reionization are encoded in the E-mode autocorrelation and temperature polarization cross-correlation power spectra at $\ell \lesssim 50$. As demonstrated in, e.g., Zaldarriaga, Furlanetto & Hernquist (2004), Hu & Holder (2003), and Keating & Miller (2006), essentially all the information on τ comes from $\ell < 50$. Although most of the τ-constraint comes from $\ell < 10$, a non-negligible amount comes from $10 < \ell < 40$. Currently, only WMAP (Page et al. 2006) and BOOMERANG (T.E. Montroy, P.A.R. Ade, J.J. Bock, J.R. Bond, J. Borrill, submitted) have detected E-mode polarization at angular scales greater than $1°$ ($\ell < 200$), and only WMAP has detected the E-mode polarization for $\ell < 100$. CBI (Readhead et al. 2004a) and CAPMAP (Barkats et al. 2005) have detected the E-mode polarization, though at smaller angular scales (higher ℓ) than are relevant for probing the EoR with CMB polarization alone.

Although no detections of the kSZ or OV effects have been reported, several groups have published detections of CMB anisotropy at scales that are relevant to both secondary temperature effects. For $3000 < \ell < 10000$, the kSZ temperature anisotropy power spectrum scales roughly as $C_l(\Omega_b h)^2 \sigma_8^p$ where $3 < p < 7$ (e.g., see Zhang et al. 2004). Dawson et al. (2002) report a detection of CMB anisotropy at subarcminute ($\ell \simeq 7000$) scales using the BIMA array and Readhead et al. (2004b) report a detection at larger (a few-arcminute) scales using CBI. The results of the two research groups are consistent, and seem to indicate significant power in excess of the CMB primary anisotropy at those scales. Both groups speculate on the likelihood that this excess structure is due to the thermal SZ effect. Although neither group has detected the kSZ or OV effects, they are valuable technological and methodological precursors for future CMB anisotropy measurements of the EoR.

6.3. Observational Challenges

A hallmark of CMB experiments since the detection of the CMB has been the mitigation and excision of systematic effects. CMB observations have gone to great lengths to combat systematic effects, including building duplicate telescopes (e.g., the COBE and WMAP satellites). Another method of improving experimental fidelity is to conduct extremely wide-band observations for foreground monitoring and removal. For example, WMAP has radiometers covering more than two octaves in frequency (Jarosik et al. 2003), and Planck's radiometers will cover nearly 5 octaves (Lamarre et al. 2003, Mennella et al. 2004).

Particularly for the large-scale E-mode polarization, measuring the CMB with multiple frequency channels alone is not sufficient. To overcome statistical (cosmic) variance on the low-ℓ C_ℓ^E peak, an experiment must observe a large survey region—at least $\sim 30° \times 30°$. Such a large area is likely to be contaminated by galactic emission, which can only be subtracted to finite precision. The most conservative approach to deal with galactic foregrounds is to exclude regions of strong foreground emission (such as the galactic plane) from the survey. For example, the WMAP team used multiple frequency channels in combination with sky-cuts that remove 15% of the sky surrounding the galactic plane (Kogut et al. 2003) and introduced correlations between multipoles at the few-percent level. Even with sophisticated foreground modeling and multiple frequency coverage the challenges of achieving polarization fidelity at the 100 nK level over $>10°$ angular scales—required, for example, to detect the signatures of noninstantaneous reionization scenarios (e.g, Holder et al. 2003, Kaplinghat et al. 2003)—are daunting. The CMB community is well aware of the challenge.

Observational challenges for secondary temperature anisotropy (kSZ and OV) measurements include point sources, primary CMB confusion, and secondary CMB confusion due to gravitational lensing and the thermal SZ effect. To detect the secondary effects, the primary CMB anisotropy must be overcome by measuring at scales smaller than the Silk damping scale, requiring large (>5-m class) telescopes. Of the foregrounds for the kSZ and OV effects, the thermal SZ effect is perhaps the most straightforward to mitigate—by measuring at the thermal SZ-null near 218 GHz. At small scales, point sources need to be excised based on ancillary measurements of their spectra and position. There are also significant hurdles to overcome in the theoretical modeling of the secondary temperature effects. In particular, separating the low-z OV effect from high-z patchy reionization effects will be challenging, though hopefully amenable to theoretical modeling and simulation.

Because galactic foregrounds and point sources contribute to spurious polarization and anisotropy with different systematics, combining the large angular CMB polarization observations with the small scale temperature anisotropy measurements represents the most promising avenue toward a faithful reconstruction of the EoR using the CMB.

6.4. Discussion and Future Prospects

WMAP's on-going measurements of CMB polarization are expected to continue until at least 2007. Planck is expected to be launched in 2008 and will start making full-sky observations of the polarization and primary temperature anisotropy of the CMB. Planck is expected to achieve a nearly cosmic variance–limited measurement of τ. A balloon-borne CMB polarimeter called EBEX (Oxley et al. 2005) will cover a large region of the sky in an attempt to measure the signature of gravitational waves imprinted on the CMB.

More reionization information will emerge from CMB B-mode polarization observations, which search for the imprint of primordial gravitational radiation on the CMB. Like the E-mode polarization reionization signature, the B-mode polarization peaks at super-degree scales so the upcoming experiments optimized to search for the B-mode signal will have appreciable sensitivity to reionization features in the E-mode spectrum (Keating & Miller 2006).

An ancillary benefit of reionization is that it boosts the primary (inflation generated) B-mode power spectrum significantly near $\ell = 10$. Due to reionization, a tighter limit on the tensor-to-scalar ratio, r, in the presence of gravitational lensing (at redshifts $z < 10$) can be obtained than that calculated in Kesden, Cooray & Kamionkowski 2002 and Knox & Song 2002. Kaplinghat, Knox & Song (2003) demonstrate that the minimum detectable inflationary energy scale behaves as $1/\tau^4$ due to the low-ℓ B-mode power spectrum peak produced by reionization. This motivates very wide-field, or full-sky, observations of CMB polarization. NASA's Beyond Einstein initiative features a CMB polarimeter called CMBPol.[1] This instrument is designed to detect the signature of inflationary gravitational waves over a wide range of inflationary energy scales. This experiment will achieve cosmic variance–limited precision on τ, which is essentially the same as Planck's sensitivity (Kaplinghat et al. 2003), but in contrast to Planck, CMBPol will be optimized to detect CMB polarization.

---

[1] http://universe.nasa.gov/program/inflation.html

Three large-format bolometric array telescopes are currently being developed to probe CMB temperature anisotropy and the kSZ and OV effects. The Atacama Cosmology Telescope (ACT, Kosowsky 2003) is a 6-m telescope operating in three frequency bands (145, 225, and 265 GHz) with $\simeq 1'$ angular resolution. With concordance cosmological parameters, ACT should be able to constrain the redshift of reionization to percent-level accuracy (Zhang, Pen & Trac 2004). The South Pole Telescope, Pen & Trac (SPT, Ruhl et al. 2004) is a 10-m telescope operating in 5 bands from 100 to 345 GHz with angular resolution ranging from $1.5'$ to $\simeq 0.5'$. Finally, the largest CMB anisotropy telescope that will be sensitive to secondary CMB temperature anisotropy is the 12-m Atacama Pathfinder EXperiment (APEX), which will use a 320-pixel array of transition edge sensor (TES) bolometers, also from a high altitude (5000 m) observatory in the Chilean Atacama desert. All of these instruments, with the exception of the CMBPol, are scheduled for "first-light" before 2010.

This section has emphasized the role of the CMB in illuminating the physics of the EoR. Both the CMB's polarization and its temperature anisotropy provide a wealth of reionization data in a fashion that is complementary to the 21-cm and Lyα observations discussed elsewhere in this review. Although the importance of large-scale CMB polarization to reionization was recognized early on, its ability to reveal more than the bulk Thomson optical depth was not initially appreciated. Now we understand that the polarization and the temperature anisotropy provide a detailed view of the EoR. In fact, they do so in completely complementary ways: the polarization reveals the EoR in a way the temperature anisotropy cannot. Likewise, the CMB's temperature anisotropy probes secondary effects to which the polarization is blind. When the new CMB polarization and temperature anisotropy observations are combined a high-fidelity image of the EoR will emerge.

7. SOURCES OF REIONIZATION

Regardless of the detailed reionization history, the IGM has been almost fully ionized since at least $z \sim 6$. This places a minimum requirement on the emissivity of UV ionizing photons per unit of comoving volume required to keep up with recombination and maintain reionization (Miralda-Escudé, Haehnelt & Rees 2000):

$$\dot{\mathcal{N}}_{ion}(z) = 10^{51.2} \mathrm{s}^{-1} \mathrm{Mpc}^{-3} \left(\frac{C}{30} \right) \times \left(\frac{1+z}{6} \right)^3 \left(\frac{\Omega_b h^2}{0.02} \right)^2, \qquad (3)$$

where $C \equiv \frac{\langle n_H^2 \rangle}{\langle n_H \rangle^2}$ is the clumping factor of the IGM. C is difficult to determine from observations and has large uncertainty when estimated from simulations ($C = 10$–100, Gnedin & Osteriker 1997). At $z < 2.5$, the UV ionizing background is dominated by quasars and AGN (e.g., Haardt & Madau 1996). At $z > 3$, the density of luminous quasars decreases faster than that of star-forming galaxies, and the ionizing background has an increased contribution from stars (e.g., Haehnelt et al. 2001). In this section we summarize what is known about the ionizing background contribution from quasars and AGN, young stars, or other sources of high energy photons (e.g., particle decay).

7.1. Quasars and AGN

Quasars and AGN are effective emitters of UV photons. The UV photon escape fraction is generally assumed to be of order unity. The UV photon emissivity from quasars and AGN can be directly measured from integrating quasar luminosity functions at high redshift. The density evolution of luminous quasars ($M_{1450} < -27$) has been well-determined from surveys such as SDSS (e.g., Warren, Hewett & Osmer 1994; Kennefick, Djorgovski & de Carvalho 1995; Schmidt, Schneider & Gunn 1995; Fan et al. 2001a,b, 2004; Richards et al. 2006) up to $z \sim 6$. Luminous quasar density declines exponentially toward high redshift: It is ~ 40 times lower at $z \sim 6$ than at its peak at $z \sim 2.5$. However, quasars have a steep luminosity function at the bright end—most of the UV photons come from the faint quasars that are currently below the detection limit at high redshift.

Fan et al. (2001b) showed that quasars could not have maintained IGM ionization at $z \sim 6$, as the shape of the luminosity function at $z \sim 6$ is not much steeper than that at $z < 3$. In fact, Fan et al. (2001a) and Richards et al. (2006) used the SDSS sample to find that at least the bright-end slope of quasars at $z > 4$ is considerably flatter than that at low redshift, consistent with the findings of the COMBO-17 survey (Wolf et al. 2003). Miralda-Escudé (2003), Yan & Windhorst (2004a), and Meiksin (2005) used different parameterizations of quasar luminosity function evolution and came to the same conclusion. Willott et al. (2005) and Mahabal et al. (2005) present surveys of a few $\deg^2$ for faint $z \sim 6$ quasars. Their detection of only one $z \sim 6$ faint QSO implies that the quasar population contribution to the ionizing background is $<30\%$ that of star-forming galaxies.

7.2. Lyman Break Galaxies

Due to the rapid decline in the AGN populations at very high z, most theoretical models assume stellar sources reionized the universe. However, despite rapid progress, there is still considerable uncertainty in estimating the total UV photon emissivity of star-forming galaxies at high redshift. Modifying the constraint from Equation 4, Bunker et al. (2004) and Bouwens et al. (2005) showed

$$\dot{\rho}_* \approx (0.026 \, M_\odot \mathrm{yr}^{-1} \mathrm{Mpc}^{-3}) \left(\frac{1}{f_{\mathrm{esc,rel}}}\right) \frac{C}{30} \left(\frac{1+z}{7}\right)^3, \qquad (4)$$

where $\dot{\rho}_*$ is the UV star-formation-rate rate, C_{30} is the IGM clumping factor, and $f_{\mathrm{esc,rel}}$ is the fraction of ionizing radiation escaping into the IGM to that escaping in the UV-continuum (~ 1500 Å). The uncertainties come from three factors: the star-formation rate, clumping factor, and UV escape factor. Current estimates of the star-formation rate at $z \sim 6$ all come from photometrically-selected i-dropout objects selected in a small number of deep fields observed by the HST: the HUDF, GOODS fields (Giavalisco et al. 2004), and UDF ACS Parallels (Thompson et al. 2005). These estimates are affected strongly by sample (or cosmic) variance at $z \sim 6$ due to the small volume of deep surveys, as well as estimates or extrapolations toward faint luminosities. The clumping factor is typically taken from cosmological simulations (e.g., $C \sim 30$, Gnedin & Osteriker 1997). The UV escape fraction remains problematic. Direct measurements of the escape fraction range from upper limits, <0.1 to <0.4 (e.g., Giallongo et al. 2002, Fernández-Soto, Lanzetta & Chen 2003; Inoue et al. 2005) to >0.5 (Steidel, Pettini & Adelberger 2001). Therefore, instead of measuring the accurate UV photon emissivity, most efforts attempt to determine whether star-forming galaxies could provide sufficient photons to at least meet the requirement in Equation 5 at $z \sim 6$.

Using HUDF data, Yan & Windhorst (2004b) found 108 plausible candidates at $5.5 < z < 6.5$ down to $m_{AB}(z) = 30.0$ mag. They estimated a steep faint-end luminosity function with a power-law index of -1.8 to -1.9, and concluded that this steep slope is sufficient to satisfy the reionization requirement and that most of the photons that ionized the universe come from dwarf galaxies. Similarly, Stiavelli, Fall & Panagia (2004) compared GOODS and HUDF observations and found that the observed mean surface brightness of galaxies at $z \sim 6$ is sufficient for reionization when the young stellar population has a top heavy IMF. Bunker et al. (2004), however, found a much lower star-formation density at $z \sim 6$ from the HUDF. Their results showed a factor of six decline in star-formation rate from $z \sim 3$ to 6. Even using the most optimistic escape fraction ($f = 1$), their results imply insufficient photons to ionize the Universe by $z \sim 6$ from stellar sources.

Bouwens et al. (2005) combined all the available datasets that include 506 i-dropout ($z \sim 6$) galaxy candidates to construct a luminosity function. **Figure 12** summarizes the determination of the evolution of star-formation rate at both the luminous and faint end. The star-formation rate begins to decline modestly at $z > 5$. The shape of the luminosity function has also evolved significantly from $z \sim 3$ to 6. However, the integrated UV luminosity (down to 0.04 $L^*_{z=3}$) did not evolve significantly. Their results support the previous claim that dwarf galaxies provide the majority of ionizing photons at $z \sim 6$, sufficient to ionize the Universe.

Most recently, Kashlinsky et al. (2005) have reported the detection of a significant excess in the Spitzer IRAC bands above that expected based on known galaxy counts. They interpret this excess as being due to population III stars forming at $z > 10$. These stars would clearly contribute to early reionization, although the magnitude of the contribution remains uncertain owing to the dificulty of the measurement, and the uncertainty of the origin of this mid-infrared background.

7.3. Dusty and Big Galaxies at High Redshift

Recent observations reveal two surprising types of galaxies at $z > 6$. First, millimeter observations of the host galaxies of the highest redshift QSOs reveal large dust masses ($\sim 10^8$ M$_\odot$) in $\sim 30\%$ of the sources (Bertoldi et al. 2003a). In one case, J1148 + 5251 at $z = 6.42$, millimeter line emission from CO has been detected, indicating a large molecular gas mass ($\sim 2 \times 10^{10}$ M$_\odot$; Walter et al. 2003, 2004; Bertoldi et al. 2003b). Second, Eyles et al. (2005) and Mobasher et al. (2005) present Spitzer observations of $z \sim 6$–6.5 galaxies. These objects show SEDs characteristic of poststarburst galaxies,

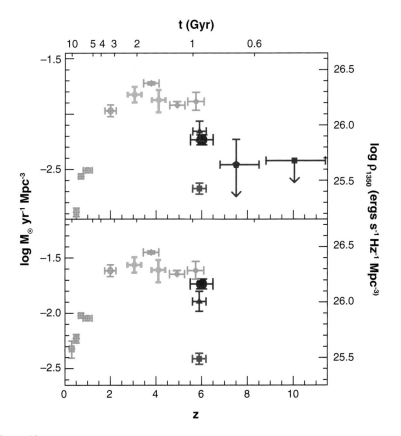

Figure 12

The cosmic star-formation history (uncorrected for extinction) integrated down to $0.3L^*_{z=3}$ (*top panel*) and $0.04L^*_{z=3}$ (*bottom panel*). These luminosities correspond to the faint-end limits for z_{850} and i_{775}-dropout probes at $z \sim 7–8$ and $z \sim 6$, respectively. The large red circle denotes determination at $z \sim 6$ from Bouwens et al. (2006) and is compared with previous determinations by Schiminovich et al. (2005) (*open squares*), Steidel et al. (1999) (*green crosses*), Giavalisco et al. (2004b) (*gray diamonds*), Bouwens et al. 2004a (*magenta triangle*), Bunker et al. (2004) (*blue squares*), Bouwens et al. 2004b (*magenta pentagon*), and Bouwens et al. (2005) (*magenta square*). The figure is divided into two panels to illustrate how much stronger the evolution is at the bright end of the luminosity function ($\gtrsim 0.3L^*_{z=3}$) than it is when integrated to the *i*-dropout faint-end limit ($0.04L^*_{z=3}$). Adapted from Bouwens et al. (2005).

with stellar masses between $10^{10}-10^{11} M_\odot$ and stellar ages of several hundred Myrs. Reddening is also infered by Chary, Stern, & Eisenhardt (2005) in one galaxy, implying a dusty galaxy ($A_v \sim 1$). These results require active star formation ($>10^2 \, M_\odot$ year $^{-1}$) starting at $z \sim 8-15$. These objects are capable of ionizing large volumes at high redshift, and could have played an important role in the early reionization of the IGM, depending on their currently ill-constrained space density (Barkana & Loeb 2005c, Panagia et al. 2005).

7.4. X-Ray Photons and Decaying Sterile Neutrinos

High-energy X-ray photons from mini-quasars or supernova remnants at high redshift could also contribute significantly to the reionization process (Oh 2001; Venkatesan, Giroux & Shull 2001; Madau et al. 2004; Ricotti & Ostriker 2004). These photons have long mean free paths, and could preionize the IGM significantly, providing the optical depth required by WMAP polarization measurement, and produce a rapid ending of reionization at $z \sim 6$ as suggested by the GP measurements. However, Dijkstra, Haiman & Loeb (2004) and Salvaterra, Haardt & Ferrara (2005) show that the hard X rays from these same sources would produce a present-day soft X-ray background. They calculated the X-ray background if such photons dominated the reionization budget, and concluded that models with accreting black holes (a combination of luminous and faint AGN) would overproduce the observed X-ray background by a large factor. A population dominated by mini-quasars could still partially ionize the IGM at $z > 6$, but its contribution could be severely constrained if the X-ray background is further resolved into discrete sources.

Finally, Hansen & Haiman (2004) proposed reionization by decaying sterile neutrinos at high redshift. However, Mapelli & Ferrara (2005) showed that by satisfying limits on the X-ray, optical, and near-IR cosmic background, the total Thomson optical depth from sterile neutrinos is small, and they could have only played a minor role in reionization. Chen & Kamionkowski (2004) and Pierpaoli (2003) show that the number of decays needed to reionize the universe at $z \sim 20$ will produce a CMB inconsistent with observations.

Furlanetto, Sokasian & Hernquist (2004) show that low-frequency observations of the power spectrum of brightness temperature fluctuations in the HI 21-cm line from the neutral IGM during reionization will be able to differentiate between uniform versus HII region–dominated reionization.

8. HI 21-cm PROBES OF REIONIZATION

The 21-cm line of neutral hydrogren presents a unique probe of the evolution of the neutral IGM and cosmic reionization. Furlanetto & Briggs (2004) point out some of the advantages of using the HI line in this regard: (*a*) unlike Lyα (i.e., the GP effect), the 21-cm line does not saturate, and the IGM remains "translucent" at large neutral fractions (Carilli et al. 2004a). And (*b*) unlike CMB polarization studies, the HI line provides full 3D information on the evolution of cosmic structure; the technique involves imaging the neutral IGM directly, and hence can easily distiguish between different reionization models (Furlanetto, Sokasian & Hernquist 2004). HI 21-cm observations can be used to study the evolution of cosmic structure from the linear regime at high redshift (i.e., density-only evolution), through the nonlinear "messy astrophysics" regime associated with luminous source formation. As such, HI measurements are sensitive to structures ranging from very large scales down to the source scale set by the cosmological Jeans mass, thereby making 21 cm the "richest of all cosmological data sets" (Barkana & Loeb 2005b).

Many programs have been initiated to study the HI 21-cm signal from cosmic reionization, and beyond. The largest near-term efforts are the Mileura Widefield

Array (MWA[2]), the Primeval Structure Telescope (PAST[3]), and the Low Frequency Array (LOFAR[4]). These telescopes are being optimized to study the power spectrum of the HI 21-cm fluctuations. The VLA-VHF[5] system is designed specifically to set limits on the cosmic Strömgren spheres around $z \sim 6$ to 6.4 SDSS QSOs. In the long term the Square Kilometer Array (SKA[6]) should have the sensitivity to perform true 3D imaging of the neutral IGM in the 21-cm line during reionization. And at the lowest frequencies (<80 MHz), the Long Wavelength Array (LWA[7]), and eventually the Lunar array (LUDAR; Corbin et al. 2005; Maccone 2004), is being designed for the higher z HI 21-cm signal from the PGM.

In this section we summarize the current theories and capabilities for detecting the neutral IGM during, and prior to, reionization using the HI 21-cm line (see also Carilli et al. 2004a; Carilli 2005; S.R. Furlanetto, S.P. Oh, & F. Briggs, in preparation).

8.1. The Physics of the Neutral IGM

The physics of radiative transfer of the HI 21-cm line through the neutral IGM have been considered in detail by many researchers (Scott & Rees 1990; Madau, Meiksin & Rees 1997; Tozzi et al. 2000; Furlanetto & Briggs 2004; Zaldarriaga, Furlanetto & Hernquist 2004; Bharadwaj & Ali 2005; Morales 2005; Santos, Cooray & Knox 2005). We review only the basic results here.

In analogy to the GP effect for Lyα absorption by the neutral IGM, the optical depth, τ, of the neutral hydrogen to 21-cm absorption for our adopted values of the cosmological parameters is

$$\tau = \frac{3c^3 \hbar A_{10} n_{HI}}{16 k_B \nu_{21}^2 T_S H(z)} \sim 0.0074 \frac{x_{HI}}{T_S}(1+\delta)(1+z)^{3/2} \left[H(z) \bigg/ \left(\frac{dv}{dr}\right) \right], \quad (5)$$

where A is the Einstein coefficient and $\nu_{21} = 1420.40575$ MHz (e.g., Santos, Cooray & Knox 2005). This equation shows immediately the rich physics involved in studying the HI 21-cm line during reionization, with τ depending on the evolution of cosmic over-densities, δ (predominantly in the linear regime), the neutral fraction, x_{HI} (i.e., reionization), the HI excitation, or spin, temperature, T_S, and the velocity structure, $\frac{dv}{dr}$, including the Hubble flow and peculiar velocities.

In the Raleigh-Jeans limit, the observed brightness temperature (relative to the CMB) due to the HI 21-cm line at a frequency of $\nu = \nu_{21}/(1+z)$ is given by

$$T_B \approx \frac{T_S - T_{CMB}}{1+z}\tau \approx 7(1+\delta)x_{HI}\left(1 - \frac{T_{CMB}}{T_S}\right)(1+z)^{1/2} \text{ mK}. \quad (6)$$

---

[2] web.haystack.mit.edu/arrays/MWA/LFD/index.html
[3] web.phys.cmu.edu/past/
[4] www.lofar.org/
[5] cfa-www.harvard.edu/dawn/
[6] www.skatelescope.org/
[7] lwa.unm.edu/index.shtml

The conversion factor from brightness temperature to specific intensity, I_ν, is given by: $I_\nu = \frac{2k_B}{(\lambda_{21}(1+z))^2} T_B = 22(1+z)^{-2} T_B$ Jy deg^{-2}. Equation 6 shows that for $T_S \sim T_{CMB}$ one expects no 21-cm signal. When $T_S \gg T_{CMB}$, the brightness temperature becomes independent of spin temperature. When $T_S \ll T_{CMB}$, we expect a strong negative (i.e., absorption) signal against the CMB.

Tozzi et al. (2000) point out that the HI excitation temperature will equilibrate with the CMB on a timescale $\sim \frac{3 \times 10^5}{(1+z)}$ year, in absence of other effects. However, collisions, and resonant scattering of Lyα photons, can drive T_S to the gas kinetic temperature, T_K (Wouthuysen 1952, Field 1959, Hirata 2005). Zygelman (2005) calculates that, for the mean IGM density, collisional coupling between T_S and T_K becomes significant for $z \geq 30$. Madau, Meiksin & Rees (1997) show that resonant scattering of Lyα photons will couple T_S and T_K when $\mathcal{J}_\alpha > 9 \times 10^{-23}(1+z)$ erg cm^{-2} s^{-1} Hz^{-1} sr^{-1}, or about one Lyα photon per every two baryons at $z = 8$.

The interplay between the CMB temperature, the kinetic temperature, and the spin temperature, coupled with radiative transfer, lead to a number of interesting physical regimes for the HI 21-cm signal (Barkana & Loeb 2004; Carilli 2005; Ali 2006): (*a*) At $z > 200$ equilibrium between T_{CMB}, T_K, and T_S is maintained by Thomson scattering off residual free electrons and gas collisions. In this case $T_S = T_{CMB}$ and there is no 21-cm signal. (*b*) At $z \sim 30$ to 200, the gas cools adiabatically, with temperature falling as $(1+z)^2$, i.e., faster than the $(1+z)$ for the CMB. However, the mean density is still high enough to couple T_S and T_K, and the HI 21-cm signal would be seen in absorption against the CMB (Sethi 2005). (*c*) At $z \sim 20$ to 30, collisions can no longer couple T_K to T_S, and T_S again approaches T_{CMB}. However, the Lyα photons from the first luminous objects (Pop III stars or mini-quasars) may induce local coupling of T_K and T_S, thereby leading to some 21-cm absorption regions (R. Cen, submitted). At the same time, X rays from these same objects could lead to local IGM warming above T_{CMB} (Chen & Miralda-Escudé 2004). Hence one might expect a patchwork of regions with no signal, absorption, and perhaps emission in the 21-cm line. (*d*) At $z \sim 6$ to 20 all the physical processes come into play. The IGM is being warmed by hard X rays from the first galaxies and black holes (Ciardi & Madau 2003, Barkana & Loeb 2004, Loeb & Zaldarriaga 2004), as well as by weak shocks associated with structure formation (Furlanetto, Zaldarriaga, & Hernquist 2004; Wang & Hu 2006), such that T_K is likely larger than T_{CMB} globally (Furlanetto, Sokasian & Hernquist 2004). Likewise, these objects are reionizing the universe, leading to a fundamental topological change in the IGM, from the linear evolution of large scale structure, to a bubble-dominated era of HII regions (Furlanetto, McQuinn & Hernquist 2005).

8.2. Capabilities to Detect HI 21-cm Signals from Reionization

The HI 21-cm signature of the neutral IGM during, and prior to, reionization can be predicted analytically using a standard Press-Schechter type of analysis of linear structure formation, plus some recipes to approximate nonlinear evolution (Gnedin & Shaver 2004; Zaldarriaga, Furlanetto & Hernquist 2004; Bharadwaj & Ali 2005; Santos, Cooray & Knox 2005; Wang & Hu 2006; Ali 2006), or through the use

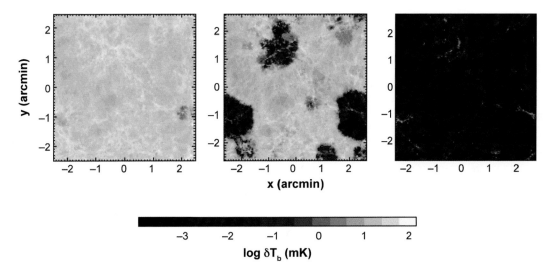

Figure 13

The simulated HI 21-cm brightness temperature distribution during reionization at $z = 12, 9,$ and 7 (Furlanetto, Sokasian & Hernquist 2004; Zaldarriaga, Furlanetto & Hernquist 2004).

of numerical simulations (Ciardi & Madau 2003; Furlanetto, Sokasian & Hernquist 2004; Wang et al. 2005).

8.2.1. Tomography. Figure 13 shows the expected evolution of the HI 21-cm signal during reionization based on numerical simulations (Furlanetto, Sokasian & Hernquist 2004; Zaldarriaga, Furlanetto & Hernquist 2004). They find that the mean HI signal is about $T_B \sim 25$ mK prior to reionization, with fluctuations of a few mK on arcmin scales due to linear density evolution. In this simulation, the HII regions caused by galaxy formation are seen in the redshift range $z \sim 8$ to 10, reaching scales up to $2'$ (frequency widths ~ 0.3 MHz, ~ 0.5 Mpc physical size). These regions have (negative) brightness temperatures up to 20 mK relative to the mean HI signal. This corresponds to 5 μ Jy beam^{-1} in a $2'$ beam at 140 MHz.

The point source rms sensitivity (dual polarization) in an image from a synthesis radio telescope is given by

$$\text{rms} = \left(\frac{1.9}{(\Delta \nu_{\text{kHz}} t_{\text{hr}})^{0.5}}\right) \left(\frac{T_{\text{sys}}}{\epsilon_{\text{eff}} A_{\text{ant}} N_{\text{ant}}}\right) \text{Jy beam}^{-1}, \quad (7)$$

where $\Delta \nu$ is the channel width in kHz, t is the integration time in hours, A_{ant} is the collecting area of each element in the array (m^2), ϵ_{eff} is the aperture efficiency, and N is the number of elements (ϵAN = total effective collecting area of the array). At low frequency, the system temperature, T_{sys}, is dominated by the synchrotron foreground, and behaves as $T_{FG} \sim 100(\frac{\nu}{200\text{MHz}})^{-2.8}$, in the coldest regions of the sky. Consider the SKA, with an effective collecting area of one square kilometer at 140 MHz, distributed over 4 km, $T_{\text{sys}} = 300$ K, with a channel width of 0.3 MHz, and integrating for one month. The rms sensitivity is then 1.3 μ Jy beam^{-1}, with a beam

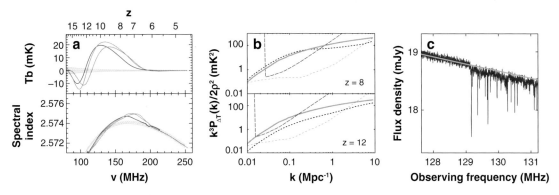

Figure 14

(*a*) Global (all sky) HI signal from reionization (Gnedin & Shaver 2004). The shaded region shows the expected thermal noise in a carefully controlled experiment. (*b*) Predicted HI 21 cm brightness temperature power spectrum (in log bins) at redshifts 8 and 12 (McQuinn et al. 2006). The gray solid line shows the signal when density fluctuations dominate. The black dotted line shows the predicted signal for $\bar{x}_i = 0.2$ at $z = 12$, and $\bar{x}_i = 0.6$ at $z = 8$, in the Furlanetto, Hernquist & Zaldarriaga (2004) semianalytic model. The blue dotted line shows the SKA sensitivity in 1000 hrs. The thick red dashed line shows the sensitivity of the pathfinder experiment LOFAR. The cutoff at low k is set by the primary beam. (*c*) The simulated SKA spectrum of a radio continuum source at $z = 10$ (Carilli, Gnedin & Owen 2002). The straight blue line is the intrinsic power law (synchrotron) spectrum of the source. The noise curve shows the effect of the 21-cm line in the neutral IGM, including noise expected for the SKA in a 100-hr integration.

FWHM $\sim 2'$. This square kilometer array will be adequate to perform 3D imaging of the average structure of the IGM during reionization.

Unfortunately, the nearer term low frequency path finder arrays will have ≤10% of the collecting area of the SKA, and will likely not be able to perform such direct 3D imaging (see table 1 in Carilli 2005). However, these near-term experiments should have enough sensitivity to probe the neutral IGM in other ways. **Figure 14** shows three possible HI 21-cm line signatures that might be observed prior to the SKA.

8.2.2. Global signal. The left panel in **Figure 14** shows the latest predictions of the global (all sky) increase in the background temperature due to the HI 21-cm line from the neutral IGM (Gnedin & Shaver 2004). The predicted HI emission signal peaks at roughly 20 mK above the foreground at $z \sim 10$. At higher redshift, prior to IGM warming, but allowing for Lyα emission from the first luminous objects, the HI is seen in absorption against the CMB. Because this is an all sky signal, the sensitivity of the experiment is independent of telescope collecting area, and the experiment can be done using small-area telescopes at low frequency, with well-controlled frequency response (Subrahmanyan, Chippendale & Ekers 2005). Note that the line signal is only $\sim 10^{-4}$ that of the mean foreground continuum emission at ~ 150 MHz.

8.2.3. Power spectra. The middle panel in **Figure 14** shows the predicted power spectrum of spatial fluctuations in the sky brightness temperature due to the HI 21-cm line (McQuinn et al. 2006). For power spectral analyses the sensitivity is greatly enhanced relative to direct imaging because the universe is isotropic, and hence one can average the measurements in annuli in the Fourier (uv) domain, i.e., the statistics of fluctuations along an annulus in the uv-plane are equivalent (Subrahmanyan et al. 1998; Zaldarriaga, Furlanetto & Hernquist 2004; Bharadwaj & Ali 2005; Bowman, Morales & Hewitt 2006; Morales 2005; Santos, Cooray & Knox 2005). Moreover, unlike the CMB, HI line studies provide spatial and redshift information, and hence the power spectral analysis can be performed in three dimensions.

Figure 14 includes the power spectrum arising from linear density fluctuations following the dark matter, plus the effect of reionization. The signature of reionization can be seen as a bump in the power spectrum above the density-only curve due to the formation of HII regions. The rms fluctuations at $z=10$ peak at about 10 mK rms on scales $\ell \sim 5000$.[8] A factor of 2 to 4 increase in the rms might be expected over this naive (Poisson) calculation due to peculiar gas velocities, i.e., infall into superclusters and along filaments (Barkana & Loeb 2005a, Ali 2006; but cf. Kuhlen, Madau & Montgomery 2005), and possible clustering of luminous sources, i.e., biased galaxy formation (Furlanetto, McQuinn & Hernquist 2005; Santos, Cooray & Knox 2005).

Also included in **Figure 14** are the noise power spectra for near- and longer-term radio arrays. The noise power spectrum, in mK^2, in standard spherical harmonic units, $\frac{\ell(\ell+1)}{2\pi}C_\ell^N$, for an array with (roughly) uniform coverage of the uv plane, is given by

$$C_\ell^N = \frac{T_{\rm sys}^2 (2\pi)^3}{\Delta \nu t f_c^2 \ell_{\rm max}^2}, \qquad (8)$$

where $\Delta \nu$ is the channel width in Hz, t is the integration time in seconds, f_c = area covering factor of the array $= N_{\rm ant}\frac{A_{\rm ant}}{A_{\rm tot}}$, $A_{\rm tot}$ is the total area of the array defined by the longest baseline, $A_{\rm tot} = \pi(\frac{b_{\rm max}}{2})^2$, and $\ell_{\rm max}$ is set by the longest baseline ($b_{\rm max}$) in the array: $\ell_{\rm max} = 2\pi b_{\rm max}/\lambda$ (Santos, Cooray & Knox 2005; for a 3D generalization, see Morales 2005). For example, the SKA will have 30% of its collecting area inside 1 km, or $f_c = 0.3$ out to 1 km. Typical noise values for near-term arrays are ~ 1 to 10 mK rms in the range $\ell = 10^3$ to 10^4, for long integrations (see **Figure 14**). This sensitivity is adequate to constrain the power spectrum of the HI 21-cm fluctuations, even if true images of the emission cannot be constructed.

A point of debate has been the fraction of HI in collapsed objects, as opposed to the diffuse IGM (Iliev et al. 2003). This fraction has a complex dependence on structure-formation history. Oh & Mack (2003) and Gnedin (2004) conclude that the majority of the HI during reionization will remain in the diffuse phase, whereas Ahn et al. (2005) argue that most of the HI will reside in mini-halos. Iliev et al. (2003) and Ahn et al. (2005) have considered the HI 21-cm power spectrum due to clustering of such mini-haloes at high redshift. For a beam size of 1′, and channel width of 0.2 MHz, they predict 3σ brightness temperature fluctuations due to clustered mini-haloes ~ 7 mK at $z = 8.5$, decreasing to 2 mK at $z = 20$.

---

[8] Or $\theta \sim \frac{180°}{\ell} = 2.2'$, or $k \sim \frac{\ell}{10^4}$ Mpc^{-1} = 0.5 Mpc^{-1} comoving.

8.2.4. Absorption toward discrete radio sources. An interesting alternative to emission studies is the possibility of studying smaller scale structure in the neutral IGM by looking for HI 21-cm absorption toward the first radio-loud objects (AGN, star-forming galaxies, GRBs) (Carilli et al. 2004b). The right panel of **Figure 14** shows the predicted HI 21-cm absorption signal toward a high redshift radio source due to the "cosmic web" prior to reionization, based on numerical simulations (Carilli, Gnedin & Owen 2002). For a source at $z = 10$, these simulations predict an average optical depth due to 21-cm absorption of about 1%, corresponding to the "radio GP effect," and about five narrow (a few km s^{-1}) absorption lines per MHz with optical depths of a few to 10 %. These latter lines are equivalent to the Lyα forest seen after reionization, and correspond to over-densities evolving in the linear regime ($\delta \leq 10$). Furlanetto & Loeb (2002) and Oh & Mack (2003) predict a similar HI 21-cm absorption line density due to gas in mini-haloes as that expected for the 21-cm forest.

Fundamental to absorption studies is the existence of radio loud sources during the EoR. This question has been considered in detail by Carilli, Gnedin & Owen (2002), Haiman, Quartaert & Bower (2004), and Jarvis & Rawlings (2005). They show that current models of radio-loud AGN evolution predict between 0.05 and 1 radio sources per square degree at $z > 6$ with $S_{150\text{MHz}} \geq 6$ mJy, adequate for EoR HI 21-cm absorption studies with the SKA.

8.2.5. Cosmic Strömgren spheres. Although direct detection of the typical structure of HI and HII regions may be out of reach of the near-term EoR 21-cm telescopes, there is a chance that even this first generation of telescopes will be able to detect the rare, very large-scale HII regions associated with luminous quasars near the end of reionization (Section 4). The expected signal is ~ 20 mK $\times x_{\text{HI}}$ on a scale $\sim 10'$ to $15'$, with a line width of ~ 1 to 2 MHz (Wyithe & Loeb 2004b). This corresponds to $0.5 \times x_{\text{HI}}$ mJy beam^{-1}, for a $15'$ beam at $z \sim 6$ to 7. Kohler et al. (2005) calculate the expected spectral dips caused by large HII regions around luminous quasars ($> 2 \times 10^{10} L_\odot$) during the EoR. In a typical spectrum from 100 to 180 MHz with a $10'$ beam they predict on average one relatively deep (–2 to –4 mK) dip per LoS on this scale. Wyithe, Loeb & Carilli (2005) perform a similar calculation using the evolution of the bright QSO luminosity function to predict the number of HII regions around active QSOs at $z > 6$. They conclude that there should be roughly one SDSS-type HII region around an active QSO (physical radius >4 Mpc) per 400 deg^2 field per 16 Mhz bandwidth at $z \sim 6$, and one $R \geq 2$ Mpc region at $z \sim 8$, and up to $\sim 100\times$ more fossil HII regions due to nonactive AGN, depending on the duty cycle.

8.2.6. Beyond reionization. Recent calculations of the brightness temperature fluctuations due to the 21-cm line have extended to redshifts higher than reionization, $z > 20$. Barkana & Loeb (2004, 2005a) predict the power spectrum of fluctuations during the era of "Lyα coupling" ($z \sim 20$ to 30; see also Kuhlen, Madau & Montgomery 2005). Brightness temperature fluctuations can be caused by emission from clustered mini-haloes, and enhanced by absorption against the CMB by the diffuse IGM, with a predicted rms ~ 10 mK for $\ell \sim 10^5$, due to a combination of linear density fluctuations, plus Poisson ("shot") noise and biasing in the Lyα source (i.e., galaxy) distribution.

Loeb & Zaldarriaga (2004) go even further in redshift, to $z > 50$ to 200. In this regime the HI generally follows linear density fluctuations, and hence the experiments are as clean as CMB studies, and $T_K < T_{\rm CMB}$, so a relatively strong absorption signal might be expected. They also point out that Silk damping, or photon diffusion, erases structures on scales $\ell > 2000$ in the CMB at recombination, corresponding to comoving scales equal to 22 Mpc. The HI 21-cm measurements can explore this physical regime at $z \sim 50$ to 300. The predicted rms fluctuations are 1 to 10 mK on scales of $\ell = 10^3$ to 10^6 ($0.2°$ to $1''$). These observations could provide the best tests of non-Gaussianity of density fluctuations, and constrain the running power-law index of mass fluctuations to large ℓ, providing important tests of inflationary structure formation. Sethi (2005) also suggests that a large global signal, up to -0.05 K, might be expected for this redshift range.

Unfortunately, the sky background is very hot $>10^4$ K at these low frequencies (<50 MHz), and the predicted signals are orders of magnitude below the expected sensitivities of even the biggest planned low frequency arrays in the coming decade.

8.3. Observational Challenges

8.3.1. Foregrounds. The HI 21-cm signal from reionization must be detected on top of a much larger synchrotron signal from foreground emission (Section 9.2). This foreground includes discrete radio galaxies, and large-scale emission from our own Galaxy, with relative contributions of about 10% and 90%, respectively. The expected HI 21-cm signal is about 10^{-4} of the foreground emission at 140 MHz.

Di Matteo et al. (2002) show that, even if point sources can be removed to the level of 1 μ Jy, the rms fluctuations on spatial scales $\leq 10'$ ($\ell \geq 1000$) due to residual radio point sources will be ≥ 10 mK just because of Poisson noise, increasing by a factor of 100 if the sources are strongly clustered (see also di Matteo, Ciardi, & Miniati 2004; Oh & Mack 2003).

A key point is that the foreground emission should be smooth in frequency, predominantly the sum of power-law, or perhaps gently curving, nonthermal spectra. A number of complimentary approaches have been presented for foreground removal (Bowman, Morales & Hewitt 2006). Gnedin & Shaver (2004) and Wang et al. (2005) consider fitting smooth spectral models (power laws or low-order polynomials in log space) to the observed visibilities or images. Morales & Hewitt (2004) and Morales (2005) present a 3D Fourier analysis of the measured visibilities, where the third dimension is frequency. The different symmetries in this 3D space for the signal arising from the noise-like HI emission, versus the smooth (in frequency) foreground emission, can be a powerful means of differentiating between foreground emission and the EoR line signal. Santos, Cooray & Knox (2005), Bharadwaj & Ali (2005), and Zaldarriaga, Furlanetto & Hernquist (2004) perform a similar analysis, only in the complementary Fourier space, meaning crosscorrelation of spectral channels. They show that the 21-cm signal will effectively decorrelate for channel separations >1 MHz, whereas the foregrounds do not. The overall conclusion of these methods is that spectral decomposition should be adequate to separate synchrotron foregrounds from the HI 21-cm signal from reionization at the mK level.

8.3.2. Ionosphere. A second potential challenge to low-frequency imaging over wide fields is phase fluctuations caused by the ionosphere. These fluctuations are due to index of refraction fluctuations in the ionized plasma, and behave as $\Delta\phi \propto \nu^{-2}$. Morever, the typical "isoplanatic patch," or angle over which a single phase error applies, is a few to 10 degrees (physical scales of tens of km in the ionosphere), depending on frequency (Cotton et al. 2004, Lane et al. 2004). Fields larger than the isoplanatic patch will have multiple phase errors across the field, and hence cannot be corrected through standard (i.e., single solution) phase self-calibration techniques. New wide field self-calibration techniques, involving multiple phase solutions over the field, or a "rubber screen" phase model (Hopkins, Doeleman & Lonsdale 2003; Cotton et al. 2004), are being developed that should allow for self-calibration over wide fields.

8.3.3. Interference. Perhaps the most difficult problem facing low-frequency radio astronomy is terrestrial (man-made) interference (RFI). The relevant frequency range corresponds to 7 to 200 MHz ($z = 200$ to 6). These are not protected frequency bands, and commercial allocations include everything from broadcast radio and television, to fixed and mobile communications.

Many groups are pursuing methods for RFI mitigation and excision (see Ellingson 2005). These include: (a) using a reference horn, or one beam of a phased array, for constant monitoring of known, strong, RFI signals, (b) conversely, arranging interferometric phases to produce a null at the position of the RFI source, and (c) real-time RFI excision, using advanced filtering techniques in time and frequency, of digitized signals both pre- and post-correlation. The latter requires very high dynamic range (many bit sampling), and very high frequency and time resolution.

In the end, the most effective means of reducing interference is to go to the remotest sites. The MWA and PAST have selected sites in remote regions of Western Australia, and China, respectively, because of known low RFI environments. Of course, the ultimate location would be the far side of the moon.

The technical challenges to HI 21-cm observations of reionization are many. Use of spectral decomposition to remove the foregrounds requires careful control of the synthesized beam as a function of frequency, with the optimal solution being a telescope design where the synthesized beam is invariant as a function of frequency (Subrahmanyan, Chippendale & Ekers 2005). High dynamic range front ends are required to avoid saturation in cases of strong interference, whereas fine spectral sampling is required to avoid Gibbs ringing in the spectral response. The polarization response must be stable and well-calibrated to remove polarized foregrounds (Haverkorn, Katgert & de Bruyn 2004). Calibration in the presence of a structured ionospheric phase screen requires new wide field calibration techniques. The very high data rate expected for many-element ($\geq 10^3$) arrays requires new methods for data transmission, cross correlation, and storage. At the lowest frequencies, ≤ 20 MHz or so, where we hope to study the PGM, phase fluctuations and the opacity of the ionosphere become problematic, leading to the proposed LUDAR project on the far side of the moon. The far side of the moon is also the best location to completely avoid terrestrial interference.

9. SUMMARY

The few years since the seminal theoretical reviews of Barkana & Loeb have seen dramatic progress in determining observational constraints on cosmic reionization. **Figure 15** shows the current limits on the cosmic neutral fraction versus redshift. The observations paint an interesting picture. On the one hand, studies of GP optical depths and variations, and the GP "gap" distribution, as well as of the thermal state of the IGM at high z, and of cosmic Strömgren spheres and surfaces around the highest redshift QSOs, suggest a qualitative change in the state of the IGM at $z \sim 6$. These data indicate a significant neutral fraction, $x_{HI} > 10^{-3}$, and perhaps as high as 0.1, at $z \geq 6$, as compared to $x_{HI} \leq 10^{-4}$ at $z < 5.5$. The IGM characteristics at this epoch are consistent with the end of the percolation stage of reionization (Gnedin & Fan 2006). On the other hand, transmission spikes in the GP trough and study of the evolution of the Lyα galaxy luminosity function indicate a neutral fraction smaller than 50% at $z \sim 6.5$. Moreover, the measurement of the large-scale polarization of the CMB implies a significant ionization fraction extending to higher redshifts, $z \sim 14$.

We emphasize that all these measurements have implicit assumptions and uncertainties, as discussed throughout this review. Indeed, the GP effect and CMB large-scale polarization studies can be considered complimentary probes of reionization, with optical depth effects limiting GP studies to the end of reionization, whereas CMB studies are weighted toward the higher redshifts, when the densities were higher. The data argue against a simple reionization history in which the IGM remains largely neutral from $z \sim 1100$ to $z \sim 6-7$, with a single phase transition at $z \sim 6$ (the "late" model in **Figure 15**) as well as against a model in which the Universe reached complete ionization at $z \sim 15-20$ and remained so ever since (the "early" model in **Figure 15**). These facts, combined with the large line-of-sight variations at the end of reionization as indicated by GP measurements, suggest a more extended reionization history. Interestingly, the latest theoretical models with reionization caused by Population II star formation are consistent with both GP optical depth and WMAP CMB polarization measurement (e.g., Gnedin & Fan 2006). Current data do not present strong evidence for a dominant contribution by metal-free Population III star formation at $z > 15$ to reionization (Haiman & Bryan 2006), and have weak constraints on, although do not require, models with multiple episodes of reionization (the double model in **Figure 15**), which were suggested by a possible high CMB Thompson optical depth based on early WMAP measurements.

Overall, we agree with the assessment of Furlanetto, Sokasian & Hernquist (2004) that reionization is less an event than a process, extended in both time and space. Much emphasis has been given to the determination of the "reionization redshift." Current observations, however, suggest that, unlike recombination, which occurs over a very narrow range in fractional redshift, $z_{recomb} = 1089 \pm 1$, cosmic reionization occurs over a fairly large fractional redshift, from $z \sim 6$ up to as high as $z \sim 14$. The redshift of HII region overlap might be quite different than the redshift at which the IGM

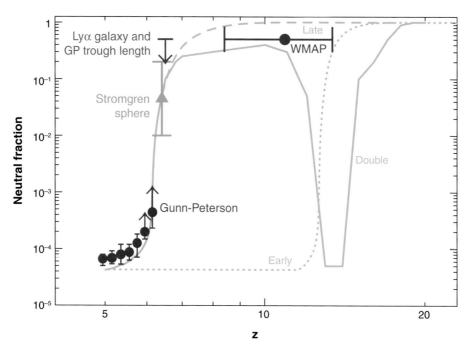

Figure 15

The volume-averaged neutral fraction of the IGM versus redshift using various techniques. The dashed line shows the fiducial model of Gnedin (2004) with late reionization at $z = 6$–7 (see also Gnedin & Fan 2006). The solid line shows an idealized model with double reionization as described in Cen (2003a), and the dotted line illustrates the model with early reionization at $z \sim 14$.

is largely (>50%) ionized. It may also differ in different parts of the universe. The implication is that the formation of the first luminous objects (stars and AGN) occurs in a "twilight zone," heavily obscured at (observed) optical wavelengths by a partially neutral IGM, and observable only at near-IR through radio wavelengths, and in the hard X rays.

We also consider the sources responsible for reionization. Current data are consistent with star-forming galaxies, in particular, relatively low-luminosity galaxies, as being the dominant sources of reionizing photons. However, again, there are a number of important, and poorly constrained, parameters in these calculations, including the IMF and the UV escape fraction, and we certainly cannot rule out a significant contribution from mini-QSOs.

Lastly, we show that low-frequency radio telescopes currently under construction should be able to make the first direct measurements of HI 21-cm emission from (and/or absorption by) the neutral IGM. These observations will present a clear picture of the reionization process, and the evolution of the neutral IGM into the dark ages.

ACKNOWLEDGMENTS

C.C. thanks the Max-Planck-Gesellschaft and the Humboldt-Stiftung for support through the Max-Planck-Forschungspreis, K. Menten, and the MPIfR for their hospitality, and S. Furlanetto for extensive comments. X.F. is grateful for support from NSF grant AST 03–07384, a Sloan Research Fellowship, and a Packard Fellowship for Science and Engineering. B.K. thanks Nathan Miller for his contributions to this review, and Steven Furlanetto, Manoj Kaplinghat, and Meir Shimon for helpful comments on this manuscript. CMBFAST was used in the preparation of this review. The NRAO is a facility of the NSF operated under cooperative agreement by AUI.

LITERATURE CITED

Aghanim M, Desert F, Puget L, Gispert R. 1996. *Astron. Astrophys.* 311:1–11

Ahn K, Shapiro P, Alvarez M, Iliev T, Martel H, Ryu D. 2005. *New Astron. Rev.* 50:179–83

Ajiki M, Taniguchi Y, Fujita S, Shioya Y, Nagao T, et al. 2003. *Astron. J.* 126:2091–2107

Ali SS. 2006. *MNRAS*. In press (astro-ph/0503237)

Anderson SF, Hogan CJ, Williams BF, Carswell RF. 1999. *Astron. J.* 117:956–62

Bahcall JN, Salpeter EE. 1965. *Ap. J.* 142:1677–80

Barkana R, Loeb A. 2001. *Phys. Rep.* 349:125–238

Barkana R, Loeb A. 2004. *Ap. J.* 601:64–69

Barkana R, Loeb A. 2005a. *Ap. J.* 626:1–11

Barkana R, Loeb A. 2005b. *Ap. J.* 624:L65–68

Barkana R, Loeb A. 2005c. *MNRAS* 363:L36

Barkats D, Bischoff C, Farese P, Gaier T, Gundersen J, et al. 2005. *Ap. J.* 619:L127–30

Barton EJ, Davé R, Smith J-DT, Papovich C, Hernquist L, Springel V. 2004. *Ap. J. Lett.* 604:L1–L4

Becker RH, Fan X, White R, Strauss M, Narayanan V, et al. 2001. *Astron. J.* 122:2850–57

Becker R, Sargent W, Rauch M, Simcoe R. 2006. *Ap. J.* 640:69–80

Bertoldi F, Carilli C, Cox P, Fan X, Strauss M, et al. 2003a. *Astron. Astrophys.* 406:L55–58

Bertoldi F, Cox P, Neri R, Carilli C, Walter F, et al. 2003b. *Astron. Astrophys.* 409:L47–50

Bharadwaj S, Ali SS. 2005. *MNRAS* 356:1519–28

Bi H. 1993. *Ap. J.* 405:479–90

Bouwens RJ, Illingworth G, Thompson R, Blakeslee J, Dickinson M, et al. 2004a. *Ap. J. Lett.* 606:L25–28

Bouwens RJ, Thompson RI, Illingworth GD, Franx M, van Dokkum P, et al. 2004b. *Ap. J. Lett.* 616:L79–82

Bouwens RJ, Illingworth GD, Blakeslee JP, Franx M. 2006. *Ap. J.* In press (astro-ph/0509641)

Bouwens RJ, Illingworth GD, Thompson RI, Franx M. 2005. *Ap. J.* 624:L5–8

Bowman J, Morales M, Hewitt J. 2006. *Ap. J.* 638:20–28

Bunker AJ, Stanway ER, Ellis RS, McMahon RG. 2004. *MNRAS* 355:374–84
Carilli C, Furlanetto S, Briggs F, Jarvis M, Rawlings S, Falcke H. 2004a. *New Astron. Rev.* 48:1029–38
Carilli C, Gnedin N, Furlanetto S, Owen F. 2004b. *New Astron. Rev.* 48:1053–61
Carilli C, Gnedin N, Owen F. 2002. *Ap. J.* 577:22–30
Carilli CL. 2005 *New Astron. Rev.* 50:162–72
Cen R. 2003a. *Ap. J.* 591:12–37
Cen R. 2003b. *Ap. J.* 591:L5–8
Cen R, Haiman Z. 2000. *Ap. J. Lett.* 542:L75–78
Cen R, Haiman Z, Mesinger A. 2005. *Ap. J.* 621:89–94
Cen R, McDonald P. 2002. *Ap. J.* 570:457–62
Cen R, Miralda-Escudé J, Ostriker JP, Rauch M. 1994. *Ap. J. Lett.* 437:L9–12
Chary R, Stern D, Eisenhardt P. 2005. *Ap. J.* 635:L5
Chen H, Prochaska JX, Bloom JS, Thompson IB. 2005. *Ap. J.* 634:L25
Chen X, Kamionkowski M. 2004. *Phys. Rev. D* 70:043502
Chen X, Miralda-Escudé J. 2004. *Ap. J.* 602:1–11
Chiu WA, Fan X, Ostriker JP. 2003. *Ap. J.* 599:759
Ciardi B, Ferrara A. 2005. *Space Sci. Rev.* 116:625–705
Ciardi B, Ferrara A, Marri S, Raimondo G. 2001. *MNRAS* 324:381–88
Ciardi B, Loeb A. 2000. *Ap. J.* 540:687–96
Ciardi B, Madau P. 2003. *Ap. J.* 596:1–8
Ciardi B, Stoehr F, White S. 2004. *MNRAS* 343:1101–9
Cool R, Eisenstein D, Johnston D, Scranton R, Brinkman J, et al. 2006. *Astron. J.* 131:736–46
Corbin M, Cen R, Windhurst R, Greeley R, Jones D, et al. 2005. Exploratory proposal to NASA
Cotton W, Condon J, Perley R, Kassim N, Lazio J, et al. 2004. *Proc. SPIE* 5489:180–89
Davidsen AF, Kriss GA, Wei Z. 1996. *Nature* 380:47–49
Dawson KS, Holzapfel WL, Carlstrom JE, Joy M, LaRoque SJ, et al. 2002. *Ap. J.* 581:86–95
Dijkstra M, Haiman Z, Loeb A. 2004. *Ap. J.* 613:646–54
Di Matteo T, Ciardi B, Miniati F. 2004. *MNRAS* 355:1053–65
Di Matteo T, Rosalba P, Abel T, Rees M. 2002. *MNRAS* 564:576–80
Djorgovski SG, Bogosavljevic M, Mahabal A. 2006. *New Astron. Rev.* 50:140–45
Djorgovski SG, Castro S, Stern D, Mahabal AA. 2001. *Ap. J. Lett.* 560:L5–8
Dodelson S. 2003. *Modern Cosmology.* Amsterdam: Academic. XIII + 440 pp.
Ellingson S. 2005. *Exp. Astron.* 17:261–67
Eyles L, Bunker A, Stanway E, Lacy M, Ellies R, Doherty M. 2005. *MNRAS* 364:443–50
Fan X, White R, Davis M, Becker R, Sttrauss M. 2000. *Astron. J.* 120:1167–74
Fan X, Strauss M, Schneider D, Gunn J, Lupton R, et al. 2001a. *Astron. J.* 121:54–65
Fan X, Narayanan V, Lupton R, Strauss M, Knapp J, et al. 2001b. *Astron. J.* 122:2833–49
Fan X, Narayanan VK, Strauss MA, White RL, Becker RH, Pentericci L, Rix H-W. 2002. *Astron. J.* 123:1247–57

Fan X, Strauss M, Schneider D, Becker R, White R, et al. 2003. *Astron. J.* 125:1649–59
Fan X, Hennawi J, Richards G, Strauss M, Schneider D, et al. 2004. *Astron. J.* 128:515–22
Fan X, Strauss M, Richards G, Hennawi J, Becker R, et al. 2006a. *Astron. J.* 131:1203–9
Fan X, Strauss MA, Becker RH, White RL, Gunn JE, et al. 2006b. *Astron. J.* In press (astro-ph/0512082)
Fernández-Soto A, Lanzetta KM, Chen H-W. 2003. *MNRAS* 342:1215–21
Field GB. 1959. *Ap. J.* 129:551–65
Furlanetto SR, Briggs FH. 2004. *New Astron. Rev.* 48:1039–52
Furlanetto SR, Hernquist L, Zaldarriaga M. 2004. *MNRAS* 354:695–707
Furlanetto SR, Loeb A. 2002. *Ap. J.* 579:1–9
Furlanetto SR, Loeb A. 2003. *Ap. J.* 588:18–34
Furlanetto SR, McQuinn M, Hernquist L. 2005. *MNRAS* 365:115–26
Furlanetto SR, Oh SP. 2005. *MNRAS* 363:1031–48
Furlanetto SR, Sokasian A, Hernquist L. 2004. *MNRAS* 347:187–95
Furlanetto SR, Zaldarriaga M, Hernquist L. 2004. *Ap. J.* 613:16–22
Giallongo E, Cristiani S, D'Odorico S, Fontana A. 2002. *Ap. J. Lett.* 568:L9–12
Giallongo E, D'Odorico S, Fontana A, McMahon RG, Savaglio S, et al. 1994. *Ap. J. Lett.* 425:L1–4
Giavalisco M, Ferguson H, Koekemoer A, Dickinson M, Alexander D, et al. 2004. *Ap. J. Lett.* 600:L93–98
Gnedin NY. 2000. *Ap. J.* 535:530–54
Gnedin NY. 2004. *Ap. J.* 610:9–13
Gnedin NY, Fan X. 2006. *Ap. J.* In press (astro-ph/0603794)
Gnedin NY, Ostriker JP. 1997. *Ap. J.* 486:581–98
Gnedin NY, Prada F. 2004. *Ap. J. Lett.* 608:L77–80
Gnedin NY, Shaver PA. 2004. *Ap. J.* 608:611–21
Gruzinov A, Hu W. 1998. *Ap. J.* 508:435
Gunn JE, Peterson BA. 1965. *Ap. J.* 142:1633–41
Haardt F, Madau P. 1996. *Ap. J.* 461:20–37
Haehnelt MG, Madau P, Kudritzki R, Haardt F. 2001. *Ap. J. Lett.* 549:L151–54
Haiman Z. 2002. *Ap. J. Lett.* 576:L1–4
Haiman Z, Bryan G. 2006. *Ap. J.* In press (astro-ph/0603541)
Haiman Z, Cen R. 2005. *Ap. J.* 623:627–31
Haiman Z, Holder GP. 2003. *Ap. J.* 595:1
Haiman Z, Knox L. 1999. ASP Conf. Ser. 181: Microwave Foregrounds, 181:227
Haiman Z, Quartaert E, Bower G. 2004. *Ap. J.* 612:698–705
Hansen SH, Haiman Z. 2004. *Ap. J.* 600:26–31
Haverkorn M, Katgert P, de Bruyn A. 2004. *Astron. Astrophys.* 427:549–59
Hernquist L, Katz N, Weinberg DH, Miralda-Escudé J. 1996. *Ap. J. Lett.* 457:L51–54
Hirata CM. 2006. *MNRAS.* 367:259–74
Holder GP, Haiman Z, Kaplinghat M, Knox L. 2003. *Ap. J.* 595:13
Hopkins P, Doeleman S, Lonsdale C. 2003. *Astron. Astrophys. Suppl.* 203:4005
Horton A, Parry I, Bland-Hawthorn J, Cianci S, King D, et al. 2004. *Proc. SPIE* 5492:1022–32

Hu EM, Cowie LL, Capak P, Kakazu Y. 2005. In *IAU Circ. 199 Conf. Proc.: "Probing Galaxies through Quasar Absorption Lines,"* ed. P Williams, C Shu, B Menard, pp. 363–68 (astro-ph/0509616)

Hu EM, Cowie LL, Capak P, McMahon RG, Hayashino T, Komiyama Y. 2004. *Astron. J.* 127:563–75

Hu EM, Cowie LL, McMahon RG, Capak P, Iwamuro F, et al. 2002. *Ap. J. Lett.* 568:L75–79

Hu W. 2000. *Ap. J.* 529:12

Hu W, Holder G. 2003. *Phys. Rev. D* 68:3001–4

Hu W, Okamoto T. 2002. *Ap. J.* 574:566

Hu W, White M. 1997. *New Astron.* 2:323

Hui L, Haiman Z. 2003. *Ap. J.* 596:9–18

Iliev I, Mellema G, Pen U, Merz H, Shapiro P, Alvarez M. 2006. *MNRAS.* In press (astro-ph/0512187)

Iliev I, Scannapieco E, Martel H, Schapiro P. 2003. *MNRAS* 341:81–90

Inoue AK, Iwata I, Deharveng J-M, Buat V, Burgarella D. 2005. *Astron. Astrophys.* 435:471–82

Jakobsen P, Boksenberg A, Deharveng JM, Greenfield P, Jedrzejewski R, Paresce F. 1994. *Nature* 370:35–39

Jarosik N, Bennett C, Halpern M, Hinshaw G, Kogut A, et al. 2003. *Ap. J. Suppl.* 145:413

Jarvis M, Rawlings S. 2005. *New Astron. Rev.* 48:1173–85

Kamionkowski M, Kosowsky A, Stebbins A. 1997. *Phys. Rev. D* 55:7368

Kaplinghat M, Chu M, Haiman Z, Holder GP, Knox L, Skordis C. 2003. *Ap. J.* 583:24

Kaplinghat M, Knox L, Song Y. 2003. *Phys. Rev. Lett.* 91:241301

Kashikawa N, Yoshida M, Shimasaku K, Nagashima M, Yahagi H, et al. 2006. *Ap. J.* 637:631

Kashlinsky A, Arendt R, Mather J, Moseley S. 2005. *Nature* 438:45–50

Kawai N, Kosugi G, Aoki K, Yamada T, Totani T, et al. 2006. *Nature.* In press (astro-ph/0512052)

Keating B, Miller N. 2006. *New Astron. Rev.* 50:184–90

Keating B, Timbie P, Polnarev A, Steinberger J. 1998. *Ap. J.* 495:580

Kennefick JD, Djorgovski SG, de Carvalho RR. 1995. *Astron. J.* 110:253–65

Kesden M, Cooray A, Kamionkowski M. 2002. *Phys. Rev. Lett.* 89:011304

Knox L. 2003. *New Astron. Rev.* 47:883

Knox L, Scoccimarro R, Dodelson S. 1998. *Phys. Rev. Lett.* 81:2004

Knox L, Song Y-S. 2002. *Phys. Rev. Lett.* 89:011303

Kodaira K, Taniguchi Y, Kashikawa N, Kaifu N, Ando H, et al. 2003. *Publ. Astron. Soc. Jpn.* 55:L17–21

Kogut A, Spergel D, Barnes C, Bennett C, Halpern M, et al. 2003. *Ap. J. Suppl.* 148:161

Kohler K, Gnedin N, Miralda-Escudé J, Shaver P. 2005. *Ap. J.* 633:552

Kosowsky A. 2003. *New Astron. Rev.* 47:939

Kriss GA, Shull JM, Oegerle W, Zheng W, Davidsen A, et al. 2001. *Science* 293:1112–16

Kuhlen M, Madau P, Montgomery R. 2005. *Ap. J.* 637:L1

Kurk JD, Cimatti A, di Serego Alighieri S, Vernet J, Daddi E, et al. 2004. *Astron. Astrophys.* 422: L13–17

Lamarre JM, Puget J, Bouchet F, Ade P, Benoit A, et al. 2003. *New Astron. Rev.* 47:1017

Lane W, Cohen A, Cotton W, Condon J, Perley R, et al. 2004. *Proc. SPIE* 5489:354–61

Lidz A, Hui L, Zaldarriaga M, Scoccimarro R. 2002. *Ap. J.* 579:491–99

Loeb A, Barkana R. 2001. *Annu. Rev. Astron. Astrophys.* 39:19–66

Loeb A, Zaldarriaga M. 2004. *Phys. Rev. Lett.* 92:1301–4

Maccone C. 2004. *35th COSPAR Assembly*, pp. 1415–19

Madau P, Ferrara A, Rees MJ. 2001. *Ap. J.* 555:92–105

Madau P, Haardt F, Rees MJ. 1999. *Ap. J.* 514:648–59

Madau P, Meiksin A, Rees MJ. 1997. *Ap. J.* 475:429–44

Madau P, Rees MJ. 2000. *Ap. J. Lett.* 542:L69–73

Madau P, Rees MJ, Volonteri M, Haardt F, Oh SP. 2004. *Ap. J.* 604:484–94

Mahabal A, Stern D, Bogosavljevic M, Djorgovski S, Thompson D. 2005. *Ap. J.* 634:L9

Malhotra S, Rhoads JE. 2004. *Ap. J. Lett.* 617:L5-L8

Malhotra S, Rhoads JE. 2005. *Ap. J.* 617:L5–8

Malhotra S, Rhoads J, Pirzkal N, Haiman Z, Xu C, et al. 2005. *Ap. J.* 626:666–79

Mapelli M, Ferrara A. 2005. *MNRAS* 364:2

Martin CL, Sawicki M. 2004. *Ap. J.* 603:414–24

McDonald P, Miralda-Escudé J. 2001. *Ap. J. Lett.* 549:L11–14

McDonald P, Miralda-Escudé J, Rauch M, Sargent WLW, Barlow TA, et al. 2000. *Ap. J.* 543:1–23

McPherson A, Born A, Sutherland W, Emerson J. 2004. *Proc. SPIE* 5489:638–49

McQuinn M, Furlanetto S, Hernquist L, Zahn O, Zaldarriaga M. 2005. *Ap. J.* 630:657

McQuinn M, Zahn O, Zaldarriaga M, Hernquist L, Furlanetto S. 2006. *Ap. J.* In press (astro-ph/0512263)

Meiksin A. 2005. *MNRAS* 356:596–606

Mennella A, Bersanelli M, Cappellini B, Maino D, Platania P, et al. 2004. *AIP Conf. Proc.* 703: Plasmas in the Laboratory and in the Universe: New Insights and New Challenges, p. 401. astro-ph/0310058

Mesinger A, Haiman Z. 2004. *Ap. J.* 611:L69–72

Miralda-Escudé J. 1998. *Ap. J.* 501:15–22

Miralda-Escudé J. 2003. *Ap. J.* 597:66–73

Miralda-Escudé J, Cen R, Ostriker JP, Rauch M. 1996. *Ap. J.* 471:582–616

Miralda-Escudé J, Haehnelt M, Rees MJ. 2000. *Ap. J.* 530:1–16

Mobasher B, Dickinson M, Ferguson H, Giavalisco M, Wiklind T, et al. 2005 *Ap. J.* 635:832

Morales M. 2005. *Ap. J.* 619:678–83

Morales M, Hewitt J. 2004. *Ap. J.* 615:7–18

Oh SP. 2001. *Ap. J.* 553:499–512

Oh SP. 2002. *MNRAS* 336:1021–29

Oh SP, Furlanetto SR. 2005. *Ap. J. Lett.* 620:L9–12
Oh SP, Mack K. 2003. *MNRAS* 346:871–77
Oxley P, Ade P, Baccigalupi C, deBernardis P, Cho H-M, et al. 2005. *Proc. SPIE* 5543(2004): 320–31
Page L, Hinshaw G, Komatsu E, Nolta D, Spergel D, et al. 2006. *Ap. J. Suppl.* In press (astro-ph/0603450)
Panagia N, Fall SM, Mobasher B, Dickinson M, Ferguson H, et al. 2005. *Ap. J.* 633:L1
Paschos P, Norman ML. 2005. *Ap. J.* 631:59–84
Pentericci L, Fan X, Rix H-W, Strauss M, Narayan V, et al. 2002. *Astron. J.* 123:2151–58
Pettini M, Madau P, Bolte M, Prochaska JX, Ellison SL, Fan X. 2003. *Ap. J.* 594:695–703
Pierpaoli E. 2003. *MNRAS* 342:L63
Rauch M. 1998. *Annu. Rev. Astron. Astrophys.* 36:267–316
Rauch M, Miralda-Escudé J, Sargent W, Barlow T, Weiberg D, et al. 1997. *Ap. J.* 489:7–20
Razoumov AO, Norman ML, Abel T, Scott D. 2002. *Ap. J.* 572: 695–704
Readhead ACS, Myers S, Pearson T, Sievers J, Mason B, et al. 2004a. *Science* 306:836
Readhead ACS, Mason B, Contaldi C, Pearson T, Bond J, et al. 2004b. *Ap. J.* 609:498
Rees MJ. 1998. *Proc. Natl. Acad. Sci. Colloq.: The Age of the Universe, Dark Matter, and Structure Formation*, pp. 47–52. Washington, DC: Natl. Acad. Sci.
Rhoads JE, Dey A, Malhotra S, Stern D, Spinrad H, et al. 2003. *Astron. J.* 125:1006–13
Richards GT, Haiman Z, Pindor B, Strauss M, Fan X, et al. 2006. *Astron. J.* 131:49–54
Ricotti M, Gnedin NY, Shull JM. 2000. *Ap. J.* 534:41–56
Ricotti M, Ostriker JP. 2004. *MNRAS* 352:547–62
Ruhl J, Ade P, Carlstrom J, Cho H, Crawford T, et al. 2004. *Proc. SPIE* 5498:11
Salvaterra R, Haardt F, Ferrara A. 2005. *MNRAS* 362:L50–54
Santos MG, Cooray A, Haiman Z, Knox L, Ma C-P. 2003. *Ap. J.* 598:756–66
Santos MG, Cooray A, Knox L. 2005. *Ap. J.* 625:575–87
Santos MR. 2004. *MNRAS* 349:1137–52
Santos MR, Ellis RS, Kneib J-P, Richard J, Kuijken K. 2004. *Ap. J.* 606:683–701
Schaye J, Aguirre A, Kim T-S, Theuns T, Rauch M, Sargent WLW. 2003. *Ap. J.* 596:768–96
Schaye J, Theuns T, Rauch M, Efstathiou G, Sargent WLW. 2000. *MNRAS* 318:817–26
Scheuer PAG. 1965. *Nature* 207:963
Schiminovich D, Ilbert O, Arnouts S, Milliard B, Tresse L, et al. 2005. *Ap. J. Lett.* 619:L47–50
Schmidt M, Schneider DP, Gunn JE. 1995. *Astron. J.* 110:68–77
Scott D, Rees MJ. 1990. *MNRAS* 247:510–16
Seljak U, Makarov A, McDonald P, Anderson S, Bahcall N, et al. 2005. *Phys. Rev. D* 71:103515
Seljak U, Zaldarriaga M. 1996. *Ap. J.* 469:437
Seshadri T, Subramanian K. 1998. *Phys. Rev. D* 58:063002

Sethi S. 2005. *MNRAS* 363:818
Shklovsky IS. 1964. *Astron. Zh.* 41:801
Shull JM, Tumlinson J, Giroux ML, Kriss GA, Reimers D. 2004. *Ap. J.* 600:570–79
Sokasian A, Abel T, Hernquist L. 2002. *MNRAS* 332:601–16
Songaila A. 2001. *Astron. J.* 561(2): L153–56
Songaila A. 2004. *Astron. J.* 127:2598–603
Songaila A. 2005. *Astron. J.* 130:1996–2005
Songaila A, Cowie LL. 2002. *Astron. J.* 123:2183–96
Songaila A, Hu EM, Cowie LL, McMahon RG. 1999. *Ap. J. Lett.* 525:L5–8
Spergel DN, Bean R, Dore O, Nolta M, Bennett C. 2006. *Ap. J. Suppl.* In press (astro-ph/0603449)
Spergel DN, Verde L, Peiris HV, Komatsu E, Nolta MR, et al. 2003. *Ap. J. Suppl.* 148:175–94
Steidel CC, Adelberger KL, Giavalisco M, Dickinson M, Pettini M. 1999. *Ap. J.* 519:1
Steidel CC, Pettini M, Adelberger KL. 2001. *Ap. J.* 546:665–71
Stern D, Yost SA, Eckart ME, Harrison FA, Helfand DJ, et al. 2005. *Ap. J.* 619:12–18
Stiavelli M, Fall SM, Panagia N. 2004. *Ap. J. Lett.* 610:L1–4
Subrahmanyan R, Chippendale A, Ekers R. 2005. *ATNF Newsl.* 56:18–19
Subrahmanyan R, Kesteven M, Ekers R, Sinclair M, Silk J. 1998. *MNRAS* 298:1189–97
Sunyaev RA, Zeldovich IB. 1980. *MNRAS* 190:413
Tagliaferri G, Antonelli L, Chincarini G, Fernandez-Soto A, Malesani D, et al. 2005. *Astron. Astrophys.* 443:L1
Taniguchi Y, Ajiki M, Nagao T, Shioya Y, Murayama T, et al. 2005. *Publ. Astron. Soc. Jpn.* 57:165–82
Tegmark M, Strauss M, Blanton M, Abazajian K, Dodelson S, et al. 2004. *Phys. Rev. D* 69:103501
Theuns T, Bernardi M, Frieman J, Hewett P, Schaye J, et al. 2002a. *Ap. J. Lett.* 574:L111–14
Theuns T, Schaye J, Zaroubi S, Kim T-S, Tzanavaris P, Carswell B. 2002b. *Ap. J. Lett.* 567:L103–6
Thompson RI, Illingworth G, Bouwens R, Dickinson M, Eisenstein D, et al. 2005. *Astron. J.* 130:1–12
Totani T, et al. 2006. *Publ. Astron. Soc. Jpn.* In press
Tozzi P, Madau P, Meiksin A, Rees M. 2000. *Ap. J.* 528:597–606
Tran K-VH, Lilly SJ, Crampton D, Brodwin M. 2004. *Ap. J. Lett.* 612:L89–92
Venkatesan A, Giroux ML, Shull JM. 2001. *Ap. J.* 563:1–8
Walter F, Bertoldi F, Carilli C, Cox P, Lo KY, et al. 2003. *Nature* 424:406–8
Walter F, Carilli C, Bertoldi F, Menten K, Cox P, et al. 2004. *Ap. J.* 615:L17–20
Wang X, Hu W. 2006. *Ap. J.* In press (astro-ph/0511141)
Wang X, Tegmark M, Santos M, Knox L. 2006. *Ap. J.* In press (astro-ph/0501081)
Warren SJ, Hewett PC, Osmer PS. 1994. *Ap. J.* 421:412–33
Webb JK, Barcons X, Carswell RF, Parnell HC. 1992. *MNRAS* 255:319–24
Weinberg DH, Miralda-Escudé J, Hernquist L, Katz N. 1997. *Ap. J.* 490:564–70

White RL, Becker RH, Fan X, Strauss MA. 2003. *Astron. J.* 126:1–14
White RL, Becker RH, Fan X, Strauss MA. 2005. *Astron. J.* 129:2102–7
Willott CJ, Percival WJ, McLure RJ, Crampton D, Hutchings JB, et al. 2005. *Ap. J.* 626:657–65
Wolf C, Wisotzki L, Borch A, Dye S, Kleinheinrich M, Meisenheimer K. 2003. *Astron. Astrophys.* 408:499–514
Wouthuysen S. 1952. *Astron. J.* 57:31–33
Wyithe JSB, Loeb A. 2003. *Ap. J.* 586:693
Wyithe JSB, Loeb A. 2004a. *Nature* 432:194–96
Wyithe JSB, Loeb A. 2004b. *Ap. J.* 610:117–27
Wyithe JSB, Loeb A. 2005. *Ap. J.* 625:1–5
Wyithe JSB, Loeb A, Barnes D. 2005. *Ap. J.* 634:715
Wyithe JSB, Loeb A, Carilli C. 2005. *Ap. J.* 628:575–82
Yan H, Windhorst RA. 2004a. *Ap. J. Lett.* 600:L1–5
Yan H, Windhorst RA. 2004b. *Ap. J. Lett.* 612:L93–96
York DG, Adelman J, Anderson J, Anderson S, Annis J, et al. 2000. *Astron. J.* 120:1579–87
Yu Q, Lu Y. 2005. *Ap. J.* 620:31
Zahn O, Zaldarriaga M, Hernquist L, McQuinn M. 2005. *Ap. J.* 630:657
Zaldarriaga M. 1997. *Phys. Rev. D* 55:1822
Zaldarriaga M, Furlanetto SR, Hernquist L. 2004. *Ap. J.* 608:622–35
Zaldarriaga M, Seljak U. 1997. *Phys. Rev. D* 55:1830
Zaldarriaga M, Seljak U. 1998. *Phys. Rev. D* 58:023003
Zhang Y, Anninos P, Norman ML. 1995. *Ap. J. Lett.* 453:L57–60
Zhang PJ, Pen UL, Trac H. 2004. *MNRAS* 347(4):1224–33
Zheng W, Kriss G, Deharvanj J, Dixon W, Kruk J, et al. 2004. *Ap. J.* 605:631–44
Zygelman B. 2005. *Ap. J.* 622:1356–62

X-Ray Emission from Extragalactic Jets

D.E. Harris[1] and Henric Krawczynski[2]

[1]Harvard-Smithsonian Center for Astrophysics, Cambridge, Massachusetts 02138; email: harris@cfa.harvard.edu

[2]Washington University, St. Louis, Missouri 63130; email: krawcz@wuphys.wustl.edu

Key Words

inverse Compton emission, relativistic jets, synchrotron emission, X-ray jets

Abstract

This review focuses on the X-ray emission processes of extragalactic jets on scales resolvable by the subarcsec resolution of the Chandra X-ray Observatory. It is divided into four parts. The introductory section reviews the classical problems for jets, as well as those associated directly with the X-ray emission. Throughout this section, we deal with the dualisms of low-powered radio sources versus high-powered radio galaxies and quasars and of synchrotron models versus inverse Compton models; and the distinction between the relativistic plasma responsible for the received radiation and the medium responsible for the transport of energy down the jet. The second section collects the observational and inferred parameters for the currently detected X-ray jets and attempts to put their relative sizes and luminosities in perspective. In the third section we first give the relevant radio and optical jet characteristics, and then examine the details of the X-ray data and how they can be related to various jet attributes. The last section is devoted to a critique of the two nonthermal emission processes and to prospects for progress in our understanding of jets.

1. THE PROBLEMS

Jets are giant collimated plasma outflows associated with some types of active galactic nuclei (AGN). The first jet was discovered in 1918 within the elliptical galaxy M87 in the Virgo cluster: "A curious straight ray lies in a gap in the nebulosity in p.a. 20°, apparently connected with the nucleus by a thin line of matter. The ray is brightest at its inner end, which is 11″ from the nucleus." (Curtis 1918). At that time, the extended feature was a mere curiosity and its nature was not understood. When radio telescopes with good angular resolution and high sensitivities became available in the 1960s, it was found that many galaxies exhibited extended radio emission consisting of a nuclear component, jets, hotspot complexes, and radio lobes. According to the standard picture, jets originate in the vicinity of a supermassive black hole (SMBH with several million to several billion solar masses) located at the center of the AGN; (c.f., the early ideas of Salpeter 1964). The jets are most likely powered by these black holes, and the jets themselves transport energy, momentum, and angular momentum over vast distances (Blandford & Rees 1974, Rees 1971, Scheuer 1974), from the "tiny" black hole of radius $r = 10^{-4} \, M_{BH}/10^9 M_\odot$ pc to radio hotspots, hotspot complexes and lobes which may be a megaparsec or more away. Thus the study of jets must address a range of scales covering a factor of 10^{10}!

Even now, after 30 years of intensive studies of radio galaxies in the radio regime, no consensus has emerged on their fundamental attributes such as composition, formation, and collimation. With the advent of the *Hubble Space Telescope* (HST) and the *Chandra X-ray Observatory* (CXO),[1] the optical and X-ray emission from jets can be studied and new tests can be evaluated that were not possible based on radio data alone. This follows because the radio, optical, and X-ray jet emissions are emitted by electrons with quite different energies (i.e., Lorentz factors,[2] γ).

This review is focused on what X-ray observations of relativistic jets can contribute to our understanding of the physical processes in jets. Although some jet detections were made with the imaging X-ray observatories *Einstein* and ROSAT, significant progress blossomed only with the CXO (Weisskopf et al. 2003) launch in 1999. For this reason, together with the limitations of space, we emphasize results obtained between the years 2000 and mid-2005. We will concentrate on spatially resolved X-ray emission from kiloparsec-scale jets. Radio observations of parsec-scale jets and broadband observations of spatially unresolved but highly variable core emission from subparsec jets of blazar-type AGN will only be discussed when they have direct implications for the inner workings of kiloparsec jets. Furthermore, we will not cover Galactic X-ray jets even though they bear many similarities to their extragalactic counterparts. Although there have been reports of thermal X-ray emission associated with jets (mainly in the context of "jet-cloud interactions"), our main concern is with the nonthermal emissions, already well established as the major process for radio through X-ray frequencies from multiple lines of argument including polarization

---

[1] NASA's first X-ray imaging satellite with subarcsecond resolution. Launched in July 1999.

[2] Lorentz factor: for relativistic electrons, $\gamma = \frac{E}{m_e \times c^2}$; for the jet's bulk velocity, $\beta = \frac{v}{c}$, $\Gamma = \frac{1}{\sqrt{1-\beta^2}}$.

data, Faraday screen parameters, X-ray spectral fitting, and the absence of emission lines.

Reviews on some aspects of jets include: *Theory of Extragalactic Radio Sources* (Begelman, Blandford & Rees 1984), *Beams and Jets in Astrophysics* (Hughes 1991), *Parsec-Scale Jets in Extragalactic Radio Sources* (Zensus 1997), and *Relativistic Jets in AGNs* (Tavecchio 2004). Among the many jet-related meetings in the last 10 years are: *Relativistic Jets in AGNs*, Cracow, 1997 (Ostrowski et al. 1997); *Ringberg Workshop on Relativistic Jets*, Ringberg Castle, 2001[3]; *The Physics of Relativistic Jets in the CHANDRA and XMM Era*, Bologna, 2002 (Brunetti et al. 2003); *Triggering Relativistic Jets*, Cozumel, 2005 (Lee & Ramirez-Ruiz 2006); and *Ultra-Relativistic Jets in Astrophysics: Observations, Theory and Simulations*, Banff, 2005[4]. Conference reviews on the unresolved core emission have been given by Coppi (1999), Krawczynski (2004, 2005), Sikora & Madejski (2001) and Tavecchio (2005).

We use the conventional definition of spectral index, α, for power-law radiation spectra: flux density, $S_\nu \propto \nu^{-\alpha}$. It is not yet known if electrons alone, or electrons and positrons, radiate the observed jet emission. We thus refer in this review to either electrons or electrons and positrons as "electrons." We use γ as the Lorentz factor of particles in the jet-frame of reference, and Γ for the bulk Lorentz factor of the jet plasma. As most X-ray emitting jets are detected on only one side of otherwise double radio sources, Γ more than a few seems likely to be generally applicable.

1.1. Jet Composition

We take the essence of a jet to be a quasi-lossless transmission line: a conduit containing relativistically moving particles and magnetic field (either of which could dominate the local energy) and/or Poynting flux. We distinguish between two substances: the "medium," which is responsible for delivering the power generated in the nucleus of the host galaxy to the end of the jet and thence to the radio lobes; and the nonthermal plasma responsible for the emission we detect in the radio, optical, and X-ray bands. Though these two substances can be one and the same for some jet models, we prefer to think of them as quite distinct. Most models explain the appearance of radio, optical, or X-ray bright hotspots in some jets as caused by the transfer of some form of energy (for example, energy associated with the medium's bulk motion or magnetic field energy) to highly relativistic emitting particles. The reader should note that we use the term medium, lacking more precise knowledge about the nature of the jet material.

Although the basic makeup of jets is still largely unknown, observations of polarized radio and optical emission show that at least some of the continuum jet emission originates as synchrotron emission from relativistic electrons gyrating in a magnetic field. Although we have this direct evidence about the emitting plasma, the jet medium responsible for delivering power to the end of the jets is largely unconstrained. The jet

---

[3] **http://www.mpa-garching.mpg.de/∼ensslin/Jets/Proceedings/**

[4] To download talks: **http://www.capca.ucalgary.ca/meetings/banff2005/index.html**

medium cannot entirely consist of the relativistic electrons that produce the observed radiation because unavoidable inverse Compton (IC) losses off the cosmic microwave background (CMB) photons would preclude the flow of high-energy electrons all the way to the end of some jets. Positing a minimal magnetic field strength of 3 μG, and ignoring the IC losses associated with starlight or quasar light, which would shorten the relevant lifetimes even more, it has been shown that electrons with γ more than a few thousand cannot survive for the time required to travel from the environs of the SMBH to the end of some jets (e.g., Harris & Krawczynski 2006).

The main contenders for the underlying jet medium are Poynting flux, electrons with $\gamma \leq 1000$, and protons. "Neutral beams" have been suggested (e.g., neutrons; Atoyan & Dermer 2004). The latter hypothesis requires that the direction into which the jet is launched changes with time to account for large-scale bending and discrete deflections such as those in 3C 120 and 3C 390.3. Real jets may be made of several components, or may involve the transition of a jet dominated by one component into a jet dominated by another; e.g., a class of models postulates that an initially electromagnetic jet transforms into a particle-dominated jet further downstream.

1.2. Jet Formation, Structure and Propagation

1.2.1. Jet formation. Jets are believed to be launched from accreting supermassive black holes and powered by either the gravitational energy of accreting matter that moves toward the black hole or, in the Blandford-Znajek process (Blandford & Znajek 1977), by the rotational energy of a rotating black hole. In the first case, jets may either be launched purely electromagnetically (Blandford 1976, Lovelace 1976), or as the result of magnetohydrodynamic processes at the inner regions of the accretion disk (Blandford & Payne 1982, Begelman, Blandford & Rees 1984, Koide, Shibata & Kudoh 1999). In the Blandford-Znajek process, the black hole rotating in the magnetic field supported by the accretion disk gives rise to a Poynting flux. Most models of jet formation face the σ-problem (σ is the ratio of electromagnetic energy density to particle energy density), namely that they predict a Poynting flux-dominated energy transport by a strongly magnetized or high-σ plasma, whereas parsec-scale observations indicate that the jets consist of particle-dominated, low-σ plasma (Celotti & Fabian 1993, Krawczynski, Coppi & Aharonian 2002, Kino, Takahara & Kusunose 2002). Understanding the launching of jets may thus require the solution of two problems: the launching of a magnetically dominated outflow, and the conversion of such an outflow into a particle-dominated jet. The latter transition is poorly understood, and requires more theoretical work.

The process of jet formation will have an impact on the steadiness of the jet flow, and will affect the amplitudes and timescales of jet luminosity variations. Modulations of the power output are believed to cause the large amplitude brightness variations of the (unresolved) X-ray and γ-ray emission from blazars (Spada et al. 2001, Tanihata et al. 2003). Large amplitude variations on timescales of thousands of years may be responsible for the radio, optical, and X-ray knots observed in many kiloparsec-scale jets (Stawarz 2004, Stawarz et al. 2004) and the bright X-ray flare of the M87 jet (Harris et al. 2006). Several recent studies show that the flaring activity of AGN can

FORWARD AND REVERSE SHOCKS

One sort of shock can arise from the interaction of a fast medium overtaking a slower medium. In the frame of the contact discontinuity separating the two media, a forward shock propagates downstream into the slower moving medium and a reverse shock propagates upstream into the faster moving medium. Particle acceleration can be associated with both. The structure of knot C in the M87 jet (with a large gradient in brightness, falling rapidly moving downstream) serves as an example of a forward shock, and the expected behavior of a reverse shock is exemplified by knot A (**Figure 3**).

be described in the language of noise processes (Uttley, McHardy & Vaughan 2005); i.e., the study of power spectra. Blazar flares show that the noise process that drives flares has a rising amplitude of the power spectrum on the relatively long timescales of a few years. If jet knots reflect nuclear variability, it would require substantial power at much longer timescales (red noise), which, of course, are not available for direct observation.

1.2.2. Transverse jet structure. In addition to the obvious uncertainties as to the identity of the jet medium and its bulk velocity, several jet models involve jet structure perpendicular to the jet axis. Radio observations of transversely resolved jets (e.g., Swain, Bridle & Baum 1998, Laing & Bridle 2004, Lara et al. 2004, Pushkarev et al. 2005) and theoretical models of the core emission of blazars (Chiaberge et al. 2000) indicate a velocity gradient across the jet. Simple models use a two-zone structure, a fast moving spine that carries most of the jet energy, surrounded by a slower sheath, each with a characteristic value of Γ (Chiaberge et al. 2000). Laing & Bridle (2004) assume a gradual decline of Γ from the jet center to the outer parts of the jet: i.e., many layers with different velocities. If the velocity difference between layers is large, the particles in some layers see the relativistically boosted photons from other layers, resulting in an increase of the IC emission (Ghisellini, Tavecchio & Chiaberge 2005). A wealth of different jet structures has been proposed and studied in the framework of explaining the prompt and afterglow emission from gamma-ray bursts (e.g., C. Graziani, D.Q. Lamb & T.Q. Donaghy, submitted) and some of these may be relevant to kiloparsec-scale jets.

The fact that jets may have a complex structure is important for interpreting the observational data. For example, the dominance of the bright jet over the dim counter-jet in a number of sources was previously thought to constrain the bulk Lorentz factor of the jets (e.g., Wardle & Aaron 1997). However, the observations may merely show that most of the radio emission comes from a slow moving plasma, and thus may not constrain the bulk Lorentz factor of the jet component that carries most of the jet energy and momentum. The boundaries between jet layers of different velocity may accelerate particles (Stawarz & Ostrowski 2002) and are of special interest for jet stability considerations.

1.2.3. Jet propagation and the occurrence of knots. The origin of jet knots (localized brightness enhancements) and the mechanism that controls the location, strength, and longevity of the shocks thought to be responsible for the existence of knots have not yet been identified unambiguously. It is important to remember, however, that there is probably more than one type of knot and that there are several suggested methods of producing brightness enhancements in addition to the conventional explanation of particle acceleration at shocks. In the case of the M87 jet, the inner knots, D, E, and F, appear to be quasi regular in size and spacing, suggesting a possible origin associated with standing waves similar to those described by Beresnyak, Istomin & Pariev (2003) or by the elliptical mode Kelvin-Helmholtz instability (Lobanov, Hardee & Eilek 2003). Quite different are the knots A and C for which steep, quasi-planar gradients in radio brightness suggest reverse and forward shocks (see side bar, *Forward and Reverse Shocks*). Note however that Bicknell & Begelman (1996) have devised a detailed model of the M87 jet. They argue that all the knots can be explained by oblique shocks, with the apparent differences ascribed to relativistic effects. Their model requires the angle between the jet axis and the line of sight to be 30° to 35°, a value substantially larger than the 10° to 20° required by the observation of fast moving blobs downstream from the leading edge of the knot HST-1 (see Section 3.1.1).

One of the alternative explanations of knots is that knots in relativistic jets could be manifestations of a change in the beaming factor. The relativistic beaming factor, δ depends both on Γ and on the viewing angle, θ (the angle between the jet axis and the line of sight in the observer's frame):

$$\delta^{-1} = \Gamma(1 - \beta \cos \theta). \tag{1}$$

If the jet medium moves in a straight line so that θ is fixed, an increase in δ requires a significant increase in Γ. Though we can imagine plausible ways to lower Γ, the critical question is, are there ways to increase Γ far from the central engine? This would entail a supply of energy such that the total power flow could decrease yet Γ could increase (e.g., by converting some power from the flow as in magnetic reconnection). Sikora et al. (2005) discuss this scenario, but deal only with the situation close to the black hole. In addition, there is circumstantial evidence for acceleration of jet features on parsec scales (e.g., Hardee, Walker & Gómez 2005) and it is generally accepted that both FR I[5] (Laing & Bridle 2002a) and quasar (Wardle & Aaron 1997) jets decelerate on parsec to kiloparsec scales; there is no indication that significant jet acceleration occurs on kiloparsec scales, which may be required for some IC models of X-ray emission.

If the jet medium is allowed to significantly change its direction, modest changes in θ can produce large changes in δ. On very long baseline interferometry (VLBI)[6] scales, there has been a long-standing debate on ballistic versus curved trajectories.

---

[5] Fanaroff-Riley class: FR I radio galaxies are of lower radio luminosity than FR IIs and quasars, and the brighter radio structures are close to the nucleus.

[6] The technique of aperture synthesis in which the component radio telescopes are not physically connected, thereby permitting the use of intercontinental baselines resulting in synthesized beam sizes of milliarcsecs.

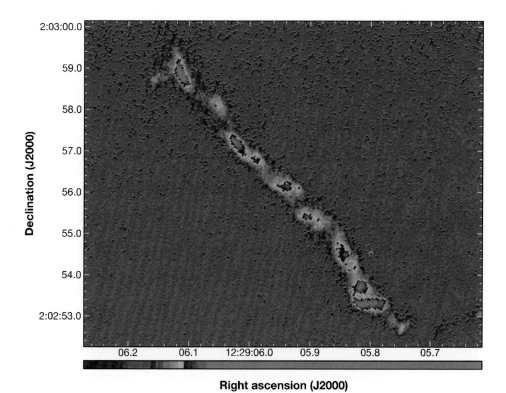

Figure 1

A *Hubble Space Telescope* image of the jet in 3C273.

On the kiloparsec scale the question arises: Does the medium move in a straight or gently curved path, or might it follow a helical pattern controlled by a field structure of the same topology? If the latter case holds, the changes in brightness along the jet could be explained by beaming effects and some of the problems for high Γ jet models, such as excessive jet length, would be mitigated. Bahcall et al. (1995) remark on the apparent helical morphology of the HST image of 3C 273 (**Figure 1**) and Nakamura, Uchida & Hirose (2001) argue for a "torsional Alvèn wave train" moving out to large distances from the central engine as a method of controlling the large-scale structure. There are of course numerous examples of large-scale bending (e.g., 3C 120, Walker, Benson & Unwin 1987) and discrete deflections (e.g., 3C 390.3, Harris et al. 1999); but in these cases we would anticipate deceleration only.

1.2.4. Terminal hotspots. Terminal hotspots, like knots, are thought to be localized volumes of high emissivity that are produced by strong shocks or a system of shocks. The somewhat hazy distinction between hotspots and knots is that downstream from a knot, the jet usually propagates much as before, whereas at the terminal hotspot, the jet itself terminates and the remaining flow is thought to create the radio lobes or tails. Thus the underlying jet medium must suffer severe deceleration and the outward

TWO-ZONE MODELS

In many areas of jet modeling, it is often the case that a simple, single power law or a simply defined emitting region is inadequate to provide all the observed emissions. Thus we are tempted to invoke another (spatial) region or a second spectral component. In almost all cases, this is done with the tacit assumption that the second component is (or can be) detected in only one channel, i.e., either synchrotron or IC. We need to realize that when we introduce a two-zone model, it precludes further analyses unless there is some hope of observing each zone in both channels. Some examples of current two-zone models are the spine/sheath jet model (e.g., Celotti, Ghisellini & Chiaberge 2001); the idea that jets contain regions of high and low magnetic field strengths, with relativistic electrons moving between these regions; and the introduction of a second spectral component to explain hard X-ray (Harris et al. 1999) or optical (Jester, Röser & Meisenheimer 2005) spectra.

flow from the hotspot is nonrelativistic and is not confined to a small angle. This is patently not true for the so-called "primary" hotspots in double or multiple systems. Instead of a terminal shock, primaries (and also aberrations such as hotspot B in 3C 390.3 North) may have oblique reflectors in essence, although the actual mechanism for bending might be more akin to refraction. For an extensive discussion of the differences between knots and hotspots, see Bridle et al. (1994).

Knots are a common property of FR I jets and generally do not lead to a total disruption of the jet, which maintains its identity downstream, be it relativistic or not. What we call knots in quasar jets may have little in common with FR I knots given their relative physical sizes.

Insofar as the X-ray emission mechanism is concerned, the initial X-ray detection of the Cygnus A hotspots (Harris, Carilli & Perley 1994) was accompanied with a demonstration that synchrotron self-Compton (SSC) emission provided a consistent explanation if the average magnetic field strength was close to the equipartition value under the assumption that the relativistic particle energy density was dominated by electrons, not protons. Essentially all the emission models for jet knots, on the other hand, have shown that SSC emission is completely inadequate to explain X-ray emission unless the magnetic field is orders-of-magnitude smaller than the equipartition value.

As the number of hotspot detections increased from CXO observations, many were found to be consistent with SSC predictions but a significant number appeared to have a larger X-ray intensity than predicted. This excess could be attributed to a field strength well below equipartition, IC emission from the decelerating jet "seeing" Doppler boosted hotspot emission (Georganopoulos & Kazanas 2003), or an additional synchrotron component (Hardcastle et al. 2004a). The last named researchers show that the strength of the excess correlates with hotspot luminosity in the sense that the strongest hotspots are consistent with SSC emission, whereas the weaker radio hotspots required the extra synchrotron component.

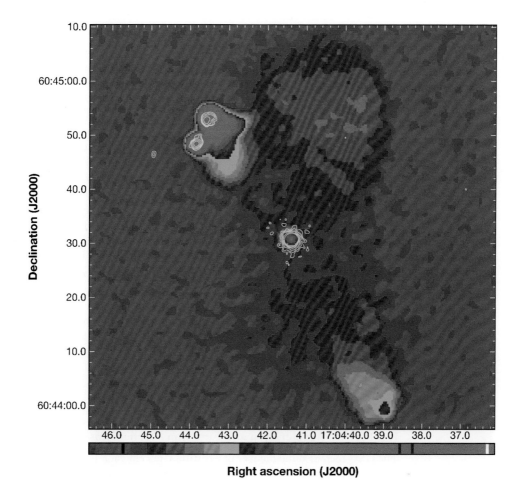

Figure 2

A radio image of the quasar 3C351 at 1.4 GHz from the Very Large Array. X-ray contours are superposed from *Chandra X-ray Observatory* data. Contour levels increase by factors of 2, from 2 to 32 in arbitrary brightness units. Note the bright north-east hotspot pair and the very weak south hotspot.

For many distant and/or faint jets, it is often difficult to be certain that a feature is a knot, a hotspot, or even a lobe. In some extreme cases, the true nature of even bright hotspots is ambiguous. An example is the double hotspot system in 3C351 shown in **Figure 2**. Displaced from the north radio lobe is a double hotspot to the NE of the core. These are bright at radio and X-ray bands. The southern radio lobe has only a weak hotspot with at most 4% of the radio intensity of the NE hotspots at 1.4 GHz. Thus the double hotspot has the hallmarks of relativistic beaming in spite of the commonly held view that hotspot radiation is not beamed (see however Dennett-Thorpe et al. 1997 for a discussion of beamed emission from hotspots). Given the fact that these bright features are not located at the outer edge of the lobe, perhaps they are knots in a jet very close to our line of sight.

1.3. Entrainment and Collimation

Long-standing problems for low-loss jets include the suppression of mixing with ambient material and the collimation and stability of jets (Hughes 1991). The process of entrainment of ambient material is closely related to the process of jet deceleration. Both processes have been studied observationally (e.g., Laing, Canvin & Bridle 2003) and numerically (e.g., Rossi et al. 2004). Possible mechanisms causing entrainment include velocity shear and Kelvin-Helmholtz instabilities (Bodo et al. 2003). Laing, Canvin & Bridle (2003) have studied FR I radio galaxies assuming that the two sides of the jets are intrinsically identical and that the observed differences in radio brightness and polarization are caused by the viewing angle and relativistic beaming effects. They find the velocity of the jet plasma decreases moving away from the jet axis and that this velocity shear decelerates the jet substantially. These arguments purport to demonstrate that there is a clear distinction between FR I and FR II radio galaxies insofar as their jet properties are concerned. Because the powerful jets of FR II radio galaxies and quasars are able to escape the high ambient density of their host galaxies and maintain their collimation out to the prominent hotspots, it is inferred that they suffer less entrainment and deceleration than FR I jets (Bicknell 1995).

Collimation of jets must be addressed both on subparsec scales during the process of launching the jet and on kiloparsec scales to explain the remarkable stability of jets. Tsinganos & Bogovalov (2002), for example, consider the former problem and demonstrate collimation for a relativistic component by a second, nonrelativistic less-collimated outflow (wind). Other collimation mechanisms include confinement by magnetic fields (Sauty, Tsinganos & Trussoni 2002), ram pressure of the ambient medium (Komissarov 1994), and radiation (Fukue, Tojyo & Hirai 2001).

1.4. Particle Acceleration and Emission Mechanisms

The CXO increased the number of jets with X-ray emission from a handful to $\approx$50 sources. Of these, 60% are classified as high-luminosity sources (quasars and FR II radio galaxies) and the remaining are low-luminosity sources (a mix of FR Is, BL Lacs, and a Seyfert galaxy). The observations indicate that the radio to X-ray emission from low-luminosity FR I sources can be explained by synchrotron models, whereas that from the high-luminosity FR II sources requires multizone synchrotron models, synchrotron and IC models, or more exotic variants.

1.4.1. Synchrotron models for FR I galaxies. For low-luminosity (FR I) radio sources, there is strong support for the synchrotron process as the dominant emission mechanism for the X-ray, optical, and, of course, radio emissions. Among the arguments supporting this view are the intensity variability found for knots in the M87 jet (Harris et al. 2006); the fact that in most cases the X-ray spectral index, α_x, is >1 and significantly larger than the radio index, α_r; and the relative morphologies in radio, optical, and X-ray emissions. For the sorts of magnetic field strengths generally ascribed to jet knots (10 to 1000 μG), synchrotron X-ray emission requires the presence of electrons of energies in the range of $10^7 < \gamma < 10^8$. As the highest energy electrons cool in equipartition magnetic fields on timescales of years, the observations

> **ENERGY LOSSES & HALFLIVES**
>
> Relativistic electrons lose energy via several processes. For both synchrotron and inverse Compton radiation, the rate of energy loss is $\propto E^2$ (E is the electron's energy). For these loss channels, the time it takes to lose half the energy (half-life) is $\propto E^{-1}$.

of single power-law spectral energy distributions extending all the way from the radio to the X-ray regime pose the problem of why there is no sign of radiative cooling. A possible solution may be that electrons escape the high-magnetic field emission region before they cool.

The radio to X-ray observations require the presence of one or more populations of high energy electrons [or protons if proton synchrotron emission is viable, Aharonian (2002)]. A common assumption is that the particles are accelerated at strong magnetohydrodynamic (MHD) shocks by the Fermi I mechanism (Bell 1978, Blandford & Ostriker 1978; see also the review by Kirk & Duffy 1999). However, there are several uncertainties. First we cannot be sure that the Fermi process is relevant because if the jet is strongly magnetized with a tangled field geometry, shock acceleration is not as effective as for strong shocks, which can exist when the field does not dominate. Next, the uncertainty of the bulk Lorentz factor of the jet medium means that we cannot be sure that Γ is large enough to allow the possibility of relativistic shocks. Finally, even if the bulk velocity of the jet is relativistic, it is still possible to have nonrelativistic shocks in the jet frame. For mildly relativistic shocks, Fermi I shock acceleration is more complicated than is the case for the nonrelativistic regime, and it is not yet well understood (Kirk & Duffy 1999).

1.4.1.1. Distributed acceleration. For FR I jets such as that in M87 (**Figure 3**), X-ray emitting electrons with $\gamma \approx 10^7$ will cool on timescales of a few years, and optical and UV emitting electrons will cool on timescales of a few decades. Thus, the interpretation of the X-ray and optical emissions from these sources as synchrotron emission implies that the emitting regions cannot be much larger than the electron acceleration regions. For bright knots that have traditionally been associated with strong shocks in the jet flow, these "life-time constraints" can easily be accommodated. However, CXO detected several jets with quasi-continuous emission along the jet (e.g., Cen A, Kataoka et al. 2006), suggesting that electron acceleration may be spatially distributed rather than restricted to a few bright knots. Wang (2002) finds that plasma turbulent waves can be a mechanism for efficient particle acceleration, producing high energy electrons in the context of blazar jets. Nishikawa et al. (2005) and Stawarz & Ostrowski (2002) propose turbulent acceleration in a jet's "boundary" or "shear" layer surrounding the jet spine. Stawarz & Ostrowski (2002) also argue that the resulting electron energy distribution should show an excess near the high energy cutoff, thereby producing a harder X-ray spectrum than would be expected based on the extrapolation of the radio and optical data.

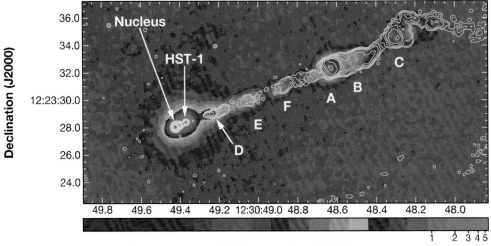

Figure 3

M87 *Chandra X-ray Observatory* image with 8 GHz contours. The X-ray image has an effective exposure of about 115 ks, consisting of 22 observations taken between 2000 and 2004. It has been smoothed with a Gaussian of FWHM = 0.25″ and the energy band is 0.2–6 keV. The color mapping is logarithmic and ranges from 0.02 (*faint green*) to a peak of 5.5 ev s^{-1} (0.049″ pixel)$^{-1}$. The radio data are from the VLA with a beam of FWHM = 0.2″. Contour levels increase by factors of 2 and start at 1 mJy/beam.

Other explanations for the quasi-continuous emission include low-level IC/CMB emission of low-energy electrons (see the discussion in the next paragraph) and synchrotron emission from electrons accelerated by magnetic reconnection. However, if the knot emission is produced by synchrotron emission from shock-accelerated electrons and the continuous emission has another origin, one might expect that the two jet regions would show markedly different spectral energy distributions. Measurement of X-ray spectral indices of the continuous emission is usually difficult, because of the fewer photons available for analysis. In the case of M87, Perlman et al. (2003) find no change of α_x between the knots and the quasi-continuous emission within a statistical accuracy of ± 0.15 in the X-ray spectral index.

1.4.1.2. Departures from power-law spectra. One of the primary reasons that IC/CMB models are preferred over synchrotron models for most FR II radio galaxies and quasars is the so-called "bow-tie problem." Conventional synchrotron spectral energy distributions call for a concave downward spectral shape, allowing for spectral breaks to steeper spectra at higher frequencies and eventual high frequency cutoffs. Thus we expect $\alpha_x \geq \alpha_{ox}$ (α_{ox} is the spectral index between optical/UV and X-ray). When this is not the case, the "bow-tie" showing the X-ray flux density and allowed range of α_x does not permit a smooth fit of a concave downward curve and instead requires a flattening of the X-ray spectrum. Examples are provided in **Figure 4**, which

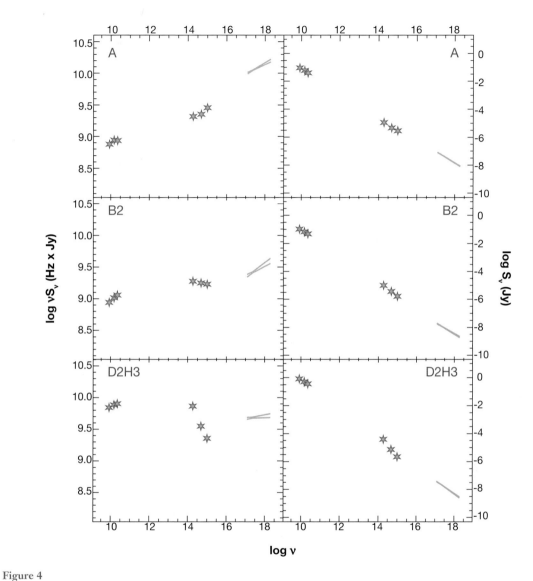

Figure 4

Examples of spectral energy distributions: three knots in the 3C273 jet. The left panels plot $\log(\nu \times S_\nu)$ as ordinate and the right panels are versions with $\log S_\nu$ vs. $\log \nu$. From top to bottom knots A, B2, and D2/H3 are shown. The X-ray data are presented as orange-colored "bow-ties," which delineate the range of acceptable power-law slopes. Note how these data preclude a fit consisting of a single-zone synchrotron spectrum, which would be a curve concave downwards. This figure was kindly provided by S. Jester. Details can be found in Jester et al. (2006).

shows the spectral energy distributions (SEDs)[7] for three knots in the 3C273 jet. The bow-tie problem is more common for FR II radio galaxies and quasars but is also found for some of the FR I radio galaxies.

In addition to the hypothesis of a second spectral component, there have been two suggestions for accommodating this behavior with synchrotron emission. The first is that mentioned above: boundary-layer acceleration (Stawarz & Ostrowski 2002) producing a flatter spectrum for the high-energy part of the electron spectrum. The other suggestion is restricted to the case where IC dominates the E^2 losses. Dermer & Atoyan (2002) argue that for the highest energy electrons, IC losses are reduced by the lower Klein-Nishina cross section so that the top end of the electron distribution experiences a reduced loss rate and, thus, an excess above the expected distribution builds up at high energies, producing a hard synchrotron spectrum at X-ray frequencies. Although this is a clever method of solving the bow-tie problem, in order to work, the photon energy density in the jet frame, $u'(\nu)$, must be larger than the magnetic field energy density, $u(B)$. To realize this, a Γ^2 boosting of the CMB is required, and the practical result is that by invoking the necessary Γ, you will already produce the observed X-ray emission by the IC/CMB process.

1.4.2. Emission models for FR II galaxies and quasars. The most pressing problem for X-ray emission from relativistic jets is the emission mechanism for the powerful jets from FR II radio galaxies and quasars. As mentioned above, the radio to X-ray spectral energy distributions of most of these sources cannot be described by a one-component synchrotron model. Such models predict a spectral energy distribution that softens at high energies. In terms of spectral indices, we expect $\alpha_x \geq \alpha_{ox}$, whereas the *Chandra* observations showed that $\alpha_x < \alpha_{ox}$ for many quasar jets. The most popular explanation is the IC model put forth by Celotti, Ghisellini & Chiaberge (2001) and Tavecchio et al. (2000). The observed large ratios of X-ray to radio luminosities are explained by postulating very fast jets with high bulk Lorentz factors Γ. Relativistic boosting increases the energy density of the CMB in the jet frame:

$$u'(\text{CMB}) = 4 \times 10^{-13}(1+z)^4 \Gamma^2 \,\text{erg cm}^{-3}. \qquad (2)$$

In this way, a single population of electrons is able to produce the radio and optical synchrotron emission in a magnetic field close to equipartition (B_{eq} generally less than 100 μG), and the IC X-ray emission by scattering off the relativistically boosted CMB.

Though we evaluate the various difficulties confronting the IC/CMB model in the final section of this review (4.2), we would like to emphasize here that the IC/CMB model requires two key ingredients for which there is at present no independent observational verification: enough low-energy electrons and highly relativistic plasma motion on kiloparsec-scales. Analysis of the SEDs of several FR II sources shows that electrons with Lorentz factors $\gamma' \approx 100$ produce the observed X rays (e.g., Harris &

---

[7]To describe the continuum spectrum of a feature, $\log(\nu \times$ flux density) is plotted against $\log \nu$. We use the term SED also for log(flux density) versus $\log \nu$.

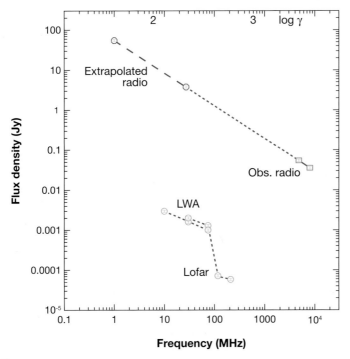

Figure 5
Segments of the synchrotron spectrum of knot WK7.8 in the jet of PKS0637-752. The observed radio flux densities are plotted toward the right edge so the short solid line allows us to determine a small section of the electron spectrum if we trust the equipartition field strength estimate; the extrapolation of this spectrum to lower frequencies is shown by the dotted line. The segment coming from the same electrons responsible for the IC/CMB X-ray emission is the long dash line to the upper left. Also shown near the bottom are sensitivity limits for low-frequency radio telescopes being designed and under construction: LWA is the Long Wavelength Array planned for a site in the southwest of the United States, and LOFAR is being built in the Netherlands. At the top are a few values of $\log \gamma$, which indicates the energy of the electrons corresponding to the emission spectrum.

Krawczynski 2002, their equation B4):

$$\gamma' = \sqrt{\frac{6.25 \times 10^{-12} \nu_{ic}(obs)}{(1 + \mu'_j)\delta\Gamma}}, \quad (3)$$

where the prime is used to denote quantities in the jet frame, μ'_j equals $\cos(\theta')$, and θ' is the angle between the jet direction and the line of sight. The uncertainties of extrapolating the electron spectrum to low energies is illustrated in **Figure 5**, which shows the spectrum of the knot in the jet of PKS0637-752. The low-energy electrons responsible for the X-ray emission produce synchrotron emission in the 1–30 MHz range, well below frequencies available from the Earth with reasonable angular resolutions. For this example we have used $\Gamma = 10$, which is the value required for the IC/CMB model (Tavecchio et al. 2000, Celotti, Ghisellini & Chiaberge 2001).

The actual electron spectrum could flatten significantly for $\gamma \leq 3000$ or even suffer a low energy cutoff. In that case there would be fewer electrons than calculated, and the required value of Γ would have to be increased to compensate. It is even conceivable that the electron spectrum could steepen at low energies and the required Γ would be much smaller than estimated. Our ignorance of the low end of the electron spectrum is very general; only in sources with very high values of magnetic field strength do ground-based radio data begin to give us the required information.

On the positive side, the IC/CMB model avoids the "far-from-equipartition" requirement of models that explain the high X-ray fluxes as synchrotron self-Compton emission from electrons up-scattering long-wavelength synchrotron photons into the X-ray band (e.g., Schwartz et al. 2000). Furthermore, it does not require the ad-hoc introduction of additional particle components, required by multizone synchrotron models (e.g., Harris et al. 1999).

2. PHYSICAL COMPARISONS OF RESOLVED X-RAY JETS

We are used to looking at images of jets that fit nicely on the page, be they Galactic microquasars, jets from relatively local FR I radio galaxies, or jets from quasars with substantial redshifts. We are struck by a number of similarities and are tempted to consider them all to share fundamental properties. In an effort to sharpen our perspective, we have devoted this part of the review to presenting the observed and deduced parameters for the X-ray jets known to us. Most of these data exist in the literature, but we have adjusted published values, where necessary, to conform to the currently standard cosmology: $H_0 = 71$ km s^{-1}; $\Omega_m = 0.3$; and $\Omega_\Lambda = 0.7$. Gathering these data will allow us to compare physical sizes and apparent luminosities.

In **Figure 6** we show the relative sizes of three jets: M87, 3C273, and PKS1127-145. Although the indicated sizes (1.6, 56, and 238 kpc, respectively) are projected sizes, it is clear that the entire jet of M87 would easily fit within a single knot of the 3C273 jet. Note also that a single 0.049″ pixel in the top panel corresponds to a few parsecs, the scale of VLBI jets, and is also comparable to the total size of jets from microquasars in our galaxy. Given the vast range of scales, can we really expect similar physical processes to operate all the way from parsec to megaparsec scales?

2.1. Gross Jet Poperties

In **Table 1** we list parameters for X-ray jets of low radio power sources from the XJET website[8] (2005.6), and **Table 2** contains the data for the more powerful sources classified as quasars or FR II radio galaxies. We have not included sources if only the terminal hotspots and/or lobes have been detected in X rays. In almost all cases, the division between the two tables corresponds to how the original investigators interpret the X-ray emission process. All the entries in **Table 1** are ascribed to synchrotron emission except for Cen B, and similarly the jets of **Table 2** are described on the basis

---

[8]http://hea-www.harvard.edu/XJET/

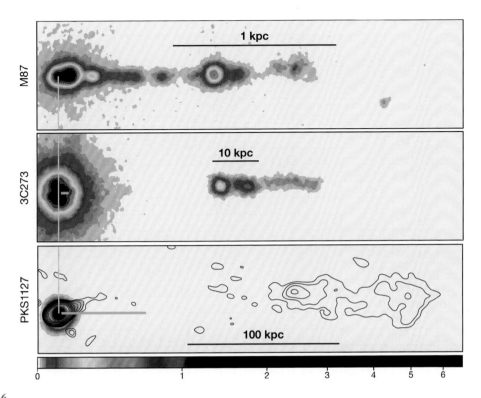

Figure 6

A comparison of three jets. The X-ray jets, M87 (*top*), 3C273 (*middle*), and PKS1127-145 (*bottom*), have been rotated for ease of comparison. All maps have had pixel randomization removed and have been smoothed with a Gaussian of FWHM = 0.25″. The absolute brightness mapping is logarithmic and the same for all three and ranges from 0.01 (*pink*) to 6.7 (the *black peak* of 3C273). The intensity units are total electron volts per second per pixel and the pixel size is 0.0492″. The overall projected size of each jet (from *top* to *bottom*) is 1.6, 56, and 238 kpc (21″, 21″, and 29″, respectively). The small cyan line overlaid on the core of 3C273 shows the length of the M87 jet if it were moved to the distance of 3C273, and the long cyan line on the bottom panel represents the total length of the 3C273 jet if it were at the distance of PKS1127-145. The X-ray jet of PKS1127-145 is too faint to be visible on the common intensity scale adopted, so radio contours are overlaid in order to show the full extent of the jet. X-ray emission is detected out to the last radio feature (refer also to **Figure 11**).

of the IC/CMB model except for PKS2152-69, for which Ly, De Young & Bechtold (2005) suggested thermal emission; Pictor A, PKS 1136-135, and 3C 273, for which both synchrotron and IC/CMB have been suggested; and 1928+738 and 3C403, which have been ascribed to synchrotron emission.

The projected jet length, both in arcsec and kiloparsec, should be accurate to about 10%; it is meant to describe the length of the X-ray jet as detected by the CXO and not the total length of the radio jet. The apparent X-ray luminosity is derived from the observed flux or flux density, assuming $\alpha_x = 1$. As pointed out by Lister (2003), such luminosities are not directly useful for correlations because we are dealing with relativistic beaming, which alters the jet-frame luminosity depending on Γ and θ.

Table 1 Parameters for Jets of Low-Power Radio Galaxies

Host name	z	Scale (kpc/as)	Length (arcsec)	Length (kpc)	log L_x (ergs^{-1})	α_x	θ (deg)	Deproj. (kpc)	Reference(s)
3C15*	0.0730	1.4	4	5.6	41.03	0.7 ± 0.4	–	–	1
NGC 315	0.0165	0.33	13	4.3	40.54	1.5 ± 0.7	–	–	2
3C31*	0.0167	0.34	8	2.7	40.56	1.1 ± 0.2	52	3.4	3
B2 0206+35	0.0369	0.72	2	1.4	41.12	–	–	–	4
3C 66B*	0.0215	0.43	7	3.0	41.03	1.3 ± 0.1	–	–	5
3C 120*	0.0330	0.65	80	52	41.95	–	–	–	6
3C 129	0.0208	0.42	2.5	1.0	39.64	–	–	–	7
PKS 0521-365*	0.055	1.06	2	2.1	41.90	1.2 ± 0.3	–	–	8
B2 0755+37*	0.0428	0.83	4	3.3	41.52	–	–	–	4
3C270	0.00737	0.15	35	5.2	39.13	–	–	–	9, 10
M84	—	0.082	3.9	0.3	38.71	0.8 ± 0.3	50	0.4	11
M87*	0.00427	0.077	20	1.5	41.32	>1	20	4.5	12, 13
Cen A	—	0.017	120	2.0	39.39	0.4 to 2.2	15	7.7	14
Cen B	0.013	0.26	8	2.1	40.13	–	—	—	15
3C296	0.0237	0.47	10	4.7	40.09	1.0 ± 0.4	–	–	16
NGC6251*	0.02488	0.49	410	200	–	1.30 ± 0.14	–	–	17
3C 346*	0.161	2.7	2	5.4	41.96	1.0 ± 0.3	20	16	18
3C 371*	0.051	0.98	4	3.9	41.87	0.7+0.4,-0.2	18	12.6	19
3C 465	0.0293	0.58	7.5	4.4	40.30	≈ 1.4	—	—	16

Notes: The scale is given in units of kiloparsec per arcsec.
All sources are classified as FR I radio galaxies except for 3C120, a Seyfert I galaxy, and the two BL Lac objects, PKS0521-365 and 3C371.
An "*" after the source name indicates that an optical detection has been reported (see **http://home.fnal.gov/~jester/optjets/**).
References: 1, Kataoka et al. (2003); 2, Worrall, Birkinshaw & Hardcastle (2003); 3, Hardcastle et al. (2002); 4, Worrall, Birkinshaw & Hardcastle (2001); 5, Hardcastle, Birkinshaw & Worrall (2001); 6, Harris, Mossman & Walker (2004); 7, Harris, Krawczynski & Taylor (2002); 8, Birkinshaw, Worrall & Hardcastle (2002); 9, Zezas et al. (2005); 10, Chiaberge et al. (2003); 11, Harris et al. (2002); 12, Wilson & Yang (2002); 13, Marshall et al. (2002); 14, Hardcastle et al. (2003); 15, Marshall et al. (2005); 16, Hardcastle et al. (2005); 17, Evans et al. (2005); 18, Worrall & Birkinshaw (2005); 19, Pesce et al. (2001).

The value of α_x given is a published value, either for the whole jet or from a brighter knot. If a reasonable estimate of the angle of the jet to the line of sight is given in the literature, it is quoted here. For a number of the quasars, θ is estimated from the IC/CMB calculation, and is thus model dependent; for others, it is estimated from VLBI studies. The resulting deprojected length suffers from similar uncertainties.

2.2. Evaluation

In **Figure 7** we show a plot of the observed parameters, jet length (projected) and observed (i.e., assuming isotropic emission) X-ray luminosity, L_x. This plot conforms to the common perception that quasars have powerful jets and are generally longer than those of FR I galaxies. Perhaps the only surprise is the gap with no jets lying between 10^{42} and 10^{43} ergs s^{-1}. The lower right is sparsely populated partly because in a large fraction of FR I jets, only the inner segment is detected in X rays. The upper left is empty because short jets at typical quasar redshifts will be difficult to resolve from the nuclear emission with arcsec resolutions. A separation of $\approx 2''$ is required to detect a jet close to a bright quasar, and at a typical redshift of 0.5, this already corresponds to 10 kpc.

The FR II radio galaxies have projected sizes comparable to those of the quasars, but are of lower apparent luminosity. The weakest jet (**Figure 7**, *lower left corner*) is M84, for which X-ray emission has been detected in the very inner part of the radio jet. 3C129 is quite similar, and joins Cen A, both points lying to the lower left of the main clump of FR Is.

This sort of plot is useful for comparative purposes, but not for interpretation because L_x is only an apparent luminosity and not the true luminosity in the jet frame and also because the length is a lower limit because of projection.

For a subset of the sources plotted in **Figure 7**, some reasonable estimate for the angle between the line of sight and the jet has been published. Because most of these jets are sensibly straight on kiloparsec scales (3C120 being a notable exception), we can obtain a deprojected length. For most of these, an estimate of the beaming factor is also available. For the majority of the quasars, the value of δ given in **Table 2** is model-dependent because it is the beaming factor required for the IC/CMB model. For the FR I and FR II radio galaxies, the δ values are derived from various lines of arguments based on geometry of the lobes, VLBI superluminal motions, and other more or less reliable methods. Assumed δs for the FR I jets are: 1.3, 3C31 and M84; 3.5, M87 and 3C371; 4, Cen A; and 3, 3C346. However, all beaming factors are suspect and the corresponding uncertainty will most likely introduce scatter in plots such as **Figure 8**, which plots $L'_x = L_x(obs)/\delta^4$ against length(obs)/$\sin \theta$.

The main purpose of **Figure 8** is to demonstrate that with the "current community interpretation" (i.e., FR I jets come from synchrotron emission whereas quasar jets are dominated by IC/CMB emission), FR I jets and quasar jets are more clearly separated on the basis of size rather than luminosity. Parameters for the smallest quasar jets (the group of 5 around log $L'_x = 42$, length = 70 kpc) are less secure because the jet emission is only of order one resolution element from the quasar core emission for these sources.

L'_x values are compromised by model dependency. If quasar jets were to come from synchrotron emission instead of IC/CMB emission, the appropriate δ could well be of order 3 or 4 (similar to that for FR I's, and adequate to explain the jet one-sidedness) instead of typical values like 10. Thus the luminosity correction when moving to the jet frame would be closer to a factor of 100 rather than 10,000 and the plot would be closer to a scaled version of **Figure 7**.

For both of these figures we need to remember that "low-power" and "high-power" sources are so divided according to their total radio luminosity. When we plot the jet luminosity we are dealing with a parameter that quantifies the jet loss, not the jet power. Because FR I jets are commonly thought of as being "lossy," the underlying assumption is that a larger fraction of the FR I jet power is radiated than is the case for FR II jets. Thus both the characteristic power and the fractional energy lost to radiation for both classes of sources are "free" parameters and the resulting

Table 2 Parameters for Jets of High-Power Radio Galaxies and Quasars

Host name	z	Scale (kpc/as)	Length (arcsec)	Length (kpc)	log L_x (ergs^{-1})	α_x	θ (deg)	Deproj. (kpc)	δ	Reference(s)
3C9	2.012	8.5	6.4	54	44.34	–	–	–	–	1
PKS 0208-512	0.999	8.04	5	40	44.47	–	8	262	7	2, 3
PKS 0413-21	0.808	7.54	2	15	43.99	–	20	44	3	2
Pictor A	0.0350	0.69	114	79	40.84	0.97 ± 0.07	>23	<201	<3	4, 5
PKS 0605-085	0.870	7.7	4	31	44.58	0.4 ± 0.7	–	–	–	6
PKS 0637-752*	0.651	6.9	12	83	44.34	0.85 ± 0.08	5.7	836	10	7, 8
3C 179	0.846	7.7	4.4	34	44.45	–	–	–	–	6
B2 0738+313	0.635	6.9	35	241	42.93	0.5 to 1.4	8	1730	7	9
0827+243	0.939	7.9	6.2	49	44.14	0.4 ± 0.2	2.5	1100	20	10
3C 207	0.68	7.1	4.6	33	43.97	0.3 ± 0.3	8	237	7	6
3C 212*	1.049	8.1	4	32	43.52	–	–	–	–	11
PKS 0903-57	0.695	7.1	3.5	25	43.90	–	20	73	3	2
PKS 0920-39	0.591	6.6	10	66	43.70	–	7	322	8	2, 3
3C 219	0.174	2.9	20	58	(41.68)	–	–	–	–	12
Q0957+561	1.41	8.5	8	68	43.69	0.9 ± 0.6	–	–	1.4	13
PKS 1030-357	1.455	8.5	12	102	44.99	–	8.6	682	9	2, 3
PKS 1046-40	0.620	6.8	4	27	43.44	–	17	93	3	2
PKS 1127-145	1.18	8.3	30	249	44.62	0.5 ± 0.2	24	612	4	14
PKS 1136-135*	0.554	6.4	6.7	43	43.92	0.4 ± 0.4	6	410	10	6
4C49.22*	0.334	4.8	5.6	27	43.62	–	6	270	14	6
PKS 1202-262	0.789	7.5	5	37	44.73	–	4.9	568	12	2, 3
3C 273*	0.1583	2.7	21	57	43.58	0.6 to 0.9	5	654	5	16, 17

Source	z	scale								Refs
4C19.44*	0.720	7.2	14.4	104	44.55	–	10	616	14	6
3C 303*	0.141	2.5	9	22	41.56	–	–	–	–	17
GB 1508+5714	4.3	6.9	2.2	15	44.96	0.9 ± 0.4	15	58	4	9, 18
PKS 1510-089	0.361	5.0	5.2	26	43.85	0.5 ± 0.4	–	–	–	6
3C 345*	0.594	6.6	2.7	18	43.65	0.7 ± 0.9	7	138	7	6
1642+690	0.751	7.3	2.7	20	43.67	–	–	–	–	6
3C 380*	0.692	7.1	1.8	13	44.68	–	13	57	–	2
1928+738*	0.302	4.4	2.6	11	43.21	0.7 ± 0.7	6	105	10	6
3C403	0.059	1.13	45	51	41.48	0.7 ± 0.4	–	–	–	19
PKS 2101-490	(1.04)	8.1	6	49	44.17	–	25	116	–	2
PKS 2152-69	0.0283	0.56	10	5.6	40.66	1.6 ± 0.4	–	–	–	20
3C 454.3*	0.859	7.7	5.2	40	44.62	–	18	129	–	2

Notes: The scale is given in units of kiloparsec per arcsec.

All sources are classified as quasars except for the four FR II radio galaxies: Pictor A, 3C219, 3C403, and PKS2152-69.

An "*" after the source name indicates that an optical detection has been reported (see also **http://home.fnal.gov/~jester/optjets/**).

The redshift for PKS2101-490 is uncertain (described as "tentative" by Marshall et al. 2005).

PKS2152-69 is odd, having a bright knot close to the core and a disparity between the radio, optical, and X-ray distributions. Ly, De Young & Bechtold (2005) argue for a thermal interpretation of the X-ray emission.

References: 1, Fabian, Celotti & Johnstone (2003); 2, Marshall et al. (2005); 3, Schwartz et al. (2006); 4, Wilson, Young & Shopbell (2001); 5, Hardcastle & Croston (2005); 6, Sambruna et al. (2004); 7, Chartas et al. (2000); 8, Schwartz et al. (2000); 9, Sieminowska et al. (2003); 10, Jorstad & Marscher (2004); 11, Aldcroft et al. (2003); 12, Comastri et al. (2003); 13, Chartas et al. (2002); 14, Sieminowska et al. (2002); 15, Marshall et al. (2001); 16, Sambruna et al. (2001); 17, Kataoka et al. (2003); 18, Cheung (2004); 19, Zezas et al. (2005); 20, Ly, De Young & Bechtold (2005).

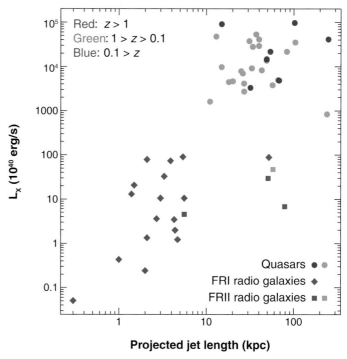

Figure 7

The observed X-ray luminosity plotted against the projected length of the jet. Quasars are plotted with filled circles; FR II radio galaxies with squares; and FR I radio galaxies (including Seyferts and BL Lac objects) with diamonds. The colors are allocated according to: red, $z > 1$; green, $1 > z > 0.1$; and blue, $0.1 > z$. The FR I (*diamond*) close to the three FR IIs is 3C120, which has a weak detection of a knot 80″ from the nucleus. The FR II (*square*) jet in the midst of the main clump of FR Is is PKS2152-69. Ly, De Young & Bechtold (2005) present evidence that the X-ray emission of the jet-related feature is thermal in origin.

luminosities (luminosity = total jet power × fractional loss to radiation) would not necessarily be expected to be similar as in **Figure 8**.

3. OBSERVATIONS OF RESOLVED JETS

3.1. Relevant Radio and Optical Considerations

In addition to the critical role of radio and optical flux densities, which complement the X-ray intensities in defining the SEDs of jet knots, these longer wavelengths provide two critical capabilities for jet observations: higher angular resolution than that of the CXO, and polarization. Moreover, in most cases, we are confident that we can interpret the data on the basis of synchrotron emission rather than being faced with the uncertainty of IC versus synchrotron emission, as is the case for the X rays.

For the SEDs, the IR-optical-UV data are usually those that determine if a synchrotron spectrum (broken power law with high energy cutoff) can be used to describe

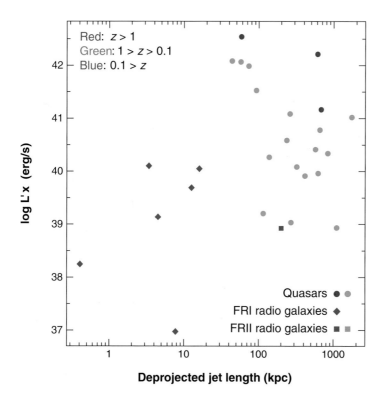

Figure 8

The best guess jet-frame luminosity corresponding to the observed X-ray luminosity, ($L'_x = L_x$ (obs)$/\delta^4$), plotted against the deprojected length of the jet. The symbols are used the same way as in **Figure 7**.

the radio to X-ray data. For example, Jester, Röser & Meisenheimer (2005) find a spectral flattening from HST data in the 3C273 jet, which is the basis for the claim that a simple synchrotron spectrum cannot fit all the data. Often the optical upper limit for a nondetection is used to preclude a single synchrotron component, whereas if no optical data were available, the radio and X-ray data could have been interpreted as a single (broken) power law.

There are at least two observational problems affecting the construction of SEDs. The first is the uncertainty that our photometry is measuring the same entity in all bands. The CXO resolution is significantly worse than that of the HST so to gather the counts for photometry, one needs at least a circle with radius of 0.5″. Thus when measuring the SED of the knots in, e.g., 3C273 (**Figure 1**), we implicitly assume that the X rays are coming from the same emitting volume as the optical/radio, and not from some additional volume such as a sheath around knots.

The second uncertainty is the absorption correction, which mainly affects the UV and soft X-ray data, and depends not only on the column density along the line of sight to the source, but also on the gas-to-dust ratio.

3.1.1. Morphology and polarization at kiloparsec scales. One of the more significant advances in understanding radio jets has been achieved by Laing & Bridle (2002a,b). For the case of FR I jets for which both sides are visible and well resolved, they have been able to use the laws of energy and momentum conservation to solve for all the physical jet parameters assuming that the observed differences in brightness and polarization between the two sides are caused by relativistic effects only (see Konigl 1980 for a general discussion of relativistic effects). For 3C31 they find $\beta = 0.8$ to 0.85 initially, then decelerating to $\beta = 0.2$ at a few kiloparsecs, with loading by entrainment being the cause of the deceleration. The solution requires cross jet velocity structure: The outside has to be going slower than the center. This is consistent with, but does not require, a simple "spine/sheath" structure.

Optical and radio polarization have also been used to study the field configuration in relation to the properties of internal shocks in jets. For example, Perlman & Wilson (2005) find that the peaks in X-ray brightness in the M87 jet coincide with minima of the optical polarization. They conclude that this is consistent with the location of internal shocks, which both produce the X-ray emission via particle acceleration and change the magnetic field direction. The observed reduction in polarized signal would then be a result of beam smearing over a region of swiftly changing field direction.

Other notable progress coming from optical data includes the discovery of features with apparent velocity of order 6 times the speed of light, moving downstream from M87/HST-1 (Biretta, Sparks & Macchetto 1999). This demonstrates that at least mildly relativistic velocities persist to kiloparsec scales. Several investigators (see, for example, Macchetto 1996) have also noted that optical emission away from bright knots requires continuous acceleration processes because the E^2 loss times are so short that the electrons responsible for the observed emission cannot travel from the shock locations; the same sort of argument was later deduced from similar morphologies observed at X-ray frequencies.

3.1.2. Parsec-scale structures. The most relevant aspect of VLBI work for X-ray jet physics is the accumulating database containing monitoring of a reasonably large sample of quasar, blazar, and BL Lac jets. The original work was the "2 cm survey," which has now become institutionalized on the web as MOJAVE[9]. These data provide a wealth of information such as the distribution of beaming factors and jet velocities (if one accepts the notion that observed proper motions of jet features reflect the underlying jet velocity and that the sources are at distances indicated by their redshifts). Kellermann et al. (2004) find apparent velocities β ranging from 0 to 15, with a tail extending up to 30 for individual features. With assumptions about brightness temperatures, this can be translated to Γ values covering a similar range. If bulk velocities of this magnitude persist to kiloparsec scales, one of the prerequisites of the IC/CMB model for X-ray jet emission will be satisfied.

Another Very Long Baseline Array (VLBA) monitoring project is described in Jorstad et al. (2005). They find a similar range for Γ using intensity variability timescales to estimate δ, with most quasar components having Γ of order 16 to 18. In

---

[9]http://www.physics.purdue.edu/astro/MOJAVE/

both of these works, there is ample evidence of nonballistic motion: velocity vectors of components misaligned with the jet vector.

Gabuzda, Murray & Cronin (2004) have used the transverse polarization structure of jets resolved with VLBI to argue for a helical structure for the magnetic field governing the emitting region; this is circumstantial evidence for nonballistic motions.

Wardle et al. (1998) argued for jet composition being a pair plasma based on circular polarization inferences, and Hirotani et al. (1999) suggested that two components in the jet of 3C279 were dominated by pair plasma on the basis of electron density arguments.

3.2. The X-ray Data

We will not cover inferences from unresolved X-ray behavior of cores, but concentrate on jet features for which we have some confidence that the radio, optical, and X-ray emission comes from the same emitting volume. We make the usual assumptions that all relativistic plasmas will emit both synchrotron and IC radiation, and because most/all X-ray jets are one sided, and these sides are the same as those that have VLBI superluminal jets, $\Gamma > 1$, but not necessarily ≥ 5 (for kiloparsec scales).

3.2.1. Jet structure. So far, there is very little transverse structure available from X-ray data. A notable exception is knot 3C120/k25 (Harris, Mossman & Walker 2004), which is resolved into three components. The jet of Cen A is well resolved because it is the nearest jet source (Kraft et al. 2002). M87 knots A, B, and C are a bit larger than the point spread function (Perlman & Wilson 2005), and Marshall (private communication) reports that several features in the jet of 3C273 are also resolved by the CXO.

In general, there is good correspondence between jet knots mapped in the radio, optical, and X-ray bands. For both IC and synchrotron emission models, this is expected in first-order approximations if there is a single relativistic electron distribution responsible for all observed emissions. Relative intensities between bands can vary depending on the relative magnitudes of energy densities in the magnetic field and in the photons, as well as on the form of the electron distribution, N(E), because different bands come from different segments of N(E).

There are, however, a few cases of gross misalignment between an X-ray feature and emissions at lower frequencies. In the M87 jet, beyond knot C (*upper right* of **Figure 3**) the radio jet makes a sudden excursion to the north downstream from a sharp gradient in radio brightness (in the opposite sense to that of the leading edge of knot A). Although there is weak radio emission downstream of this radio edge, the X ray brightens. One interpretation might be that the radio jet encounters an "obstacle" causing an internal shock and a jet deflection. The X-ray emission would then come from the obstacle, and not be associated properly with the jet itself. A similar situation occurs just downstream of knot A in the jet of 3C273 where the radio jet deflects to the south before resuming its principal direction, whereas the X-ray emission, and in

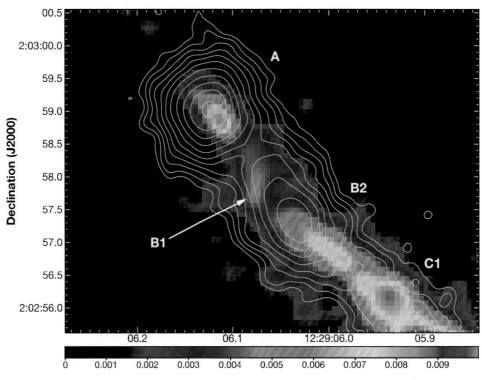

Figure 9

The first part of the jet in 3C273. The quasar itself is well off this picture, up and to the left. The false color image is from a VLA map at 22 GHz, kindly supplied by R. Perley. The radio beamsize is 0.35″ FWHM. The color scale is below the image and given in Jy/beam. The contours are from recent CXO data, smoothed with a Gaussian of FWHM = 0.25″. The lowest contour is 0.008 ev s^{-1} per 0.049″ pixel and successive contours increase by factors of $\sqrt{2}$. Note how the radio ridge line heads directly south right after knot A; thus knot B1 lies on the northern edge of the jet in the UV and X ray, but on the southern side of the jet in the radio. Although B1 and B2 are not well resolved with the CXO data, they are clearly separate in the radio and HST images (Jester, Röser & Meisenheimer 2005).

this case also the optical emission, continues further north along the main jet vector defined by knot A and the rest of the jet (**Figure 9**).

Another example of discrepant correspondence between radio and X rays is Cen A (Kraft et al. 2002, Hardcastle et al. 2003, 2004b), shown in **Figure 10**. Though many radio and X-ray features align well, albeit with quite different relative intensities, there are a few X-ray knots that have no obvious corresponding radio enhancements. For PKS1136-135 (Sambruna et al. 2002, 2004), the radio emission associated with the first bright X-ray knot ("A") is extremely weak; this is another example of the range of relative intensities between radio and X-ray emissions.

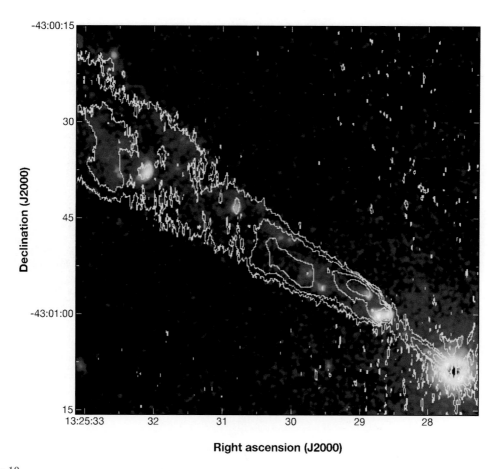

Figure 10

The jet in Cen A. The colors show the X-ray image smoothed with an 0.5″ FWHM Gaussian and the contours are from the VLA at 8 GHz with a beamsize of 1.1″ × 0.25″ (PA of major axis ≈0). The first contour level is 0.12 mJy/beam and successive intervals increase by factors of 4. This figure was provided by M. Hardcastle.

3.2.1.1. Offsets. Another common, although not universal, effect is the offset between peak X-ray, optical, and radio brightness distributions when mapped with similar angular resolutions. It is generally the case that when this occurs, the higher frequency brightness peaks at the upstream end of the knot, and the underlying cause seems to be a steepening of the spectra moving downstream (Hardcastle et al. 2003). A few examples are Cen A, **Figure 10** for knot A2 at RA = 13h 25m 29s; M87 knots D and F (**Figure 3**); and knot B in PKS1127-145 (**Figure 11**). For the nearby sources Cen A and M87, the magnitude of the projected offsets are of order tens of parsecs, whereas for PKS1127-145 at $z = 1.18$ the observed offset is of order 10 kpc. Additional examples from the FR I category are listed in Bai & Lee (2003), a paper devoted to the offset effect.

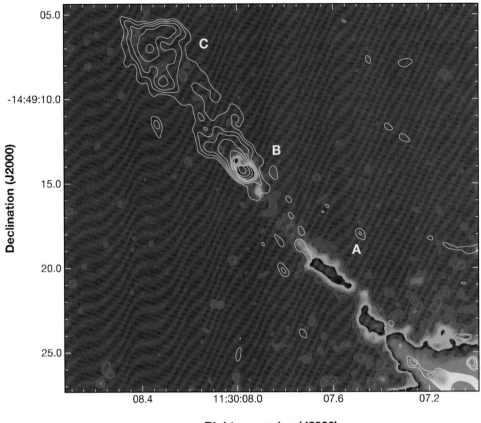

Figure 11

The jet of PKS1127-145. The colors show a *Chandra* X-ray image smoothed with an 0.5″ FWHM Gaussian, and the contours are from the VLA at 8 GHz with a beamsize of 0.78″ × 0.58″ in PA = 62°. The first contour level is 0.2 mJy/beam and successive intervals increase by factors of $\sqrt{2}$. These new data will be published by A. Siemiginowska et al. (in preparation).

For the simplest synchrotron scenario, if electrons were to be accelerated at a single location (i.e., a shock) and then be advected down the jet, all synchrotron bands should coincide insofar as peak brightness goes, even though downstream we would expect to lose the highest energy electrons sooner than the lower energy electrons responsible for the radio emission. This statement assumes perfect angular resolution, whereas normally, our beamsizes are not adequate to discern these structural differences. Thus even with the same angular resolution, the peak brightness of the X-ray emission, being centered on the shock, can occur upstream of the radio centroid, which has been shifted downstream a bit because the downstream plasma will continue to produce radio emission (but not X-ray emission).

Another possibility to explain the observed offsets that are not fine-tuned to the beamsize is an increasing magnetic field strength downstream from the shock, thereby enhancing the radio emissivity. Though the physical scales for the nearby sources are reasonably consistent with these models (travel time matching E^2 halflives), the 10 kpc offset for PKS1127-145 (again assuming synchrotron emission) would most likely rely on the second explanation. It is also the case, as emphasized earlier, that whereas we expect the electrons responsible for the X-ray emission to travel no further than some tens of light years, there is emission between the bright knots in many jets, and this supports the presence of a distributed, quasi-continuous acceleration mechanism (see Section 1.4).

For the IC/CMB model, currently there is no reasonable explanation as to why the X-ray emission should drop off more rapidly than the radio/optical synchrotron emission because the X-ray producing electrons have $\gamma \leq 200$, and thus much longer halflives than the electrons responsible for the radio (and optical) emission. There are, of course, *ad hoc* possibilities such as a precursor shock system (or some other mechanism) that would accelerate copious numbers of electrons only up to some small value of γ like 1000.

3.2.1.2. Progressions. Another effect that is closely associated with offsets is what we call progressions. This term is applied to those jets for which the X-ray intensity is highest at the upstream end, thereafter generally decreasing down the jet, whereas the radio intensity increases along the jet. Progressions are rather common; this effect is shown for seven quasars in figure 5 of Sambruna et al. (2004). The most striking example is 3C273 and profiles are shown in **Figure 12**. Note that the optical knots are of relatively constant brightness. If this jet were to be observed with a single resolution element, there would be a clear offset between peak brightnesses in the radio and X-ray emissions, which would most likely be comparable to the length of the bright part of the jet, $\approx 6''$ (15 kpc). Referring to **Figure 6**, we see that the 3C273 jet is about the size of a single "knot" in the PKS1127-145 jet. Thus we see that progressions and offsets can be considered to be two manifestations of an underlying spectral behavior. Both offsets and progressions are observed in FR I and in quasar jets although the common explanations for the two classes differ. For synchrotron models an increasing magnetic field strength is posited, thereby increasing the synchrotron emissivity. For IC models, a gradual decrease in jet bulk velocity is assumed, leading to a diminishing $u'(\nu)$ in the jet frame (Georganopoulos & Kazanas 2004).

3.2.1.3. Emission between the knots. Although the conventional view of synchrotron jets is that electrons responsible for X-ray emission cannot propagate more than a few light years from their acceleration site, lower brightness emission is often detected between the brighter knots. In their survey of quasar jets, Sambruna et al. (2004) [their section 4.3] remark on this attribute for PKS0605-085 and possibly also for 3C207 and PKS1136-135. Because of the lower brightness levels in both radio and X-ray bands, it has been difficult to obtain the data necessary to perform spectral tests for the emission mechanism. Quasi-continuous jet emission is expected for the IC/CMB

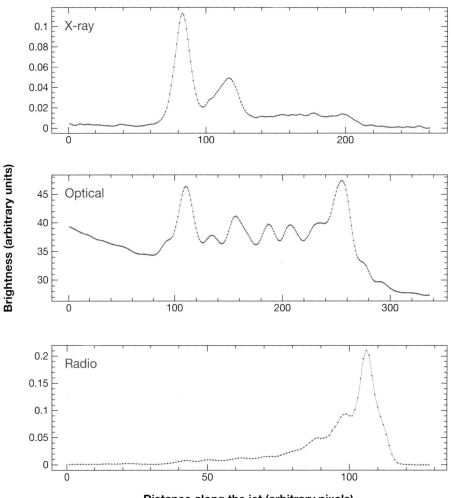

Figure 12

Profiles along the jet in 3C273. The quasar is off the plots to the left. The length shown is 16.7″ and the width used was 2″. Top panel: the X-ray data smoothed with a Gaussian of FWHM = 0.25″. Middle: a profile from an archival HST exposure (F622W), smoothed with a Gaussian of 0.5″. Bottom: a profile from an 8 GHz VLA map kindly supplied by R. Perley. The clean beam is 0.5″. The vertical scales are linear, in arbitrary units.

process, whereas a synchrotron hypothesis would require continuous acceleration processes, as outlined in Section 1.4.

3.2.2. Relative intensities and SEDs. Combining radio, optical, and X-ray photometry to create a broad band spectrum is the standard method to discriminate between synchrotron and IC emission. Resting on the common assumptions of a power-law distribution for the relativistic electrons and E^2 losses affecting the highest energy electrons more severely than the lower energy electrons, the notion that a one-zone synchrotron source must have a concave downwards spectrum has been generally accepted. This approach can work even if only three flux densities are available (i.e., one radio, one optical, and the X-ray measurement), but can become stronger with more data, permitting estimates of the spectral index to be obtained within each band.

There are several variations of this test such as plotting α_{ro} against α_{ox} (e.g., Sambruna et al. 2004, figure 4), or simply demonstrating that an optical upper limit lies below the line connecting the radio and X-ray data (e.g., Schwartz et al. 2006). Of the 34 quasars and FR II radio galaxies listed in **Table 2**, 9 have been shown by Sambruna et al. (2004) to have knots with $\alpha_{ro} > \alpha_{ox}$ and another 8 have optical upper limits that preclude a simple synchrotron fit. Twelve of the sources do not yet have useful optical data available, and the remaining 5 consist of a few FR II radio galaxies and a couple of quasars for which some knots have spectra consistent with a synchrotron fit while others do not.

Although it is thus fairly easy to demonstrate that a simple (i.e., "single zone") synchrotron spectrum fails to apply to most knots in quasar jets, once high quality data are available, serious problems arise also for the single zone IC/CMB models. Examples are spine-sheath models devised to benefit both from high and low Γ effects, and 3C273 for which some knots seem to have $\alpha_x > \alpha_r$, contrary to the expectation that electron spectra will most likely flatten at low energies (Jester et al. 2006).

3.2.3. Variability.
Because the CXO has been providing X-ray photometry of jet components for only six years, to detect variability we require a small physical size abetted by a significant value of δ to compress the elapsed time in our frame.

Thus, clear intensity variability has so far only been found in Cen A (Hardcastle et al. 2003) and M87 (Harris et al. 2006). This is not meant to preclude the possibility of detection of variability in larger structures that could well contain small scale structure. If one were to ascribe the factor of 100 difference in apparent luminosities between 3C273 and M87 to a factor of 3 difference in δ, then an event such as the flare in knot HST-1 of M87 (Harris et al. 2006) would be easily detected. The factor of 50 increase in the X-ray flux from HST-1 means that what was once an inconspicuous X-ray knot for a time outshone the remainder of the jet plus the unresolved core of M87.

Proper motion has been observed for radio features in Cen A (Hardcastle et al. 2003) and optical features moving at up to 6c downstream of HST-1 in M87 (Biretta, Sparks & Macchetto 1999). In both cases, the associated X-ray features align with stationary radio or optical components.

3.3. Jet Detection Statistics

In the standard picture of AGN unification (Urry & Padovani 1995), there are two main classes of AGN, low-power AGN (BL Lac objects and low-power radio galaxies) and high-power AGN (quasars and high-power radio galaxies). The differences of AGN within each class are explained with a different degree of alignment between the line of sight and the symmetry axis of the AGN (assumed to be parallel with the AGN jets). BL Lac objects are interpreted to be the aligned versions of FR I radio galaxies, and steep spectrum radio quasars (SSRQ, sources with a radio spectral index α_r, >0.5 at a few GHz) and flat spectrum radio quasars (FSRQ, α_r, <0.5) are increasingly aligned versions of the FR II parent population. Urry & Padovani (1995) derive the luminosity functions of the beamed AGN from the luminosity function of the parent populations. For their specific source samples, their analysis indicates that FR II radio galaxies have jets with bulk Lorentz factors of between 5 and 40, and that SSRQs and FSRQs are FR IIs with jets aligned to within $\sim38°$ and $\sim14°$ to the line of sight, respectively. The bulk Lorentz factors of the FR I jets are less constrained but seem to be somewhat lower than those of the FR II jets, and the jets of their radio selected BL Lacs seem to be aligned to within $12°$ to the line of sight.

In this context, we now consider the sources with X-ray jets (see **Tables 1** and **2**). The low-power sources with X-ray jets are mostly FR I radio galaxies, except for 3C120, a Seyfert I galaxy and the two BL Lacs PKS0521-365 and 3C371. The high-power sources with X-ray jets are all classified as quasars except for the four FR II radio galaxies Pictor A, 3C219, 3C403, and PKS2152-69. Remarkably, almost all X-ray jets from the nonaligned FR I and FR II sources can be explained as synchrotron emission from mildly relativistic jets; Lorentz factors of a few are needed to explain the nondetection of counterjets. Most sources for which the simple synchrotron picture does not work are quasars. In this case, explaining the X-ray emission requires Lorentz factors on the order of 10 and viewing angles on the order of $10°$. The IC/CMB interpretation of the X-ray emission thus indicates that the X-ray jets detected so far are similarly closely aligned to the line of sight as the average radio-selected FSRQs used in the FR II/quasar unification analysis described above.

Sambruna et al. (2004) and Marshall et al. (2005) used the CXO to study the fraction of sources with X-ray jet emission for certain source samples. Sambruna et al. (2004) studied sources with bright 1.4 GHz radio emission and a radio knot detection more than $3''$ away from the core. Out of a sample of 17 sources, X-ray jets were detected for 10 sources. Marshall et al. (2005) studied two samples of flat spectrum radio sources. One sample consisted of sources selected for their high 5 GHz flux density. The other sample consisted of sources with one-sided linear radio jet morphology. Out of 19 sources of the first sample, 16 were detected with short CXO observations. The detection probability in the second sample was lower, but this finding was not statistically significant.

The samples used in these "survey-type studies" were biased toward beamed sources. As the alignment of the sources is poorly constrained by the longer wavelength data, the high detection fraction with CXO cannot be used to argue for or against the IC/CMB model for those sources for which simple synchrotron models do not account for the X-ray emission.

As mentioned above, the radio spectral index can be used as an indicator of the jet orientation relative to the line of sight. A similar indicator is the lobe over core dominance at intermediate radio frequencies like 5 GHz. A test of the IC/CMB model is to check that the orientation parameters indicate an aligned jet for all quasars that exhibit the bow-tie problem. Indeed only for one source (PKS 1136-135) do we find at the same time $\alpha_r \gg 0.5$, lobe over core dominance and a SED that indicates a bow-tie problem. However, this source does not make a strong case against the IC/CMB model. Only knot B exhibits the bow-tie problem and close inspection of the radio-X-ray morphology shows that it may well be a hotspot rather than a jet knot. We conclude that the IC/CMB scenario is not grossly inconsistent with other orientation indicators.

4. DISCUSSION & SUMMARY

4.1. Critique of the Synchrotron Emission Model

There seems to be little doubt that the X-ray emission from most or all jets of FR I sources is dominated by the synchrotron process. When SEDs are available, they are consistent with concave downwards fits. There are no problems with the synchrotron parameters such as magnetic field strength or energy requirements. Light curves for variable knots also support the synchrotron model even if the predictions for behavior at lower frequencies still need to be verified. The alternative of IC/CMB emission requires unreasonable beaming parameters such as angles to the line of sight, which are too small compared to a host of other estimates.

Perhaps the most important implication to be deduced from FR I jets is the necessity for distributed emission rather than a finite number of shocks. While we don't doubt the evidence for strong, discrete shocks (e.g., a large gradient in radio brightness, often facing upstream), some additional process is required.

The most likely alternatives for the "additional process" are the aforementioned "distributed acceleration" and IC/CMB emission. As outlined in Section 1.1, one of the candidates for the underlying jet "medium" is electrons with $\gamma <$ a few thousand. If that option were correct, then it could well be the case that even very modest values of Γ, δ, and θ would suffice for an IC/CMB model of inter-knot emission, and many of the problems of this process for knots would not be present. It is likely that sufficient data have now accumulated in the relevant archives that this test could be performed for a number of the brighter jets with well-defined knots.

To explain jet segments devoid of detectable emission, this scenario would indicate that jets are inherently intermittent. Aside from testing this suggestion by careful photometry, spectral analysis, and calculation of beaming parameters, it would be a somewhat unbelievable coincidence if the energy spectral index, p, $[N(\gamma) \propto \gamma^{-p}]$ was the same for the postulated low-energy electrons responsible for the jet's energy transport and for the highest energy electrons with $\gamma \approx 10^7$ responsible for the knot X-ray emission. Thus one could reasonably expect to see a marked change in α_x moving from knot to inter-knot regions.

The largest hurdle for the application of synchrotron emission to jets of quasars comes from those cases for which the optical intensity is so low (or undetected) that it precludes a concave downwards spectral fit from radio to X-ray emissions. The associated bow-tie problem (Section 1.4) has been reported also for FR I sources (e.g., 3C120, knot 'k25'; Harris, Mossman & Walker 2004). None of the possible solutions has been accepted by the community and progress on this issue depends on a demonstration that a key ingredient of spectral hardening at high energy is indicated by some independent means. Examples would be the confirmation of a prediction from the two-zone model or finding independent support for the shear layer acceleration model.

4.2. Critique of the IC/CMB Emission Model

Although the idea of augmenting $u(\nu)$ compared to $u(B)$ in the jet frame had been used for jets close to black holes where $u(\nu)$ was thought to be dominated by UV radiation (e.g., Dermer & Schlickeiser 1994, Blandford & Levinson 1995, Sikora et al. 1997). Celotti, Ghisellini & Chiaberge (2001) and Tavecchio et al. (2000) applied this concept to kiloparsec-scale jets for which the CMB dominates $u(\nu)$. By positing that the X-ray knots of PKS0637-752 had a value of $\Gamma \approx 10$, similar to the values deduced from superluminal proper motions for the parsec-scale jet (Tingay et al. 2000), they were able to show that IC/CMB can explain the observed X-ray intensities while still maintaining equipartition conditions between $u(B)$ and $u(p)$ (where $u(p)$ is the energy density in relativistic particles). This idea was quickly adopted by the community because it was already realized that the preponderance of one-sided X-ray jets requires $\Gamma \geq 3$ or 4 and it provided a solution to the vexing problem of too little optical intensity to provide a reasonable synchrotron fit to the spectrum.

Additional support for this model is supplied by the jets that show the progression, discussed above, with a decreasing ratio of X-ray-to-radio intensity moving away from the core. Under the IC/CMB model, all that is required is a general deceleration of the jet, thereby reducing $u'(\nu)$ in the jet frame (e.g., for 3C273, Sambruna et al. 2001). There is, of course the problem of explaining why the IC/CMB X-ray intensity of 3C273/knot A happens to fall so close to the extrapolation of the radio/optical synchrotron spectrum (Marshall et al. 2001).

There are a number of additional uncertainties and problems for the IC/CMB model although none of these represents a definitive refutation.

4.2.1. Offsets and lifetime considerations.
In Section 3.2.1 we discussed offsets between X-ray and radio brightness distributions of jet knots. The low-energy ($\gamma \approx 100$) electrons responsible for the X-ray emission will have E^2 lifetimes in excess of 10^6 yr, which is sufficient to travel to the end of even a megaparsec jet. Thus when we are confronted with a knotty X-ray jet, the question arises: Once a copious supply of these electrons are generated (e.g., at knot A in 3C273), why does the emission fade to a low level and then rise again for the next knot instead of forming a continuous or cumulatively brightening jet? One might devise a rather contrived scenario by having the beaming factor decrease to end one knot, and then either increase at the location of the next knot, or posit the injection of enough new electrons to produce a bright knot even though the beaming factor is less than that enjoyed by the first knot. This explanation is unsatisfactory when the radio emission is considered because it should

follow the X-ray behavior if δ is the controlling factor. The same sort of problem affects the observed offsets (Atoyan & Dermer 2004, Stawarz et al. 2004); the X-ray brightness should persist further downstream than the optical and radio brightnesses, the opposite of what is observed.

4.2.2. Energetics. Dermer & Atoyan (2004) and Atoyan & Dermer (2004) have made a comprehensive review of the X-ray emission processes and emphasize that the original formulation of IC/CMB emission worked on the equipartition assumption based on the radio data. When the electron spectrum is extended down to the low energies required by IC/CMB, the particle energy density [and hence also $u(B)$] increase significantly. This leads them to conclude that excessive energies for the jet are required, even under the "optimistic" assumption that the jet is made of an electron/positron plasma without cold protons. For PKS0637-752, they find the kinetic luminosity is $\geq 7 \times 10^{46}$ erg s^{-1} for $\delta = 27, \theta \leq 2°$, and increases for more reasonable beaming parameters. The associated total energy is $\geq 10^{57}$ ergs for the case of $\delta = \Gamma \approx 10, \theta \approx 5°$.

4.2.3. Uncertainty of extrapolation of the electron spectrum. One of the implicit assumptions of every IC/CMB calculation (i.e., to determine the required beaming parameters to explain the observed radio and X-ray intensities) is that the electron spectrum extends to very low energies with a slope $p = 2\alpha_r + 1$. If that were to be the case, then α_x should have the same value as α_r. However, as demonstrated in **Figure 5**, we currently have no knowledge that this condition holds. If α_x is less than α_r, it would indicate a low frequency break to a flatter spectrum and the estimated beaming parameters would be wrong. With fewer low-energy electrons than assumed by the extrapolation, Γ and δ would have to be larger and θ correspondingly smaller, exacerbating some of the problems listed above. It is, of course, conceivable that the electron spectrum takes an upturn at low energies, in which case the error goes in the opposite direction.

Another assumption often, but not always, present is that of equipartition. Because every calculation requires a value of the magnetic field in order to move from the observed segment of the synchrotron spectrum to obtain the corresponding segment of the electron spectrum, the usual method is to assume equipartition. When that constraint is removed as in the case of arguing for a field strength well below equipartition (e.g., Kataoka & Stawarz 2005), the electron spectrum can be considered undefined, and one can conjure up whatever number of low energy electrons are needed to explain the X rays for a given beaming factor. In the case of Kataoka & Stawarz (2005), a small value of Γ was invoked based on radio asymmetry arguments (Wardle & Aaron 1997). The initial analysis of PKS0637-752 (Schwartz et al. 2000) also suggested substantial dominance of $u(p)$ because the IC/CMB scenario with beaming was not widely known at that time.

Finally, not only do IC/CMB models require a substantial extrapolation of the electron energy distribution to low energies with a fixed power law, they also require some fine tuning of a strict cutoff in the distribution at some slightly lower γ in order not to over-produce the optical emission (e.g., Sambruna et al. 2004, their table 7).

4.2.4. Small angles to the line of sight and physical length of jets.
From recent quasar surveys with the CXO (Sambruna et al. 2004, Marshall et al. 2005, Schwartz et al. 2006), fitting IC/CMB models yield δ values that range from 3 to 11. For $\Gamma = \delta$, this means θ is most commonly between 4° and 11°. Because most X-ray jets are reasonably straight, the physical length of jets sometimes exceeds 1 Mpc (n.b.: the jet lengths given in **Tables 1** and **2** refer primarily to the X-ray extent; the radio jet is often longer).

Many workers (e.g., Dermer & Atoyan 2004) find megaparsec-scale quasar jets uncomfortably long, and it is certainly the case that most FR II radio galaxies, the "face-on" counterparts of quasars under the unified scheme, are much smaller. However, there are a small number of "giant radio galaxies," and even a few quasars with sizes considerably greater than 1 Mpc (e.g., Riley & Warner 1990).

In some cases source morphology inferences are in conflict with small θ. Wilson, Young & Shopbell (2001) argue that if the IC/CMB model with equipartition is applied to the jet in Pictor A, $\Gamma = \delta = 7.2$ and $\theta = 8°$. Such an angle to the line of sight would mean that the total extent of the source would be on the order of 3 Mpc and the hotspots at the outer end of each lobe should be seen projected onto the radio lobes instead of protruding beyond the lobes as they are actually situated. Although Wilson, Young & Shopbell (2001) conclude that an IC/CMB model at a more reasonable $\theta \approx 23°$ would require $B < B_{eq}$, Hardcastle & Croston (2005) subsequently have made a strong case that the X-ray emission from the Pictor A jet is synchrotron emission, not IC/CMB.

4.2.5. Expectations for jets with $z > 1$.
Schwartz (2002) has argued that at higher redshifts there should be more jet detections because the increase in $u(\nu)$ of the CMB by the factor $(1 + z)^4$ will compensate for the usual redshift dimming of surface brightness. In addition to this effect, we might expect to see more of the lower Γ jets with larger beaming cones because the $(1 + z)^4$ factor already will statistically increase the ratio of $u(\nu)/u(B)$ regardless of the Γ^2 factor from the jet's bulk velocity. So far, these predictions have not been realized (Bassett et al. 2004), and Kataoka & Stawarz (2005) [see their figure 10] have emphasized that the required δ values for the IC/CMB model generally decrease with redshift. At this stage, the only quasar jet detection with z substantially greater than or equal to 2 is GB 1508+5714 ($z = 4.4$).

4.3. Tests to Differentiate between Synchrotron and IC/CMB Models

The basic tenet of the IC/CMB model for jet knots is that the X-ray emission is sampling the low-energy end of the power-law electron distribution. Therefore, the IC emission must continue to higher frequencies, unlike the synchrotron spectrum, which is already relatively steep and most likely will show an exponential cutoff at somewhat harder X-ray energies than available with the CXO. If we could measure the X-ray spectrum of quasar knots at much higher frequencies, and found a smooth continuation, it would be a clear confirmation of the IC/CMB model. If on the other hand, we were to find a cutoff in the X-ray spectrum, that would indicate synchrotron emission. Unfortunately, there are no real prospects of convincingly performing this test because it is so difficult to reach the required sensitivity and angular resolution above 10 keV. The CXO band is too narrow to define the expected cutoff, which may well be smeared over a wide frequency band by internal source structure.

Another option for discriminating these emission mechanisms will become available as new radio telescopes with unprecedented sensitivity and resolution at low frequencies come on line in the next several years. Both LOFAR in the Netherlands and the LWA (Long Wavelength Array) in the United States will have the capability to resolve jet knots and determine the characteristics of the electron distribution at the low energies of interest (**Figure 5**). Each of these instruments will have reasonably wide frequency coverage so that not only the amplitude, but also the slope of the low-frequency emission, can be measured. If we find that the low-frequency radio data indicate that the spectrum flattens significantly or has a low-frequency cutoff, then the IC/CMB model will have serious problems.

Optical and IR telescopes can be used to achieve detections of jet knots that currently have only upper limits. This band plays a crucial role for the IC/CMB model because there is still substantial uncertainty as to the origin of the currently detected optical features: Is this emission from the top end of the synchrotron spectrum or the bottom end of the IC spectrum? Robust detections and photometry at several wavelengths should clarify this problem, which impacts on the general "fine tuning" of the low-energy end of the electron spectrum (Section 4.2.3).

4.4. Detectability of the Extended Jet Emission by Gamma-Ray Telescopes

The *EGRET* detector on board the Compton Gamma-Ray Observatory established that blazars, AGNs with their jets aligned with the line of sight, are strong sources of gamma rays. The EGRET experiment (approximately 20 MeV to 30 GeV, or 5×10^{21} to 7×10^{24} Hz) detected a total of 66 blazars with redshifts up to $z \sim 2$ (Hartman, Bertsch & Bloom 1999). A small number of blazars (currently 10) with redshifts between 0.031 and 0.186 have been detected at even higher energies (GeV to TeV, frequencies above 10^{25} Hz) with ground-based Cherenkov telescopes (Krawczynski 2005). Rapid gamma-ray flux variability on timescales between 15 min and a few hours, together with assumptions about IR to UV emission co-spatially emitted with the gamma rays, have been used to derive a lower limit on the Doppler factor $\delta \gtrsim 10$ of the emitting plasma based on gamma-ray opacity arguments (Gaidos et al. 1996, Mattox, Wagner & Malkan 1997). All of these observations refer to very small physical scales, resulting in completely unresolved data from the nuclear regions.

If the extended jet emission detected by *Chandra* indeed originates from the IC/CMB process, the IC component should in principle be detectable in the MeV/GeV energy range with the Gamma-ray Large Area Space Telescope (GLAST) to be launched in 2007 (McEnery, Moskalenko & Ormes 2004), and possibly also in the GeV/TeV energy regime with ground-based telescopes like H.E.S.S., VERITAS,

MAGIC, and CANGAROO III (Aharonian 2004, Weekes 2003). GLAST has a sensitivity for the flux above 100 MeV of 3×10^{-13} ergs cm^{-2} s^{-1} for 5 yrs of sky-survey observations. Cherenkov telescopes like VERITAS and H.E.S.S. have a 100-GeV sensitivity of 9×10^{-13} ergs cm^{-2} s^{-1} for 100-hrs integration. These estimates are derived by the instrument teams for photon indices of 2. For harder photon spectra with indices of 1.5, the $\nu \times f_\nu$ sensitivities are about a factor of two better. IC/CMB models predict gamma-ray fluxes between 10^{-13} and a few times 10^{-12} ergs cm^{-2} s^{-1} (Dermer & Atoyan 2004, Tavecchio et al. 2004) so these new observatories should have sufficient sensitivity for detection.

The angular resolution of GLAST for a single photon will be 3.4° at 100 MeV, and 0.1° at 10 GeV; typical source localization accuracies will be tens of arcminutes near detection threshold and 0.5 arcmin for very strong sources.[10] For most sources, the angular distance between the core and the kiloparsec-scale jet is only a few arcseconds and GLAST will not be able to distinguish between core and jet emission on the basis of the spatial information. Furthermore, variability studies will be limited to rather long timescales and large fractional flux variations.

Cherenkov telescopes have better angular resolutions ($\approx 0.1°$) and source localization accuracies ($\approx 20''$). For $\sim$100 GeV photons however, the transparency of the Universe is limited to redshifts on the order of 0.5 owing to the gamma-rays pair-producing on IR background photons (1 to 40 microns) from galaxies. Detection and identification of gamma rays from kiloparsec-scale jets would thus require very strong sources with very extended X-ray jets at low redshifts; the chances for obtaining unambiguous results are not promising.

4.5. Prospects

4.5.1. Synchrotron emission.
In general, synchrotron emission is a powerful diagnostic of relativistic plasmas, and in the particular case of X-ray frequencies, informs us as to the location of acceleration sites. The major problem is the unknown magnetic field strength, which precludes a direct determination of the electron energy distribution.

Because the X-ray emitting electrons have such a high energy and, consequently, short lifetime, we expect variability in jets will continue to offer new insights. With multifrequency monitoring, it should be possible to disentangle light travel times from E^2 halflives and thus obtain a different estimate of the magnetic field strength and/or $u'(\nu)$ as well as δ (Harris et al. 2006).

As more jets are studied with greater sensitivity, we believe the chances are good that we should find a few objects that display the effects of a high energy cutoff in the CXO band. Though we assume that all synchrotron plasmas have cutoffs, few if any have actually been observed in radio, optical, or X-ray bands. This result would impact the acceleration scenario by providing an estimate of the extent in energy of the electron distribution.

---

[10]Refer to **http://www-glast.slac.stanford.edu/** for more information.

On the theoretical front, we need additional ideas of how deviations from a power-law electron spectrum can occur. The two proposals currently available are rather restricted in applicability and should be further developed.

4.5.2. IC emission. If the jet X-ray emission from powerful sources is indeed from the IC/CMB process, we can study different attributes of the underlying relativistic plasma than those involved in synchrotron emission. In particular, we can obtain vital information about the low-energy part of the electron spectrum. Both the amplitude and slope for $\gamma \leq 1000$ are germane to the injection problem for shock acceleration as well as permitting greatly improved estimates of the total particle energy density and hence the energetics of the emitting plasma.

As is well known, estimates of the photon energy density are amenable to direct observational input, and this permits us to pass more confidently from the emission spectrum to the electron spectrum. Once the electron spectrum is known, then the observed synchrotron component will provide the magnetic field strength. The basic physics is understood and IC emission is mandatory in all relativistic plasmas. The only questions are, how much emission is there and what is the frequency range of the emission?

For the beaming IC/CMB model applied to jets, some "paradigm shifts" will be in order. If current estimates of beaming parameters are correct, many of the relatively bright X-ray knots are, in their own frame, rather unimpressive: Luminosities of order 10^{38} to 10^{39} erg s^{-1} would be common and the canonical 10^{44} erg s^{-1} would no longer be relevant.

Another effect means that our view of jets close to the line of sight is actually a stretched-out version of the time history of a very small fraction of the "current jet length" (by which we mean the distance from the outermost knot or hotspot to the core, at the time we observe the jet tip). This can be quickly grasped by reversing time and sending a signal from the earth to the quasar. As the wavefront of our signal passes the jet tip, the jet is moving relativistically toward the quasar. For example, take a 100,000-l.y. jet at 5° to our line of sight. If the jet has a bulk velocity of 0.99c, by the time our wavefront reaches the quasar about 98.6% of the jet (as it existed when our wavefront first reached the tip) has now been swallowed by the black hole, and is thus not observable by us. What we see, which appears to be 100,000 l.y. in length, is actually just the 1,400 l.y. long tip of the "current jet," as it was at progressively earlier times as we move back from the tip. The most important aspect of this effect is to make the necessary adjustments when comparing quasar jets to those lying closer to the plane of the sky. What might we actually be studying if all we see is 1% of the current jet length? The hotspot? If so, what we call knots in the jets would actually be bits of the hotspot brightening and fading over its 100,000-year-long journey to its "present" location.

4.6. Summary

Within a few years, the uncertainty as to the X-ray emission process for quasar jets should be eliminated and then we will either have a method of measuring the low-energy end of the relativistic electron distribution (if the IC/CMB model applies) or we will have new insights into the behavior and loss mechanisms affecting the highest-energy electrons (if synchrotron models apply). If we are convinced that the IC/CMB model is correct, then a number of conclusions are already clear: Detected quasar jets lie close to the line of sight and have large Lorentz factors. That in turn means we can solve for some of the basic jet parameters such as energy flux, and most likely we will improve our understanding of cross-jet velocity structure: Many different lines of argument point to the necessity of some sort of "spine-sheath" structure.

In Section 2 of this review, we examined the differences between the jets of FR I radio galaxies and those of quasars. Will the distinctions in jet length and luminosity translate to differences in X-ray emission process? If so, why are there so many similarities between low-power and high-power sources such as offsets and progressions? We may also expect to better understand the underlying reasons for brightness fluctuations along jets, and if the small knots of FR I jets have the same genesis as the kiloparsec-scale knots in quasar jets. All of these lines of investigation will hopefully elucidate the dichotomy between the plasma that emits the radiation we observe and the medium that transports the energy and momentum over such vast distances.

ACKNOWLEDGMENTS

We thank C. Cheung, S. Jester, M. Hardcastle, and many other colleagues for useful discussions. C. Cheung and L. Stawarz kindly gave us helpful comments on the manuscript and the editor of this series, R. Blandford, provided valuable advice. This work has made use of NASA's Astrophysics Data System Bibliographic Services and the XJET website. Partial support was provided by NASA contract NAS8-03060 and grant GO3-4124A. H.K. thanks the Department of Energy for support in the framework of the Outstanding Junior Investigator program.

LITERATURE CITED

Aharonian FA. 2002. *MNRAS* 332:215–30

Aharonian FA. 2004. *Very High Energy Cosmic Gamma Radiation: A Crucial Window on the Extreme Universe*. River Edge, NJ: World Sci. 495 pp.

Aldcroft T, Siemiginowska A, Elvis M, Mathur S, Nicastro F, Murray S. 2003. *Ap. J.* 597:751

Atoyan A, Dermer CD. 2004. *Ap. J.* 613:151–58

Bahcall JN, Kirhakos S, Schneider DP, Davis RJ, Muxlow TWB, et al. 1995. *Ap. J. Lett.* 452:L91

Bai JM, Lee MG. 2003. *Ap. J. Lett.* 585:L113–16

Bassett LC, Brandt WN, Schneider DP, Vignali C, Chartas G, Garmire GP. 2004. *Astron. J.* 128:523–33

Begelman MC, Blandford RD, Rees MJ. 1984. *Rev. Mod. Phys.* 56:255

Bell AR. 1978. *MNRAS* 182:147

Beresnyak AR, Istomin YN, Pariev VI. 2003. *Astron. Astrophys.* 403:793–804

Bicknell GV. 1995. *Ap. J. Suppl.* 101:29

Bicknell GV, Begelman MC. 1996. *Ap. J.* 467:597
Biretta JA, Sparks WB, Macchetto F. 1999. *Ap. J.* 520:621
Birkinshaw M, Worrall DM, Hardcastle MJ. 2002. *MNRAS* 335:142–50
Blandford R, Payne DG. 1982. *MNRAS* 199:883
Blandford R, Rees M. 1974. *MNRAS* 169:395
Blandford RD. 1976. *MNRAS* 176:465–81
Blandford RD, Levinson A. 1995. *Ap. J.* 441:79
Blandford RD, Ostriker JP. 1978. *Ap. J.* 221:L29
Blandford RD, Znajek RL. 1977. *MNRAS* 179:433–56
Bodo G, Rossi P, Mignone A, Massaglia S, Ferrari A. 2003. *New Astron. Rev.* 47:557–59
Bridle AH, Hough DH, Lonsdale CJ, Burns JO, Laing RA. 1994. *Astron. J.* 108:766–820
Brunetti G, Harris D, Sambruna R, Setti G, eds. 2003. *New Astron. Rev.* 47:(6–7):411–712
Celotti A, Fabian AC. 1993. *MNRAS* 264:228
Celotti A, Ghisellini G, Chiaberge M. 2001. *MNRAS* 321:L1–5
Chartas G, Gupta V, Garmire G, Jones C, Falco EE, et al. 2002. *Ap. J.* 565:96–104
Chartas G, Worrall DM, Birkinshaw M, Cresitello-Dittmar M, Cui W, et al. 2000. *Ap. J.* 542:655–66
Cheung CC. 2004. *Ap. J. Lett.* 600:L23–26
Chiaberge M, Celotti A, Capetti A, Ghisellini G. 2000. *Astron. Astrophys.* 358:104–12
Chiaberge M, Gilli R, Macchetto F, Sparks W, Capetti A. 2003. *Ap. J.* 582:645
Comastri A, Brunetti G, Dallacasa D, Bondi M, Pedani M, Setti G. 2003. *MNRAS* 340:L52
Coppi PS. 1997. In *Relativistic Jets in AGNs*, eds. M Ostrowski, M Sikora, G Madejski, M Begelman, pp. 333–52. Kraków, Poland: Jagellonian Univ.
Curtis HD. 1918. *Pub. Lick. Obs.* 13:31
Dennett-Thorpe J, Bridle A, Scheuer P, Laing R, Leahy J. 1997. *MNRAS* 289:753
Dermer CD, Atoyan A. 2004. *Ap. J.* 611:L9
Dermer CD, Atoyan AM. 2002. *Ap. J. Lett.* 568:L81–84
Dermer CD, Schlickeiser R. 1994. *Ap. J. Suppl.* 90:945–48
Evans DA, Hardcastle MJ, Croston JH, Worrall DM, Birkinshaw M. 2005. *MNRAS* 359:363–82
Fabian A, Celotti A, Johnstone R. 2003. *MNRAS* 338:L7
Fukue J, Tojyo M, Hirai Y. 2001. *Publ. Astron. Soc. Jpn.* 53:555–63
Gabuzda DC, Murray É, Cronin P. 2004. *MNRAS* 351:L89–93
Gaidos JA, Akerlof CW, Biller SD, Boyle PJ, Breslin AC, et al. 1996. *Nature* 383:319
Georganopoulos M, Kazanas D. 2003. *Ap. J. Lett.* 589:L5–8
Georganopoulos M, Kazanas D. 2004. *Ap. J. Lett.* 604:L81–84
Ghisellini G, Tavecchio F, Chiaberge M. 2005. *Astron. Astrophys.* 432:401–10
Hardcastle M, Birkinshaw M, Worrall D. 2001. *MNRAS* 326:1499
Hardcastle M, Croston J. 2005. *MNRAS* 363:649
Hardcastle MJ, Harris DE, Worrall DM, Birkinshaw M. 2004a. *Ap. J.* 612:729–48
Hardcastle MJ, Worrall DM, Birkinshaw M, Laing RA, Bridle AH. 2002. *MNRAS* 334:182–92

Hardcastle MJ, Worrall DM, Birkinshaw M, Laing RA, Bridle AH. 2005. *MNRAS* 358:843–50

Hardcastle MJ, Worrall DM, Kraft RP, Forman WR, Jones C, Murray SS. 2003. *Ap. J.* 593:169–83

Hardcastle MJ, Worrall DM, Kraft RP, Forman WR, Jones C, Murray SS. 2004b. *Nuc. Phys. B Proc. Suppl.* 132:116–21

Hardee PE, Walker RC, Gómez JL. 2005. *Ap. J.* 620:646–64

Harris D, Mossman A, Walker R. 2004. *Ap. J.* 615:161

Harris DE, Cheung CC, Biretta JA, Junor W, Perlman ES, Sparks WB, Wilson AS. 2006. *Ap. J.* 640:211–18

Harris DE, Carilli CL, Perley RA. 1994. *Nature* 367:713

Harris DE, Finoguenov A, Bridle AH, Hardcastle MJ, Laing RA. 2002. *Ap. J.* 580:110–13

Harris DE, Hjorth J, Sadun AC, Silverman JD, Vestergaard M. 1999. *Ap. J.* 518:213–18

Harris DE, Krawczynski H. 2002. *Ap. J.* 565:244–55

Harris DE, Krawczynski H. 2006. See Lee & Ramirez-Ruiz 2006, In press

Harris DE, Krawczynski H, Taylor GB. 2002. *Ap. J.* 578:60–63

Hartman RC, Bertsch DL, Bloom SD. 1999. *Ap. J. Suppl.* 123:79

Hirotani K, Iguchi S, Kimura M, Wajima K. 1999. *Publ. Astron. Soc. Jpn.* 51:263–67

Hughes PA. 1991. *Beams and Jets in Astrophysics*. Cambridge, UK: Cambridge Univ. Press 583, pp.

Jester S, Harris DE, Marshall H, Meisenheimer K, Perley R. 2006. *Ap. J.* In press

Jester S, Röser HJ, Meisenheimer K. 2005. *Astron. Astrophys.* 431:477–502

Jorstad SG, Marscher AP. 2004. *Ap. J.* 614:615–25

Jorstad SG, Marscher AP, Lister ML, Stirling AM, Cawthorne TV, et al. 2005. *Astron. J.* 130:1418–65

Kataoka J, Edwards P, Georganopoulos M, Takahara F, Wagner S. 2003. *Astron. Astrophys.* 399:91

Kataoka J, Stawarz L. 2005. *Ap. J.* 622:797–810

Kataoka J, Stawarz L, Aharonian F, Takahara F, Ostrowski M, Edwards PG. 2006. *Ap. J.* 641:158–68

Kellermann KI, Lister ML, Homan DC, Vermeulen RC, Cohen MH, et al. 2004. *Ap. J.* 609:539–63

Kino M, Takahara F, Kusunose M. 2002. *Ap. J.* 564:97

Kirk JG, Duffy P. 1999. *J. Phys. G* 25:163

Koide S, Shibata K, Kudoh T. 1999. *Ap. J.* 522:727

Komissarov SS. 1994. *MNRAS* 266:649

Konigl A. 1980. *Relativistic Effects in Extragalactic Radio Sources*, PhD thesis, Calif. Inst. Tech., Pasadena, CA

Kraft RP, Forman WR, Jones C, Murray SS, Hardcastle MJ, Worrall DM. 2002. *Ap. J.* 569:54–71

Kraft RP, Hardcastle MJ, Worrall DM, Murray SS. 2005. *Ap. J.* 622:149–59

Krawczynski H. 2004. *New Astron. Rev.* 48:367

Krawczynski H. 2005. *ASP Conf. Ser. 350* In press (astro-ph/050862)

Krawczynski H, Coppi PS, Aharonian FA. 2002. *MNRAS* 336:721
Laing RA, Bridle AH. 2002a. *MNRAS* 336:1161–80
Laing RA, Bridle AH. 2002b. *MNRAS* 336:328–52
Laing RA, Bridle AH. 2004. *MNRAS* 348:1459–72
Laing RA, Canvin JR, Bridle AH. 2003. *New Astron. Rev.* 47:577–79
Lara L, Giovannini G, Cotton WD, Feretti L, Venturi T. 2004. *Astron. Astrophys.* 415:905–13
Lee WH, Ramirez-Ruiz E, eds. 2006. *Triggering Relativistic Jets, Cozumel, 2005, Rev. Mex. Astron. Astrofis., Ser. Conf.* In press
Lister ML. 2003. *Ap. J.* 599:105–15
Lobanov A, Hardee P, Eilek J. 2003. *New Astron. Rev.* 47:629–32
Lovelace RVE. 1976. *Nature* 262:649–52
Ly C, De Young DS, Bechtold J. 2005. *Ap. J.* 618:609–17
Macchetto FD. 1996. In *IAU Symp. 175: Extragalactic Radio Sources*
Marshall HL, Harris DE, Grimes JP, Drake JJ, Fruscione A, et al. 2001. *Ap. J. Lett.* 549:L167–71
Marshall HL, Miller BP, Davis DS, Perlman ES, Wise M, et al. 2002. *Ap. J.* 564:683–87
Marshall HL, Schwartz DA, Lovell JEJ, Murphy DW, Worrall DM, et al. 2005. *Ap. J. Suppl.* 156:13–33
Mattox JR, Wagner SJ, Malkan M. 1997. *Ap. J.* 476:692
McEnery JE, Moskalenko IV, Ormes JF. 2004. In *Cosmic Gamma Ray Sources*, eds. K Cheng, G Romero. Kluwer ASSL Series. (astro-ph/0406250)
Nakamura M, Uchida Y, Hirose S. 2001. *New Astron.* 6:61–78
Nishikawa KI, Hardee P, Richardson G, Preece R, Sol H, Fishman GJ. 2005. *Ap. J.* 622:927
Ostrowski M, Sikora M, Madejski G, Begelman M, eds. 1997. *Relativistic Jets in AGNs*, Konfederacka 6, 30–306, Krakow. Poligrafia Inspecktoratu Towarzystwa Salezjanskiego
Perlman ES, Harris DE, Biretta JA, Sparks WB, Macchetto FD. 2003. *Ap. J. Lett.* 599:L65–68
Perlman ES, Wilson AS. 2005. *Ap. J.* 627:140–55
Pesce JE, Sambruna RM, Tavecchio F, Maraschi L, Cheung CC, et al. 2001. *Ap. J. Lett.* 556:L79–82
Pushkarev AB, Gabuzda DC, Vetukhnovskaya YN, Yakimov VE. 2005. *MNRAS* 356:859–71
Rees MJ. 1971. *Nature* 229:312
Riley JM, Warner PJ. 1990. *MNRAS* 246:1P
Rossi P, Bodo G, Massaglia S, Ferrari A, Mignone A. 2004. *Astrophys. Space Sci.* 293:149–55
Salpeter EE. 1964. *Ap. J.* 140:796–800
Sambruna R, Urry C, Tavecchio F, Maraschi L, Scarpa R, et al. 2001. *Ap. J.* 549:L161
Sambruna RM, Gambill JK, Maraschi L, Tavecchio F, Cerutti R, et al. 2004. *Ap. J.* 608:698–720
Sambruna RM, Maraschi L, Tavecchio F, Urry CM, Cheung CC, et al. 2002. *Ap. J.* 571:206–17

Sauty C, Tsinganos K, Trussoni E. 2002. *LNP Vol. 589: Relativistic Flows in Astrophysics* 589:41
Scheuer P. 1974. *MNRAS* 166:513
Schwartz DA. 2002. *Ap. J.* 569:L23
Schwartz DA, Marshall HL, Lovell JEJ, Murphy DW, Bicknell GV, et al. 2006. *Ap. J.* 640:592–602
Schwartz DA, Marshall HL, Lovell JEJ, Piner BG, Tingay SJ, et al. 2000. *Ap. J. Lett.* 540:L69
Siemiginowska A, Bechtold J, Aldcroft TL, Elvis M, Harris DE, Dobrzycki A. 2002. *Ap. J.* 570:543–56
Siemiginowska A, Stanghellini C, Brunetti G, Fiore F, Aldcroft T, et al. 2003. *Ap. J.* 595:643
Sikora M, Begelman MC, Madejski GM, Lasota JP. 2005. *Ap. J.* 625:72–77
Sikora M, Madejski G. 2001. *AIP* 558:275
Sikora M, Madejski G, Moderski R, Poutanen J. 1997. *Ap. J.* 484:108
Spada M, Ghisellini G, Lazzati D, Celotti A. 2001. *MNRAS* 325:1559–70
Stawarz L. 2004. *Ap. J.* 613:119–28
Stawarz L, Ostrowski M. 2002. *Ap. J.* 578:763–74
Stawarz L, Sikora M, Ostrowski M, Begelman MC. 2004. *Ap. J.* 608:95–107
Swain MR, Bridle AH, Baum SA. 1998. *Ap. J. Lett.* 507:L29–33
Tanihata C, Takahashi T, Kataoka J, Madejski GM. 2003. *Ap. J.* 584:153
Tavecchio F. 2005. In *Proc. Tenth Marcel Grossmann Meet. Gen. Relativity, Rio de Janeiro, Brazil, July 2003*, ed. M Novello, S Perez-Bergliaffa, R. Ruffini. Singapore: World Sci.
Tavecchio F. 2004. *Mem. Soc. Astron. Ital. Suppl.* 5:211
Tavecchio F, Maraschi L, Sambruna RM, et al. 2004. *Ap. J.* 614:64
Tavecchio F, Maraschi L, Sambruna RM, Urry CM. 2000. *Ap. J. Lett.* 544:L23–26
Tingay SJ, Jauncey DL, Reynolds JE, Tzioumis AK, King EA, et al. 2000. *Adv. Space Res.* 26:677–80
Tsinganos K, Bogovalov S. 2002. *MNRAS* 337:553–58
Urry CM, Padovani P. 1995. *Publ. Astron. Soc. Pac.* 107:803
Uttley P, McHardy IM, Vaughan S. 2005. *MNRAS* 359:345
Walker R, Benson J, Unwin S. 1987. *Ap. J.* 316:546
Wang JC. 2002. *Chin. J. Astron. Astrophys.* 2:1–7
Wardle JFC, Aaron SE. 1997. *MNRAS* 286:425
Wardle JFC, Homan DC, Ojha R, Roberts DH. 1998. *Nature* 395:457–61
Weekes T. 2003. *Very High Energy Gamma-Ray Astronomy*. Bristol, UK: Inst. Physics Pub.
Weisskopf MC, Aldcroft TL, Bautz M, Cameron RA, Dewey D, et al. 2003. *Exper. Astron.* 16:1–68
Wilson A, Young A, Shopbell P. 2001. *Ap. J.* 547:740
Wilson AS, Yang Y. 2002. *Ap. J.* 568:133
Worrall D, Birkinshaw M. 2005. *MNRAS* 360:926
Worrall D, Birkinshaw M, Hardcastle M. 2001. *MNRAS* 326:L7
Worrall DM, Birkinshaw M, Hardcastle MJ. 2003. *MNRAS* 343:L73–78
Zensus JA. 1997. *Annu. Rev. Astron. Astrophys.* 35:607–36
Zezas A, Birkinshaw M, Worrall DM, Peters A, Fabbiano G. 2005. *Ap. J.* 627:711–20

The Supernova–Gamma-Ray Burst Connection

S.E. Woosley[1] and J.S. Bloom[2]

[1]Department of Astronomy and Astrophysics, University of California, Santa Cruz, California 95064; email: woosley@ucolick.org

[2]Department of Astronomy, University of California, Berkeley, California 94720; email: jbloom@astro.berkeley.edu

Key Words

gamma-ray astronomy, gamma-ray bursts, stellar evolution, supernovae

Abstract

Observations show that at least some gamma-ray bursts (GRBs) happen simultaneously with core-collapse supernovae (SNe), thus linking by a common thread nature's two grandest explosions. We review here the growing evidence for and theoretical implications of this association, and conclude that most long-duration soft-spectrum GRBs are accompanied by massive stellar explosions (GRB-SNe). The kinetic energy and luminosity of well-studied GRB-SNe appear to be greater than those of ordinary SNe, but evidence exists, even in a limited sample, for considerable diversity. The existing sample also suggests that most of the energy in the explosion is contained in nonrelativistic ejecta (producing the supernova) rather than in the relativistic jets responsible for making the burst and its afterglow. Neither all SNe, nor even all SNe of Type Ibc produce GRBs. The degree of differential rotation in the collapsing iron core of massive stars when they die may be what makes the difference.

1. INTRODUCTION

Gamma-ray bursts (GRBs), discovered by Klebesadel, Strong & Olson (1973), are brief (~seconds), intense flashes of electromagnetic radiation with typical photon energies ~100 keV that arrive at Earth from unpredictable locations several times daily (e.g., Fishman & Meegan 1995). They are isotropically distributed on the sky and, so far as we know, not one has ever repeated.[1] The production of GRBs is believed to require some small amount of matter accelerated to ultrarelativistic speeds (e.g., Mészáros 2002) and beamed to a small fraction of the sky (Frail et al. 2001). In many of the longer lasting events the total energy in γ rays, corrected for this beaming, is $\sim 10^{51}$ erg.

Core-collapse supernovae (SNe), on the other hand, are the explosive deaths of massive stars that occur when their iron cores collapse to neutron stars or black holes (e.g., Woosley & Janka 2005). They are, in general, not accompanied by highly relativistic mass ejection, but are visible from all angles and last from weeks to months. They may be either of Type Ib or Ic, if their hydrogen envelopes are lost, or Type II if they are not (Filippenko 1997). The total kinetic energy here is also $\sim 10^{51}$ erg, roughly the same as the energy of the jet that makes a GRB.

SNe are the most powerful explosions in the modern universe, rivaling in pure wattage the rest of the observable universe combined, but most of the emission, $\sim 10^{53}$ erg s^{-1}, is in neutrinos, which are, unless one happens to be very nearby, unobservable. GRBs are the brightest explosions in the universe, in terms of electromagnetic radiation per unit solid angle, sometimes as bright as if the rest mass of the sun, 2×10^{54} erg, had been turned to γ rays in only 10 seconds. The light per solid angle from SNe is about 10 orders of magnitude fainter.

Despite the similarity in kinetic energy scale, it was thought for decades that these two phenomena had no relation to one another (though see Colgate 1968; Paczyński 1986). This was largely because, until the late 1990s, no one knew just how far away—and hence how luminous—the GRBs really were. Indeed, the consensus in the mid-1990s was that even if GRBs were ever found to be of cosmological origin, they were likely the result of a merger in a degenerate binary system, such as double neutron stars (Blinnikov et al. 1984; Paczyński 1986; Eichler et al. 1989; Narayan, Paczyński & Piran 1992) or a neutron star and black hole (Narayan, Piran & Shemi 1991; Paczyński 1991) and would not be found in conjunction with SNe.

Beginning in 1997 with the localization of long-wavelength counterparts (Costa et al. 1997; van Paradijs et al. 1997) and the confirmation of the cosmological distance scale (Metzger et al. 1997),[2] it became increasingly clear that those well-studied GRBs were associated with young stars in distant actively star-forming galaxies, and not with

---

[1]Not included in our review here are the "soft-gamma repeaters" (Woods & Thompson 2005), related phenomenologically to classic GRBs but believed to be associated with highly magnetic neutron stars in the local group of galaxies (though see Tanvir et al. 2005). There is no known direct temporal connection of "soft-gamma repeaters" (SGRs) to supernovae.

[2]By 1997, most of the community had already come to accept a cosmological distance scale for GRBs because of the isotropic distribution of locations found by the Burst and Transient Source Experiment (Fishman & Meegan 1995; Paczyński 1995).

old stars in mature galaxies, as expected of the merger hypothesis. Within a few years, evidence mounted against the merger hypothesis and implicated GRBs, to the surprise of many, as due to the death of massive stars (Section 2.1.2).

This association with star formation did not require a causal connection though. Perhaps the young stars resulted in SNe that produced neutron stars or black holes, which in turn made the bursts some time later. Nor was it clear that any supernova accompanying a GRB would be especially bright.[3] The watershed event that brought the SN-GRB connection to the forefront was the discovery of a GRB on April 25, 1998 (GRB 980425), in conjunction with one of the most unusual SNe ever seen, SN 1998bw (Galama et al. 1998b). The SN and the GRB were coincident both in time and place. Theory had anticipated that event (Woosley 1993), but not its brilliance. Though the physical connection was initially doubted by some, and GRB 980425 was very subenergetic compared with most GRBs, the in-depth study of the "SN-GRB connection" began in earnest on that day (Section 2.1.4).

Aspects of the SN-GRB connection have been reviewed previously (Wang & Wheeler 1998; van Paradijs, Kouveliotou & Wijers 2000; van Paradijs 2001; Wheeler 2001; Woosley, MacFadyen & Heger 2001; Mészáros 2002; Weiler et al. 2002; Nomoto et al. 2004; Bloom 2005; Della Valle 2005; Höflich et al. 2005; Matheson 2005; Piran 2005), but usually as cursory overviews of the connection, or parts of larger reviews of GRBs in general. Here we review specifically the observations, history, and theory relating to the SN-GRB connection. Since, until recently, the only accurately known counterparts were for GRBs of the so-called long-soft variety,[4] our discussion centers on these events (though see Sections 2.2 and 2.3).

What we know now is that at least one other supernova besides SN 1998bw, namely SN 2003dh (Section 2.1.6), has happened nearly simultaneously with a GRB. This time the GRB (030329) was of a more normal energy. A compelling spectroscopic case can also be made for SN associations with GRB 031203 (SN 2003lw) and possibly GRB 021211 (SN 2002lt). There have also been "bumps" in the optical afterglows of many GRBs (Section 2.1.5) consistent in color, timing, and brightness with what is expected from Type I SNe of luminosity comparable to SN 1998bw. Indeed, given the difficulties in making the key observations, the data and models we review (Sections 2.1, 2.4, and 2.5) are consistent with, though not conclusive proof of, the hypothesis that *ALL* long-soft GRBs are accompanied by SNe of Type Ic.[5] Still, there is evidence for considerable diversity in the brightness, rise times, and evolution of these events. The well-studied SNe that accompany GRBs (GRB-SNe) also show

---

[3] Here we distinguish "supernova," the subrelativistic explosion of a stellar mass object, from the "optical afterglow" produced when the relativistic material responsible for a GRB impacts the surrounding medium (Mészáros & Rees 1997a; Piran 2005).

[4] Though the light curves and spectra of cosmic GRBs are very diverse, they can be broadly categorized on the basis of duration and power spectrum into two groups (Kouveliotou et al. 1993; Fishman & Meegan 1995)—"short-hard" bursts with a median duration of 0.3 s, and "long-soft" bursts with a median duration of 20 s. The average peak energy of the shorter class is about 50% greater than the long class—360 keV vs 220 keV (Hakkila et al. 2000).

[5] A Type Ic SN has no hydrogen in its spectrum and also lacks strong lines of He I or Si II that would make it Type Ib or Ia. See Filippenko (1997) for review.

evidence for broad lines, indicative of high-velocity ejecta. This suggests a subclassification of Type Ic SNe, called "Type Ic-BL" (whether they are associated with GRBs or not). Type Ic-BL is a purely observational designation, and makes no reference to a specific progenitor model (e.g., "collapsars") nor to the model-specific energetics or brightness of the explosions. In the latter respect, the GRB modeling community has come to view the label of "hypernova" for typing GRB-SNe as somewhat narrow and subjective.

Though all GRBs of the long-soft variety may be accompanied by SNe, not all SNe, or even all SNe of Type Ic-BL, make GRBs. But why should some stars follow one path to death and others another? As described in Section 3.3, rotation is emerging as the distinguishing ingredient. Though it is a conjecture still to be proven, GRBs may only come from the most rapidly rotating and most massive stars, possibly favored in regions of low metallicity. Ordinary SNe, on the other hand, which comprise about 99% of massive star deaths, may come from stars where rotation plays a smaller role or no role at all. Indeed, the SN-GRB connection is forcing a re-evaluation of the role of rotation in the deaths of all sorts of massive stars.

The continued observation of GRBs and the SNe that accompany them should yield additional diagnostics that will help the community gain deeper insight into both phenomena. Some of these diagnostics, especially those that might shed light on the prime mover, or "central engine," that drives all these explosions, are discussed in Section 3.4, and we end with a discussion of future directions in the field in Section 4.

2. OBSERVATIONS

2.1. Observational Evidence for a SN-GRB Association

2.1.1. Early indications. Colgate (1968), in the only paper to predict the existence of GRBs before their discovery, associated them with the breakout of relativistic shocks from the surfaces of SNe. This motivated the discoverers (Klebesadel, Strong & Olson 1973) to search for SN-GRB coincidences, but none were found. We know now that the transients from breakout itself are too faint to be GRBs at cosmological distances.[6] Even with thousands more GRBs and hundreds of SNe localized by 1997, no clear observational connection could be established (e.g., Hartmann & Woosley 1988).

Bohdan Paczyński for years (e.g., Paczyński 1986; though see also Usov & Chibisov 1975) suggested a cosmological origin for GRBs, and pointed out that the requisite energy (in γ rays) would be comparable to the (kinetic) energy of a supernova. When the first redshifts of GRBs were determined (Metzger et al. 1997; Kulkarni et al. 1998c), the implied energies were up to $\sim 10^3$ times larger than 10^{51} erg. In fact, the largest inferred γ-ray energy exceeded the rest mass energy of a neutron star (Kulkarni et al. 1999). Later, however, with the inclusion of a geometric correction for beaming

---

[6]Aspects of the Colgate model remain viable. A relativistic blast wave interacting with a circumstellar medium might be a reasonable model for some events (Woosley, Eastman & Schmidt 1999; Matzner & McKee 1999; Tan, Matzner & McKee 2001), especially if the ejecta are beamed.

(Rhoads 1997; Halpern et al. 2000), the total energy release in γ rays came down to around 10^{51} erg (Kumar & Piran 2000; Frail et al. 2001; Freedman & Waxman 2001; Bloom, Frail & Kulkarni 2003; Friedman & Bloom 2005), with a small, but significant number of bursts at lower energies (e.g., Soderberg et al. 2004b). That the actual energy release in long-duration GRBs and SNe is comparable is consistent with an association, but does not require a common origin. Similar energetics are expected from a variety of viable cosmological progenitors. Yet, even before direct confirmation, several independent lines of evidence pointed tantalizingly to a direct SN-GRB connection. On length scales spanning parsecs ("circumburst"), to galactic, to cosmic distances, GRBs revealed their origin.

2.1.2. Location in and around distant galaxies. The various scenarios for making GRBs (Section 3.2) have implications for their observed locations. Because neutron stars experience "kicks" at birth, the long delay before coalescence would lead to bursts farther from star-forming regions (Livio et al. 1998; Paczyński 1998; Bloom, Sigurdsson & Pols 1999; Fryer, Woosley & Hartmann 1999; Belczyński, Bulik & Zbijewski 2000) than very massive stars. Therefore, subarcsecond localizations of afterglows with respect to distant galaxies provided, early on, an indirect means for testing hypotheses about the progenitors. GRB 970228, for example, was localized on the outskirts of a faint galaxy (van Paradijs et al. 1997), essentially ruling out (Sahu et al. 1997) disruptive events around a central massive black hole (Carter 1992). Unfortunately, the imaging capabilities of current instruments cannot resolve the immediate environment ($\lesssim 100$ pc) of GRBs that originate beyond ~ 100 Mpc. Hence the location of individual bursts (for instance, in an H II region) cannot be used as a definitive test of their nature. However, statistical studies reveal a strong correlation of the locations with the blue light of galaxies (Bloom, Kulkarni & Djorgovski 2002; Fruchter et al. 2006). It seems that long-soft GRBs happen preferentially in the regions where the most massive stars die.

No host galaxy stands out as exceptional, yet, in the aggregate, they are faint and blue (Mao & Mo 1998; Le Floc'h et al. 2003), systematically smaller, dimmer, and more irregular than M_* galaxies at comparable redshifts (C. Wainwright, E. Berger & B.E. Penprase, unpublished data; see also Mao & Mo 1998; Hogg & Fruchter 1999; Djorgovski et al. 2003; Conselice et al. 2005). Of the more than 60 GRB hosts known, only one (GRB 990705) appears to be associated with a normal spiral (Masetti et al. 2000), and that could be a coincidence (given the large solid angle of a big face-on spiral, the GRB has a greater probability of having occurred at higher redshift than its putative host). No long-duration GRB has ever been definitively associated with an early-type galaxy.[7]

There is convincing spectroscopic evidence that typical GRB host galaxies are forming stars actively, perhaps at a higher rate per unit mass than field galaxies

---

[7] One group's photometry of the host of GRB 970508 (Chary, Becklin & Armus 2002) indicated a significant old population, but this was not confirmed by other groups. Furthermore, morphological analysis showed that the galaxy surface brightness fitted an exponential profile better than an $r^{-1/4}$ profile (Fruchter et al. 2000b).

(Djorgovski et al. 2001; Christensen, Hjorth & Gorosabel 2004). Submillimeter observations of some GRB hosts once appeared to indicate the presence of large amounts (∼hundreds $M_\odot$ yr^{-1}) of obscured star formation (Berger et al. 2003a), but recent mid-infrared observations of the same galaxies suggest significantly smaller star-formation rates (Le Floc'h et al. 2006). There are indications (e.g., Bloom, Djorgovski & Kulkarni 2001), in particular from the line ratios of [Ne III] to [O II], that the H II regions are especially hot, indicating a propensity of GRB hosts to make more massive stars. More recently, a growing body of absorption-line spectroscopic evidence suggests that GRB hosts—or more precisely, the regions through which GRB afterglows are viewed—are low in metallicity (Vreeswijk et al. 2001; Savaglio, Fall & Fiore 2003; Prochaska et al. 2004) though it remains to be seen whether metallicities are significantly different from field galaxies and damped Lyman α systems at comparable redshifts. As a class, the hosts of long-duration GRBs thus appear to favor progenitors that are closely connected with metal-poor massive stars (see Section 3.3.3).

The locations of long duration GRBs on the largest scale—in redshift space—confirm the expectations from smaller spatial scales. The GRB rate appears to track the global star-formation rate (Loredo & Wasserman 1998; Bloom 2003; Firmani et al. 2004; Price & Schmidt 2004; Natarajan et al. 2005; Jakobsson et al. 2006). This too implicates a progenitor that makes a GRB without appreciable ($\gtrsim$Gyr) delay following a starburst (Totani 1997; Wijers et al. 1998). Though bursts from massive stars could, in principle, be observable to redshifts of ∼30 (Mészáros & Rees 2003), bursts from merger remnants with long delay times since starburst (such as, perhaps, short-hard GRBs) should not be observed beyond $z \approx 6$: there would simply not be enough time since the formation of the first stars.

2.1.3. Absorption spectroscopy. Absorption line spectroscopy of GRB afterglows can also constrain the environments around GRB progenitors (Perna & Loeb 1998), on both galactic and stellar scales. The columns of neutral hydrogen and metals seen in absorption in the highest redshift GRBs are generally significantly larger than those seen through quasar sight lines (e.g., Savaglio, Fall & Fiore 2003). This is not necessarily an effect local to the GRB, but possibly a consequence of the location of GRBs in the inner regions of their host.

Spectroscopy of the afterglow of GRB 021004 revealed significantly blueshifted ($\gtrsim$100 km s^{-1}) absorption features relative to the highest z system (Chornock & Filippenko 2002; Møller et al. 2002; Vreeswijk et al. 2004). Though some have claimed that the high velocities could be due to radiative acceleration of the circumburst material (e.g., Schaefer et al. 2003), a more natural explanation appears to be that the absorption is occurring in the fast moving Wolf-Rayet (WR) winds from the progenitor. Indeed, detailed modeling of WR winds and interactions with the interstellar medium appears to accommodate significant column densities at a variety of blueshifts extending out to ∼2000 km/s (van Marle, Langer & Garcia-Segura 2005; although see Mirabal et al. 2002). High-resolution spectra of more recent bursts also appear to show WR features (e.g., Prochaska, Chen & Bloom 2006). The presence of fine structure lines of, for example, Fe$^+$ indicates a warm, dense medium that has only been observed in Galactic WR winds and never been seen in any extragalactic sightlines. Interestingly, because of the redshifting of UV lines that would be otherwise inaccessible to groundbased spectroscopy, high resolution spectra of GRBs are now offering unique detailed diagnostics of WR winds. Circumburst diagnostics are discussed further in Section 3.4.4.

2.1.4. GRB 980425 and SN 1998bw. GRB 980425 triggered detectors on board both BeppoSAX and BATSE (Kippen 1998). At high energies, it was seemingly unremarkable (Galama et al. 1998b; Kippen 1998) with a typical soft spectrum ($E_{peak} \approx 150$ keV) and moderate duration ($\Delta T \approx 23$ sec). Within the initial 8-arcmin-radius error circle was an underluminous (0.02 L_*) late-type galaxy (ESO184–G82; $z = 0.0085$, Tinney et al. 1998), found 2.5 days after the GRB to host a young supernova, designated as SN 1998bw (Galama et al. 1998a; Galama et al. 1998b; Lidman et al. 1998; Sadler et al. 1998). On temporal and spatial grounds, the physical association of the GRB with the young SN was initially controversial (Galama et al. 1998b; Pian et al. 1998), but after a careful reanalysis of the X-ray data (Pian et al. 2000) the association of the GRB with the SN was confirmed: Consistent with the location of the SN was a slowly variable X-ray source. This transient X-ray source provided an improved spatial and temporal connection between the GRB and the SN. It is now widely accepted that GRB 980425 was coincident SN 1998bw (Kouveliotou et al. 2004).

The evolution of SN 1998bw was unusual at all wavelengths. The discovery of prompt radio emission just a few days after the GRB (Kulkarni et al. 1998b) (**Figure 1**) was novel. Almost irrespective of modeling assumptions, the rapid rise of radio emission from SN 1998bw showed that the time of the SN explosion was the same as the GRB to about one day. The brightness temperature several days after the GRB

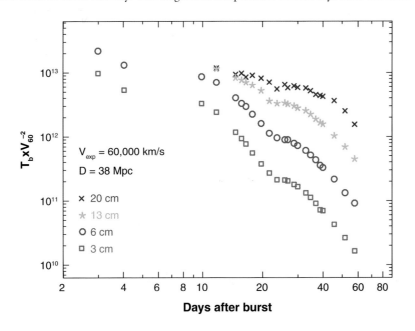

Figure 1

The evolution of the brightness temperature of radio SN 1998bw, the most luminous radio SN ever recorded. The brightness temperature is computed under the assumption that the radio photosphere expanded with the same velocity inferred from optical spectroscopy ($\sim 0.2c$). In order for the true brightness temperature to be less than the "Compton Catastrophe" value ($T_{CC} \approx 10^{12}$ K), relativistic motion in the first week after GRB 980425 is required. From Kulkarni et al. (1998b).

suggested that the radio photosphere moved relativistically with $\Gamma \gtrsim 3$. Still, the total energy in relativistic ejecta was small, $\lesssim 3 \times 10^{50}$ erg (Li & Chevalier 1999), about two orders of magnitude less than in the SN explosion itself.

The early optical spectrum of the SN stymied astronomers. Lidman et al. (1998) wrote

> The relative intensity of the different regions of the spectrum is changing from day to day. The absence of H lines suggests that the object is not a Type-II supernova; the lack of Si at 615 nm indicates that it is not a regular Type-Ia supernova. The nature of this puzzling object still evades identification....

The initial IAUC-designated Type was that of a Ib (Sadler et al. 1998), but a reclassification to a "peculiar Type Ic" was suggested (Filippenko 1998; Patat & Piemonte 1998) when no He nor Si II $\lambda 6355$ was found. SN 1998bw peaked in the V-band 16.2 days (rest frame) after the GRB with $M_V = -19.16 \pm 0.05 + 5 \log h_{71}$ mag (Galama et al. 1998a). Given the unknown peculiar velocity of the host and the uncertainty in the extinction along the line of sight, this absolute peak brightness is uncertain at the 10% level.

If GRB 980425 arose from the $z = 0.0085$ galaxy, then it must have been a very underluminous burst. Assuming isotropic emission, the energy in γ rays was $E_\gamma = 8.5 \pm 0.1 \times 10^{47}$ erg, more than three orders of magnitude fainter than the majority of long-duration GRBs (Frail et al. 2001; Bloom, Frail & Kulkarni 2003). Any collimation would imply an even smaller energy release in γ rays. Still, many held, on purely phenomenological grounds, that GRB 980425 and SN 1998bw had to be physically associated. The SN was simply was too unusual not be connected with the GRB.

Accepting the connection and given the very low redshift, the burst site of at least one (albeit faint) GRB was studied in unprecedented detail. The SN position was found with the *Hubble Space Telescope* (HST) to have been in an apparent star-forming region within 100 pc of several young stars (Fynbo et al. 2000), as expected of core-collapse SNe. Likewise, late-time *Chandra* X-ray imaging revealed an X-ray point source (**Figure 2**) consistent with the optical and radio SN position (Kouveliotou et al. 2004) (and, curiously, a variable ultraluminous X-ray source in the same spiral arm). Even at such low redshifts, the late-time HST imaging was incapable of resolving any star cluster or companion. This fact serves as a bleak reminder that GRBs that occur at even higher redshifts may not have their progenitors nor immediate progenitor environments directly observed.

The unusual properties of the event led some to suggest that GRB 980425 represented not only a phenomenological subclass of GRBs, but one physically distinct from other long-duration GRBs (Bloom et al. 1998; Matzner & McKee 1999; Norris, Bonnell & Watanabe 1999; Woosley, Eastman & Schmidt 1999; Tan, Matzner & McKee 2001). The consensus is now that both GRB 980425 and SN 1998bw, in particular the energetics of such, represent the extreme in a continuum of events all with the same underlying physical model. Indeed GRB 031203 and its associated SN

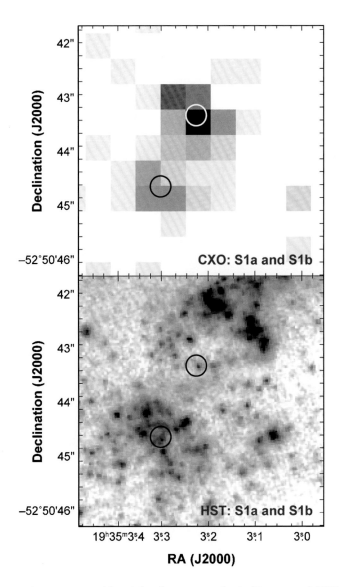

Figure 2
The location of SN 1998bw as viewed at late times by *Chandra* (*top*) and HST (*bottom*). The circles represent the 1σ astrometric position from *Chandra*. The fading X-ray source ("S1a") consistent with SN 1998bw is to the southeast (*bottom left*). From Kouveliotou et al. (2004).

(Section 2.1.6) are now considered the closest cosmological instance of GRB 980425 and SN 1998bw (Soderberg et al. 2004b).

2.1.5. Late-time bumps. Viewing GRB 980425, and its origin, as distinct from the "cosmological" set of GRBs was the norm in 1998. Though the detection of a contemporaneous SN would be a natural consequence of a massive star origin (e.g., Woosley 1993; Hansen 1999), no other GRB had obvious late-time emission that resembled a SN. A report of a red emission "bump" following GRB 980326 (Bloom & Kulkarni 1998) was interpreted as being caused by a coincident SN at about a

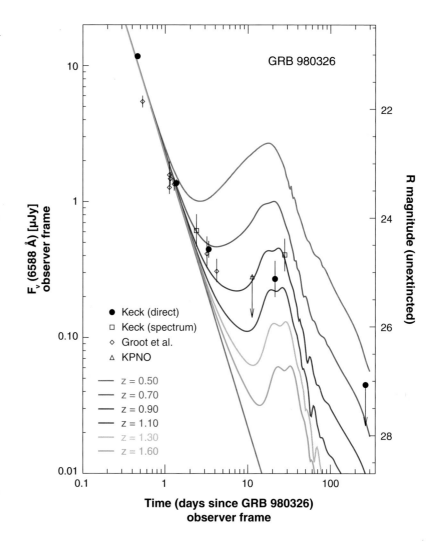

Figure 3

An optical supernova-like bump superimposed on the afterglow of GRB 980326. Models of SN 1998bw at different redshifts are shown. The color and light curve of the bump was found to be consistent with 1998bw at redshift of unity. Without a measured redshift, however, there is a degeneracy between the absolute brightness, distance, and extinction due to dust in the host galaxy. From Bloom et al. (1999).

redshift of unity (Castro-Tirado & Gorosabel 1999; Bloom et al. 1999) (**Figure 3**). Without a spectroscopic redshift for GRB 980326 and multiband photometry around the peak of the bump, the absolute peak brightness and type of the purported SN could not be known. The available data were also consistent with a dust echo (Esin & Blandford 2000; Reichart 2001) or dust re-radiation (Waxman & Draine 2000) from material surrounding the GRB. A subsequent reanalysis of the afterglow of GRB 970228 revealed evidence for a bump that appeared to be the same absolute magnitude as SN 1998bw with similar rise times (Reichart 1999; Galama et al. 2000; Reichart, Lamb & Castander 2000). Similar reports of bumps were made (Fruchter et al. 2000a; Sahu et al. 2000; Berger et al. 2001; Björnsson et al. 2001; Castro-Tirado et al. 2001; Lazzati et al. 2001; Sokolov 2001; Covino et al. 2003; Gorosabel et al.

2005a; Masetti et al. 2005), but none as significant and with as clear-cut connection to SNe as GRB 980326 and GRB 970228 (see Bloom 2005).

Concerted multiepoch ground-based and space-based observing campaigns following several GRBs strengthened the notion that late-time bumps were indeed SNe (Bloom et al. 2002; Garnavich et al. 2003b; Price et al. 2003; Stanek et al. 2005). The SN of GRB 011121 showed a spectral rollover during peak at around 4000 Å, nominally expected of core-collapse SNe in the photospheric phase. The typing of the SN associated with GRB 011211 was controversial, with Garnavich et al. (2003b) showing evidence that the brightness and color evolution resembled 1998S (a Type IIn; see also Meurs & Rebelo 2004) and Bloom et al. (2002) showing consistency with a Ic-like curve interpolated between the faint and fast 1994I and the bright and slow 1998bw. We discuss the overall census of GRB-SNe photometry in Section 2.5.

2.1.6. Spectroscopic evidence and a clear case. Though several bumps were found with characteristics remarkably similar to Type I SNe, the first truly solid evidence for a connection between ordinary GRBs and SNe came with the detection of the low-redshift ($z = 0.1685$; Greiner et al. 2003c) GRB 030329 and its associated supernova, SN 2003dh (for recent reviews, see Matheson 2004 and Della Valle 2005). Shortly after its discovery (the brightest burst HETE-2 ever saw), the afterglow of the GRB (Peterson & Price 2003; Torii 2003) was very bright ($R \sim 13$ mag). It faded slowly, undergoing several major rebrightening events in the first few days (Burenin et al. 2003; Greiner et al. 2003a; Matheson et al. 2003c; Bloom et al. 2004; Lipkin et al. 2004). Given the low redshift, several spectroscopic campaigns were initiated. Spectra of the afterglow (Chornock et al. 2003; Garnavich et al. 2003a; Hjorth et al. 2003; Kawabata et al. 2003; Matheson et al. 2003a,b; Stanek et al. 2003), 6.6 and 7.7 days after the GRB, showed a deviation from a pure power law and the emergence of broad SN spectral features (**Figure 4**).

As the afterglow faded, the SN became more prominent and showed remarkable similarity to SN 1998bw (**Figure 5**). Spectrapolarimetric observations at later times showed that the SN light was somewhat polarized ($P < 1\%$) indicating mild asymmetry in the subrelativistic ejecta. Given the broad spectral features, indicating high velocities ($\gtrsim 25000$ km s^{-1}; Stanek et al. 2003; Hjorth et al. 2003; Mazzali et al. 2003) and apparent absence of hydrogen, helium, and strong Si II $\lambda 6355$ absorption, a classification as Type Ic-BL was natural (Matheson et al. 2003c). We leave the discussion of the modeling of SN 2003dh to Section 3, but we emphasize here that even what should be the easiest-to-measure "observable" of a SN-GRB can be highly model dependent: the reported peak magnitude of SN 2003dh differed by more than 1 mag (from 0.6 mag fainter to 0.5 mag brighter than SN 1998bw) (Hjorth et al. 2003; Mazzali et al. 2003; Matheson et al. 2003c; Bloom et al. 2004; Lipkin et al. 2004). In part, the differences can be due to the quality of the observations near peak, but much of the difference can be ascribed to different assumptions regarding the extinction toward SN 1998bw, the modeled brightness of the afterglow at the time of peak, and the modeled k-corrections of the SN (1998bw) template.

There have been a few other reports of spectroscopic identifications of a SN associated with a cosmological GRB. A SN-like brightening was first reported by

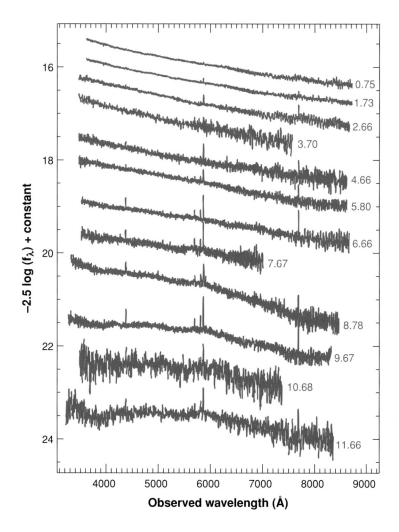

Figure 4
The discovery spectra of the emergence of SN 2003dh from the glare of the afterglow of GRB 030329. Shown is the observed spectra, a combination of afterglow and supernova. Days since the GRB are noted at right. The narrow emission lines are from the host galaxy and do not change in intensity throughout. From Matheson (2004).

Bailyn et al. (2003) at the position of the low redshift ($z = 0.1055$; Prochaska et al. 2004) GRB 031203. The presence of SN 2003lw was confirmed photometrically in multiple optical and infrared bands (Bersier et al. 2004; Cobb et al. 2004; Malesani et al. 2004; Gal-Yam et al. 2004). Spectra 17 and 27 days after the GRB exhibited broad spectral features reminiscent of SN 1998bw at similar epochs (Malesani et al. 2004), but the light-curve behavior was more of a broad plateau around peak than 1998bw (Cobb et al. 2004; Gal-Yam et al. 2004; Thomsen et al. 2004). Della Valle et al. (2003) reported both a bump and a low resolution spectrum at the position of GRB 021211 ($z = 1.006$). A recent claim of a SN associated with Swift burst GRB 050525a has been made based upon a photometric bump and a low signal-to-noise spectrum near peak (Della Valle et al. 2006).

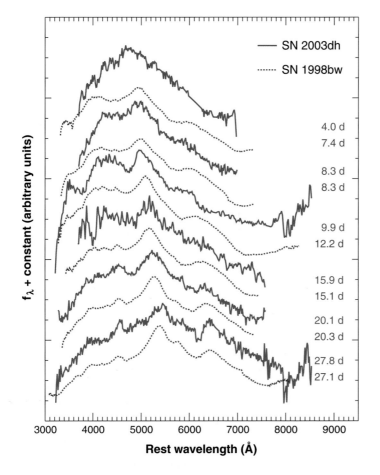

Figure 5
A comparison of the rest frame optical spectrum of SN 1998bw and SN 2003dh (associated with GRB 030329). The spectral features and evolution show a remarkable similarity suggesting that the SN explosions of cosmological GRBs are connected to the mechanisms inferred for 1998bw. The afterglow component has been modeled and removed from the observed spectra. From Hjorth et al. (2003).

2.2. Short-Hard Gamma-Ray Bursts

"Short-hard bursts" constitute ~30% of the BATSE sample (Kouveliotou et al. 1993) and, if, as now appears likely, they typically are sampled from a smaller redshift than long-soft bursts, they could be the most frequent form of GRB in the universe. Some models (Waxman & Mészáros 2003; Zhang, Woosley & MacFadyen 2003) predict an association of short-hard bursts with massive star death, and hence with SNe.

Recently, however, the counterparts of several short-hard bursts have been discovered (Fox et al. 2005; Gehrels 2005; Hjorth 2005a; Villasenor 2005; Bloom et al. 2006; Soderberg et al. 2006a). These GRBs have been found at lower redshifts than typical long bursts, but it is not yet conclusively established that the true bursting rate is significantly skewed toward lower redshift (Fox et al. 2005; Gal-Yam et al. 2005; Bloom & Prochaska 2006; Prochaska et al. 2006). Short bursts tend to have prompt burst energy releases much smaller than that of long bursts, and this may limit their detectability beyond redshifts of unity. Short bursts have not been found in regions of obvious active star formation (though two are on the outskirts of a starburst galaxy)

(Bloom & Prochaska 2006; Prochaska et al. 2006; Soderberg et al. 2006a). In fact, the hosts of three of the five well-localized short GRBs are elliptical galaxies (see also Berger et al. 2005 and Bloom et al. 2006). This strongly implicates old stars or stellar remnants as the progenitors of short-hard GRBs. This claim is buttressed by the fact that in at least two cases (GRB 050509b: Bloom et al. 2006 and Hjorth et al. 2005b; GRB 050709: Fox et al. 2005) the limits on any accompanying SN are very tight, stronger than any limits placed on long-duration GRB counterparts. No SN is present that is more than 1% as bright as SN 1998bw ($M_R > -12$ at 16 days in GRB 050709; Hjorth et al. 2005b). Still, the internal-external shock model for short-duration GRBs appears to accommodate the data (Fox et al. 2005; Lee, Ramirez-Ruiz & Granot 2005; Panaitescu 2005; Bloom et al. 2006).

With such a small sample, it would be prudent to wait (e.g., Bloom & Prochaska 2006) before claiming that all short-hard bursts are the result of merging compact objects, but the data so far are certainly consistent with that hypothesis. This raises the interesting need to define yet another class of "peculiar supernova" (mini-SN?), if the radioactive ejecta of the mergers prove capable of powering a brief optical and X-ray display (S.R. Kulkarni, unpublished data; see also Li & Paczyński 1998).

If short-hard bursts are merging compact objects, the duration of the bursts also has some interesting implications for GRBs in general. Associating the event duration with the operation of the central engine implies that the viscous lifetime of any accretion disk created in the merger, typically 0.1 $M_\odot$ (Rosswog, Ramirez-Ruiz & Davies 2003; Setiawan, Ruffert & Janka 2004), is ∼0.1 s. This timescale is far too short for disk accretion to be the power source of long soft GRBs unless there is an additional mechanism for mass accretion onto the disk over longer timescales (Section 3).

2.3. Cosmic X-Ray Flashes

Cosmic X-ray flashes (XRFs; Heise et al. 2001) are observationally similar to classic GRBs, only softer, with a similar distribution of durations (Sakamoto et al. 2004). An intermediary in the spectral continuum between XRFs and classic GRBs is the so-called X-ray rich (XRR) GRB. Although the cosmological distance scale was well established (Bloom et al. 2003), it was not until Soderberg et al. (2004a) determined a spectroscopic redshift ($z = 0.251$; XRF 020903) that the energetics of any XRF was firmly determined. Though the brightness of XRFs implies a similar energy release (per solid angle) to GRBs, the internal-external shock model for the prompt and afterglow emission of XRFs is not as well established. No broadband study of the light curve of an XRF afterglow has been carried out, so the synchrotron origin, though consistent with the data, is uncertain.

Because of their similar characteristics to long-soft GRBs, it is generally thought that the underlying cause is the same (Section 3), i.e., the explosive death of a massive star. The emission could be softer because one is just outside the edge of an ordinary GRB jet (Yamazaki, Yonetoku & Nakamura 2003; Granot, Ramirez-Ruiz & Perna 2005); because an ordinary GRB jet had a cocoon of relativistic matter directed toward us with a moderate Lorentz factor; or because the jet itself had a larger baryonic loading and hence lower Lorentz factor (Zhang, Woosley & Heger 2004). The latter two explanations implicitly assume that the XRF is produced by an external shock where lower Lorentz factor correlates with softer spectra. The opposite behavior is expected in the internal shock model.

An important clue to the origin of XRFs came with the discovery of a SN-like bump associated with XRF 020903 (Soderberg et al. 2004a). There was a clear rise and decay and, when a low S/N spectrum was obtained near peak, the galaxy-subtracted spectrum was a reasonably good match to the spectrum of SN 1998bw at a similar epoch. This suggests that at least one XRF originates from the death of a massive star.

However, aside from XRF 020903, a concerted search for SN signatures in XRFs 011030, 020427, 030723, 040701, 040812, and 040916 (Levan et al. 2005; Soderberg et al. 2005) turned up no clear evidence for associated SNe. A bump peaking in the R-band ∼16 days after XRF 030723 has been interpreted as a SN at redshift ∼0.5 (Fynbo et al. 2004; Tominaga et al. 2004). However, the optical spectrum showed no clear evidence for features and, more importantly, both the K-band (A.M. Soderberg, private communication) and X-ray light curve (Butler et al. 2005) appeared to track the R-band light curve. This is contrary to the expectation from a SN, where the IR-optical colors evolve and the IR light luminosity peaks after the optical light. In other XRFs, no bump was seen. Most constraining is that any SN in XRF 040701 ($z = 0.21$) would have to have been over 3 mag fainter than SN 1998bw (Soderberg et al. 2005), fainter than all GRB-SNe known to date.[8] Though fewer bump searches have been conducted for XRFs than for GRBs, the nondetections are significant because of the low average redshift of XRFs. The absolute magnitudes probed by deep (mostly HST) imaging rival all the bump searches in GRBs. Whereas all GRBs less than redshift 0.7 have claimed bump detections, six XRFs (one with redshift and two more with inferred redshifts less than unity) show no evidence for a SN-like bump. The search for SNe from XRR GRBs has been more sucessful, with at least two (041006 and 040924) showing strong evidence for late-time bumps (Soderberg et al. 2006b). Both appear at peak to have been fainter than SN 1998bw.

It may be that the SNe in XRFs are inherently faint (or absent), which would have important implications for the models, but the numbers are still small. Was XRF 020903 truly an XRF or an outlier in the classic GRB population? Could the optical extinction for XRF 040701 have been greater than estimated? Could the XRFs with no SN bump and no well-determined redshift be farther away than we think? The study of XRF-related SNe will be a subject of great interest in the coming years.

2.4. Characteristics of Supernovae Associated with Gamma-Ray Bursts

The distinguishing feature of a GRB-SN that sets it apart from all other SNe is the concentration of significant kinetic energy in relativistic ejecta ($\beta\Gamma \gtrsim 2$). Here, β is the velocity of the ejecta divided by the speed of light and the Lorentz factor

---

[8]Note that no optical afterglow of any sort was seen in XRF 040701 so the optical extinction could not be measured as with GRB 011121 (Price et al. 2002). Therefore this quoted limit for an XRF-SN relies upon a rather uncertain estimate of the optical extinction based on the X-ray spectrum.

$\Gamma = (1 - \beta^2)^{-1/2}$. This does not necessarily require that the SN be bright, or even exceptionally energetic, though GRB-SNe often are. It also does not preclude the existence of SNe without GRBs, powered by the same energy source (Section 3.4.2). But to produce a GRB, one needs at least as much energy in relativistic ejecta as is observed in γ-ray and afterglow emission. That is, $E_{Rel} \gtrsim E_\gamma$. The value of E_γ is difficult to measure directly because of the effects of beaming, but in typical bursts, it is around 10^{51} erg (Frail et al. 2001; Bloom, Frail & Kulkarni 2003). E_{Rel} can be inferred from radio observations at such late times that beaming is no longer important, and is $\sim 5 \times 10^{51}$ erg (Berger et al. 2003c; Berger, Kulkarni & Frail 2004). Of course, there can be considerable variation in both these numbers.

The SNe accompanying GRBs also differ from common SNe (Filippenko 1997) in other ways (**Table 1**), most obviously the absence of hydrogen in their spectra: GRB-SNe appear to be Type I SNe. Indeed, where spectra of sufficient quality exist to be sure, the SN is of Type Ic-BL. Some of the broad peaks seen in the GRB-SNe spectra are likely due to low opacity, rather than due to emission from a single ion spread over large velocity ranges (e.g., Iwamoto et al. 2003). Near maximum light, GRB-SNe do appear to show broad absorption lines of O I, Ca II, and Fe II

Table 1 Properties of good candidate supernovae associated with γ-ray bursts, X-ray flashes, and X-ray rich γ-ray bursts

Name Burst/SN	z	Peak [mag]	T^a_{peak} [day]	SN likeness/ designation	References
GRB 980425/1998bw	0.0085	$M_V = -19.16 \pm 0.05$	17	Ic-BL	b
GRB 030329/2003dh	0.1685	$M_V = -18.8$ to -19.6	$10-13$	Ic-BL	c
GRB 031203/2003lw	0.1005	$M_V = -19.0$ to -19.7	$18-25$	Ibc-BL	d
XRF 020903	0.25	$M_V = -18.6 \pm 0.5$	~ 15	Ic-BL	e
GRB 011121/2001dk	0.365	$M_V = -18.5$ to -19.6	$12-14$	I (IIn?)	f
GRB 050525a	0.606	$M_V \approx -18.8$	12	I	g
GRB 021211/2002lt	1.00	$M_U = -18.4$ to -19.2	~ 14	Ic	h
GRB 970228	0.695	$M_V \sim -19.2$	~ 17	I	i
XRR 041006	0.716	$M_V = -18.8$ to -19.5	$16-20$	I	j
XRR 040924	0.859	$M_V = -17.6$	~ 11	?	k
GRB 020405	0.695	$M_V \sim -18.7$	~ 17	I	l

[a]The time of peak brightness is reported in the rest frame if the redshift is known, observed frame otherwise.
[b]Galama et al. 1998a.
[c]Hjorth et al. 2003; Stanek et al. 2003; Bloom et al. 2004; Lipkin et al. 2004.
[d]Malesani et al. 2004; Cobb et al. 2004; Thomsen et al. 2004; Gal-Yam et al. 2004.
[e]Soderberg et al. 2005.
[f]Bloom et al. 2002; Garnavich et al. 2003b; Greiner et al. 2003b.
[g]Della Valle et al. 2006.
[h]Della Valle et al. 2004.
[i]Galama et al. 2000; Reichart 1999.
[j]Stanek et al. 2005; Soderberg et al. 2006b.
[k]Soderberg et al. 2006b.
[l]Price et al. 2003.

(Iwamoto et al. 1998). About seven days before maximum, the width of a weak Si II line in SN 1998bw suggested expansion speeds in excess of 30,000 km s^{-1} (Patat et al. 2001). There has never been a photospheric spectrum of a confirmed GRB-SN that indicated the presence of H and no optical He I lines have been seen (e.g., $\lambda 6678$, $\lambda 7065$, and $\lambda 7281$), leading to a classification as a Type Ic.[9] The late time nebular phases (at least in the case of SN 1998bw) show lines of [O I], Ca II, Mg I, and Na I D (Sollerman et al. 2002).

As detailed in Section 2.5, the median peak magnitude of the observed GRB-SNe sample is comparable to that of Type Ia SNe. This large apparent brightness could be misleading, however, because of the stringent requirements—rapidly declining optical afterglow, low redshift, faint host galaxy—needed for detection (Section 2.5). In one case, GRB 010921, the upper limit on any SN is absolute magnitude -17.7 and, as noted previously, some of the SNe with XRFs may be fainter still. Indeed, when the nondetections of GRB-SNe are accounted for, the inferred "true" mean of the sample ($M_V = -18.2 \pm 0.4 + 5 \log h_{71}$) is considerably fainter than the mean of normal Type Ia SNe.

At late times, the decay of ^{56}Co often leads to a steady exponential decline in the light curve of Type Ib SNe, providing all decay energy remains trapped. SN 1998bw initially declined somewhat faster than this, presumably because of γ-ray escape (McKenzie & Schaefer 1999; Sollerman et al. 2002). At very late times ($\gtrsim 500$ days), a flattening seen in the light curve could be interpreted as greater retention of the energy from radioactive decay as well as the contribution of species other than ^{56}Co (Sollerman et al. 2002), but this could have other explanations. Other GRB-SNe could not be followed with sufficient sensitivity to see the exponential tail.

The radio emission of GRB 980425 and SN 1998bw showed no evidence for polarization (Kulkarni et al. 1998b), which suggests that the mildly relativistic ejecta were not highly asymmetric, at least in projection. Still, internal Faraday dispersion in the ejecta would serve to suppress radio polarization (A.M. Soderberg, private communication). The optical light of SN 1998bw showed significant evidence for polarization at the 0.5% level (Patat et al. 2001), which is consistent with polarization inferred in other core-collapse SNe (Wang et al. 1996; Leonard et al. 2002). This implies some degree of asphericity in the nonrelativistic ejecta (e.g., Höflich, Wheeler & Wang 1999) but, as Patat et al. (2001) noted, there is a degeneracy between the viewing angle and the level of asymmetry. Significant polarization was observed for the afterglow of GRB 030329 over many epochs, but by the time SN 2003dh dominated the optical light, two epochs of observations revealed only marginally significant polarization (Greiner et al. 2003a; Kawabata et al. 2003) (even then, the afterglow could have contaminated the polarization signal). Though polarization is surely a critical ingredient toward understanding the SN explosion geometry, Klose et al. (2004) have pointed out that interstellar dust for mildly extinguished lines of sight

---

[9]An infrared feature observed in SN 1998bw may have been due to He (Patat et al. 2001), but that is not a secure identification and, more importantly, the distinction between Ib and Ic depends on the observed presence or absence, respectively, of He I in the optical waveband (T. Matheson, private communication).

Table 2 Physical properties of γ-ray burst-supernovae

GRB/SN	E_{SN} (10^{52} erg)	E_{Rel} (10^{49} erg)	$E_{iso}(\gamma)$[a] (10^{49} erg)	E_γ (10^{49} erg)	$M(^{56}Ni)$ ($M_\odot$)	Refs.
980425/1998bw	2–3[b]	1–30	0.06–0.08	<0.08	0.5–0.7	c
030329/2003dh	2–5	≈50	1070	7–46	0.3–0.55	d
031203/2003lw	2–3	2	2.94 ± 0.11	<3	0.5–0.7	e

[a]The absence of increasing energy inferred in the afterglow blast wave at late times suggests that these sources were not off-axis GRBs (e.g., Soderberg et al. 2004b) and therefore that $E_{iso}(\gamma)$ is indeed an upper limit to the true energy released in γ rays.
[b]Höflich, Wheeler & Wang (1999) inferred a kinetic energy for SN 1998bw one order of magnitude smaller based upon a highly asymmetric model.
[c]Iwamoto et al. 1998; Woosley, Eastman & Schmidt 1999; Li & Chevalier 1999; Galama et al. 1998b; Friedman & Bloom 2005.
[d]Mazzali et al. 2003; Berger et al. 2003c; Woosley & Heger 2006; Deng et al. 2005; Frail et al. 2005; Friedman & Bloom 2005.
[e]Malesani et al. 2004; Gal-Yam et al. 2004; Lipkin et al. 2004; Soderberg et al. 2004b; Sazonov, Lutovinov & Sunyaev 2004; Friedman & Bloom 2005.

should artificially induce polarization at the 1% level. Thus, even if polarization is detected in GRB-SNe, the interpretation is anything but straightforward.

In the case where the brightness is as great as Type Ia, theory demands a large production of ^{56}Ni (Section 3.4.1). In order to make so much ^{56}Ni, large shock energies and stellar masses are required, at least in one-dimensional models. The high energy is consistent with the rapid expansion velocities inferred from the spectra. The combination—high mass, large velocity, and big ^{56}Ni mass—implies kinetic energies $E_{SN} \sim 10^{52}$ erg. Because the collapse of the iron core to a neutron star, plus any accretion into a black hole, must release a total energy in neutrinos $E_\nu \sim 10^{53}$ erg, one has the interesting energy relation, $E_\nu \gg E_{SN} \gtrsim E_{Rel} \gtrsim E_\gamma$. There may be significant gravitational radiation as well, with $E_{GW} \sim E_\nu$ (van Putten et al. 2004b). The energy in γ rays is only a small fraction, $\sim 1\%$ of the total energy released in the explosion.

Table 2 gives the energetics and theoretical masses of ^{56}Ni produced in each of the three GRBs with definite spectroscopic SNe. For those GRBs that exhibit jet breaks and are thought to be observed nearly pole on, an actual E_γ can be inferred. In other cases, E_{Rel} probably provides an upper limit to E_γ, but E_γ could be much greater than $E_{\gamma,iso}$ if the event was observed off axis. SN 2003lw had a light curve and spectrum similar to SN 1998bw, and its energy and ^{56}Ni production are assumed here to be the same (Gal-Yam et al. 2004). It is noteworthy that E_{Rel} and E_γ for all three bursts in **Table 2** are much less than the canonical 5×10^{51} and 10^{51} erg, respectively, mentioned above for common GRBs. This reflects, at least in part, the greater likelihood of discovering a SN if the optical afterglow is faint.

So far, no clear correlation has been found between GRB properties, or even E_{Rel}, and the brightness or energy of the SN, though the simplest theory would suggest that more energy input by the central engine would make both more ^{56}Ni and more relativistic ejecta.

The demographics of local Ibc SNe compared with the GRB rate can be used to constrain the frequency with which SNe accompany GRBs (whether the GRB is detected or otherwise). At optical wavebands, Type Ic-BL SNe like 1998bw comprise about $\sim 5\%$ of all Type Ibc SNe, which is to simply assert the unusual nature of

1998bw-like SNe. Radio surveys of local Ibc SNe at early times reveal that no more than 3% (Berger et al. 2003b) harbor SNe with relativistic ejecta like SN 1998bw. At late times, the radio emission from GRBs initially directed off axis should become observable as Γ of the shock slows to unity. Yet none of the local Ibc SNe studied show evidence for an off-axis jet with large relativistic energy. Specifically, at the 90%-confidence level less than 10% of all Ibc harbor an ordinary off-axis GRB (Soderberg et al. 2006c). Moreover, the class of optical BL SNe cannot all be related to GRBs at the 84% confidence level (Soderberg et al. 2006c).

2.5. Do All Gamma-Ray Bursts Have Supernova Counterparts?

Though 2003 was a banner year for the SN-GRB connection, there has since been a noticeable lack of new GRBs with spectroscopic SNe. Indeed, most long-duration GRBs do not have a detected associated SN. Yet the indirect evidence (Section 2.1.2) supports the consensus view that most long-duration GRBs arise from the death of massive stars. These two statements can be reconciled by invoking observational biases that are either extrinsic or intrinsic to the explosions. Extrinsic biases, those that result in the decreased probability for discovery even if the SN is bright, are straightforward to enumerate:

1. *Localization*—Poor localizations of bursts from their afterglows dramatically hamper the ability for large aperture telescopes to discover emerging SNe. These poor localizations can be endemic to the detection scheme (e.g., BATSE) or because bursts are found in regions of the sky that are unfavorable for optical or X-ray followup (e.g., near the Sun). Afterglows are less likely to be found near full moon because of the brighter sky and less sensitive IR detectors. Likewise, if the bump peaks near full moon then the point source sensitivity is reduced.
2. *Dust*—Fainter SNe are expected from bursts that occur near the line of sight through the Galaxy, or in especially extinction-riddled regions of their hosts. GRB 021211 and GRB 031203 were behind significant Galactic columns, which diminished the sensitivity of the SN observations.
3. *Redshift and luminosity distance*—With increasing luminosity distance comes higher distance modulus, but the k corrections at optical wavelengths are particularly unkind beyond redshifts of $z \approx 0.5$. Absorption line blanketing due to metals (e.g., Fe) supresses the emissivity below the blackbody luminosity blueward of $\sim$4000 Å making Type Ibc SNe especially difficult to detect at increasing redshift (Bloom et al. 1999). We expect essentially no optical flux from GRB SNe at $z \gtrsim 1.2$. Because most Swift bursts have been found above redshift of unity, it is disappointing, but not surprising, that only one SN (associated with GRB 050525a) was found in the first 14 months of Swift observations.
4. *Host galaxies*—The host galaxies of GRBs, though generally fainter than L_*, can still contaminate the light of GRB SNe at late times. If all hosts were 0.1 L_*, then the integrated light of hosts would be M_V(host) ≈ -19 mag, comparable to the brightest SNe. Thus, without high-resolution imaging to resolve out diffuse host light, or high signal-to-noise image differencing, finding the SN point source is all but impossible for SNe that peak at magnitudes fainter than

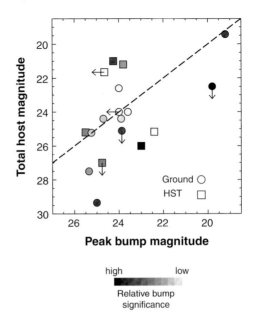

Figure 6

Illustration of the difficulty in finding γ-ray burst-supernovae: claimed bump peak magnitudes versus the integrated magnitude of the host galaxy. Relative significances of the detections are based on our subjective analysis of the believability of the bumps. Many bumps were claimed by differencing catalog magnitudes—small systematic errors in measuring the true host magnitudes artificially reveal bumps. High-resolution imaging ameliorates some of the endemic problems of bump detections from catalogs.

the integrated light of the hosts. **Figure 6** makes this point rather dramatically. An important demonstration is that the SN associated with GRB 031203 was only 0.22 mag brighter than the host galaxy at peak. Had this burst originated from higher redshift, the photometric sensitivity would be diminished, and the SN might not have been discovered.

The biases against detection introduced by intrinsic properties of the explosions are less tractable. Local observations of Type Ibc SNe show a more than 5 mag spread in peak brightness, likely related to the diversity of dust extinction, the spread in explosion energy, and ^{56}Ni production. Rise times also range from about one to several weeks. SNe that make the same mass of ^{56}Ni but which peak later are fainter. An intrinsically faint SN obviously has less of a chance of being detected. The GRB afterglow brightnesses at late times may also be comparable to or brighter than the peak of a SN. It is again sobering to note that the SN associated with GRB 030329 might never have been recognized at higher redshift. SN 2003dh was discovered spectroscopically when the SN contributed less than 5% of the total flux (Matheson et al. 2003c). If there is no obvious bump in the light curve when the SN peaks (e.g., Lipkin et al. 2004) a photometry-only campaign could miss the SN altogether. Still, the SN might be recovered with precision color photometry.

Zeh, Klose & Hartmann (2004) published an important photometric study of bumps in GRBs, fitting 21 of the best sampled afterglows and finding evidence for nine bumps. Statistically significant evidence for bumps is found in 4 GRB afterglows (990712, 991208, 011121, and 020405), while 5 have marginal significance (970228, 980703, 000911, 010921, and 021211). All of these GRBs had bumps claimed prior to the Zeh analysis. Zeh, Klose & Hartmann (2004) emphasize that all GRBs with $z \lesssim 0.7$ appear to have bumps, which, of course, would be expected if all long-duration GRBs have associated SNe. However, bump detection does not necessarily imply a SN detection. There are, in fact, important cases where multiband photometry has shown that late-time bumps may not be due to a SN. The late-time light curves of GRB 990712 and XRF 030723, for example (see Section 2.3), do not appear consistent with a SN. **Figure 7** shows the results of our compilation of bumps and upper-limits from the literature.

If GRB-SNe are a particular subset of Type Ibc SNe, then it is useful to ask if the GRB-SNe sample draws only from the bright end of the Type Ibc distribution

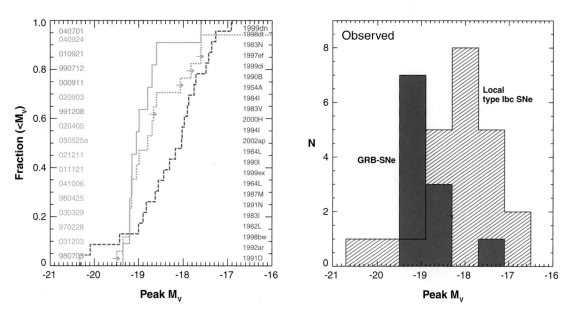

Figure 7

Comparison of the peak M_V of γ-ray burst-supernovae (GRB-SNe) to the local Ibc population. (*Left*) Cumulative distribution of the Ibc SNe from Richardson, Branch & Baron (2006) shown as a dashed line. The distribution of the observed GRB-SNe is shown as a solid line; we include SNe from XRFs and XRR GRBs. The dotted line includes those GRBs with no significant bump detection or bumps that do not resemble SNe. The names of the associated GRB are shown on the left and the names of the Ib (*blue*) and Ic (*mauve*) are shown on the right. (*Right*) Histogram of the secure bumps as compared to the local Ibc population. GRB-SNe populate the "bright" subclass of the Richardson, Branch & Baron (2006) sample, but, as we show in the text, the bump population is consistent with being drawn from the entire local Ibc population.

(Bloom 2005; Soderberg et al. 2006b). Richardson, Branch & Baron (2006) report the detailed modeling of 24 local Type Ibc SNe and derive the distribution of peak absolute M_V accounting for Galactic and host extinction.[10] **Figure 7**, comparing the observed GRB-SNe population with the Richardson sample, makes clear that the observed GRB-SNe are at the bright end of the observed Type Ibc population. Here we use the GRB-SNe compilation from **Table 1**; in the case where a range of M_V was reported in the literature, we assume that M_V is the average of the two measurements that define that range (following from 1998bw, we also assume that $M_U - M_V = 0.19$ mag for GRB 021211). The median of the observed GRB-SNe sample is $M_V = -19.1 + 5 \log h_{71}$ mag, whereas the median peak magnitude in the local Ibc sample is $M_V = -17.9 + 5 \log h_{71}$ mag.

A more subtle but important question is whether the true GRB-SNe distribution in peak magnitude, when nondetections are folded in, is consistent within having been drawn from the local Type Ibc population. Treating lower limits and insignificant bump measurements from the literature as nondetections of GRB-SNe, we find that the mean of the true underlying peak magnitude distribution (from the Kaplan-Meier estimator) is $M_V = -18.2 \pm 0.4 + 5 \log h_{71}$, whereas the mean peak magnitude in the local Ibc sample is $M_V = -18.3 \pm 0.2 + 5 \log h_{71}$ mag. By all relevant "survival analysis" tests (Lavalley, Isobe & Feigelson 1992), we conclude that the GRB-SNe population is statistically consistent with having been drawn from the same population as the local Ibc from Richardson, Branch & Baron.[11] Moreover, although the GRB-SNe population is consistent with both the Ib and Ic sample, the connection with Ic SNe is favored.[12]

3. MODELS

3.1. Core-Collapse Supernova Models

Models for ordinary core-collapse SNe have been extensively reviewed by Woosley & Weaver (1986); Bethe (1990); Burrows (2000); Woosley, Heger & Weaver (2002); Buras et al. (2003, 2006); Woosley & Janka (2005); Janka et al. (2005); Mezzacappa (2005). The current "standard model," by no means universally accepted (Wheeler, Akiyama & Williams 2005), begins with the collapse of the iron core of a highly evolved star that had a main sequence mass of over 10 $M_\odot$. The collapse, triggered by electron capture and the partial photodisintegration of the iron at temperatures

---

[10] In what follows, the Richardson, Branch & Baron magnitudes have been recalibrated to $H_0 = 71$ km s^{-1} Mpc^{-1}. We have not included the peak M_V from SN 1999cq because that value is degenerate with 1994I (T. Matheson, private communication). The sample compiled by Soderberg et al. (2006b) contains a subset (13) Ibc SNe.

[11] That is, the probability that the observed deviation is due to random chance is greater than 0.3 for all tests (e.g., the Generalized Wilcoxon Tests). We have assumed no censoring of the local Ibc data; because bright Ibc is systematically detected over faint Ibc, this clearly biases the Ibc sample to brighter magnitudes.

[12] Specifically, P (Gehan's Generalized Wilcoxon Test) $= 0.14$ when comparing the GRB-SNe sample to local Ib SNe and P (Gehan's Generalized Wilcoxon Test) $= 0.96$ when comparing to local Ic SNe. Other tests show similar improvements for the Ic comparison.

$T \sim 10^{10}$ K and densities $\rho \sim 10^{10}$ g cm^{-3}, continues until the center of the central core exceeds nuclear density by a factor of about two. The rebound, generated by this overshoot and the short range repulsive component of the nuclear force, launches a shock wave, but this "prompt" shock wave quickly loses all outward velocity owing to photodisintegration and neutrino losses. By ~ 0.1 sec after the onset of the collapse, one has a "proto-neutron star" with radius ~ 30 km and mass 1.4 $M_\odot$ with a standing accretion shock at ~ 150 km through which matter is falling at about 0.1–0.3 $M_\odot$ s^{-1}.

Over the next tenth of a second or so, the neutron star radiates a small fraction of its binding energy as neutrinos, $L_\nu \sim 10^{53}$ erg s^{-1}. Approximately 10% of these capture on nucleons in the region between the neutron star and accretion shock. This energy deposition drives vigorous convection, which helps transport energy to the shock and also keeps the absorbing region cool enough that it does not efficiently re-radiate the neutrino energy it absorbed. If $\sim 10^{51}$ erg can be deposited in a few tenths of a second, the accretion can be shut off. The continuing neutrino energy deposition then inflates a bubble of pairs and radiation that pushes off the rest of the star making the SN. If not, accretion continues until a black hole is formed. In this standard model, rotation and magnetic fields are assumed to have negligible effect.

The problem with this scenario, as many have noted, is that it is not robust in the computer simulations. More often than not, the neutrino energy deposition, by itself, fails to launch and sustain an outbound shock of greater than about 10^{51} erg, as is required by analysis of SN 1987A (e.g., Bethe 1990), observations of the light curves of ordinary Type IIp SNe (Woosley & Weaver 1986; Chieffi et al. 2003; Elmhamdi, Chugai & Danziger 2003), SN remnants and interstellar medium heating (Dickel, Eilek & Jones 1993; Thornton et al. 1998), nucleosynthesis constraints (Woosley & Weaver 1995), and neutron star masses (Timmes, Woosley & Weaver 1996). If the explosion energy is too weak, large amounts of matter fall back producing a black hole and robbing the Galaxy of the necessary iron and other heavy elements. As of this writing it remains unclear whether the problem with the models is "simply" one of computational difficulty[13] or whether key physics is lacking. If additional physics is required, rotation and magnetic fields are the leading candidates. However, it remains possible that the answer is being affected by uncertainties in the high density equation of state, changes in fundamental particle physics (especially neutrino flavor mixing), or the possible role of vibrational energy (Burrows et al. 2006).

3.2. The Gamma-Ray Burst Central Engine

The general theory of GRBs, with some discussion of models, has been recently reviewed by Mészáros (2002) and Piran (2005). Unlike a model for SNe alone, a viable GRB model must deliver, far away from the progenitor star, focused jets with at least 200 times as much energy in motion and fields as in rest mass. The jet typically

---

[13] The correct calculation must be done in three dimensions to capture the complex fluid flow in the convective region where the neutrinos deposit their energy, and the neutrino transport itself must be followed in great detail and coupled to the hydrodynamics.

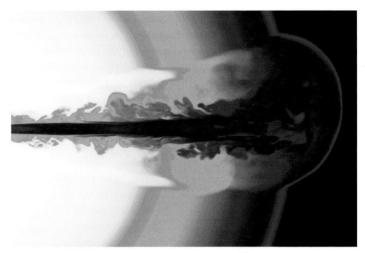

Figure 8

Break out of a relativistic γ-ray burst jet with energy 3×10^{50} erg s^{-1} 8 s after it is launched from the center of a 15 M$_\odot$ WR star. The radius of the star is 8.9×10^{10} cm and the core jet, at infinity, will have a Lorentz factor $\Gamma \sim 200$. Note the cocoon of mildly relativistic material that surrounds the jet and expands to larger angles. Once it has expanded and converted its internal energy this cocoon material will have Lorentz factor $\Gamma \sim 15$–30. An off-axis observer may see a softer display dominated by this cocoon ejecta. If the star were larger or the jet stayed on a shorter time, the relativistic core would not emerge, though there would still be a very energetic, highly asymmetric explosion. (Zhang, Woosley & Heger 2004.)

must have an opening angle ~ 0.1 radian and a power $\sim 10^{50}$ erg s^{-1}. In addition, at least occasionally, the model must deliver $\sim 10^{52}$ erg of kinetic energy to a much larger solid angle (~ 1 radian) to produce SNe like SN 2003dh and SN 1998bw. This is 10 times more than an ordinary SN.

Except in the supranova model (Section 3.2.3), the fact that a relativistic jet must escape the host star without losing too much of its energy severely constrains the model for the central engine. Zhang, Woosley & Heger (2004) have shown that the jet head travels significantly slower than the speed of light and requires 8–25 s to reach the surface for jet energies from 3×10^{48} erg s^{-1} to 3×10^{50} erg s^{-1} (**Figure 8**). If the jet is interrupted or changes its orientation significantly in that time, the flow rapidly degrades to subrelativistic energies and is incapable of making a GRB. Thus acceptable models must provide $\gtrsim 10^{50}$ erg s^{-1} of relativistic, beamed energy for $\gtrsim 10$ s. If, for example, one accepts that the duration of short-hard bursts (~ 0.3 s) reflects the activity of a central engine, the energy source for short-hard bursts and long-soft ones cannot be the same. Similarly, any model that delivers its energy impulsively, on a timescale much less than 10 s, will not make a GRB, even if that energy is initially large and highly focused.

3.2.1. The millisecond magnetar model. Building on earlier models of electromagnetic explosions (Usov 1992; Thompson 1994; Meszaros & Rees 1997b), many groups (Wheeler et al. 2000; Lyutikov & Blackman 2001; Drenkhahn & Spruit 2002; Lyutikov & Blandford 2003) have developed models in which the energy source for GRBs is the rotation of a highly magnetized neutron star with an initial period of about one millisecond (i.e., rotating near breakup). For a rotational velocity $\Omega \sim 5000$ rad s^{-1} and a dynamo-generated magnetic field, $B \sim 2 \times 10^{15}$ G, the rotational energy, $E \sim I\Omega^2/2 \sim 10^{52}$ erg[14] and the dipole spin-down luminosity, $L \sim B^2 R^6 \Omega^4/c^3 \sim 10^{50}$ erg s^{-1}, are typical of GRBs and the SNe that accompany them. Thompson, Chang & Quataert (2004) have considered the coupling between the neutrino-powered wind that must accompany any protoneutron star going through its Kelvin-Helmholtz contraction (Duncan, Shapiro & Wasserman 1986; Qian & Woosley 1996) and the strong magnetic field of a millisecond magnetar. Large powers up to 10^{52} erg s^{-1} can, in principle, be extracted by a centrifugally driven wind. The strength of these models is that they relate GRBs to the birth of an object known to exist, the magnetar, with an energy scale that is about right for a neutron star rotating near break up. The Poynting flux models further offer the possibility of highly energetic outflows with essentially no limit on the Lorentz factor (Blandford 2002). The fields required $\gtrsim 10^{15}$ G are large, but no larger than in other models.

So far, however, all these models ignore the accretion, $\gtrsim 0.1$ M$_\odot$ s^{-1}, that occurs onto the proto-neutron star for several seconds before it contracts to its final radius and develops its full rotation rate. This accretion must be reversed before the neutron star becomes a black hole. The models are also characterized by a monotonically declining power. Though they do not, so far, consider the two events separately, an initial blast to a large solid angle is probably necessary to explode the star and make the necessary ^{56}Ni (Section 3.4.1). A declining power would be available 10 s later to make the GRB itself. It may be that the neutron star models require the initial operation of a successful neutrino-powered explosion before they can function (Fryer & Warren 2004).

3.2.2. Collapsars. The necessary conditions to make a collapsar are black hole formation in the middle of a massive star and sufficient angular momentum to make a disk around that hole (Woosley 1993; MacFadyen & Woosley 1999). The angular momentum needed is at least the value of the last stable orbit around a black hole of several solar masses, $j = 2\sqrt{3} GM/c = 4.6 \times 10^{16} M_{BH}/3$ cm^2 s^{-1} for a nonrotating hole (where M_{BH} is the black-hole mass in solar units) and $j = 2/\sqrt{3} GM/c = 1.5 \times 10^{16} M_{BH}/3$ cm^2 s^{-1} for a Kerr hole with $a = 1$. This compares with an angular momentum in the millisecond magnetar model of $j = R^2 \Omega \sim 5 \times 10^{15}$ cm^2 s^{-1} if $\Omega \sim 5000$ rad s^{-1} and $R = 10$ km. Because the black holes in the collapsar model are typically very rapidly rotating and because the specific angular momentum at 3 M$_\odot$ is always greater than that at 1.5 M$_\odot$, the minimum angular momentum requirements of the collapsar and millisecond

---

[14]The moment of inertia of a neutron star for an appropriate range of masses and radii is 0.35 MR^2 (Lattimer & Prakash 2001).

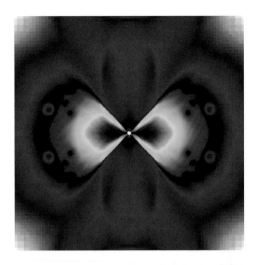

Figure 9
Collapse of the core of a rapidly rotating 14 $M_\odot$ Wolf-Rayet star. Twenty seconds after collapse, a black hole of 4.4 $M_\odot$ has formed and has accreted at ~ 0.1 $M_\odot$ s^{-1} for the last 15 seconds. The figure is 1800 km across and the inner boundary is at 13 km. Colors indicate density on a logarithmic scale, with the highest density in the equatorial plane near the black hole being 9×10^8 g cm^{-3} (MacFadyen & Woosley 1999; W. Zhang, S.E. Woosley & A.I. MacFadyen, in preparation).

magnetar model are similar. Interestingly, there may also be a maximum value of j for the collapsar to work (MacFadyen & Woosley 1999; Narayan, Piran & Kumar 2001; Lee & Ramirez-Ruiz 2006). If j is too great, the disk forms at too great a radius to effectively dissipate its binding energy as neutrino emission and photodisintegration. The collapsar model additionally invokes the formation of a black hole, which seems likely above some critical mass (Section 3.3.1).

Provided a disk and hole form (**Figure 9**), the greatest uncertainty in this model is the mechanism for turning disk binding energy or black-hole rotation energy into directed relativistic outflows. Three possible mechanisms are discussed: (*a*) neutrinos (Woosley 1993; Popham, Woosley & Fryer 1999; Narayan, Piran & Kumar 2001; Di Matteo, Perna & Narayan 2002); (*b*) magnetic instabilities in the disk (Blandford & Payne 1982; Proga et al. 2003); and (*c*) magnetohydrodynamic (MHD) extraction of the rotational energy of the black hole (Blandford & Znajek 1977; Lee, Brown & Wijers 2000; Mizuno et al. 2004). In the first case, neutrino pairs are generated in the hot disk and impact one another with the greatest angle along the rotational axis. Efficient energy deposition is favored by large-angle collisions and the small volume of the region, especially for Kerr black holes, but probably the efficiency is no greater than $\sim 1\%$ of the total neutrino emission, or $\sim 10^{51}$ erg. As the estimate of the total energy in the relativistic component of a GRB has come down in recent years, the neutrino version of the collapsar model has become more attractive. However, up to three orders of magnitude greater energy is available from methods that more directly tap the gravitational potential of the disk or the rotation of the black hole. The neutrino version produces a hot, high entropy jet, whereas some versions of the MHD models produce colder, Poynting flux jets (Section 3.2.4).

In the collapsar model, the SN and the GRB derive their energies from different sources. The SN and the ^{56}Ni that makes it bright are produced by a disk "wind" (MacFadyen & Woosley 1999; MacFadyen 2003; Kohri, Narayan & Piran 2005). This wind is subrelativistic with a speed comparable to the escape velocity of the inner disk, or about 0.1 c (**Figure 10**). If 1 M$_\odot$ accretes to make the GRB and half of this is lost to the wind, this is 10^{52} erg. The wind begins as neutrons and protons in nearly equal proportions and thus ends up, after cooling, as ^{56}Ni. This nickel probably

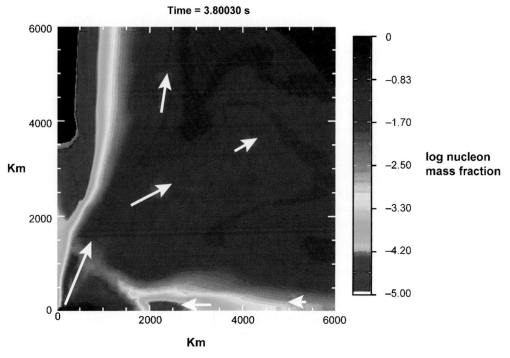

Figure 10

A "wind" of nucleons blows off the black-hole accretion disk. In the collapsar model, this wind is responsible both for blowing up the star and producing the ^{56}Ni that makes it bright. In this numerical simulation, the action of magnetohydrodynamical instabilities in the disk is represented by a simple "alpha-disk" viscosity ($\alpha \approx 0.3$). The highest wind velocity (*long white arrow*) is ~20,000 km s^{-1}, and the mass in the wind is a fraction (~50%) of the 0.1 M$_\odot$ s^{-1} accretion rate. Matter flows in at the equator, is photodisintegrated, and ejected as neutrons and protons. Farther out these nucleons cool and assemble to form ^{56}Ni. This is a separate phenomenon from the core jet that makes the γ-ray burst. The figure is color coded by the logarithm of the mass fraction of nucleons with the dark brown red disk being pure nucleons with a temperature $\sim 4 \times 10^{10}$ K. The outer radius of the figure is 6000 km cm, so the inner disk of 300 km is the dark part, in the lower left corner (MacFadyen & Woosley 1999; MacFadyen 2003).

comes out in a large cone (polar angle ∼1 rad) surrounding the GRB jet though it might get mixed to other angles during the explosion.

In numerical simulations so far, the explicit MHD processes that drive the wind have not been followed. An α-viscosity is used instead and the wind is driven by thermal dissipation. Thus the larger energy (10^{52} erg) of the SN definitely has a MHD origin [if only the disk instabilities responsible for its viscosity (Balbus & Papaloizou 1999)], while that of the GRB remains ambiguous. There are also versions of the collapsar model in which the black hole is not made promptly by the failure of the initial shock to truncate accretion, but by fall back in a SN that explodes with insufficient energy for all its matter to escape (MacFadyen, Woosley & Heger 2001). Still other versions place the black hole in pair instability SNe at high red shift (Fryer, Woosley & Heger 2001). Both of these variations probably give transients that last longer than the typical value, 20 s, for long-soft GRBs, but if they are to work, the energy source must be MHD. Neutrino annihilation is too inefficient for the low accretion rate in the fallback model and large black hole radius in the pair-instability model.

It is a prediction of the collapsar (though possibly the other models as well) that the central engine remains active for a long time after the principal burst is over, potentially contributing to the GRB afterglow (Burrows et al. 2005). This is because the jet and disk wind are inefficient at ejecting all the matter in the equatorial plane of the pre-collapse star and some continues to fall back and accrete (MacFadyen, Woosley & Heger 2001).

Finally, the collapsar model attempts to explain the time structure of GRBs and to produce the variable Lorentz factor necessary for the internal shock model to function (Piran 2005). The jet, as it passes through the star, is modulated by its interaction with the surrounding matter (Zhang, Woosley & MacFadyen 2003). That is, even a jet introduced with constant power in the star's center emerges with a highly variable density and energy at the surface. However, this interaction happens far from the central engine and would be present in any model where a relativistic jet of radiation and matter must penetrate the star.

3.2.3. Supranovae. The maximum mass for a differentially rotating neutron star can be up to ∼50% larger than in the nonrotating case (T. Gold as cited in Blandford & Rees 1972; Morrison, Baumgarte & Shapiro 2004). This gives baryonic maximum masses around 3–3.5 solar masses, well above the largest iron core masses expected in massive stars (Woosley, Heger & Weaver 2002). Uniform rotation causes less of an increase, ∼20%. The simplest version of the supranova model (Vietri & Stella 1998) assumes that such a "hypermassive" neutron star initially forms with a mass above the critical value for a slowly rotating neutron star. The external star is blown away in an initial SN that makes the neutron star. Some time later, years in the original model, dipole radiation slows the neutron star to the point that it collapses to a black hole. For a soft equation of state with an adiabatic index $\Gamma - 4/3 \ll 1$, as much as 10–20% of the mass of the collapsing neutron star avoids capture and goes into a disk about the central hole (Duez et al. 2004; Shapiro 2004). This disk accretes and a GRB jet is produced by MHD processes. The SN evacuates the near region of troublesome baryons that might contaminate the jet and also provides a shell of heavy elements about 10^{16} cm from the burst. The discovery of X-ray lines in the afterglows of some GRBs provided support for this model (Vietri et al. 2001), but to the extent that the lines themselves have become questionable (Section 3.4.5), that motivation is less compelling. The disk, neutron-star combination produced here is similar to that in models for short-hard bursts (Section 2.2) and, to the extent that those are inherently less energetic than long-soft bursts, one wonders if there is sufficient energy and duration for a long-soft burst.

It is clear in the case of events like GRB 980425 and GRB 030329 that the SN and the GRB happened nearly simultaneously—within a few days of each other at most. Delays between several hours and several months are also ruled out because neither the GRB jet nor its emissions would escape the still compact SN. Nevertheless the model is not ruled out for those GRBs in which no SN has been observed, nor is it ruled out for situations in which the delay is seconds rather than years. Because the critical mass for differential rotation is considerably larger than for rigid rotation, there is a range of masses for which simply enforcing rigid rotation (while conserving angular momentum) will lead to collapse. Baumgarte, Shapiro & Shibata (2000) estimate the magnetic braking time to be

$$\tau_B \sim \frac{R}{v_a} \sim 1 \left(\frac{B}{10^{14}\,\mathrm{G}}\right)^{-1} \left(\frac{R}{15\,\mathrm{km}}\right)^{-1/2} \left(\frac{M}{3\,M_\odot}\right)^{1/2} \mathrm{s}. \quad (1)$$

It is unclear whether the outcome of an object experiencing such a collapse is deformation, accompanied by gravitational radiation, complete collapse, or collapse to a black hole plus a disk. In any case, this might be a transition object along the way to the collapsar model. Indeed, for the angular momentum that is invoked, a hypermassive neutron star rotating at break up is a likely, though often ignored, initial stage in the collapsar model.

3.2.4. Poynting flux or fireball?

Related to the uncertainty in the birth of the GRB jet in the above models is a key uncertainty in the nature of the jet itself. Does it consist of hot baryons, thermally loaded with pairs and radiation greatly exceeding the rest mass (Piran 1999; Mészáros 2002), or is the "jet" characterized by large-scale magnetic fields, dynamically dominant and present from start (at the central engine) to finish (in the afterglow) (Meszaros & Rees 1997b; Lyutikov & Blackman 2001; Blandford 2002; Lyutikov, Pariev & Blandford 2003)? In the latter case the baryons play little role and may as well be absent.

Numerical simulations (Aloy et al. 2000; Zhang, Woosley & Heger 2004) show that relativistic fireballs, given mild initial collimation, can pass through stars of solar radius and exit the star with their large energy per baryon intact. No such calculations exist yet for Poynting flux jets to show that their ordered electromagnetic energy does not become thermalized on the way out. This difficulty might be overcome once the jet has bored a hole so that there is a low-density (albeit optically thick) line of sight to the center of the star. That is, a jet could initially be a fireball and make a transition early in the burst to being Poynting flux dominated.

The most important characteristic, observationally, of Poynting flux models is that there are no internal or reverse shocks. In the model developed by Lyutikov &

Blandford, the GRB emission comes from 10^{16} cm not 10^{13} cm. Poynting flux models have the capability of producing large polarization (Lyutikov, Pariev & Blandford 2003), which might be a diagnostic of the model. A strong early optical afterglow, as in GRB 990123, may also be easier to accommodate in Poynting flux models (Zhang & Mészáros 2002; Fan, Wei & Wang 2004). Poynting flux models also predict no high energy neutrino flux accompanying the GRB, but do predict that GRBs could be the site of ultrahigh energy cosmic rays (Blandford 2002).

3.3. Progenitor Stars

All GRB progenitors must lose their hydrogen envelope prior to death. The radius of even a blue supergiant is several hundred light seconds and the head of the jet which makes the GRB travels significantly slower than light while inside the star (Zhang, Woosley & MacFadyen 2003; Zhang, Woosley & Heger 2004). One might envision situations where a very asymmetric SN might occur, powered by the same central engine as a GRB, but a ∼20 s GRB of the common variety is very unlikely. This is consistent with the observation that the limited set of spectroscopic SNe associated with GRBs are, so far, of Type I. The progenitors must also be massive enough and occur frequently enough to explain the observed statistics. Finally, because only a small fraction of massive stars make GRBs when they die, special circumstances must be involved.

3.3.1. Mass. Single stars over about 10 $M_\odot$ on the main sequence are required to make an iron core and collapse to a compact remnant. Stars of still higher mass produce more massive iron cores and have greater accretion rates onto that core once it collapses. Higher mass stars also have greater gravitational binding energy outside the iron core (Woosley, Heger & Weaver 2002). Both effects make an explosion more difficult. It has been speculated that above some critical mass, a black hole forms before an outgoing shock is launched (Fryer 1999), setting the stage for the collapsar model. For a somewhat smaller mass, a black hole could still form from fallback (MacFadyen, Woosley & Heger 2001), though the accretion goes on at a much lower rate. For stars that do not make black holes, the faster rotation associated with neutron stars resulting from the death of the most massive stars (Heger, Woosley & Spruit 2005) still makes them more promising GRB progenitors. The necessary helium core mass to make a GRB is probably near 10 $M_\odot$, corresponding to a main sequence star of at least 25–30 $M_\odot$, but there are production channels that involve appreciably lighter single stars (Woosley & Heger 2006).

There are also binary channels for making long-soft GRBs including (*a*) the merger, by common envelope, of two massive stars, both of which are burning helium in their centers (Fryer & Heger 2005); (*b*) the merger of a black hole with the helium core of a massive star (Fryer & Woosley 1998; Zhang & Fryer 2001); or (*c*) the merger of a black hole and a white dwarf (Fryer 1999). If the common envelope is completely dispersed and the mass of the merged helium core is ∼10 $M_\odot$ or more, model (*a*) gives a resulting WR star similar to the single star models. Cases (*b*) and (*c*) result in black-hole accretion with higher angular momentum and produce longer bursts, probably longer than typical GRBs. The white dwarf merger model would not produce a SN like SN 1998bw or 2003dh and might not give the observed degree of concentration of GRBs in star-forming regions.

3.3.2. Rotation. The role of rotation in ordinary SNe has long been debated (Hoyle 1946; Fowler & Hoyle 1964; Leblanc & Wilson 1970; Ostriker & Gunn 1971). GRBs aside, current models assign a role ranging from dominant (Akiyama et al. 2003; Ardeljan, Bisnovatyi-Kogan & Moiseenko 2005; Wheeler, Akiyama & Williams 2005), to important (Thompson, Quataert & Burrows 2005), and unimportant (Fryer & Warren 2002, 2004; Scheck et al. 2004). No such ambiguity surrounds the role of rotation in making GRBs. It is crucial in all current models (Section 3.2).

Attempts to model the evolution of angular momentum in massive stars to the point where their iron cores become unstable to collapse have yielded uncertain results. There is general agreement that the omission of magnetic torques leads to cores that do indeed rotate rapidly enough to make a GRB (Heger, Langer & Woosley 2000; Hirschi, Meynet & Maeder 2004). Indeed, one could easily end up with the converse problem—too great a fraction of massive stars would make GRBs. However, incorporation of approximate magnetic torques (Spruit 2002) in single stars that evolve through a red giant phase gives too little rotation for GRBs (Heger, Woosley & Spruit 2005), though just about the right amount for pulsars (Ott et al. 2006). This suggests either that the estimated torques are wrong or that special circumstances are required to make a GRB.

Recently, Woosley & Heger (2006) and Yoon & Langer (2005) have discussed the possibility that single massive stars on the high velocity tail of the rotational velocity distribution function might experience "homogeneous evolution" (Maeder 1987), bypassing red giant formation altogether. Such stars can die with very rapid rotation rates and large core masses but only if the metallicity is low.

The possibility that GRBs are a consequence of binary evolution is frequently discussed (Smartt et al. 2002; Podsiadlowski et al. 2004; Tutukov & Cherepashchuk 2003, 2004; Fryer & Heger 2005; Petrovic et al. 2005), but the same caveats apply. Unless the merger occurs well into helium burning, stars with solar metallicity end up with iron cores that rotate too slowly to make GRBs if the estimated magnetic torques are applied (Petrovic et al. 2005; Woosley & Heger 2006).

In general, the rotation rate for WR stars is not well determined observationally, but is expected, on theoretical grounds, to be rapid, at least for low metallicity (Meynet & Maeder 2005).

3.3.3. Metallicity. Both the mass and rotation rate of potential GRB progenitors are strongly influenced by mass loss. Expansion of deeper layers to replenish what has been lost from the stellar surface causes them to rotate more slowly and ultimately this torque is communicated to the core. Mass loss also makes the star lighter and easier to explode, hence no black hole. Thus if one wants a GRB, it is helpful if the mass-loss rate, especially during the WR phase of the evolution leading up to the GRB, is small. WR stars of low metallicity are known to have smaller mass-loss rates (Vink & de Koter 2005), scaling approximately as $Z^{0.86}$ down to metallicities of 1% solar (where Z refers here to the primordial iron abundance, not the abundances of carbon and oxygen on the surfaces of WC and WO stars). GRBs will therefore be favored in regions of low metallicity (MacFadyen & Woosley 1999) as has been observed in several cases (Fynbo et al. 2003; Prochaska et al. 2004; Gorosabel et al. 2005b; Sollerman et al. 2005). Reducing the metallicity to below 10% solar will therefore possibly increase both the frequency and violence of the outbursts. In the collapsar model, more stars will make black holes and those holes will accrete more matter. In the magnetar model, more rapidly rotating neutron stars will be made. This does not preclude the possibility of GRBs in solar metallicity stars by some rare channel of binary evolution, or the estimates of magnetic torques used in the stellar evolution models may be too big.

3.3.4. Frequency. Madau, Della Valle & Panagia (1998) estimate that the core-collapse rate of all massive stars from redshift 1 to 4 amounts to an observed event rate of 20 per 4 arcmin square on the sky. Over the full sky this corresponds to about 5 SNe per second, a number that should approximately characterize the observable universe. GRBs, however, are thought to occur throughout the universe at a rate of about 3 per day. Correcting for an average factor of 300 in beaming, this means that, universe-wide, the GRB rate is only about 0.2% of the SN rate. Even allowing for a large number of events in which a GRB-like central engine might produce a SN or an XRF without a bright GRB, this suggests that GRBs are a rare branch of stellar evolution requiring unusual conditions. If, however, the GRB rate is strongly metallicity sensitive this fraction might increase with redshift. This is consistent with observations that restrict the GRB rate to be no greater than about 1% of the rate of core collapse SNe (Gal-Yam et al. 2006).

3.4. Model Diagnostics

3.4.1. The supernova light curve and properties. The fact that bright SNe sometimes accompany GRBs offers a powerful insight into the explosion mechanism. Type I SNe of all sorts are believed to be powered, at peak, by the decay of radioactive ^{56}Ni, and its daughter ^{56}Co, to ^{56}Fe. ^{56}Ni is only made when matter with near neutron-proton equality (e.g., ^{28}Si, ^{16}O, etc.) is heated to high temperature ($\gtrsim 4 \times 10^9$ K). These high temperatures and the 6.077-day half-life of ^{56}Ni require that it be made in the explosion, not long before. Neither merging neutron stars nor supranovae with a long delay are able to do this.

In a spherically symmetric explosion, the production of ^{56}Ni is limited by the amount of ejected matter interior to a radius given by $4/3\pi r^3 a(4 \times 10^9)^4 E_{\rm exp}$, where $E_{\rm exp}$ is 10^{51} erg in an ordinary SN, and perhaps 10^{52} erg in a GRB SN. To make the 0.5 M$_\odot$ of ^{56}Ni inferred from some explosions, 10^{52} erg must therefore be deposited in a time equal to that required by the shock to go 10,000 km. Typical SN shocks at this radius move faster than 10,000 km s^{-1}, so the energy source must radiate a power of $\sim 10^{52}$ erg s^{-1} during the first second of the explosion. This is far more power, at least during the first second, than is required to make the GRB itself.

The GRB jet itself is relatively inefficient at making ^{56}Ni itself if it starts out with a small solid angle. This is both because a small amount of matter is intercepted by the jet and because at very high explosion energies the rapid expansion results in the production of α-particles, not just ^{56}Ni. Still, one expects the distribution of ^{56}Ni in the ejecta to be very asymmetric and concentrated along the rotational axis (Maeda & Nomoto 2003). Because a GRB observer is also situated along this same axis, high velocities and large blue shifts should be seen. The ^{56}Ni may be rapidly mixed far out in the explosion making it brighter, early on, than a more spherically symmetric explosion. The deformation may also make the SN light mildly polarized.

An important issue is whether the amount of ^{56}Ni and the expansion rate are likely to be the same in all SNe with GRBs and XRFs, i.e., is the SN a standard candle? Despite the similarities between SN 2003dh and SN 1998bw (Section 2.1.6), the answer probably is "no." One expects a large variation in the masses and rotation rates of the progenitor stars, especially when metallicity effects are folded in. Different stars will give different rotation rates to their neutron cores, accrete different amounts of mass into black holes of varying sizes, present different density structures to the outgoing blast, etc.; SNe expanding at a slower rate will be fainter, even if they make the same amount of ^{56}Ni because their light will peak later after more decay and adiabatic degradation. A "supernova" more than 10 times fainter than SN 1998bw would be surprising, especially in the collapsar model, which needs to accrete about a solar mass just to get the jet out of the star. But it still might be explained either by pulsar models or collapsars in which most of the energy came from black-hole rotation instead of disk accretion. On the other hand, a SN more than twice as bright as SN 1998bw would also be surprising. Excepting pair instability SNe, no such large ^{56}Ni mass has ever been verified in any SN and it is probable that a collapsar with such a vigorous wind would shut off its own accretion. These limits are consistent with the observed spread in SN luminosities in XRFs (Soderberg et al. 2005).

3.4.2. Unusual supernovae and transition objects. SNe are visible at all angles, last a long time, and do not require relativistic mass ejection. There could therefore be a large number of "orphan" SNe in which the same central engine acts, but which, for some reason, does not give an observable GRB. Perhaps the jet had too much baryon loading or died prematurely. Perhaps a GRB went off in another direction, or for 20 seconds we simply were not looking at the right place. Still the fraction of all SNe with the unusual characteristics of GRB-SNe is small (Section 3.3.4). It is also much easier to see the GRB from a great distance, hence many GRB-SNe go undetected. A volume-limited sample of GRBs and broad line, Type I SNe would give interesting constraints on the beaming and jet break out, but so far the data is too limited.

Specific cases include Type Ic supernovae SN 1997ef (Iwamoto et al. 2000), SN 1997dq (Mazzali et al. 2004) and SN 2002ap (Mazzali et al. 2002; Yoshii et al. 2003). All show evidence for peculiarity and high energy. Despite high velocity, a photosphere persists until late times. SN 2003jd, also a Type Ic, shows evidence for two components of oxygen, one at high velocity, one at low (Mazzali et al. 2005). These observations are best explained in a two component model (Maeda et al. 2002, 2003; S.E. Woosley & A. Heger, unpublished data) in which the SN has a high velocity along its rotational axes, reflecting the activity of some jet-like energy source, and low velocities in the equator.

Perhaps the clearest case so far of a transition SN is SN 2005bf (Tominaga et al. 2005; Folatelli et al. 2006; Maeda et al. 2006), a SN that (*a*) contained broad lines of Fe II and Ca II, but only a trace of hydrogen, (*b*) had two distinct velocity components, and (*c*) had two luminosity peaks, the second being almost as bright as a Ia but occurring 40 days after the explosion. No spherically symmetric model with a monotonically declining distribution of ^{56}Ni is capable of explaining the observations. The explosion had to be an exceptionally massive helium core to peak so late and yet make an unusually large amount of ^{56}Ni to be so bright then. The explosion energy was also well over 10^{51} erg. But the spectrum did not resemble so much that of SN 1998bw as an ordinary Ib or Ic SN.

All these SNe so far are the explosion of WR stars of one sort or another. It is expected on theoretical grounds that stripped-down helium cores will retain a higher angular momentum than those embedded in red giant envelopes (Section 3.3.2). Also, the asymmetry of the explosion tends to be damped out in stars with extended envelopes (Wang et al. 2001) and the high velocities in the helium core are tamped. Still hyperactivity has been reported in Type II SNe. SN 1997cy (Germany et al. 2000), the brightest SN ever and a Type II, was possibly associated with short-hard burst GRB 970514. However with four month's uncertainty in the explosion date and difficulty making short-hard bursts in massive stars, this association is probably coincidental. The high luminosity, attributed by Germany et al. to 2.6 M$_\odot$ of ^{56}Ni, may have been due to circumstellar interaction (Turatto et al. 2000). Similar caveats apply to the identification of Type II SN 1999E with GRB 980910 (Rigon et al. 2003).

3.4.3. Energetics. As was discussed in Section 2.4, the energy budget of a GRB-SN can be broadly partitioned according to the Lorentz factor of its ejecta. The SN itself, as determined by its spectral line widths, its nucleosynthesis, and some estimate of its mass, is a measure of the nonrelativistic ($1 < \Gamma < 1.005$) kinetic energy, E_{SN}. The afterglow measures the energy in $\beta\Gamma \gtrsim 2$ ejecta, E_{Rel}, and the GRB measures the energy of matter with $\Gamma \gtrsim 200$, E_{GRB}. In general, one expects $E_{SN} > E_{Rel}$ and by definition, $E_{Rel} > E_{GRB}$, but there is no reason that the ratio, E_{SN}/E_{Rel}, should be a constant from event to event, and it probably is not (Malesani et al. 2004; Soderberg et al. 2004b).

In the collapsar model, E_{SN}/E_{Rel} measures the energy in the disk wind compared to that in the jet. In the pulsar model, it measures the ratio of prompt large-angle energy input to late-time input with small angles. In both cases, the answer could vary with accretion rate, angular momentum, magnetic field strength, and stellar mass. Conversely, observations of this ratio will ultimately constrain these uncertainties in the models.

3.4.4. Optical spectroscopy of the afterglow. As light from the GRB and, later, the SN passes through the wind of the progenitor star, distinctive lines may be created that are informative of the star's mass-loss rate, wind speed, and composition (Mirabal et al. 2003; Schaefer et al. 2003; van Marle, Langer & Garcia-Segura 2005). Multiple components are seen with speeds $\sim$500 km s^{-1} and $\sim$3000 km s^{-1}. The highest velocity lines probably originate near the progenitor and reflect the WR wind speed when it died, but lower speed lines are produced by a nebula farther out that is either the fractured residual of a collision between the WR wind and a previous red supergiant wind or radiatively accelerated by the burst. Observations so far are consistent with the wind of a WR star, perhaps of class WC, but constrain the lifetime of the WR progenitor to be shorter than expected (van Marle, Langer & Garcia-Segura 2005).

Some have suggested an origin as due to radiative acceleration of dense clumpy nebular material by the GRB afterglow (Mirabal et al. 2002; Schaefer et al. 2003; Ramirez-Ruiz et al. 2005). Then the absence of detectable photoionization or deceleration places constraints on the circumburst radii of the absorbing material (Mirabal et al. 2002; Schaefer et al. 2003) and possible companions to the progenitor (Fiore et al. 2005; Starling et al. 2005). Other (less locally based) origins are possible, such as SNe remnants (Lazzati et al. 2002), quasar absorption systems (Mirabal et al. 2002), or galactic superwinds. Given the alternatives, and because only a handful of GRBs have exhibited such blueshifted absorption, absorption spectroscopy has yet to establish a definitive connection with the circumburst environment. Still, with more frequent spectroscopic observations at early times after a GRB, there is hope that the measurement of time-dependent metal columns and velocities could provide important diagnostics of the progenitors.

3.4.5. X-ray lines. X-ray lines are potentially a powerful diagnostic of the SN-GRB combination. Lines from various elements have been reported in the afterglows of at least seven GRBs using a variety of instruments (e.g., Piro et al. 2000; Reeves et al. 2002; Piran 2005, and references therein). They are sometimes seen in emission about 10 hours after the burst with a luminosity of 10^{44}–10^{45} erg s^{-1} for at least several hours. Typically the lines are Fe-Kα, though the Kα lines of Si, S, Ar, and Ca have also been reported. Unfortunately, the statistical significance of these signals is not universally accepted (Rutledge & Sako 2003). If real, the lines might be expected in the supranova model, though other alternatives have been discussed (Mészáros & Rees 2001; Kallman, Mészáros & Rees 2003; Kumar & Narayan 2003) in which the SN and GRB are simultaneous.

3.4.6. Gamma-ray burst-supernovae remnants. The SNe that accompany GRBs may be hyperenergetic compared with ordinary SNe and are certainly accompanied by jets. These peculiarities might manifest themselves in the remnant a long time after the explosion is over. However, the jet energy is probably less than the SN energy, and initial asymmetries are eroded once the explosion has swept up several times the initial mass of the progenitor star. Ayal & Piran (2001) estimate a time $\sim$5000 years for this to occur, and, given the low event rate for GRBs, estimate a remnant population $\sim$0.05 per galaxy. On the other hand, Perna, Raymond & Loeb (2000) and Roberts et al. (2003) discuss unusual supernova remnants that may have involved unusual energy or asymmetry. These observations may have other explanations though, e.g., multiple SNe.

3.4.7. Compact remnants. The black hole left in the collapsar model would be very rapidly rotating (Kerr parameter about 1), but the spin of an isolated black hole is unobservable from far away. If the GRB occurred in a binary system that somehow remained bound (admittedly a big if) and later became an accreting X-ray source (Brown et al. 2000), measurement of the mass and Kerr parameter (M. Middleton, C. Done, M. Gierlinski & S. Davis, submitted; Shafee et al. 2006; Psaltis 2004) of the black hole would both show large values. Indeed, measurements of Kerr parameters in the range 0.5 to 1, even for black holes that may not have made GRBs, but do not seem to have been spun up by later accretion, would strongly suggest that rotation is an important component of some SNe.

The object left in the millisecond magnetar model would be a very magnetic neutron star (B $\sim 10^{15}$ G), with a still rapid rotation rate and high luminosity, though probably not visible unless the burst were relatively nearby. This would essentially be the birth of a magnetar. Because the magnetic activity of such objects has been associated with SGRs (e.g., Woods & Thompson 2005) and because SGR activity might possibly be visible out to a distance of 70 Mpc (Tanvir et al. 2005), nearby GRB sites, specifically that of GRB 980425 at 40 Mpc, might be monitored for repeated bursting activity.

3.4.8. Nucleosynthesis. Pruet, Thompson & Hoffman (2004) and Surman & McLaughlin (2005) find that the nucleosynthesis in the disk wind of collapsars consists mostly of ^{56}Ni for the relevant range of accretion rates. The winds do not preserve the large neutron excess characteristic of the inner disk because the outgoing nucleons capture electron-positron pairs and neutrinos. If "bubbles" remove disk material at an unusually low density and rapid rate, heavier nuclei can be produced, even the r-process, but this is highly uncertain.

In the ideal case GRBs would greatly overproduce one or more nuclear species not made elsewhere. Pruet, Surman & McLaughlin (2004) have found that the wind from collapsar disks can synthesize interesting large abundances of ^{42}Ca, ^{45}Sc, ^{46,49}Ti, and ^{63}Cu, but these same species can be produced in ordinary SNe (Woosley, Heger & Weaver 2002).

3.4.9. Afterglows and density gradients. The afterglow of a GRB in radio and X-ray bands is generally regarded as coming from the external shock of the GRB producing jet as it decelerates in the external medium. Breaks in the light curves of this emission can yield information on the opening angle of the jet and therefore on the actual total relativistic energy in the event (Frail et al. 2001; Piran et al. 2001; Panaitescu & Kumar 2002). In addition, the afterglow offers unique insight into the mass-loss history of the star just before it exploded. If the metallicity is low, one expects mass-loss rates much smaller than for typical WR stars in our galaxy (Section 3.3.3).

It is important to note that radio emission from GRBs and Type Ibc SNe samples the mass loss during an epoch of stellar evolution that is otherwise unobserved (and therefore not tightly constrained). During the past several hundred years of their lives WR stars over 8 $M_\odot$ are burning carbon and heavier elements in their cores (Woosley, Heger & Weaver 2002). For a wind speed of 10^8 cm s^{-1}, this mass loss determines the distribution of mass out to 10^{18} cm wherein all the afterglow is formed. The mass inside 10^{15} cm, where the burst itself gets made, reflects the last few months in the star's life when it was burning oxygen and silicon. So long as the mass-loss rate depends only on the surface luminosity of the star, it will not change much, for a WR star of given mass and metallicity, from helium burning until explosion. The luminosity varies by only about 50%. But if these late stages are pulsationally unstable with a short growth time, the mass loss could be quite different—perhaps higher. The mass loss of WR stars is also known to be clumpy (Hamann & Koesterke 1998), and that could complicate the modeling.

In general, though, unless the mass loss is rapidly varying, which seems doubtful in carbon burning, the density should scale as r^{-2}. This scaling is consistent with radio observations of some GRBs (Chevalier & Li 2000; Li & Chevalier 2001; Panaitescu & Kumar 2002; Price et al. 2002; Greiner et al. 2003b), but inconsistent with others (Chevalier & Li 1999; Kumar & Panaitescu 2003). The latter is difficult to reconcile with the otherwise successful paradigm that long-soft GRBs originate from the deaths of massive stars, but the complex interplay between the winds and the interstellar medium could mask global wind signatures and even mimic a constant density environment (Wijers 2001; Chevalier, Li & Fransson 2004; Ramirez-Ruiz et al. 2005).

3.4.10. Gravitational radiation and neutrinos. All models produce compact objects and require a lot of rotation and thus predict a gravitational radiation signature of some sort (Fryer, Woosley & Heger 2001; Davies et al. 2002; Kobayashi & Mészáros 2003; van Putten et al. 2004a,b). However, most of the models are cylindrically symmetric. Perhaps the best opportunity would be from the initial collapse that leads to the collapsar. The proto-neutron star has more angular momentum than even a neutron star rotating at break up and thus might pass through a highly deformed stage before collapsing to a black hole (Baumgarte, Shapiro & Shibata 2000). But the cylindrically symmetric exclusion of the excess angular momentum in a disk is also a possibility that could greatly diminish the gravitational radiation.

The neutrino burst from core collapse is not much brighter than in ordinary SNe and may even be fainter. Given the large distances and soft spectrum, these neutrinos are probably not visible above the background. However, very energetic neutrinos can be produced by a relativistic jet traversing a massive star (Mészáros & Waxman 2001; Razzaque, Mészáros & Waxman 2004, 2005).

4. THE FUTURE—A MYSTERY UNSOLVED

Although the observations of the past seven years have revealed an exciting link between SNe and the long-soft GRBs, it would be premature to think that we understand either one of them very well. No complete physical model currently exists for even the most common variety of SN. Indeed, one of the most important consequences of the SN-GRB connection may be a better understanding of how massive stars die.

Some specific diagnostics that might help with this were given in Section 3.4. Here we mention a few places where we think significant progress could happen in the next decade. Some progress will come simply from a larger sample of GRB-SN and from codes of increased realism running on more powerful computers. Other advances may require the development of space missions beyond Swift and ground-based facilities that are only now in the planning stages. We restrict our list to science specifically related to the SN-GRB connection, not everything we want to know about, or can do with GRBs.

- How variable—in energy, mass, and luminosity—is the class of SNe that accompany XRFs and GRBs? We have taken the position here that all of these high-energy transients, except perhaps the short bursts, are accompanied by stellar explosions of some sort. Is that true? Were SN 1998bw and SN 2003dh unusually bright? Are there any systematic differences in the SNe that accompany long-soft GRBs of different duration, energy, spectral hardness, etc.?
- Are GRBs favored by low metallicity? Do the average properties of GRBs vary significantly (in their rest frame) with redshift? Because mass loss decreases with metallicity, GRBs from high redshift might preferentially come from more massive and more rapidly rotating stars. This might be reflected in the properties of the bursts and their afterglows.
- Pushing this to the extreme, can we use GRBs to study Population III stars at very high redshift, including stars of much higher mass than those that die as SNe today? Bursts from redshift 10–20 would be both highly time-dilated and severely reddened. A new mission or mission strategy may be necessary that combines observations in the infrared and hard X-ray bands.
- What is the most common form of GRB in the universe? It is possible that observations so far have been selectively biased to more luminous events. Are events like GRB 980425 actually more frequent than "ordinary" GRBs? More sensitive studies over a long period could eventually give, at least, a volume-limited local sample.
- What is the relationship between XRFs and GRBs? Are XRFs the result of GRBs seen off axis, the result of jets that have lower Lorentz factors at all angles, or something else? Observationally, it will be important to see if the distribution of properties of XRFs and GRBs is continuous from one extreme to the other. Theoretical models are still primitive. Are both phenomena due to internal shocks or is a mixture of internal and external shocks involved?
- Similarly, is there a continuum of events between core collapse SNe and GRBs, or are they two discrete classes of phenomena? Rapid differential rotation in the core of a massive star when it dies has been implicated as a necessary ingredient for GRBs. Rotation may play a role in producing all manner of unusual SNe like those mentioned in Section 3.4.2, even those that have no GRBs. But is it important in ordinary Type IIp SNe?
- How is the jet launched in a long-soft GRB? Is the jet a fireball or Poynting flux? This is largely an ongoing issue for theory and simulation with important implications for active galactic nuclei and pulsars as well as GRBs. There may

be observational diagnostics, however, in the polarization of afterglows and the strength of the optical afterglow.

- How long does the central engine operate? Is its power at late times continuous or episodic? Recent studies with Swift have shown some evidence in some bursts for substantial energy input continuing long after the main burst is over. Variable late-time energy input could be a consequence of incomplete ejection of all mass in the SN, which leads to continued accretion in the collapsar model, though pulsar-based explanations are not ruled out.
- Do SN-like displays ever occur with short-hard bursts? Present data suggest that they do not, but the exceptions should continue to be sought.
- In the longer time frame, neutrino bursts and gravitational radiation may possibly yield the greatest insight into the nature and activity of the central engine, as it is only in these emissions that the central engine is directly observable.

ACKNOWLEDGEMENTS

This review has greatly benefited from discussions with many people and presentations at conferences too many to mention by name. We are particularly grateful, however, to Roger Blandford for a critical reading of the manuscript and many useful comments; Alex Filippenko for discussions of the characteristics of GRB-SNe; Alex Heger for helping us understand the role of rotation in the advanced stages of stellar evolution; Thomas Janka for critical comments on SN models; Chryssa Kouveliotou for discussions of GRB 980425 and GRBs in general; Andrew MacFadyen and Weiqun Zhang for discussions of the collapsar model and three of the figures used in the text; Tom Matheson for his careful reading and comments, especially on GRB-SNe properties; Bethany Cobb for comments; and Alicia Soderberg for many helpful comments on SNe and XRFs and a detailed critique of an earlier draft of this manuscript. J.S.B. also offers special thanks to Andrew Friedman, Dale Frail, Shrinivas Kulkarni, and Robert Kirshner.

LITERATURE CITED

Akiyama S, Wheeler JC, Meier DL, Lichtenstadt I. 2003. *Ap. J.* 584:954–70
Aloy MA, Müller E, Ibáñez JM, Martí JM, MacFadyen A. 2000. *Ap. J. Lett.* 531:L119–22
Ardeljan NV, Bisnovatyi-Kogan GS, Moiseenko SG. 2005. *MNRAS* 359:333–44
Ayal S, Piran T. 2001. *Ap. J.* 555:23–30
Bailyn C, Dokkum PV, Buxton M, Cobb B, Bloom JS. 2003. *GCN Circ.* 2486
Balbus SA, Papaloizou JCB. 1999. *Ap. J.* 521:650–58
Baumgarte TW, Shapiro SL, Shibata M. 2000. *Ap. J. Lett.* 528:L29–32
Belczyński K, Bulik T, Zbijewski W. 2000. *Astron. Astrophys.* 355:479–84
Berger E, Cowie LL, Kulkarni SR, Frail DA, Aussel H, Barger AJ. 2003a. *Ap. J.* 588:99–112
Berger E, Diercks A, Frail DA, Kulkarni SR, Bloom JS, et al. 2001. *Ap. J.* 556:556–61
Berger E, Kulkarni SR, Frail DA. 2004. *Ap. J.* 612:966–73

Berger E, Kulkarni SR, Frail DA, Soderberg AM. 2003b. *Ap. J.* 599:408–18

Berger E, Kulkarni SR, Pooley G, Frail DA, McIntyre V, et al. 2003c. *Nature* 426:154–57

Berger E, Price PA, Cenko SB, Gal-Yam A, Soderberg AM, et al. 2005. *Nature* 438:988–90

Bersier D, Rhoads J, Fruchter A, Cerón JMC, Strolger L, et al. 2004. *GCN Circ.* 2544

Bethe HA. 1990. *Rev. Mod. Phys.* 62:801–66

Björnsson G, Hjorth J, Jakobsson P, Christensen L, Holland S. 2001. *Ap. J. Lett.* 552:L121–24

Blandford R. 2002. In *Lighthouses of the Universe*, ed. M Gilfanov, R Sunyaev, E Churazov, pp. 381–404. Berlin: Springer

Blandford RD, Payne DG. 1982. *MNRAS* 199:883–903

Blandford RD, Rees MJ. 1972. *Astrophys. Lett.* 10:77

Blandford RD, Znajek RL. 1977. *MNRAS* 179:433–56

Blinnikov SI, Novikov ID, Perevodchikova TV, Polnarev AG. 1984. *Pis ma Astron. Zhurnal* 10:422–28

Bloom JS. 2003. *Astron. J.* 125:2865–75

Bloom JS. 2005. In *IAU Colloq. 192: Cosmic Explosions, on the 10th Anniv. SN1993J*, p. 411

Bloom JS, Djorgovski SG, Kulkarni SR. 2001. *Ap. J.* 554:678

Bloom JS, Fox D, van Dokkum PG, Kulkarni SR, Berger E, et al. 2003. *Ap. J.* 599:957

Bloom JS, Frail DA, Kulkarni SR. 2003. *Ap. J.* 594:674–83

Bloom JS, Kulkarni SR. 1998. *GCN Circ.* 161

Bloom JS, Kulkarni SR, Djorgovski SG. 2002. *Astron. J.* 123:1111–48

Bloom JS, Kulkarni SR, Djorgovski SG, Eichelberger AC, Cote P, et al. 1999. *Nature* 401:453–56

Bloom JS, Kulkarni SR, Harrison F, Prince T, Phinney ES, Frail DA. 1998. *Ap. J. Lett.* 506:L105–8

Bloom JS, Kulkarni SR, Price PA, Reichart D, Galama TJ, et al. 2002. *Ap. J. Lett.* 572:L45–49

Bloom JS, Prochaska JX. 2006. In *Proc. 16th Ann. Oct. Astrophys. Conf.: GRBs in the SWIFT Era*, ed. S Holt, N Gehrels, J Nousek. In press (astro-ph/0602058)

Bloom JS, Prochaska JX, Pooley D, Blake CH, Foley RJ, et al. 2006. *Ap. J.* 638:354–68

Bloom JS, Sigurdsson S, Pols OR. 1999. *MNRAS* 305:763–69

Bloom JS, van Dokkum PG, Bailyn CD, Buxton MM, Kulkarni SR, Schmidt BP. 2004. *Astron. J.* 127:252–63

Brown GE, Lee C-H, Wijers RAMJ, Lee HK, Israelian G, Bethe HA. 2000. *New Astron. Rev.* 5:191–210

Buras R, Rampp M, Janka H-T, Kifonidis K. 2003. *Phys. Rev. Lett.* 90(24):241101

Buras R, Rampp M, Janka H-T, Kifonidis K. 2006. *Astron. Astrophys.* 447:1049–92

Burenin RA, Sunyaev RA, Pavlinsky MN, Denisenko DV, Terekhov OV, et al. 2003. *Astron. Lett.* 29:573

Burrows A. 2000. *Nature* 403:727–33

Burrows A, Livne E, Dessart L, Ott CD, Murphy J. 2006. *Ap. J.* 640:878

Burrows DN, Romano P, Falcone A, Kobayashi S, Zhang B, et al. 2005. *Science* 309:1833–35

Butler NR, Sakamoto T, Suzuki M, Kawai N, Lamb DQ, Graziani C, et al. 2005. *Ap. J.* 621:884–93
Carter B. 1992. *Ap. J. Lett.* 391:L67–70
Castro-Tirado AJ, Gorosabel J. 1999. *Astron. Astrophys. Suppl.* 138:449–50
Castro-Tirado AJ, Sokolov VV, Gorosabel J, Castro Cerón JM, Greiner J, et al. 2001. *Astron. Astrophys.* 370:398–406
Chary R, Becklin EE, Armus L. 2002. *Ap. J.* 566:229–38
Chevalier RA, Li Z-Y. 1999. *Ap. J. Lett.* 520:L29–32
Chevalier RA, Li Z-Y. 2000. *Ap. J.* 536:195
Chevalier RA, Li Z-Y, Fransson C. 2004. *Ap. J.* 606:369–80
Chieffi A, Domínguez I, Höflich P, Limongi M, Straniero O. 2003. *MNRAS* 345:111–22
Chornock R, Filippenko AV. 2002. *GCN Circ.* 1605
Chornock R, Foley RJ, Filippenko AV, Papenkova M, Weisz D, Garnavich P. 2003. *IAU Circ.* 8114
Christensen L, Hjorth J, Gorosabel J. 2004. *Astron. Astrophys.* 425:913–26
Cobb BE, Bailyn CD, van Dokkum PG, Buxton MM, Bloom JS. 2004. *Ap. J. Lett.* 608:L93–96
Colgate SA. 1968. *Can. J. Phys.* 46:476
Conselice CJ, Vreeswijk PM, Fruchter AS, Levan A, Kouveliotou C, Fynbo JPU, et al. 2005. *Ap. J.* 633:29–40
Costa E, Frontera F, Heise J, Feroci M, in't Zand J, et al. 1997. *Nature* 387:783–85
Costa E, Frontera F, Hjorth J, eds. 2001. *Gamma-Ray Bursts in the Afterglow Era*. Berlin: Springer-Verlag
Covino S, Lazzati D, Ghisellini G, Fugazza D, Campana S, et al. 2003. In *AIP Conf. Proc. 662: Gamma-Ray Burst and Afterglow Astronomy 2001: A Workshop Celebrating the First Year of the HETE Mission*, pp. 393–95
Davies MB, King A, Rosswog S, Wynn G. 2002. *Ap. J. Lett.* 579:L63–66
Della Valle M. 2005. *Nuovo Cimento C* 28:563–73
Della Valle M, Malesani D, Benetti S, Testa V, Hamuy M, et al. 2003. *Astron. Astrophys.* 406:L33–37
Della Valle M, Malesani D, Benetti S, Testa V, Hamuy M, et al. 2004. In *AIP Conf. Proc. 727: Gamma-Ray Bursts: 30 Years of Discovery*, pp. 403–7
Della Valle M, Malesani D, Bloom JS, Benetti S, Chincarini G, et al. 2006. *Ap. J. Lett.* 642:L103–6
Deng J, Tominaga N, Mazzali PA, Maeda K, Nomoto K. 2005. *Ap. J.* 624:898–905
Di Matteo T, Perna R, Narayan R. 2002. *Ap. J.* 579:706–15
Dickel JR, Eilek JA, Jones EM. 1993. *Ap. J.* 412:648–63
Djorgovski SG, Kulkarni SR, Bloom JS, Frail DA, Harrison FA, et al. 2001. See Costa et al. 2001, p. 218
Djorgovski SG, Kulkarni SR, Frail DA, Harrison FA, Bloom JS, et al. 2003. *Proc. SPIE* 4834:238–47
Drenkhahn G, Spruit HC. 2002. *Astron. Astrophys.* 391:1141–53
Duez MD, Liu YT, Shapiro SL, Stephens BC. 2004. *Phys. Rev. D* 69(10):104030
Duncan RC, Shapiro SL, Wasserman I. 1986. *Ap. J.* 309:141–60

Eichler D, Livio M, Piran T, Schramm DN. 1989. *Nature* 340:126–28
Elmhamdi A, Chugai NN, Danziger IJ. 2003. *Astron. Astrophys.* 404:1077–86
Esin AA, Blandford R. 2000. *Ap. J. Lett.* 534:L151–54
Fan YZ, Wei DM, Wang CF. 2004. *Astron. Astrophys.* 424:477–84
Filippenko AV. 1997. *Annu. Rev. Astron. Astrophys.* 35:309–55
Filippenko AV. 1998. *IAU Circ.* No. 6896
Fiore F, D'Elia V, Lazzati D, Perna R, Sbordone L, et al. 2005. *Ap. J.* 624:853–67
Firmani C, Avila-Reese V, Ghisellini G, Tutukov AV. 2004. *Ap. J.* 611:1033–40
Fishman GJ, Meegan CA. 1995. *Annu. Rev. Astron. Astrophys.* 33:415–58
Folatelli G, Contreras C, Phillips MM, Woosley SE, Blinnikov S, et al. 2006. *Ap. J.* 641:1039–50
Fowler WA, Hoyle F. 1964. *Ap. J. Suppl.* 9:201
Fox DB, Frail DA, Price PA, Kulkarni SR, Berger E, et al. 2005. *Nature* 437:845–50
Frail DA, Kulkarni SR, Sari R, Djorgovski SG, Bloom JS, et al. 2001. *Ap. J. Lett.* 562:L55–58
Frail DA, Soderberg AM, Kulkarni SR, Berger E, Yost S, et al. 2005. *Ap. J.* 619:994–98
Freedman DL, Waxman E. 2001. *Ap. J.* 547:922–28
Friedman AS, Bloom JS. 2005. *Ap. J.* 627:1–25
Fruchter AS, Levan AJ, Strolger L, Vreeswijk PM, Thorsett SE, et al. 2006. *Nature* 441:463–68
Fruchter AS, Pian E, Gibbons R, Thorsett SE, Ferguson H, et al. 2000b. *Ap. J.* 545:664–69
Fruchter A, Vreeswijk P, Hook R, Pian E. 2000a. *GCN Rep.* 752
Fryer CL. 1999. *Ap. J.* 522:413–18
Fryer CL, Heger A. 2005. *Ap. J.* 623:302–13
Fryer CL, Warren MS. 2002. *Ap. J. Lett.* 574:L65–68
Fryer CL, Warren MS. 2004. *Ap. J.* 601:391–404
Fryer CL, Woosley SE. 1998. *Ap. J. Lett.* 502:L9
Fryer CL, Woosley SE, Hartmann DH. 1999. *Ap. J.* 526:152–77
Fryer CL, Woosley SE, Heger A. 2001. *Ap. J.* 550:372–82
Fynbo JPU, Holland S, Andersen MI, Thomsen B, Hjorth J, et al. 2000. *Ap. J. Lett.* 542:L89–93
Fynbo JPU, Jakobsson P, Möller P, Hjorth J, Thomsen B, et al. 2003. *Astron. Astrophys.* 406:L63–66
Fynbo JPU, Sollerman J, Hjorth J, Grundahl F, Gorosabel J, et al. 2004. *Ap. J.* 609:962–71
Galama TJ, Tanvir N, Vreeswijk PM, Wijers RAMJ, Groot PJ, et al. 2000. *Ap. J.* 536:185–94
Galama TJ, Vreeswijk PM, Pian E, Frontera F, Doublier V, Gonzalez J-F. 1998a. *IAU Circ.* No. 6895
Galama TJ, Vreeswijk PM, Van Paradijs J, Kouveliotou C, Augusteijn T, et al. 1998b. *Nature* 395:670–72
Gal-Yam A, Moon D-S, Fox DB, Soderberg AM, Kulkarni SR, et al. 2004. *Ap. J. Lett.* 609:L59–62
Gal-Yam A, Nakar E, Ofek E, Fox DB, Cenko SB, et al. 2005. *Ap. J.* 639:331–39

Gal-Yam A, Ofek EO, Poznanski D, Levinson A, Waxman E, et al. 2006. *Ap. J.* 639:331–39

Garnavich P, Matheson T, Olszewski EW, Harding P, Stanek KZ. 2003a. *IAU Circ.* 8114:1

Garnavich PM, Stanek KZ, Wyrzykowski L, Infante L, Bendek E, et al. 2003b. *Ap. J.* 582:924–32

Gehrels N, Sarazin CL, O'Brien PT, Zhang B, Barbier L, et al. 2005. *Nature* 437:851–54

Germany LM, Reiss DJ, Sadler EM, Schmidt BP, Stubbs CW. 2000. *Ap. J.* 533:320–28

Gorosabel J, Fynbo JPU, Fruchter A, Levan A, Hjorth J, et al. 2005a. *Astron. Astrophys.* 437:411–18

Gorosabel J, Pérez-Ramírez D, Sollerman J, de Ugarte Postigo A, Fynbo JPU, et al. 2005b. *Astron. Astrophys.* 444:711–21

Granot J, Ramirez-Ruiz E, Perna R. 2005. *Ap. J.* 630:1003–14

Greiner J, Klose S, Reinsch K, Martin Schmid H, Sari R, et al. 2003a. *Nature* 426:157–59

Greiner J, Klose S, Salvato M, Zeh A, Schwarz R, et al. 2003b. *Ap. J.* 599:1223–27

Greiner J, Peimbert M, Estaban C, Kaufer A, Vreeswijk P, et al. 2003c. *GCN Circ.* 2020

Hakkila J, Haglin DJ, Pendleton GN, Mallozzi RS, Meegan CA, Roiger RJ. 2000. *Ap. J.* 538:165–80

Halpern JP, Uglesich R, Mirabal N, Kassin S, Thorstensen J, et al. 2000. *Ap. J.* 543:697–703

Hamann W-R, Koesterke L. 1998. *Astron. Astrophys.* 335:1003–8

Hansen BMS. 1999. *Ap. J.* 512:L117–20

Hartmann D, Woosley SE. 1988. In *Multiwavelength Astrophysics*, ed. F Córdova, pp. 189–233. Cambridge, UK: Cambridge Univ. Press

Heger A, Langer N, Woosley SE. 2000. *Ap. J.* 528:368–96

Heger A, Woosley SE, Spruit HC. 2005. *Ap. J.* 626:350–63

Heise J, in't Zand J, Kippen RM, Woods PM. 2001. See Costa et al. 2001, p. 16

Hirschi R, Meynet G, Maeder A. 2004. *Astron. Astrophys.* 425:649–70

Hjorth J, Sollerman J, Gorosabel J, Granot J, Klose S, et al. 2005b. *Ap. J. Lett.* 630:L117–20

Hjorth J, Sollerman J, Møller P, Fynbo JPU, Woosley SE, et al. 2003. *Nature* 423:847–50

Hjorth J, Watson D, Fynbo JPU, Price PA, Jensen BL, et al. 2005a. *Nature* 437:859–61

Höflich P, Baade D, Khokhlov A, Wang L, Wheeler JC. 2005. In *IAU Colloq. 192: Cosmic Explosions, on the 10th Anniv. SN1993J*, p. 403

Höflich P, Wheeler JC, Wang L. 1999. *Ap. J.* 521:179–89

Hogg DW, Fruchter AS. 1999. *Ap. J.* 520:54–58

Hoyle F. 1946. *MNRAS* 106:343

Iwamoto K, Mazzali PA, Nomoto K, Umeda H, Nakamura T, et al. 1998. *Nature* 395:672–74

Iwamoto K, Nakamura T, Nomoto K, Mazzali PA, Danziger J, et al. 2000. *Ap. J.* 534:660–69

Iwamoto K, Nomoto K, Mazzali PA, et al. 2003. In *LNP Vol. 598: Supernovae and Gamma-Ray Bursters*, ed. K Weiler, pp. 243–81

Jakobsson P, Levan A, Fynbo JPU, Priddey R, Hjorth J, et al. 2006. *Astron. Astrophys.* 447:897–903

Janka HT, Scheck L, Kifonidis K, Müller E, Plewa T. 2005. In *ASP Conf. Ser. 332: The Fate of the Most Massive Stars*, p. 372

Kallman TR, Mészáros P, Rees MJ. 2003. *Ap. J.* 593:946–60

Kawabata KS, Deng J, Wang L, Mazzali P, Nomoto K, et al. 2003. *Ap. J. Lett.* 593:L19–22

Kippen RM. 1998. *GCN Circ.* 67

Klebesadel RW, Strong IB, Olson RA. 1973. *Ap. J. Lett.* 182:L85–88

Klose S, Palazzi E, Masetti N, Stecklum B, Greiner J, et al. 2004. *Astron. Astrophys.* 420:899–903

Kobayashi S, Mészáros P. 2003. *Ap. J.* 589:861–70

Kohri K, Narayan R, Piran T. 2005. *Ap. J.* 629:341–61

Kouveliotou C, Meegan CA, Fishman GJ, Bhat NP, Briggs MS, et al. 1993. *Ap. J. Lett.* 413:L101–4

Kouveliotou C, Woosley SE, Patel SK, Levan A, Blandford R, et al. 2004. *Ap. J.* 608:872–82

Kulkarni SR, Bloom JS, Frail DA, Ekers R, Wieringa M, et al. 1998a. *IAU Circ.* No. 6903

Kulkarni SR, Djorgovski SG, Odewahn SC, Bloom JS, Gal RR, et al. 1999. *Nature* 398:389–94

Kulkarni SR, Djorgoski SG, Ramaprakash AN, Goodrich R, Bloom JS, et al. 1998c. *Nature* 393:35–39

Kulkarni SR, Frail DA, Wieringa MH, Ekers RD, Sadler EM, et al. 1998b. *Nature* 395:663–69

Kumar P, Narayan R. 2003. *Ap. J.* 584:895–903

Kumar P, Panaitescu A. 2003. *MNRAS* 346:905–14

Kumar P, Piran T. 2000. *Ap. J.* 535:152–57

Lattimer JM, Prakash M. 2001. *Ap. J.* 550:426–42

Lavalley M, Isobe T, Feigelson E. 1992. In *ASP Conf. Ser. 25: Astron. Data Anal. Software Syst. I*, p. 245

Lazzati D, Covino S, Ghisellini G, Fugazza D, Campana S, et al. 2001. *Astron. Astrophys.* 378:996–1002

Lazzati D, Rossi E, Covino S, Ghisellini G, Malesani D. 2002. *Astron. Astrophys.* 396:L5–9

Le Floc'h E, Charmandaris V, Forrest WJ, Mirabel F, Armus L, Devost D. 2006. *Ap. J.* 642:636–52

Le Floc'h E, Duc P-A, Mirabel IF, Sanders DB, Bosch G, et al. 2003. *Astron. Astrophys.* 400:499–510

Leblanc JM, Wilson JR. 1970. *Ap. J.* 161:541

Lee HK, Brown GE, Wijers RAMJ. 2000. *Ap. J.* 536:416–19

Lee WH, Ramirez-Ruiz E. 2006. *Ap. J.* 641:961–71

Lee WH, Ramirez-Ruiz E, Granot J. 2005. *Ap. J. Lett.* 630:L165–68

Leonard DC, Filippenko AV, Chornock R, Foley RJ. 2002. *Publ. Astron. Soc. Pac.* 114:1333–48
Levan A, Patel S, Kouveliotou C, Fruchter A, Rhoads J, et al. 2005. *Ap. J.* 622:977–85
Li L-X, Paczyński B. 1998. *Ap. J. Lett.* 507:L59–62
Li Z-Y, Chevalier RA. 1999. *Ap. J.* 526:716–26
Li Z-Y, Chevalier RA. 2001. *Ap. J.* 551:940–45
Lidman C, Doublier V, Gonzalez J-F, Augusteijn T, Hainaut OR, et al. 1998. *IAU Circ.* No. 6895
Lipkin YM, Ofek EO, Gal-Yam A, Leibowitz EM, Poznanski D, et al. 2004. *Ap. J.* 606:381–94
Livio M, Sahu K, Panagia N, eds. 2001. *Supernovae and Gamma-Ray Bursts: The Greatest Explosions Since the Big Bang*. Cambridge, UK: Cambridge Univ. Press
Livio M, Sahu KC, Petro L, Fruchter AS, Pian E, et al. 1998. In *Gamma Ray Bursts: 4th Huntsville Symp.*, Vol. 428, ed. CA Meegan, R Preece, T Koshut, p. 483. Woodbury, NY: AIP
Loredo TJ, Wasserman IM. 1998. *Ap. J.* 502:75
Lyutikov M, Blackman EG. 2001. *MNRAS* 321:177–86
Lyutikov M, Blandford R. 2003. *MNRAS* Submitted (astro-ph/0312347)
Lyutikov M, Pariev VI, Blandford RD. 2003. *Ap. J.* 597:998–1009
MacFadyen AI. 2003. In *From Twilight to Highlight: The Physics of Supernovae, ESO Astrophys. Symp.*, ed. W Hillebrandt, B Leibundgut, p. 97
MacFadyen AI, Woosley SE. 1999. *Ap. J.* 524:262–89
MacFadyen AI, Woosley SE, Heger A. 2001. *Ap. J.* 550:410–25
Madau P, Della Valle M, Panagia N. 1998. *MNRAS* 297:L17
Maeda K, Mazzali PA, Deng J, Nomoto K, Yoshii Y, et al. 2003. *Ap. J.* 593:931–40
Maeda K, Nakamura K, Nomoto K, Mazzali PA. 2002. *Ap. J.* 565:405–12
Maeda K, Nomoto K. 2003. *Ap. J.* 598:1163
Maeda K, Nomoto K, Mazzali PA, Deng J. 2006. *Ap. J.* 640:854–77
Maeder A. 1987. *Astron. Astrophys.* 178:159–69
Malesani D, Tagliaferri G, Chincarini G, Covino S, Della Valle M, et al. 2004. *Ap. J. Lett.* 609:L5–8
Mao S, Mo HJ. 1998. *Astron. Astrophys.* 339:L1–4
Masetti N, Palazzi E, Pian E, Hunt L, Fynbo JPU, et al. 2005. *Astron. Astrophys.* 438:841–53
Masetti N, Palazzi E, Pian E, Hunt LK, Méndez M, et al. 2000. *Astron. Astrophys.* 354:473–79
Matheson T. 2004. In *Cosmic Explosions in Three Dimensions: Asymmetries in Supernovae and Gamma-Ray Bursts*, ed. P Höflich, P Kumar, JC Wheeler, p. 351. Cambridge, UK: Cambridge Univ. Press
Matheson T. 2005. In *ASP Conf. Ser. 332: The Fate of the Most Massive Stars*, p. 416
Matheson T, Garnavich P, Hathi N, Jansen R, Windhorst R, et al. 2003a. *GCN Circ.* 2107
Matheson T, Garnavich P, Olszewski EW, Harding P, Eisenstein D, et al. 2003b. *GCN Circ.* 2120
Matheson T, Garnavich PM, Stanek KZ, Bersier D, Holland ST, et al. 2003c. *Ap. J.* 599:394–407

Matzner CD, McKee CF. 1999. *Ap. J.* 510:379–403

Mazzali PA, Deng J, Maeda K, Nomoto K, Filippenko AV, Matheson T. 2004. *Ap. J.* 614:858–63

Mazzali PA, Deng J, Maeda K, Nomoto K, Umeda H, et al. 2002. *Ap. J. Lett.* 572:L61–65

Mazzali PA, Deng J, Tominaga N, Maeda K, Nomoto K, et al. 2003. *Ap. J. Lett.* 599:L95–98

Mazzali PA, Kawabata KS, Maeda K, Nomoto K, Filippenko AV, et al. 2005. *Science* 308:1284–87

McKenzie EH, Schaefer BE. 1999. *Publ. Astron. Soc. Pac.* 111:964–68

Mészáros P. 2002. *Annu. Rev. Astron. Astrophys.* 40:137–69

Mészáros P, Rees MJ. 1997a. *Ap. J.* 476:232–37

Mészáros P, Rees MJ. 1997b. *Ap. J. Lett.* 482:L29–32

Mészáros P, Rees MJ. 2001. *Ap. J. Lett.* 556:L37–40

Mészáros P, Rees MJ. 2003. *Ap. J. Lett.* 591:L91–94

Mészáros P, Waxman E. 2001. *Phys. Rev. Lett.* 87(17):171102

Metzger MR, Djorgovski SG, Kulkarni SR, Steidel CC, Adelberger KL, et al. 1997. *Nature* 387:879

Meurs EJA, Rebelo MCA. 2004. *Nucl. Phys. B Proc. Suppl.* 132:324–26

Meynet G, Maeder A. 2005. *Astron. Astrophys.* 429:581–98

Mezzacappa A. 2005. *Annu. Rev. Nucl. Part. Sci.* 55:467–515

Mirabal N, Halpern JP, Chornock R, Filippenko AV, Terndrup DM, et al. 2003. *Ap. J.* 595:935–49

Mirabal N, Halpern JP, Kulkarni SR, Castro S, Bloom JS, et al. 2002. *Ap. J.* 578:818–32

Mizuno Y, Yamada S, Koide S, Shibata K. 2004. *Ap. J.* 615:389–401

Møller P, Fynbo JPU, Hjorth J, Thomsen B, Egholm MP, et al. 2002. *Astron. Astrophys.* 396:L21–24

Morrison IA, Baumgarte TW, Shapiro SL. 2004. *Ap. J.* 610:941–47

Narayan R, Paczyński B, Piran T. 1992. *Ap. J. Lett.* 395:L83

Narayan R, Piran T, Kumar P. 2001. *Ap. J.* 557:949–57

Narayan R, Piran T, Shemi A. 1991. *Ap. J. Lett.* 379:L17–20

Natarajan P, Albanna B, Hjorth J, Ramirez-Ruiz E, Tanvir N, Wijers R. 2005. *MNRAS* 364:L8–12

Nomoto K, Maeda K, Mazzali PA, Umeda H, Deng J, Iwamoto K. 2004. In *Stellar Collapse*, ed. C Fryer, p. 277. Amsterdam: Kluwer Academic

Norris JP, Bonnell JT, Watanabe K. 1999. *Ap. J.* 518:901–8

Ostriker JP, Gunn JE. 1971. *Ap. J. Lett.* 164:L95

Ott CD, Burrows A, Thompson TA, Livne E, Walder R. 2006. *Ap. J. Suppl.* 164:130–55

Paczyński B. 1986. *Ap. J.* 308:L43–46

Paczyński B. 1991. *Acta Astron.* 41:257–67

Paczyński B. 1995. *Publ. Astron. Soc. Pac.* 107:1167–75

Paczyński B. 1998. *Ap. J.* 494:L45–48

Panaitescu A. 2005. *MNRAS Lett.* 367:L42–46

Panaitescu A, Kumar P. 2002. *Ap. J.* 571:779–89

Patat F, Cappellaro E, Danziger J, Mazzali PA, Sollerman J, et al. 2001. *Ap. J.* 555:900–17

Patat F, Piemonte A. 1998. *IAU Circ.* 6918

Perna R, Loeb A. 1998. *Ap. J.* 501:467

Perna R, Raymond J, Loeb A. 2000. *Ap. J.* 533:658–69

Peterson BA, Price PA. 2003. *GCN Circ.* 1985

Petrovic J, Langer N, Yoon S-C, Heger A. 2005. *Astron. Astrophys.* 435:247–59

Pian E, Amati L, Antonelli LA, Butler RC, Costa E, et al. 2000. *Ap. J.* 536:778–87

Pian E, Antonelli LA, Daniele MR, Rebecchi S, Torroni V, et al. 1998. *GCN Circ.* 61

Piran T. 1999. *Phys. Rep.* 314:575–667

Piran T. 2005. *Rev. Mod. Phys.* 76:1143–210

Piran T, Kumar P, Panaitescu A, Piro L. 2001. *Ap. J. Lett.* 560:L167–69

Piro L, Garmire G, Garcia M, Stratta G, Costa E, et al. 2000. *Science* 290:955–58

Podsiadlowski P, Mazzali PA, Nomoto K, Lazzati D, Cappellaro E. 2004. *Ap. J. Lett.* 607:L17–20

Popham R, Woosley SE, Fryer C. 1999. *Ap. J.* 518:356–74

Price PA, Berger E, Reichart DE, Kulkarni SR, Yost SA, et al. 2002. *Ap. J. Lett.* 572:L51–55

Price PA, Kulkarni SR, Berger E, Fox DW, Bloom JS, et al. 2003. *Ap. J.* 589:838–43

Price PA, Schmidt BP. 2004. In *AIP Conf. Proc. 727: Gamma-Ray Bursts: 30 Years of Discovery*, p. 503

Prochaska JX, Bloom JS, Chen H-W, Foley RJ, Perley DA, et al. 2006. *Ap. J.* 642:989–94

Prochaska JX, Bloom JS, Chen H-W, Hurley KC, Melbourne J, et al. 2004. *Ap. J.* 611:200–7

Prochaska JX, Chen HW, Bloom JS. 2006. *Ap. J.* In press

Proga D, MacFadyen AI, Armitage PJ, Begelman MC. 2003. *Ap. J. Lett.* 599:L5–8

Pruet J, Surman R, McLaughlin GC. 2004. *Ap. J. Lett.* 602:L101–4

Pruet J, Thompson TA, Hoffman RD. 2004. *Ap. J.* 606:1006–18

Psaltis D. 2004. In *X-Ray Timing 2003: Rossi and Beyond*, ed. P Kaaret, FK Lamb, JH Swank, 714:29–35. Meville, NY: AIP

Qian Y-Z, Woosley SE. 1996. *Ap. J.* 471:331

Ramirez-Ruiz E, García-Segura G, Salmonson JD, Pérez-Rendón B. 2005. *Ap. J.* 631:435–45

Razzaque S, Mészáros P, Waxman E. 2004. *Phys. Rev. Lett.* 93(18):181101

Razzaque S, Mészáros P, Waxman E. 2005. *Phys. Rev. Lett.* 94(10):109903

Reeves JN, Watson D, Osborne JP, Pounds KA, O'Brien PT, et al. 2002. *Nature* 416:512–15

Reichart DE. 1999. *Ap. J. Lett.* 521:L111–15

Reichart DE. 2001. *Ap. J.* 554:643–59

Reichart DE, Lamb DQ, Castander FJ. 2000. In *Gamma Ray Bursts: 5th Huntsville Symp.*, ed. GJ Fishman, RM Kippen, RS Mallozzi, 526:414. Meville, NY: AIP

Rhoads JE. 1997. *Ap. J. Lett.* 487:L1–4

Richardson D, Branch D, Baron E. 2006. *Astron. J.* 131:2233–44

Rigon L, Turatto M, Benetti S, Pastorello A, Cappellaro E, et al. 2003. *MNRAS* 340:191–96

Roberts TP, Goad MR, Ward MJ, Warwick RS. 2003. *MNRAS* 342:709–14

Rosswog S, Ramirez-Ruiz E, Davies MB. 2003. *MNRAS* 345:1077–90

Rutledge RE, Sako M. 2003. *MNRAS* 339:600–6

Sadler EM, Stathakis RA, Boyle BJ, Ekers RD. 1998. *IAU Circ.* 6901

Sahu KC, Livio M, Petro L, Macchetto FD, van Paradijs J, et al. 1997. *Nature* 387:476

Sahu KC, Vreeswijk P, Bakos G, Menzies JW, Bragaglia A, et al. 2000. *Ap. J.* 540:74–80

Sakamoto T, Lamb DQ, Graziani C, Donaghy TQ, Suzuki M, et al. 2004. *Ap. J.* 602:875–85

Savaglio S, Fall SM, Fiore F. 2003. *Ap. J.* 585:638–46

Sazonov SY, Lutovinov AA, Sunyaev RA. 2004. *Nature* 430:646–48

Schaefer BE, Lamb DQ, Graziani C, Donaghy TQ, Suzuki M, et al. 2003. *Ap. J.* 588:387–99

Scheck L, Plewa T, Janka H-T, Kifonidis K, Müller E. 2004. *Phys. Rev. Lett.* 92(1):011103

Setiawan S, Ruffert M, Janka H-T. 2004. *MNRAS* 352:753–58

Shafee R, McClintock JE, Narayan R, Davis SW, Li L-X, Remillard RA. 2006. *Ap. J. Lett.* 636:L113–16

Shapiro SL. 2004. *Ap. J.* 610:913–19

Smartt SJ, Vreeswijk PM, Ramirez-Ruiz E, Gilmore GF, Meikle WPS, et al. 2002. *Ap. J. Lett.* 572:L147–51

Soderberg AM, Berger E, Kasliwal M, Frail DA, Price PA, et al. 2006a. In press (astro-ph/0601455)

Soderberg AM, Kulkarni SR, Berger E, Fox DB, Price PA, et al. 2004a. *Ap. J.* 606:994–99

Soderberg AM, Kulkarni SR, Berger E, Fox DW, Sako M, et al. 2004b. *Nature* 430:648–50

Soderberg AM, Kulkarni SR, Fox DB, Berger E, Price PA, et al. 2005. *Ap. J.* 627:877–87

Soderberg AM, Kulkarni SR, Price PA, Fox DB, Berger E, et al. 2006b. *Ap. J.* 636:391–99

Soderberg AM, Nakar E, Berger E, Kulkarni SR. 2006c. *Ap. J.* 638:930–37

Sokolov VV. 2001. See Costa et al. 2001, p. 136

Sollerman J, Holland ST, Challis P, Fransson C, Garnavich P, et al. 2002. *Astron. Astrophys.* 386:944–56

Sollerman J, Ostlin G, Fynbo JPU, Hjorth J, Fruchter A, Pedersen K. 2005. *New Astron. Rev.* 11:103–15

Spruit HC. 2002. *Astron. Astrophys.* 381:923–32

Stanek KZ, Garnavich PM, Nutzman PA, Hartman JD, Garg A, et al. 2005. *Ap. J. Lett.* 626:L5–9

Stanek KZ, Matheson T, Garnavich PM, Martini P, Berlind P, et al. 2003. *Ap. J. Lett.* 591:L17–20

Starling RLC, Wijers RAMJ, Hughes MA, Tanvir NR, Vreeswijk PM, et al. 2005. *MNRAS* 360:305–13

Surman R, McLaughlin GC. 2005. *Ap. J.* 618:397–402
Tan JC, Matzner CD, McKee CF. 2001. *Ap. J.* 551:946–72
Tanvir NR, Chapman R, Levan AJ, Priddey RS. 2005. *Nature* 438:991–93
Thompson C. 1994. *MNRAS* 270:480
Thompson TA, Chang P, Quataert E. 2004. *Ap. J.* 611:380–93
Thompson TA, Quataert E, Burrows A. 2005. *Ap. J.* 620:861–77
Thomsen B, Hjorth J, Watson D, Gorosabel J, Fynbo JPU, et al. 2004. *Astron. Astrophys.* 419:L21–25
Thornton K, Gaudlitz M, Janka H-T, Steinmetz M. 1998. *Ap. J.* 500:95–119
Timmes FX, Woosley SE, Weaver TA. 1996. *Ap. J.* 457:834–43
Tinney C, Stathakis R, Cannon R, Wieringa M, Frail DA, et al. 1998. *IAU Circ.* No. 6896
Tominaga N, Deng J, Mazzali PA, Maeda K, Nomoto K, et al. 2004. *Ap. J. Lett.* 612:L105–8
Tominaga N, Tanaka M, Nomoto K, Mazzali PA, Deng J, et al. 2005. *Ap. J. Lett.* 633:L97–100
Torii K. 2003. *GCN Circ.* 1986
Totani T. 1997. *Ap. J. Lett.* 486:L71
Turatto M, Suzuki T, Mazzali PA, Benetti S, Cappellaro E, et al. 2000. *Ap. J. Lett.* 534:L57–61
Tutukov AV, Cherepashchuk AM. 2003. *Astron. Rep.* 47:386–400
Tutukov AV, Cherepashchuk AM. 2004. *Astron. Rep.* 48:39–44
Usov VV. 1992. *Nature* 357:472–74
Usov VV, Chibisov GV. 1975. *Sov. Astron.* 19:115
van Marle A-J, Langer N, Garcia-Segura G. 2005. *Astron. Astrophys.* 444:837–47
van Paradijs JA. 2001. In *Black Holes in Binaries and Galactic Nuclei*, ed. L Kaper, EPJ Van Den Heuvel, PA Woudt, p. 316. Berlin: Springer
van Paradijs J, Groot PJ, Galama T, Kouveliotou C, Strom RG, et al. 1997. *Nature* 386:686–89
van Paradijs J, Kouveliotou C, Wijers RAMJ. 2000. *Annu. Rev. Astron. Astrophys.* 38:379–425
van Putten MH, Lee HK, Lee CH, Kim H. 2004a. *Phys. Rev. D* 69:104026
van Putten MH, Levinson A, Lee HK, Regimbau T, Punturo M, Harry GM. 2004b. *Phys. Rev. D* 69:044007
Vietri M, Ghisellini G, Lazzati D, Fiore F, Stella L. 2001. *Ap. J. Lett.* 550:L43–48
Vietri M, Stella L. 1998. *Ap. J. Lett.* 507:L45–46
Villasenor JS, Lamb DQ, Ricker GR, Atteia JL, Kawai N, et al. 2005. *Nature* 437:855–58
Vink JS, de Koter A. 2005. *Astron. Astrophys.* 442:587–96
Vreeswijk PM, Ellison SL, Ledoux C, Wijers RAMJ, Fynbo JPU, et al. 2004. *Astron. Astrophys.* 419:927
Vreeswijk PM, Fruchter A, Kaper L, Rol E, Galama TJ, et al. 2001. *Ap. J.* 546:672–80
Wang L, Howell DA, Höflich P, Wheeler JC. 2001. *Ap. J.* 550:1030–35
Wang L, Wheeler JC. 1998. *Ap. J. Lett.* 504:L87
Wang L, Wheeler JC, Li Z, Clocchiatti A. 1996. *Ap. J.* 467:435

Waxman E, Draine BT. 2000. *Ap. J.* 537:796–802

Waxman E, Mészáros P. 2003. *Ap. J.* 584:390–98

Weiler KW, Panagia N, Montes MJ, Sramek RA. 2002. *Annu. Rev. Astron. Astrophys.* 40:387–438

Wheeler JC. 2001. See Livio et al 2001, pp. 356–76

Wheeler JC, Akiyama S, Williams PT. 2005. *Astrophys. Space Sci.* 298:3–8

Wheeler JC, Yi I, Höflich P, Wang L. 2000. *Ap. J.* 537:810–23

Wijers RAMJ. 2001. In *Gamma-Ray Bursts in the Afterglow Era, Proc. Int. Workshop, Rome, CNR Hqrs., 17–20 Oct. 2000*, ed. E Costa, F Frontera, J Hjorth, p. 306. Berlin: Springer-Verlag

Wijers RAMJ, Bloom JS, Bagla J, Natarajan P. 1998. *MNRAS* 294:L17–21

Woods PM, Thompson C. 2005. In *Compact Stellar X-Ray Sources*, ed. WHG Lewin, M van der Klis. Cambridge, UK: Cambridge Univ. Press, Chapter 14

Woosley S, Heger A. 2006. *Ap. J.* 637:914–21

Woosley SE. 1993. *Ap. J.* 405:273–77

Woosley SE, Eastman RG, Schmidt BP. 1999. *Ap. J.* 516:788–96

Woosley SE, Heger A, Weaver TA. 2002. *Rev. Mod. Phys.* 74:1015–71

Woosley SE, Janka T. 2005. *Nat. Phys.* 3:147–54

Woosley SE, MacFadyen AI, Heger A. 2001. See Livio et al. 2001, pp. 171–83

Woosley SE, Weaver TA. 1986. *Annu. Rev. Astron. Astrophys.* 24:205

Woosley SE, Weaver TA. 1995. *Ap. J. Suppl.* 101:181–235

Yamazaki R, Yonetoku D, Nakamura T. 2003. *Ap. J. Lett.* 594:L79–82

Yoon S-C, Langer N. 2005. *Astron. Astrophys.* 443:643–48

Yoshii Y, Tomita H, Kobayashi Y, Deng J, Maeda K, et al. 2003. *Ap. J.* 592:467–74

Zeh A, Klose S, Hartmann DH. 2004. *Ap. J.* 609:952–61

Zhang B, Mészáros P. 2002. *Ap. J.* 566:712–22

Zhang W, Fryer CL. 2001. *Ap. J.* 550:357–67

Zhang W, Woosley SE, Heger A. 2004. *Ap. J.* 608:365–77

Zhang W, Woosley SE, MacFadyen AI. 2003. *Ap. J.* 586:356–71

Subject Index

A

A 1524-617, 53
Above horizontal branch (AHB) stars, 125–28, 130–31, 134
Absolute magnitude calibrations
　of Cepheids, 93–134
　of RR Lyrae stars, 113–25, 133
Abundances
　globular clusters (GCs) and, 225–26
Active galactic nuclei (AGN)
　jets and, 464, 493–94
　mapping the evolution of, 178, 180
　star formation and, 183–84
Advection-dominated accretion flow (ADAF), 78
Age-metallicity
　globular clusters (GCs) and, 226–27
AGN
　See Active galactic nuclei (AGN)
Angular momentum
　black hole (BH), 50
　Cepheids and, 110
Anisotropy
　cosmic microwave background (CMB), 438
Asteroids, 310

B

B1757-24, 37, 38
B1957+20, 26, 37, 42
Beyond Einstein initiative, 439
Binary systems
　pulsars, 27
　pulsar winds in, 40–42
Black-hole binaries (BHBs), 49–86
　astrophysics and, 80–81
　census of, 51–54
　Fe emission line, 78
　Fe-Kα line, 83–84
　hardness-intensity diagram (HID), 60–62
　physics and, 80–81
　power-density spectrum (PDS) and, 55–56, 59
　quasi-periodic oscillations (QPOs) and, 56, 59, 60
　quiescent state, 62
　radio jets and, 60–62
　steep power law (SPL), 78–79
　three-state description, 57–60
　X-ray light curves of, 55
　X-ray observations of, 55–56
　X-ray overviews of, 63–72
　X-ray spectra, 56, 57
　X-ray states, 63–72
Black-hole candidates (BHC), 51–54
Black holes (BHs)
　astrophysics and, 80–81
　importance of, 50
　jets and, 464
　magnetohydrodynamics (MHD) simulations of, 77–78
　outburst states of, 57–60
　physics and, 80–81
　as probes of strong gravity, 80–85
　spin, 81
　　measuring, 80–85
　　quasi-periodic oscillations (QPOs) and, 84–85
　steep power law (SPL) and, 59, 60
　X-ray quasi-periodic oscillations (QPOs) and, 72–76

Bow shocks
    pulsars and, 34–38, 42
Brown dwarf(s), 296–98, 313–14
    formation, 281, 283–84
    spectroscopy of, 296–98
    and white dwarf binary systems, 298

C

3C 58
    pulsar wind nebulae (PWN), 29–31, 38
    X-ray observations of, 33
    X-ray spectra, 34
Canada-France Redshift Survey (CFRS), 166, 175–76
Carbon
    C2, 387–88
    C3, 387–88
        diffuse cloud models and, 395
    diatomic
        interstellar, 378–79
    diffuse clouds and, 381–82, 389–91
    diffuse interstellar bands (DIBs) and, 398
    diffuse molecular clouds and, 371
    translucent clouds and, 372
Carbon dioxide
    diffuse clouds and, 382
Carbon monoxide
    dense molecular clouds and, 373
    diffuse clouds and, 376–78, 380, 386–87
Cepheid(s)
    absolute magnitude calibrations, 93–134
    angular momentum measurements, 110
    anomalous, 130–33
    dwarf spheroidals and, 131–34
    extragalactic distance scale and, 111–13
    instability strip, 96–103
    light curve shape, 126
    period-color relations and, 98–103
    population II, 128–30
    RR Lyrae variables and, 115
Chandra X-ray Observatory (CXO), 324–25
    pulsar wind nebulae and, 19
    spectra and time variability from, 353–55
    X-ray emission from jets and, 464
Clouds
    classification for types of, 369–73
    dense molecular, 373
    diffuse atomic, 369–71
    diffuse interstellar, 367–406
    diffuse molecular, 371–72
    translucent, 369, 372–74
Cluster(s)
    disk evolution in, 286–87
    ellipticals
        color-magnitude relation, 159–61
        versus field ellipticals, 166–72
    formation, 195
    galaxies
        formation, 210
    high redshift
        fundamental plane, 162–65
    young, 276–79
Color bimodality
    globular clusters (GCs) and, 195–207, 215–16
Color dispersion
    globular clusters (GCs) and, 204
Color distributions
    globular clusters (GCs) and, 199, 205–6
    globular clusters (GCs) subpopulations and, 258
Color-magnitude (C-M) relation
    cluster ellipticals and, 159–61
    elliptical galaxies and, 147–49
Color-metallicity relations
    globular clusters (GCs) and, 198, 204–5
Coma ellipticals, 156–57
COMBO-17, 166–67, 177–79
Comets
    Spitzer and, 310–11
Copernicus orbital observatory, 374
Cosmic microwave background (CMB), 384
    future study of, 439–40
    jets and, 466, 474, 476–80, 491–99, 501, 502
    observational challenges, 438–39
    reionization and, 415–16, 431–40, 453
    Wilkinson Microwave Anisotropy Probe (WMAP) and, 436–39
COSMOS project, 178
Crab Nebula, 18–19, 28, 30–34, 38, 40
    X-ray spectra, 34
Cyg X-1, 55
    soft state of, 56–57
    X-ray reflection component, 78

D

Dark-matter halos, 254
Debris disks, 313, 314
    direct imaging, 301–4
    evolution, 299–308
    spectral properties, 299–308
Deep Groth Strip Survey, 170
DEEP2 survey, 177–78, 182

Delta Scuti
  stars, 122–23
  variables, 95, 98
dEs
  color distributions of, 246–47
  kinematics of globular clusters (GCs) in, 247
  metal-rich globular clusters (GCs) populations and, 205
Diffuse clouds
  chemical models of, 388–95
  classification of, 369–73
  defined, 369
  diatomics in, 376–80
  dust grains and, 368
  formation of large molecules in, 399–400
  H_3, 381
  HD in, 376
  hydrogen and, 384–86
  indicators of physical conditions, 384–88
  key transitions in, 389–91
  molecule-bearing, 400–2
  molecules in, 373–83
  physical and chemical processes, 388–89
  polyatomic molecules in, 380–83
  those of special interest, 400–5
  very large molecules in, 395–400
Diffuse interstellar bands (DIBs), 396–400
Diffuse interstellar clouds, 367–406
Diffuse molecular clouds, 371–72
  three sightlines dominated by, 402–5
Disk evolution, 286–92
  spectroscopic studies of, 287–90
Disk galaxies
  globular clusters (GCs) and, 236
Disks
  galaxy formation and, 143
  star formation and, 183
Distant red galaxies (DRG), 176, 182
Dust grains
  diffuse clouds and, 368
  young stellar objects (YSOs) and, 290–92
Dwarf ellipticals
  globular clusters luminosity functions (GCLF), 245–46
  specific frequencies, 243–45
Dwarf galaxies
  globular clusters (GCs) and, 242–47
Dwarfs
  galaxies and, 142
Dwarf spheroidals
  Cepheids and, 131–34

E

Early-type galaxies (ETG), 141–84
  color evolution of, 159–61
  evolution of the line indices, 165
  line strength diagnostics and, 151–56
  luminosity function (LF), 161–62
  number density of, 175
Einstein Observatory, 324
Elliptical galaxies
  color-magnitude (C-M) relation and, 147–49
  formation of, 141–84, 207–8
  fundamental plane and, 149–51
  at high redshift, 159–80
  in the local universe, 146–59
Elliptical galaxy formation, 141–84
Ellipticals
  field *versus* cluster, 166–72
  formation of, 180–83
  globular clusters (GCs) kinematics and, 233–36
  versus spiral bulges, 156–58
Epoch of reionization (EoR), 415, 416, 431–36, 438–40
  radio loud source during, 450
Extragalactic distance scale
  Cepheids and, 111–13
Extragalactic globular clusters
  galaxy formation and, 193–258
Extremely red objects (ERO), 176–77, 181

F

Faint Fuzzies (FF), 214–15
Far Ultraviolet Spectroscopic Explorer (FUSE) observatory, 374
51 Pegasi, 308
Filaments
  pulsar wind nebulae (PWN) and, 33
FORS Deep Field, 183
Fossil evidence
  early-type galaxies (ETG) and, 158
4U 1543-47
  X-ray overview of, 65, 66
4U 1630-47, 53
Fundamental plane (FP)
  elliptical galaxies and, 149–51
  galaxies, 162–65
  high redshift cluster, 162–65

G

G21.5-0.9, 31
  X-ray spectra, 34
G54.1+0.3, 31

G292.0+1.8, 31
G359.23-0.82, 36–38
Galaxies
- accretion-powered X-ray binaries (XRBs) in, 325
- cosmic reionization and, 442–43
- distant red galaxies (DRG), 176, 182
- FR1, 472–78, 480–81, 485–86, 494, 502
- FRII radio galaxies, 476–78, 481, 493, 494
- Lyα, 429–31
- low-mass X-ray binaries (LMXBs) and, 330–31
- Lyman break, 441–42
- merging
  - globular clusters (GCs) and, 206
- star-formation history of, 158–59
- X-ray populations of old systems, 329–42
- X-ray source populations of, 344–46
- X-ray sources in, 323–59
- *See also* Elliptical galaxies; Galaxy

Galaxy
- Cepheids, 94
- evolution, 143
  - modeling of, 184
  - X-ray emission and, 358–59
- formation
  - elliptical, 141–84
  - extragalactic globular clusters and, 193–258
- globular clusters (GCs) in, 236
- period-color relations and, 98–103
- period-luminosity (P-L) relations calibrations of, 103, 105–11, 133
- See also Galaxies

Gamma-ray bursts (GRBs)
- absorption spectroscopy and, 512–13
- afterglow and density gradients, 542–43
- afterglow optical spectroscopy, 540–41
- core-collapse supernovae and, 507–45
- cosmic X-ray flashes (XRFs) and, 520–21, 544
- future study of, 543–45
- intergalactic medium (IGM) and, 427–28
- models, 529–36
- nucleosynthesis, 542
- progenitor stars, 536–38
- short-hard, 519–20
- X-ray lines, 541

Gamma-ray telescopes
- jets and, 499–500

Gemini Deep Deep Survey (GDDS), 172
Globular cluster(s) (GCs), 193–258
- age-dating of, 215–23
- azimuthal distributions, 210–12
- color bimodality, 195–207, 215–16
- color dispersion, 204
- color distributions of, 246–47
- color for, 199, 204
- color/metallicity of, 196–97
- color-metallicity relations, 204–5
- dissipationless accretion, 249–50
- dwarf galaxies and, 242–47
- formation, 210, 247–58
  - dark-matter halos, 254–55
  - efficiency, 208
  - metal-poor, 254–57
  - recent scenarios, 250–54
- future work on, 257–58
- global properties, 207–15
- kinematics, 124–25, 233–36, 247
- low-mass X-ray binaries (LMXBs), 336–38
- luminosity for, 199, 204
- luminosity functions, 237–39
- major merges, 248–49
- metallicity *versus* luminosity, 197–98
- models for formation, 206–7
- near-infrared (NIR) photometry and, 226–27
- properties of color distributions, 200–3
- proto, 206
- radial distributions, 210–12
- sizes, 239–41
- specific frequencies, 243–45
- specific frequency, 207–8
- spectroscopy and, 215–26
- subpopulations, 208–10
- variations with galaxy morphology, 212–14

Globular star clusters, 193–258
Gravitational wave astronomy, 80
GRB 980425
- SN 1998bw and, 509, 513–15, 517

Great Observatories Origins Deep Survey (GOODS), 169–70, 172, 175, 181, 183
GRO J1655-40, 64–65, 84–85
- spectra of, 58–60
- spectral observations of, 79

GRS 1739-278, 53
GRS 1758-258, 53
GRS 1915+105, 53, 84–85
- black-hole binaries (BHBs), 70, 72

Gunn-Peterson (GP) absorption, 415, 418–26
Gunn-Peterson (GP) effect
- observations of, 418–26, 453

GX 339-4, 53, 57, 62, 70, 71, 83

H

H 1743-322, 53, 64, 74
  X-ray overview of, 65, 68
Hardness-intensity diagram (HID)
  black-hole binaries (BHBs) and, 60–62, 70, 72
HD 62542
  diffuse interstellar bands (DIBs), 403
HD 204827, 403, 405
HD 210121, 405
Helium
  reionization, 416–17
Hierarchical Merging model, 143–44
High-frequency quasi-periodic oscillations (HFQPOs), 75–76
High-mass X-ray binaries (HMXBs), 326
  evolution of, 358
  X-ray luminosity functions (XLFs), 346–49
Horizontal branch stars
  globular clusters (GCs) ages and, 222–23
HR diagram
  Cepheid instability strip in, 96–103
  white dwarf sequence fittings in, 124
Hydrogen
  diffuse atomic clouds and, 371
  diffuse clouds and, 373–76, 384–86, 389, 390
  diffuse molecular clouds and, 372
  H_3^+, 394–95
    in diffuse molecular gas, 406
  in the intergalactic medium (IGM), 418
  reionization, 416
Hydrogen chloride
  diffuse clouds and, 380

I

IC 4051, 196
Instability strip
  Cepheid, 96–103
  Large Magellanic Clouds (LMC), 104
Intergalactic medium (IGM)
  evolution of, 416–54
  physics of, 445–46
  reionization, 416–54
  thermal state evolution, 428–29
Interstellar medium (ISM)
  molecules in, 368
Inverse Compton (IC)
  jets and, 466, 467, 474, 476–81, 486, 491–99, 501, 502
Iron
  early-type galaxies (ETG) and, 151–56, 158
  emission line
    black-hole binaries (BHBs), 78
  star-formation history and, 223–25

J

J1119-6127, 39
J1846-0258, 39
Jets
  in black holes (BH), 57
  collimation and, 472
  composition, 465–66
  cosmic microwave background (CMB) and, 466, 474, 476–80, 491–99, 501, 502
  detection statistics, 493–95
  distributed acceleration and, 473–74
  emission between knots and, 491–92
  emission mechanisms, 472
  energies, 497
  entrainment and, 472
  extragalactic, 463–502
  formation of, 466–67
  FR II radio galaxies and, 476–78, 481, 493, 494
  gamma-ray telescopes and, 499–500
  gross properties, 478–80
  Inverse Compton (IC) and, 466, 467, 474, 476–81, 486, 491–99, 501, 502
  kiloparsec scales and, 485–86
  knots and, 468–72, 502
  observations of, 484–95
  offsets, 489–92, 496–97
  parsec scale structures and, 486–87
  particle acceleration and, 472
  physical length of, 498
  power-law spectra, 474–76
  problems with, 464–65
  progressions, 491
  propagation, 468–69
  pulsar wind nebulae (PWN) and, 31–33
  quasars and, 476–78, 480–83, 494, 501–2
  relative intensities, 492–93
  structure, 487–89
  synchrotron emission and, 498–502
  synchrotron emission model, 495–96
  synchrotron models for, 472–78
  terminal hotspots, 469–71
  transverse structure, 467
  two-zone models, 470
  variability, 493
  X-ray, 478–84, 487–93

K

Kerr metric
    black holes (BH) and, 81
Kormendy relation, 162
Kuiper Belt Objects (KBOs), 311–14

L

Large Magellanic Clouds (LMC)
    Cepheids, 94
    instability strip, 104
    period-color relations and, 98–103
    period-luminosity (P-L) relations calibrations
        of, 103, 105–11, 133
LDSS redshift survey, 176
LMC X-1, 53
LMC X-3, 53
Low-frequency quasi-periodic oscillations
        (LFQPOs), 73–75
    models, 79
Low-mass X-ray binaries (LMXBs), 326
    constraints on, 340–42
    early-type galaxies and, 330–31
    evolution of, 358
    formation, 338–40
    globular clusters and, 336–38
    spectra and variability, 332
    X-ray luminosity functions and, 333–36, 338
Luminosity
    calibration
        of Cepheids, 95
        of pulsators, 93–134
    galaxy, 250–52
        metal-rich globular clusters (GCs) and, 205
    globular clusters (GCs) and, 199, 204
    luminosity-metallicity relation
        RR Lyrae (RRL), 95
    simple stellar population, 145
    spin-down, 21–22
Luminosity function(s) (LF)
    early-type galaxies (ETG), 161–62
    globular clusters (GCs)
        dwarf ellipticals and, 245–46
    globular clusters (GCs) and, 237–39
Lyman break galaxies (LBGs), 181

M

M31
    globular clusters (GCs), 225, 227–30, 236, 242
M33, 236
    globular clusters (GCs) in, 223
M51, 213, 215
M81, 213
    globular clusters (GCs) ages, 218
M87, 204, 209, 240
    color-magnitude diagrams (CMDs), 199
    first jet discovered in, 464
    globular clusters (GCs) in, 218, 223
    kinematics of globular clusters (GCs) in, 234
M101, 213
Magnesium
    early-type galaxies (ETG), 151–56, 158
Magnetars
    spin down, 39–40
Magnetic field
    neutron star, 21
Magnetohydrodynamics (MHD) simulations
    of black holes (BH), 77–78
Magnetorotational instability (MRI), 77
Main sequence fitting
    HB calibration and, 121–22
Mass
    black hole (BH), 50
    function
        black hole (BH), 51
    mass-metallicity relation
        globular clusters (GCs) and, 199
    mass-to-light (M/L) ratios
        galaxy formation and, 207
        stellar, 145, 146
Metal enrichment
    cosmic reionization and, 427
Metallicities
    early-type galaxies (ETG) and, 151–56
    of globular clusters (GCs) in dwarfs, 250–52
Metallicity
    age-metallicity
        globular clusters (GCs) and, 226–27
    calibrating Cepheid, 109, 110
    Cepheid, 111, 133
    globular clusters (GCs) and, 215–21
    globular clusters (GCs) color bimodality and,
        195
    globular clusters (GCs) formation and,
        250–52
    luminosity-metallicity relation
        RR Lyrae (RRL), 95
    main sequence and, 122
    mass-metallicity relation
        globular clusters (GCs) and, 199
    M31 bulge, 230–32
    stellar, 145–46
    zero age horizontal branch (ZAHB) and,
        115–21

Milky Way
  black-hole binaries (BHBs), 53–54
  globular clusters (GCs) and, 218
  globular clusters (GCs) subpopulations and, 196
Mills, Bernard
  on his life and work, 1–14
Molecules
  in space, 368
  interstellar, 368
Monolithic Collapse model, 143
MUNICS survey, 173

N

Near-IR (NIR) photometry
  globular clusters (GCs) and, 226–27
Nebulae
  pulsar wind, 17–41
Neutrinos
  cosmic reionization and, 444
Neutron star(s)
  defined, 19
  SN of 1054 CE and, 19
  spin evolution of, 20–21
  winds from, 39–40
NGC 253
  kinematics, 236
NGC 524
  globular clusters (GCs) ages, 218
NGC 1023
  Faint Fuzzies (FF), 214–15
  globular clusters (GCs) ages, 218
NGC 1052, 209
  globular clusters (GCs) ages, 218
NGC 1316, 227
NGC 1399, 204, 226, 233–35
  globular clusters (GCs) ages, 218
  luminosity, 208
  radial distributions, 211
NGC 1407, 199
  CN anomaly, 225
  globular clusters (GCs), 219
    ages, 218
NGC 3115, 218, 226, 234, 240–41
  CN anomaly, 225
NGC 3311, 196
NGC 3379, 209
  globular clusters (GCs) kinematics in, 236
NGC 3384, 215
NGC 3610
  CN anomaly, 225–26
  globular clusters (GCs) ages, 218
NGC 4258, 110–11
NGC 4365, 226, 227, 232
  globular clusters (GCs) ages, 218
NGC 4406
  surface density of globular clusters (GCs) in, 212
NGC 4472, 199, 204, 209, 218, 234
  color gradients in, 211
  globular clusters (GCs)
    ages, 218
    color bimodality, 195
    surface density of, 212
NGC 4565
  number of globular clusters (GCs), 213
NGC 4594, 199, 209
  globular clusters (GCs)
    ages, 218
    sizes, 241
    systems, 196
NGC 4636, 204, 208
NGC 4649
  color-magnitude diagrams (CMDs), 199
NGC 4874, 196
NGC 5128, 209, 226, 230–32, 234
  globular clusters (GCs), 233
    color bimodality, 195
    and field stars, 230–32
NGC 5195, 215
NGC 5466, 131
NGC 5846, 227
NGC 6388, 95, 123–24, 129–30, 133
NGC 6441, 95, 123–24, 129–30, 133
NGC 6822, 243
NGC 6946, 213
NGC 7192, 227
NGC 7814, 213
Nitrogen
  diffuse clouds and, 379–80
Nitrogen hydride (NH)
  diffuse clouds and, 379

O

1E 1740.7-2942, 53
Ophiuchi, 400–2
  detection of CO toward, 377, 378
  molecules detected toward, 382
Outbursts, 284, 286
Outflows, 284, 286

P

Penrose process
  black hole spin and, 81
Period-color relations
  of Cepheids, 98–103
  period-luminosity (P-L) relations and, 94
Period-luminosity (P-L) relation(s), 126–27
  for anomalous Cepheids, 132
  calibrations of, 103, 105–11, 133
  of Cepheids, 94, 96–98, 101, 112–13
    metallicity and, 111, 133
  of classical Cepheid variables, 94
  color differences and, 102
  period-color relations and, 94
  population II Cepheids and, 130
Planck
  cosmic microwave background (CMB) study and, 439
Planet(s)
  orbiting a solar-type star in 51 Pegasi, 308
  transiting, 308–10
Planetary system
  formation and evolution, 299–308
Pleiades
  error in main sequence placement, 109
Polarimetry
  black hole (BH) spin and, 82
Power density spectrums (PDSs)
  black-hole binaries (BHBs) and black-hole candidate, 76
Pregalactic medium (PGM)
  three phases of, 416
Protoplanetary disks, 275–76, 313
  evolution of, 286–92
PSR B1509-58, 32–33, 40, 41
PSR B1853+01, 26–27, 37
PSR J1747-2958, 36–38
Pulsar(s)
  binary systems, 27
  bow shocks and, 34–38, 42
  defined, 19
  in interstellar gas, 27
  SN of 1054 CE and, 19
  spin down, 20–22
  winds, 27–28
    binary systems and, 40–42
Pulsar wind nebulae (PWN), 17–41
  bow-shock, 27
  defined, 19
  evolution, 23–27, 42
  filamentary, 33
  jets and, 31–33
  knot-like structures in, 34
  overall properties, 20–22
  radio emission from, 22
  Sedov supernova remnant and, 24–27
  shocks and, 35–38
  spectra, 34
  supernova remnant(s) (SNR) and, 23–24, 38–39
  TeV gamma rays, 40
  tori and, 31–33
  wind termination shock and, 28, 30–31, 42
  wisps and, 31–33
  X-ray emission from, 22
  young, 27–34, 42
    observed properties, 28–30
Pulsators
  luminosity calibration of, 93–134

Q

Quasars
  cosmic Strömgren spheres (CSS) and, 426
  Gunn-Peterson effect and, 418–26, 453
  jets and, 476–78, 480–83, 494, 501–2
  reionization and, 427
Quasi-periodic oscillations (QPOs)
  black holes (BH) and, 72–76
  mechanisms, 79
Quasi-soft sources (QSSs), 350–51

R

Radio emission
  pulsar wind nebulae (PWN) and, 22
Radio frequency interference (RFI), 452
Radio jet(s)
  black holes (BH) and, 57
  hard state and, 78
  unified model for, 61–62
Radius-surface brightness relation, 162
Reionization
  basic model, 417
  cosmic, 415–54
    metal enrichment and, 427
  cosmic microwave background (CMB) and, 415–16, 431–40, 453
  HI 21-cm probes of, 444–52, 454
  sources of, 440–44, 454
Relativity
  general relativity (GR), 50
ROSAT, 324
Rosat Deep Cluster Survey, 160, 161

Rossi X-Ray Timing Explorer (RXTE), 50, 57, 59, 63, 86
RR Lyrae (RRL)
    as distance indicators, 95
    luminosity calibrations of variables, 95
    stars
        absolute magnitude calibrations of, 113–25, 133
    variables
        Cepheids and, 115

S

Sedov supernova remnant
    pulsar wind nebulae (PWN) and, 24–27
Shocks
    forward and reverse, 467
    pulsar wind nebulae (PWN) and, 35–38
    supernova remnants (SNR) and, 23–24
Simple stellar populations (SSPs), 145, 146
Small Magellanic Clouds (SMC)
    Cepheids, 94
    period-color relations and, 98–103
    period-luminosity (P-L) relations calibrations of, 103, 105–11, 133
SN 1986, 39
SN 1987A, 38
SN 1998bw
    GRB 980425 and, 509, 513–15, 517
SNR G327.1-1.1, 25–26
SNR W44, 26–27
Spectroscopy
    absorption
        gamma-ray bursts (GRBs) and, 512–13
    diffuse interstellar bands (DIBs), 399
    globular clusters (GCs) and, 215–26
    rocket-based, 374
Spheroids
    galaxy formation and, 143
Spiral galaxies
    bulges of, 156–58
    globular clusters (GCs) and, 212–14
Spitzer Space Telescope, 269–314
    architecture of, 271–74
    cryo-thermal system, 271
    instruments, 273–74
    Legacy Science Program, 276
    optics, 273
    orbit, 270–71
    pointing and control system, 273
Star(s)
    clusters, 194–95
    evolution of, 274–98
    formation of, 275–76
    spheroid, 142, 143
    See also Above horizontal branch (AHB) stars; Neutron stars; Star formation
Star formation, 275–76
    disks and, 183
    early-type galaxies (ETG) and, 183
    gamma-ray bursts (GRBs) and, 509
    globular clusters (GCs) and, 194
    high-mass, 284
    history
        of early-type galaxies (ETG), 155–56
        of galaxies, 158–59
    local universe, 183
    models, 183–84
Star-formation rate(s) (SFR), 195, 208
    early-type galaxies (ETG) and, 180
Steep power law (SPL)
    black-hole binaries (BHBs) and, 70, 72, 78–79
    black holes (BH) and, 59, 60, 70, 72
    quasi-periodic oscillations (QPOs) and, 59, 60
Stellar evolution, 274–98
    post-main sequence, 292–93
Stellar populations
    radius-surface brightness relation, 162
    synthetic, 145–46
Strömgren spheres
    cosmic, 426, 450
Substellar objects, 275–76, 313
Supernova, 313
    core-collapse
        models, 528–29
    explosion in 1054 CE, 18–19
Supernovae
    described, 508
    and gamma-ray bursts (GRBs) connection, 510–19
    infrared echoes, 294–95
    observations of, 293–96
Supernova remnant(s) (SNR), 18
    high velocity pulsars and, 34–35
    pulsar wind nebulae (PWN) and, 23–24, 38–39
Super soft sources (SSSs), 350–51
SX Phoenicis stars, 122–23
Synchrotron emission
    jets and, 499–500
Synchrotron emission model
    jets and, 495–96
Synchrotron models
    FR 1 galaxies and, 472–78

T

Telescopes
   to probe the cosmic microwave background (CMB), 440
Tori
   pulsar wind nebulae (PWN) and, 31–33

U

Ultra-compact dwarf (UCD) galaxies, 204
Ultraluminous infrared galaxies (ULIRG), 181, 182
Ultraluminous X-ray sources (ULXs), 324, 325, 351–58
   models of formation and evolution, 356–58
   star-forming stellar populations and, 352–53
Unidentified infrared bands (UIBs)
   in diffuse clouds, 395–96, 400

V

VLT VIMOS Deep Survey (VVDS), 178

W

Water
   diffuse clouds and, 382
White dwarf
   and brown dwarf binary systems, 298
   sequence fitting, 124
Wilkinson Microwave Anisotropy Probe (WMAP) and, 436–39, 453
Wind(s)
   from neutron stars, 39–40
   pulsar, 27–28
   pulsar wind nebulae, 17–41
   termination shock
      pulsar wind nebulae (PWN) and, 28, 30–31, 42
Wisps
   pulsar wind nebulae (PWN) and, 31–33

X

XMM-Newton
   spectra and time variability from, 353–55
X Persei, 402
X-ray
   quasi-periodic oscillations (QPOs), 72–76
X-ray binaries (XRBs), 49–86, 324–26
   old populations, 329–42
   young populations, 342–49
X-ray continuum
   to measure black hole spin, 82–83
X-ray emission
   extragalactic jets and, 463–502
   galaxy evolution and, 358–59
   pulsar wind nebulae (PWN) and, 22
X-ray light curves
   black-hole binaries (BHBs) and, 55
X-ray luminosity functions (XLFs), 328–29
   low-mass X-ray binaries (LMXBs) and, 333–36, 338
   star-forming populations and, 346–49
X-ray photometry, 327–28
X-ray photons
   cosmic reionization and, 444
X-ray rich (XRR) gamma-ray bursts (GRBs), 520
X-rays
   population studies in, 326–29
X-ray sources
   galaxies and, 323–59
X-ray state(s)
   black holes (BH) and, 65
   physical models for, 77–79
X-ray timing
   black holes (BH) and, 55–56
XTE J1118+480
   Fe emission line, 78
XTE J1550-564, 65, 67, 74, 84–85
   quiescent spectrum, 62
   spectral observations of, 79
XTE J1650-500, 83
XTE J1655-40, 83
XTE J1748-288, 53
XTE J1859+226, 65, 69

Y

Young massive star clusters (YMCs), 206, 208
Young stellar objects (YSOs), 275–83
   dust grains in, 290–92

Z

Zero age horizontal branch (ZAHB), 115–21, 126–28
   metallicity, 115–21

Cumulative Indexes

Contributing Authors, Volumes 33–44

A

Alexander DR, 35:137–77
Allard F, 35:137–77
Aller LH, 33:1–17
Angel JR, 36:507–37
Aparicio A, 43:387–434
Arnett D, 33:115–32
Aschwanden MJ, 39:175–210
Asplund M, 43:481–530

B

Bachiller R, 34:111–54
Bailyn CD, 33:133–62
Balbus SA, 41:555–97
Balick B, 40:439–86
Bally J, 39:403–55
Baraffe I, 38:337–77
Barkana R, 39:19–66
Basri G, 38:485–519
Bastian TS, 36:131–88
Beck R, 34:153–204
Beers TC, 43:531–79
Beiersdorfer P, 41:343–90
Belloni T, 42:317–64
Benz AO, 36:131–88
Bertschinger E, 36:599–654
Bessell MS, 43:293–336
Bethe HA, 41:1–14
Bhattacharjee A, 42:365–84
Bignami GF, 34:331–81
Blaauw A, 42:1–37
Black DC, 33:359–80

Blake GA, 36:317–68
Blanco VM, 39:xiv, 1–18
Bland-Hawthorn J, 40:487–537; 43:769–826
Bloom JS, 44:507–56
Bodenheimer P, 33:199–238
Bolte M, 34:461–510
Borgani S, 40:539–77
Bothun G, 35:267–307
Bowyer S, 38:231–88
Branch D, 36:17–55
Brandenburg A, 34:153–204
Brandt WN, 43:827–59
Brighenti F, 41:191–239
Brodie JP, 44:193–267
Bromm V, 42:79–118
Burrows A, 40:103–36
Busso M, 37:239–309
Butler RP, 36:57–97

C

Cameron AGW, 37:1–36
Canup RM, 42:441–75
Caraveo PA, 34:331–81
Carilli CL, 40:319–48; 44:415–62
Carlstrom JE, 40:643–80
Carretta E, 42:385–440
Catling D, 41:429–63
Cecil G, 43:769–826
Cesarsky CJ, 38:761–814
Chabrier G, 38:337–77
Charnley SB, 38:427–83

Christensen-Dalsgaard J, 41:599–643
Christlieb N, 43:531–79
Churchwell E, 40:27–62
Chyba CF, 43:31–74
Clayton DD, 42:39–78
Cox DP, 43:337–85
Crenshaw DM, 41:117–67

D

Davidson K, 35:1–32
de Vegt C, 37:97–125
Dickinson M, 38:667–715
Dodelson S, 40:171–216
Dole H, 43:727–68
Draine BT, 41:241–89
Drake JJ, 38:231–88
Dultzin-Hacyan D, 38:521–71
Dwek E, 39:249–307

E

Ehrenfreund P, 38:427–83
Elmegreen BG, 42:211–73, 275–316
Ellis RS, 35:389–443
Evans NJ II, 37:311–62

F

Fabbiano G, 44:323–66
Falcke H, 39:309–52

Fan X, 44:415–62
Fazio G, 44:269–321
Feigelson ED, 37:363–408
Fender R, 42:317–64
Ferguson HC, 38:667–715
Ferland G, 37:487–531
Ferland GJ, 41:517–54
Ferrari A, 36:539–98
Filippenko AV, 35:309–55
Fishman GJ, 33:415–58
Frank A, 40:439–86
Franklin FA, 39:581–631
Franx M, 35:637–75
Freeman K, 40:487–537
Friel ED, 33:381–414

G

Gaensler BM, 44:17–47
Gallart C, 43:387–434
Gallino R, 37:239–309
Gary DE, 36:131–88
Gautschy A, 33:75–113; 34:551–606
Gawiser E, 43:861–918
Genzel R, 38:761–814
George IM, 41:117–67
Giacconi R, 43:1–30
Giavalisco M, 40:579–641
Gilmore G, 35:637–75
Glassgold AE, 34:241–78
Goldreich P, 42:549–601
Goldstein ML, 33:283–325
Gosling JT, 34:35–73
Gosnell TR, 43:139–94
Gratton R, 42:385–440
Güdel M, 40:217–61

H

Hamann F, 37:487–531
Hand KP, 43:31–74
Hansen BMS, 41:465–515
Harris DE, 44:463–506
Hartmann L, 34:205–39
Hasinger G, 43:827–59
Hauschildt PH, 35:137–77
Hauser MG, 39:249–307
Haxton WC, 33:459–503
Herbig GH, 33:19–73
Herwig F, 43:435–79

Hickson P, 35:357–88
Hillebrandt W, 38:191–230
Holder GP, 40:643–80
Hollenbach DJ, 35:179–215
Holman MJ, 39:581–631
Horányi M, 34:383–418
Howard RF, 34:75–109
Hu W, 40:171–216
Hubbard WB, 40:103–36
Hudson H, 33:239–82
Humphreys RM, 35:1–32

I

Impey C, 35:267–307

J

Jewitt DC, 40:63–101
Johnston KJ, 37:97–125

K

Kahabka P, 35:69–100
Kahn SM, 41:291–342
Kasting JF, 41:429–63
Keating B, 44:415–62
Kellermann KI, 39:457–509
Kennicutt RC Jr, 36:189–231; 42:603–83
Kenyon SJ, 34:205–39
Kirkpatrick JD, 43:195–245
Knapp GR, 36:369–433
Kormendy J, 33:581–624; 42:603–83
Kouveliotou C, 38:379–425
Kovalevsky J, 36:99–129
Kraemer SB, 41:117–67
Krawczynski H, 44:463–506
Kudritzki RP, 38:613–66
Kulsrud RM, 37:37–64

L

Lada CJ, 41:57–115
Lada EA, 41:57–115
Lagache G, 43:727–68
Larson RB, 42:79–118
Lean J, 35:33–67
Lebreton Y, 38:35–77
Lecar M, 39:581–631

Leibundgut B, 39:67–98
Liebert J, 41:465–515
Lin DNC, 33:505–40; 34:703–47
Lithwick Y, 42:549–601
Lo KY, 43:625–76
Loeb A, 39:19–66
Low BC, 43:103–37
Lunine JI, 40:103–36
Luu JX, 40:63–101

M

Maeder A, 38:143–90
Maraschi L, 35:445–502
Marcy GW, 36:57–97
Marsden BG, 43:75–102
Marziani P, 38:521–71
Massey P, 41:15–56
Mateo M, 34:511–50
Mateo ML, 36:435–506
Mathews WG, 41:191–239
Matthaeus WH, 33:283–325
McCall BJ, 44:367–414
McCarthy PJ, 42:477–515
McClintock JE, 44:49–92
McGaugh SS, 40:263–317
McWilliam A, 35:503–56
Meegan CA, 33:415–58
Melia F, 39:309–52
Mellier Y, 37:127–89
Mészáros P, 40:137–69
Meynet G, 38:143–90
Miesch MS, 41:599–643
Mills B, 44:1–15
Mirabel IF, 34:749–92; 37:409–43
Montes MJ, 40:387–438
Montmerle T, 37:363–408
Moran JM, 39:457–509
Morris M, 34:645–701
Moss D, 34:153–204
Mulchaey JS, 38:289–335
Murray NW, 39:581–631

N

Narlikar JV, 39:211–48; 41:169–89
Niemeyer JC, 38:191–230
Nittler LR, 42:39–78
Norman C, 40:539–77

O

O'Connell RW, 37:603–48
O'Dell CR, 39:99–136
Oke JB, 38:79–111
Olszewski EW, 34:511–50
Osterbrock DE, 38:1–33

P

Paczynski B, 34:419–59
Padmanabhan T, 39:211–48
Paerels FBS, 41:291–342
Panagia N, 40:387–438
Papaloizou JCB, 33:505–40;
 34:703–47
Peale SJ, 37:533–602
Pethick CJ, 42:169–210
Pinsonneault M, 35:557–605
Poland AI, 39:175–210
Prochaska JX, 43:861–918
Puetter RC, 43:139–94
Puget J-L, 43:727–68
Puls J, 38:613–66

Q

Quirrenbach A, 39:353–401

R

Rabin DM, 39:175–210
Rauch M, 36:267–316
Reese ED, 40:643–80
Refregier A, 41:645–67
Reid IN, 37:191–237;
 43:247–92
Reipurth B, 39:403–55
Remillard RA, 44:49–92
Renzini A, 44:141–92
Rephaeli Y, 33:541–79
Richstone D, 33:581–624
Rieke G, 44:269–321
Roberts DA, 33:283–325
Robinson B, 37:65–96
Rodriguez LF, 37:409–43
Roellig TL, 44:269–321
Rosati P, 40:539–77
Rubin V, 39:137–74
Ryan J, 33:239–82

S

Saio H, 33:75–113; 34:551–606
Salpeter EE, 40:1–25
Sandage A, 37:445–86;
 43:581–624; 44:93–140
Sanders DB, 34:749–92
Sanders RH, 40:263–317
Sandquist EL, 38:113–41
Sari R, 42:549–601
Savage BD, 34:279–329
Scalo J, 42:211–73, 275–316
Schatzman E, 34:1–34
Sembach KR, 34:279–329
Serabyn E, 34:645–701
Shibazaki N, 34:607–44
Shukurov A, 34:153–204
Slane PO, 44:17–47
Sneden C, 42:385–440
Snow TP, 44:367–414
Sofue Y, 39:137–74
Sokoloff D, 34:153–204
Solomon PM, 43:677–725
Song I, 42:685–721
Sramek RA, 40:387–438
Starrfield S, 35:137–77
Stetson PB, 34:461–510
Strader J, 44:193–267
Sulentic JW, 38:521–71
Suntzeff NB, 34:511–50

T

Taam RE, 38:113–41
Tammann GA, 44:93–140
Tanaka Y, 34:607–44
Tarter J, 39:511–48
Taylor GB, 40:319–48
Teerikorpi P, 35:101–36
Thomas JH, 42:517–48
Thompson MJ, 41:599–643
Tielens AGGM, 35:179–215
Tohline JE, 40:349–85
Toomre J, 41:599–643
Townes CH, 35:xiii–xliv

U

Ulrich M-H, 35:445–502
Urry CM, 35:445–502

V

van de Hulst HC, 36:1–16
VandenBerg DA, 34:461–510
Vanden Bout PA, 43:677–725
van den Heuvel EPJ, 35:69–100
van der Klis M, 38:717–60
van Dishoeck EF, 36:317–68;
 42:119–67
van Paradijs J, 38:379–425
Van Winckel H, 41:391–427
van Woerden H, 35:217–66
Veilleux S, 43:769–826
Vennes S, 38:231–88

W

Waelkens C, 36:233–66
Wagner SJ, 33:163–97
Wakker BP, 35:217–66
Wallerstein G, 36:369–433;
 38:79–111
Wasserburg GJ, 37:239–309
Waters LBFM, 36:233–66
Watson DM, 44:269–321
Weiler KW, 40:387–438
Weiss NO, 42:517–48
Weissman PR, 33:327–57
Werner M, 44:269–321
Wijers RAMJ, 38:379–425
Williams R, 38:667–715
Willson LA, 38:573–611
Witzel A, 33:163–97
Wolfe AM, 43:861–918
Woolf N, 36:507–37
Woosley SE, 44:507–56
Wyse RFG, 35:637–75

Y

Yahil A, 43:139–94
Yakovlev DG, 42:169–210

Z

Zensus JA, 35:607–36
Zhang M, 43:103–37
Zoccali M, 43:387–434
Zuckerman B, 39:549–80;
 42:685–721

Chapter Titles, Volumes 33–44

Prefatory Chapters

An Astronomical Rescue	LH Aller	33:1–17
The Desire to Understand the World	E Schatzman	34:1–34
A Physicist Courts Astronomy	CH Townes	35:xiii–xliv
Roaming Through Astrophysics	HC van de Hulst	36:1–16
Adventures in Cosmogony	AGW Cameron	37:1–36
A Fortunate Life in Astronomy	DE Osterbrock	38:1–33
Telescopes, Red Stars, and Chilean Skies	VM Blanco	39:1–18
A Generalist Looks Back	EE Salpeter	40:1–25
My Life in Astrophysics	HA Bethe	41:1–14
My Cruise Through the World of Astronomy	A Blaauw	42:1–37
An Education in Astronomy	R Giacconi	43:1–30
An Engineer Becomes an Astronomer	B Mills	44:1–15

Sun

High-Energy Particles in Solar Flares	H Hudson, J Ryan	33:239–82
Magnetohydrodynamic Turbulence in the Solar Wind	ML Goldstein, DA Roberts, WH Matthaeus	33:283–325
The Solar Neutrino Problem	WC Haxton	33:459–503
Solar Active Regions as Diagnostics of Subsurface Conditions	RF Howard	34:75–109
The Sun's Variable Radiation and Its Relevance for Earth	J Lean	35:33–67
Radio Emission from Solar Flares	TS Bastian, AO Benz, DE Gary	36:131–88
The New Solar Corona	MJ Aschwanden, AI Poland, DM Rabin	39:175–210

The Internal Rotation of the Sun	MJ Thompson, J Christensen-Dalsgaard, MS Miesch, J Toomre	41:599–643
Implusive Magnetic Reconnection in the Earth's Magnetotail and the Solar Corona	A Bhattacharjee	42:365–84
Fine Structure in Sunspots	JH Thomas, NO Weiss	42:517–48
The Hydromagnetic Nature of Solar Coronal Mass Ejections	M Zhang, BC Low	43:103–37

Solar System and Extrasolar Planets

The Kuiper Belt	PR Weissman	33:327–57
Completing the Copernican Revolution: The Search for Other Planetary Systems	DC Black	33:359–80
Corotating and Transient Solar Wind Flows in Three Dimensions	JT Gosling	34:35–73
Charged Dust Dynamics in the Solar System	M Horányi	34:383–418
Detection of Extrasolar Giant Planets	GW Marcy, RP Butler	36:57–97
Astronomical Searches for Earth-Like Planets and Signs of Life	N Woolf, JR Angel	36:507–37
Origin and Evolution of the Natural Satellites	SJ Peale	37:533–602
Organic Molecules in the Interstellar Medium, Comets, and Meteorites: A Voyage from Dark Clouds to the Early Earth	P Ehrenfreund, SB Charnley	38:427–83
Chaos in the Solar System	M Lecar, FA Franklin, MJ Holman, NW Murray	39:581–631
Kuiper Belt Objects: Relics from the Accretion Disk of the Sun	JX Luu, DC Jewitt	40:63–101
Theory of Giant Planets	WB Hubbard, A Burrows, JI Lunine	40:103–36
Evolution of a Habitable Planet	JF Kasting, D Catling	41:429–63
Astrophysics with Presolar Stardust	DD Clayton, LR Nittler	42:39–78
Planet Formation by Coagulation: A Focus on Uranus and Neptune	P Goldreich, Y Lithwick, R Sari	42:549–601
Astrobiology: The Study of the Living Universe	CF Chyba, KP Hand	43:31–74
Sungrazing Comets	BG Marsden	43:75–102
First Fruits of the *Spitzer Space Telescope*: Galactic and Solar System Studies	M Werner, G Fazio, G Rieke, TL Roellig, DM Watson	44:269–321

Stars

Stellar Pulsations Across the HR Diagram: Part 1	A Gautschy, H Saio	33:75–113
Explosive Nucleosynthesis Revisited: Yields	D Arnett	33:115–32
Blue Stragglers and Other Stellar Anomalies: Implications for the Dynamics of Globular Clusters	CD Bailyn	33:133–62
Angular Momentum Evolution of Young Stars and Disks	P Bodenheimer	33:199–238
The Old Open Clusters of the Milky Way	ED Friel	33:381–414
Gamma-Ray Bursts	GJ Fishman, CA Meegan	33:415–58
Theory of Accretion Disks I: Angular Momentum Transport Processes	JCB Papaloizou, DNC Lin	33:505–40
Bipolar Molecular Outflows from Young Stars and Protostars	R Bachiller	34:111–54
The FU Orionis Phenomenon	L Hartmann, SJ Kenyon	34:205–39
Circumstellar Photochemistry	AE Glassgold	34:241–78
Geminga, Its Phenomenology, Its Fraternity, and Its Physics	GF Bignami, PA Caraveo	34:331–81
The Age of the Galactic Globular Cluster System	DA VandenBerg, M Bolte, PB Stetson	34:461–510
Old and Intermediate-Age Stellar Populations in the Magellanic Cloud	EW Olszewski, NB Suntzeff, M Mateo	34:511–50
Stellar Pulsations Across the HR Diagram: Part 2	A Gautschy, H Saio	34:551–606
X-Ray Novae	Y Tanaka, N Shibazaki	34:607–44
Theory of Accretion Disks II: Application to Observed Systems	DNC Lin, JCB Papaloizou	34:703–47
Eta Carina and Its Environment	K Davidson, RM Humphreys	35:1–32
Luminous Supersoft X-Ray Sources	P Kahabka, EPJ van den Heuvel	35:69–100
Model Atmospheres of Very Low Mass Stars and Brown Dwarfs	PH Hauschildt, S Starrfield, F Allard, DR Alexander	35:137–77
Optical Spectra of Supernovae	AV Filippenko	35:309–55
Abundance Ratios and Galactic Chemical Evolution	A McWilliam	35:503–56
Mixing in Stars	M Pinsonneault	35:557–605
Type Ia Supernovae and the Hubble Constant	D Branch	36:17–55

Herbig Ae/Be Stars	C Waelkens, LBFM Waters	36:233–66
Carbon Stars	G Wallerstein, GR Knapp	36:369–433
The HR Diagram and the Galactic Distance Scale After Hipparcos	IN Reid	37:191–237
Nucleosynthesis in Asymptotic Giant Branch Stars: Relevance for Galactic Enrichment and Solar System Formation	M Busso, R Gallino, GJ Wasserburg	37:239–309
High-Energy Processes in Young Stellar Objects	ED Feigelson, T Montmerle	37:363–408
Sources of Relativistic Jets in the Galaxy	IF Mirabel, LF Rodriguez	37:409–43
Stellar Structure and Evolution: Deductions from Hipparcos	Y Lebreton	38:35–77
Common Envelope Evolution of Massive Binary Stars	RE Taam, EL Sandquist	38:113–41
The Evolution of the Rotating Stars	A Maeder, G Meynet	38:143–90
Type Ia Supernova Explosion Models	W Hillebrandt, JC Niemeyer	38:191–230
Theory of Low-Mass Stars and Substellar Objects	G Chabrier, I Baraffe	38:337–77
Gamma-Ray Burst Afterglows	J van Paradijs, C Kouveliotou, RAMJ Wijers	38:379–425
Observations of Brown Dwarfs	G Basri	38:485–519
Mass Loss from Cool Stars: Impact on the Evolution of Stars and Stellar Populations	LA Willson	38:573–611
Winds from Hot Stars	R-P Kudritzki, J Puls	38:613–66
Millisecond Oscillations in X-Ray Binaries	M van der Klis	38:717–60
Cosmological Implications from Observations of Type Ia Supernovae	B Leibundgut	39:67–98
The Orion Nebula and Its Associated Population	CR O'Dell	39:99–136
Herbig-Haro Flows: Probes of Early Stellar Evolution	B Reipurth, J Bally	39:403–55
Dusty Circumstellar Disks	B Zuckerman	39:549–80
Theories of Gamma-Ray Bursts	P Mészáros	40:137–69
Stellar Radio Astronomy: Probing Stellar Atmospheres from Protostars to Giants	M Güdel	40:217–61
The Origin of Binary Stars	JE Tohline	40:349–85
Radio Emission from Supernovae and Gamma-Ray Bursters	KW Weiler, N Panagia, MJ Montes, RA Sramek	40:387–438
Shapes and Shaping of Planetary Nebulae	B Balick, A Frank	40:439–86

Massive Stars in the Local Group: Implications for Stellar Evolution and Star Formation	P Massey	41:15–56
Embedded Clusters in Molecular Clouds	CJ Lada, EA Lada	41:57–115
Post-AGB Stars	H Van Winckel	41:391–427
Cool White Dwarfs	BMS Hansen, J Liebert	41:465–515
Quantitative Spectroscopy of Photoionized Clouds	GJ Ferland	41:517–54
Enhanced Angular Momentum Transport in Accretion Disks	SA Balbus	41:555–97
Neutron Star Cooling	DG Yakovlev, CJ Pethick	42:169–210
GRS 1915 + 105 and The Disc-Jet Coupling in Accreting Black Hole Systems	R Fender, T Belloni	42:317–64
Abundance Variations within Globular Clusters	R Gratton, C Sneden, E Carretta	42:385–440
Young Stars Near the Sun	B Zuckerman, I Song	42:685–721
New Spectral Types L and T	JD Kirkpatrick	43:195–245
High-Velocity White Dwarfs and Galactic Structure	IN Reid	43:247–92
The Adequacy of Stellar Evolution Models for the Interpretation of the Color-Magnitude Diagrams of Resolved Stellar Populations	C Gallart, M Zoccali, A Aparicio	43:387–434
Evolution of Asymptotic Giant Branch Stars	F Herwig	43:435–79
New Light on Stellar Abundance Analyses: Departures from LTE and Homogeneity	M Asplund	43:481–530
The Discovery and Analysis of Very Metal-Poor Stars in the Galaxy	TC Beers, N Christlieb	43:531–79
The Evolution and Structure of Pulsar Wind Nebulae	BM Gaensler, PO Slane	44:17–47
X-Ray Properties of Black-Hole Binaries	RA Remillard, JE McClintock	44:49–92
Absolute Magnitude Calibrations of Population I and II Cepheids and Other Pulsating Variables in the Instability Strip of the Hertzsprung-Russell Diagram	A Sandage, GA Tammann	44:93–140
First Fruits of the *Spitzer Space Telescope*: Galactic and Solar System Studies	M Werner, G Fazio, G Rieke, TL Roellig, DM Watson	44:269–321
Populations of X-Ray Sources in Galaxies	G Fabbiano	44:323–66
The Supernova–Gamma-Ray Burst Connection	SE Woosley, JS Bloom	44:507–56

Interstellar Medium

The Diffuse Interstellar Bands	GH Herbig	33:19–73

Gamma-Ray Bursts	GJ Fishman, CA Meegan	33:415–58
Bipolar Molecular Outflows from Young Stars and Protostars	R Bachiller	34:111–54
Galactic Magnetism: Recent Developments and Perspectives	R Beck, A Brandenburg, D Moss, A Shukurov, D Sokoloff	34:153–204
Circumstellar Photochemistry	AE Glassgold	34:241–78
Interstellar Abundances from Absorption-Line Observations with the *Hubble Space Telescope*	BD Savage, KR Sembach	34:279–329
Eta Carina and Its Environment	K Davidson, RM Humphreys	35:1–32
Dense Photodissociation Regions (PDRs)	DJ Hollenbach, AGGM Tielens	35:179–215
High-Velocity Clouds	BP Wakker, H van Woerden	35:217–66
Chemical Evolution of Star-Forming Regions	EF van Dishoeck, GA Blake	36:317–68
A Critical Review of Galactic Dynamos	RM Kulsrud	37:37–64
Physical Conditions in Regions of Star Formation	NJ Evans II	37:311–62
Gamma-Ray Burst Afterglows	J van Paradijs, C Kouveliotou, RAMJ Wijers	38:379–425
Organic Molecules in the Interstellar Medium, Comets, and Meteorites: A Voyage from Dark Clouds to the Early Earth	P Ehrenfreund, SB Charnley	38:427–83
Herbig-Haro Flows: Probes of Early Stellar Evolution	B Reipurth, J Bally	39:403–55
Ultra-Compact HII Regions and Massive Star Formation	E Churchwell	40:27–62
Theories of Gamma-Ray Bursts	P Mészáros	40:137–69
Radio Emission from Supernovae and Gamma-Ray Bursters	KW Weiler, N Panagia, MJ Montes, RA Sramek	40:387–438
Shapes and Shaping of Planetary Nebulae	B Balick, A Frank	40:439–86
Interstellar Dust Grains	BT Draine	41:241–89
ISO Spectroscopy of Gas and Dust: From Molecular Clouds to Protoplanetary Disks	EF van Dishoeck	42:119–67
Interstellar Turbulence I: Observations and Processes	BG Elmegreen, J Scalo	42:211–73
Interstellar Turbulence II: Implications and Effects	J Scalo, BG Elmegreen	42:275–316

The Three-Phase Interstellar Medium Revisited	DP Cox	43:337–85
Molecular Gas at High Redshift	PM Solomon, PA Vanden Bout	43:677–725
The Evolution and Structure of Pulsar Wind Nebulae	BM Gaensler, PO Slane	44:17–47
Diffuse Atomic and Molecular Clouds	TP Snow, BJ McCall	44:367–414
The Supernova–Gamma-Ray Burst Connection	SE Woosley, JS Bloom	44:507–56

Galaxy

Blue Stragglers and Other Stellar Anomalies: Implications for the Dynamics of Globular Clusters	CD Bailyn	33:133–62
The Old Open Clusters of the Milky Way	ED Friel	33:381–414
Gravitational Microlensing in the Local Group	B Paczynski	34:419–59
The Age of the Galactic Globular Cluster System	DA VandenBerg, M Bolte, PB Stetson	34:461–510
The Galactic Center Environment	M Morris, E Serabyn	34:645–701
Abundance Ratios and Galactic Chemical Evolution	A McWilliam	35:503–56
First Results from Hipparcos	J Kovalevsky	36:99–129
Reference Frames in Astronomy	KJ Johnston, C de Vegt	37:97–125
The HR Diagram and the Galactic Distance Scale After Hipparcos	IN Reid	37:191–237
The Supermassive Black Hole at the Galactic Center	F Melia, H Falcke	39:309–52
The New Galaxy: Signatures of its Formation	K Freeman, J Bland-Hawthorn	40:487–537
Enhanced Angular Momentum Transport in Accretion Disks	SA Balbus	41:555–97
First Fruits of the *Spitzer Space Telescope*: Galactic and Solar System Studies	M Werner, G Fazio, G Rieke, TL Roellig, DM Watson	44:269–321

Galaxies

Old and Intermediate-Age Stellar Populations in the Magellanic Cloud	EW Olszewski, NB Suntzeff, M Mateo	34:511–50
Luminous Infrared Galaxies	DB Sanders, IF Mirabel	34:749–92
Low Surface Brightness Galaxies	C Impey, G Bothun	35:267–307
Compact Groups of Galaxies	P Hickson	35:357–88
Faint Blue Galaxies	RS Ellis	35:389–443

Galactic Bulges	RF Wyse, G Gilmore, M Franx	35:637–75
Star Formation in Galaxies Along the Hubble Sequence	RC Kennicutt Jr.	36:189–231
Dwarf Galaxies of the Local Group	ML Mateo	36:435–506
Far-Ultraviolet Radiation from Elliptical Galaxies	RW O'Connell	37:603–48
X-ray Properties of Groups of Galaxies	JS Mulchaey	38:289–335
The Hubble Deep Fields	HC Ferguson, M Dickinson, R Williams	38:667–715
Extragalactic Results from the Infrared Space Observatory	R Genzel, CJ Cesarsky	38:761–814
Rotation Curves of Spiral Galaxies	Y Sofue, V Rubin	39:137–74
Cluster Magnetic Fields	GB Taylor, CL Carilli	40:319–48
The Evolution of X-ray Clusters of Galaxies	P Rosati, S Borgani, C Norman	40:539–77
Lyman Break Galaxies	M Giavalisco	40:579–641
Hot Gas in and Around Elliptical Galaxies	WG Mathews, F Brighenti	41:191–239
EROs and Faint Red Galaxies	PJ McCarthy	42:477–515
Secular Evolution and the Formation of Pseudobulges in Disk Galaxies	J Kormendy, RC Kennicutt, Jr.	42:603–83
The Classification of Galaxies: Early History and Ongoing Developments	A Sandage	43:581–624
Mega-Masers and Galaxies	KY Lo	43:625–76
Molecular Gas at High Redshift	PM Solomon, PA Vanden Bout	43:677–725
Dusty Infrared Galaxies: Sources of the Cosmic Infrared Background	G Lagache, J-L Puget, H Dole	43:727–68
Galactic Winds	S Veilleux, G Cecil, J Bland-Hawthorn	43:769–826
Deep Extragalactic X-Ray Surveys	WN Brandt, G Hasinger	43:827–59
Stellar Population Diagnostics of Elliptical Galaxy Formation	A Renzini	44:141–92
Extragalactic Globular Clusters and Galaxy Formation	JP Brodie, J Strader	44:193–267
Populations of X-Ray Sources in Galaxies	G Fabbiano	44:323–66

Active Galactic Nuclei

Intraday Variablity in Quasars and BL Lac Objects	SJ Wagner, A Witzel	33:163–97
Theory of Accretion Disks I: Angular Momentum Transport Processes	JCB Papaloizou, DNC Lin	33:505–40
Inward Bound: The Search for Supermassive Black Holes in Galactic Nuclei	J Kormendy, D Richstone	33:581–624

Theory of Accretion Disks II: Application to Observed Systems	DNC Lin, JCB Papaloizou	34:703–47
Variability of Active Galactic Nuclei	M-H Ulrich, L Maraschi, CM Urry	35:445–502
Parsec-Scale Jets in Extragalactic Radio Sources	JA Zensus	35:607–36
Modeling Extragalactic Jets	A Ferrari	36:539–98
Element Abundances in Quasistellar Objects: Star Formation and Galactic Nuclear Evolution at High Redshifts	F Hamann, G Ferland	37:487–531
Phenomenology of Broad Emission Lines in Active Galactic Nuclei	JW Sulentic, P Marziani, D Dultzin-Hacyan	38:521–71
Extragalactic Results from the Infrared Space Observatory	R Genzel, CJ Cesarsky	38:761–814
The Reionization of the Universe by the First Stars and Quasars	A Loeb, R Barkana	39:19–66
Mass Loss from the Nuclei of Active Galaxies	DM Crenshaw, SB Kraemer, IM George	41:117–67
Mega-Masers and Galaxies	KY Lo	43:625–76
Galactic Winds	S Veilleux, G Cecil, J Bland-Hawthorn	43:769–826
Deep Extragalactic X-Ray Surveys	WN Brandt, G Hasinger	43:827–59
X-Ray Emission from Extragalactic Jets	DE Harris, H Krawczynski	44:463–506

Cosmology

Comptonization of the Cosmic Microwave Background: The Sunyaev-Zeldovich Effect	Y Rephaeli	33:541–79
Observational Selection Bias Affecting the Determination of the Extragalactic Distance Scale	P Teerikorpi	35:101–36
Type Ia Supernovae and the Hubble Constant	D Branch	36:17–55
The Lyman Alpha Forest in the Spectra of Quasistellar Objects	M Rauch	36:267–316
Simulations of Structure Formation in the Universe	E Bertschinger	36:599–654
Probing the Universe with Weak Lensing	Y Mellier	37:127–89
The Hubble Deep Fields	HC Ferguson, M Dickinson, R Williams	38:667–715
Extragalactic Results from the Infrared Space Observatory	R Genzel, CJ Cesarsky	38:761–814
The Reionization of the Universe by the First Stars and Quasars	A Loeb, R Barkana	39:19–66

Cosmological Implications from Observations of Type Ia Supernovae	B Leibundgut	39:67–98
Standard Cosmology and Alternatives: A Critical Appraisal	JV Narlikar, T Padmanabhan	39:211–48
The Cosmic Infrared Background: Measurements and Implications	MG Hauser, E Dwek	39:249–307
Cosmic Microwave Background Anisotropies	W Hu, S Dodelson	40:171–216
Modified Newtonian Dynamics as an Alternative to Dark Matter	RH Sanders, SS McGaugh	40:263–317
Cluster Magnetic Fields	GB Taylor, CL Carilli	40:319–48
The Evolution of X-ray Clusters of Galaxies	P Rosati, S Borgani, C Norman	40:539–77
Lyman Break Galaxies	M Giavalisco	40:579–641
Cosmology with the Sunyaev-Zeldovich Effect	JE Carlstrom, GP Holder, ED Reese	40:643–80
Action at a Distance and Cosmology: A Historical Perspective	JV Narlikar	41:169–89
Weak Gravitational Lensing by Large-Scale Structure	A Refregier	41:645–67
The First Stars	V Bromm, RB Larson	42:79–118
Dusty Infrared Galaxies: Sources of the Cosmic Infrared Background	G Lagache, J-L Puget, H Dole	43:727–68
Damped Ly(alpha) Systems	AM Wolfe, E Gawiser, JX Prochaska	43:861–918
Observational Constraints on Cosmic Reionization	X Fan, CL Carilli, B Keating	44:415–62

Instrumentation and Techniques

Frequency Allocation: The First Forty Years	B Robinson	37:65–96
The First 50 Years at Palomar: 1949-1999: The Early Years of Stellar Evolution, Cosmology, and High-Energy Astrophysics	A Sandage	37:445–86
The First 50 Years at Palomar, 1949–1999 Another View: Instruments, Spectroscopy, and Spectrophotometry and the Infrared	G Wallerstein, JB Oke	38:79–111
Extreme Ultraviolet Astronomy	S Bowyer, JJ Drake, S Vennes	38:231–88
Optical Interferometry	A Quirrenbach	39:353–401
The Development of High Resolution Imaging in Radio Astronomy	KI Kellermann, JM Moran	39:457–509
The Search for Extraterrestrial Intelligence (SETI)	J Tarter	39:511–48
High-Resolution X-Ray Spectroscopy with *Chandra* and *XMM-Newton*	FBS Paerels, SM Kahn	41:291–342
Laboratory X-Ray Astrophysics	P Beiersdorfer	41:343–90

Digital Image Reconstruction: Deblurring and Denoising	RC Puetter, TR G A	43:139–94
Standard Photometric Systems	MS	43:293–336

ANNUAL REVIEWS
Intelligent Synthesis of the Scientific Literature

Annual Reviews – Your Starting Point for Research Online
http://arjournals.annualreviews.org

- Over 900 Annual Reviews volumes—more than 25,000 critical, authoritative review articles in 32 disciplines spanning the Biomedical, Physical, and Social sciences— available online, including all Annual Reviews back volumes, dating to 1932
- Current individual subscriptions include seamless online access to full-text articles, PDFs, Reviews in Advance (as much as 6 months ahead of print publication), bibliographies, and other supplementary material in the current volume and the prior 4 years' volumes
- All articles are fully supplemented, searchable, and downloadable — see http://astro.annualreviews.org
- Access links to the reviewed references (when available online)
- Site features include customized alerting services, citation tracking, and saved searches

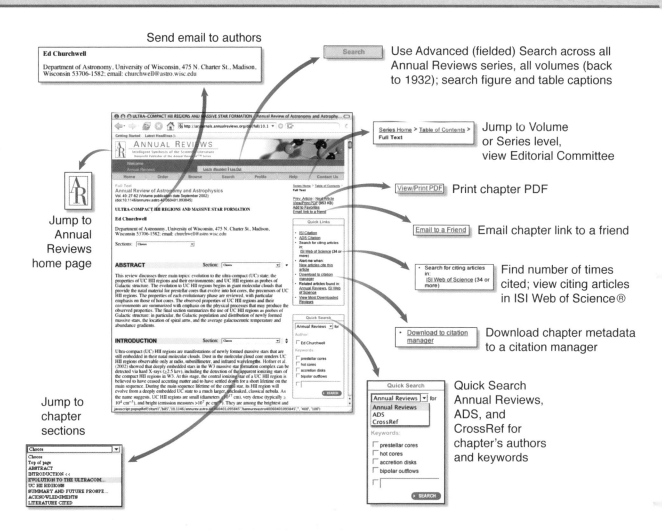

Copyright ® 2006 Annual Reviews, Nonprofit Publisher of the *Annual Review of* Series